Lecture Notes in Computer Science 4627

Commenced Publication in 1973
Founding and Former Series Editors:
Gerhard Goos, Juris Hartmanis, and Jan van Leeuwen

T0180859

Moses Charikar   Klaus Jansen
Omer Reingold   José D.P. Rolim (Eds.)

Approximation, Randomization, and Combinatorial Optimization

Algorithms and Techniques

10th International Workshop, APPROX 2007
and 11th International Workshop, RANDOM 2007
Princeton, NJ, USA, August 20-22, 2007
Proceedings

 Springer

Volume Editors

Moses Charikar
Princeton University
Computer Science Dept.
35 Olden Street, Princeton, NJ 08540-5233, USA
E-mail: moses@cs.princeton.edu

Klaus Jansen
University of Kiel
Institute for Computer Science
Olshausener Str. 40, 24098 Kiel, Germany
E-mail: kj@informatik.uni-kiel.de

Omer Reingold
The Weizmann Institute of Science
Faculty of Mathematics and Computer Science
Rehovot 76100, Israel
E-mail: omer.reingold@weizmann.ac.il

José D.P. Rolim
Centre Universitaire d'Informatique
Battelle bâtiment A
route de Drize 7, 1227 Carouge, Geneva, Switzerland
E-mail: rolim@cui.unige.ch

Library of Congress Control Number: 2007932294

CR Subject Classification (1998): F.2, G.2, G.1

LNCS Sublibrary: SL 1 – Theoretical Computer Science and General Issues

ISSN   0302-9743
ISBN-10   3-540-74207-7 Springer Berlin Heidelberg New York
ISBN-13   978-3-540-74207-4 Springer Berlin Heidelberg New York

Springer is a part of Springer Science+Business Media

springer.com

© Springer-Verlag Berlin Heidelberg 2007
Printed in Germany

Typesetting: Camera-ready by author, data conversion by Scientific Publishing Services, Chennai, India
Printed on acid-free paper   SPIN: 12106011   06/3180   5 4 3 2 1 0

Preface

This volume contains the papers presented at the 10th International Workshop on Approximation Algorithms for Combinatorial Optimization Problems (APPROX 2007) and the 11th International Workshop on Randomization and Computation (RANDOM 2007), which took place concurrently at Princeton University, on August 20–22, 2007. APPROX focuses on algorithmic and complexity issues surrounding the development of efficient approximate solutions to computationally difficult problems, and this was the 10th in the series after Aalborg (1998), Berkeley (1999), Saarbrücken (2000), Berkeley (2001), Rome (2002), Princeton (2003), Cambridge (2004), Berkeley (2005), and Barcelona (2006). RANDOM is concerned with applications of randomness to computational and combinatorial problems, and this was the 11th workshop in the series following Bologna (1997), Barcelona (1998), Berkeley (1999), Geneva (2000), Berkeley (2001), Harvard (2002), Princeton (2003), Cambridge (2004), Berkeley (2005), and Barcelona (2006).

Topics of interest for APPROX and RANDOM are: design and analysis of approximation algorithms, hardness of approximation, small space and data streaming algorithms, sub-linear time algorithms, embeddings and metric space methods, mathematical programming methods, coloring and partitioning, cuts and connectivity, geometric problems, game theory and applications, network design and routing, packing and covering, scheduling, design and analysis of randomized algorithms, randomized complexity theory, pseudorandomness and derandomization, random combinatorial structures, random walks/Markov chains, expander graphs and randomness extractors, probabilistic proof systems, random projections and embeddings, error-correcting codes, average-case analysis, property testing, computational learning theory, and other applications of approximation and randomness.

The volume contains 21 contributed papers, selected by the APPROX Program Committee out of 49 submissions, and 23 contributed papers, selected by the RANDOM Program Committee out of 50 submissions.

We would like to thank all of the authors who submitted papers and the members of the Program Committees:

APPROX 2007

Nikhil Bansal, IBM T J Watson
Moses Charikar, Princeton University (chair)
Chandra Chekuri, University of Illinois, Urbana-Champaign
Julia Chuzhoy, IAS
Venkatesan Guruswami, University of Washington
Howard Karloff, AT&T Labs Research
Guy Kortsarz, Rutgers, Camden

Robert Krauthgamer, IBM Almaden
Claire Mathieu, Brown University
Seffi Naor, Microsoft Research and Technion
Chaitanya Swamy, University of Waterloo
Lisa Zhang, Bell Labs

RANDOM 2007

Irit Dinur, Hebrew University
Thomas Hayes, Toyota Technological Institute at Chicago
Piotr Indyk, MIT
Russell Martin, University of Liverpool
Dieter van Melkebeek, University of Wisconsin, Madison
Michael Mitzenmacher, Harvard University
Michael Molloy, University of Toronto
Cristopher Moore, University of New Mexico
Sofya Raskhodnikova, Penn State University
Omer Reingold, Weizmann Institute (chair)
Ronen Shaltiel, University of Haifa
Asaf Shapira, Microsoft Research
Aravind Srinivasan, University of Maryland
Angelika Steger, ETH Zürich
Emanuele Viola, IAS

We would also like to thank the external subreferees: Scott Aaronson, Dimitris Achlioptas, Nir Ailon, Gorjan Alagic, Christoph Ambühl, Matthew Andrews, Arash Asadpour, Ziv Bar-Yossef, Petra Berenbrink, Nicla Bernasconi, Andrej Bogdanov, Niv Buchbinder, Sourav Chakraborty, Ho-Leung Chan, Aaron Clauset, Anne Condon, Artur Czumaj, Varsha Dani, Aparna Das, Roee Engelberg, Eldar Fischer, Abraham Flaxman, Tom Friedetzky, Rajiv Gandhi, Leszek Gąsieniec, Daniel Golovin, Ronen Gradwohl, Anupam Gupta, Dan Gutfreund, Magnus M. Halldorsson, Prahladh Harsha, Gabor Ivanyos, Mark Jerrum, Valentine Kabanets, Shiva Kasiviswanathan, Jonathan Katz, Neeraj Kayal, David Kempe, Rohit Khandekar, Sanjeev Khanna, Lefteris Kirousis, Phil Klein, Jochen Konemann, Roman Kotecky, Robert Krauthgamer, Michael Krivelevich, Fabian Kuhn, Eyal Kushilevitz, Emmanuelle Lebhar, Michael Landberg, Ron Lavi, Liane Lewin-Eytan, Avner Magen, Martin Marciniszyn, Frank McSherry, Julian Mestre, Vahab S. Mirrokni, Torsten Muetze, Ashwin Nayak, Ofer Neiman, Ilan Newman, Zeev Nutov, Konstantinos Panagiotou, Rina Panigrahy, Konstantin Pervyshev, Boris Pittel, Yuval Rabani, Prasad Raghavendra, Rajmohan Rajaraman, Dror Rawitz, Daniel Reichman, Dana Ron, Eyal Rozenman, Alex Russell, Jared Saia, Mohammad R Salavatipour, Alex Samorodnitsky, Rahul Santhanam, Warren Schudy, Justus Schwartz, Roy Schwartz, Amir Shpilka, Nikolai Simonov, Adam Smith, Joel Spencer, Reto Spöhel, Venkatesh Srinivasan, Mukund Sundararajan, Maxim Sviridenko, Jan Vondrak, John Watrous, Ryan Williams, David

Williamson, Prudence Wong, Orly Yahalom, Sergey Yekhanin, Neal Young, and Michele Zito.

We gratefully acknowledge the support from the Department of Computer Science at the Princeton University, the Faculty of Mathematics and Computer Science of the Weizmann Institute of Science, the Institute of Computer Science of the Christian-Albrechts-Universität zu Kiel and the Department of Computer Science of the University of Geneva.

August 2007   Moses Charikar and Omer Reingold, Program Chairs
                               Klaus Jansen and José D. P. Rolim, Workshop Chairs

Table of Contents

Contributed Talks of APPROX

Contributed Talks of RANDOM

Approximation Algorithms and Hardness for Domination with Propagation

Ashkan Aazami[1] and Michael D. Stilp[2]

[1] Department of Combinatorics and Optimization, University of Waterloo,
Waterloo, Ontario N2L3G1, Canada
aaazami@uwaterloo.ca
[2] School of Industrial and Systems Engineering, Georgia Institute of Technology,
Atlanta, Georgia 30332, USA
mstilp3@isye.gatech.edu

Abstract. The power dominating set (PDS) problem is the following extension of the well-known dominating set problem: find a smallest-size set of nodes S that power dominates all the nodes, where a node v is power dominated if (1) v is in S or v has a neighbor in S, or (2) v has a neighbor w such that w and all of its neighbors except v are power dominated. Note that rule (1) is the same as for the dominating set problem, and that rule (2) is a type of propagation rule that applies iteratively. We use n to denote the number of nodes. We show a hardness of approximation threshold of $2^{\log^{1-\epsilon} n}$ in contrast to the logarithmic hardness for dominating set. This is the first result separating these two problems. We give an $O(\sqrt{n})$ approximation algorithm for planar graphs, and show that our methods cannot improve on this approximation guarantee. We introduce an extension of PDS called ℓ-round PDS; for $\ell = 1$ this is the dominating set problem, and for $\ell \geq n - 1$ this is the PDS problem. Our hardness threshold for PDS also holds for ℓ-round PDS for all $\ell \geq 4$. We give a PTAS for the ℓ-round PDS problem on planar graphs, for $\ell = O(\frac{\log n}{\log \log n})$. We study variants of the greedy algorithm, which is known to work well on covering problems, and show that the approximation guarantees can be $\Theta(n)$, even on planar graphs. Finally, we initiate the study of PDS on directed graphs, and show the same hardness threshold of $2^{\log^{1-\epsilon} n}$ for directed acyclic graphs.

Keywords: Approximation Algorithms, Hardness of Approximation, PTAS, Dominating Set, Power Dominating Set, Planar Graphs, Integer Programming, Greedy Algorithms.

1 Introduction

A DOMINATING SET of an (undirected) graph $G = (V, E)$ is a set of nodes S such that every node in the graph is in S or has a neighbor in S. The problem of finding a dominating set of minimum size is an important problem that has been extensively studied, especially in the last twenty years, see the books by Haynes et al. [16,17]. Our focus is on an extension called the POWER DOMINATING SET

M. Charikar et al. (Eds.): APPROX and RANDOM 2007, LNCS 4627, pp. 1–15, 2007.

(abbreviated as PDS) problem. Power domination is defined by two rules; the first rule is the same as for the DOMINATING SET problem, but the second rule allows a type of indirect propagation. More precisely, given a set of nodes S, the set of nodes that are *power dominated* by S, denoted $\mathcal{P}(S)$, is obtained as follows

(Rule 1) if node v is in S, then v and all of its neighbors are in $\mathcal{P}(S)$;
(Rule 2) (propagation) if node v is in $\mathcal{P}(S)$, one of its neighbors w is not in $\mathcal{P}(S)$, and all other neighbors of v are in $\mathcal{P}(S)$, then w is inserted into $\mathcal{P}(S)$.

It can be seen that the set $\mathcal{P}(S)$ is independent of the sequence in which nodes are inserted into $\mathcal{P}(S)$ by Rule 2. The PDS problem is to find a node-set S of minimum size that power dominates all the nodes (i.e., find $S \subseteq V$ with $|S|$ minimum such that $\mathcal{P}(S) = V$). We use $\mathsf{Opt}(G)$ to denote the size of an optimal solution for graph G. We use n to denote $|V(G)|$, the number of nodes in the input graph.

For example, consider the planar graph in Figure 1; the graph has t disjoint triangles, and three (mutually disjoint) paths such that each path has exactly one node from each triangle; note that $|V| = 3t$. The minimum dominating set has size $\Theta(|V|)$, since the maximum degree is 4. The minimum power dominating set has size one – if S has any one node of the innermost (first) triangle (like v), then $\mathcal{P}(S) = V^1$.

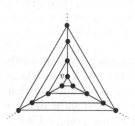

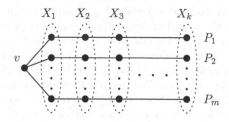

Fig. 1. Example for PDS   **Fig. 2.** Example for ℓ-round PDS

The PDS problem arose in the context of the monitoring of electric power networks. A power network contains a set of nodes and a set of edges connecting the nodes. A power network also contains a set of *generators*, which supply power, and a set of *loads*, where the power is directed to. In order to monitor a power network

---

[1] In more detail, we apply Rule 1 to see that all the nodes of the innermost (first) triangle and one node of the second triangle are in $\mathcal{P}(S)$; then by two applications of Rule 2 (to each of the nodes in the first triangle not in S), we see that the other two nodes of the second triangle are in $\mathcal{P}(S)$; then by three applications of Rule 2 (to each of the nodes in the second triangle) we see that all three nodes of the third triangle are in $\mathcal{P}(S)$; etc.

we need to measure all the state variables of the network by placing measurement devices. A Phasor Measurement Unit (PMU) is a measurement device placed on a node that has the ability to measure the voltage of the node and the current phase of the edges connected to the node; PMUs are expensive devices. The goal is to install the minimum number of PMUs such that the whole system can be monitored. These units have the capability of monitoring remote elements via propagation (as in Rule 2); see Brueni [8], Baldwin et al. [5], and Mili et al. [24]. Most measurement systems require one measurement device per node, but this does not apply to PMUs; hence, PMUs give a huge advantage. To see this in more detail consider a power network $G = (V, E)$, and assume that the resistances of the edges in the power network are known, and the goal is to measure the voltages of all nodes. For simplicity, assume that there are no generators and loads. By placing a PMU at node v we can measure the voltage of v and the electrical current on each edge incident to v. Next, by using Ohm's law we can compute the voltage of any node in the neighborhood of v (Rule 1). Now assume that the voltage on v and all of its neighbors except w is known. By applying ohm's law we can compute the current on the edges incident to v except $\{v, w\}$. Next by using Kirchhoff's law we compute the current on the edge $\{v, w\}$. Finally, applying Ohm's law on the edge $\{v, w\}$ gives us the voltage of w (Rule 2).

PMUs are used to monitor large system stability and to give warnings of system-wide failures. PMUs have become increasingly popular for monitoring power networks, and have been installed by several electricity companies since 1988 [23,6]. For example, the USA Western States Coordinating Council (WSCC) had installed around 34 PMUs by 1999 [6]. By now, several hundred PMUs have been installed world wide [26]. Some researchers in electrical engineering regard PMUs as the most important device for the future of power systems [25].

Our motivation comes from the area of approximation algorithms and hardness results. The DOMINATING SET problem is a so-called *covering problem*; we wish to cover all the nodes of the graph by choosing as few node neighborhoods as possible. In fact, the DOMINATING SET problem is a special case of the well-known SET COVERING[2] problem. Such covering problems have been extensively investigated. One of the key positive results dates from the 1970's, when Johnson [18] and Lovász [21] showed that the greedy method achieves an approximation guarantee of $O(\log n)$ where n denotes the size of the ground set.

Several negative results (on the hardness of approximation) have been discovered over the last few years: Lund and Yannakakis [22] showed that SET COVERING is hard to approximate within a ratio of $\Omega(\log n)$, modulo some variants of the P $\neq$ NP assumption. Later, Feige [12] showed that SET COVERING is hard to approximate within $(1 - \epsilon)\ln n$, modulo some variants of the P $\neq$ NP assumption. A natural question is what happens to covering problems (in the setting of approximation algorithms and hardness results) when we augment the covering rule with a propagation rule. PDS seems to be a key problem of this

---

[2] Given a family of sets on a groundset, find the minimum number of sets whose union equals the groundset.

type, since it is obtained from the DOMINATING SET problem by adding a simple propagation rule.

Previous literature: Apparently, the earliest theoretical publications on PDS are Brueni [8], Baldwin et al. [5], Mili set al. [24]. Later, Haynes et al. [15] showed that the problem is NP-complete even when the input graph is bipartite; they presented a linear-time algorithm to solve PDS optimally on trees. Kneis et al. [19] generalized this result to a linear-time algorithm that finds an optimal solution for graphs that have bounded tree-width, relying on earlier results of Courcelle et al. [10]. Kneis et al. [19] also showed that PDS is a generalization of DOMINATING SET. Guo et al. [14] developed a combinatorial algorithm based on dynamic programming for optimally solving PDS on graphs with bounded tree-width in linear-time. Even for planar graphs, the DOMINATING SET problem is NP-hard [13], and the same holds for PDS [14]. Brueni and Heath [9] have more results on PDS, including the NP-completeness of PDS on planar bipartite graphs.

Our contributions: Our results substantially improve on the understanding of PDS in the context of approximation algorithms.

- For general graphs, we show that PDS cannot be approximated better than $2^{\log^{1-\epsilon} n}$, unless $\mathsf{NP} \subseteq \mathsf{DTIME}(n^{polylog(n)})$. This is a substantial improvement over the previous logarithmic hardness result. This seems to be the first known "separation" result between PDS and DOMINATING SET.
- We introduce an extension of PDS called the ℓ-round PDS problem by adding a parameter ℓ to PDS which restricts the number of "parallel" rounds of propagation that can be applied (see Section 3.2 for formal definitions). The rules are the same as PDS, except we apply the propagation rule in parallel, in contrast to PDS where we apply the propagation rule sequentially. This problem has applications to monitoring power networks when there is a time constraint that should be met; deducing information through a parallel round of propagation takes one unit of time and in some applications we want to detect a failure in the network after at most ℓ units of time. Moreover, the ℓ-round PDS problem has theoretical significance because it allows one to study the spectrum of problems between DOMINATING SET and PDS, and see how the hardness threshold changes as we change the ℓ parameter. We show that ℓ-round PDS for $\ell \geq 4$ cannot be approximated better than $2^{\log^{1-\epsilon} n}$, unless $\mathsf{NP} \subseteq \mathsf{DTIME}(n^{polylog(n)})$.
- We focus on planar graphs, and give a PTAS for ℓ-round PDS for $\ell = O(\frac{\log n}{\log \log n})$. Baker's PTAS [4] for the DOMINATING SET problem on planar graphs is a special case of our result with $\ell = 1$, and no similar result of this type was previously known for $\ell > 1$. Note that the ℓ-round PDS problem remains NP-hard on planar graphs (for all $\ell \geq 1$). Also, note that our PTAS does not apply to PDS in general, because the running time is super-polynomial for $\ell = \omega(\frac{\log n}{\log \log n})$.
- We introduce the notion of strong regions and weak regions as a means of obtaining lower bounds on the size of an optimal solution for PDS. Based on

this, we develop an approximation algorithm[3] for PDS that gives a guarantee of $O(k)$ given a (not necessarily optimal) tree decomposition of the input graph with width k (see Definition 1). Note that the (optimal) tree-width of the input graph is some integer, denoted by tw, between 1 and $n - 1$ and is NP-hard to compute, but there are efficient approximation algorithms that find a tree decomposition of width $O(\text{tw} \cdot \log n)$ [7]. Our algorithm requires the tree decomposition as part of the input. Guo et al. [14] developed a dynamic programming algorithm for optimally solving PDS on graphs of tree-width k in time exponential in k^2. In contrast, our algorithm runs in time polynomial in n (independent of k). So our approximation algorithm runs in polynomial time on graphs with arbitrarily large tree-width, unlike the algorithm by Guo et al.[14]. Our algorithm provides an $O(\sqrt{n})$-approximation algorithm for PDS on planar graphs because a tree decomposition of a planar graph with width $O(\sqrt{n})$ can be computed efficiently [3]. Moreover, we show that our methods (specifically, the lower bounds used in our analysis) cannot improve on the $O(\sqrt{n})$ approximation guarantee for planar graphs.
- Greedy algorithms provide a successful approach (sometimes the best) for lots of problems including the SET COVERING problem and the DOMINATING SET problem. We focus on different variations of the greedy algorithm for PDS and we show that they perform very poorly, even on planar graphs.
- We study an integer programming formulation for ℓ-round PDS; this extends to PDS.
- We extend PDS in a natural way to directed graphs and prove that even for directed acyclic graphs, PDS is hard to approximate within the same threshold as for undirected graphs, modulo the same complexity assumption. We give a linear-time algorithm based on dynamic programming for directed PDS when the underlying undirected graph has bounded tree-width. This builds on results and methods of Guo et al. [14].

2 Hardness of PDS and ℓ-Round PDS

In this section we prove that PDS (and ℓ-round PDS for $\ell \geq 4$) cannot be approximated better than $2^{\log^{1-\epsilon} n}$ ratio, for any $\epsilon > 0$. We will prove this result by a reduction from the MINREP problem.

Theorem 1. *The PDS (and ℓ-round PDS for $\ell \geq 4$) problem cannot be approximated within ratio $2^{\log^{1-\epsilon} n}$, for any $\epsilon > 0$, unless $\mathsf{NP} \subseteq \mathsf{DTIME}(n^{polylog(n)})$.*

---

[3] An approximation algorithm for a (minimization) optimization problem means an algorithm that runs in polynomial time and computes a solution whose cost is within a guaranteed factor of the optimal cost; the *approximation guarantee* is the worst-case ratio, over all inputs of a given size, of the cost of the solution computed by the algorithm to the optimal cost.

In the MINREP problem we are given a bipartite graph $G = (A, B, E)$ with a partitioning of A and B into (equal size) subsets, say $A = \bigcup_{i=1}^{q_A} A_i$ and $B = \bigcup_{i=1}^{q_B} B_i$, where $|A_i| = m_A = \frac{|A|}{q_A}$ and $|B_i| = m_B = \frac{|B|}{q_B}$. This partitioning naturally defines a super bipartite graph $\mathcal{H} = (\mathcal{A}, \mathcal{B}, \mathcal{E})$. The super nodes of H are $\mathcal{A} = \{A_1, A_2, \cdots, A_{q_A}\}$ and $\mathcal{B} = \{B_1, B_2, \cdots, B_{q_B}\}$, and the super edges are $\mathcal{E} = \{A_i B_j | \exists a \in A_i, b \in B_j : ab \in E(G)\}$. We say that super edge $A_i B_j$ is covered by $ab \in E(G)$ if $a \in A_i$ and $b \in B_j$. The goal in MINREP is to pick the minimum number of nodes, $A' \cup B' \subseteq V(G)$, from G such that all the super edges in $\mathcal{H}$ are covered. The following theorem states the hardness of the MINREP problem [20].

Theorem 2. *[20] MINREP cannot be approximated within ratio $2^{\log^{1-\epsilon} n}$, for any $\epsilon > 0$, unless $\mathsf{NP} \subseteq \mathsf{DTIME}(n^{polylog(n)})$, where $n = |V(G)|$.*

The reduction: Theorem 1 is proved by a reduction from the MINREP problem. The reduction works for PDS and ℓ-round PDS (for $\ell \geq 4$) at the same time. In the following we create an instance of PDS (ℓ-round PDS), $\overline{G} = (\overline{V}, \overline{E})$, from an instance $G = (A, B, E)(\mathcal{H} = (\mathcal{A}, \mathcal{B}, \mathcal{E}))$ of the MINREP problem.

1. Add a new node w^* (master node) to the graph G, and add an edge between w^* and all the nodes in G. Also add new nodes w_1^*, w_2^*, w_3^* and connect them by an edge to w^*.
2. $\forall i \in \{1, \ldots, q_A\}, j \in \{1, \ldots, q_B\}$ do the following:
   (a) Let $E_{ij} = \{e_1, e_2, \ldots, e_\kappa\}$ be the set of edges between $A_i = \left\{a_{i_1}, \ldots, a_{i_{m_A}}\right\}$ and $B_j = \left\{b_{j_1}, \ldots, b_{j_{m_B}}\right\}$ in G, where κ is the number of edges between A_i and B_j.
   (b) Remove E_{ij} from G.
   (c) Let the edge $e_q \in E_{i,j}$ be incident to a_{i_q} and b_{j_q} (in G). In this labeling for simplicity the same node might get different labels. Let D_{ij} be the graph in Figure 3 (A dashed line shows an edge between the master node w^* and that node). Make $k = 4$ new copies of the graph D_{ij} and then identify nodes a_{i_q}'s, b_{j_q}'s with the corresponding nodes in A_i and B_j (in G). Note that the k copies are sharing the same set of nodes, A_i and B_j, but other nodes are disjoint.
3. Let $\overline{G} = (\overline{V}, \overline{E})$ be the obtained graph.

The analysis: The key part of the analysis is to show that the size of an optimal solution in ℓ-round PDS (for $\ell \geq 4$) is equal to the size of an optimal solution for PDS, and they are exactly one more than the size of an optimal solution in the MINREP instance (Lemma 1). The number of nodes in the constructed graph is at most $|V(\overline{G})| \leq 4 + |V(G)| + 10k |E(G)|$. This shows that the above reduction is a *gap preserving reduction* from MINREP to PDS and ℓ-round PDS (for $\ell \geq 4$) with the same gap (hardness ratio) as the MINREP problem.

Lemma 1. *(A^*, B^*) is an optimal solution to the instance $G = (A, B, E)$ of the MINREP problem if and only if $\Pi^* = A^* \cup B^* \cup \{w^*\} \subseteq V(\overline{G})$ is an optimal solution to the instance $\overline{G}$ of PDS (and ℓ-round PDS for all $\ell \geq 4$).*

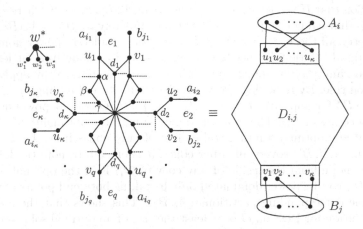

Fig. 3. D_{ij} graph

Proof. The node w^* should be in any optimal solution in order to power dominate w_1^*, w_2^*, w_3^*, since otherwise we need to have at least 2 nodes from $\{w_1^*, w_2^*, w_3^*\}$ to get a feasible solution. By taking w^* all the nodes in $A \cup B$ (and also the nodes inside D_{ij}'s that are the neighbors of w^*) will be power dominated.

First assume that $A^* \cup B^*$ is an optimal solution for the MINREP instance G. Now let us show that $\Pi = A^* \cup B^* \cup \{w^*\}$ is a feasible solution to the PDS instance $\overline{G}$. As described above the nodes in $A \cup B$ and some nodes inside D_{ij}'s are power dominated by w^*. Consider a pair of subsets (A_i, B_j) in G. The set $A^* \cup B^*$ covers all the super edges in $\mathcal{H}$. So there exists an edge $e_q = \{a_{i_q}, b_{j_q}\} \in E(G)$ such that $a_{i_q} \in A^*, b_{j_q} \in B^*$, where it covers the super edge (A_i, B_j). Since a_{i_q} and b_{j_q} are in Π they will power dominates their neighbors, u_q and v_q, in all the $k = 4$ copies of D_{ij} in $\overline{G}$. After u_q and v_q become power dominated, the node d_q will power dominate the center node in D_{ij}. It is easy to check that after the center node become power dominated all of the nodes in D_{ij} will be power dominated (through the gadgets on α, β, γ nodes). This shows that Π is a feasible solution for PDS in $\overline{G}$. Also it is straightforward to check that such a solution will power dominate the entire graph in at most 4 parallel rounds. Therefore, $\mathtt{Opt}(\overline{G}) \leq \mathtt{Opt}_\ell(\overline{G}) \leq |A^* \cup B^*| + 1$. The size of an optimal solution for PDS is a lower bound on the size of an optimal solution for the ℓ-round PDS; this follows from the fact the an optimal solution for ℓ-round PDS is a feasible solution for PDS. So it is enough to prove that the above upper bound is also a lower bound for PDS in $\overline{G}$. And this also proves that the size of an optimal solution for PDS and ℓ-round PDS (for $\ell \geq 4$) is equal on $\overline{G}$.

Let $\Pi^* \subseteq V(\overline{G})$ be an optimal solution for PDS. As we saw above w^* should be in any optimal solution for PDS. Now define $A' = A \cap \Pi^*$ and $B' = B \cap \Pi^*$. First we prove that any optimal solution of PDS only contains nodes from $A \cup B \cup \{w^*\}$, and then we show that (A', B') covers all the super edges. Suppose for the contradiction that Π^* contains some nodes not in $A \cup B \cup \{w^*\}$. So there are some D_{ij} graphs that cannot be power dominated completely by $\Pi^* \cap (A \cup B \cup \{w^*\})$. By symmetry all the $k = 4$ copies of D_{ij} are not completely power dominated. So the optimal solution Π^* needs to have at least one node from at least 3 of the 4 copies, and the remaining one might be power dominated by applying the propagation rule. By removing these 3 nodes from Π^* and adding $a_{i_q} \in A_i$ and $b_{j_q} \in B_j$ to Π^* for some arbitrary $e_q = \{a_{i_q}, b_{j_q}\} \in E(G)$ we can power dominate all of the 4 copies of D_{ij}. This is a contradiction with the optimality of Π^*. This proves that any optimal solution will consist of only nodes from $A \cup B \cup \{w^*\}$. To show that (A', B') covers all super edges, it is enough to note the following: suppose no node from the inside of any copy of D_{ij} is in the optimal solution; then any D_{ij} can be power dominated only by taking both end points of an edge between the corresponding partitions (A_i, B_j). This shows that the size of an optimal solution for PDS on $\overline{G}$ is at least the size of an optimal solution for the MINREP problem on G plus 1. This completes the proof of the lemma. □

3 Approximation Algorithms on Planar Graphs

There has been a lot of research previously on designing *polynomial time approximation schemes*[4] (PTAS) for problems on planar graphs. Some of the most important results are the outerplanar layering technique by Baker [4], and the bidimensionality theory by Demaine and Hajiaghayi [11]. Baker [4] showed that the dominating set problem in planar graphs has a PTAS. Demaine and Hajiaghayi obtained PTASs for some variants of the DOMINATING SET problem on planar graphs (such as connected dominating sets). We have examples showing that none of these successful approaches applies to PDS and ℓ-round PDS.

In the Baker method we first partition the graph into smaller graphs. Then we solve the problem optimally on the subgraphs and finally we return the union of the solutions as a solution for the original graph. Consider the graph shown in Figure 1. The size of an optimal solution for PDS is 1 but if we apply the Baker method the size of the output solution will be at least as large as the number of small graphs in the partition which can be $\Theta(n)$. An important property of bidimensionality is that the size of an optimal solution should not increase when we contract any edge. We have examples showing that this property does not apply to PDS nor to ℓ-round PDS [2].

3.1 PDS on Planar Graphs

In this section we will describe a $(k+1)$-approximation algorithm for PDS in graphs with tree-width k. This will imply that PDS on planar graphs can be approximated within ratio $O(\sqrt{n})$, since the tree-width of a planar graph G with n nodes is $O(\sqrt{n})$ and in $O(n^{\frac{3}{2}})$ time such a tree decomposition can be

---

[4] A *polynomial time approximation scheme* (PTAS) is an algorithm that for any fixed $\epsilon > 0$ can provide a solution with cost within $(1 + \epsilon)$ ratio of the optimal solution in polynomial time.

found [3]. The algorithm makes one bottom-to-top pass over the tree T of the tree decomposition of the graph G (see Definition 1) and constructs a solution Π for PDS (initially, $\Pi = \emptyset$). At each node r_j of T we check whether every solution of PDS that contains Π has at least one node from the bags of subtree rooted at r_j; if yes, then X_{r_j} (the bag corresponding to r_j) is added to Π ($\Pi := \Pi \cup X_{r_j}$), otherwise Π is not updated. The key point in the analysis is to show that $\mathrm{Opt}(G) \geq m$, where m is the number of nodes of T where we updated Π. To get this lower bound, we introduce the notion of strong regions.

Definition 1. *A tree decomposition of a graph $G = (V, E)$ is a pair $\langle \{X_i \subseteq V \mid i \in I\}, T \rangle$, where $T = (I, F)$ is a tree, satisfying the following three properties: (1) $\bigcup_{i \in I} X_i = V$; (2) For all edges $\{u, v\} \in E$ there is an $i \in I$ such that $\{u, v\} \subseteq X_i$; (3) For all $i, j, k \in I$, if j is on the unique path from i to k in T, then we have: $X_i \cap X_k \subseteq X_j$. The width of $\langle \{X_i \mid i \in I\}, T \rangle$ is the $\max_{i \in I} |X_i| - 1$. The tree-width of G is defined as the minimum width over all tree decompositions. The nodes of the tree T are called T-nodes and each X_i is called a* bag.

Let us define some concepts needed for the analysis of our algorithm. Consider the given graph $G = (V, E)$, the neighborhood of $R \subseteq V$ is $nbr(R) = \{v \in V | \exists uv \in E, u \in R, v \notin R\}$, and the exterior of R is defined by $ext(R) = nbr(V \setminus R)$, i.e., $ext(R)$ consists of the nodes in R that are adjacent to a node in $V \setminus R$.

Definition 2. *Given a graph $G = (V, E)$ and a set $\Pi \subseteq V$, the subset $R \subseteq V$ is called Π-strong region if $R \not\subseteq \mathcal{P}(\Pi \cup nbr(R))$, otherwise the set R is called Π-weak region. The region R is called minimal Π-strong if it is a Π-strong region and $\forall r \in R, R - r$ is a Π-weak region.*

It is easy to check from the definition that a Π-strong region is also $\emptyset$-strong (or shortly strong) region. Also it can be checked that the subset $R \subseteq V$ is a Π-strong region iff for every feasible solution $\Pi \cup \Pi^*$ for G, where $\Pi \cap \Pi^* = \emptyset$, we have $R \cap \Pi^* \neq \emptyset$.

Theorem 3. *[2] Given a graph $G = (V, E)$ with its tree decomposition of width k as an input, Algorithm 1 is a $k + 1$-approximation algorithm.*

As we mention earlier, the planar graphs have tree-width $O(\sqrt{n})$, and such a tree-width decomposition can be found in $O(n^{\frac{3}{2}})$ time [3]. This fact together with the above theorem provides a $O(\sqrt{n})$-approximation algorithm for PDS on planar graphs.

Corollary 1. *Algorithm 1 is an $O(\sqrt{n})$-approximation algorithm for the PDS problem on planar graphs.*

The analysis of our algorithm is tight; moreover, we can prove that any approximation algorithm for planar graphs that uses the number of disjoint strong regions as a lower bound has the approximation guarantee of $\Omega(\sqrt{n})$ (the details are given in [2]). This is shown by proving that the $\sqrt{n} \times \sqrt{n}$ grid has no two

---

Algorithm 1. $(k + 1)$-approximation Algorithm

1: Given a tree decomposition $\langle \{X_i | i \in I\}, T \rangle$, take an arbitrary node, r, as a root of T.
2: Let I_ℓ be the set of T-nodes of distance ℓ from the root, and let d be the maximum distance from r.
3: $\Pi \leftarrow \emptyset, \quad a \leftarrow 0$
4: **for** $i = d$ to 0 **do**
5:     Let $I_i = \{r_1, \ldots, r_k\}$ and let T_j be the subtree in T rooted at r_j.
6:     Let Y_j be the set of nodes in G corresponding to the T-nodes in T_j.
7:     **for all** induced subgraph $G_j = G[Y_j]$ **do**
8:         **if** G_j is Π-strong **then**
9:             $\Pi \leftarrow \Pi \cup X_{r_j}, a \leftarrow a + 1, ST_a \leftarrow Y_j \setminus \bigcup_{s=1}^{a-1} ST_s$; where ST_a is the a-th strong region found.
10:         **end if**
11:     **end for**
12: **end for**
13: Output $\Pi_O = \Pi$

---

disjoint strong regions, while having an optimal solution of size $\Theta(\sqrt{n})$. This means that we cannot improve the approximation guarantee without finding a better lower bounding technique.

3.2 ℓ-Round PDS on Planar Graphs

In this section we present a PTAS for the ℓ-round PDS problem on planar graphs when $\ell = O(\frac{\log n}{\log \log n})$. Baker's PTAS [4] for the DOMINATING SET problem on planar graphs is a special case of our result with $\ell = 1$, but there are no previous results of this type for $\ell > 1$. Our PTAS works in the same fashion as Baker's PTAS, but our analysis and proofs are novel contributions of this paper. It is known that PDS remains NP-hard even on planar graphs [19,14]. The same reduction used in [19,14] also proves that ℓ-round PDS is NP-hard on planar graphs [1].

First let us formally define the ℓ-round PDS problem. Given a graph $G = (V, E)$ and a subset of nodes $S \subseteq V$, the set of nodes, $\mathcal{P}^k(S)$, that can be power dominated by applying at most k rounds of parallel propagation is defined recursively as follows:

$$\mathcal{P}^k(S) = \begin{cases} \bigcup_{v \in S} N[v] & k = 1 \\ \mathcal{P}^{k-1}(S) \bigcup \{v : (u, v) \in E, N[u] \setminus \{v\} \subseteq \mathcal{P}^{k-1}(S)\} & k \geq 2 \end{cases}$$

where $N[u]$ is the closed neighborhood of u in G, i.e. $N[u] = \{u\} \cup \{w : \{u, w\} \in E(G)\}$. The goal in the ℓ-round PDS problem is to find a minimum size subset of nodes $S \subseteq V$, such that $\mathcal{P}^\ell(S) = V$. We denote by $\mathtt{Opt}_\ell(G)$ the size of an optimal solution for the ℓ-round PDS problem. For example consider the G graph given in Figure 2; note that the graph has $k \cdot m + 1$ nodes. It is easy to check that the node v can power dominate the entire graph in exactly k parallel rounds; the set X_i will be power dominated at the i-th parallel round.

Hence $\mathtt{Opt}_\ell(G) = 1$ for all $\ell \geq k$. Also it is easy to prove that $\mathtt{Opt}_\ell(G) = m$ for $\ell = k - 1$; we need to take a node other than v from each path P_i to power dominate G in $k - 1$ parallel rounds.

Our PTAS needs to solve instances of the "generalized" ℓ-round PDS problem, where the goal is to power dominate at least all nodes of a given subset $V' \subseteq V$; we denote the instance of generalized ℓ-round PDS of G with respect to the subset V' by $\langle G, V' \rangle$. The following result is proved by designing a dynamic programming algorithm (the details are given in [1]).

Proposition 1. *Given a pair $\langle G, V' \rangle$ where $G = (V, E)$ is a planar graph with tree-width k and $V' \subseteq V$, a minimum size set $S \subseteq V$ such that $V' \subseteq \mathcal{P}^\ell(S)$ can be obtained in time $O(c^{k \log \ell} \cdot |V|)$, for a constant c.*

Now we describe our PTAS for solving ℓ-round PDS on planar graphs for small values of ℓ. First we provide some useful definitions and notations. Consider an embedding of a planar graph G, we define the nodes at level i, L_i, as follows [4]. Let L_1 be the set of nodes on the exterior face in the given embedding of G. For $i > 1$, the set L_i is defined as the set of nodes on the exterior face of the graph induced on $V \setminus \cup_{j=1}^{i-1} L_j$. We denote by $L(a,b) = \cup_{i=a}^{b} L_i$ the set of nodes at levels a through level b. A planar graph is called d-outerplanar if it has an embedding with all nodes of level $\leq d$. Given a graph $G = (V, E)$ and $X \subseteq V$, we denote by $G[X]$ the subgraph induced on the set of nodes X.

Now let us describe our PTAS informally. Consider a parameter k, which is a function of the parameter ℓ and the approximation factor $(1 + \epsilon)$, that will be defined later in the formal description of our algorithm. Consider a fixed embedding of G. We decompose the graph in k different ways, $\mathcal{D}_1, \ldots, \mathcal{D}_k$. In each decomposition the graph is decomposed into blocks of $k + 4\ell - 2$ consecutive levels. The block j in $\mathcal{D}_i$ is defined as $B_{i,j} = L(jk+i-2\ell+1, (j+1)k+i-1+2\ell-1)$. We denote the k middle levels of the block $B_{i,j}$ by $C_{i,j} = L(jk+i, (j+1)k+i-1)$. In our PTAS, for each decomposition $\mathcal{D}_i$, we optimally solve instances of the generalized ℓ-round PDS problem $(\mathcal{I}_{i,j} = \langle G[B_{i,j}], C_{i,j} \rangle)$ defined for each block in $\mathcal{D}_i$. Note that each instance $\mathcal{I}_{i,j}$ can be optimally solved by using the dynamic programming algorithm from Proposition 1. Let $\mathcal{O}_{i,j}$ denote the optimal solution for this instance. Then we take the union of the solutions corresponding to blocks in $\mathcal{D}_i$, $\Pi_i = \cup_{j \geq 0} \mathcal{O}_{i,j}$. By doing this for all k decompositions we get k different feasible solutions for the original graph G. Finally, we chose the solution with minimum size among these k solutions. The $2\ell - 1$ extra levels around the k middle levels plays an important role in the feasibility and the near optimality of the final output of the algorithm.

Theorem 4. *[1] Let ℓ be a given parameter, where $\ell = O(\frac{\log n}{\log \log n})$. Then Algorithm 2 is a PTAS for the ℓ-round PDS problem on planar graphs.*

4   Greedy Algorithm for PDS

In this section we consider some variations of greedy algorithms for PDS and show that they perform very poorly even on planar graphs. In contrast, for other problems such as DOMINATING SET and SET COVERING, greedy algorithms perform

---

Algorithm 2. PTAS for ℓ-round PDS

1: Given a planar embedding of G, and the parameter $0 < \epsilon \leq 1$.
2: Let $k = 4 \cdot \lceil \frac{\ell}{\epsilon} \rceil$.
3: **for** $i = 1$ to k **do**
4:    **for all** $j \geq 0$ **do**
5:       Solve "generalized" ℓ-round PDS on $\langle G[B_{i,j}], C_{i,j} \rangle$
6:       Let $\mathcal{O}_{i,j}$ be an optimal solution for $\langle G[B_{i,j}], C_{i,j} \rangle$
7:    **end for**
8:    $\Pi_i = \cup_{j \geq 0} \mathcal{O}_{i,j}$
9: **end for**
10: $r \leftarrow argmin\{|\Pi_i| : i = 1, \cdots, k\}$
11: Output $\Pi_O = \Pi_r$.

---

well since they achieve a logarithmic approximation guarantee, and no substantial improvement is possible by any polynomial-time algorithm, under complexity assumptions like $\mathsf{P} \neq \mathsf{NP}$. The most natural greedy algorithm is the one that in each step, adds a new node v to the current solution Π such that v power dominates the maximum number of new nodes ($v = argmax\{|\mathcal{P}(\Pi \cup \{v\}) \backslash \mathcal{P}(\Pi)| : v \in V \backslash \Pi\}$). Unfortunately, this greedy algorithm might find a solution of size $\Theta(n)$ on planar graphs that have an optimal solution of size $\Theta(1)$.

Proposition 2. *The greedy algorithm may find a solution Π such that $|\Pi| \geq \Theta(n) \cdot Opt(G)$.*

Proof. Consider a graph G that is obtained from a $9\ell \times 9m$ grid by subdividing all row-edges, except with minor changes in the four corners as shown in the Fig4(a). Partition the graph G into $(\ell \cdot m)$ 9×9 grids (ignoring the nodes introduced by subdivision), Fig. 4(b). It is easy to check that any single node can power dominate at most 7 nodes, and center nodes of any one of the 9×9 grids achieve this maximum. So the greedy algorithm at the first iteration may take the center node of any one of the 9×9 grids. Assuming all nodes taken by the algorithm so far have been these center nodes, we see that taking another center node maximizes $|\mathcal{P}(\Pi \cup \{v\}) - \mathcal{P}(\Pi)|$ over all $v \in V$. So the greedy algorithm could continue taking all center nodes and after that taking possibly other nodes until it finds a feasible solution. Let Π be the output of the greedy algorithm that contains all the center nodes. The size of the output is at least $m \cdot \ell = \Theta(n)$. By taking all of the nodes in the first column it is easy to check that they will power dominate G. Hence $Opt(G) \leq \ell$. Now by fixing $\ell = \Theta(1)$ we get: $|\Pi| \geq \ell \cdot m \geq Opt(G) \cdot m = Opt(G) \cdot \Theta(n)$. $\qquad\square$

We considered two other variations of the greedy algorithm, namely *Proximity* and *Cleanup*, and show that they are not promising [2]. In each step of the *Proximity* algorithm we chose a node such that the set of all power dominated nodes induces a connected subgraph, and subject to this, the number of newly power dominated nodes is maximized. Informally, this is to escalate the use of the propagation rule. In the *Cleanup* algorithm, we first run the greedy algorithm, to find a solution (node set) Π, then we repeatedly remove nodes from Π, until Π is an inclusionwise minimal power dominating set.

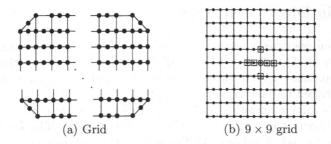

(a) Grid   (b) 9 × 9 grid

Fig. 4. Bad example for the greedy algorithm

5 Integer Programming Formulation for ℓ-Round PDS

In this section we present an integer programming (IP) formulation for the ℓ-round PDS. Then we consider its LP relaxation, and we show that it has integrality ratio of $\Omega(n)$.

Given an undirected graph $G = (V, E)$ and parameter $1 \leq \ell \leq n$, where $n = |V|$. Define the set of parallel rounds $\mathbf{T} = \{1, \ldots, \ell\}$. The variables in the IP formulation are as follows. Let S^* be an optimal solution. We have a binary variable x_v for each node v, where it is equal to 1 if and only if node v is in S^* ($S^* = \{v \in V : x_v = 1\}$). The goal is to minimize $\sum_v x_v$ subjects to the following constraints. For each node v and "parallel" round $t \in \mathbf{T}$ we have a binary variable z_v^t, where $z_v^t = 1$ means that node v is power dominated on (or before) parallel round t. Also we have binary variables $Y_{u \to v}^t$ and $Y_{v \to u}^t$ for each edge $\{u, v\} \in E$ and each parallel round $t \in \mathbf{T}$, where $Y_{u \to v}^t = 1$ means that node u can power dominate node v at the parallel round $t + 1$. We have a set of constraints for the termination condition saying that every node should be power dominated at the end of the last round ($\forall v \in V : 1 \leq z_v^\ell$). The second set of constraints are for power dominating all the nodes in the closed neighborhood of a node that is in the optimal solution ($\forall v \in V : z_v^1 \leq \sum_{u \in N[v]} x_u$). We have another set of constraints for each edge $\{u, v\} \in E$ that checks if the propagation rule (Rule 2) can be applied to u to power dominate v at the next round; node u is ready only if u and all of its neighbors except v are already power dominated ($\forall(u, v) : \{u, v\} \in E, \forall w \in N[u] \setminus \{v\}, \forall t \in \mathbf{T} : Y_{u \to v}^t \leq z_w^t$). The last set of constraints are for checking if node v is power dominated at time t; node v is power dominated only if either it is in the optimal solution or at least one of its neighbors can power dominate v at time $t-1$ ($\forall v \in V, \forall t \in \mathbf{T} \setminus \{1\} : z_v^t \leq \sum_{u \in N(v)} Y_{u \to v}^{t-1} + x_v$). When a node v becomes power dominated, it should stay power dominated. Although it may seem redundant, we need to have the second term, $+x_v$, in the right hand side of the last constraint to have the monotonicity of the z variables. To get an LP relaxation, we relax the variables to be non-negative instead of being binary. We can show that this LP relaxation has integrality ratio of $\Omega(n)$ provided $\ell = \Omega(\log n)$, even for planar graphs (the proof is given in [1]).

6   PDS in Directed Graphs

In this section we extend power domination to directed graphs to get the DI-
RECTED POWER DOMINATING SET (DPDS) problem. We prove that DPDS is
hard to approximate within the same ratio as the undirected case even when the
input is restricted to directed acyclic graphs.

Let G be a directed graph, and $S \subseteq V(G)$. For any node u let $N^+(u)$ denote
the out-neighbors of u, i.e. $N^+(u) := \{w : (u,w) \in E(G)\}$. The *DPDS problem*
has the following set of rules: (D1) The node v is power dominated if $v \in S$ or
$\exists u \in S : v \in N^+(u)$, and (D2) The node $v \in N^+(u)$ is power dominated if u and
all out-neighbors of u except v are power dominated. The set of nodes that can
be power dominated by the above two rules is denoted by $\mathcal{P}(S)$. We say S power
dominates G if $\mathcal{P}(S) = V(G)$. The goal in the DPDS problem is to find a node
set S with minimum size that power dominates all the nodes, i.e. $\mathcal{P}(S) = V$.
Our main results on DPDS are as follows (the proofs are given in [2]).

Theorem 5. *DPDS even when restricted to directed acyclic graphs can-
not be approximated within ratio $2^{\log^{1-\epsilon} n}$, for any $\epsilon > 0$, unless $NP \subseteq DTIME(n^{polylog(n)})$.*

Guo et al. [14] designed a dynamic programming algorithm for optimally solving PDS on graphs with bounded tree-width. The key idea in their dynamic
programming algorithm is a new formulation of PDS in terms of a valid orientation of edges. We extend the valid orientation in [14] to valid colorings for
directed graphs [2]. This gives a reformulation for directed PDS and leads to our
linear-time algorithm for directed PDS when the underlying undirected graph
has bounded tree-width.

Theorem 6. *Given a directed graph G and tree width-k decomposition (not necessarily an optimal decomposition) of its underlying undirected graph, DPDS can
be optimally solved in $O(c^{k^2} \cdot n)$ time for a global constant c.*

References

1. Aazami, A.: Domination in graphs with bounded propagation: algorithms, formulations and hardness results (manuscript, submitted to a journal)
2. Aazami, A., Stilp, M.D.: Approximation algorithms and hardness for domination
   with propagation (manuscript, submitted to a journal)
3. Alber, J., Bodlaender, H.L., Fernau, H., Kloks, T., Niedermeier, R.: Fixed parameter algorithms for dominating set and related problems on planar graphs.
   Algorithmica 33(4), 461–493 (2002)
4. Baker, B.S.: Approximation algorithms for NP-complete problems on planar
   graphs. J. ACM 41(1), 153–180 (1994)
5. Baldwin, T.L., Mili, L., Boisen, M.B.J., Adapa, R.: Power system observability with
   minimal phasor measurement placement. IEEE Trans. Power Syst 8(2), 707–715
   (1993)
6. Bhargava, B.: Synchronized phasor measurement system project at Southern California Edison Co. Power Engineering Society Summer Meeting 1, 16–22 (1999)

7. Bodlaender, H.L., Gilbert, J.R., Hafsteinsson, H., Kloks, T.: Approximating treewidth, pathwidth, frontsize, and shortest elimination tree. J. Algorithms 18(2), 238–255 (1995)
8. Brueni, D.J.: Minimal PMU placement for graph observability, a decomposition approach. Masters thesis, Virginia Polytechnic Institute and State University, Blacksburg, VA (1993)
9. Brueni, D.J., Heath, L.S.: The PMU placement problem. SIAM J. Discret. Math. 19(3), 744–761 (2005)
10. Courcelle, B., Makowsky, J.A., Rotics, U.: Linear time solvable optimization problems on graphs of bounded clique width. In: Workshop on Graph-Theoretic Concepts in Computer Science, pp. 1–16 (1998)
11. Demaine, E.D., Hajiaghayi, M.: Bidimensionality: new connections between FPT algorithms and PTASs. In: SODA, Philadelphia, PA, USA, pp. 590–601 (2005)
12. Feige, U.: A threshold of ln n for approximating set cover. J. ACM 45(4), 634–652 (1998)
13. Garey, M.R., Johnson, D.S.: Computers and Intractability: A Guide to the Theory of NP-Completeness. W. H. Freeman & Co., New York, NY, USA (1979)
14. Guo, J., Niedermeier, R., Raible, D.: Improved algorithms and complexity results for power domination in graphs. In: Liśkiewicz, M., Reischuk, R. (eds.) FCT 2005. LNCS, vol. 3623, pp. 172–184. Springer, Heidelberg (2005)
15. Haynes, T.W., Hedetniemi, S.M., Hedetniemi, S.T., Henning, M.A.: Domination in graphs applied to electric power networks. SIAM J. Discrete Math. 15(4), 519–529 (2002)
16. Haynes, T.W., Hedetniemi, S.T., Slater, P.J.: Domination in graphs. Marcel Dekker (1998)
17. Haynes, T.W., Hedetniemi, S.T., Slater, P.J.: Fundamentals of Domination in Graphs. Marcel Dekker (1998)
18. Johnson, D.S.: Approximation algorithms for combinatorial problems. J. Comput. Syst. Sci. 9(3), 256–278 (1974)
19. Kneis, J., Mölle, D., Richter, S., Rossmanith, P.: Parameterized power domination complexity. Inf. Process. Lett. 98(4), 145–149 (2006)
20. Kortsarz, G., Krauthgamer, R., Lee, J.R.: Hardness of approximation for vertex-connectivity network design problems. SIAM J. Comput. 33(3), 704–720 (2004)
21. Lovász, L.: On the ratio of optimal integral and fractional covers. Discrete Mathematics 13, 383–390 (1975)
22. Lund, C., Yannakakis, M.: On the hardness of approximating minimization problems. J. ACM 41(5), 960–981 (1994)
23. Martin, K.E.: Phasor measurements at the Bonneville Power Administration. Power Systems and Communications Infrastructures for the Future (September 2002)
24. Mili, L., Baldwin, T.L., Phadke, A.: Phasor measurements for voltage and transient stability monitoring and control. In: Proceedings of the EPRI-NSF Workshop on Application of Advanced Mathematics to Power Systems (1991)
25. Nuqui, R.F.: State Estimation and Voltage Security Monitoring Using Synchronized Phasor Measurements. PhD thesis, Virginia Polytechnic Institute and State University(July 2001)
26. Phadke, A.: Synchronized phasor measurements – a historical overview. Transmission and Distribution Conference and Exhibition 1, 476–479 (2002)

A Knapsack Secretary Problem with Applications

Moshe Babaioff[1,*], Nicole Immorlica[2], David Kempe[3,**],
and Robert Kleinberg[4,***]

[1] UC Berkeley School of Information
moshe@ischool.berkeley.edu
[2] Microsoft Research
nickle@microsoft.com
[3] University of Southern California, Department of Computer Science
dkempe@usc.edu
[4] Cornell University, Department of Computer Science
rdk@cs.cornell.edu

Abstract. We consider situations in which a decision-maker with a fixed budget faces a sequence of options, each with a cost and a value, and must select a subset of them online so as to maximize the total value. Such situations arise in many contexts, e.g., hiring workers, scheduling jobs, and bidding in sponsored search auctions.

This problem, often called the *online knapsack problem*, is known to be inapproximable. Therefore, we make the enabling assumption that elements arrive in a *random* order. Hence our problem can be thought of as a weighted version of the classical *secretary problem*, which we call the *knapsack secretary problem*. Using the random-order assumption, we design a constant-competitive algorithm for arbitrary weights and values, as well as a e-competitive algorithm for the special case when all weights are equal (i.e., the *multiple-choice secretary problem*). In contrast to previous work on online knapsack problems, we do not assume any knowledge regarding the distribution of weights and values beyond the fact that the order is random.

1 Introduction

Allocation of resources under uncertainty is a very common problem in many real-world scenarios. Employers have to decide whether or not to hire candidates, not knowing whether future candidates will be stronger or more desirable. Machines need to decide whether to accept jobs without knowledge of the importance or profitability of future jobs. Consulting companies must decide which

---

* Supported by NSF ITR Award ANI-0331659.
** Work supported in part by NSF CAREER Award 0545855.
*** Partially supported by an NSF Mathematical Sciences Postdoctoral Research Fellowship. Portions of this work were completed while the author was a postdoctoral fellow at UC Berkeley.

M. Charikar et al. (Eds.): APPROX and RANDOM 2007, LNCS 4627, pp. 16–28, 2007.

jobs to take on, not knowing the revenue and resources associated with potential future requests.

More recently, online auctions have proved to be a very important resource allocation problem. Advertising auctions in particular provide the main source of monetization for a variety of Internet services including search engines, blogs, and social networking sites. Additionally, they are the main source of customer acquisition for a wide array of small online businesses, the so-called "mom and pop shops" of the networked world. In bidding for the right to appear on a web page (such as a search engine), advertisers have to trade off between large numbers of parameters, including keywords and viewer attributes. In this scenario, an advertiser may be able to estimate accurately the bid required to win a particular auction, and the benefit either in direct revenue or name recognition to be gained, but may not know about the trade off for future auctions.

All of these problems involve an online scenario, wherein an algorithm has to make decisions on whether to accept an offer (such as a candidate, job, or a bidding opportunity), based solely on the required resource investment (or *weight*) w and projected *value* v of the current offer, without knowledge of the weights or values of future offers. The total weight of all selected offers may not exceed a given budget W. Thus, the problem we are concerned with is an online knapsack problem. In general, this problem does not permit any good competitive ratio, as evidenced by trivial bad examples. Instead, we focus on the case where the offers arrive in a uniformly *random* order.

Summary of Results: In this model, we prove two results: for the case of general weights and values, we give a constant-competitive online algorithm (specifically, it is $10e$-competitive). For the special case where all the weights are uniform, and the weight constraint thus poses a constraint on the total *number* of offers that can be accepted, we improve the approximation factor to e, via two simple and natural algorithms.

Secretary Problems: When the weights are uniform and equal to the weight constraint, our problem reduces to the famous *secretary problem*, or the problem of selecting online an element of maximum value in a randomly-ordered sequence. This problem was first introduced by Dynkin [9] in 1963. His paper gives an algorithm which selects the maximum value element with probability that tends to $1/e$ as n tends to infinity and hence is e-competitive. Many generalizations of this problem have been studied in the literature. In one natural generalization, Kleinberg [11] considers the multiple-choice secretary problem in which k elements need to be selected and the goal is to maximize the combined value (sum) of the selected elements. Kleinberg presents an asymptotically optimal $1/(1 - 5/\sqrt{k})$-competitive algorithm for this problem. Another closely related generalization considered in the literature is the *matroid secretary problem*, introduced by Babaioff et al. [2], in which the elements of a weighted matroid arrive in a random order. As each element is observed, the algorithm makes an irrevocable decision to choose it or skip it, with the constraint that the chosen elements must constitute an independent set. Again, the objective is to maximize the combined weight of the chosen elements. Babaioff et al. give an $O(\log k)$-competitive algorithm for the matroid secretary problem, where k is the rank of the matroid, as well as constant-competitive algorithms for several specific matroids.

In this paper, we study both the multiple-choice secretary problem and a weighted generalization, which we call the *knapsack secretary problem*. The multiple-choice secretary problem is a special case of the matroid secretary problem (for the truncated uniform matroid). We show how to apply an intuitive algorithmic idea proposed by Babaioff et al. [2] to get a e-competitive algorithm for this problem for any k. Hence, our result improves upon the competitive ratio of the algorithm by Kleinberg [11] for small k and is significantly simpler. The knapsack secretary problem, on the other hand, can not be interpreted as a matroid secretary problem, and hence none of the previous results apply. In this paper, we give the first constant-competitive algorithm for this problem, using intuition from the standard 2-approximation algorithm for the offline knapsack problem.

Knapsack Problems: Our work builds upon the literature for knapsack problems. It is well known that the NP-complete (offline) knapsack problem admits an FPTAS as well as a simple 2-approximation, whereas the online knapsack problem is inapproximable to within any non-trivial multiplicative factor. Assuming that the density (value to weight ratio) of every element is in a known range $[L, U]$, and that each weight is much smaller than the capacity of the knapsack (or that the packing is allowed to be fractional), Buchbinder and Naor [4,5] give an algorithm with a multiplicative competitive ratio of $O(\log(U/L))$ for *online knapsack* based on a general online primal-dual framework. They also show an $\Omega(\log(U/L))$ lower bound on the competitive ratio of any algorithm under such assumptions.

Several papers have also considered a *stochastic online knapsack* problem [12,13] in which the value and/or weight of elements are drawn according to a known distribution. These papers provide algorithms with an additive approximation ratio of $\Theta(\log n)$ as well as showing that no online algorithm can achieve a constant additive approximation. Dean et al. [7,8] consider a *stochastic offline knapsack* problem where the algorithm knows the values and the distribution of the weights of the elements. They present an involved way for choosing the order of the elements so as to achieve a constant-competitive outcome in the multiplicative sense. The main difficulty in their model is that the weight of an element is not revealed until it is actually selected.

Our results show that a constant-competitive algorithm exists for any sequence *when elements arrive in a random order*. The random order assumption allows us to eliminate all assumptions from previous papers, e.g., that elements have small weights [4,5], and densities are bounded [4,5] or drawn according to a known distribution [7,8,12,13].[1] In return, we are able to design a constant-competitive online algorithm for our setting. In contrast, for the online setting

---

[1] In contrast to the Dean et al. [7,8] models, our model and the others mentioned make the stronger assumption that the weights of elements are learned *before* deciding whether or not to select them.

of Buchbinder and Naor, there is a super-constant lower bound of $\Omega(\ln(U/L))$ for a worst-case order of arrivals [4,5].

Sponsored Search: Several recent papers have considered applications of the knapsack problem to auction design. Aggarwal and Hartline [1] design truthful auctions which are revenue competitive when the auctioneer is constrained to choose agents with private values and publicly known weights that fit into a knapsack. Knapsack algorithms have also been used to design bidding strategies for budget-constrained advertisers in sponsored search auctions. That the bidding problem in such settings is similar to knapsack was first noted by Borgs et al. [3] (who considered using knapsack to model slot selection) and Rusmevichientong and Williamson [16] (who considered using stochastic knapsack to model keyword selection). The bidding problem was further studied in papers by Feldman et al. [10] and Muthukrishnan et al. [15] which consider the problem of slot selection in more complicated settings, including interactions between keywords and stochastic information. All these papers assume that the set of keywords and distributional information regarding values and weights are known upfront by the algorithm; hence the algorithms they develop are inspired by offline knapsack problems. Recently, Chakrabarty et al. [6] modeled the bidding problem using online knapsack. Under the same assumptions as the paper of Buchbinder and Naor [4,5] mentioned above, Chakrabarty et al. design a $(\ln(U/L) + 1)$-competitive online algorithm for a worst case sequence of keywords.

Outline of paper: In Section 2, we introduce a formal model for the knapsack secretary problem. We then give a pair of e-competitive algorithms for the unweighted knapsack secretary problem in Section 3. Finally, in Section 4, we design a constant-competitive algorithm for the general case.

2   Model

In formalizing the resource allocation problem, we will adopt the terminology of the *secretary problem*, and think of our problem as a *weighted secretary problem*. A set $U = \{1, \ldots, n\}$ of n elements or *secretaries* each have non-negative *weight* $w(i)$ and *value* $v(i)$. We extend the notation to sets by writing $w(S) := \sum_{i \in S} w(i)$ and $v(S) := \sum_{i \in S} v(i)$.

The algorithm will be given a *weight bound* W, and must select, in an online fashion, a set $S \subseteq U$ of secretaries (approximately) solving the following knapsack problem:

$$\text{Maximize} \sum_{i \in S} v(i) \quad \text{subject to} \quad \sum_{i \in S} w(i) \leq W. \tag{1}$$

We assume that the secretaries in U are presented to the algorithm in a uniformly random order. In order to be able to number the elements by their arrival order, we assume that the actual weights and values are obtained as $v = v_0 \circ \pi$, $w = w_0 \circ \pi$, where π is a uniformly random permutation of n elements, and v_0, w_0 are arbitrary initial weight and value functions. For simplicity, we also assume that no two secretaries have the same values $v(i), v(j)$. This is easy to ensure, by fixing a random (but consistent) tie-breaking between elements of the same value, based for instance on the identifier of the element.[2]

The algorithm is online in the following sense: initially, the algorithm knows only n, the total number of secretaries, but knows nothing about the distribution of weights or values. Whenever a secretary i arrives, the algorithm learns its weight $w(i)$ and value $v(i)$. It must then irrevocably decide whether to *select i* or *pass*: a selected secretary cannot later be discarded, nor can a passed secretary be added. Thus, the algorithm maintains a set S of currently selected secretaries, which grows over the course of the execution, but must always satisfy $w(S) \leq W$.

Clearly, this setting does not permit the design of an optimal algorithm. Hence, we look for algorithms which are constant-competitive in that the expected value of the selected set S is within a constant of the optimum value. More precisely, we say an algorithm is α-competitive for the weighted secretary problem if for any initial weight and value functions v_0, w_0

$$\alpha \cdot \mathrm{E}\left[v(S)\right] \geq v(S^*),$$

where S^* is the optimal solution to Program 1 and the expectation is over all permutations π.

Note that this is a generalization of the classical secretary problem of Dynkin [9]. In the classical secretary problem, all weights are one (i.e., $w(i) = 1$ for all i) and the weight bound W is also one; thus, the algorithm is to select exactly one secretary. Dynkin gives a e-competitive algorithm for this special case. Our formulation can also be used to capture the k-secretary problem by setting all weights equal to one and the weight bound W equal to k. This case has been studied by Kleinberg [11], who gave a $1/(1 - 5/\sqrt{k})$-competitive algorithm.

In the following sections, we first present two algorithms for the k-secretary problem. Our algorithms are simpler than those of Kleinberg and show that there is a e-competitive algorithm for *all* k (Kleinberg's result is strictly worse than e for small k). We then present a constant-competitive algorithm for the general case of weighted secretaries, although the constant is worse than that of k-secretaries.

3   The Unweighted Case

In this section we present two simple algorithms for the unweighted case (i.e., the multiple-choice secretary problem), in which all weights $w(i)$ are equal to 1 and the knapsack capacity W is equal to k. Both algorithms will achieve a competitive guarantee no worse than e. While the second algorithm, called the "optimistic algorithm" is perhaps more natural (and our analysis is almost certainly not tight), the first algorithm, called the "virtual algorithm", has a significantly simpler analysis, yielding essentially a tight bound on its performance.

Both algorithms are based on the same idea of a sampling period of $t \in \{k + 1, \ldots, n\}$ steps (during which the algorithm passes on all candidates), followed

---

[2] Note that such a tie-breaking can be accomplished in polynomial time.

by hiring some of the secretaries for the remaining $n - t$ steps. We call t the *threshold time* of the algorithms, and denote the set of sampled elements by T. We leave t unspecified for now; after analyzing the algorithm, we will specify the optimal value of t, which will be approximately n/e.

Both algorithms use the first t time steps to assemble a *reference set* R, consisting of the k elements with the largest $v(i)$ values seen during the first t steps. These elements are kept for comparison, but *not selected*. Subsequently, when an element $i > t$ with value $v(i)$ is observed, a decision of whether to select i into the set S is made based on $v(i)$ and R, and the set R is possibly updated. At any given time, let $j_1, j_2, \ldots, j_{|R|}$ be the elements of R, sorted by decreasing $v(j_i)$.

Virtual: In the virtual algorithm, i is selected if and only if $v(i) > v(j_k)$, and $j_k \le t$ (j_k is in the sample). In addition, whenever $v(i) > v(j_k)$ (regardless of whether $j_k \le t$), element i is added to R, while element j_k is removed from R.

  Thus, R will always contain the best k elements seen so far (in particular, $|R| = k$), and i is selected if and only if its value exceeds that of the kth best element seen so far, and the kth best element was seen during the sampling period.

Optimistic: In the optimistic algorithm, i is selected if and only if $v(i) > v(j_{|R|})$. Whenever i is selected, $j_{|R|}$ is removed from the set R, but no new elements are ever added to R. Thus, intuitively, elements are selected when they beat one of the remaining reference points from R.

  We call this algorithm "optimistic" because it removes the reference point $j_{|R|}$ even if $v(i)$ exceeds, say, $v(j_1)$. Thus, it implicitly assumes that it will see additional very valuable elements in the future, which will be added when their values exceed those of the remaining, more valuable, j_i.

We first observe that neither algorithm ever selects more than k secretaries. Each selection involves the removal of a sample $j_i \in R \cap T$ from R, and no elements from T are ever added to R by either algorithm after time t. Since R starts with only k samples, no more than k elements can be selected.

Next, we prove that both the virtual and the optimistic Algorithm are e-competitive, if $t = \lfloor n/e \rfloor$ elements are sampled.

Theorem 1. *The competitive ratio of both the Virtual and the Optimistic Algorithm approaches e as n tends to infinity, when the algorithms sample $t = \lfloor n/e \rfloor$ elements.*

The proof of the theorem for both algorithms follows from stronger lemmas, establishing that *each* of the top k elements is selected with probability at least $1/e$. Specifically, let $v_1^*, v_2^*, \ldots, v_k^*$ denote the k largest elements of the set $\{v(1), v(2), \ldots, v(n)\}$, and for $a = 1, 2, \ldots, k$ let $i_a^* = v^{-1}(v_a^*)$ be the index in the sequence $v(i)$ at which v_a^* appeared. We will then establish the following lemmas:

Lemma 1. *For all $a \leq k$, the probability that the virtual algorithm selects element v_a^* is*
$$\text{Prob}[i_a^* \in S] \geq \tfrac{t}{n} \ln(n/t).$$

Lemma 2. *For all $a \leq k$, the probability that the optimistic algorithm selects element v_a^* is*
$$\text{Prob}[i_a^* \in S] \geq \tfrac{t}{n} \ln(n/t).$$

Proof of Theorem 1. The theorem follows immediately from these two lemmas, as the expected gain of the algorithm is
$$\text{E}\left[v(S)\right] \geq \textstyle\sum_{a=1}^{k} \text{Prob}[i_a^* \in S] \cdot v_a^* > \tfrac{t}{n} \ln(n/t) \cdot v(S^*).$$

$\tfrac{t}{n} \ln(n/t)$ is maximized for $t = n/e$, and setting $t = \lfloor n/e \rfloor$ gives us that $t/n \to 1/e$ as $n \to \infty$. Thus, the algorithms' competitive ratios approach e as n tends to infinity. ∎

The proof of Lemma 1 turns out to be surprisingly simple and elegant, while the proof of Lemma 2 for the optimistic algorithm is significantly more complex, and will be given in the full version of this paper.

Proof of Lemma 1. If v_a^* is observed at time $i_a^* = i > t$, it will be selected if and only if the kth smallest element of R at that time was sampled at or before time t. Because the permutation is uniformly random, this happens with probability $t/(i-1)$. Each i is equally likely to be the time at which v_a^* is observed, so the probability of selecting v_a^* is
$$\text{Prob}[i_a^* \in S] = \textstyle\sum_{i=t+1}^{n} \tfrac{1}{n} \cdot \tfrac{t}{i-1} = \tfrac{t}{n} \sum_{i=t+1}^{n} \tfrac{1}{i-1} > \tfrac{t}{n} \int_t^n \tfrac{dx}{x} = \tfrac{t}{n} \ln\left(\tfrac{n}{t}\right). \qquad ∎$$

Notice that the proof of Lemma 1 is essentially tight. Each of the top k elements is selected with probability approaching $1/e$ in the limit for our choice of t.

4   The Weighted Case

In this section, we present an algorithm for the weighted case, with a competitive ratio of $10e$. The algorithm is based on the familiar paradigm of sampling a constant fraction of the input and using the sample to define a selection criterion which is then applied to the subsequent elements observed by the algorithm. One complication which arises in designing algorithms for the weighted case is the need to address at least two cases: either there is a single element (or, more generally, a bounded number of elements) whose value constitutes a constant fraction of the optimal knapsack solution, or there is no such element[3]. In the former case(s), we use a selection criterion based on the values of elements but ignoring their sizes. In the latter case, we use a selection criterion based on the *value density*, i.e., the ratio of value to weight. To incorporate both cases, we randomize the selection criterion.

---

[3] This type of case analysis is reminiscent of the case analysis which underlies the design of polynomial-time approximation schemes for the offline version of the knapsack problem.

4.1 Notation

For $i \in U$, we define the *value density* (or simply "density") of i to be the ratio

$$\rho(i) = \frac{v(i)}{w(i)}.$$

We will assume throughout this section that distinct elements of U have distinct densities; this assumption is justified for the same reason our assumption of distinct values is justified. (See Section 2.) If $Q \subseteq U$ and $x > 0$, it will be useful to define the "optimum fractional packing of elements of Q into a knapsack of size x." This is defined to be a vector of weights $(y_Q^{(x)}(i))_{i=1}^n$ which is a solution of the following linear program (that is, $y_Q^{(x)}(i) = y(i)$).

$$\begin{aligned}
\max \quad & \sum_{i=1}^n v(i)y(i) \\
\text{s.t.} \quad & \sum_{i=1}^n w(i)y(i) \leq x \\
& y(i) = 0 \qquad \forall i \notin Q \\
& y(i) \in [0,1] \quad \forall i.
\end{aligned} \tag{2}$$

The reader may verify the following easy fact about $y_Q^{(x)}(i)$: there exists a *threshold density* $\rho_Q^{(x)}$ such that $y_Q^{(x)}(i) = 1$ for all $i \in Q$ such that $\rho(i) > \rho_Q^{(x)}$ and $y_Q^{(x)}(i) = 0$ for all $i \in Q$ such that $\rho(i) < \rho_Q^{(x)}$. Finally, for a set $R \subseteq U$ we will define $v_Q^{(x)}(R), w_Q^{(x)}(R)$ by

$$v_Q^{(x)}(R) = \sum_{i \in R} v(i)y_Q^{(x)}(i)$$

$$w_Q^{(x)}(R) = \sum_{i \in R} w(i)y_Q^{(x)}(i).$$

4.2 The Algorithm

For convenience, we assume in this section that $W = 1$. (To reduce from the general case to the $W = 1$ case, simply rescale the weight of each element by a factor of $1/W$.) Our algorithm begins by sampling a random number $a \in \{0,1,2,3,4\}$ from the uniform distribution. The case $a = 4$ is a special case which will be treated in the following paragraph. If $0 \leq a \leq 3$, then the algorithm sets $k = 3^a$ and runs the k-secretary algorithm from Section 3 (with $t = \lfloor n/e \rfloor$) to select at most k elements. If the k-secretary algorithm selects an element i whose weight $w(i)$ is greater than $1/k$, we override this decision and do not select the element.

If $a = 4$, our algorithm operates as follows. It samples a random $t \in \{1, 2, \ldots, n\}$ from the binomial distribution $B(n, 1/2)$, i.e. the distribution of the number of heads observed when a fair coin is tossed n times. Let $X = \{1, 2, \ldots, t\}$ and $Y = \{t+1, t+2, \ldots, n\}$. For every element $i \in X$, the algorithm observes $v(i)$ and $w(i)$ but does not select i. It then sets $\hat{\rho} = \rho_X^{(1/2)}$ and selects every element $i \in Y$ which satisfies $w(i) \leq 3^{-4}$, $\rho(i) \geq \hat{\rho}$, and $w(S_{<i} \cup \{i\}) \leq 1$, where $S_{<i}$ denotes the set of elements which were already selected by the algorithm before observing i.

4.3   Analysis of the Algorithm

Theorem 2. *The algorithm in Section 4.2 is (10e)-competitive.*

Proof. Let $\mathsf{OPT} \subseteq U$ denote the maximum-value knapsack solution, and suppose that $i_1, i_2, \ldots, i_m$ are the elements of OPT arranged in decreasing order of weight. Partition OPT into five sets $B_0, B_1, \ldots, B_4$. For $0 \le j \le 3$,

$$B_j = \{i_\ell \mid 3^j \le \ell < 3^{j+1}\},$$

while for $j = 4$, $B_4 = \{i_{81}, i_{82}, \ldots, i_m\}$. Let $b_j = v(B_j)$ for $0 \le j \le 4$.

Let S denote the set of elements selected by the algorithm. For $0 \le j \le 4$, define

$$g_j = E\left[v(S) \mid a = j\right]$$

where a denotes the random element of $\{0, 1, 2, 3, 4\}$ sampled in the first step of the algorithm. In Lemmas 3 and 4 below, we prove that $b_j \le 2eg_j$ for $0 \le j \le 4$. Summing over j, we obtain:

$$
\begin{aligned}
v(\mathsf{OPT}) &= b_0 + b_1 + b_2 + b_3 + b_4 \\
&\le 2e(g_0 + g_1 + g_2 + g_3 + g_4) \\
&= (10e) \sum_{j=0}^{4} \mathrm{Prob}[a = j]g_j \\
&= 10e\, E[v(S)].
\end{aligned}
$$

This establishes the theorem.   □

Lemma 3. *For $0 \le j \le 3$, $b_j \le 2eg_j$.*

Proof. Let $k = 3^j$. Recall that every element $i \in B_j$ appears in at least the kth position on a list of elements of OPT arranged in decreasing order of weight. Since the sum of the weights of all elements of OPT is at most 1, we have that $w(i) \le 1/k$ for every $i \in B_j$. Let $Q = \{i \in U \mid w(i) \le 1/k\}$, and let R be the maximum-value k-element subset of Q. Since $B_j \subseteq Q$ and $|B_j| \le 2k$, we have $v(B_j) \le 2v(R)$. On the other hand, Theorem 1 implies that $g_j \ge v(R)/e$. The lemma follows by combining these two bounds.   □

Lemma 4. $b_4 \le 2eg_4$.

Proof. Assuming the algorithm chooses $a = 4$, recall that it splits the input into a "sample set" $X = \{1, 2, \ldots, t\}$ and its complement $Y = \{t+1, \ldots, n\}$, where t is a random sample from the binomial distribution $B(n, 1/2)$. Recall that in the case $a = 4$, the algorithm aims to fill the knapsack with multiple items of weight at most $1/81$, and value density at least equal to the value density of the optimal solution for the sample (and a knapsack of size $1/2$). Thus, let $Q \subseteq U$ consist of all elements $i \in U$ such that $w_0(i) \le 1/81$. We will show that with sufficiently high constant probability, the algorithm obtains a "representative" sample, in the sense that the optimal value density estimated from X is bounded from above and below in terms of the optimal value density for all of Q (with different knapsack sizes). This in turn will imply that each element of Q is picked by the algorithm with constant probability, more specifically, with probability at least 0.3.

To obtain sufficiently high probability, we rely on the independence of membership in X between elements, which in turn allows us to apply Chernoff Bounds. Recall that we encoded the random ordering of the input by assuming that there exists a fixed pair of functions v_0, w_0 and a uniformly random permutation π on U, such that $v = v_0 \circ \pi$, $w = w_0 \circ \pi$. This implies that, conditional on the value of t, $\pi^{-1}(X)$ is a uniformly-random t-element subset of U. Since t itself has the same distribution as the cardinality of a uniformly-random subset of U, it follows that $\pi^{-1}(X)$ is a uniformly-random subset of U. For each $i \in U$, if we define

$$\zeta_i = \begin{cases} 1 \text{ if } \pi(i) \in X \\ 0 \text{ otherwise,} \end{cases}$$

then the random variables ζ_i are mutually independent, each uniformly distributed in $\{0, 1\}$.

Since $B_4 \subseteq Q$ and $w(B_4) \leq 1$,

$$b_4 \leq v_Q^{(1)}(Q) \leq \frac{4}{3} v_Q^{(3/4)}(Q). \tag{3}$$

For every j such that $y_{\pi(Q)}^{(3/4)}(\pi(j)) > 0$ we will prove that $\text{Prob}[\pi(j) \in S \mid a = 4] > 0.3$. This implies the first inequality in the following line, whose remaining steps are clear from the definitions.

$$v_{\pi(Q)}^{(3/4)}(\pi(Q)) < E\left[\tfrac{10}{3} v_{\pi(Q)}^{(3/4)}(S) \mid a = 4\right] \leq \tfrac{10}{3} E[v(S) \mid a = 4] = \tfrac{10}{3} g_4. \tag{4}$$

Combining (3) and (4) we will have derived $b_4 \leq (40/9)g_4 < 2eg_4$, thus establishing the lemma.

Note that for all $i \in U$, $x > 0$, the number $y_{\pi(Q)}^{(x)}(\pi(i))$ does not depend on the random permutation π, since it is the i-th component of the solution of linear program (2) with v_0 and w_0 in place of v and w, and the solution to the linear program does not depend on π. We will use the notation $y(i, x)$ as shorthand for $y_{\pi(Q)}^{(x)}(\pi(i))$. Fix any $j \in Q$. We will show that j will be picked by the algorithm with probability at least 0.3. To prove this, we will upper and lower bound the total weight of $\pi(Q)$ (scaled by the fractional solutions for knapsacks of different sizes) seen in X and Y. This will allow us to reason that j will have density exceeding $\hat{\rho}$, and there will still be room in S by the time j is encountered.

We will reason about the expected fractional weight of items other than j in X in a knapsack of size $3/4$, and of items other than j in Y in a knapsack of size $3/2$. Formally, we define the random variables

$$Z_1 = w_{\pi(Q)}^{(3/4)}(X \setminus \{\pi(j)\}) = \sum_{i \in Q \setminus \{j\}} w_0(i) y(i, 3/4) \zeta_i \tag{5}$$

$$Z_2 = w_{\pi(Q)}^{(3/2)}(Y \setminus \{\pi(j)\}) = \sum_{i \in Q \setminus \{j\}} w_0(i) y(i, 3/2)(1 - \zeta_i) \tag{6}$$

Since Z_1, Z_2 are sums of independent random variables taking values in the interval $[0, 1/81]$, we can use the following form of the Chernoff bound, obtained from standard forms [14] by simple scaling: If $z_1, z_2, \ldots, z_n$ are independent random variables taking values in an interval $[0, z_{\max}]$ and if $Z = \sum_{i=1}^n z_i$, $\mu = E[Z]$, then for all $\delta > 0$,

$$\text{Prob}[Z \geq (1 + \delta)\mu] < \exp\left(-\frac{\mu}{z_{\max}}[(1 + \delta)\ln(1 + \delta) - \delta]\right).$$

Because the expectations of Z_1 and Z_2 are

$$E[Z_1] = \tfrac{1}{2} w_{\pi(Q)}^{(3/4)}(\pi(Q) \setminus \{\pi(j)\}) = \tfrac{1}{2}\left(\tfrac{3}{4} - w_0(j)y(j, 3/4)\right) \in \left[\tfrac{3}{8} - \tfrac{1}{162}, \tfrac{3}{8}\right],$$
$$E[Z_2] = \tfrac{1}{2} w_{\pi(Q)}^{(3/2)}(\pi(Q) \setminus \{\pi(j)\}) = \tfrac{1}{2}\left(\tfrac{3}{2} - w_0(j)y(j, 3/2)\right) \in \left[\tfrac{3}{4} - \tfrac{1}{162}, \tfrac{3}{4}\right],$$

applying the Chernoff Bound to Z_1 and Z_2 with $z_{\max} = 1/81$, $\delta = \tfrac{1}{3} - \tfrac{8}{243}$ yields $\text{Prob}[Z_1 \geq 1/2 - 1/81] < 0.3$ and $\text{Prob}[Z_2 \geq 1 - 2/81] < 0.1$.

Let $\mathcal{E}$ denote the event that $Z_1 < \tfrac{1}{2} - \tfrac{1}{81}$ and $Z_2 < 1 - \tfrac{2}{81}$. By a union bound, $\text{Prob}[\mathcal{E} \mid a = 4] > 0.6$. Conditional on the event $\mathcal{E}$ (and on the event that $a = 4$), the element $\pi(j)$ can add no more than $1/81$ to the weight of X or Y (whichever one it belongs to). Hence, $w_{\pi(Q)}^{(3/4)}(X) < 1/2$ and $w_{\pi(Q)}^{(3/2)}(Y) < 1 - \tfrac{1}{81}$, which in turn implies $w_{\pi(Q)}^{(3/2)}(X) > 1/2 > w_{\pi(Q)}^{(3/4)}(X)$, since every element of $\pi(Q)$ belongs to either X or Y and $w_{\pi(Q)}^{(3/2)}(\pi(Q)) = 3/2$. Because the threshold density for a fractionally packed knapsack with larger capacity cannot be larger than for a knapsack with smaller capacity, the above bounds on the weight imply that

$$\rho_{\pi(Q)}^{(3/4)} \geq \rho_X^{(1/2)} \geq \rho_{\pi(Q)}^{(3/2)}. \tag{7}$$

Let S^+ denote the set of all elements of $Y \setminus \{\pi(j)\}$ whose value density is greater than or equal to $\hat{\rho} = \rho_X^{(1/2)}$. (Note that the algorithm will pick every element of S^+ that it sees until it runs out of capacity, and it will not pick any element which does not belong to S^+ except possibly $\pi(j)$.) We claim that the combined size of the elements of S^+ is at most $1 - \tfrac{1}{81}$. This can be seen from the fact that for all but at most one $i \in S^+$, the coefficient $y_{\pi(Q)}^{(3/2)}(i)$ is equal to 1. Hence the combined size of all the elements of S^+ is bounded above by

$$\tfrac{1}{81} + w_{\pi(Q)}^{(3/2)}(Y \setminus \{\pi(j)\}) = \tfrac{1}{81} + Z_2 < 1 - \tfrac{1}{81},$$

from which it follows that the algorithm does not run out of room in its knapsack before encountering $\pi(j)$. If $y(j, 3/4) > 0$, then $\rho(\pi(j)) \geq \rho_{\pi(Q)}^{(3/4)}$ and (7) implies that $\rho(\pi(j)) \geq \hat{\rho}$. Thus, the algorithm will select $\pi(j)$ if $\pi(j) \in Y$. Finally, note that the event $\pi(j) \in Y$ is independent of $\mathcal{E}$, so

$$\text{Prob}[\pi(j) \in S \mid \mathcal{E} \wedge (a = 4)] = \text{Prob}[\pi(j) \in Y \mid \mathcal{E} \wedge (a = 4)] = \tfrac{1}{2}.$$

Combining this with the bound $\text{Prob}[\mathcal{E} \mid a = 4] > 0.6$ established earlier, we obtain

$$\text{Prob}[\pi(j) \in S \mid a = 4] > 0.3,$$

which completes the proof of the lemma. $\square$

5   Conclusion

In this paper, we have presented algorithms for a knapsack version of the secretary problem, in which an algorithm has to select, in an online fashion, a maximum-value subset from among the randomly ordered items of a knapsack problem. We gave a constant-competitive algorithm in this model, as well as a e-approximation for the k-secretary problem, in which all items have identical weights.

The competitive ratios we obtain are certainly not tight, and it appears that the analysis for the "optimistic algorithm" is not tight, either. Determining the exact competitive ratio for this algorithm, as well as improving the algorithm for the knapsack problem, are appealing directions for future work.

Furthermore, many natural variants of the secretary problem remain to be studied. How general a class of set systems admits a constant-factor (or even a e) approximation in the random ordering model? An appealing conjecture of Babaioff et al. [2] states that a e approximation should be possible for all matroids. We have shown that there is an interesting class of non-matroid domains - knapsack secretary problems - that admits a constant-factor approximation. Are there other natural classes of non-matroid domains that admit a constant-factor approximation?

An interesting question is how the random ordering model relates with other models of stochastic optimization. In particular, the "sample-and-optimize" approach taken in all algorithms in this paper bears superficial similarity to the standard techniques in multi-stage stochastic optimization. It would be interesting to formalize this similarity, and perhaps derive new insights into both classes of problems.

References

1. Aggarwal, G., Hartline, J.: Knapsack auctions. In: SODA, pp. 1083–1092 (2006)
2. Babaioff, M., Immorlica, N., Kleinberg, R.: Matroids, secretary problems, and online mechanisms. In: SODA, pp. 434–443 (2007)
3. Borgs, C., Chayes, J., Etesami, O., Immorlica, N., Jain, K., Mahdian, M.: Dynamics of bid optimization in online advertisement auctions. In: Proceedings of the 16th International World Wide Web Conference, (to appear 2007)

4. Buchbinder, N., Naor, J.: Online primal-dual algorithms for covering and packing problems. In: Brodal, G.S., Leonardi, S. (eds.) ESA 2005. LNCS, vol. 3669, Springer, Heidelberg (2005)
5. Buchbinder, N., Naor, J.: Improved bounds for online routing and packing via a primal-dual approach. In: FOCS (2006)
6. Chakrabarty, D., Zhou, Y., Lukose, R.: Budget constrained bidding in keyword auctions and online knapsack problems. In: WWW2007, Workshop on Sponsored Search Auctions (2007)
7. Dean, B., Goemans, M., Vondrák, J.: Approximating the stochastic knapsack problem: The benefit of adaptivity. In: FOCS, pp. 208–217 (2004)
8. Dean, B., Goemans, M., Vondrák, J.: Adaptivity and approximation for stochastic packing problems. In: SODA, pp. 395–404 (2005)
9. Dynkin, E.B.: The optimum choice of the instant for stopping a Markov process. Sov. Math. Dokl. 4 (1963)
10. Feldman, J., Muthukrishnan, S., Pal, M., Stein, C.: Budget optimization in search-based advertising auctions. In: Proceedings of the 8th ACM Conference on Electronic Commerce, (to appear)
11. Kleinberg, R.: A multiple-choice secretary problem with applications to online auctions. In: SODA, pp. 630–631 (2005)
12. Lueker, G.: Average-case analysis of off-line and on-line knapsack problems. In: SODA, pp. 179–188 (1995)
13. Marchetti-Spaccamela, A., Vercellis, C.: Stochastic on-line knapsack problems. Mathematical Programming 68, 73–104 (1995)
14. Motwani, R., Raghavan, P.: Randomized Algorithms. Cambridge University Press, Cambridge (1995)
15. Muthukrishnan, S., Pal, M., Svitkina, Z.: Stochastic models for budget optimization in search-based. manuscript (2007)
16. Rusmevichientong, P., Williamson, D.P.: An adaptive algorithm for selecting profitable keywords for search-based advertising services. In: Proceedings of the 7th ACM Conference on Electronic Commerce, ACM Press, New York (2006)

An Optimal Bifactor Approximation Algorithm for the Metric Uncapacitated Facility Location Problem

Jaroslaw Byrka*

Centrum voor Wiskunde en Informatica,
Kruislaan 413, NL-1098 SJ Amsterdam, Netherlands
J.Byrka@cwi.nl

Abstract. We consider the metric uncapacitated facility location problem(UFL). In this paper we modify the $(1 + 2/e)$-approximation algorithm of Chudak and Shmoys to obtain a new (1.6774,1.3738)-approximation algorithm for the UFL problem. Our linear programing rounding algorithm is the first one that touches the approximability limit curve $(\gamma_f, 1 + 2e^{-\gamma_f})$ established by Jain et al. As a consequence, we obtain the first optimal approximation algorithm for instances dominated by connection costs.

Our new algorithm - when combined with a (1.11,1.7764)-approximation algorithm proposed by Jain, Mahdian and Saberi, and later analyzed by Mahdian, Ye and Zhang - gives a 1.5-approximation algorithm for the metric UFL problem. This algorithm improves over the previously best known 1.52-approximation algorithm by Mahdian, Ye and Zhang, and it cuts the gap with the approximability lower bound by 1/3.

The algorithm is also used to improve the approximation ratio for the 3-level version of the problem.

1 Introduction

The Uncapacitated Facility Location (UFL) problem is defined as follows. We are given a set $\mathcal{F}$ of n_f *facilities* and a set $\mathcal{C}$ of n_c *clients*. For every facility $i \in \mathcal{F}$, there is a nonnegative number f_i denoting the *opening cost* of the facility. Furthermore, for every client $j \in \mathcal{C}$ and facility $i \in \mathcal{F}$, there is a *connection cost* c_{ij} between facility i and client j. The goal is to open a subset of the facilities $\mathcal{F}' \subseteq \mathcal{F}$, and connect each client to an open facility so that the total cost is minimized. The UFL problem is NP-complete, and max SNP-hard (see [8]). A UFL instance is *metric* if its *connection cost* function satisfies a kind of *triangle inequality*, namely if $c_{ij} \leq c_{ij'} + c_{i'j'} + c_{i'j}$ for any $i, i' \in \mathcal{C}$ and $j, j' \in \mathcal{F}$.

The UFL problem has a rich history starting in the 1960's. The first results on approximation algorithms are due to Cornuéjols, Fisher, and Nemhauser [7] who

---

* Supported by the EU Marie Curie Research Training Network ADONET, Contract No MRTN-CT-2003-504438.

M. Charikar et al. (Eds.): APPROX and RANDOM 2007, LNCS 4627, pp. 29–43, 2007.

considered the problem with an objective function of maximizing the "profit" of connecting clients to facilities minus the cost of opening facilities. They showed that a greedy algorithm gives an approximation ratio of $(1 - 1/e) = 0.632\ldots$, where e is the base of the natural logarithm. For the objective function of minimizing the sum of connection cost and opening cost, Hochbaum [9] presented a greedy algorithm with an $O(\log n)$ approximation guarantee, where n is the number of clients. The first approximation algorithm with constant approximation ratio for the minimization problem where the connection costs satisfy the triangle inequality, was developed by Shmoys, Tardos, and Aardal [14]. Several approximation algorithms have been proposed for the metric UFL problem after that, see for instance [8,4,5,6,15,10,12]. Up to now, the best known approximation ratio was 1.52, obtained by Mahdian, Ye, and Zhang [12]. Many more algorithms have been considered for the UFL problem and its variants. We refer an interested reader to survey papers by Shmoys [13] and Vygen [16].

We will say that an algorithm is a λ-approximation algorithm for a minimization problem if it computes, in polynomial time, a solution that is at most λ times more expensive than the optimal solution. Specifically, for the UFL problem we consider the notion of *bifactor approximation* studied by Charikar and Guha [4]. We say that an algorithm is a (λ_f,λ_c)-approximation algorithm if the solution it delivers has total cost at most $\lambda_f \cdot F^* + \lambda_c \cdot C^*$, where F^* and C^* denote, respectively, the facility and the connection cost of an optimal solution.

Guha and Khuller [8] proved by a reduction from Set Cover that there is no polynomial time λ-approximation algorithm for the metric UFL problem with $\lambda < 1.463$, unless $NP \subseteq DTIME(n^{\log \log n})$. Sviridenko showed that the approximation lower bound of 1.463 holds, unless $P = NP$ (see [16]). Jain et al. [10] generalized the argument of Guha and Khuller to show that the existence of a (λ_f,λ_c)-approximation algorithm with $\lambda_c < 1 + 2e^{-\lambda_f}$ would imply $NP \subseteq DTIME(n^{\log \log n})$.

1.1   Our Contribution

We modify the $(1+2/e)$-approximation algorithm of Chudak [5], see also Chudak and Shmoys [6], to obtain a new $(1.6774, 1.3738)$-approximation algorithm for the UFL problem. Our linear programing (LP) rounding algorithm is the first one that achieves an optimal bifactor approximation due to the matching lower bound of $(\lambda_f, 1 + 2e^{-\lambda_f})$ established by Jain et al. In fact we obtain an algorithm for each point $(\lambda_f, 1 + 2e^{-\lambda_f})$ such that $\lambda_f \geq 1.6774$, which means that we have an optimal approximation algorithm for instances dominated by connection cost (see Figure 1).

Our main technique is to modify the support graph corresponding to the LP solution before clustering, and to use various average distances in the fractional solution to bound the cost of the obtained solution. Modifying the solution in such a way was introduced by Lin and Vitter [11] and is called *filtering*. Throughout this paper we will use the name *sparsening technique* for the combination of filtering with our new analysis.

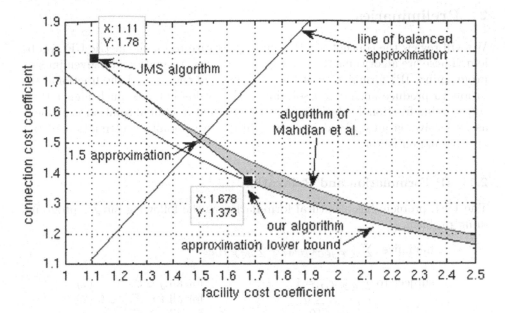

Fig. 1. Bifactor approximation picture. The gray area corresponds to the improvement due to our algorithm.

One could view our contribution as an improved analysis of a minor modification of the algorithm by Sviridenko [15], which also introduces filtering to the algorithm of Chudak and Shmoys. The filtering process that is used both in our algorithm and in the algorithm by Sviridenko is relatively easy to describe, but the analysis of the impact of this technique on the quality of the obtained solution is quite involved in each case. Therefore, we prefer to state our algorithm as an application of the sparsening technique to the algorithm of Chudak and Shmoys, which in our opinion is relatively easy do describe and analyze.

The motivation for the sparsening technique is the "irregularity" of instances that are potentially tight for the original algorithm of Chudak and Shmoys. We propose a way of measuring and controlling this irregularity. In fact our clustering is the same as the one used by Sviridenko in his 1.58-approximation algorithm [15], but we continue our algorithm in the spirit of Chudak and Shmoys' algorithm, which leads to an improved bifactor approximation guaranty.

Our new algorithm may be combined with the $(1.11, 1.7764)$-approximation algorithm of Jain et al. to obtain a 1.5-approximation algorithm for the UFL problem. This is an improvement over the previously best known 1.52-approximation algorithm of Mahdian et al., and it cuts of a $1/3$ of the gap with the approximation lower bound by Guha and Khuler [8].

We also note that the new $(1.6774, 1.3738)$-approximation algorithm may be used to improve the approximation ratio for the 3-level version of the UFL problem to 2.492.

2   Preliminaries

We will review the concept of LP-rounding algorithms for the metric UFL problem. These are algorithms that first solve the linear relaxation of a given integer programing (IP) formulation of the problem, and then round the fractional solution to produce an integral solution with a value not too much higher than the starting fractional solution. Since the optimal fractional solution is at most as expensive as an optimal integral solution, we obtain an estimation of the approximation factor.

2.1   IP Formulation and Relaxation

The UFL problem has a natural formulation as the following integer programming problem.

$$\text{minimize } \sum_{i \in \mathcal{F}, j \in \mathcal{C}} c_{ij} x_{ij} + \sum_{i \in \mathcal{F}} f_i y_i$$

$$\begin{aligned}
\text{subject to } \sum_{i \in \mathcal{F}} x_{ij} &= 1 & \text{for all } j \in \mathcal{C} & \quad (1)\\
x_{ij} - y_i &\leq 0 & \text{for all } i \in \mathcal{F}, j \in \mathcal{C} & \quad (2)\\
x_{ij}, y_i &\in \{0, 1\} & \text{for all } i \in \mathcal{F}, j \in \mathcal{C} & \quad (3)
\end{aligned}$$

A linear relaxation of this IP formulation is obtained by replacing Condition (3) by the condition $x_{ij} \geq 0$ for all $i \in \mathcal{F}, j \in \mathcal{C}$. The value of the solution to this LP relaxation will serve as a lower bound for the cost of the optimal solution. We will also make use of the following dual formulation of this LP.

$$\text{maximize } \sum_{j \in \mathcal{C}} v_j$$

$$\begin{aligned}
\text{subject to } \sum_{j \in \mathcal{C}} w_{ij} &\leq f_i \text{ for all } i \in \mathcal{F} & (4)\\
v_j - w_{ij} &\leq c_{ij} \text{ for all } i \in \mathcal{F}, j \in \mathcal{C} & (5)\\
w_{ij} &\geq 0 \quad \text{ for all } i \in \mathcal{F}, j \in \mathcal{C} & (6)
\end{aligned}$$

2.2   Clustering

The first constant factor approximation algorithm for the metric UFL problem by Shmoys et al., but also the algorithms by Chudak and Shmoys, and by Sviridenko are based on the following clustering procedure. Suppose we are given an optimal solution to the LP relaxation of our problem. Consider the bipartite graph G with vertices being the facilities and the clients of the instance, and where there is an edge between a client j and a facility i if the corresponding variable x_{ij} in the optimal solution to the LP relaxation is positive. We call G a *support graph* of the LP solution. If two clients are both adjacent to the same facility in graph G, we will say that they are *neighbors* in G.

The clustering of this graph is a partitioning of clients into clusters together with a choice of a leading client for each of the clusters. This leading client is called a *cluster center*. Additionally we require that no two cluster centers are neighbors in the support graph. This property helps us to open one of the

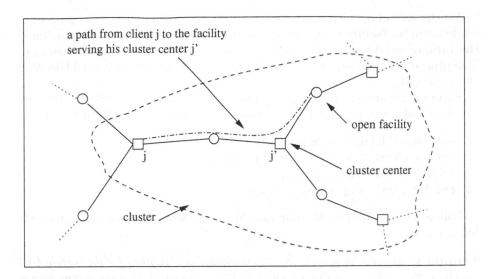

a path from client j to the facility
serving his cluster center j'

open facility

cluster center

j

j'

cluster

Fig. 2. A cluster. If we make sure that at least one facility is open around a cluster center j' , then any other client j from the cluster may use this facility. Because the connection costs are assumed to be metric, the distance to this facility is at most the length of the shortest path from j to the open facility.

adjacent facilities for each cluster center. Formally we will say that a clustering is a function $g : C \rightarrow C$ that assigns each client to the center of his cluster. For a picture of a cluster see Figure 2.

All the above mentioned algorithms use the following procedure to obtain the clustering. While not all the clients are clustered, choose greedily a new cluster center j, and build a cluster from j and all the neighbors of j that are not yet clustered. Obviously the outcome of this procedure is a proper clustering. Moreover, it has a desired property that clients are close to their cluster centers. Each of the mentioned LP-rounding algorithms uses a different greedy criterion for choosing new cluster centers. In our algorithm we will use the clustering with the greedy criterion of Sviridenko [15].

2.3 Scaling and Greedy Augmentation

The techniques described here are not directly used by our algorithm, but they help to explain why the algorithm of Chudak and Shmoys is close to optimal. We will discuss how scaling facility opening costs before running an algorithm, together with another technique called *greedy augmentation* may help to balance the analysis of an approximation algorithm for the UFL problem.

The greedy augmentation technique introduced by Guha and Khuller [8] (see also [4]) is the following. Consider an instance of the metric UFL problem and a feasible solution. For each facility $i \in \mathcal{F}$ that is not opened in this solution, we may compute the impact of opening facility i on the total cost of the solution, also called the *gain* of opening i, denoted by g_i. The greedy augmentation procedure, while there is a facility i with positive gain g_i, opens a facility i_0 that maximizes the ratio of saved cost to the facility opening cost $\frac{g_i}{f_i}$, and updates values of g_i. The procedure terminates when there is no facility whose opening would decrease the total cost.

Suppose we are given an approximation algorithm A for the metric UFL problem and a real number $\delta \geq 1$. Consider the following algorithm $S_\delta(A)$.

1. scale up all facility opening costs by a factor δ;
2. run algorithm A on the modified instance;
3. scale back the opening costs;
4. run the greedy augmentation procedure.

Following the analysis of Mahdian, Ye, and Zhang [12] one may prove the following lemma.

Lemma 1. *Suppose A is a (λ_f, λ_c)-approximation algorithm for the metric UFL problem, then $S_\delta(A)$ is a $(\lambda_f + ln(\delta), 1 + \frac{\lambda_c - 1}{\delta})$-approximation algorithm for this problem.*

This method may be applied to balance an (λ_f, λ_c)-approximation algorithm with $\lambda_f \ll \lambda_c$. However, our 1.5-approximation algorithm is balanced differently. It is a composition of two algorithms that have opposite imbalances.

3 Sparsening the Graph of the Fractional Solution

In this section we describe a technique that we use to control the expected connection cost of the obtained solution. It is based on modifying a fractional solution in a way introduced by Lin and Vitter [11] and called *filtering*.

The filtering technique has been successfully applied to the facility location problem, also in the algorithms of Shmoys, Tardos, and Aardal [14] and of Sviridenko [15]. We will give an alternative analysis of what is the effect of applying filtering on a fractional solution to the LP relaxation of the UFL problem.

Suppose that for a given UFL instance we have solved its LP relaxation, and that we have an optimal primal solution (x^*, y^*) and the corresponding optimal dual solution (v^*, w^*). Such a fractional solution has facility cost $F^* = \sum_{i \in \mathcal{F}} f_i y_i^*$ and connection cost $C^* = \sum_{i \in \mathcal{F}, j \in \mathcal{C}} c_{ij} x_{ij}^*$. Each client j has its share v_j of the total cost. This cost may again be divided into a client's fractional connection cost $C_j^* = \sum_{i \in \mathcal{F}} c_{ij} x_{ij}^*$, and his fractional facility cost $F_j^* = v_j^* - C_j^*$.

3.1 Motivation and Intuition

The idea behind the sparsening technique is to make use of some irregularities of an instance if they occur. We call an instance *regular* if the facilities that fractionally serve a client j are all at the same distance from j. For such an instance the algorithm of Chudak and Shmoys produces a solution whose cost is bounded by $F^* + (1 + \frac{2}{e})C^*$, which also follows from our analysis in Section 4. It remains to use the technique described in section 2.3 to obtain an optimal 1.463...-approximation algorithm for such regular instances.

The instances that are not regular are called *irregular*. Difficult to understand are the irregular instances. In fractional solutions for these instances particular clients are fractionally served by facilities at different distances. Our approach is to divide facilities serving a client into two groups, namely *close* and *distant* facilities. We will remove links to distant facilities before the clustering step, so that if there are irregularities, distances to cluster centers should decrease.

We measure the local irregularity of an instance by comparing a fractional connection cost of a client to the average distance to his distant facilities. In the case of a regular instance, the sparsening technique gives the same results as technique described in section 2.3, but for irregular instances sparsening also takes some advantage of the irregularity.

3.2   Details

We will start by modifying the primal optimal fractional solution (x^*, y^*) by scaling the y-variables by a constant $\gamma > 1$ to obtain a suboptimal fractional solution $(x^*, \gamma \cdot y^*)$. Now suppose that the y-variables are fixed, but that we now have a freedom to change the x-variables in order to minimize the total cost. For each client j we change the corresponding x-variables so that he uses his closest facilities in the following way. We choose an ordering of facilities with nondecreasing distances to client j. We connect client j to the first facilities in the ordering so that among facilities fractionally serving j only the latest one in the chosen ordering may be opened more then it serves j. Formally, for any facilities i and i' such that i' is later in the ordering, if $x_{ij} < y_i$ then $x_{i'j} = 0$.

Without loss of generality, we may assume that this solution is complete (i.e. there are no $i \in \mathcal{F}, j \in \mathcal{C}$ such that $0 < x_{ij} < y_i$). Otherwise we may split facilities to obtain an equivalent instance with a complete solution - see [15][Lemma 1] for a more detailed argument.

Let $(\overline{x}, \overline{y})$ denote the obtained complete solution. For a client j we say that a facility i is one of *his close facilities* if it fractionally serves client j in $(\overline{x}, \overline{y})$. If $\overline{x}_{ij} = 0$, but facility i was serving client j in solution (x^*, y^*), then we say, that i is a *distant* facility of client j.

Definition 1. *Let*

$$r_\gamma(j) = \begin{cases} \frac{\frac{\gamma}{\gamma-1}\sum_{i \in \{i \in \mathcal{F} | \overline{x}_{ij}=0\}} c_{ij}x_{ij}^* - C_j^*}{F_j^*} & \text{for } F_j^* > 0 \\ 0 & \text{for } F_j^* = 0. \end{cases}$$

The value $r_\gamma(j)$ is a measure of the irregularity of the instance around client j. It is the average distance to a distant facility minus the fractional connection cost C_j^* (C_j^* is the general average distance to both close and distant facilities) divided by the fractional facility cost of a client j; or it is equal 0 if $F_j^* = 0$. Observe, that $r_\gamma(j)$ takes values between 0 and 1. $r_\gamma(j) = 0$ means that client j

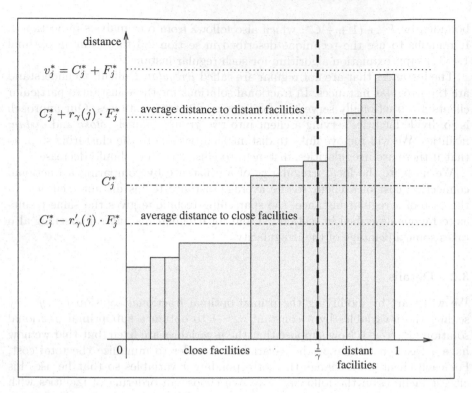

Fig. 3. Distances to facilities serving client j; the width of a rectangle corresponding to facility i is equal to x^*_{ij}. Figure explains the meaning of $r_\gamma(j)$.

is served in the solution (x^*, y^*) by facilities that are all at the same distance. In the case of $r_\gamma(j) = 1$ the facilities are at different distances and the distant facilities are all so far from j that j is not willing to contribute to their opening. In fact, for clients j with $F^*_j = 0$ the value of $r_\gamma(j)$ is not relevant for our analysis.

To get some more intuition for the F^*_j and $r_\gamma(j)$ values, imagine that you know F^*_j and C^*_j, but the adversary is constructing the fractional solution and he decided about distances to particular facilities fractionally serving client j. One could interpret F^*_j as a measure of freedom the adversary has when he chooses those distances. In this language, $r_\gamma(j)$ is a measure of what fraction of this freedom is used to make distant facilities more distant than average facilities.

Let $r'_\gamma(j) = r_\gamma(j) * (\gamma - 1)$. For client j with $F^*_j > 0$ we have $r'_\gamma(j) = \frac{C^*_j - \sum_{i \in \mathcal{F}} c_{ij} \overline{x}_{ij}}{F^*_j}$ which is the fractional connection cost minus the average distance to a close facility, divided by the fractional facility cost of a client j.

Observe, that for every client j the following hold (see Figure 3):

- his average distance to a close facility equals $D^C_{av}(j) = C^*_j - r'_\gamma(j) \cdot F^*_j$,
- his average distance to a distant facility equals $D^D_{av}(j) = C^*_j + r_\gamma(j) \cdot F^*_j$,
- his maximal distance to a close facility is at most the average distance to a distant facility, $D^C_{max}(j) \le D^D_{av}(j) = C^*_j + r_\gamma(j) \cdot F^*_j$.

Consider the bipartite graph G obtained from the solution $(\overline{x}, \overline{y})$, where each client is directly connected to his close facilities. We will greedily cluster this graph in each round choosing the cluster center to be an unclustered client j with the minimal value of $D^C_{av}(j) + D^C_{max}(j)$. In this clustering, each cluster center has a minimal value of $D^C_{av}(j) + D^C_{max}(j)$ among clients in his cluster.

4   Our New Algorithm

Consider the following algorithm $A1(\gamma)$:

1. Solve the LP relaxation of the problem to obtain a solution (x^*, y^*).
2. Scale up the value of the facility opening variables y by a constant $\gamma > 1$, then change the value of the x-variables so as to use the closest possible fractionally open facilities (see Section 3.2).
3. If necessary, split facilities to obtain a complete solution $(\overline{x}, \overline{y})$.
4. Compute a greedy clustering for the solution $(\overline{x}, \overline{y})$, choosing as cluster centers unclustered clients minimizing $D^C_{av}(j) + D^C_{max}(j)$.
5. For every cluster center j, open one of his close facilities randomly with probabilities $\overline{x}_{ij}$.
6. For each facility i that is not a close facility of any cluster center, open it independently with probability $\overline{y}_i$.
7. Connect each client to an open facility that is closest to him.

In the analysis of this algorithm we will use the following result:

Lemma 2. *Given n independent events $e_1, e_2, \ldots, e_n$ that occur with probabilities $p_1, p_2, \ldots, p_n$ respectively, the event $e_1 \cup e_2 \cup \ldots \cup e_n$ (i.e. at least one of e_i) occurs with probability at least $1 - \frac{1}{e^{\sum_{i=1}^{n} p_i}}$, where e denotes the base of the natural logarithm.*

Theorem 1. *Algorithm $A1(\gamma = 1.67736)$ produces a solution with expected cost $E[cost(SOL)] \leq 1.67736 \cdot F^* + 1.37374 \cdot C^*$.*

Proof. The expected facility opening cost of the solution is
$$E[F_{SOL}] = \sum_{i \in \mathcal{F}} f_i \overline{y}_i = \gamma \cdot \sum_{i \in \mathcal{F}} f_i y^*_i = \gamma \cdot F^*.$$
To bound the expected connection cost we show that for each client j there is an open facility within a certain distance with a certain probability. If j is a cluster center, one of his close facilities is open and the expected distance to this open facility is $D^C_{av}(j) = C^*_j - r'_\gamma(j) \cdot F^*_j$.

If j is not a cluster center, he first considers his close facilities (see Figure 4). If any of them is open, the expected distance to the closest open facility is at most $D^C_{av}(j)$. From Lemma 2, with probability $p_c \geq (1 - \frac{1}{e})$, at least one close facility is open.

Suppose none of the close facilities of j is open, but at least one of his distant facilities is open. Let p_d denote the probability of this event. The expected distance to the closest facility is then at most $D^D_{av}(j)$.

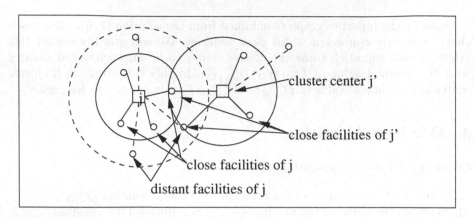

Fig. 4. Facilities that client j may consider: his close facilities, distant facilities, and close facilities of cluster center j'

If neither any close nor any distant facility of client j is open, then he connects himself to the facility serving his cluster center $g(j) = j'$. Again from Lemma 2, such an event happens with probability $p_s \leq \frac{1}{e^\gamma}$. In the following we will show that if $\gamma < 2$ then the expected distance from j to the facility serving j' is at most $D_{av}^D(j) + D_{max}^C(j') + D_{av}^C(j')$. Let $\mathcal{C}_j$ ($\mathcal{D}_j$) be the set of close (distant) facilities of j. For any set of facilities $X \subset \mathcal{F}$, let $d(j, X)$ denote the weighted average distance from j to $i \in X$ (with values of opening variables y_i as weights).

If the distance between j and j' is at most $D_{av}^D(j) + D_{av}^C(j')$, then the remaining $D_{max}^C(j')$ is enough for the distance from j' to any of his close facilities. Suppose now that the distance between j and j' is bigger than $D_{av}^D(j) + D_{av}^C(j')$ (*). We will bound $d(j', \mathcal{C}_{j'} \setminus (\mathcal{C}_j \cup \mathcal{D}_j))$, the average distance from cluster center j' to his close facilities that are neither close nor distant facilities of j (since the expected connection cost that we compute is on the condition that j was not served directly). The assumption(*) implies that $d(j', \mathcal{C}_j \cap \mathcal{C}_{j'}) > D_{av}^C(j')$. Therefore, if $d(j', \mathcal{D}_j \cap \mathcal{C}_{j'}) \geq D_{av}^C(j')$, then $d(j', \mathcal{D}_j \setminus (\mathcal{C}_j \cup \mathcal{D}_j)) \leq D_{av}^C(j')$ and the total distance from j is small enough.

The remaining case is that $d(j', \mathcal{D}_j \cap \mathcal{C}_{j'}) = D_{av}^C(j') - z$ for some positive z (**). Let $\hat{y} = \sum_{i \in (\mathcal{C}_{j'} \cap \mathcal{D}_j)} \overline{y}_i$ be the total fractional opening of facilities in $\mathcal{C}_{j'} \cup \mathcal{D}_j$ in the modified fractional solution $(\overline{x}, \overline{y})$. From (*) we conclude, that $d(j, \mathcal{D}_j \cap \mathcal{C}_{j'}) \geq D_{av}^D(j) + z$, which implies $d(j, \mathcal{D}_j \setminus \mathcal{C}_{j'}) \leq D_{av}^D(j) - z \cdot \frac{\hat{y}}{\gamma - 1 - \hat{y}}$ (note that (**) implies $(\mathcal{D}_j \setminus \mathcal{C}_{j'}) \neq \emptyset$ and $\gamma - 1 - \hat{y} > 0$), hence $D_{max}^C(j) \leq D_{av}^D(j) - z \cdot \frac{\hat{y}}{\gamma - 1 - \hat{y}}$. Combining this with assumption (*) we conclude that the minimal distance from j' to a facility in $\mathcal{C}_j \cap \mathcal{C}_{j'}$ is at least $D_{av}^D(j) + D_{av}^C(j') - D_{max}^C(j) \geq D_{av}^C(j') + z \cdot \frac{\hat{y}}{\gamma - 1 - \hat{y}}$. Assumption (**) implies $d(j', \mathcal{C}_{j'} \setminus \mathcal{D}_j) = D_{av}^C(j') + z \cdot \frac{\hat{y}}{1 - \hat{y}}$. Concluding, if $\gamma < 2$ then $d(j', \mathcal{C}_{j'} \setminus (\mathcal{D}_j \cup \mathcal{C}_j)) \leq D_{av}^C(j') + z \cdot \frac{\hat{y}}{\gamma - 1 - \hat{y}}$.

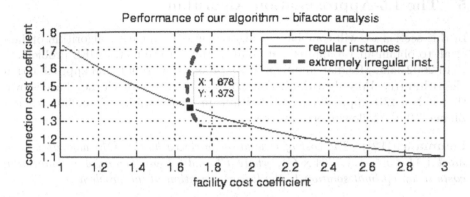

Fig. 5. Figure presents performance of our algorithm for different values of parameter γ. The solid line corresponds to regular instances with $r_\gamma(j) = 0$ for all j and it coincides with the approximability lower bound curve. The dashed line corresponds to instances with $r_\gamma(j) = 1$ for all j. For a particular choice of γ we get a horizontal segment connecting those two curves; for $\gamma \approx 1.67736$ the segment becomes a single point. Observe that for instances dominated by connection cost only a regular instance may be tight for the lower bound.

Therefore, the expected connection cost from j to a facility in $\mathcal{C}_{j'} \setminus (\mathcal{D}_j \cup \mathcal{C}_j)$ is at most

$$
\begin{aligned}
& D^C_{max}(j) + D^C_{max}(j') + d(j', \mathcal{C}_{j'} \setminus (\mathcal{D}_j \cup \mathcal{C}_j)) \\
& \leq D^D_{av}(j) - z \cdot \frac{\hat{y}}{\gamma - 1 - \hat{y}} + D^C_{max}(j') + D^C_{av}(j') + z \cdot \frac{\hat{y}}{\gamma - 1 - \hat{y}} \\
& = D^D_{av}(j) + D^C_{max}(j') + D^D_{av}(j').
\end{aligned}
$$

Putting all the cases together, the expected total connection cost is

$$
\begin{aligned}
E[C_{SOL}] & \leq \sum_{j \in \mathcal{C}} \left(p_c \cdot D^C_{av}(j) + p_d \cdot D^D_{av}(j) + p_s \cdot (D^D_{av}(j) + D^C_{max}(j') + D^C_{av}(j')) \right) \\
& \leq \sum_{j \in \mathcal{C}} \left((p_c + p_s) \cdot D^C_{av}(j) + (p_d + 2p_s) \cdot D^D_{av}(j) \right) \\
& = \sum_{j \in \mathcal{C}} \left((p_c + p_s) \cdot (C^*_j - r'_\gamma(j) \cdot F^*_j) + (p_d + 2p_s) \cdot (C^*_j + r_\gamma(j) \cdot F^*_j) \right) \\
& = ((p_c + p_d + p_s) + 2p_s) \cdot C^* \\
& \quad + \sum_{j \in \mathcal{C}} \left((p_c + p_s) \cdot (-r_\gamma(j) \cdot (\gamma - 1) \cdot F^*_j) + (p_d + 2p_s) \cdot (r_\gamma(j) \cdot F^*_j) \right) \\
& = (1 + 2p_s) \cdot C^* + \sum_{j \in \mathcal{C}} \left(F^*_j \cdot r_\gamma(j) \cdot (p_d + 2p_s - (\gamma - 1) \cdot (p_c + p_s)) \right) \\
& \leq (1 + \tfrac{2}{e^\gamma}) \cdot C^* + \sum_{j \in \mathcal{C}} \left(F^*_j \cdot r_\gamma(j) \cdot (\tfrac{1}{e} + \tfrac{1}{e^\gamma} - (\gamma - 1) \cdot (1 - \tfrac{1}{e} + \tfrac{1}{e^\gamma})) \right).
\end{aligned}
$$

By setting $\gamma = \gamma_0 \approx 1.67736$ such that $\frac{1}{e} + \frac{1}{e^{\gamma_0}} - (\gamma_0 - 1) \cdot (1 - \frac{1}{e} + \frac{1}{e^{\gamma_0}}) = 0$, we obtain $E[C_{SOL}] \leq (1 + \frac{2}{e^{\gamma_0}}) \cdot C^* \leq 1.37374 \cdot C^*$. $\qquad \square$

The algorithm A1 with $\gamma = 1 + \epsilon$ (for a sufficiently small positive ϵ) is essentially the algorithm of Chudak and Shmoys.

5 The 1.5-Approximation Algorithm

In this section we will combine our algorithm with an earlier algorithm of Jain et al. to obtain an 1.5-approximation algorithm for the metric UFL problem.

In 2002 Jain, Mahdian and Saberi [10] proposed a primal-dual approximation algorithm (the JMS algorithm). Using a dual fitting approach they have shown that it is a 1.61-approximation algorithm. In a later work of Mahdian, Ye and Zhang [12] the following was proven.

Lemma 3 ([12]). *The cost of a solution produced by the JMS algorithm is at most* $1.11 \times F^* + 1.7764 \times C^*$*, where* F^* *and* C^* *are facility and connection costs in an optimal solution to the linear relaxation of the problem.*

Theorem 2. *Consider the solutions obtained with the A1 and JMS algorithms. The cheaper of them is expected to have a cost at most* 1.5 *times the cost of the optimal fractional solution.*

Proof. Consider the algorithm A2 that with probability $p = 0.313$ runs the JMS algorithm and with probability $1 - p$ runs the A1 algorithm. Suppose that you are given an instance, and F^* and C^* are facility and connection costs in an optimal solution to the linear relaxation of the problem for this instance. Consider the expected cost of the solution produced by algorithm A2 for this instance. $E[cost] \le p \cdot (1.11 \cdot F^* + 1.7764 \cdot C^*) + (1-p) \cdot (1.67736 \cdot F^* + 1.37374 \cdot C^*) = 1.4998 \cdot F^* + 1.4998 \cdot C^* < 1.5 * (F^* + C^*) \le 1.5 * OPT.$ □

Instead of the JMS algorithm we could take the algorithm of Machdian et al. [12] - the MYZ(δ) algorithm that scales the facility costs by δ, runs the JMS algorithms, scales back the facility costs and finally runs the greedy augmentation procedure. With a notation introduced in Section 2.3, the MYZ(δ) algorithm is the $S_\delta(JMS)$ algorithm. The MYZ(1.504) algorithm was proven [12] to be a 1.52-approximation algorithm for the metric UFL problem. We may change the value of δ in the original analysis to observe that MYZ(1.1) is a (1.2053,1.7058)-approximation algorithm. This algorithm combined with our A1 (1.67736,1.37374)-approximation algorithm gives a 1.4991-approximation algorithm, which is even better than just using JMS and A1, but it gets more complicated and the additional improvement is tiny.

6 Multilevel Facility Location

In the k-level facility location problem the clients need to be connected to open facilities on the first level, and each open facility ,except on the last, k-th level, needs to be connected to an open facility on the next level. Aardal, Chudak, and Shmoys [1] gave a 3-approximation algorithm for the k-level problem with arbitrary k. Ageev, Ye, and Zhang [2] proposed a reduction of a k-level problem to a $(k - 1)$-level and a 1-level problem, which results in a recursive algorithm.

This algorithm uses an approximation algorithm for the single level problem and has a better approximation ratio, but only for instances with small k. Using our new (1.67736,1.37374)-approximation algorithm instead of the JMS algorithm within this framework improves approximation for each level. In particular, in the limit as k tends to ∞ we get 3.236-approximation which is the best possible for this construction.

By a slightly different method, Zhang [17] obtained a 1.77-approximation algorithm for the 2-level problem. By reducing to a problem with smaller number of levels, he obtained 2.523[1] and 2.81 approximation algorithms for the 3-level and the 4-level version of the problem. If we modiffy the algorithm by Zhang for the 3-level problem, and use the new (1.67736,1.37374)-approximation algorithm for the single level part, we obtain a 2.492-approximation, which improves on the previously best known approximation by Zhang. Note, that for $k > 4$ the best known approximation factor is still due to Aardal et al. [1].

7 Concluding Remarks

The presented algorithm was described as a procedure of rounding a particular fractional solution to the LP relaxation of the problem. In the presented analysis we compared the cost of the obtained solution with the cost of the starting fractional solution. If we appropriately scale the cost function in the LP relaxation before solving the relaxation, we easily obtain an algorithm with a bifactor approximation guaranty in a stronger sense. Namely, we get a comparison of the produced solution with any feasible solution to the LP relaxation of the problem. Such a stronger guaranty was, however, not necessary to construct the 1.5-approximation algorithm for the metric UFL problem.

With the 1.52-approximation algorithm of Mahdian et al. it was not clear for the authors if a better analysis of the algorithm could close the gap with the approximation lower bound of 1.463 by Guha and Khuler. Byrka and Aardal [3] have recently given a negative answer to this question by constructing instances that are hard for the MYZ algorithm. Similarly, we now do not know if our new algorithm $A1(\gamma)$ could be analyzed better to close the gap. Construction of hard instances for our algorithm remains an open problem.

The technique described in Section 2.3 enables to move the bifactor approximation guaranty of an algorithm along the approximability lower bound of Jain et al. (see Figure 1) towards higher facility opening costs. If we developed a technique to move the analysis in the opposite direction, together with our new algorithm, it would imply closing the approximability gap for the metric UFL problem. It seems that with such an approach we would have to face the difficulty of analyzing an algorithm that closes some of the previously opened facilities.

---

[1] This value deviates slightly from the value 2.51 given in the paper. The original argument contained a minor calculation error.

Acknowledgments

The author would like to thank Karen Aardal for all her support and many helpful comments on earlier drafts of this paper. The author also thanks David Shmoys, Steven Kelk, Evangelos Markakis and anonymous referees for their advice and valuable remarks.

References

1. Aardal, K., Chudak, F., Shmoys, D.B.: A 3-approximation algorithm for the k-level uncapacitated facility location problem Information Processing Letters. 72, 161–167 (1999)
2. Ageev, A., Ye, Y., Zhang, J.: Improved combinatorial Approximation algorithms for the k-level facility location problem. In: Baeten, J.C.M., Lenstra, J.K., Parrow, J., Woeginger, G.J. (eds.) ICALP 2003. LNCS, vol. 2719, pp. 145–156. Springer, Heidelberg (2003)
3. Byrka, J., Aardal, K.: The approximation gap for the metric facility location problem is not yet closed. Operations Research Letters (ORL) 35(3), 379–384 (2007)
4. Charikar, M., Guha, S.: Improved combinatorial algorithms for facility location and k-median problems. In: Proc. of the 40th IEEE Symp. on Foundations of Computer Science (FOCS), pp. 378–388. IEEE Computer Society Press, Los Alamitos (1999)
5. Chudak, F.A.: Improved approximation algorithms for uncapacitated facility location. In: Proc. of the 6th Integer Programing and Combinatorial Optimization (IPCO), pp. 180–194 (1998)
6. Chudak, F.A., Shmoys, D.B.: Improved approximation algorithms for the uncapacitated facility location problem. SIAM J. Comput. 33(1), 1–25 (2003)
7. Cornuéjols, G., Fisher, M.L., Nemhauser, G.L.: Location of bank accounts to optimize float: An analytic study of exact and approximate algorithms. Management Science 8, 789–810 (1977)
8. Guha, S., Khuller, S.: Greedy strikes back: Improved facility location algorithms. In: Proc. of the 9th ACM-SIAM Symp. on Discrete Algorithms (SODA), pp. 228–248. ACM Press, New York (1998)
9. Hochbaum, D.S.: Heuristics for the fixed cost median problem. Mathematical Programming 22, 148–162 (1982)
10. Jain, K., Mahdian, M., Saberi, A.: A new greedy approach for facility location problems. In: Proc. of the 34th ACM Symp. on Theory of Computing (STOC), pp. 731–740. ACM Press, New York (2002)
11. Lin, J.-H., Vitter, J.S.: ϵ-approximations with minimum packing constraint violation. In: Proc. of the 24th ACM Symp. on Theory of Computing (STOC), pp. 771–782. ACM Press, New York (1992)
12. Mahdian, M., Ye, Y., Zhang, J.: Improved approximation algorithms for metric facility location problems. In: Jansen, K., Leonardi, S., Vazirani, V.V. (eds.) APPROX 2002. LNCS, vol. 2462, pp. 229–242. Springer, Heidelberg (2002)
13. Shmoys, D.B.: Approximation algorithms for facility location problems. In: Jansen, K., Khuller, S. (eds.) APPROX 2000. LNCS, vol. 1913, pp. 265–274. Springer, Heidelberg (2000)
14. Shmoys, D.B., Tardos, É., Aardal, K.: Approximation algorithms for facility location problems (extended abstract). In: Proc. of the 29th ACM Symp. on Theory of Computing (STOC), pp. 265–274. ACM Press, New York (1997)

15. Sviridenko, M.: An improved approximation algorithm for the metric uncapacitated facility location problem. In: Proc. of the 9th Integer Programming and Combinatorial Optimization (IPCO), pp. 240–257 (2002)
16. Vygen, J.: Approximation algorithms for facility location problems (Lecture Notes), Report No. 05950-OR, Research Institute for Discrete Mathematics, University of Bonn(2005), http://www.or.uni-bonn.de/~vygen/fl.pdf
17. Zhang, J.: Approximating the two-level facility location problem via a quasi-greedy approach. Mathematical Programming, Ser. A 108, 159–176 (2006)

Improved Approximation Algorithms for the Spanning Star Forest Problem

Ning Chen[1,*], Roee Engelberg[2,**], C. Thach Nguyen[1],
Prasad Raghavendra[1,***], Atri Rudra[1,†], and Gyanit Singh[1]

[1] Department of Computer Science & Engineering,
University of Washington, Seattle, USA
{ning,ncthach,prasad,atri,gyanit}@cs.washington.edu
[2] Department of Computer Science,
Technion, Haifa, Israel
roee@cs.technion.ac.il

Abstract. A *star* graph is a tree of diameter at most two. A *star forest* is a graph that consists of node-disjoint star graphs. In the spanning star forest problem, given an unweighted graph G, the objective is to find a star forest that contains all the vertices of G and has the maximum number of edges. This problem is the complement of the dominating set problem in the following sense: On a graph with n vertices, the size of the maximum spanning star forest is equal to n minus the size of the minimum dominating set.

We present a 0.71-approximation algorithm for this problem, improving upon the approximation factor of 0.6 of Nguyen et al. [9]. We also present a 0.64-approximation algorithm for the problem on node-weighted graphs. Finally, we present improved hardness of approximation results for the weighted versions of the problem.

1 Introduction

A *star* graph is a tree of diameter at most two. Equivalently, a star graph consists of a vertex designated *center* along with a set of *leaves* adjacent to it. In particular, a singleton vertex is a star as well. Given an undirected graph, a spanning star forest consists of a set of node-disjoint stars that cover all the nodes in the graph. In the spanning star forest problem, the objective is to maximize the number of edges (or equivalently, leaves) present in the forest.

A dominating set of a graph is a subset of the vertices such that every other vertex is adjacent to a vertex in the dominating set. Observe that in a spanning star forest solution, each vertex is either a center or adjacent to a center. Hence the set of centers form a dominating set of the graph. Therefore, the size of the

---

* Supported in part by NSF CCF-0635147.
** This work was done while the author was visiting University of Washington.
*** Supported in part by NSF CCF-0343672.
† Supported in part by NSF CCF-0343672.

M. Charikar et al. (Eds.): APPROX and RANDOM 2007, LNCS 4627, pp. 44–58, 2007.

maximum spanning star forest is the number of vertices minus the size of the minimum dominating set. Computing the maximum spanning star forest of a graph is NP-hard because computing the minimum dominating set is NP-hard.

The spanning star forest problem has found applications in computational biology. Nguyen et al. [9] use the spanning star forest problem to give an algorithm for the problem of aligning multiple genomic sequences, which is a basic bioinformatics task in comparative genomics. The spanning star forest problem and its directed version have found applications in the comparison of phylogenetic trees [3] and the diversity problem in the automobile industry [1].

Surprisingly, even though the maximum spanning star forest is a natural NP-hard problem, there is not much literature on approximation algorithms for this problem. In fact, the first approximation algorithms for this problem appeared recently in the work of Nguyen et al. [9]. They gave a number of approximation algorithms: the most general one being a 0.6-approximation algorithm on an unweighted graph. This should be contrasted with the complementary problem of minimizing the size of the dominating set of the graph which is known to be hard to approximate within a factor of $(1 - \epsilon)\ln n$ for any $\epsilon > 0$ unless NP is in $DTIME(n^{\log \log n})$ [4,8]. This disparity in approximability of complementary problems is fairly commonplace (for example the maximum independent set is not approximable to within any polynomial factor while its complement problem of minimum vertex cover can be approximated to within a factor of 2). Nguyen et al. [9] also showed that the spanning star forest problem is hard to approximate to within a factor of $\frac{545}{546} + \epsilon$ unless P=NP. The paper also gave algorithms with better approximation factors for special graphs such as planar graphs and trees (in fact, for trees the optimal spanning star forest can be computed in linear time).

There are some natural weighted generalizations of the spanning star forest problem. The first generalization is when edges have weights and the objective is to maximize the weights of the edges in the spanning star forest solution. There is a simple 0.5-approximation algorithm for this case [9]. Note that the edge-weighted version is no longer the complement of the (weighted) dominating set problem. Another generalization is the case when nodes have weights. The objective now is to maximize the weights of nodes that are leaves in the spanning star forest solution. This problem is the natural complement of the weighted minimum dominating set problem. To the best of our knowledge, the approximability of the node-weighted spanning star forest problem has not been considered before.

1.1   Our Results and Techniques

We prove the following results in this paper. First, we improve the result of [9] by giving a 0.71-approximation algorithm for the unweighted spanning star forest problem. Second, we give a 0.64-approximation algorithm for the node-weighted

spanning star forest problem. Finally, we prove better hardness of approximation results for the weighted versions of the problem. In particular, we show that
the node and edge-weighted spanning star forest problem cannot be approxi-
mated to within a factor of $\frac{31}{32} + \epsilon$ and $\frac{19}{20} + \epsilon$, respectively, for any $\epsilon > 0$ unless
P=NP.

Our algorithms are based on an LP relaxation of the spanning star forest problem and randomized rounding. For each vertex we have a variable x_i which is 1 if
x_i is a leaf. However, the natural rounding scheme of making vertex i a leaf with
probability x_i does not give a good approximation ratio. Instead, we make vertex i a leaf with probability $f(t, x_i) = e^{-t(1-x_i)}$, where the value of t is carefully
chosen. Note that for fixed t, the function $f(t, x_i)$ is non-linear in x_i. Non-linear
rounding schemes used in ([5,7]) round with probability x_i^c, where c is a fixed
constant or is a value that depends on the input[1]. An interesting point about the
rounding is that the function $f(t, x_i)$ is nonzero even for $x_i = 0$, so with some low
probability, the rounding can round a variable $x_i = 0$ to 1.

The nonlinear rounding algorithm, obtains an approximation factor of $\ln \frac{n}{OPT}$
$+ O(1)$ for the dominating set problem, where n is the number of vertices in the
graph and OPT is the value of the optimal (fractional) dominating set. This
almost matches the best known approximation factor due to Slavík (for the
more general set cover problem) [10].

However, the LP rounding only provides a 0.5 approximation, when the dominating set is large (say $0.5n$). To get the claimed factor of 0.71 for unweighted
graphs, we use the LP algorithm in conjunction with another algorithm. The
idea is to divide the input graph G into the union of a subgraph G' and some
trees, where in G' the minimum degree is at least 2. Given a spanning star forest
solution for G', we can "lift" back the solution to the original graph G. Then we
use as a black box the algorithm from [9] that produces a spanning star forest
of size at least $\frac{3}{5}n$ on a n-vertex graph of minimum degree 2.

We now turn to the node-weighted spanning star forest problem. Our LP
rounding algorithm can be easily generalized to the node-weighted case. As in
the unweighted case, the LP rounding algorithm by itself does not give us the
stated factor of 0.64. To get the claimed approximation factor, we combine our
rounding algorithm with the following trivial factor 0.5 algorithm: Compute any
spanning tree, designate an arbitrary vertex as root. Divide the tree in to levels
based on distance from the root. Make nodes at alternate levels as centers. It is
easy to check that one of the two solutions will have weight at least $\frac{1}{2}$ times the
sum of the weights of all nodes.

Finally, we turn to our hardness of approximation results. The hardness results
are obtained by gadget reductions from the result of Håstad [6] that states that
$MAX3SAT$ is NP-hard to approximate to within a factor of $\frac{7}{8} + \epsilon$, for any $\epsilon > 0$,
unless P=NP.

---

[1] For the problem of maximum k-densest subgraph, a randomized rounding using $c = 0.5$ appears to be a folklore result that is attributed to Goemans [5].

2 Preliminaries

In this paper, we will consider undirected simple graphs that can be unweighted, node-weighted (where weights are on the nodes) or edge-weighted (where the weights are on the edges). Without loss of generality, assume that G is connected, otherwise we can consider each connected components separately. We say a graph is a *star* if there is one vertex (called the *center*) incident to all edges in the graph (all other vertices are called *leaves*). The *size* of a star is the number of edges in the star (for weighted case, it is the sum of weights of the edges or the sum of the weights of the leaves in the star, for edge-weighted and node-weighted stars respectively). In particular, a singleton vertex is a star of size 0.

A *spanning star forest* of a graph G is a collection of node disjoint stars that covers all vertices of G. The problem we are interested in is to find a spanning star forest that maximizes the sum of the sizes of its constituent stars. The unweighted, node-weighted and edge-weighted versions of the problem are denoted by UNWEIGHTED SPANNING STAR FOREST, NODE-WEIGHTED SPANNING STAR FOREST and EDGE-WEIGHTED SPANNING STAR FOREST, respectively.

We will now fix some notation. Unless mentioned otherwise, a graph $G = (V, E)$ will be an unweighted graph. For a node-weighted graph, for any vertex $v_i \in V$, its weight will be denoted by $w_i \geq 0$. For an edge-weighted graph, for any edge $e \in E$, its weight will be denoted by $w_e \geq 0$. Further, for a vertex $v_i \in V$, $N(i)$ will denote the neighbor set of v_i in G, that is, $N(i) = \{v_j \mid (v_i, v_j) \in E\}$. We will usually denote $|V|$ by n. By abuse of notation, we will use $OPT(G)$ to denote the optimal spanning star forest for G as well as its the total size. Given a maximization problem, we say that an algorithm is an α-approximation for $0 < \alpha \leq 1$, if for every input instance the algorithm produces a solution whose objective value is at least α times that of the optimal solution for that instance.

3 An LP-Based Algorithm

In this section we will present a linear programming based algorithm for the NODE-WEIGHTED SPANNING STAR FOREST problem. Towards this, we define the following linear programming relaxation. For every vertex i, the variable x_i has the following meaning: $x_i = 1$ if v_i is a leaf in the spanning star forest and is 0 otherwise. For a vertex v_i, it is not possible to have all vertices in $N(i) \cup \{v_i\}$ as leaves. These constraints have been included in the linear program.

$$\max \sum_{v_i \in V} w_i \cdot x_i$$

$$s.t. \ x_i + \sum_{v_j \in N(i)} x_j \leq |N(i)|, \ \forall \, v_i \in V$$

$$0 \leq x_i \leq 1, \ \forall \, v_i \in V$$

Let $LP_{OPT}(G)$ be the value of the optimal solution of the LP. For the rest of the section, fix an optimal solution $\{x_i\}_{i \in V}$. Let $W = \sum_{i=1}^{n} w_i$ be the sum of the weights of all the nodes in G. Define

$$a = \frac{\sum_{i=1}^{n} w_i x_i}{\sum_{i=1}^{n} w_i} = \frac{\sum_{i=1}^{n} w_i x_i}{W}. \tag{1}$$

Notice that this implies that the optimal objective value is aW. Note that setting all $x_i = 1/2$ gives a feasible solution with value $W/2$. Thus, $a \geq 1/2$. We will round the given optimal LP solution using the following rounding algorithm.

ROUNDING-ALG.

1. Make vertex v_i a leaf with probability $e^{-t(a)(1-x_i)}$,
   where $t(a) = \frac{1}{a} \ln \left(\frac{1}{1-a} \right)$. (Note that as $1/2 \leq a < 1$, $t(a) \geq 0$.)
2. Let L_1 denote the set of vertices declared leaves in the first step.
3. Let $L_2 = \{v_i \in V \mid v_i \cup N(i) \subseteq L_1\}$. Declare all vertices in $L_1 \setminus L_2$ as leaves.
4. Assign every leaf vertex to one of its neighbors that is not declared a leaf. Ties are broken arbitrarily.

We have the following approximation guarantee for the above rounding algorithm.

Lemma 1. *Given an LP solution $\{x_i\}_{i \in V}$, ROUNDING-ALG outputs a spanning star forest with expected size at least $aW(1-a)^{\frac{1}{a}-1}$. That is, it is a $(1-a)^{\frac{1}{a}-1}$ factor approximation algorithm for the spanning star forest problem.*

Proof. It is easy to verify that ROUNDING-ALG does indeed generate a valid spanning star forest. For notational convenience let $t = t(a)$ where a is as defined in (1). Now the expected total weight of all leaves after step 2 of ROUNDING-ALG is

$$\mathbb{E}(\ell_1) = \sum_{i=1}^{n} w_i e^{-t(1-x_i)} = e^{-t} W \left(\frac{\sum_{i=1}^{n} w_i e^{t x_i}}{W} \right)$$
$$\geq e^{-t} W (e^{t a W})^{\frac{1}{W}} = W e^{-t(1-a)}$$

The inequality above is obtained by the fact that the arithmetic mean is larger than the geometric mean, and then using $\sum_{i=1}^{n} w_i x_i = aW$. Now after step 3, a vertex v_i can cease to be a leaf with probability exactly

$$e^{-t(1-x_i)} \prod_{j \in N(i)} e^{-t(1-x_j)}.$$

Thus, if ℓ_2 is the total weight of vertices that were leaves after step 2 but ceased to be leaves after step 3, then its expectation is given by

$$
\mathbb{E}(\ell_2) = \sum_{i=1}^{n} w_i \left(e^{-t(1-x_i)} \prod_{j \in N(i)} e^{-t(1-x_j)} \right)
$$

$$
= \sum_{i=1}^{n} w_i e^{-t} \left(e^{-t(|N(i)| - \sum_{j \in N(i)} x_j - x_i)} \right)
$$

$$
\leq \sum_{i=1}^{n} w_i e^{-t} = W e^{-t}
$$

The inequality follows from the fact that the x_i's form a feasible solution. Now the expected value of the solution produced by ROUNDING-ALG is the expected total weight of leaves at the end of step 3. In other words, the expected value is given by

$$
\mathbb{E}(\ell_1) - \mathbb{E}(\ell_2) \geq W \left(\frac{e^{at} - 1}{c^t} \right)
$$

Now substituting the value $t = \frac{1}{a} \ln \left(\frac{1}{1-a} \right)$ completes the proof.   ∎

We have the following remarks concerning ROUNDING-ALG.

- The integrality gap of the LP is at most 3/4: consider a 4-cycle. Note that setting all $x_i = 2/3$ is a valid solution, giving an LP optimal value of 8/3. However, the integral optimum value is 2.
- The randomized rounding algorithm can easily be derandomized using the method of conditional expectations [2]. In fact, exact formulas for $\mathbb{E}(\ell_1)$ and $\mathbb{E}(\ell_2)$ are presented in the proof and the conditional expectations are easy to compute from these formulas.
- In the worst case where $a = 1/2$, the approximation ratio of ROUNDING-ALG for spanning star forest is rather bad (equal to 0.5). However, as we will see in the next two sections, we will take advantage of ROUNDING-ALG to get good approximation algorithms.

3.1   Application of ROUNDING-ALG to Dominating Set

Observe that the approximation ratio in Lemma 1 improves as the value of a increases. In particular, the approximation ratio tends to 1 as a approaches 1. This suggests that the above rounding scheme yields an approximation algorithm for the complementary objective of minimizing the dominating set. In fact, by analyzing the behavior of the function as a approaches 1, we obtain the following result.

Theorem 1. *The* ROUNDING-ALG *computes a* $\left(\ln\frac{W}{OPT_f} + 1 + 2\frac{OPT_f}{W}\ln\frac{W}{OPT_f}\right)$ *approximation ratio solution for the weighted dominating set problem, where* OPT_f *is the total weight of the optimal fractional dominating set solution.*

Proof. Let the optimal LP value for the spanning star forest be given by aW, where W is the sum of all the node weights. This implies that the optimal (fractional) dominating set has size $OPT_f = (1-a)W$.

Now, the dominating set returned by ROUNDING-ALG has size

$$W - aW(1-a)^{\frac{1}{a}-1} = OPT_f \cdot \frac{1 - (1-a)^{\frac{1}{a}-1}a}{1-a}$$

Let $a = 1 - \epsilon$. We have

$$\frac{1-(1-a)^{\frac{1}{a}-1}a}{1-a} = \frac{1 - \epsilon^{\frac{1}{1-\epsilon}-1}(1-\epsilon)}{\epsilon} = \frac{1-\epsilon^{\frac{\epsilon}{1-\epsilon}}}{\epsilon} + \epsilon^{\frac{\epsilon}{1-\epsilon}}$$

As $\epsilon < 1$, $\epsilon^{\frac{\epsilon}{1-\epsilon}} \leq 1$. Thus, the approximation ratio (for the dominating set problem) is at most:

$$\frac{1-\epsilon^{\frac{\epsilon}{1-\epsilon}}}{\epsilon} + 1 = \frac{1 - e^{\frac{\epsilon}{1-\epsilon}\ln\epsilon}}{\epsilon} + 1 \leq \frac{1 - \left(1 - \frac{\epsilon}{1-\epsilon}\ln\epsilon\right)}{\epsilon} + 1$$

$$= \frac{\frac{\epsilon}{1-\epsilon}\ln\frac{1}{\epsilon}}{\epsilon} + 1 \leq \ln\frac{1}{\epsilon}(1 + 2\epsilon) + 1 = \ln\frac{1}{\epsilon} + 2\epsilon\ln\frac{1}{\epsilon} + 1,$$

where in the above we have used that since $0 < \epsilon \leq 1$, $\frac{\epsilon}{1-\epsilon}\ln\frac{1}{\epsilon} < 1$. Further, for any $0 < y < 1$ and $0 < x \leq 1/2$, we have the following inequalities: $e^{-y} \geq 1 - y$ and $\frac{1}{1-x} \leq 1 + 2x$. Note that for our case we can always find a dominating set of size at most $W/2$, that is, $\epsilon \leq 1/2$. The proof is complete by noting that $\epsilon = OPT_f/W$. ∎

We remark that $\epsilon = \frac{OPT_f}{W}\ln\frac{W}{OPT_f}$ in general is at most 1. However, if $OPT_f = o(W)$, then $\epsilon = o(1)$. This result is close to the best known bound of $\left(OPT_f - \frac{1}{2}\right)\ln\frac{n}{OPT_f} + OPT_f$ from the analysis of greedy algorithm for set cover (and hence, applicable to dominating set too) in [10].

4   An Approximation Algorithm for the UNWEIGHTED SPANNING STAR FOREST Problem

In this section, we will describe a 0.71-approximation algorithm for the UN-WEIGHTED SPANNING STAR FOREST problem. We will use the following two known results.

Theorem 2 ([9]). *For any connected unweighted graph G of minimum degree at least 2, if the number of vertices $n \geq 8$, there is a polynomial time algorithm (denoted by* ORACLE-ALG*) to compute a spanning star forest of G of size at least $3n/5$.*

Theorem 3 ([9]). *For any tree T rooted at r, let $OPT_{ct}(T)$ and $OPT_{lf}(T)$ be the optimal value of spanning star forest of T given the condition that r is declared a center and leaf, respectively. Then $OPT_{ct}(T)$ and $OPT_{lf}(T)$ can be computed in polynomial time.*

Starting with the given connected graph G, we will generate a subgraph from G recursively as follows: Whenever there is a vertex in the current graph of degree 1, remove the vertex and the edge incident to it from the graph. Denote the final resulting subgraph to be G'. Note that G' is connected and every vertex in it has degree at least 2. Let

$S = \{v_i \in G' \mid \text{at least one edge incident to } v_i \text{ is dropped in the above process}\}.$

For simplicity, assume $S = \{v_1, \ldots, v_h\}$ and let $(G \setminus G') \cup S$ denote the induced subgraph on the vertex set $(V(G) \setminus V(G')) \cup S$.

Consider the subgraph $(G \setminus G') \cup S$: it is easy to verify that $(G \setminus G') \cup S$ is composed of h disconnected trees rooted at vertices in S. Denote these trees by $T_1, \ldots, T_h$, where the root of T_j is v_j. Let $OPT_{ct}(T_j)$ and $OPT_{lf}(T_j)$ be the optimal value of spanning star forest for T_j with the condition that v_j is declared a center and leaf, respectively. According to Theorem 3, $OPT_{ct}(T_j)$ and $OPT_{lf}(T_j)$ can be computed in polynomial time. Define

$$S_1 = \{v_j \in S \mid OPT_{ct}(T_j) < OPT_{lf}(T_j)\}$$
$$S_2 = \{v_j \in S \mid OPT_{ct}(T_j) \geq OPT_{lf}(T_j)\}$$

Let $N'(S_2)$ be the set of neighbors of S_2 in G'. Observe that $|N'(S_2)| \geq 2$ (otherwise, all vertices in S_2 would have been removed earlier). Consider the subgraph $G' \setminus S_2$ and assume that there are k vertices in $G' \setminus S_2$. We add two extra vertices u and v and connect u and v to all vertices in $N'(S_2)$. Let the resulting graph be G^* (see Figure 1 for an example). Note that G^* is a connected graph of minimum degree at least 2. Thus by Theorem 2, we can compute a spanning star forest of G^* of size at least $\frac{3}{5} \cdot (k + 2)$ in polynomial time.

Now we are ready to describe our algorithm.

TREECUTTING-ALG.

1. For each $i \in S_2$, declare i a center.
2. If the number of vertices in $G' \setminus S_2$ is smaller than (say) 1000,
3.   compute the optimal spanning star forest of G' given vertices in S_2 are centers.
4. Else,
5.   compute spanning star forests of G^* by ORACLE-ALG and ROUNDING-ALG.
6.   declare each $v_i \in G' \setminus S_2$ either a center of leaf according to $\max\{\text{ORACLE-ALG}(G^*), \text{ROUNDING-ALG}(G^*)\}$.
7. Given the choices made for the vertices in S, compute the best possible spanning star forest for $T_1, \ldots, T_h$.

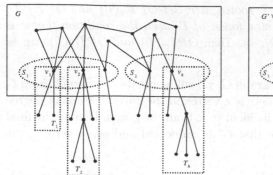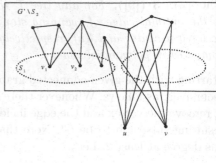

Fig. 1. Illustration of Graph G (left) and G^* (right)

Note that all vertices in S_2 are declared centers. Thus, in Step 6, the declaration of each vertex $v_i \in G' \setminus S_2$ is feasible (it is either covered by another vertex in $G' \setminus S_2$ or by a vertex in S_2). Therefore, the algorithm outputs a feasible spanning star forest solution.

In the following discussions, let $\alpha(G)$ and $\beta(G)$ be the value returned by Oracle-Alg(G) and Rounding-Alg(G), respectively. It can be seen that

$$
\begin{aligned}
&\text{TreeCutting-Alg}(G) \\
&\geq \max\{\alpha(G^*), \beta(G^*)\} - 2 + \sum_{v_i \in S_1} OPT(T_i \setminus v_i) + \sum_{v_j \in S_2} OPT_{ct}(T_j). \quad (2)
\end{aligned}
$$

where "-2" is because in the worst case, both u and v are leaves in the output of Oracle-Alg(G^*) or Rounding-Alg(G^*), but they do not contribute to the solution of $G' \setminus S_2$.

Observe that for any graph G'' and any vertex $w \in G''$, given a spanning star forest solution where w is a leaf, we can easily get a solution where w is a center by switching the declaration of w from leaf to center. Thus,

$$
OPT(G'' \mid w \text{ is a center}) \geq OPT(G'' \mid w \text{ is a leaf}) - 1.
$$

For any $v_j \in S_2$, note that

$$
\begin{aligned}
&OPT(G' \mid v_j \text{ is a center}) + OPT_{ct}(T_j) \\
&\geq OPT(G' \mid v_j \text{ is a leaf}) - 1 + OPT_{ct}(T_j) \\
&\geq OPT(G' \mid v_j \text{ is a leaf}) - 1 + OPT_{lf}(T_j),
\end{aligned}
$$

where the second inequality follows from the definition of S_2. Therefore,

$$
\begin{aligned}
OPT(G) = \max\{ &OPT(G' \mid v_j \text{ is a center}) + OPT_{ct}(T_j), \\
&OPT(G' \mid v_j \text{ is a leaf}) + OPT_{lf}(T_j) - 1 \} \\
= &OPT(G' \mid v_j \text{ is a center}) + OPT_{ct}(T_j)
\end{aligned}
$$

In other words, in the optimal solution of G, we can always assume vertices in S_2 are declared centers.

For any $v_i \in S_1$, we know essentially $OPT_{ct}(T_i) = OPT_{lf}(T_i) - 1$. Note that the root v_i contributes zero to $OPT_{ct}(T_i)$ and one to $OPT_{lf}(T_i)$. That is, regardless of the contribution of v_i, the contribution of vertices in $T_i \setminus \{v_i\}$ in $OPT_{ct}(T_i)$ and $OPT_{lf}(T_i)$ is the same. In other words, for any declaration of v_i (either center or leaf), we can always get the same optimal value for $T_i \setminus \{v_i\}$.

Therefore,

$$OPT(G) = OPT(G \mid \text{every } v_j \in S_2 \text{ is a center})$$
$$= OPT(G' \mid \text{every } v_j \in S_2 \text{ is a center})$$
$$+ \sum_{v_i \in S_1} OPT(T_i \setminus v_i) + \sum_{v_j \in S_2} OPT_{ct}(T_j). \qquad (3)$$

Thus, when k is small (i.e., TREECUTTING-ALG goes through Step 2,3), where recall that k is the number of vertices in $G' \setminus S_2$, TREECUTTING-ALG$(G) = OPT(G)$. Hence, we can assume that k is large (i.e., TREECUTTING-ALG goes through Step 4,5,6).

Assume that the optimal LP value satisfies $LP_{OPT}(G^*) = a \cdot (k+2)$, where recall that $G^* = (G' \setminus S_2) \cup \{u, v\}$. Hence,

$$\frac{\text{TREECUTTING-ALG}(G)}{OPT(G)}$$

$$\geq \frac{\max\{\alpha(G^*), \beta(G^*)\} - 2 + \sum_{v_i \in S_1} OPT(T_i \setminus v_i) + \sum_{v_j \in S_2} OPT_{ct}(T_j)}{OPT(G' \mid v_j \text{ is a center}, v_j \in S_2) + \sum_{v_i \in S_1} OPT(T_i \setminus v_i) + \sum_{v_j \in S_2} OPT_{ct}(T_j)}$$
$$\qquad (4)$$

$$\geq \frac{\max\{\alpha(G^*), \beta(G^*)\} - 2}{OPT(G' \mid v_j \text{ is a center}, v_j \in S_2)} \qquad (5)$$

$$\geq \frac{\max\{\alpha(G^*), \beta(G^*)\} - 2}{LP_{OPT}(G^*)} \qquad (6)$$

$$\geq \max\left\{ \frac{\frac{3}{5}(k+2)}{a \cdot (k+2)}, \frac{\beta(G^*)}{LP_{OPT}(G^*)} \right\} - \frac{2}{a \cdot (k+2)} \qquad (7)$$

$$= \max\left\{ \frac{0.6}{a}, (1-a)^{\frac{1}{a}-1} \right\} - \frac{2}{a \cdot (k+2)} \qquad (8)$$

$$> 0.71 \qquad (9)$$

where (4) follows from (2) and (3), (5) follows from the fact that the summations are non negative, (6) follows from the fact that the LP optimal is larger than the integral optimal value, (7) follows from Theorem 2, (8) follows from Lemma 1, and (9) follows by an estimation using a computer aided numerical analysis (Figure 2).

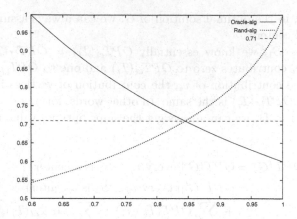

Fig. 2. The approximation ratios for ORACLE-ALG and ROUNDING-ALG. The horizontal line is 0.71.

In conclusion, we have the following result.

Theorem 4. TREECUTTING-ALG *gives a 0.71-approximation ratio solution for the* UNWEIGHTED SPANNING STAR FOREST *problem*.

5   An Approximation Algorithm for the NODE-WEIGHTED SPANNING STAR FOREST Problem

In this section, we present a 0.64-approximation algorithm for the node-weighted spanning star forest problem. Consider the following simple algorithm.

TRIVIAL-ALG

1. Compute a spanning tree T of the graph G, and pick an arbitrary vertex r as its root. Let h denote the height T rooted at r. For each integer k, let N_k denote the set of vertices at a distance of k (in the tree) from the root r.
2. Output the spanning star forest with the higher weight of the following:
   – centers: $N_0 \cup N_2 \cup \dots$, leaves: $N_1 \cup N_3 \cup \dots$
   – centers: $N_1 \cup N_3 \cup \dots$, leaves: $N_0 \cup N_2 \cup \dots$
   Essentially, the two spanning star forests are obtained by picking alternate levels in the spanning tree T.

It is easy to see that the following holds for TRIVIAL-ALG.

Proposition 1. TRIVIAL-ALG *always outputs a solution with value at least* $W/2$.

Theorem 5. *There exists a polynomial time algorithm that solves the* NODE-WEIGHTED SPANNING STAR FOREST *problem with an approximation factor of*

$$\min_{a \in [1/2,1)} \max \left(\frac{1}{2a}, (1-a)^{\frac{1}{a}-1} \right) > 0.64$$

Proof. Consider the algorithm that runs TRIVIAL-ALG and ROUNDING-ALG and picks the better of the two solutions– this algorithm obviously has polynomial running time. Let aW denote the value of the LP optimum. From Proposition 1, the TRIVIAL-ALG produces a spanning star forest with weight at least $W/2$, and hence an approximation ratio of at least $\frac{W/2}{aW} = \frac{1}{2a}$. Clearly this also implies that $a > \frac{1}{2}$. The claim on the approximation ratio follows from Lemma 1. The lower bound on the ratio follows by an estimation using a computer aided numerical analysis (Figure 3). ∎

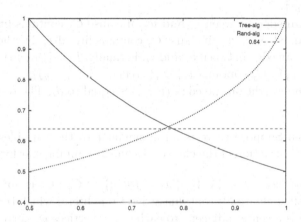

Fig. 3. The approximation ratios for TRIVIAL-ALG and ROUNDING-ALG. The horizontal line is 0.64.

6 Hardness of Approximation

The hardness results are obtained by a reduction from the following strong hardness for $MAX3SAT$.

Theorem 6 ([6]). *For every $\epsilon > 0$, given a 3-CNF formula ϕ it is NP-hard to distinguish between the following two cases:*

– *There exists an assignment satisfying $1 - \epsilon$ fraction of the clauses in ϕ*
– *No assignment satisfies more than $\frac{7}{8} + \epsilon$ fraction of the clauses in ϕ.*

Further, the hardness result holds even if each variable x_i is constrained to appear positively and negatively an equal number of times, i.e the literals $x_i, \bar{x}_i$ appear in equal number of clauses.

Theorem 7. *For any $\eta > 0$, it is NP-hard to approximate the* EDGE-WEIGHTED SPANNING STAR FOREST *problem within $\frac{19}{20} + \eta$.*

Proof. Let ϕ be a 3-CNF formula on n variables $\{x_1, x_2, \ldots, x_n\}$. Further let $C_1, C_2, \ldots, C_m$ be the set of clauses in ϕ. From Theorem 6, we can assume that each literal appears positively and negatively an equal number of times. For each i, let d_i denote the number of clauses containing x_i (respectively $\bar{x}_i$). Without loss of generality, we assume that $d_i \geq 2$ for all i. This can be achieved by just repeating the formula ϕ three times. A simple counting argument shows that $\sum_{i=1}^{n} d_i = \frac{3m}{2}$.

Create an edge-weighted graph G_ϕ as follows:

- Introduce one vertex u_i for each literal x_i and v_i for literal $\bar{x}_i$, and one vertex w_j for each clause C_j. Formally $V = \{u_1, \ldots, u_n\} \cup \{v_1, \ldots, v_n\} \cup \{w_1, \ldots, w_m\}$.
- Introduce an edge between u_i and w_j, if clause C_j contains literal x_i. Similarly, add an edge (v_i, w_j) if clause C_j contains literal $\bar{x}_i$. Furthermore, for all i, introduce an edge between u_i and v_i. Formally, $E = \{(u_i, w_j) \mid C_j \text{ contains } x_i\} \cup \{(v_i, w_j) \mid C_j \text{ contains } \bar{x}_i\} \cup \{(u_1, v_1), \ldots, (u_n, v_n)\}$.
- For all i, the weight on the edge (u_i, v_i) is equal to d_i. The rest of the edges have weight 1.

Completeness: Suppose there is an assignment to the variables $\{x_1, \ldots, x_n\}$ that satisfies $1 - \epsilon$ fraction of the clauses. Define a spanning star forest as follows:

- Centers: $\{u_i \mid x_i = true\} \cup \{v_i \mid x_i = false\} \cup \{C_j \mid C_j \text{ is not satisfied}\}$.
- Every satisfied clause C_j contains at least one literal which is assigned true. Thus there is a center adjacent to each of the vertices w_j corresponding to a satisfied clause. Since for each i, one of u_i or v_i is a center, the other vertex can be a leaf. Thus the set of leaves is given by: $\{u_i \mid x_i = false\} \cup \{v_i \mid x_i = true\} \cup \{w_j \mid C_j \text{ is satisfied}\}$.

Therefore, the total edge weight of the spanning star forest is given by

$$\sum_{i=1}^{n} d_i + |\{w_j \mid C_j \text{ is satisfied}\}| = \sum_{i=1}^{n} d_i + (1-\epsilon)m = \frac{3m}{2} + (1-\epsilon)m = \left(\frac{5}{2} - \epsilon\right)m.$$

Soundness: Consider the optimal spanning star forest solution OPT of G_ϕ. Without loss of generality, we can assume that for each i, exactly one of $\{u_i, v_i\}$ is a center, and the other is a leaf attached to it. This is because:

- If both u_i and v_i are centers, then modify the spanning star forest by deleting all the leaves attached to v_i, and making v_i a leaf of u_i. The total weight of the spanning star forest solution does not decrease, since we delete at most d_i edges of weight 1 and introduce an edge of weight d_i.

- If one of u_i and v_i is a center (say u_i) and the other (i.e. v_i) is a leaf but not attached to u_i, then we can disconnect v_i from its center and attach it to u_i. This operation increases the weight of the spanning star forest by $d_i - 1$, which contradicts to the optimality of the solution.
- If both u_i and v_i are leaves, then making u_i a center and attaching v_i to it will increase the weight of the solution by $d_i - 2$, again a contradiction.

From the spanning star forest solution OPT, obtain an assignment to ϕ as follows: $x_i = true$ if u_i is a center in OPT and $x_i = false$ otherwise. If vertex w_j is a leaf in OPT, then there is a center (say u_i) adjacent to it, which implies that clause C_j is satisfied by the assignment of x_i. A similar argument applies when the vertex w_j is adjacent to a center v_i. Therefore, the total weight of OPT is given by

$$\sum_{i=1}^{n} d_i + |\{w_j \mid C_j \text{ is satisfied}\}| = \frac{3m}{2} + |\{w_j \mid C_j \text{ is satisfied}\}|$$

In particular, if at most $(\frac{7}{8} + \epsilon)$-fraction of the clauses in ϕ can be satisfied, then the weight of OPT is at most $\frac{3m}{2} + (\frac{7}{8} + \epsilon)m - (\frac{19}{8} + c)m$.

From the completeness and soundness arguments, it is NP-hard to distinguish whether G_ϕ has a spanning star forest of weight $(\frac{5}{2} - \epsilon)m$ or $(\frac{19}{8} + \epsilon)m$. Thus it is NP-hard to approximate the EDGE-WEIGHTED SPANNING STAR FOREST problem within a factor of $(\frac{19}{8} + \epsilon)/(\frac{5}{2} - \epsilon)$. The claim follows by picking a small enough ϵ. ∎

The proof of the next theorem is similar to the previous one.

Theorem 8. *For any $\eta > 0$, it is NP-hard to approximate the* NODE-WEIGHTED SPANNING STAR FOREST *problem within $\frac{31}{32} + \eta$.*

Proof. Let ϕ be a 3-CNF formula on n variables $\{x_1, x_2, \ldots, x_n\}$ and m clauses $C_1, C_2, \ldots, C_m$. From Theorem 6, we can assume that each literal appears positively and negatively an equal number of times. For each i, let d_i denote the number of clauses containing x_i (respectively $\bar{x}_i$).

Create a node-weighted graph G_ϕ as follows:

- Introduce three vertices a_i, u_i, v_i for each variable x_i, and one vertex w_j for each clause C_j. Formally $V = \{a_1, \ldots, a_n\} \cup \{u_1, \ldots, u_n\} \cup \{v_1, \ldots, v_n\} \cup \{w_1, \ldots, w_m\}$.
- Introduce an edge between u_i and w_j, if clause C_j contains literal x_i. Similarly, add an edge (v_i, w_j) if clause C_j contains the literal $\bar{x}_i$. Furthermore, for all i, introduce edges $(a_i, u_i), (u_i, v_i), (v_i, a_i)$. Formally, $E = \{(u_i, w_j) \mid C_j$ contains $x_i\} \cup \{(v_i, w_j) \mid C_j$ contains $\bar{x}_i\} \cup \{(a_1, u_1), (u_1, v_1), (v_1, a_1), \ldots, (a_n, u_n), (u_n, v_n), (v_n, a_n)\}$
- For all i, the weight of nodes a_i, u_i, v_i is equal to d_i. The weight of the rest of nodes is 1.

Completeness: Suppose there is an assignment to the variables $\{x_1, \ldots, x_n\}$ that satisfies $1 - \epsilon$ fraction of the clauses. Define a spanning star forest solution as follows:

- Centers: $\{u_i \mid x_i = true\} \cup \{v_i \mid x_i = false\} \cup \{C_j \mid C_j$ is not satisfied$\}$.
- Every satisfied clause C_j contains at least one literal which is assigned true. Thus there is a center adjacent to each of the vertex w_j corresponding to a satisfied clause. Since for each i, one of u_i or v_i is a center, the other remaining two in $\{a_i, u_i, v_i\}$ can be leaves. Thus the set of leaves is given by : $\{u_i \mid x_i = false\} \cup \{v_i \mid x_i = true\} \cup \{w_j \mid C_j$ is satisfied$\} \cup \{a_i\}$.

The total node weight of the spanning star forest solution is given by

$$\sum_{i=1}^{n} 2d_i + |\{w_j \mid C_j \text{ is satisfied}\}| = \sum_{i=1}^{n} 2d_i + (1-\epsilon)m = 3m + (1-\epsilon)m = (4 - \epsilon)\, m.$$

The rest of the proof is similar to that of Theorem 7 and is omitted due to space considerations.   ∎

References

1. Agra, A., Cardoso, D., Cerfeira, O., Rocha, E.: A spanning star forest model for the diversity problem in automobile industry. In: ECCO XVIII, Minsk (May 2005)
2. Alon, N., Spencer, J.: The Probabilistic Method. John Wiley and Sons, Inc. New York, NY (1992)
3. Berry, V., Guillemot, S., Nicholas, F., Paul, C.: On the approximation of computing evolutionary trees. In: Proceedings of the Eleventh Annual International Computing and Combinatorics Conference, pp. 115–123 (2005)
4. Fiege, U.: A threshold of $\ln n$ for approximating set cover. Journal of the ACM 45(4), 634–652 (1998)
5. Goemans, M.X.: Mathematical programming and approximation algorithms. Lecture at Udine School, Udine, Italy (1996)
6. Håstad, J.: Some optimal inapproximability results. Journal of the ACM 48(4), 798–859 (2001)
7. Krauthgamer, R., Mehta, A., Rudra, A.: Pricing commodities, or how to sell when buyers have restricted valuations, (Manuscript, 2007)
8. Lund, C., Yannakakis, M.: On the hardness of approximating minimization problems. Journal of the ACM 41(5), 960–981 (1994)
9. Nguyen, C.T., Shen, J., Hou, M., Sheng, L., Miller, W., Zhang, L.: Approximating the spanning star forest problem and its applications to genomic sequence alignment. In: SODA, pp. 645–654 (2007)
10. Slavík, P.: A tight analysis of the greedy algorithm for set cover. In: STOC, pp. 435–441 (1996)

Packing and Covering δ-Hyperbolic Spaces by Balls*

Victor Chepoi and Bertrand Estellon

Laboratoire d'Informatique Fondamentale de Marseille,
Faculté des Sciences de Luminy, Universitée de la
Méditerranée, F-13288 Marseille Cedex 9, France
{chepoi,estellon}@lif.univ-mrs.fr

Abstract. We consider the problem of covering and packing subsets of δ-hyperbolic metric spaces and graphs by balls. These spaces, defined via a combinatorial Gromov condition, have recently become of interest in several domains of computer science. Specifically, given a subset S of a δ-hyperbolic graph G and a positive number R, let $\gamma(S, R)$ be the minimum number of balls of radius R covering S. It is known that computing $\gamma(S, R)$ or approximating this number within a constant factor is hard even for 2-hyperbolic graphs. In this paper, using a primal-dual approach, we show how to construct in polynomial time a covering of S with at most $\gamma(S, R)$ balls of (slightly larger) radius $R + \delta$. This result is established in the general framework of δ-hyperbolic geodesic metric spaces and is extended to some other set families derived from balls. The covering algorithm is used to design better approximation algorithms for the augmentation problem with diameter constraints and for the k-center problem in δ-hyperbolic graphs.

Keywords: covering, packing, ball, metric space, approximation algorithm.

1 Introduction

The *set cover problem* is a classical question in computer science [39] and combinatorics [9]. In this problem, given a collection $\mathcal{S}$ of subsets of a domain U with n elements, the task is to find a subcollection of $\mathcal{S}$ of minimum size $\gamma(\mathcal{S})$ whose union is U. It was one of Karp's 21 NP-complete problems. More recently, it has been shown that, under the assumption $P \neq NP$, set cover cannot be approximated in polynomial time to within a factor of $c \cdot \ln n$, where c is a small constant; see [3] and the references cited therein. On the other hand, set cover can be approximated in polynomial time to within a factor of $\ln n + 1$ using several algorithms [39], in particular, using the greedy algorithm. The *set packing problem* asks to find a maximum number $\pi(\mathcal{S})$ of pairwise disjoint subsets of $\mathcal{S}$. Another problem closely related to set cover is the hitting set problem. A subset

---

* This research was partly supported by the ANR grant BLAN06-1-138894 (projet OPTICOMB).

M. Charikar et al. (Eds.): APPROX and RANDOM 2007, LNCS 4627, pp. 59–73, 2007.

T is called a *hitting set* of $\mathcal{S}$ if $T \cap S \neq \emptyset$ for any $S \in \mathcal{S}$. The *minimum hitting set problem* asks to find a hitting set of $\mathcal{S}$ of smallest cardinality $\tau(\mathcal{S})$.

Numerous algorithmic and optimization problems can be formulated as set cover or set packing problems for structured set families. For example, many papers consider cover and packing problems with set families like intervals and unions of intervals of a line, subtrees of a tree, or cliques, cuts, paths, and balls of a graph. For example, in case of covering with balls, one can expect that the specific metric properties of graphs in question yield better algorithmic results in comparison with the general set cover. Although the set cover problem can be viewed as a particular instance of covering with unit balls of rather special graphs, for several graphs classes polynomial time algorithms have been designed. These algorithms resides on the treelike structure of those graphs and on the equality between ball covering and packing numbers of such graphs.

In this note, we consider the problem of covering and packing by balls and union of balls of hyperbolic metric spaces and graphs. The *ball* $B(x, R)$ of center x and radius $R \geq 0$ consists of all points of a metric space (X, d) at distance at most R from x. In our paper, we will consider covering and packing problems of the following type: given a finite subset S of points of X, a radius R, and a slack parameter δ, find a good covering of S with balls of radius at most $R + \delta$. We show that if the metric space (X, d) is δ-hyperbolic, then in polynomial time we can construct a covering of S with balls of radius $R + \delta$ and a set of the same size of pairwise disjoint balls of radius R centered at points of S. This type of results is obtained for arbitrary subfamilies of balls and for set-families consisting of unions of κ balls of (X, d). We apply these results to design better approximation algorithms for the k-center problem and the augmentation problem with diameter constraints in δ-hyperbolic graphs.

1.1   Geodesic and δ-Hyperbolic Metric Spaces

Let (X, d) be a metric space. A *geodesic segment* joining two points x and y from X is a map ρ from the segment $[a, b]$ of length $|a - b| = d(x, y)$ to X such that $\rho(a) = x, \rho(b) = y$, and $d(\rho(s), \rho(t)) = |s - t|$ for all $s, t \in [a, b]$. A metric space (X, d) is *geodesic* if every pair of points in X can be joined by a geodesic. We will denote by $[x, y]$ any geodesic segment connecting the points x and y. Every graph $G = (V, E)$ equipped with its standard distance d_G can be transformed into a (network-like) geodesic space (X, d) by replacing every edge $e = (u, v)$ by a segment $[u, v]$ of length 1. These segments may intersect only at their commons ends. Then (V, d_G) is isometrically embedded in a natural way in (X, d).

Introduced by Gromov [29], δ-hyperbolicity measures, to some extent, the deviation of a metric from a tree metric. Recall that a metric space (X, d) embeds into a tree network (with positive real edge lengths), that is, d is a *tree metric*, if and only if for any four points u, v, w, x the two larger ones of the distance sums $d(u, v) + d(w, x), d(u, w) + d(v, x), d(u, x) + d(v, w)$ are equal. Now, a metric space (X, d) is called δ-*hyperbolic* if the two larger distance sums differ by at most δ. A connected graph $G = (V, E)$ equipped with standard graph metric d_G is δ-hyperbolic if (V, d_G) is a δ-hyperbolic metric space.

In case of geodesic metric spaces, there exist several equivalent definitions of δ-hyperbolic metric spaces involving different but comparable values of δ [5,28,29]. In this paper, we will use the definition employing δ-thin geodesic triangles. A *geodesic triangle* $\Delta(x, y, z)$ with vertices $x, y, z \in X$ is a union $[x, y] \cup [x, z] \cup [y, z]$ of three geodesic segments connecting these vertices. Let m_x be the point of the geodesic segment $[y, z]$ located at distance $\alpha_y := (d(y, x) + d(y, z) - d(x, z))/2$ from y. Then m_x is located at distance $\alpha_z := (d(z, y) + d(z, x) - d(y, x))/2$ from z because $\alpha_y + \alpha_z = d(y, z)$. Analogously, define the points $m_y \in [x, z]$ and $m_z \in [x, y]$ both located at distance $\alpha_x := (d(x, y) + d(x, z) - d(y, z))/2$ from x; see Fig. 1 for a construction. There exists a unique isometry φ which maps the geodesic triangle $\Delta(x, y, z)$ to a star $\Upsilon(x', y', z')$ consisting of three solid segments $[x', m']$, $[y', m']$, and $[z', m']$ of lengths α_x, α_y, and α_z, respectively. This isometry maps the vertices x, y, z of $\Delta(x, y, z)$ to the respective leaves x', y', z' of $\Upsilon(x', y', z')$ and the points m_x, m_y, and m_z to the center m of this tripod. Any other point of $\Upsilon(x', y', z')$ is the image of exactly two points of $\Delta(x, y, z)$. A geodesic triangle $\Delta(x, y, z)$ is called δ-*thin* [5] if for all points $u, v \in \Delta(x, y, z)$, $\varphi(u) = \varphi(v)$ implies $d(u, v) \leq \delta$. A geodesic metric space (X, d) is called δ-*hyperbolic* if all geodesic triangles of X are δ-thin. Note that our δ-hyperbolic metric spaces will be 2δ-hyperbolic if we will use the first definition of δ-hyperbolicity; see the proof of Proposition 2.1 of [5].

Throughout this paper, we will suppose that all metric spaces are either geodesic or graphic with δ-thin geodesic triangles. Additionally, in case of geodesic spaces (X, d), we will assume the following computational assumption: *there exists an oracle which, given two points $x, y \in X$, it returns a geodesic segment $[x, y]$.* In case of graph-distance d_G or of geodesic spaces derived from graphs, the role of this oracle is played by any shortest path algorithm.

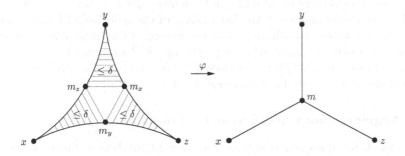

Fig. 1. A geodesic triangle $\Delta(x, y, z)$, the points m_x, m_y, m_z, and the tripod $\Upsilon(x', y', z')$

1.2   r-Domination and r-Packing

Now, we will formulate the r-*domination* and r-*packing* problems, which correspond to covering and packing by balls. Let S be a subset of not necessarily distinct points of a metric space (X, d) and let $r : S \to \mathbb{R}_+$ be a map associating to each point $s \in S$ a positive number $r(s)$. We say that a subset C of X r-*dominates* S if for each point $s \in S$ there exists a point $c \in C$ such that $d(s, c) \leq r(s)$. In other words, C is a hitting set for the family of balls $\mathcal{B}_{S,r} = \{B(s, r(s)) : s \in S\}$. A subset P of S is called an r-*packing* of S, if for each pair x, x' of points of P we have $r(x) + r(x') < d(x, x')$ (in other words, the family $B(x, r(x)), x \in P$, consists of pairwise disjoint balls). The r-*domination problem* is to find an r-dominating set with minimum size $\gamma(S, r)$ and the r-*packing problem* is to find an r-packing set with maximum size $\pi(S, r)$. Then $\gamma(S, r) = \tau(\mathcal{B}_{S,r})$ and $\pi(S, r)$ are called the r-*domination* and the r-*packing numbers* of (X, d) (these numbers are well-defined when S is finite). If S is a subset of vertices of a graph $G = (V, E)$ and the r-dominating set C is also contained in V, then we denote the respective r-domination and r-packing numbers by $\gamma_G(S, r)$ and $\pi_G(S, r)$. If $r(s) \equiv R$ for all $s \in S$, then we obtain the problem of covering S with a minimum number of balls of radius R; in the particular case $r(s) \equiv 1$ and $S \subseteq V$, we obtain the well-known domination problem of a graph.

The r-domination problem is closely related with the k-center clustering problem [10,30,31,39]. In the k-*center problem,* given a set S of n points in a metric space (X, d), the goal is to find the smallest R^* and the position of k centers, such that any point of S is at distance of at most R^* from one of those centers (in other words, R^* is the least radius such that S can be covered with at most k balls of radius R^*).

A κ-*ball* $^\kappa B$ of a metric space (X, d) is the union of κ balls $B(x_1, r_1), \ldots, B(x_\kappa, r_\kappa)$, i.e., $^\kappa B = \bigcup_{j=1}^{\kappa} {}^\kappa B(x_j, r_j)$. It extends the notions of d-intervals of [1,7]; these are unions of d closed intervals of $\mathbb{R}$. Indeed, each interval $[a, b]$ can be viewed as a closed ball of $\mathbb{R}$ of radius $(b - a)/2$ centered at the point $(a + b)/2$. As in the case of r-domination, any finite family $^\kappa \mathcal{B}_{S,\mathbf{r}}$ of κ-balls can be defined via the set S of centers of all balls and a multi-valued map $\mathbf{r} : S \to \mathbb{R}_+$ which associates to each point $s \in S$ the list of radii of the balls from $\bigcup {}^\kappa \mathcal{B}_{S,\mathbf{r}}$ centered at s. Thus two κ-balls may have two balls centered at the same point. The $(\kappa, \mathbf{r})$-*domination problem* consists in finding a hitting set C for a family $^\kappa \mathcal{B}_{S,\mathbf{r}}$ of κ-balls of minimum cardinality $\gamma(S, \mathbf{r})$. Analogously, the $(\kappa, \mathbf{r})$-*packing problem* is to find a maximum number $\pi(S, \mathbf{r})$ of pairwise disjoint κ-balls of $^\kappa \mathcal{B}_{S,\mathbf{r}}$.

1.3   Augmentation Under Diameter Constraints

In Section 4, we apply our results on covering δ-hyperbolic graphs with balls to the following *augmentation problem under diameter constraints* (problem ADC): Given a graph $G = (V, E)$ with n vertices and a positive integer D, add a minimum number OPT(D) of new edges E' such that the augmented graph $G' = (V, E \cup E')$ has diameter at most D. ADC can be viewed as a network improvement problem where G is the initial communication network and a minimum number of additional communication links must be added so that the upgraded network G' ensures a low communication delay.

1.4   Our Results

Using the notation established in previous subsections, the main algorithmic results of our paper can be formulated in the following way: for geodesic δ-hyperbolic spaces and δ-hyperbolic graphs, $\gamma(S, r + \delta) \leq \pi(S, r)$ and $\gamma(S, \mathbf{r} + 2\delta) \leq 2\kappa^2\pi(S, \mathbf{r})$ hold. Moreover, it is possible to construct in polynomial time an $(r + \delta)$-dominating set C and an r-packing P such that $|C| = |P|$, and a $(\kappa, \mathbf{r}+2\delta)$-dominating set C and a $(\kappa, \mathbf{r})$-packing P such that $|C| \leq 2\kappa^2|P|$. Using these results, we show that one can augment in polynomial time a δ-hyperbolic graph $G = (V, E)$ to a graph of diameter $2R + 2\delta$ using at most $2\text{OPT}(2R)$ new edges. These results also show that for δ-hyperbolic graphs, the well-known 2-approximation algorithm [31] for the k-center problem returns a solution of radius at most $\text{OPT} + \delta$. Notice also that the problem of approximating $\gamma(S, r)$ within a constant is hard already for 2-hyperbolic graphs and $r(s) \equiv 1$ for all $s \in S$, because the split graphs, which encode the general set cover problem (the elements x of the domain U form a clique and the sets S of $\mathcal{S}$ form a stable set so that the vertices x and S are adjacent if and only if $x \in S$), are chordal, and therefore, 2-hyperbolic.

1.5   Related Work

We briefly review the known results related with the subject of our paper. The inequality $\gamma(S, r) \geq \pi(S, r)$ holds in any metric space (X, d), because two points of an r-packing cannot be r-dominated by the same point. On the other hand, the equality $\gamma_G(S, r) = \pi_G(S, r)$ holds for trees [13,14], strongly chordal graphs [15,25], dually chordal graphs [11], and it is at the heart of linear-time algorithms for r-covering and r-packing problems for these graphs. The paper [21] proposes an exact fixed-parameter algorithm for the (NP-hard) problem of covering planar graphs with a minimum number of balls of radius R. Finally, [17] shows that every planar graph of diameter $2R$ can be covered with a fixed number of balls of radius R. Covering and packing problems for special families of subtrees of a tree have been considered in [8]. Alon [1,2] established that if $^\kappa\mathcal{I}$ is a family of κ-intervals of the line (or a family consisting of unions of κ subtrees of a tree), then $\tau(^\kappa\mathcal{I}) \leq 2\kappa^2\pi(^\kappa\mathcal{I})$. In case of κ-intervals, Bar-Yehuda et al. [7] presented a factor 2κ algorithm for approximating $\pi(^\kappa\mathcal{I})$. Their algorithm is based on rounding a fractional solution of the linear relaxation of the problem and construction of a respective packing using the local ratio technique.

The k-center problem is a well-studied k-clustering and facility location problem [10,30,39]. The general problem is NP-hard to approximate with a factor smaller than 2 (see Theorem 5.7 of [39]). The analogous problem in Euclidean spaces is NP-hard to approximate with a factor smaller than 1.822 [26]. Hochbaum and Shmoys [31] present a (best possible) factor 2 approximation algorithm for the general k-center problem.

The augmentation problem of graphs with diameter constraints has been introduced in [20]. It is already non-trivial when the input graph is a path [4,24]. Approximation algorithms for this augmentation problem has been designed in

[18,19,22,32]. In particular, [18,19] propose factor 2 approximation algorithms for the augmentation problem of trees and dually chordal graphs with even and odd diameters $2R$ and $2R + 1$ based on particular coverings of trees with balls of radius $R - 1$ and R.

δ-Hyperbolic metric spaces play an important role in geometric group theory and in geometry of negatively curved spaces [5,28,29]. δ-Hyperbolicity captures the basic common features of "negatively curved" spaces like the hyperbolic space $\mathbb{H}^k$, Riemannian manifolds of strictly negative sectional curvature, and of discrete spaces like trees and the Caley graphs of word-hyperbolic groups. It is remarkable that a strikingly simple concept leads to such a rich general theory [5,28,29]. More recently, the concept of δ-hyperbolicity emerged in discrete mathematics, algorithms, and networking. For example, it has been shown empirically in [38] that the internet topology embeds with better accuracy into a hyperbolic space than into a Euclidean space of comparable dimension. A few algorithmic problems in hyperbolic spaces and hyperbolic graphs have been considered in recent papers [23,27,33,35]. 0-Hyperbolic metric spaces are exactly the tree metrics. On the other hand, the Poincaré half space in $\mathbb{R}^k$ with the hyperbolic metric is δ-hyperbolic with $\delta = \log 3$. A full characterization of 1-hyperbolic graphs has been given in [6]; see also [34] for a partial characterization. Chordal graphs (graphs in which all induced cycles have length 3) are 2-hyperbolic [34]. For chordal graphs as well as dually chordal and strongly chordal graphs one can construct trees approximating the graph-distances within a constant 2 or 3 [12], from which follows that those graphs have low δ-hyperbolicity (this result has been extended in [16] to all graphs in which the largest induced cycle is bounded by some constant δ; this result implies that those graphs are δ-hyperbolic). In general, the distance of a δ-hyperbolic space on n points can be approximated within a factor of $2\delta \log n$ by a tree metric [29,28] and this approximation is sharp.

2 r-Domination and r-Packing

Let (X, d) be a geodesic δ-hyperbolic space. Given an instance (S, r) of the r-domination and r-packing problems, denote by $r + \delta$ the function defined by setting $(r + \delta)(x) := r(x) + \delta$ for all $x \in S$. For each point $x \in S$, define the set $S_x := \{y \in S : r(x) + r(y) \geq d(x, y)\}$ of all points which cannot belong to the same r-packing set as x. Next auxiliary result shows that in any compact subset S of X one can always find a point x such that x and all points of the set S_x can be $(r + \delta)$-dominated by a common point $c \in X$.

Lemma 1. *For any compact subset S of X, there exist two points $x \in S$ and $c \in X$ such that $d(c, y) \leq r(y) + \delta$ for any point $y \in S_x$, i.e., S_x is $(r + \delta)$-dominated by c.*

Proof. Let x, z be a pair of points of S maximizing the value $M := d(x, z) - r(x)$ (such a pair exists because S is compact). If $M \leq \delta$, then the point z $(r + \delta)$-dominates all points of S and we can set $c := z$. Suppose now that $M > \delta$. Pick a geodesic segment $[x, z]$ between x and z, and let c be the point of $[x, z]$ located at distance $r(x)$ from x. Consider any point $y \in S$ such that $r(x) + r(y) \geq d(x, y)$. We assert that $d(y, c) \leq r(y) + \delta$. For this pick any two geodesic segments $[x, y]$ and $[y, z]$ between the pairs x, y and y, z. Let $\Delta(x, y, z) := [x, y] \cup [x, z] \cup [y, z]$ be the geodesic triangle formed by the three geodesic segments and let m_x, m_y, and m_z be the three points on these geodesics as defined above. We distinguish two cases. First suppose that c belongs to the portion of $[x, z]$ comprised between the points x and m_y. In this case, since $d(x, m_y) = d(x, y) - \alpha_y$, we obtain

$$d(c, m_y) = d(x, y) - \alpha_y - r(x) \leq r(x) + r(y) - \alpha_y - r(x) = r(y) - \alpha_y.$$

Since $\Delta(x, y, z)$ is δ-thin, the triangle condition yields

$$d(c, y) \leq d(c, m_y) + d(m_y, m_x) + d(m_x, y) \leq d(c, m_y) + \delta + \alpha_y \leq r(y) + \delta.$$

On the other hand, if c belongs to the portion of $[x, z]$ comprised between z and m_y, then the choice of the points x and z yields $d(y, z) - r(y) \leq d(x, z) - r(x)$. Since $d(x, z) = \alpha_x + \alpha_z$ and $d(y, z) = \alpha_y + \alpha_z$, we conclude that $\alpha_y - r(y) \leq \alpha_x - r(x)$. Thus $d(c, m_y) = r(x) - \alpha_x \leq r(y) - \alpha_y$. As a result, we deduce that $d(c, y) \leq d(c, m_y) + \delta + \alpha_y < r(y) + \delta$. □

The following result can be viewed as the variant for δ-hyperbolic spaces of the classical Jung theorem asserting that each subset S of the Euclidean space $\mathbb{E}^m$ with finite diameter D is contained in a ball of radius at most $\sqrt{\frac{m}{2(m+1)}} D$.

Corollary 1. *If the diameter of a compact geodesic δ-hyperbolic metric space (X, d) is $D := 2R$, then X can be covered by single ball of radius $R + \delta$, i.e., the radius of X is at most $R + \delta$.*

Proof. Let $S := X$ and $r(x) \equiv R$. Since $d(x, y) \leq 2R = r(x) + r(y)$ for any pair $x, y \in X$, we conclude that $S_x = X$ for any point $x \in X$. Since X is compact, by Lemma 1, there exist a point $x \in S = X$ and a point $c \in X$ such that $X = S_x \subseteq B(c, r(x) + \delta) = B(c, R + \delta)$. □

The following result, generalizing Corollary 1, can be viewed as the analogy of the classical Helly property for balls.

Corollary 2. *If $B(x_i, r_i), i \in I$, is a collection of pairwise intersecting balls of a geodesic δ-hyperbolic metric space (X, d) with a compact set $S := \{x_i : i \in I\}$ of centers, then the balls $B(x_i, r_i + \delta), i \in I$, have a nonempty intersection.*

Proof. Set $r(x_i) := r_i$. Then, as in previous result, since $d(x_i, x_j) \leq r_i + r_j = r(x_i) + r(x_j)$, the equality $S_x = S$ holds for any point x_i of S. By Lemma 1, S is $(r + \delta)$-dominated by a single point c. Obviously this point belongs to all balls $B(x_i, r_i + \delta)$, establishing the result. □

Now, we present the main result of this paper. It generalizes the equality $\gamma(S, r) = \pi(S, r)$ for trees to all δ-hyperbolic spaces in the following way:

Theorem 1. *Let S be a finite subset of a geodesic δ-hyperbolic metric space (X, d). Then $\gamma(S, r + \delta) \leq \pi(S, r)$. Moreover, a set C $(r + \delta)$-dominating the set S and an r-packing P of S such that $|C| = |P|$ can be constructed using a polynomial in $|S|$ number of calls of the oracle for computing a geodesic segment in (X, d).*

Proof. The proof of this result is algorithmic: we construct the r-packing P and the $(r + \delta)$-dominating set C step by step taking care that the following properties hold: (i) each time when a new point is inserted in C, then a new point is also inserted in P, and (ii) at the end, the set P is an r-packing and C is an $(r + \delta)$-dominating set for S.

The algorithm starts with $S' := S$, $C := \emptyset$, and $P := \emptyset$. While the set S' is nonempty, the algorithm applies Lemma 1 to the current set S' in order to obtain a point $x \in S'$ and a point $c \in X$ which $(r+\delta)$-dominates the set S'_x. The algorithm adds the point x to P and the point c to C, and then it updates the set S' by removing from S' all points which are $(r + \delta)$-dominated by c, and so on. The algorithm terminates in at most $|S|$ rounds. Notice also that $|P| = |C|$, because when the point x is inserted in P, then at the same step x is removed from S' because x is $(r + \delta)$-dominated by the point c which is included at that step in C.

We assert that at the end, P is an r-packing of S and C is an $(r + \delta)$-dominating set for S. Indeed, C $(r + \delta)$-dominates S because S' is empty when the algorithm halts and that each point $s \in S$ is $(r + \delta)$-dominated by the point which is inserted in C at the iteration when s is removed from S'. To show that P is an r-packing it suffices to show that after each iteration the updated set P is an r-packing. So, suppose that at the current iteration the point y has been inserted in the set P, which before this insertion was an r-packing. We must show that $P \cup \{y\}$ is an r-packing as well. Suppose by way of contradiction that $d(x, y) \leq r(x) + r(y)$ for some point $x \in P$. Consider the iteration at which the point x was inserted in P and suppose that at this iteration the point c was inserted in C. Since $y \in S'_x$ and all points of S'_x are $(r + \delta)$-dominated by c, the algorithm will remove at this iteration y from S', thus y cannot be inserted in P at a later stage, contrary to our assumption. This ensures that P is an r-packing during all execution of the algorithm. □

Consider now the case of r-domination and r-packing for graphs $G = (V, E)$ such that the underlying geodesic metric space (X, d) is δ-hyperbolic. More precisely, let S be a subset of vertices of G, let r be a map from S to $\mathbb{N}_+$, and we are searching for a subset of vertices $C \subseteq V$ which $(r + \delta)$-dominates S. Now, if we will run in (X, d) the algorithm described in Theorem 1 with S and r as an input, then the r-dominating set C returned by this algorithm must be a subset of V. For this, it suffices to notice that each vertex $c \in C$ is defined according to the choice of Lemma 1. The point c in this lemma is located at distance $r(x)$ from x on a geodesic segment $[x, z]$. Since $r(x)$ and $d(x, z)$ are integers, we conclude that c is a vertex of G. In case of graphs, we can specify the oracle computing geodesic segments: it suffices to use any shortest-path algorithm in G. Finally, notice that if $r(x) := R$ for all $x \in S$, then an $(r + \delta)$-dominating set

C corresponds to the set of centers of balls of radius $R + \delta$ covering the set S. Summarizing, we obtain the following observation.

Corollary 3. *Let S be a subset of vertices of a finite δ-hyperbolic graph $G = (V, E)$. Then $\gamma_G(S, r + \delta) \leq \pi_G(S, r)$. Moreover a set $C \subseteq V$ $(r + \delta)$-dominating the set S and an r-packing P of S such that $|C| = |P|$ can be constructed in polynomial time.*

3 $(\kappa, \mathbf{r})$-Domination and $(\kappa, \mathbf{r})$-Packing

Let ${}^\kappa\mathcal{B}_{S,\mathbf{r}}$ be a finite family of κ-balls of a δ-hyperbolic geodesic metric space. For $\varepsilon > 0$, denote by ${}^\kappa\mathcal{B}_{S,\mathbf{r}+\varepsilon}$ the family of balls obtained by "inflating" each ball $B(s, r(s))$ of $\bigcup {}^\kappa\mathcal{B}_{S,\mathbf{r}}$ until its radius becomes $r(s) + \varepsilon$, i.e., obtained by replacing $\mathbf{r}$ by the function $\mathbf{r} + \varepsilon$. We call the κ-balls of ${}^\kappa\mathcal{B}_{S,\mathbf{r}+\varepsilon}$ ε-*inflated κ-balls*. This section is devoted to the proof of the following result:

Theorem 2. *Let ${}^\kappa\mathcal{B}_{S,\mathbf{r}} = \{{}^\kappa B_1, \ldots, {}^\kappa B_m\}$ be a family of κ-balls of a δ-hyperbolic geodesic metric space. Then $\gamma(S, \mathbf{r} + 2\delta) \leq 2\kappa^2 \pi(S, \mathbf{r})$. Moreover a hitting set C for ${}^\kappa\mathcal{B}_{S,\mathbf{r}+2\delta}$ and a packing P of ${}^\kappa\mathcal{B}_{S,\mathbf{r}}$ such that $|C| \leq 2\kappa^2 |P|$ can be constructed with a polynomial in $|S|$ number of calls of the oracle for computing geodesic segments in (X, d).*

For each κ-ball ${}^\kappa B_i$, denote by S_i the set of centers of the balls constituting ${}^\kappa B_i$. For each $s \in S_i$, let $r_i(s)$ be the radius of the ball of ${}^\kappa B_i$ centered at s. Clearly, $S = \bigcup_{i=1}^m S_i$. For a point $v \in X$, let $N[v] := \{i : d(v, s) \leq r_i(s) \text{ for some } s \in S_i\}$ be the set of indices of all κ-balls ${}^\kappa B_i$ covering v. For any $i = 1, \ldots, m$, let $N[i]$ be the set of indices of all κ-balls which cannot be included in a common $(\kappa, \mathbf{r})$-packing with ${}^\kappa B_i$, i.e., $N[i] = \bigcup\{N[v] : v \in {}^\kappa B_i\}$. Clearly, if $j \in N[i]$, then $i \in N[j]$. Notice also that $i \in N[i]$.

Now we can formulate a pair of dual linear programs whose optimal solutions $\pi_f(S, \mathbf{r})$ and $\gamma_f(S, \mathbf{r})$ are an optimal fractional packing and an optimal fractional covering for ${}^\kappa\mathcal{B}_{S,\mathbf{r}}$, respectively. For this, we introduce a variable x_i for each κ-ball ${}^\kappa B_i$ and a dual variable y_v for each point $v \in X$.

$$\begin{cases} \max & \sum_{i=1}^m x_i \\ \text{s.t.} & \sum_{i \in N[v]} x_i \leq 1 \quad \forall v \in X \\ & x_i \qquad\qquad \geq 0 \quad \forall i = 1, \ldots, m \end{cases} \qquad \Pi(S, \mathbf{r})$$

$$\begin{cases} \min & \int_{v \in X} y_v \\ \text{s.t.} & \int_{v \in {}^\kappa B_i} y_v \geq 1 \quad \forall i = 1, \ldots, m \\ & y_v \qquad\quad\ \geq 0 \quad \forall v \in X. \end{cases} \qquad \Gamma(S, \mathbf{r})$$

Notice that the first linear program contains as many constraints as points in the space X, while the second linear program assumes that we can integrate over the balls of X. In fact, one can easily rewrite $\Pi(S, \mathbf{r})$ using only a finite number of constraints: since there exists only a finite number of patterns of intersections of balls in $\bigcup {}^\kappa\mathcal{B}_{S,\mathbf{r}}$, we can pick a point v in each type of intersection and write the constraints $\sum_{i \in N[v]} x_i \leq 1$ only for such v. Denote the resulting finite set by V^*. We can also rewrite $\Gamma(S, \mathbf{r})$ by replacing the integration by a sum over all points $v \in V^*$ belonging to $^\kappa B_i$. The resulting linear programs have respectively m variables, $|V^*|$ constraints and, vice-versa, $|V^*|$ variables and m constraints.

Now, we will construct in polynomial time a set V of size $\kappa \cdot m$ and formulate the linear programs on V instead of X or V^*. Then we relate the admissible and optimal solutions of resulting linear programs with those of $\Pi(S, \mathbf{r})$ and $\Gamma(S, \mathbf{r})$.

Let $\mathcal{B}$ denote the set of all balls participating in κ-balls of the family $^\kappa \mathcal{B}_{S, \mathbf{r}}$. Denote the radius function of these balls by r and by $S = \bigcup_i^m S_i$ the multi-set of centers of the balls from $\mathcal{B}$. The set V is constructed iteratively, starting with $V := \emptyset$, $S' = S$, and $\mathcal{B}' := \mathcal{B}$. At each iteration, given the current set of balls $\mathcal{B}'$ and the set S' of their centers, we apply Lemma 1 to find a point $s \in S'$, a ball $B \in \mathcal{B}'$ centered at s, and a point $c_s \in X$ such that the set S'_s is $(r + \delta)$-dominated by c_s. Then the point c_s is inserted in V and the ball B is removed from $\mathcal{B}'$. The algorithm halts when $\mathcal{B}'$ becomes empty. Clearly, the returned set V has cardinality $\kappa \cdot m$. Denote by $\pi'_f(S, \mathbf{r})$ and $\gamma'_f(S, \mathbf{r})$ the optimal solutions of the following linear programs:

$$\begin{cases} \max \ \sum_{i=1}^m x_i \\ \text{s.t.} \ \sum_{i \in N[v]} x_i \leq 1 \quad \forall v \in V \\ \quad\quad x_i \quad\quad\quad \geq 0 \quad \forall i = 1, \dots, m \end{cases} \qquad \Pi'(S, \mathbf{r})$$

$$\begin{cases} \min \ \sum_{v \in C} y_v \\ \text{s.t.} \ \sum_{v \in {}^\kappa B_i} y_v \geq 1 \quad \forall i = 1, \dots, m \\ \quad\quad y_v \quad\quad\quad \geq 0 \quad \forall v \in V. \end{cases} \qquad \Gamma'(S, \mathbf{r})$$

Lemma 2. *Any admissible solution $\{x_i : i = 1, \dots, m\}$ of $\Pi'(S, \mathbf{r} + \delta)$ is also an admissible solution of $\Pi(S, \mathbf{r})$. Moreover, $\gamma'_f(S, \mathbf{r} + \delta) \leq \gamma_f(S, \mathbf{r})$.*

Proof. Notice that it suffices to check the inequality $\sum_{i \in N[v]} x_i \leq 1$ only for points $v \in X$ for which the set $N[v]$ is nonempty. Then v belongs to at least one ball from the set $\mathcal{B}$. Among such balls, let B be the first ball considered by the algorithm constructing the set V. Let s be the center of B and c_s be the point included in V when the ball B is removed from $\mathcal{B}'$. Notice that the set $S(v)$ of centers of all balls of $\mathcal{B}$ containing v belongs to S'_s. The definition of c_s yields that $S(v)$, as a part of S'_s, is $(r + \delta)$-dominated by the point c_s of V. Writing down the constraint of $\Pi'(S, \mathbf{r} + \delta)$ defined by the point c_s, we conclude that the sum of x_i's over all δ-inflated κ-balls containing c_s is at most 1. Since the δ-inflations of all κ-balls containing v all contain c_s, we conclude that $\sum_{i \in N[v]} x_i \leq 1$ holds in $\Pi(S, \mathbf{r})$. $\square$

Lemma 3. *If $\mathbf{x} = \{x_i : i = 1, \dots, m\}$ is an admissible solution of $\Pi'(S, \mathbf{r} + \delta)$, then there exists a κ-ball $^\kappa B_i$ such that $\sum_{j \in N[i]} x_j \leq 2\kappa$.*

Proof. The proof of this result is inspired by the averaging argument used in the proof of Lemma 4.1 of [7]. Define a graph $\mathbf{N}$ with $1, \dots, m$ as the set of vertices and in which ij is an edge if and only if $j \in N[i]$ (and consequently $i \in N[j]$).

For each edge ij of $\mathbf{N}$, set $z(i,j) = x_i \cdot x_j$. Since $i \in N[i]$, define $z(i,i) = x_i^2$. In the sum $\sum_{i=1}^{m} \sum_{j \in N[i]} z(i,j)$ every $z(i,j)$ is counted twice. On the other hand, an upper bound on this sum can be obtained in the following way. For a point $s \in S$, let $N^{\delta}[c_s]$ be the set of indices of all δ-inflated κ-balls which contain the point c_s. Now, for each κ-ball ${}^{\kappa}B_i$ consider its set of centers S_i, and for each $s \in S_i$, add up $z(i,j)$ for all $j \in N^{\delta}[c_s]$, and then multiply the total sum by 2. This way we computed the sum $2 \sum_{i=1}^{m} \sum_{s \in S_i} \sum_{j \in N^{\delta}[c_s]} z(i,j)$. We assert that this suffices. Indeed, pick any $z(i,j)$ for an edge ij of the graph $\mathbf{N}$. Thus the κ-balls ${}^{\kappa}B_i$ and ${}^{\kappa}B_j$ contain two intersecting balls B and B', say B is centered at $s \in S_i$. Suppose without loss of generality that the algorithm for constructing the set V considers B before B'. Then necessarily $j \in N^{\delta}[c_s]$, because c_s hits the δ-inflation of the ball B'. Hence the term $z(i,j)$ will appear at least once in the triple sum, establishing the required inequality

$$\sum_{i=1}^{m} \sum_{j \in N[i]} z(i,j) \leq 2 \sum_{i=1}^{m} \sum_{s \in S_i} \sum_{j \in N^{\delta}[c_s]} z(i,j).$$

Taking into account that $z(i,j) = x_i \cdot x_j = z(j,i)$, this inequality can be rewritten in the following way:

$$\sum_{i=1}^{m} x_i \sum_{j \in N[i]} x_j \leq 2 \sum_{i=1}^{m} x_i \sum_{s \in S_i} \sum_{j \in N^{\delta}[c_s]} x_j.$$

Now, since c_s hits all δ-inflated κ-balls from $N^{\delta}[c_s]$ and $\mathbf{x}$ is an admissible solution of $\Pi'(S, \mathbf{r} + \delta)$, we conclude that $\sum_{j \in N^{\delta}[c_s]} x_j \leq 1$. Thus $\sum_{s \in S_i} \sum_{j \in N^{\delta}[c_s]} x_j \leq |S_i|$. Since $|S_i| \leq \kappa$, we deduce that $\sum_{i=1}^{m} x_i \sum_{j \in N[i]} x_j \leq 2\kappa \sum_{i=1}^{m} x_i$. Hence, there exists ${}^{\kappa}B_i$ such that $x_i \sum_{j \in N[i]} x_j \leq 2\kappa x_i$, yielding $\sum_{j \in N[i]} x_j \leq 2\kappa$. $\square$

Lemma 4. *It is possible to construct in polynomial time an integer admissible solution* $\mathbf{x}^*$ *of the linear program* $\Pi(S, \mathbf{r})$ *of size at least* $\pi'_f(S, \mathbf{r} + \delta)/(2\kappa)$.

Proof. Let $\mathbf{x} = \{x_1, \ldots, x_m\}$ be an optimal (fractional) solution of the linear program $\Pi'(S, \mathbf{r} + \delta)$ (it can be found in polynomial time). We will iteratively use Lemma 3 to $\mathbf{x}$ to derive an integer solution $\mathbf{x}^* = \{x_1^*, \ldots, x_m^*\}$ for the linear program $\Pi(S, \mathbf{r})$. The algorithm starts with the set ${}^{\kappa}\mathcal{B}' := {}^{\kappa}\mathcal{B}_{S,\mathbf{r}}$ of m κ-balls. By Lemma 3 there exists a κ-ball ${}^{\kappa}B_i \in {}^{\kappa}\mathcal{B}'$ such that $\sum_{j \in N[i]} x_j \leq 2\kappa$. We set $x_i^* := 1$ and $x_j^* := 0$ for all $j \in N[i] \setminus \{i\}$, then we remove all κ-balls ${}^{\kappa}B_j$ with $j \in N[i]$ from ${}^{\kappa}\mathcal{B}'$. The algorithm continues with the current set ${}^{\kappa}\mathcal{B}'$ of κ-balls until it becomes empty. Notice that at all iterations of the algorithm the restriction of $\mathbf{x}$ on the κ-balls of ${}^{\kappa}\mathcal{B}'$ remains an admissible solution of the linear program $\Pi'(S', \mathbf{r} + \delta)$ defined by ${}^{\kappa}\mathcal{B}'$. This justifies the use of Lemma 3 at all iterations of the algorithm.

To show that $\mathbf{x}^*$ is an admissible solution of $\Pi(S, \mathbf{r})$, suppose by way of contradiction that there exist two intersecting κ-balls ${}^{\kappa}B_i$ and ${}^{\kappa}B_j$ with $x_i^* = 1 = x_j^*$. Suppose that the algorithm selects ${}^{\kappa}B_i$ before ${}^{\kappa}B_j$. Consider the iteration when x_i^* becomes 1. Since $j \in N[i]$, at this iteration x_j^* becomes 0 and ${}^{\kappa}B_j$ is removed from $^\kappa\mathcal{B}'$. Thus x_j^* cannot become 1 at a later stage. This shows that the κ-balls $^\kappa B_i$ with $x_i^* = 1$ indeed constitute a packing for $^\kappa\mathcal{B}_{S,\mathbf{r}}$.

It remains to compare the costs of the solutions $\mathbf{x}$ and $\mathbf{x}^*$. For this, notice that according to the algorithm, for each κ-ball $^\kappa B_i$ with $x_i^* = 1$ we can define a subset $N'[i]$ of $N[i]$ such that $i \in N'[i]$, $x_j^* = 0$ for all $j \in N'[i] \setminus \{i\}$, and $\sum_{j \in N'[i] \cup \{i\}} x_j \leq 2\kappa$. Hence, the κ-balls of $^\kappa\mathcal{B}_{S,\mathbf{r}}$ can be partitioned into groups, such that each group contains a single κ-ball selected in the integer solution and the total cost of the fractional solutions of the balls from each group is at most 2κ. This shows that $\sum_{i=1}^m x_i^* \geq (\sum_{i=1}^m x_i)/(2\kappa)$. □

Lemma 5. *It is possible to construct in polynomial time an integer solution* $\mathbf{y}^*$ *of the linear program* $\Gamma(S, \mathbf{r} + \delta)$ *of size at most* $\kappa\gamma_f'(S, \mathbf{r})$.

Proof. Let $\mathbf{y} = \{y_v : v \in V\}$ be an optimal (fractional) solution of the linear program $\Gamma'(S, \mathbf{r})$. Since $\sum_{v \in ^\kappa B_i} y_v \geq 1$ for all $i = 1, \ldots, m$, each κ-ball $^\kappa B_i$ contains a ball B_i such that $\kappa\sum_{v \in B_i} y_v \geq 1$. Let s_i be the center of the ball B_i and let $r(s_i)$ be its radius. Set $S = \{s_1, \ldots, s_m\}$. Notice that $\mathbf{y}' = \{y_v' : v \in V\}$ defined by setting $y_v' = \kappa \cdot y_v$ if $v \in \bigcup_{i=1}^m B_i$ and $y_v' = 0$ otherwise, is a fractional covering for the family of balls $B_1, \ldots, B_m$. Thus the cost of $\mathbf{y}'$ is at least $\gamma_f(S, r) = \pi_f(S, r)$. Notice also that the cost of $\mathbf{y}'$ is at most κ times the cost of $\mathbf{y}$. By Theorem 1, we can construct in polynomial time a set C of size at most $\pi(S, r)$ which $(r + \delta)$-dominates the set S. Let $\mathbf{y}^* = \{y_v : v \in V\}$ be defined by setting $y_v^* = 1$ if $v \in C$ and $y_v^* = 0$ otherwise. Since $\pi(S, r) \leq \pi_f(S, r)$, putting all things together, we obtain:

$$\sum_{v \in V} y_v^* = |C| \leq \pi(S, r) \leq \pi_f(S, r) = \gamma_f(S, r) \leq \sum_{v \in V} y_v' \leq \kappa \sum_{v \in V} y_v = \kappa\gamma_f'(S, \mathbf{r}). \square$$

Now, we are ready to complete the proof of Theorem 2. According to Lemma 4 we can construct in polynomial time an integer solution $\mathbf{x}^*$ for $\Pi(S, \mathbf{r})$ of size at least $\pi_f'(S, \mathbf{r} + \delta)/(2\kappa)$. Let $P = \{^\kappa B_i : x_i^* = 1\}$. On the other hand, applying Lemma 5 for the radius function $\mathbf{r} + \delta$ instead of $\mathbf{r}$, we can construct in polynomial time an integer solution $\mathbf{y}^*$ of the linear program $\Gamma(S, \mathbf{r} + 2\delta)$ of size at most $\kappa\gamma_f'(S, \mathbf{r} + \delta)$. Let $C = \{v \in V : y_v^* = 1\}$. Since, by duality, $\gamma_f'(S, \mathbf{r} + \delta) = \pi_f'(S, \mathbf{r} + \delta)$, we deduce that $|C| \leq 2\kappa^2|P|$, as required.

4   Augmentation Under Diameter Constraints

Denote by $\mathrm{OPT}(D)$ the minimum number of edges necessary to decrease the diameter of the input δ-hyperbolic graph $G = (V, E)$ until D. First suppose that the resulting diameter D is even, say $D = 2R$. We recall the relationship between the augmentation problem ADC and the r-domination problem established in [19] for trees. Let E^* be an optimal augmentation, i.e. $|E^*| = \mathrm{OPT}(D)$ and the graph $G^* = (V, E \cup E^*)$ has diameter D. Denote by C the set of end-vertices of the edges of E^* and let V' the set of all vertices which are located at distance less than or equal to $R - 1$ from a vertex of C (in other words, V' is the union of all balls or radius $R - 1$ centered at vertices of C). Since E^* is a solution for the problem ADC, it can be easily shown (see [19] for missing details) that the diameter in G of the set $Q := V \setminus V'$ is at most $2R$. Since G is δ-hyperbolic, from Corollary 1 and the discussion preceding Corollary 3 we infer that Q can be covered by a single ball $B(c^*, R+\delta)$ of radius $R+\delta$. Let Q' be the set of vertices of G located outside $B(c, R+\delta)$. Since $Q' \subseteq V' = \bigcup \{B(x, R - 1) : x \in C\}$ and each edge of E^* has both ends in C, we conclude that $\gamma_G(Q', R - 1) \leq \gamma_G(V', R - 1) \leq |C| \leq 2|E^*| = 2\mathrm{OPT}(D)$.

Now, we turn this analysis of an optimal solution (which we do not know how to construct) into a polynomial time algorithm which instead will find a set E' of new $2\mathrm{OPT}(D)$ edges so that the resulting graph $G' = (V, E \cup E')$ will have diameter at most $D + 2\delta$ (instead of D, as required). As for trees [19], the algorithm will try every vertex c' of G as a center of a ball of radius $R + \delta$ and it covers the set $V \setminus B(c', R + \delta)$ with at most $\pi_G(V \setminus B(c', R + \delta), R - 1)$ balls of radius $R - 1 + \delta$. This is done using the procedure described in Theorem 1. Among $|V|$ such coverings, the algorithm selects the one with a minimum number of balls. Let c' be the center of the ball of radius $R + \delta$ providing this covering C'. Then the algorithm returns as the set E' of new edges all pairs of the form $c'c$, where c is a center of a ball of radius $R - 1 + \delta$ from C'. Notice that the graph obtained from G after adding the new edges has diameter at most $2R + 2\delta$. Finally notice that since the algorithm tested the vertex c^* described above as the center of the ball of radius $R + \delta$, by Theorem 1 we conclude that $|C'| \leq \pi_G(Q', R - 1)$, showing that $|C'| \leq 2\mathrm{OPT}(2R)$. We obtain the following result:

Proposition 1. *Given a δ-hyperbolic graph $G = (V, E)$ and $R \geq 1$, one can construct in polynomial time an admissible solution for the problem ADC with $D = 2R + 2\delta$ which contains at most $2\mathrm{OPT}(2R)$ edges.*

5 k-Center Problem

Let $G = (V, E)$ be a δ-hyperbolic graph and S be a set of n input vertices of the k-center problem. Then, as we noticed already, the k-center problem consists in finding the smallest radius R^* such that the set S can be covered with at most k balls of radius R^*. The value of R^* belongs to the list Δ of size $O(|V| \cdot |S|)$ consisting of all possible distinct values of distances from the vertices of G to the set S. As in some other minmax problems [30,31,39], the approximation algorithm tests the entries of Δ, using a parameter R, which is the "guess" of the optimal radius. For current $R \in \Delta$, instead of running the algorithm of Hochbaum and Shmoys [31], we use the algorithm described in Theorem 1 and Corollary 3 with $r(x) = R$ for all $x \in S$. This algorithm either finds a covering of S with at most k balls of radius $R + \delta$ or it returns an r-packing P of size greater than k. In the second case, we conclude that $\gamma_G(S, r) \geq \pi_G(S, r) > k$, therefore the tested value R is too small, yielding $R < R^*$. Now, if R is the least value for which the algorithm does not return the negative answer, then $R \leq R^*$, and we obtain a solution for the k-center problem of radius $R + \delta \leq R^* + \delta$.

Proposition 2. *Given a δ-hyperbolic graph $G = (V, E)$, one can construct in polynomial time an admissible solution for the k-center problem having radius at most* OPT $+ \delta$.

References

1. Alon, N.: Piercing d-intervals. Discrete and Computational Geometry 19, 333–334 (1998)
2. Alon, N.: Covering a hypergraph of subgraphs. Discrete Mathematics 257, 249–254 (2002)
3. Alon, N., Moshkovitz, D., Safra, M.: Algorithmic construction of sets for k-restrictions. ACM Transactions on Algorithms (to appear)
4. Alon, N., Gyárfás, A., Ruszinkó, M.: Decreasing the diameter of bounded degree graphs. J. Graph Theory 35, 161–172 (2000)
5. Alonso, J.M., Brady, T., Cooper, D., Ferlini, V., Lustig, M., Mihalik, M., Shapiro, M., Short, H.: Notes on word hyperbolic groups. In Group Theory from a Geometrical Viewpoint, ICTP Trieste 1990. In: Ghys, E., Haefliger, A., Verjovsky, A. (eds.), World Scientific, pp. 3–63 (1991)
6. Bandelt, H.-J., Chepoi, V.: 1-Hyperbolic graphs. SIAM J. Discrete Mathematics 16, 323–334 (2003)
7. Bar-Yehuda, R., Halldorsson, M.M., Naor, J., Shachnai, H., Shapira, I.: Scheduling split intervals. SIAM Journal on Computing 36, 1–15 (2006) (and SODA' 2002)
8. Bárány, I., Edmonds, J., Wolsey, L.A.: Packing and covering a tree by subtrees. Combinatorica 6, 221–233 (1986)
9. Berge, C.: Hypergraphs. North-Holland, Amsterdam (1989)
10. Bern, M., Eppstein, D.: Approximation algorithms for geometric problems. In: Hochbaum, D.S. (ed.) Approximation Algorithms for NP-Hard Problems, pp. 296–345. PWS, Boston, MA (1997)
11. Brandstädt, A., Chepoi, V., Dragan, F.: Clique r-domination and clique r-packing problems on dually chordal graphs. SIAM J. Discrete Mathematics 10, 109–127 (1997)
12. Brandstädt, A., Chepoi, V., Dragan, F.: Distance approximating trees for chordal and dually chordal graphs. J. Algorithms 30, 166–184 (1999) (and ESA'97)
13. Chandrasekaran, R., Dougherty, A.: Location on tree networks: p-center and q-dispersion problems. Mathematics Operations Research 6, 50–57 (1981)
14. Chang, G.J.: Labeling algorithms for domination problems in sun-free chordal graphs. Discrete Applied Mathematics 22, 21–34 (1988/1989)
15. Chang, G.J., Nemhauser, G.L.: The k-domination and k-stability problems on sun-free chordal graphs. SIAM J. Alg. Disc. Meth. 5, 332–345 (1986)
16. Chepoi, V., Dragan, F.: A note on distance approximating trees in graphs. European J. Combinatorics 21, 761–766 (2000)
17. Chepoi, V., Estellon, B., Vaxès, Y.: On covering planar graphs with a fixed number of balls. Discrete and Computational Geometry 37, 237–244 (2007)
18. Chepoi, V., Estellon, B., Nouioua, K., Vaxès, Y.: Mixed covering of trees and the augmentation problem with odd diameter constraints. Algorithmica 45, 209–226 (2006)
19. Chepoi, V., Vaxès, Y.: Augmenting trees to meet biconnectivity and diameter constraints. Algorithmica 33, 243–262 (2002)

20. Chung, F.R.K., Garey, M.R.: Diameter bounds for altered graphs. J. Graph Theory 8, 511–534 (1984)
21. Demaine, E.D., Fomin, F.V., Hajiaghayi, M., Thilikos, D.M.: Fixed-parameter algorithms for (k, r)-center in planar graphs and map graphs. In: Baeten, J.C.M., Lenstra, J.K., Parrow, J., Woeginger, G.J. (eds.) ICALP 2003. LNCS, vol. 2719, pp. 33–47. Springer, Heidelberg (2003)
22. Dodis, Y., Khanna, S.: Designing networks with bounded pairwise distance. In: Proc. 31th Annual ACM Symposium on the Theory of Computing STOC' 1999, pp. 750–759 (1999)
23. Eppstein, D.: Squarepants in a tree: sum of subtree clustering and hyperbolic pants decomposition. In: Proc. 18th ACM-SIAM Symposium on Discrete Algorithms SODA' 2007, ACM Press, New York (2007)
24. Erdös, P., Gyárfás, A., Ruszinkó, M.: How to decrease the diameter of triangle–free graphs. Combinatorica 18, 493–501 (1998)
25. Farber, M.: Domination, independent domination and duality in strongly chordal graphs. Discrete Applied Mathematics 7, 115–130 (1984)
26. Feder, T., Greene, D.H.: Optimal algorithms for approximate clustering. In: Proc. 20th ACM Symp. Theory of Computing, STOC' 1988, pp. 434–444 (1988)
27. Gavoille, C., Ly, O.: Distance labeling in hyperbolic graphs. In: Deng, X., Du, D.-Z. (eds.) ISAAC 2005. LNCS, vol. 3827, pp. 1071–1079. Springer, Heidelberg (2005)
28. Ghys, E., de la Harpe, P. (eds.) Les groupes hyperboliques d'après M. Gromov. Progress in Mathematics vol. 83 Birkhäuser (1990)
29. Gromov, M.: Hyperbolic Groups. In: Essays in group theory Gersten,S.M.(ed.), MSRI Series 8, pp. 75–263 (1987)
30. Hochbaum, D.S.: Various notions of approximations: good, better, best, and more. In: Hochbaum, D.S. (ed.) Approximation Algorithms for NP-Hard Problems, pp. 346–398. PWS, Boston MA (1997)
31. Hochbaum, D.S., Shmoys, D.B.: A unified approach to approximation algorithms for bottleneck problems. Journal of the ACM 33, 533–550 (1986)
32. Ishii, T., Yamamoto, S., Nagamochi, H.: Augmenting forests to meet odd diameter requirements. Discrete Optimization 3, 154–164 (2006)
33. Kleinberg, R.: Geographic routing using hyperbolic space. In: Proc. 26th Annual IEEE International Conference on Computer Communications, INFOCOM' 2007, IEEE Computer Society Press, Los Alamitos (2007)
34. Koolen, J., Moulton, V.: Hyperbolic bridged graphs. European J. Combinatorics 23, 683–699 (2002)
35. Krauthgamer, R., Lee, J.R.: Algorithms on negatively curved spaces. In: Proc. 47th Annual IEEE Symposium on Foundations of Computer Science, FOCS' 2006, IEEE Computer Society Press, Los Alamitos (2006)
36. Li, C.-L., McCormick, S.T., Simchi–Levi, D.: On the minimum-cardinality-bounded-diameter and the bounded-cardinality-minimum-diameter edge addition problems. Operations Research Letters 11, 303–308 (1992)
37. Schoone, A.A., Bodlaender, H.L., van Leeuwen, J.: Diameter increase caused by edge deletion. J. Graph Theory 11, 409–427 (1987)
38. Shavitt, Y., Tankel, T.: On internet embedding in hyperbolic spaces for overlay construction and distance estimation. IEEE/ACM Transactions on Networking and INFOCOM' 2004. ACM Press, New York (to appear)
39. Vazirani, V.: Approximation Algorithms. Springer, Berlin (2001)

Small Approximate Pareto Sets for Bi-objective Shortest Paths and Other Problems

Ilias Diakonikolas* and Mihalis Yannakakis**

Department of Computer Science
Columbia University
New York, NY 10027, USA
{ilias,mihalis}@cs.columbia.edu

Abstract. We investigate the problem of computing a minimum set of solutions that approximates within a specified accuracy ϵ the Pareto curve of a multiobjective optimization problem. We show that for a broad class of bi-objective problems (containing many important widely studied problems such as shortest paths, spanning tree, and many others), we can compute in polynomial time an ϵ-Pareto set that contains at most twice as many solutions as the minimum such set. Furthermore we show that the factor of 2 is tight for these problems, i.e., it is NP-hard to do better. We present further results for three or more objectives, as well as for the dual problem of computing a specified number k of solutions which provide a good approximation to the Pareto curve.

1 Introduction

In many decision making situations it is typically the case that more than one criteria come into play. For example, when purchasing a product (car, tv, etc.) we care about its cost, quality, etc. When choosing a route we may care about the time it takes, the distance travelled, etc. When designing a network we may care about its cost, its capacity (the load it can carry), its coverage. This type of *multicriteria* or *multiobjective* problems arise across many diverse disciplines, in engineering, in economics and business, healthcare, and others. The area of multiobjective optimization has been (and continues to be) extensively investigated in the management science and optimization communities with many papers, conferences and books (see e.g. [Cli, Ehr, EG, FGE, Miet]).

In multiobjective problems there is typically no uniformly best solution in all objectives, but rather a trade-off between the different objectives. This is captured by the *trade-off* or *Pareto curve*, the set of all solutions whose vector of objective values is not dominated by any other solution. The trade-off curve represents the range of reasonable "optimal" choices in the design space; they are precisely the optimal solutions for all possible global "utility" functions that depend monotonically on the different objectives. A decision maker, presented with the trade-off curve, can select a solution that corresponds best to his/her preferences; of course different users generally may have

---

* Research partially supported by NSF grant CCF-04-30946 and an Alexander S. Onassis Foundation Fellowship.

** Research partially supported by NSF grant CCF-04-30946.

M. Charikar et al. (Eds.): APPROX and RANDOM 2007, LNCS 4627, pp. 74–88, 2007.
© Springer-Verlag Berlin Heidelberg 2007

different preferences and select different solutions. The problem is that the trade-off curve has typically exponential size (for discrete problems) or is infinite (for continuous problems), and hence we cannot construct the full curve. Thus, we have to contend with an approximation of the curve: We want to compute efficiently and present to the decision makers a small set of solutions (as small as possible) that represents as well as possible the whole range of choices, i.e. that provides a good approximation to the Pareto curve. Indeed this is the underlying goal in much of the research in the multiobjective area, with many heuristics proposed, usually however without any performance guarantees or complexity analysis, as we do in theoretical computer science.

In recent years we initiated a systematic investigation [PY1, VY] to develop the theory of multiobjective approximation along similar rigorous lines as the approximation of single objective problems. The approximation to the Pareto curve is captured by the concept of an ϵ-*Pareto set*, a set P_ϵ of solutions that approximately dominates every other solution; that is, for every solution s, the set P_c contains a solution s' that is within a factor $1 + \epsilon$ of s, or better, in all the objectives. (As usual in approximation, it is assumed that all objective functions take positive values.) Such an approximation was studied before for certain problems, e.g. multiobjective shortest paths, for which Hansen [Han] and Warburton [Wa] showed how to construct an ϵ-Pareto set in polynomial time (for fixed number of objectives). Note that typically in most real life multiobjective problems the number of objectives is small. In fact, the great majority of the multiobjective literature concerns the case of two objectives.

Consider a multiobjective problem with d objectives, for example shortest path with cost and time objectives. For a given instance, and error tolerance ϵ, we would like to compute a smallest set of solutions that form an ϵ-Pareto set. Can we do it in polynomial time? If not, how well can we approximate the smallest ϵ-Pareto set? Note that an ϵ-Pareto set is not unique: in general there are many such sets, some of which can be very small and some very large. First, to have any hope we must ensure that there exists at least a polynomial size ϵ-Pareto set. Indeed, in [PY1] it was shown that this is the case for every multiobjective problem with a fixed number of polynomially computable objectives. Second we must be able to construct at least one such set in polynomial time. This is not always possible. A necessary and sufficient condition for polynomial computability for all $\epsilon > 0$ is the existence of a polynomial algorithm for the following *Gap problem*: Given a vector of values b, either compute a solution that dominates b, or determine that no solution dominates b by at least a factor $1 + \epsilon$ (in all the objectives). Many multiobjective problems were shown to have such a routine for the Gap problem (and many others have been shown subsequently).

Construction of a polynomial-size approximate Pareto set is useful, but not good enough in itself: For example, if we plan a trip, we want to examine just a few possible routes, not a polynomial number in the size of the map. More generally, in typical multicriteria situations, the selected representative solutions are investigated more thoroughly by the decision maker (designer, physician, corporation, etc.) to assess the different choices and pick the most preferable one, based possibly on additional factors that are perhaps not formalized or not even quantifiable. We thus want to select as small a set as possible that achieves a desired approximation. In [VY] the problem of constructing a minimum ϵ-Pareto set was raised formally and investigated in a general framework. It was shown that for all bi-objective problems with a polynomial-time Gap routine, one can construct an ϵ-Pareto set that contains at most 3 times the number of points of the smallest such set; furthermore, the factor 3 is best possible in the sense that for some problems it is NP-hard to do better. Further results were shown for 3 and more objectives, and for other related questions. Note that although the factor 3 of [VY] is best possible in general for two objectives, one may be able to do better for specific problems.

We show in this paper, that for an important class of bi-objective problems (containing many widely studied natural ones such as shortest paths, spanning tree, knapsack, scheduling problems and others) we can obtain a 2-approximation, and furthermore the factor of 2 is tight for them, i.e., it is NP-hard to do better. Our algorithm is a general algorithm that relies on a routine for a stronger version of the Gap problem, namely a routine that solves approximately the following *Restricted problem*: Given a (hard) bound b_1 for one objective, compute a solution that optimizes approximately the second objective subject to the bound. Many problems (e.g. shortest paths, etc.) have a polynomial time approximation scheme for the Restricted problem. For all such problems, a 2-approximation to the minimum ϵ-Pareto set can be computed in polynomial time. Furthermore, the number of calls to the Restricted routine (and an associated equivalent dual routine) is linear in the size OPT_ϵ of the optimal ϵ-Pareto set.

The bi-objective shortest path problem is probably the most well-studied multiobjective problem. It is the paradigmatic problem for dynamic programming (thus can express a variety of problems), and arises itself directly in many contexts. One area is network routing with various QoS criteria (see e.g. [CX2, ESZ, GR+, VV]). For example, an interesting proposal in a recent paper by Van Mieghen and Vandenberghe [VV] is to have the network operator advertise a portfolio of offered QoS solutions for their network (a trade-off curve), and then users can select the solutions that best fit their applications. Obviously, the portfolio cannot include every single possible route, and it would make sense to select carefully an "optimal" set of solutions that cover well the whole range. Other applications include the transportation of hazardous materials (to minimize risk of accident, and population exposure) [EV], and many others; we refer to the references, e.g. [EG] contains pointers to the extensive literature on shortest paths, spanning trees, knapsack, and the other problems. Our algorithm applies not only to the above standard combinatorial problems, but more generally to any bi-objective problem for which we have available a routine for the Restricted problem; the objective functions and the routine itself could be complex pieces of software without a simple mathematical expression.

After giving the basic definitions and background in Sect. 2, we present in Sect. 3 our general lower and upper bound results for bi-objective problems, as well as applications to specific problems. In Sect. 4 we present some results for $d = 3$ and more objectives. Here we assume only a Gap routine; i.e. these results apply to all problems with a polynomial time constructible ϵ-Pareto set. It was shown in [VY] that for $d = 3$ it is in general impossible to get a constant factor approximation to the optimal ϵ-Pareto set, but one has to relax ϵ. Combining results from [VY] and [KP] we show that for any $\epsilon' > \epsilon$ we can construct an ϵ'-Pareto set of size $cOPT_\epsilon$, i.e. within a (large) constant factor c of the size OPT_ϵ of the optimal ϵ-Pareto set. For general d, the problem can be reduced to a Set Cover problem whose VC dimension and codimension are at most d, and we can construct an ϵ'-Pareto set of size $O(d \log(dOPT_\epsilon))OPT_\epsilon$.

We discuss also the *Dual* problem: For a specified number k of points, find k points that provide the best approximation to the Pareto curve, i.e. that form an ϵ-Pareto set with the minimum possible ϵ. In [VY] it was shown that for $d = 2$ objectives the problem is NP-hard, but we can approximate arbitrarily well (i.e. there is a PTAS) the minimum approximation ratio $\rho^* = 1 + \epsilon^*$. As we'll see, for $d = 3$ this is not possible, in fact one cannot get any multiplicative approximation (unless P=NP). We use a relationship of the Dual problem to the asymmetric k-center problem and techniques from the latter problem to show that the Dual problem can be approximated (for $d = 3$) within a constant power, i.e. we can compute k points that cover every point on the Pareto curve within a factor $\rho' = (\rho^*)^c$ or better in all objectives, for some constant c. For small ρ^*, i.e. when there is a set of k points that provides a good approximation to the Pareto curve, constant factor and constant power are related, but in general of course they are not.

2 Definitions and Background

A multiobjective optimization problem Π has a set $\mathcal{I}_\Pi$ of *valid instances*, every instance $I \in \mathcal{I}_\Pi$ has a set of solutions $\mathcal{S}(I)$. There are d objective functions, $f_1, \ldots, f_d$, each of which maps every instance I and solution $s \in \mathcal{S}(I)$ to a value $f_j(I, s)$. The problem specifies for each objective whether it is to be maximized or minimized. We assume as usual in approximation that the objective functions have positive rational values, and that they are polynomial-time computable. We use m to denote the maximum number of bits in numerator and denominator of the objective function values.

We say that a d-vector u *dominates* another d-vector v if it is at least as good in all the objectives, i.e. $u_j \geq v_j$ if f_j is to be maximized ($u_j \leq v_j$ if f_j is to be minimized). Similarly, we define domination between any solutions according to the d-vectors of their objective values. Given an instance I, the *Pareto set* $P(I)$ is the set of undominated d-vectors of values of the solutions in $\mathcal{S}(I)$. Note that for any instance, the Pareto set is unique. (As usual we are also interested in solutions that realize these values, but we will often blur the distinction and refer to the Pareto set also as a set of solutions that achieve these values. If there is more than one undominated solution with the same objective values, $P(I)$ contains one of them).

We say that a d-vector u *c-covers* another d-vector v if u is at least as good as v up to a factor of c in all the objectives, i.e. $u_j \geq v_j/c$ if f_j is to be maximized ($u_j \leq cv_j$ if f_j is to be minimized). Given an instance I and $\epsilon > 0$, an ϵ-*Pareto set* $P_\epsilon(I)$ is a set of d-vectors of values of solutions that $(1 + \epsilon)$-cover all vectors in $P(I)$. For a given instance, there may exist many ϵ-Pareto sets, and they may have very different sizes. It is shown in [PY1] that for every multiobjective optimization problem in the aforementioned framework, for every instance I and $\epsilon > 0$, there exists an ϵ-Pareto set of size $O((4m/\epsilon)^{d-1})$, i.e. polynomial for fixed d.

An approximate Pareto set always exists, but it may not be constructible in polynomial time. We say that a multiobjective problem Π has a polynomial time approximation scheme (respectively a fully polynomial time approximation scheme) if there is an algorithm, which, given instance I and a rational number $\epsilon > 0$, constructs an ϵ-Pareto set $P_\epsilon(I)$ in time polynomial in the size $|I|$ of the instance I (respectively, in time polynomial in $|I|$, the representation size $|\epsilon|$ of ϵ, and in $1/\epsilon$). Let MPTAS (resp. MFPTAS) denote the corresponding class of problems. There is a simple necessary and sufficient condition [PY1], which relates the efficient computability of an ϵ-Pareto set for a multi-objective problem Π to the following *GAP Problem*: given an instance I of Π, a (positive rational) d-vector b, and a rational $\delta > 0$, either return a solution whose vector dominates b or report that there does not exist any solution whose vector is better than b by at least a $(1+\delta)$ factor in all of the coordinates. As shown in [PY1], a problem is in MPTAS (resp. MFPTAS) if and only if there is a subroutine GAP that solves the GAP problem for Π in time polynomial in $|I|$ and $|b|$ (resp. in $|I|$, $|b|$, $|\delta|$ and $1/\delta$).

We say that an algorithm that uses a routine as a black box to access the solutions of the multiobjective problem is *generic*, as it is not geared to a particular problem, but applies to all of the problems for which the particular routine is available. All that such an algorithm needs to know about the input instance is bounds on the minimum and maximum possible values of the objective functions. (For example, if the objective functions are positive rational numbers whose numerators and denominators have at most m bits, then an obvious lower bound on the objective values is 2^{-m} and an obvious upper bound is 2^m; however, for specific problems better bounds may be available.) Based on the bounds, the algorithm calls the given routine for certain values of its parameters, and uses the returned results to compute an approximate Pareto set.

For a given instance, there may exist many ϵ-Pareto sets, and they may have very different sizes. We want to compute one with the smallest possible size, which we'll denote OPT_ϵ. [VY] gives generic algorithms that compute small ϵ-Pareto sets and are applicable to all multiobjective problems in M(F)PTAS, i.e. all problems possessing a (fully) polynomial GAP routine. They consider the following "dual" problems: Given an instance and an $\epsilon > 0$, construct an ϵ-Pareto set of as small size as possible. And dually, given a bound k, compute an ϵ-Pareto set with at most k points that has as small an ϵ value as possible. In the case of two objectives, they give an algorithm that computes an ϵ-Pareto set of size at most $3 \cdot OPT_\epsilon$; they show that no algorithm can be better than 3-competitive in this setting. For the dual problem, they show that the optimal ϵ-value can be approximated arbitrarily closely. For three objectives, they show that no algorithm can be c-competitive for any constant c, unless it is allowed to use a larger ϵ value. They also give an algorithm that constructs an ϵ'-Pareto set of cardinality at most $4 \cdot OPT_\epsilon$, for any $\epsilon' > (1 + \epsilon)^2 - 1$.

In a general multiobjective problem we may have both minimization and maximization objectives. In the remainder, we will assume for convenience that all objectives are minimization objectives; this is without loss of generality, since we can simply take the reciprocals of maximization objectives.

Due to space constraints, most proofs are deferred to the full version of this paper.

3   Two Objectives

We use the following notation in this section. Consider the plane whose coordinates correspond to the two objectives. Every solution is mapped to a point on this plane. We use x and y as the two coordinates of the plane. If p is a point, we use $x(p)$, $y(p)$ to denote its coordinates; that is, $p = (x(p), y(p))$.

We consider the class of bi-objective problems Π for which we can approximately minimize one objective (say the y-coordinate) subject to a "hard" constraint on the other (the x-coordinate). Our basic primitive is a (fully) polynomial time routine for the following *Restricted problem* (for the y-objective): Given an instance $I \in \mathcal{I}_\Pi$, a (positive rational) bound C and a parameter $\delta > 0$, either return a solution point $\tilde{s}$ satisfying $x\,(\tilde{s}) \leq C$ and $y\,(\tilde{s}) \leq (1 + \delta) \cdot \min \{y$ over all solutions $s \in \mathcal{S}(I)$ having $x(s) \leq C\}$ or report that there does not exist any solution s such that $x\,(s) \leq C$. For simplicity, we will drop the instance from the notation and use Restrict$_\delta\,(y, x \leq C)$ to denote the solution returned by the corresponding routine. If the routine does not return a solution, we will say that it returns NO. We say that a routine Restrict$_\delta\,(y, x \leq C)$ runs in polynomial time (resp. fully polynomial time) if its running time is polynomial in $|I|$ and $|C|$ (resp. $|I|$, $|C|$, $|\delta|$ and $1/\delta$). The Restricted problem for the x-objective is defined analogously. We will also use the Restricted routines with strict inequality bounds; it is easy to see that they are polynomially equivalent.

Note that in general the two objectives could be nonlinear and completely unrelated. Moreover, it is possible that a bi-objective problem possesses a (fully) polynomial Restricted routine for the one objective, but not for the other.

The considered class of bi-objective problems is quite broad and contains many well-studied natural ones. Applications include the shortest path problem [Han, Wa] and generalizations [EV, GR+, CX2, VV], cost-time trade-offs in query evaluation [PY2], spanning tree [GR, HL] and related problems [CX1]. The aforementioned problems possess a polynomial Restricted routine for *both* objectives. For several other problems [ABK1, ABK2, CJK, DJSS], the Restricted routine is available for one objective *only*, and it is NP-hard to even separately optimize the other objective. An example is the following scheduling problem: We are given a set of n jobs and a fixed number m of machines. Executing job j on machine i requires time p_{ij} and incurs cost c_{ij}. We are interested in the trade-off between makespan and cost. Minimizing the makespan is NP-hard even for $m = 2$, but there is an FPTAS for the Restricted problem for the makespan objective [ABK1].

In Sect. 3.1, we show that, even if the given bi-objective problem possesses a (fully) polynomial Restricted routine *for both objectives*, no generic algorithm can guarantee an approximation ratio better than 2. (This lower bound applies *a fortiori* if the Restricted routine is available for one objective only.) Furthermore, we show that for two such natural problems, namely, the bi-objective shortest path and spanning tree problems, it is NP-hard to do better than 2. In Sect. 3.2 we give a matching upper bound: we present an algorithm that is 2-competitive and applies to all of the problems that possess a polynomial Restricted routine for one of the two objectives.

3.1 Lower Bound

To prove a lower bound for a generic procedure, we present two Pareto sets which are indistinguishable from each other using the Restricted routine as a black box, yet whose smallest ϵ-Pareto sets are of different sizes. We omit the proof.

Proposition 1. *Consider the class of bi-objective problems that possess a fully polynomial Restricted routine for both objectives. Then, for any $\epsilon > 0$, there is no polynomial time generic algorithm that approximates the size of the smallest ϵ-Pareto set P_ϵ^* to a factor better than 2.*

In fact, we can prove something stronger (assuming P $\neq$ NP) for the bi-objective short-est path (*BSP*) and spanning tree (*BST*) problems. In the *BSP* problem, we are given a graph, positive rational "costs" and "delays" for each edge and two specified nodes s and t. The set of feasible solutions is the set of $s - t$ paths. The *BST* problem is defined analogously. These problems are well-known to possess polynomial Restricted routines for *both* objectives [LR, GR].

Theorem 1

a. For the bi-objective Shortest Path problem, for any k from $k = 1$ to a polynomial, it is NP-hard to distinguish the case that the minimum size OPT_ϵ of the optimal ϵ-Pareto set is k from the case that it is $2k - 1$.

b. The same holds for the bi-objective Spanning Tree problem for any fixed k.

The proof is omitted, due to lack of space. For $k = 1$ the theorem says that it is NP-hard to tell if one point suffices or we need at least 2 points for an ϵ approximation. We proved that the theorem holds also for more general k to rule out additive and asymptotic approximations as well. Similar hardness results can be shown for several other related problems.

3.2   Two Objectives Algorithm

Before we give the algorithm, we remark that if we have *exact* (not approximate) Restricted routines for both objectives then we can compute the *optimal* ϵ-Pareto set by a simple greedy algorithm. The algorithm is similar to the one given in [KP, VY] for the (special) case where all the solution points are given explicitly in the input. The Greedy algorithm proceeds by iteratively selecting points $q_1, \ldots, q_k$ in decreasing x (increasing y) as follows: We start by computing a point q_1' having minimum y coordinate (among all feasible solutions); q_1 is then selected to be the *leftmost* solution point satisfying $y(q_1) \leq (1 + \epsilon)y(q_1')$. During the jth iteration ($j \geq 2$) we initially compute the point q_j' with minimum y-coordinate among all solution points s having $x(s) < x(q_{j-1})/(1+\epsilon)$ and select as q_j the leftmost point which satisfies $y(q_j) \leq (1 + \epsilon)y(q_j')$. The algorithm terminates when the last point selected $(1 + \epsilon)$-covers the leftmost solution point. It follows by an easy induction that the set $\{q_1, q_2, \ldots, q_k\}$ is an ϵ-Pareto set of minimum cardinality. This exact algorithm is applicable to biobjective linear programming (and all problems reducible to it, for example biobjective flows), the biobjective *global* min-cut problem [AZ] and several scheduling problems [CJK]. For these problems we can compute an ϵ-Pareto set of minimum cardinality.

If we have approximate Restricted routines, one may try to modify the Greedy algorithm in a straightforward way to take into account the fact that the routines are not exact. However, it can be shown that this modified Greedy algorithm is suboptimal, in particular it does not improve on the factor 3 that can be obtained from the general GAP routine (details omitted). More care is required to achieve a factor 2, matching the lower bound. We will describe now how to accomplish this.

Assume that we have an approximate Restricted routine for the y-objective. Our generic algorithm will also use a polynomial routine for the following *Dual Restricted problem* (for the x-objective): Given an instance, a bound D and $\delta > 0$, either return a solution $\tilde{s}$ satisfying $y(\tilde{s}) \leq (1 + \delta)D$ and $x(\tilde{s}) \leq \min\{x(s)$ over all solutions s

having $y(s) \leq D$} or report that there does not exist any solution s such that $y(s) \leq D$. We use the notation DualRestrict$_\delta$ $(x, y \leq D)$ to denote the solution returned by the corresponding routine. The following proposition establishes the fact that any bi-objective problem that possesses a (fully) polynomial Restricted routine for the one objective, also possesses a (fully) polynomial *Dual Restricted* routine for the other.

Proposition 2. *For any bi-objective optimization problem, the problems Restrict$_\delta$ $(y, \cdot)$ and DualRestrict$_\delta$ $(x, \cdot)$ are polynomially equivalent.*

Algorithm Description: We first give a high-level overview of the 2-competitive algorithm. The algorithm iteratively selects a set of solution points $\{q_1, \ldots, q_r\}$ (in decreasing x) by judiciously combining the two routines. The idea is, in addition to the Restricted routine (for the y-coordinate), to use the Dual Restricted routine (for the x-coordinate) in a way that circumvents the problems previously identified for the greedy algorithm. More specifically, after computing the point q_i' in essentially the same way as the greedy algorithm, we proceed as follows: We select as q_i a point that: (i) has y-coordinate at most $(1 + \epsilon)y(q_i')/(1 + \delta)$ and (ii) has x-coordinate *at most* the minimum x over all solutions s with $y(s) \leq (1 + \epsilon)y(q_i')/(1 + \delta)^2$ for a suitable δ. This can be done by a call to the Dual Restricted routine for the x-objective. Intuitively this selection means that we give some "slack" in the y-coordinate to "gain" some slack in the x-coordinate. Also notice that, by selecting the point q_i in this manner, there may exist solution points with y-values in the interval $((1+\epsilon)y(q_i')/(1+\delta)^2, (1+\epsilon)y(q_i')/(1+\delta)]$ whose x-coordinate is *arbitrarily* smaller than $x(q_i)$. In fact, the optimal point $(1 + \epsilon)$-covering q_i can be such a point. However, it turns out that this is sufficient for our purposes and, if δ is chosen appropriately, this scheme can guarantee that the point q_{2i} lies to the left (or has the same x-value) of the i-th rightmost point of the optimal solution. We now proceed with the formal description of the algorithm. In what follows, the error tolerance is set to $\delta \doteq \sqrt[3]{1 + \epsilon} - 1$ ($\approx \epsilon/3$ for small ϵ). If $\sqrt[3]{1 + \epsilon}$ is not rational, we let δ be a rational that approximates $\sqrt[3]{1 + \epsilon} - 1$ from below, i.e. $(1 + \delta)^3 \leq (1 + \epsilon)$, and which has representation size (number of bits) $|\delta| = O(|\epsilon|)$.

Algorithm 2-Competitive
If Restrict$_{\delta_0 \leftarrow 1}(y, x \leq 2^m) = $ NO **then** halt.
$q_1' = $ Restrict$_\delta(y, x \leq 2^m)$;
$q_{\text{left}} = $ DualRestrict$_{\delta_0 \leftarrow 1}(x, y \leq 2^m)$; $x_{\min} = x(q_{\text{left}})$;
$\bar{y}_1 = y(q_1')(1 + \delta)$;
$q_1 = $ DualRestrict$_\delta(x, y \leq \bar{y}_1)$;
$\bar{x}_1 = x(q_1)/(1 + \epsilon)$;
$Q = \{q_1\}; i = 1$;
While ($\bar{x}_i > x_{\min}$) **do**
$\{$ $q_{i+1}' = $ Restrict$_\delta(y, x < \bar{x}_i)$;
    $\bar{y}_{i+1} = [(1 + \epsilon)/(1 + \delta)] \cdot \max\{\bar{y}_i, y(q_{i+1}')/(1 + \delta)\}$;
    $q_{i+1} = $ DualRestrict$_\delta(x, y \leq \bar{y}_{i+1})$;
    $\bar{x}_{i+1} = x(q_{i+1})/(1 + \epsilon)$;
    $Q = Q \cup \{q_{i+1}\}$;
    $i = i + 1; \}$
Return Q.

Theorem 2. *The above algorithm computes a 2-approximation to the smallest ϵ-Pareto set in time $O(OPT_\epsilon)$ subroutine calls, where $1/\delta = O(1/\epsilon)$.*

Sketch of Proof: We can assume that the solution set is non-empty. In this case, (i) the solution point q_{left} has minimum x-value among all feasible solutions and (ii) q'_1 has y-value *at most* $(1 + \delta)y_{\text{min}}$. (We denote by $x_{\text{min}}, y_{\text{min}}$ the minimum values of the objectives in each dimension.) It is also easy to see that each subroutine call returns a point; so, all the points are well-defined. Let $Q = \{q_1, q_2, \ldots, q_r\}$ be the set of solution points produced by the algorithm. We will prove that the set Q is an ϵ-Pareto set of size at most $2OPT_\epsilon$. We note the following simple properties.

Fact. 1. For each $i \in [r - 1]$ it holds (i) $x(q'_{i+1}) < x(q_i)/(1 + \epsilon)$ and (ii) for each solution point t with $x(t) < x(q_i)/(1 + \epsilon)$, we have $y(t) \geq y(q'_{i+1})/(1 + \delta)$.
2. For each $i \in [r]$ it holds (i) $y(q_i) \leq (1 + \delta)\bar{y}_i$ and (ii) for each solution point t with $y(t) \leq \bar{y}_i$ we have $x(t) \geq x(q_i)$.

We first show that Q is an ϵ-Pareto set. It is not hard to see that the x coordinates of the points $q_1, q_2, \ldots, q_r$ of Q form a strictly decreasing sequence, while this is not necessarily the case with the y coordinates. We claim that the point q_1 $(1 + \epsilon)$-covers all of the solution points that have x-coordinate at least $x(q_1)/(1 + \epsilon)$. To wit, let t be a solution point with $x(t) \geq x(q_1)/(1+\epsilon)$. It suffices to argue that $y(t) \geq y(q_1)/(1+\epsilon)$. By property 2-$(ii)$ we have $y(q_1) \leq (1 + \delta)\bar{y}_1 = (1 + \delta)^2 y(q'_1)$ and the definition of q'_1 implies that $y(t) \geq y(q'_1)/(1 + \delta)$, *for any solution point t.* By combining these facts we get that for any solution point t it holds $y(t) \geq y(q_1)/(1 + \delta)^3 \geq y(q_1)/(1 + \epsilon)$. Moreover, for each $i \in [r] \setminus \{1\}$ the point q_i $(1 + \epsilon)$-covers all of the solution points that have their x-coordinate in the interval $[x(q_i)/(1 + \epsilon), x(q_{i-1})/(1 + \epsilon))$. Let t be a solution point satisfying $x(q_i)/(1+\epsilon) \leq x(t) < x(q_{i-1})/(1+\epsilon)$. Suppose (for the sake of contradiction) that there exists such a point t with $y(t) < y(q_i)/(1 + \epsilon)$. By property 2-$(i)$ and the definition of $\bar{y}_i$ this implies $y(t) < \max\{\bar{y}_{i-1}, y(q'_i)/(1 + \delta)\}$. Now since $x(t) < x(q_{i-1})/(1 + \epsilon)$, property 1-$(ii)$ gives $y(t) \geq y(q'_i)/(1 + \delta)$. Furthermore, since $x(t) < x(q_{i-1})$, by property 2-(ii) it follows that $y(t) > \bar{y}_{i-1}$. Finally note that there are no solution points with x-coordinate smaller than $x(q_r)/(1 + \epsilon)$.

We now bound the size of the set of points Q in terms of the size of the optimal ϵ-Pareto set. Let $P^*_\epsilon = \{p^*_1, p^*_2, \ldots, p^*_k\}$ be the optimal ϵ-Pareto set, where its points p^*_i, $i \in [k]$, are ordered in increasing order of their y- and decreasing order of their x-coordinate. We will show that $|Q| = r \leq 2k$. This follows from the following claim, whose proof is omitted.

Claim. If the algorithm selects a solution point q_{2i-1} (i.e. if $2i - 1 \leq r$), then there must exist a point p^*_i in P^*_ϵ (i.e. it holds $i \leq k$) and if the algorithm selects a point q_{2i}, then $x(p^*_i) \geq x(q_{2i})$.

We now analyze the running time of the algorithm. Let k be the number of points in the smallest ϵ-Pareto set, $k = OPT_\epsilon$. The algorithm involves $r \leq 2k$ iterations of the while loop; each iteration involves two calls to the subroutines. Therefore, the total running time is bounded by $4k$ subroutine calls. $\square$

In the case of the bi-objective shortest path problem, the Restricted problem can be solved in time $O(en/\epsilon)$ for acyclic (directed) graphs [ESZ], and in time $O(en(\log\log n + 1/\epsilon))$ for general graphs [LR] with n nodes and e edges. The Dual Restricted problem can be solved with the same complexity. Thus, our algorithm runs in time $O(en(\log\log n + 1/\epsilon)OPT_\epsilon)$ for general graphs and $O(enOPT_\epsilon/\epsilon)$ for acyclic graphs. The time complexity is comparable or better than previous algorithms, which furthermore do not provide any guarantees on the size.

4 d Objectives

The results in this section use the GAP routine and thus apply to all problems in M(F)PTAS.

4.1 Approximation of the Optimal ϵ-Pareto Set

Recall that for $d \geq 3$ objectives we are forced to compute an ϵ'-Pareto set, where $\epsilon' > \epsilon$, if we are to have a guarantee on its size [VY]. For any $\epsilon' > \epsilon$, a logarithmic approximation for the problem is given in [VY], by a simple reduction to the Set Cover problem. We can sharpen this result, by exploiting additional properties of the corresponding set system.

Theorem 3
1. For any $\epsilon' > \epsilon$ there exists a polynomial time generic algorithm that computes an ϵ'-Pareto set Q such that $|Q| \leq O\big(d\log(dOPT_\epsilon)\big) \cdot OPT_\epsilon$.
2. For $d = 3$, the algorithm outputs an ϵ'-Pareto set Q satisfying $|Q| \leq cOPT_\epsilon$, where c is a constant.

Consider the following problem $\mathcal{Q}(P, \epsilon)$: Given a set of n points $P \subseteq \mathbf{R}_+^d$ as input and $\epsilon > 0$, compute the smallest ϵ-Pareto set of P. It should be stressed that, by definition, the set of points P is given *explicitly* in the input. (Note the major difference with our setting: for a typical multiobjective problem there are exponentially many solution points and they are not given explicitly.) This problem can be solved in linear time for $d = 2$ by a simple greedy algorithm. For $d = 3$ it is NP-hard and can be approximated within some (large) constant factor c [KP]. If d is arbitrary (i.e. part of the input, e.g. $d = n$), the problem is hard to approximate better than within a $\Omega(\log n)$ factor [VY].

The following fact relates the approximability of $\mathcal{Q}$ with the problem of computing a small ϵ'-Pareto set for a multiobjective problem Π, given the GAP primitive. Let $\epsilon > 0$ be a given rational number. For any $\epsilon' > \epsilon$, we can find a $\delta > 0$ such that $1/\delta = O(1/(\epsilon' - \epsilon))$ satisfying $1 + \epsilon' \geq (1 + \epsilon)(1 + \delta)^2$.

Proposition 3 (implicit in [VY]). *Suppose that there exists an r-factor approximation algorithm for $\mathcal{Q}$. Then, for any $\epsilon' > \epsilon$, we can compute an ϵ'-Pareto set Q, such that $|Q| \leq rOPT_\epsilon$ using $O((m/\delta)^d)$ GAP$_\delta$ calls.*

Sketch of Proof: First compute a δ-Pareto set R, by using the original algorithm of [PY1]. Then apply the r-approximation algorithm for $\mathcal{Q}$ to $(1 + \epsilon)(1 + \delta)$-cover R. It is easy to see that the computed set of points is an ϵ'-Pareto set of the desired cardinality. $\qquad\qquad\square$

Part 2 of Theorem 3 follows immediately from the fact that Q is constant factor approximable for $d = 3$ [KP] and the above proposition. We consider the case of general d in the remainder. The problem $Q(P, \epsilon)$ can be phrased as a set cover problem as follows: For each input point $q \in P$ and $\epsilon > 0$, define $S_{q,\epsilon} = \{x \in \mathbf{R}^d \mid q \leq (1 + \epsilon) \cdot x\}$ (the subset of $\mathbf{R}^d$ $(1 + \epsilon)$-covered by q). For each point $r \in P$, r is $(1 + \epsilon)$-covered by q iff $r \in S_{q,\epsilon}$. Now consider the set system $\mathcal{F}(P, \epsilon) = (P, \mathcal{S}(P, \epsilon))$, where $\mathcal{S}(P, \epsilon) = \{P_{q,\epsilon} \equiv P \cap S_{q,\epsilon} \mid q \in P\}$. Clearly, there is a bijection between set covers of $\mathcal{F}(P, \epsilon)$ and ϵ-Pareto sets of P. For $q \in P$ and $\epsilon > 0$, define $S_{q,\epsilon}^D = \{x \in \mathbf{R}^d \mid x \leq (1 + \epsilon) \cdot q\}$. A point r $(1 + \epsilon)$-covers q iff $r \in S_{q,\epsilon}^D$. The "dual" set system of $\mathcal{F}(P, \epsilon)$ is defined as $\mathcal{F}^D(P, \epsilon) = (P, \mathcal{S}^D(P, \epsilon))$, where $\mathcal{S}^D(P, \epsilon) = \{P_{q,\epsilon}^D \equiv P \cap S_{q,\epsilon}^D \mid q \in P\}$. In words, the elements are the points of P and for each point $q \in P$ we have a set consisting of the points $r \in P$ that $(1 + \epsilon)$-cover q. An ϵ-Pareto set of P is equivalent to a hitting set of $\mathcal{F}^D$.

For a set system $(U, \mathcal{R})$, we say that $X \subseteq U$ is *shattered* by $\mathcal{R}$ if for any $Y \subseteq X$, there exists a set $R \in \mathcal{R}$ with $X \cap R = Y$. The *VC-dimension* of the set system is the maximum size of any set shattered by $\mathcal{R}$. We can show the following:

Proposition 4. *The VC-dimension of the set systems $\mathcal{F}(P, \epsilon)$ and $\mathcal{F}^D(P, \epsilon)$ is upper bounded by d.*

As shown in [BG], for any (finite) set system with VC-dimension d, there exists a polynomial time $O(d \log(dOPT))$-factor approximation algorithm for the minimum *hitting set* problem, where OPT is the cost of the optimal solution. If we apply this result to the dual set system $\mathcal{F}^D(P, \epsilon)$, we conclude:

Proposition 5. *Problem Q can be approximated within a factor of $O(d \log(dOPT_\epsilon))$.*

Part 1 of Theorem 3 follows by combining Propositions 3 and 5.

4.2 The Dual Problem

For a given number k, we want to find k points that provide the best approximation to the Pareto curve, i.e. such that every Pareto point is ρ^*-covered by one of the k selected points for the minimum possible ratio $\rho^* = 1 + \epsilon^*$. It was shown in [VY] that for $d = 2$ the problem is NP-hard but has a PTAS. We show now that for $d = 3$ any multiplicative factor for the dual problem is impossible, even for explicitly given points; we can only hope for a constant power, and only above a certain constant. We omit the proof.

Theorem 4. *Consider the Dual problem for $d = 3$ objectives and explicitly given points.*

1. It is NP-hard to approximate the minimum ratio ρ^* within any polynomial multiplicative factor.
2. It is NP-hard to compute k points that approximate the Pareto curve with ratio better than $(\rho^*)^{3/2}$.

Of course if ρ^* is upperbounded by a constant, i.e. if k points suffice to provide a good approximation to the Pareto curve, then a constant power of ρ^* is also bounded by a constant. In [VY] the Dual problem was related to the Asymmetric k-center problem, and this was used to show that (i) for any d, a set of k points can be computed that approximates the Pareto curve with ratio $(\rho^*)^{O(\log^* k)}$, and (ii) for unbounded d and explicitly given points, it is hard to do much better. Since the metric ρ for the dual problem is a ratio (multiplicative coverage) versus distance (additive coverage) in the k-center problem, in some sense the analogue of constant factor approximation for the Dual problem is constant power.

Can we achieve a constant power $(\rho^*)^c$ for all problems in MPTAS with a fixed number d of objectives? We show that the answer is Yes for $d = 3$ and provide a conjecture that implies it for general d.

Consider the following *generalization* $Q'(A, P, \epsilon)$ of problem Q: Given a set of n points $P \subseteq \mathbf{R}_+^d$, a subset $A \subseteq P$ and $\epsilon > 0$, compute the smallest subset $P_\epsilon^*(A) \subseteq P$ that $(1 + \epsilon)$-covers A. It is easy to see that for $d = 3$ the arguments of [KP] for Q can be applied to Q' as well showing that it admits a constant factor approximation. We believe that in fact for all fixed d there may well be a constant factor approximation. Proving (or disproving) this for $d > 3$ seems quite challenging.

The following weaker statement seems more manageable:

Conjecture 1. For any fixed d, there exists a polynomial time *bicriterion* approximation algorithm for $Q'(A, P, \epsilon)$, that outputs an $(1 + \epsilon)^{f(d)}$-cover $C \subseteq P$ of A, satisfying $|C| \le g(d) \cdot |P_\epsilon^*(A)|$, for some functions $f, g : \mathbf{N} \to \mathbf{N}$.

For $d = 3$, conjecture 1 holds with $f(3) \le 2$ and $g(3) \le 4$. This can be shown by a technical adaptation of the 3-objectives algorithm in [VY], that is omitted here.

For general multiobjective problems with a polynomial GAP routine, we formulate the following conjecture:

Conjecture 2. For any fixed d, there exists a polynomial time generic algorithm, that outputs an $(1 + \epsilon)^{f(d)}$-cover C, whose cardinality is $|C| \le g(d) \cdot OPT_\epsilon$, for some functions $f, g : \mathbf{N} \to \mathbf{N}$.

The case of $d = 3$ is proved in [VY] with $f(3) =$ any constant $c' > 2$ and $g(3) = 4$. Note that, by (a variant of) Proposition 3, Conjecture 1 implies Conjecture 2. The converse is also partially true: Conjecture 2 implies Conjecture 1, if in the statement of the latter, problem Q' is substituted with problem Q.—

In the following theorem, we show that a constant factor bicriterion approximation for Q' implies a constant power approximation for the problem of computing k solutions that cover the feasible set with minimum ratio, given the GAP routine.

Theorem 5. *Consider a (implicitly represented) d-objective problem in MPTAS and suppose that the minimum achievable ratio with k points is ρ^*.*

1. For $d = 3$ objectives we can compute k points which approximate the Pareto set with ratio $O((\rho^*)^c)$ for some constant c.
2. If Conjecture 1 holds, then the same result holds for any fixed number d of objectives.

Sketch of Proof: Part 1 follows from 2 since Conjecture 1 holds for $d = 3$. To show 2, consider first the following ("dual") problem $\mathcal{D}(P, k)$: We are given *explicitly* a set P of n points in $\mathbf{R}_+^d$ and a positive integer k and we want to compute a subset of P of cardinality (at most) k that ρ-covers P with minimum ratio ρ. Let $\rho^* = 1 + \epsilon^*$ denote the optimal value of the ratio. As shown in [VY], this problem can be reduced to the asymmetric k-center problem. A c-factor approximation algorithm for the asymmetric k-center problem implies a $(\rho^*)^c$-factor approximation algorithm for the problem at hand.

We claim that, if problem $Q'(A, P, \epsilon)$ admits a $((1 + \epsilon)^{f(d)}, g(d))$-bicriterion approximation, then problem $\mathcal{D}(P, k)$ admits a $(\rho^*)^{h(d)}$ approximation for some function h. This is implied by the above reduction and the following more general fact: If we have an instance of the asymmetric k-center problem (problem $\mathcal{D}(P, k)$ in our setting) such that a certain collection of associated set cover subproblems (which are instances of problem $Q'(A, P, \epsilon)$ here) admits a constant factor bicriterion approximation (an algorithm that blows up both criteria by a constant factor), then this instance admits a constant factor *unicriterion* approximation. This implication is not stated (or proved) in [PV], but follows easily from their work. One way to prove it is to apply Lemma 5 of [PV] in a recursive manner. This is the approach we follow. We defer the technical details to the full version. Aaron Archer recently informed us [Ar2] that he has an alternative method that yields better constants.

For a general multiobjective problem where the solution points are not given explicitly, we impose a geometric $\sqrt{1 + \delta}$ grid for a suitable δ, call GAP at the grid points, and then apply the above algorithm to the set of points returned. Then the set of k points computed by the algorithm provide a $(1 + \epsilon')^{h(d)}$-cover of the Pareto curve, where $1 + \epsilon' = (1 + \epsilon)(1 + \delta)^2$.   □

We should remark that the algorithms of this section are less satisfactory that the bi-objective algorithm of the previous section (and the 2d and 3d algorithms of [VY]) in several respects. One weakness is that the constants c obtained (for $d = 3$) are quite large: in the case of Theorem 3, c is the same constant as in [KP] (which is around 200), and in the case of Theorem 5 the constant that comes out of the recursive method applied to the [VY] algorithm is around 100 (using Archer's new technique instead would reduce it to around 20). A second weakness of the algorithms is that they start by applying the general method of [PY1] calling the GAP routine on a grid, and thus incur always the worst-case time complexity even if there is a very small ϵ-Pareto set. Thus, we view our algorithms in this section mainly as theoretical proofs of principle, i.e. that certain (constant) approximations can be computed in polynomial time, but it would be very desirable and important to improve both the constants and the time.

5   Conclusion

We investigated the problem of computing a minimum set of solutions for a multiobjective optimization problem that represents approximately the whole Pareto curve within a desired accuracy ϵ. We developed tight approximation algorithms for the bi-objective shortest path problem, spanning tree, and a host of other bi-objective problems. Our algorithms compute efficiently an approximate Pareto set that contains at most twice as many solutions as the minimum one; furthermore improving on the factor 2 for these specific problems is NP-Hard. The algorithm works in general for all bi-objective problems for which we have a routine for the Restricted problem of approximating one objective subject to a (hard) bound on the other. The algorithm calls this Restricted routine and a dual one as black boxes and makes quite effective use of them: for every instance, the number of calls is linear (at most 4 times) in the number of points in the optimal solution for that instance.

We presented also results for three and more objectives, both for the problem of computing an optimal ϵ-Pareto set and for the dual problem of selecting a specified number k of points that provide the best approximation of the full Pareto curve. As we indicated at the end of the last section, there is still a lot of room for improvement both in the time complexity and the constants of the approximations achieved. We would like especially to resolve Conjecture 2, hopefully positively. It would be great to have a general efficient method for any (small) fixed number d of objectives that computes for every instance a succinct approximate Pareto set with small constant loss in accuracy and in the number of points, and do it in time proportional to the number of computed points, i.e., the optimal approximate Pareto set for the instance in hand.

References

[ABK1]   Angel, E., Bampis, E., Kononov, A.: A FPTAS for Approximating the Unrelated Parallel Machines Scheduling Problem with Costs. In: Meyer auf der Heide, F. (ed.) ESA 2001. LNCS, vol. 2161, Springer, Heidelberg (2001)

[ABK2]   Angel, E., Bampis, E., Kononov, A.: On the approximate trade-off for bicriteria batching and parallel machine scheduling problems. Theoretical Computer Science 306(1-3), 319–338 (2003)

[ABV1]   Aissi, H., Bazgan, C., Vanderpooten, D.: Approximation Complexity of Min-Max (Regret) Versions of Shortest Path, Spanning Tree and Knapsack. In: Brodal, G.S., Leonardi, S. (eds.) ESA 2005. LNCS, vol. 3669, Springer, Heidelberg (2005)

[ABV2]   Aissi, H., Bazgan, C., Vanderpooten, D.: Complexity of Min-Max (Regret) Versions of Cut problems. In: Deng, X., Du, D.-Z. (eds.) ISAAC 2005. LNCS, vol. 3827, Springer, Heidelberg (2005)

[AN+]   Ackermann, H., Newman, A., Röglin, H., Vöcking, B.: Decision Making Based on Approximate and Smoothed Pareto Curves. In: Deng, X., Du, D.-Z. (eds.) ISAAC 2005. LNCS, vol. 3827, Springer, Heidelberg (2005)

[Ar1]   Archer, A.F.: Two $O(\log^* k)$-approximation algorithms for the asymmetric k-center problem. In: Proc. 8th Conf. on Integer programming and Combinatorial Optimization, pp. 1–14 (2001)

[Ar2]   Archer, A.F.: Personal communication (2007)

[AZ]   Armon, A., Zwick, U.: Multicriteria Global Minimum Cuts. Algorithmica 46, 15–26 (2006)

[BG]   Brönnimann, H., Goodrich, M.T.: Almost Optimal Set Covers in Finite VC-Dimension. Discrete and Computational Geometry 14(4), 463–479 (1995)

[CG+]   Chuzhoy, J., Guha, S., Halperin, E., Khanna, S., Kortsartz, G., Naor, S.: Asymmetric k-center is $\log^* n$-hard to Approximate. In: Proc. 36th ACM STOC, pp. 21–27. ACM Press, New York (2004)

[CJK]   Cheng, T.C.E., Janiak, A., Kovalyov, M.Y.: Bicriterion Single Machine Scheduling with Resource Dependent Processing Times. SIAM J. Optimization 8(2), 617–630 (1998)

[Cli]   Climaco, J. (ed.): Multicriteria Analysis. Springer, Heidelberg (1997)

[CX1]   Chen, G., Xue, G.: A PTAS for weight constrained Steiner trees in series-parallel graphs. Theoretical Computer Science 1-3(304), 237–247 (2003)

[CX2]   Chen, G., Xue, G.: K-pair delay constrained minimum cost routing in undirected networks. In: Proc. SODA, pp. 230–231 (2001)

[DJSS]   Dongarra, J., Jeannot, E., Saule, E., Shi, Z.: Bi-objective Scheduling Algorithms for Optimizing Makespan and Reliability on Heterogeneous Systems. In: Proc. SPAA (2007)

[Ehr]   Ehrgott, M.: Multicriteria optimization, 2nd edn. Springer, Heidelberg (2005)

[EG]   Ehrgott, M., Gandibleux, X.: An annotated bibliography of multiobjective combinatorial optimization problems. OR Spectrum 42, 425–460 (2000)

[ESZ]   Ergun, F., Sinha, R., Zhang, L.: An improved FPTAS for Restricted Shortest Path. Information Processing Letters 83(5), 237–239 (2002)

[EV]   Erkut, E., Verter, V.: Modeling of transport risk for hazardous materials. Operations Research 46, 625–642 (1998)

[FGE]   Figueira, J., Greco, S., Ehrgott, M. (eds.): Multiple Criteria Decision Analysis: State of the Art Surveys. Springer, Heidelberg (2005)

[GJ]   Garey, M.R., Johnson, D.S.: Computers and Intractability. W. H. Freeman, New York (1979)

[GR]   Goemans, M.X., Ravi, R.: The Constrained Minimum Spanning Tree Problem. In: Proc. SWAT, pp. 66–75 (1996)

[GR+]   Goel, A., Ramakrishnan, K.G., Kataria, D., Logothetis, D.: Efficient computation of delay-sensitive routes from one source to all destinations. In: Proc. IEEE INFOCOM (2001)

[Han]   Hansen, P.: Bicriterion Path Problems. In: Bekic, H. (ed.) Programming Languages and their Definition. LNEMS, vol. 177, pp. 109–127. Springer Verlag, Heidelberg (1979)

[Has]   Hassin, R.: Approximation schemes for the restricted shortest path problem. Mathematics of Operations Research 17(1), 36–42 (1992)

[HL]   Hassin, R., Levin, A.: An efficient polynomial time approximation scheme for the constrained minimum spanning tree problem. SIAM J. Comput. 33(2), 261–268 (2004)

[KP]   Koltun, V., Papadimitirou, C.H.: Approximately dominating representatives. Theoretical Computer Science 371, 148–154 (2007)

[LR]   Lorenz, D.H., Raz, D.: A simple efficient approximation scheme for the restricted shortest path problem. Operations Research Letters 28(5), 213–219 (2001)

[Miet]   Miettinen, K.M.: Nonlinear Multiobjective Optimization. Kluwer Academic Publishers, Dordrecht (1999)

[PV]   Panigrahy, R., Vishwanathan, S.: An $O(\log^* n)$ approximation algorithm for the asymmetric p-center problem. J. of Algorithms 27(2), 259–268 (1998)

[PY1]   Papadimitriou, C.H., Yannakakis, M.: On the Approximability of Trade-offs and Optimal Access of Web Sources. In: Proc.41st IEEE FOCS, IEEE Computer Society Press, Los Alamitos (2000)

[PY2]   Papadimitriou, C.H., Yannakakis, M.: Multiobjective Query Optimization. In: Proc. ACM PODS, ACM Press, New York (2001)

[TZ]   Tsaggouris, G., Zaroliagis, C.D.: Multiobjective Optimization: Improved FPTAS for Shortest Paths and Non-linear Objectives with Applications. In: Asano, T. (ed.) ISAAC 2006. LNCS, vol. 4288, pp. 389–398. Springer, Heidelberg (2006)

[VV]   Van Mieghen, P., Vandenberghe, L.: Trade-off Curves for QoS Routing. In:Proc. INFOCOM (2006)

[VY]   Vassilvitskii, S., Yannakakis, M.: Efficiently computing succinct trade-off curves. Theoretical Computer Science 348, 334–356 (2005)

[Wa]   Warburton, A.: Approximation of Pareto Optima in Multiple-Objective Shortest Path Problems. Operations Research 35, 70–79 (1987)

Two Randomized Mechanisms for Combinatorial Auctions

Shahar Dobzinski

The School of Computer Science and Engineering,
The Hebrew University of Jerusalem

Abstract. This paper discusses two advancements in the theory of designing truthful randomized mechanisms.

Our first contribution is a new framework for developing truthful randomized mechanisms. The framework enables the construction of mechanisms with polynomially small failure probability. This is in contrast to previous mechanisms that fail with constant probability. Another appealing feature of the new framework is that bidding truthfully is a *strongly* *dominant* strategy. The power of the framework is demonstrated by an $O(\sqrt{m})$-mechanism for combinatorial auctions that succeeds with probability $1 - O(\frac{\log m}{\sqrt{m}})$.

The other major result of this paper is an $O(\log m \log \log m)$ randomized truthful mechanism for combinatorial auction with *subadditive* bidders. The best previously-known truthful mechanism for this setting guaranteed an approximation ratio of $O(\sqrt{m})$. En route, the new mechanism also provides the best approximation ratio for combinatorial auctions with *submodular* bidders currently achieved by truthful mechanisms.

1 Introduction

1.1 Background

The field of Algorithmic Mechanism Design [14] has received much attention in recent years. The main goal of research in this field is to design efficient algorithms for computationally-hard problems, in a way that handles the strategic behavior of the participants. In particular, we are interested in mechanisms that are *truthful*, i.e., where the dominant strategy of each bidder is to report his true preferences[1]. For a recent excellent introduction to the basics of mechanism design the reader is referred to [13].

Arguably the most important problem in algorithmic mechanism design is the design of truthful efficient mechanisms for combinatorial auctions. In a combinatorial auction we have n bidders and a set M of items, $|M| = m$. Each bidder i has a valuation function v_i. The common assumptions are that each valuation function is normalized ($v_i(\emptyset) = 0$) and monotone (for every

---

[1] One could also consider algorithms that lead to another arbitrary equilibrium, but the revelation principle tells us that the restriction to truthful mechanisms is essentially without loss of generality.

M. Charikar et al. (Eds.): APPROX and RANDOM 2007, LNCS 4627, pp. 89–103, 2007.

$S \subseteq T \subseteq M, v_i(T) \geq v_i(S)$). The goal is to find a partition $S_1, ..., S_n$ of the items, such that the total social welfare, $\Sigma_i v_i(S_i)$, is maximized. Notice that a naive representation of each valuation function requires exponential number of bits in n and m, the parameters we require our algorithms to be polynomial in. This paper therefore assumes that the valuations are represented as black boxes which can answer a specific natural type of queries, *demand queries*[2].

Combinatorial auctions demonstrate the clash between the computational and economic aspects that stands in the heart of algorithmic mechanism design: on one hand, obtaining a truthful mechanism is easy, ignoring computational limitations: find the optimal solution and use VCG payments, probably the main technique of mechanism design. Unfortunately, combinatorial auctions are hard to exactly solve and even to approximate: in the general case, an approximation ratio of $m^{\frac{1}{2}-\epsilon}$ cannot be obtained in polynomial time, for any constant $\epsilon > 0$ [12]. It is known that in general using VCG together with an approximation algorithm (in contrast to the optimal algorithm) does not result in a truthful mechanism [15], hence other techniques are required.

Recently, a series of papers [10,3,4] used *randomization* to construct truthful mechanisms with good approximation ratios. Referring to randomization seems necessary, as we have some evidence that *deterministic* mechanisms do not have much power [9,1]. Two types of randomized mechanisms were considered: mechanisms that are truthful *in expectation*, and mechanisms that are truthful *in the universal sense*. Mechanisms that are truthful in expectation are truthful only with respect to bidders that maximize their *expected* utility (risk-neutral bidders), and do not know the outcome of the random coin flips. Mechanisms that are truthful in the universal sense are stronger, as they are truthful for every type of bidders, and even if the outcome of the random coins is known in advance. For a more thorough discussion of the differences the reader is referred to [3]; This paper considers only mechanisms that are truthful in the universal sense.

1.2 Our Results

An Improved Framework for Designing Truthful Mechanisms. In [3] a general framework for designing truthful mechanisms was presented. The framework uses random-sampling methods (introduced in [6]), and was quite successful in enabling the design of mechanisms that provide good approximation ratios and are truthful in the universal sense (rather than mechanisms that are truthful in expectation [10,4]). However, the framework of [3] suffers from two major drawbacks:

- **The probability of success:** In general, given a randomized mechanism, running the mechanism again (in order to increase the success probability) makes the mechanism no longer truthful. Thus, the main goal of the framework of [3] was to achieve good approximation ratios with high probability.

---

[2] In a demand query the bidder is presented with a price of p_j for each item, and the answer is the bundle that maximizes the profit, $v(S) - \Sigma_{j \in s} p_j$.

Unfortunately, this was achieved by trading success probability with approximation ratio. For example, in its application to combinatorial auction with general bidders, the framework provided an approximation ratio of $O(\frac{\sqrt{m}}{\epsilon^3})$ with probability $1 - \epsilon$, for any $\epsilon > 0$. In particular, this means that in order to obtain a probability of success that is better than a constant, we have to compromise on a non-optimal approximation guarantee (by more than a constant factor).

- **The motivation of bidders to participate:** The framework of [3] uses a randomly selected group of bidders that cannot win any items at all. Yet, bidders in it are still required to provide information about their valuations. In other words, bidding truthfully is only a *weakly* dominant strategy for these bidders, as their utility will always be 0 regardless of the information they provide.

One contribution of the current paper is an improved framework that overcomes both limitations. The improved framework enables the design of mechanisms that provide good approximation ratios with high probability, without compromising on the approximation ratio. In addition, for every outcome of the random coins bidding truthfully can sometime strictly improve the utility of each bidder. The main application of the improved framework is the following theorem, that guarantees the optimal approximation ratio possible for combinatorial auctions with general bidders with high success probability:

Theorem: There exists a truthful randomized $O(\sqrt{m})$-approximation mechanism for combinatorial auctions with general bidders. The approximation ratio is achieved with probability of at least $1 - O(\frac{\log m}{\sqrt{m}})$. Bidding truthfully is a strongly dominant strategy of each bidder.

The analysis of the framework of [3] was quite straightforward. The improved framework presented in this paper is more involved, and its analysis requires some effort.

A Mechanism for Combinatorial Auctions with Subadditive Bidders. Another line of research that has gained popularity recently is determining the approximation ratios possible when the bidders' valuations are restricted (e.g., [11,2,4,5,8]). Most of the research concentrated in the case when the bidders' valuations are known to be either subadditive (for every $S, T \subseteq M$, $v(S)+v(T) \geq v(S \cup T)$) or submodular (for every $S, T \subseteq M$, $v(S)+v(T) \geq v(S \cup T)+v(S \cap T)$). Between these two classes lies the syntactic class of XOS valuations (see [3] for a definition), that strictly contains the class of submodular valuations, but is strictly contained in the class of subadditive valuations.

In [3] another application of the framework is presented, an $O(\frac{\log^2 m}{\epsilon^3})$ mechanism for combinatorial auctions with XOS bidders that succeeds with probability of $1 - \epsilon$. The improved framework presented in this paper can be applied to construct an $O(\log^2 m)$ mechanism for the same setting that succeeds with probability $1 - O(\frac{1}{\log m})$. Yet, for the case of combinatorial auctions with *subadditive* bidders, the best mechanism until now guaranteed an approximation ratio of only $O(\sqrt{m})$ [2]. Feige [4] presents an $O(\frac{\log m}{\log \log m})$-mechanism that is truthful in expectation for this case[3]. Ignoring incentives issues, a ratio of 2 can be achieved [4].

This paper presents the first truthful mechanism that provides a poly-logarithmic approximation ratio for combinatorial auctions with subadditive bidders. En route, this mechanism also improves upon the best mechanism known for combinatorial auctions with submodular bidders [3]. Another important improvement over [3] is this paper's mechanism uses the simple and natural demand queries, and not, as in [3], the syntactically defined XOS queries.

Theorem: There exists a truthful $O(\log m \log \log m)$-mechanism for combinatorial auctions with subadditive bidders, that succeeds with probability of $1 - O(\frac{1}{\log m})$.

We point out that unlike the improved framework that this paper describes, where our main contribution is game-theoretic, the main improvement of the mechanism for subadditive bidders is of algorithmic nature: we define a new combinatorial property of valuation functions, namely, α-supporting prices, and show that if a class of valuations exhibit this property, then we can essentially derive an $O(\alpha)$-approximation algorithm. Finally, we prove that every subadditive valuation has an $O(\log m)$-supporting prices, and show how to use the framework to derive a truthful mechanism that achieves almost this approximation ratio.

1.3   Open Questions

The main question we leave open is to determine the exact approximation ratio that truthful mechanisms can guarantee for combinatorial auctions with subadditive bidders. Obtaining mechanisms with constant approximation ratios is of particular interest. It will be also very interesting to understand the power of randomization; For example, for combinatorial auctions with general valuations this paper presents an optimal $O(\sqrt{m})$-randomized mechanism. Yet, the best deterministic mechanism achieves a ratio of $O(\frac{m}{\sqrt{\log m}})$ [7]. For combinatorial auctions with subadditive bidders, the gap is also large: the best deterministic mechanism achieves a ratio of $O(\sqrt{m})$ [2], while the best randomized mechanism is this paper's $O(\log m \log \log m)$ mechanism. A recent result [1] suggests that this gap cannot be bridged, at least with the current techniques.

A second open question is to develop mechanisms that fail not only with polynomially small probability, but with exponentially small probability, as common in traditional algorithm design. This seems beyond the power of the framework, and any other random-sampling based techniques, due to the need to "gather statistics", which fails with polynomially small probability. Developing such mechanisms seems very challenging.

---

[3] In fact the mechanism of [4] uses an even weaker notion than truthfulness in expectation, as truthfulness maximizes the expected utility only approximately.

2   An Improved Framework for Truthful Mechanisms

The new framework presented in this section uses two types of auctions as sub-routines: a second-price auction with non-anonymous reserve prices, and a fixed price auction.

A *second-price auction with non-anonymous reserve prices* is a variation of the well-known second-price auction. Here each bidder is presented with a (different) reserve price. If the winner (the bidder with the highest bid) bids above his reserve price, he wins the item and pays the maximum price of the second-highest bid and his reserve price. If the winner bids below his reserve price, then he pays nothing and the item is not allocated. The auction is truthful if each bidder's reserve price is independent of his bid.

In a *fixed-price auction with price p* the bidders are ordered arbitrarily, and we assign the first bidder his most demanded set when the price of each item is p. Then, we assign the second bidder his most demanded set from the set of remaining items (the items that were not allocated to the first bidder), again, when the price of each remaining item is p and so on. If p does not depend on the valuations of the bidders, then a fixed-price auction is truthful.

Let us now present the framework itself. To create a specific mechanism, we have to specify **four parameters**. These parameters are marked in **bold**.

The Framework

1. Add each bidder to exactly one of the following groups of bidders: with probability $\frac{1}{2}$ to STAT, and with probability $\frac{1}{2}$ to FIXED.
2. For each set of bidders S, we denote by $\overline{OPT}_S$ **an estimation of the optimal solution** with the participation of bidders in S only.
3. Conduct a second-price auction with non-anonymous reserve prices for selling the bundle of all items with the participation of bidders in STAT only. Set the reserve price of the ith bidder that is in STAT to be $\frac{\overline{OPT}_{STAT\setminus\{i\}}}{\alpha}$, **for some α.**
   If there was a winner in the second-price auction, continue to Case 1, otherwise continue to Case 2.
   Case 1: There was a winner.
4. Conduct a second-price auction with non-anonymous reserve prices for selling the bundle of all items with the participation of *all* bidders. Set the reserve price of each bidder $i \in STAT$ to $\frac{\overline{OPT}_{STAT\setminus\{i\}}}{\alpha}$. Set the reserve price of each bidder $i \in FIXED$ to 0. Let w be the winning bidder. Assign w the bundle of all items, and let w pay $\max(\max_{w \neq i \in N} v_i(M), r_w)$, where r_w denotes the reserve price of bidder w. Finish the mechanism.
   Case 2: There was no winner.
5. Use the bidders in STAT to **calculate a price per item of p** to be used in a fixed-price auction among the bidders in FIXED. Conduct a fixed-price auction with this price. Let S_i denote the bundle bidder i is assigned in this auction.

6. Select, uniformly at random, a price r from the set $[\frac{\overline{OPT}_{STAT}}{\alpha}, \frac{\overline{OPT}_{STAT}}{\alpha}, \ldots,$ $\frac{\overline{OPT}_{STAT}}{\beta}]$, **for some** β. Conduct a second-price auction with a reserve price of r for selling the bundle of all items among the bidders of FIXED. Let j denote the winner in auction (if one exists). If there are at least two bids above r, allocate the bundle to the bidder with the highest bid, let him pay the second-most highest bid, and finish the mechanism. Otherwise, continue to the next step.

7. If there is no bid above r, then assign each bidder i the bundle S_i and let him pay $|S_i| \cdot p$. If there is *exactly* one bid above r, let this bidder j choose his maximum-profit allocation of the following two, and assign accordingly:
   - Assign bidder j all items, and let him pay r.
   - Assign bidder j the bundle S_j and let him pay $|S_j| \cdot p$. Assign the rest of the bidders no items at all.

We will first see that mechanisms constructed using the framework are truthful. As an application, we will see an improved mechanism for combinatorial auctions with general bidders.

Theorem 1. *Mechanisms constructed using the framework are truthful. Moreover, for each bidder i bidding truthfully is a strongly dominant strategy.*

Proof. We first prove that the (strongly) dominant strategy of each bidder in STAT is to bid truthfully. Then, we show that this is true for bidders in FIXED too.

Look at some bidder in STAT. This bidder can only win the bundle of all items, by participating a second-price auction with non-anonymous reserve prices. Observe that the reserve price does not depend on the declaration of the bidder. Clearly, such an auction is truthful, hence the truthfulness of the mechanism with respect to bidders in STAT. Also observe that for every input there is a declaration that makes the bidder win and gain a positive utility, thus truthfully reporting his valuation is a strongly dominant strategy.

We now handle bidders in FIXED. If there is a winner in the second-price auction, then bidders in FIXED essentially participate in a second-price auction, which is truthful. Otherwise, there was no winner in the second-price auction. Observe that if the value of all items for some bidder is below r, then, independently of his declaration, he either wins some items via the fixed-price auction, that is truthful, or does not win any items at all. If his value of all items is above r and there is another bidder with a value above r, then this bidder participates in a second-price auction which is truthful. Finally, if this bidder is the only one that bids above r then he chooses between getting the bundle of all items with a price of r, or getting the bundle he won in the fixed-price auction in the corresponding price. In particular, he cannot be hurt by declaring above r. Notice that in any case no other bidder gets any items at all, thus all other bidders have no incentive to bid untruthfully in the fixed-price auction[4].   □

---

[4] Interestingly enough, in this case even if the *other* bidders do lie regarding their preferences in the fixed-price auction, this will not hurt the approximation ratio!

We do note that in the two applications of the framework presented in the paper we use iterative methods to compute an estimation of the optimal solution (namely, solving the corresponding linear program). Thus it might be the case that a bidder will have no incentive to truthfully answer *some* queries (but will have no incentive to answer untruthfully too). This problem do not arise if the valuations can be succinctly described, or in any other situation where the estimation of the optimal solution can be done in a non-iterative way.

2.1 The Solution Concept: Dominant Strategies vs. Ex-Post Nash

It is well known in the economic literature that in iterative mechanisms the revelation principle do not hold anymore and the solution concept becomes ex-post Nash[5]. Consider an iterative variant of a second-price auction where the first bidder bids, and the second bidder bids *after* him. If the strategy of the first bidder is: "if the first bidder bids 5 then I bid 3, otherwise I bid 4", then bidding truthfully is no longer a dominant strategy of the first bidder. As this example demonstrate, in order to prove that in an iterative mechanism is bidding truthfully is a dominant strategy, we have to make sure, roughly speaking, that bidders do not get "extra information" from the previous rounds (i.e., the bid of the first player in the example). See, e.g., [10] for a discussion. Fortunately, this is not the case with mechanisms constructed using the framework, but showing it is subtle. Let us now explicitly explain why bidding truthfully is a dominant strategy in mechanisms constructed using the framework. We will have to be careful about the implementation details, though.

In the first stage, we ask each bidder i to declare $v_i(M)$. *We hide bidder i's bid from the rest of the bidders.* After determining the winner, we use the rest of the bidders in STAT to calculate the winner's reserve price. The crucial point is that the only new information that bidders in STAT gained is the identity of the bidder. Thus, the only way the winner's bid can influence this calculation is by bidding lower and losing, but then the winner's profit can only decrease to zero.

If the winner bids above his reserve price, we use the bids of the bidders in FIXED to determine the identity of the bidder that will be allocated the bundle of all items. Otherwise, we calculate, using bidders in STAT, the price per item p for the fixed price auction (again, the bids of the bidders in FIXED are hidden from bidders in STAT), and perform the fixed price auction. We now allocate the items to bidders in FIXED according to the framework. Notice that when the bidders participate in the fixed-price auction, they do not know the bids on the bundle of all items, and thus they are not influenced by these bids. The mechanism now have all the information it needs to decide the resulting allocation.

---

[5] In an *ex-post Nash* equilibrium bidding truthfully is a (weakly) dominant strategy of each bidder if all other bidders bid truthfully.

2.2   Combinatorial Auctions with General Bidders

We now present the main application of the improved framework, an $O(\sqrt{m})$ approximation mechanism that succeeds with probability $1 - O(\frac{\log m}{\sqrt{m}})$. To make the mechanism concrete, we need to specify four parameters:

- **The value of α:** we let $\alpha = 100\sqrt{m}$.
- **The value of β:** we let $\beta = 100 \log m$.
- **The value of p:** we let the price per item in the fixed-price auction to be $p = \frac{\overline{OPT_{STAT}}}{100m}$.
- **The way we estimate the optimal solution restricted to a set of bidders:** we use the value of the optimal fractional allocation (details in the full version).

Let us start the proof of the approximation ratio, and in the same time provide some intuition regarding the framework. Some notation first: we denote by OPT_S^* the optimal fractional solution with the participation of bidders in S only. Bidder i is called a *dominant* bidder if $v_i(M) \geq \frac{OPT^*}{\sqrt{m}}$, and *super dominant* if $v_i(M) \geq \frac{OPT^*}{10 \log m}$. The analysis involves considering three different cases; the first two are similar in all mechanisms developed using the framework, while the third one is much more algorithmically challenging. The main difficulty (but not the only one) in this case is to show that a fixed-price auction provides a good approximation ratio. See the mechanism for subadditive bidders for a much more complicated example than the one presented in this section.

Case 1: There is a Super-Dominant Bidder. We first prove that if there is at least one super-dominant bidder then the mechanism always ends with a good approximation ratio. There are two possibilities: there might be a super-dominant bidder in STAT, or all super-dominant bidders are in FIXED. In the first case we observe that the reserve price offered to each of the bidders is at most $\frac{OPT^*}{100\sqrt{m}}$, since $OPT_S^* \leq OPT^*$ for any set S. Thus, there must be a winner in the second-price auction. Allocating the bundle of all item to the bidder that values it the most, which must be a super-dominant bidder, is a good approximation to the optimal welfare, as required.

Assume that all super-dominant bidders are in FIXED. If there was a winner in the second-price auction of Step 4, then all bidders essentially participate in a second-price auction. As in the previous case, we are guaranteed to get a good approximation ratio. If there was no winner, then the analysis is divided into cases once again, depending on the number of super-dominant bidders. If there are at least two super-dominant bidders, we will have a second-price auction for selling the bundle of all items, and we are guaranteed to get a good approximation. The case where there is exactly one super-dominant bidder is a bit more tricky. This super-dominant bidder i has to choose between getting the bundle of all items with a price of r, or getting the bundle S_i he won in the fixed-price and pay $|S_i| \cdot p$. If i takes M we get a good approximation ratio. If i takes the bundle S_i, observe that the profit of this bidder from taking M is at least $v_i(M) - r \geq \frac{OPT^*}{10\log m} - \frac{OPT^*}{100\log m} \geq \frac{OPT^*}{2\log m}$, thus the profit from taking S_i must be bigger, and in particular $v_i(S_i) \geq \frac{OPT^*}{2\log m}$, which gives a good approximation ratio even if i takes S_i.

Case 2: Many Dominant Bidders. If there are more than $\log m$ dominant bidders, then with probability of at least $1 - \frac{1}{2^{\log m}} = 1 - \frac{1}{m}$ at least one dominant bidder will be in STAT, since each bidder is selected to STAT with a probability of $\frac{1}{2}$. If this happens, then using similar arguments to the previous claim, we are guaranteed that the bundle of all items will be sold to some dominant bidder. Thus, we get a good approximation ratio with high probability.

Case 3: A Few Dominant Bidders. The last case we need to handle is the case where there are at most $\log m$ dominant bidders, and no super-dominant bidders. For most mechanisms developed using the framework this is the non-standard part that requires a deeper combinatorial understanding of the problem. For the mechanism presented in this section, we borrow ideas from [3]. The second mechanism we present (for subadditive valuations) requires much more effort to handle this third case.

First notice that dominant bidders contribute together a value of at most $\frac{OPT^*}{10\log m} \cdot \log m = \frac{OPT^*}{10}$ to the optimal welfare. Let ND be the set of bidders that are not dominant. Clearly, $OPT^*_{ND} \geq \frac{9OPT^*}{10}$. The proof of the next lemma can be found in the full version of this paper:

Lemma 1. *With probability of at least $O(\frac{1}{\sqrt{m}})$, we have that both events hold together: $OPT^*_{ND \cap STAT} \geq \frac{OPT^*}{8}$, and $OPT^*_{ND \cap FIXED} > \frac{OPT^*}{8}$.*

With probability of at most $O(\frac{1}{\sqrt{m}})$ the conclusion of the lemma does not hold, and we assume that the approximation ratio is 0. Otherwise, if there was a winner in the second-price auction of Step 4, we get a good approximation ratio, since the reserve price is at least $\frac{OPT^*}{400\sqrt{m}}$. If there was no winner, and there are at least two bidders that value the bundle of all items above r, we also get a good approximation since one of them will get the bundle, and $r \geq \Omega(\frac{OPT^*}{\sqrt{m}})$. The next lemma handles the case where no bidder bids above r (see the full version for a proof):

Lemma 2. *Let $A = FIXED \cap ND$. Then, a fixed-price auction with the participation of bidders in $FIXED$ with a price per item $p = \frac{OPT^*_A}{c \cdot m}$, for some $c > 2$, returns an allocation that has a value of $O(\frac{OPT^*_A}{c\sqrt{m}})$.*

We have to handle an additional single case where there is only one bidder i that bids above r. Our concern is that the profit from taking M, $v_i(M) - r$, is smaller than $\frac{OPT}{100\sqrt{m}}$, and the profit from taking the bundle S_i that i won in the fixed-price auction is larger, but $v_i(S_i)$ is low. We will see that this case occurs with very small probability. First, observe that this problem does not arise for the next (if possible) value of r, $r' = r + \frac{OPT}{100\sqrt{m}}$, since $v_i(M) < r'$. Also notice that for smaller values of r, like $r'' = r - \frac{OPT}{100\sqrt{m}}$, we have that $v_i(M) - r' > \frac{OPT}{100\sqrt{m}}$. Thus, if bidder i chooses to take S_i in this case, it holds that $v_i(S_i) \geq \frac{OPT}{100\sqrt{m}}$. (For smaller values we may have more than one bidder that bids above r', but this is a "good" case, similarly to before.) We conclude that there exists *at most* one value of r that gives a bad approximation ratio. However, there are $O(\frac{\sqrt{m}}{\log m})$ possible values, and we choose one uniformly at random. Hence, we are not guaranteed to get a good approximation ratio with probability of at most $O(\frac{\log m}{\sqrt{m}})$. In conclusion, we have the following theorem:

Theorem 2. *There exists an $O(\sqrt{m})$-approximation mechanism for combinatorial auction with general bidders that succeeds with probability $1 - O(\frac{\log m}{\sqrt{m}})$*

3   A Truthful Mechanism for Subadditive Valuations

This section introduces a new truthful mechanism for combinatorial auctions with subadditive bidders. The mechanism uses the framework of the previous section. However, the main novelty of the mechanism is combinatorial: roughly speaking, we prove that for every subadditive valuation and bundle S, one can set a price for each item such that every subset of S is profitable in these prices, yet the sum of prices is high (comparing to the value of S). We use this property to show that if all valuations are subadditive, then a fixed-price auction, with a well-chosen price per item, returns an allocation that is a good approximation to the optimal welfare. This will enable the construction of the mechanism.

3.1   On A Combinatorial Property of Subadditive Valuations

Definition 1. *A vector of non-negative prices $\overrightarrow{p} = (p_1, ..., p_m)$ α-supports a set of items S in a valuation v if the following two conditions hold together:*

- *S **is strongly profitable**: for every $T \subseteq S$ we have that $v(T) \geq \Sigma_{j \in T} p_j$.*
- *The prices are high: $\Sigma_{j \in S} p_j \geq \frac{v(S)}{\alpha}$.*

In [2] it is proved that the class of XOS valuations is *exactly* the class of valuations for which every subset of items has 1-supporting prices. The main combinatorial property we prove is that if v is subadditive, then for every set of items S there are $O(\log m)$-supporting prices.

Lemma 3. *Let v be a subadditive valuation. Then, for every $S \subseteq M$ there exists a vector $\overrightarrow{p}$ that $\frac{\log m}{2e}$-supports it. Furthermore, the price of each item is either 0 or p, for some $p > 0$. These prices can be found by using polynomially many demand queries.*

Proof. The lemma will be proved by constructing a series of sets $S = S_0 \supseteq S_1 \supseteq \cdots \supseteq S_k = T$. Each set S_i is defined to be the most demanded set when the price of each item $j \in S_{i-1}$ is $\frac{v(S_{i-1})}{|S_{i-1}| \log m}$, and the price of the rest of the items is ∞ (that is, S_i is the answer to the demand query with these prices). The construction stops at the first step k where $|S_k| \geq \frac{|S_{k-1}|}{2}$. Clearly, $k \leq \log m$.

Let us first prove that $v(T) \geq \frac{v(S)}{e}$. Denote by p_j^i the price of item j in the i'th demand query. In each step i we have that $\Sigma_{j \in S_{i-1}} p_j^i \leq \frac{v(S_{i-1})}{\log m}$. Thus the profit from the set S_{i-1} in prices $\{p_j^i\}_j$ is at least $v(S_{i-1})(1 - \frac{1}{\log m})$. The most demanded set S_i must be at least as profitable, hence at least as valuable: $v(S_i) \geq v(S_{i-1})(1 - \frac{1}{\log m})$. After $k \leq \log m$ steps we have that $v(T) = v(S_k) \geq v(S)(1 - \frac{1}{\log m})^{\log m} \geq \frac{v(S)}{e}$.

We now prove that T is strongly profitable in prices $\{p_j^i\}_j$.

Proposition 1. *Let v be a subadditive valuation. Let S be the most demanded set (the most profitable set) when the price of each item j is p_j. Then S is strongly profitable in these prices.*

Proof. We use the following fact: for each two subsets, S and T, $S \subseteq T$, and a subadditive valuation v, we have that $v(S) \geq v(T) - v(T \setminus S)$. In other words, the value of S is at least as large as the marginal value of S given T. Given this fact, assume, for a contradiction, that the most demanded set T in prices $(p_1, ..., p_m)$ in v is not strongly profitable. Thus, there exists a set $S \subseteq T$ such that $v(S) < \Sigma_{j \in S} p_j$. But then it follows by the fact that $v(T \setminus S) - \Sigma_{j \in T \setminus S} p_j > v(T) - \Sigma_{j \in T} p_j$, a contradiction to our assumption that S is the most demanded set.

All that is left is to prove the correctness of the fact. Since v is subadditive we have that $v(S) + v(T \setminus S) \geq v(T)$. By subtracting $v(T \setminus S)$ from both sides of the equation, we get $v(S) \geq v(T) - v(T \setminus S)$, as needed. $\square$

To finish the lemma, we explicitly describe the $\frac{\log m}{2e}$-supporting prices of S: we set the price of each item $j \in T$ to p_j^k, and the price of each item $j \notin T$ to 0. $\square$

It will be interesting to determine if subadditive valuations exhibit c-supporting prices for every bundle, for some constant c. The following example rules out the possibility that $c < 2$: let the set of items M consist of m homogeneous items. The subadditive valuation v is defined as follows: $v(M) = 2$, and $v(S)=1$, for all $S \neq M$. We will show that M does not have 2-supporting prices. By the definition of a strongly profitable set for each set S, $|S| = m - 1$, we have that $\Sigma_{j \in S} p_j \leq 1$. Thus there must exist some item j' with $p_j \leq \frac{1}{m-1}$. But then $\Sigma_{j \in M} p_j = \Sigma_{j \in M \setminus \{j'\}} p_j + p_{j'} \leq 1 + \frac{1}{m-1}$, while $v(M) = 2$.

Towards developing the mechanism itself, we require a generalization of the α-supporting prices property:

Definition 2. *Let $v_1, ..., v_n$ be subadditive valuations, and let $A = (S_1, ..., S_n)$ be an allocation. Let $\overrightarrow{p}$ be a vector of non-negative prices. We say the $\overrightarrow{p}$ α-supports an allocation A if both the following conditions hold:*

- *For each i, and for each $T \subseteq T_i$ we have that $v_i(T) \geq \Sigma_{j \in T} p_j$.*
- *$\Sigma_i v_i(S_i) \geq \frac{\Sigma_j p_j}{\alpha}$*

We will say that $\overrightarrow{p}$ supports A if the second condition does not necessarily hold.

We will sometime abuse the notation and say that an allocation A is α-supported by a price p if there exists a vector $\overrightarrow{p} \in \{0,p\}^m$ that α-supports an allocation A.

Lemma 4. *Let $A = (S_1, ..., S_n)$ be an allocation. If all valuations are subadditive, then there exists a vector of non-negative prices $\overrightarrow{p} \in \{0,p\}^m$ that $\frac{\log m}{2e}$-supports A. These prices can be found using polynomially many demand queries.*

The proof is easy given Lemma 3, and can be found in the full version of this paper. The next lemma turns this combinatorial property into an algorithmic result:

Lemma 5. *Let $A = (S_1, ..., S_n)$ be an allocation, and let $\overrightarrow{p} \in \{0,p\}^m$ support it, where $p > 0$. If all valuations are subadditive, then a fixed-price auction with a price of $\frac{p}{2}$ returns an allocation that has a value of $\Omega(\Sigma_i \Sigma_{j \in S_i} p_j)$.*

Proof. Let $(ALG_1, ..., ALG_n)$ be the allocation produced by the fixed-price auction. Let LB_i denote the sum of prices of items held in A by bidders $i+1, ..., n$ immediately after bidder i is queried and allocated a set of items in the fixed-price auction. I.e., $LB_i = \Sigma_{t=i+1}^{n} \Sigma_{j \in S_t \setminus (\cup_{t=1}^{i} ALG_t)} p_j$. Let $LB_0 = \Omega(\Sigma_i \Sigma_{j \in S_i} p_j)$. To prove the lemma it suffices to show that for every $i \geq 1$, $v_i(ALG_i) = \Omega(LB_{i-1} - LB_i)$. We can therefore assume that for all i and each item $j \in S_i$, $p_j = p$ (otherwise we remove each item j with $p_j = 0$ from the S_i's).

Observe that $LB_{i-1} - LB_i$ is the sum of two terms: the sum of prices of items in S_i that are available but were not allocated to bidder i (i.e., $\Sigma_{j \in S_i \setminus (\cup_{t=1}^{i} ALG_t)} p_j$), and the sum of prices of items allocated in the fixed-price auction to bidder i but belong to bidders $i+1, .., n$ in $(S_1, ..., S_n)$ (i.e., $\Sigma_{j \in ALG_i \cap (\cup_{t=i+1}^{n} S_t)} p_j$).

To bound the first term, let $T = S_i \setminus \cup_{t=1}^{i} ALG_t$. Notice that $T \subseteq S_i$ and S_i is strongly profitable in a price of p per item (recall that $(S_1, ..., S_n)$ is supported by $\overrightarrow{p}$). Hence, if the price of each item is p, then the profit from T is at least $\frac{v(T)}{2}$. We get that the profit from the most demanded set in a price of p per item, ALG_i, is at least $\frac{v(T)}{2}$. However, ALG_i is the most demanded set, so it must be at least as profitable, and in particular $v_i(ALG_i) \geq \frac{v(T)}{2}$.

As for the second term, for any set $U \in ALG_i$, by Proposition 1 we have that $v(U) \geq \Sigma_{j \in U} p_j / 2$. In particular, the last inequality is true for $ALG_i \cap (\cup_{t=i+1}^{n} S_t)$. The lemma follows.   □

As evident from the last two lemmas, for every allocation there exists a price p such that a fixed-price auction with a price of p per item returns an allocation that is an $O(\log m)$ approximation to the welfare of the allocation.

3.2   The $O(\log m \log \log m)$-Truthful Mechanism for Subadditive Valuations

Let us now specify the four parameters of the framework needed for the mechanism: let $\alpha = 500 \log m \log \log m$, and $\beta = 100 \log \log m$. To estimate the value of the optimal solution restricted to a set of bidders we use the 2-approximation algorithm of Feige [4]. To find the price to be used in the fixed-price auction, we obtain an allocation A that is a 2 approximation to OPT_{STAT} [4]. We then find a vector $\{0, p\}^m$ that $O(\log m)$-supports A (Lemma 4), and let p be the price used in the fixed-price auction.

A bidder i is called *super dominant* if $v_i(M) \geq \frac{OPT}{100 \log \log m}$, and *dominant* if $v_i(M) \geq \frac{OPT}{500 \log m \log \log m}$. As in the mechanism for general bidders, if there is a super-dominant bidder, then the mechanism will surely end with a good approximation ratio. If there are more than $\log \log m$ dominant bidders, then with probability $\frac{1}{\log m}$ one of them will be selected to the STAT group, and we are guaranteed to get a good approximation in this case too.

As usual, the hard case is when there are at most $\log \log m$ dominant bidders, and no super-dominant bidders. The main effort is to show that we can get a good price p for the fixed-price auction using bidders in STAT. Observe that in this case dominant bidders contribute a value of at most $\log \log m \cdot \frac{OPT}{100 \log \log m} = \frac{OPT}{100}$. Hence we have that $OPT_{ND} \geq \frac{99 OPT}{100}$, where ND is the set of bidders that are not dominant. The next lemma proves that if there is an price that $O(\log m)$-supports an allocation with a good value, then this price also supports an allocation with a good value restricted to bidders in FIXED (proof in the full version):

Lemma 6. *Let $A = (A_1, ..., A_n)$ be an allocation where $\Sigma_i v_i(A_i) \geq \frac{OPT}{t}$. Let p be a price that $10 \log m$-supports it[6]. Let S be a set of bidders where each bidder is selected independently at random to A with probability $\frac{1}{2}$. If $v_i(M) \leq \frac{OPT}{500 \cdot t \cdot \log m \log \log m}$ then, with probability $1 - \frac{1}{\log^{1.5} m}$, there is an allocation B restricted to bidders in S only, where p supports B, and $\Sigma_i |B_i| \cdot p \geq \frac{OPT}{2 \cdot t \log m}$.*

By itself the lemma is not enough to finish this case: the allocation A from the lemma involves all bidders, while we only have information regarding allocations to bidders in STAT. Moreover, we do not know a-priori the price p from the lemma, and the conclusion of the lemma must hold for each one of the many possibilities. To this end, observe that only prices that are larger than $\frac{OPT}{500 m \log m}$ can $O(\log m)$ support an allocation with a good value, since otherwise the sum of prices of the supported allocation $(S_1, ..., S_n)$ is not high enough: $\Sigma_i |S_i| \cdot p \leq m \cdot p \leq \frac{OPT}{500 \log m}$. We therefore restrict our attention to the set of prices $P = \{ \frac{OPT}{1000 \log m}, \frac{OPT}{500 \log m}, \frac{OPT}{250 \log m}, \cdots \}$. Observe that $|P| = O(\log m)$.

We claim that with probability of at least $1 - O(\frac{1}{\log m})$ the following conditions hold together (again, otherwise we assume the algorithm provides an approximation ratio of 0):

1. $OPT_{STAT} \geq OPT/8$.
2. $OPT_{FIXED} \geq OPT/8$.
3. For each price $p \in P$, if there is an allocation that hold a constant fraction of the welfare, and is $O(\log m)$-supported by p, then there is an allocation to bidders in $FIXED$ that holds a constant fraction of the welfare and is $O(\log m)$-supported by p.

---

[6] Notice that the existence of such a price is guaranteed by Lemma 4.

In the full version of the paper we show that the first two conditions do not hold with probability of $O(\frac{1}{\log m})$. The third condition alone does not hold with probability of at most $O(\frac{1}{\log^2 m})$, as there are $O(\log m)$ "sub-conditions", one for each price, each sub-condition does not hold with probability of at most $O(\frac{1}{\log^2 m})$ (by Lemma 6). The claim follows by using the union bound.

With high probability and by the first condition, if there was a winner in the second-price auction, then we get a good approximation ratio. Next we assume that there was no winner in the second-price auction.

Recall that bidders in FIXED face a randomly selected price r. As in the previous mechanism, if two bidders bid above r then we get a good approximation ratio. The problematic case is when there is one bidder that bids above r and chooses to take the bundle he won in the fixed-price auction. Again, there is exactly one value of r that causes this. We choose a value of r uniformly at random from a set of $O(\log m)$ possible values, hence the mechanism fails to provide a good approximation ratio in this case with probability $O(\frac{1}{\log m})$.

We now claim that the price p that $O(\log m)$-supports the allocation to bidders in STAT also supports some allocation with a good value restricted to bidders in FIXED: an allocation with a good value restricted to bidders in STAT is also an allocation with the same value to all bidders. This allocation is $O(\log m)$-supported by some price $p \in P$. By the third condition, there must be an allocation with a good value to bidders in FIXED that is $O(\log m)$ supported by this price. Finally, by Lemma 5 we get that the fixed-price auction returns an allocation that is an $O(\log m)$-approximation to the optimal welfare. We have the following theorem:

Theorem 3. *There exists a truthful randomized $O(\log m \log\log m)$ approximation mechanism for combinatorial auctions with subadditive bidders that obtains this ratio with probability of at least $1 - O(\frac{1}{\log m})$.*

Acknowledgements. I thank Noam Nisan, Ariel Procaccia, Michael Schapira, and the anonymous reviewers for helpful discussions and comments. I thank Liad Blumrosen for a discussion that lead to this paper. This research was supported by grants from the Israel Science Foundation and the USA-Israel Bi-national Science Foundation. The full version of this extended abstract can be found in the author's homepage.

References

1. Dobzinski, S., Nisan, N.: Limitations of vcg-based mechanisms. In: STOC'07 (2007)
2. Dobzinski, S., Nisan, N., Schapira, M.: Approximation algorithms for combinatorial auctionss with complement-free bidders. In: STOC'05 (2005)
3. Dobzinski, S., Nisan, N., Schapira, M.: Truthful randomized mechanisms for combinatorial auctions. In: STOC'06, (2006)
4. Feige, U.: On maximizing welfare where the utility functions are subadditive. In: STOC'06 (2006)

5. Feige, U., Vondrak, J.: Approximation algorithms for allocation problems: Improving the factor of 1-1/e. In: FOCS'06 (2006)
6. Goldberg, A., Hartline, J., Karlin, A., Saks, M., Wright, A.: Competitive auctions. Games and Economic Behaviour (2006)
7. Holzman, R., Kfir-Dahav, N., Monderer, D., Tennenholtz, M.: Bundling equilibrium in combinatrial auctions. Games and Economic Behavior 47, 104–123 (2004)
8. Khot, S., Lipton, R.J., Markakis, E., Mehta, A.: Inapproximability results for combinatorial auctions with submodular utility functions. In: Deng, X., Ye, Y. (eds.) WINE 2005. LNCS, vol. 3828, Springer, Heidelberg (2005)
9. Lavi, R., Mu'alem, A., Nisan, N.: Towards a characterization of truthful combinatorial auctions. In: The 44th Annual IEEE Symposium on Foundations of Computer Science (FOCS) (2003)
10. Lavi, R., Swamy, C.: Truthful and near-optimal mechanism design via linear programming. In: FOCS(2005)
11. Lehmann, B., Lehmann, D., Nisan, N.: Combinatorial auctions with decreasing marginal utilities. In: ACM conference on electronic commerce (2001)
12. Nisan, N.: The communication complexity of approximate set packing and covering. In: Widmayer, P., Triguero, F., Morales, R., Hennessy, M., Eidenbenz, S., Conejo, R. (eds.) ICALP 2002. LNCS, vol. 2380, Springer, Heidelberg (2002)
13. Nisan, N.: ntroduction to Mechanism Design (for Computer Scientists). In: Nisan, N., Roughgarden, T., Tardos, E., Vazirani, V. (eds.) "Algorithmic Game Theory" (2007)
14. Nisan, N., Ronen, A.: Algorithmic mechanism design. In: STOC (1999)
15. Nisan, N., Ronen, A.: Computationally feasible vcg-based mechanisms. In: ACM Conference on Electronic Commerce (2000)

Improved Approximation Ratios for Traveling Salesperson Tours and Paths in Directed Graphs

Uriel Feige[1] and Mohit Singh[2,*]

[1] Microsoft Research and Weizmann Institute
uriel.feige@weizmann.ac.il
[2] Tepper School of Business, Carnegie Mellon University
mohits@andrew.cmu.edu

Abstract. In metric asymmetric traveling salesperson problems the input is a complete directed graph in which edge weights satisfy the triangle inequality, and one is required to find a minimum weight walk that visits all vertices. In the asymmetric traveling salesperson problem (ATSP) the walk is required to be cyclic. In asymmetric traveling salesperson path problem (ATSPP), the walk is required to start at vertex s and to end at vertex t.

We improve the approximation ratio for ATSP from $\frac{4}{3}\log_3 n \simeq 0.84\log_2 n$ to $\frac{2}{3}\log_2 n$. This improvement is based on a modification of the algorithm of Kaplan et al [JACM 05] that achieved the previous best approximation ratio. We also show a reduction from ATSPP to ATSP that loses a factor of at most $2 + \epsilon$ in the approximation ratio, where $\epsilon > 0$ can be chosen to be arbitrarily small, and the running time of the reduction is polynomial for every fixed ϵ. Combined with our improved approximation ratio for ATSP, this establishes an approximation ratio of $(\frac{4}{3} + \epsilon)\log_2 n$ for ATSPP, improving over the previous best ratio of $4\log_e n \simeq 2.76\log_2 n$ of Chekuri and Pal [Approx 2006].

1 Introduction

One of the most well studied NP-hard problems in combinatorial optimization is the minimum Traveling Salesperson (TSP) problem [8]. The input to this problem is a graph with edge weights, and the goal is to find a cyclic tour of minimum weight that visits every vertex exactly once. In the symmetric version of the problem, the graph is undirected, whereas in the asymmetric version the graph is directed. In the metric version of the problem the input graph is a complete graph (with anti-parallel edges in the directed case), and the edge weights (denoted by w) satisfy the triangle inequality $w(u, v) + w(v, w) \geq w(u, w)$. (Most often, not all edge distances are given explicitly, but rather they can be computed efficiently. For example, they may be shortest path distances between the given points in some input graph, or the distances between points in some normed space.) In the non-metric version a cyclic tour might not exist at all,

---

* This work was done when the author was visiting Microsoft Research. Supported by NSF Grant CCF-0430751.

M. Charikar et al. (Eds.): APPROX and RANDOM 2007, LNCS 4627, pp. 104–118, 2007.

and deciding whether such a tour exists is NP-hard (being equivalent to Hamiltonicity). In the metric version of the problem a cyclic tour always exists, and we shall be interested in polynomial time approximation algorithms that find short cyclic tours. The performance measure of an algorithm is its approximation ratio, namely, the maximum (taken over all graphs) of the ratio between the weight of the cyclic tour output by the algorithm (or the expected weight, for randomized algorithms) and the weight of the shortest cyclic tour in the given graph. Throughout, we use n to denote the number of vertices in the input graph, and the approximation ratio is often expressed as a function of n.

In this paper we shall be dealing only with metric instances of TSP. In this case, every tour that visits every vertex at least once can be converted into one that visits every vertex exactly once (by skipping over redundant copies of vertices), without increasing the weight of the tour. A cyclic tour that visits every vertex at least once will simply be called a *tour*, and the TSP problem is equivalent to that of finding the shortest tour.

For symmetric TSP, the well known algorithm of Christofides [6] achieves an approximation ratio of 3/2. Despite much effort, no better approximation algorithm is known, except for some special cases [1,2]. Considerable efforts have been made to improve over the 3/2 approximation ratio using approaches based on linear programming relaxations of TSP. Specifically, a linear programming bound of Held and Karp [9] is conjectured to provide a 4/3 approximation ratio. In terms of negative results, it is known that there is some (small) ϵ such that symmetric TSP is NP-hard to approximate within a ratio of $1 + \epsilon$ (see [13] for explicit bounds on ϵ).

The asymmetric TSP (ATSP) problem includes the symmetric version as a special case (when anti-parallel edges have the same weight), and hence, is no easier to approximate. The known hardness of approximation results are of the form $1 + \epsilon$, with a slightly larger ϵ than for the symmetric case (see [13]). There are known examples for which the Held-Karp lower bound for ATSP is a factor of 2 away from the true optimum [4] (whereas for symmetric TSP this lower bound is at most a factor of 3/2 from the optimum [15,14]).

Frieze, Galbiati and Maffioli [7] designed an approximation algorithm for ATSP with approximation ratio $O(\log n)$. Blaser [3] notes that the approximation ratio proved in [7] is precisely $\log_2 n$ (with leading constant 1), and then designs an algorithm for which he shows an approximation ratio of $0.999 \log_2 n$. Subsequently, Kaplan et al [10] designed an algorithm with approximation ratio $4/3 \log_3 n \simeq 0.842 \log_2 n$ (using a technique that they apply to other related problems as well). In this paper, we provide a modest improvement in the leading constant of the approximation ratio. We show that the analysis of the algorithm of Kaplan et al [10] is not tight and it achieves a better ratio of $0.79 \log_2 n$. We then give an improved algorithm which returns a solution with approximation ratio of $\frac{2}{3} \log_2 n$. This result is summarized in the following theorem.

Theorem 1. *Given a complete directed graph $G = (V, E)$ with a weight function w satisfying triangle inequality, there exists a polynomial time algorithm which* *returns a Hamiltonian cycle of weight at most $\frac{2}{3}\log_2 n \cdot OPT$ where OPT is the weight of the minimum weight Hamiltonian cycle.*

Another interesting variant of the ATSP problem is the asymmetric traveling salesman path problem in which we are not required to find a Hamiltonian cycle of minimum weight but a Hamiltonian path between two specified vertices s and t. Lam and Newman [12] gave an $O(\sqrt{n})$-approximation algorithm to the problem. Chekuri and Pal [5] used the $2H_n$-approximation algorithm of Kleinberg and Williamson [11] for the ATSP problem in combination with an *augmentation lemma* to obtain a $4H_n$-approximation algorithm for the AT-SPP problem. We show that ATSPP problem can be approximated nearly as well as the ATSP problem by showing a general reduction which converts an α-approximation algorithm for the ATSP problem into a $(2+\epsilon)\alpha$-approximation algorithm for the ATSPP problem. This involves generalizing and strengthening the *augmentation lemma* of [5] and using it to obtain a dynamic programming based algorithm for the ATSPP problem which we show in the following result.

Theorem 2. *Given a complete directed graph $G = (V, E)$ with a weight function w satisfying triangle inequality, vertices s and t and an α-approximation algorithm to the ATSP problem, there exists an algorithm which returns a Hamiltonian path from s to t of weight at most $(2+\epsilon)\alpha \cdot OPT$ where OPT is the weight of the minimum weight Hamiltonian path from s to t. The running time of the algorithm is polynomial in the size of the graph for any fixed $\epsilon > 0$.*

Observe that it is trivial to obtain an α-approximation for the ATSP from an α-approximation to ATSPP problem. The above theorem shows that both these problems can be approximated to nearly the same factor. Along with the Theorem 1 and Theorem 2, we have the following corollary.

Corollary 1. *Given a complete directed graph $G = (V, E)$ with a weight function w satisfying triangle inequality, vertices s and t and a fixed $\epsilon > 0$, there is a polynomial time algorithm which returns a Hamiltonian path from s to t of weight at most $(\frac{4}{3} + \epsilon)\log_2 n \cdot OPT$ where OPT is the weight of the minimum weight Hamiltonian path from s to t.*

In Section 2 we prove Theorem 2 and in Section 3 we prove Theorem 1.

2   From ATSP to ATSPP

In this section we show that an α-approximation algorithm $AlgTSP$ for the ATSP problem with metric weights can be used to obtain a $(2 + \epsilon)\alpha$-approximation algorithm for the ATSPP problem with metric weights.

First a few definitions. Given a graph $G = (V, E)$ we call a (s, t)-walk in G *spanning* if it visits every vertex of G at least once. Vertices and edges can appear more than once on a walk. A tour is an (s, s)-walk which is spanning. Observe that a tour is independent of vertex s. Given a directed path P and vertices u and v on P such that v occurs after u on P, we denote $P(u, v)$ to be the sub-path of P

starting at u and ending at v. Given two paths P and Q, we say that Q respects the ordering of P, if Q contains all vertices of P, and for every two vertices u and v in P, u appears before v in Q iff u appears before v in P.

In an instance $\mathcal{I} = (G, w, s, t)$ of the $ATSPP$ problem we are given a directed graph G with a weight function w on edges which satisfies the triangle inequality and the task is to find a minimum weight Hamiltonian path from s to t. In an instance $\mathcal{I} = (G, w)$ of $ATSP$ the task is to find a minimum weight Hamiltonian cycle.

Observe that as the weights satisfy triangle inequality, any spanning (s, t)-walk can be "shortcutted" to obtain a Hamiltonian path from s to t of no greater weight, and every tour can be "shortcutted" into Hamiltonian Cycle of no greater weight. Hence, it is enough to find a spanning (s, t)-walk for the ATSPP problem and a tour for the ATSP problem.

2.1   Overview

Here we present an overview of our reduction from ATSPP to ATSP. For every fixed $\epsilon > 0$, this reduction works in polynomial time, and transforms a factor α approximation algorithm for ATSP into a factor $(2 + \epsilon)\alpha$ approximation for ATSPP.

Let s denote the starting vertex and t denote the ending vertex for ATSPP, and let OPT denote the weight of the minimum weight spanning path from s to t. Assume for simplicity that the value of OPT is known. Without loss of generality, we assume that for every pair of vertices (u, v) the graph contains an edge (u, v) whose weight is the shortest distance from u to v.

Let $d(t, s)$ denote the distance from t to s in the input graph (this distance might be infinite). The difficult case is when $OPT < d(t, s)$, and this is the case that we will address in this overview. In the first phase of the reduction, we modify the input graph as follows. We remove all edges entering s and all edges exiting t, and put in an edge (t, s) of weight $\min[d(t, s), OPT]$. We update the shortest path distance between all pairs of vertices not involving s and t to reflect the existence of this new edge.

Observe that the new graph has a ATSP tour of weight at most $2OPT$. In the second stage we use the approximation ratio for ATSP to find a simple tour (with no repeated vertices) of weight at most $2\alpha OPT$. Observe that in this tour s follows immediately after t, because the only edge leading out of t leads into s. Remove the edge (t, s) from the tour, which now becomes a spanning (s, t) path of weight at most $(2\alpha - 1)OPT$.

Unfortunately, we are not done at this point. The problem is that the weight of an edge (u, v) of the path might be shorter than its corresponding weight in the original graph, due to the fact that the shortest path distance between (u, v) decreased when we added the edge (t, s). We replace every such problematic edge (u, v) with the path $u - t - s - v$. Now the edge (t, s) reappears in our spanning path. Since the edge (t, s) might have weight more than OPT in the original graph, the spanning path that we have does not correspond to a spanning path of the same weight in the original graph.

In the next phase of our reduction, we remove all copies of (t, s) from the spanning path. This results in breaking the path into a collection of paths from s to t, such that every vertex (other than s and t) appears on exactly one of these paths. If the number of paths is r, then the sum of weights of all the paths is at most $(2\alpha - r)OPT$, because altogether we removed r copies of (t, s).

The last stage of our reduction uses the following structural lemma which generalizes and strengthens the augmentation lemma of [5].

Lemma 1. *For every collection of k paths $P_1, \ldots, P_k$ between s and t such that no vertex appears in more than one path (some vertices might not appear on any of the paths), there is a single path between s and t that visits all vertices in $\cup_{i=1}^k P_i$, respects the order of each of the original paths, and weighs no more than the weight of the original paths plus k times the weight of minimum ATSPP.*

The proof of this lemma appears in section 2.3.

Having established the lemma, we can limit ourselves to finding an ATSPP that respects the order of the vertices on the paths, and then get a $(2\alpha - r + r)OPT = 2\alpha OPT$ approximation ratio. Such a path can be found by dynamic programming in time roughly n^r. If r is constant, this results in a polynomial time 2α approximation for ATSPP.

To make the algorithm polynomial also when r is not constant, we lose $(1 + \epsilon)$ in the approximation ratio (the running time will be exponential in $1/\epsilon$). Rather than merging all paths simultaneously, merge only k paths at a time, where $k = 1/\epsilon$. Doing so using dynamic programming takes time roughly n^k, costs k times OPT, and decreases the number of paths by $k - 1$.

Now, we expand on the overview given above. Before giving the algorithm and proof of Theorem 2 we first prove the structural result in Lemma 1.

2.2   Proof of Lemma 1

Proof of Lemma 1: Let P denote the optimal ATSPP from s to t. We maintain a path Q starting from s and prefix paths Q_i of paths P_i with the property that Q visits the vertices of $\cup_i Q_i$ and respects the order of each Q_i. In each iteration we will extend Q and at least one of Q_i maintaining the above property. For each path P_i, we maintain a vertex $front_i$ which is the next vertex to be put in order, that is, the successor of Q_i in P_i. We maintain an invariant that all front vertices, except possibly $front_j$, occurs on $P(v, t)$ where v is the last vertex on Q and $front_j$ is the front of path P_j which contains v. We initialize $Q = (s)$ and $Q_j = (s)$ and $front_j$ to be the second vertex in P_j for each j. The invariant is trivially satisfied at initialization.

Now, we describe an iteration. Let v be the last vertex of Q and P_j be the path containing v. Let $u = front_i$ be the first vertex on path $P(v, t)$ (sub-path of P starting at v and ending at t) among all front vertices. First we assume that $i \neq j$ and describe the updates. Let w be the last vertex on P_j which occurs on $P(s, u)$, i.e., each vertex occurring after w on P_j occurs after u on P. Now, extend $Q \leftarrow Q$-$P_j(v, w)$-$P(w, u)$. We update $Q_j = Q_j$-$P_j(v, w)$, $Q_i = Q_i$-(u). We also update front$_j$ to be vertex succeeding w in P_j and front$_i$ to be the vertex succeeding old front$_i$ in P_i. In this case we say that we jumped out of path P_j to P_i using the sub-path $P(w, u)$ of the optimal path P. Observe that last vertex of Q is u. The front vertices of all paths, except for P_i and P_j, do not change and each of them occur after u by the choice of u as the first front vertex on path $P(v, t)$. Hence, the invariant is satisfied for all paths except possibly for path P_j (we do not need to check for path P_i as it contains u). The new front$_j$ cannot occur on $P(s, v)$ else it would have chosen instead of w and hence it occurs after v on P proving the invariant in this case as well.

Now if $i = j$, we do not use any sub-path of P and do not jump out of P_i. Let w be the last vertex on P_i occurring on $P(s, u)$. We extend Q by using a sub-path of P_i as follows: $Q \leftarrow Q$-$P_i(v, w)$. We update $Q_i = Q_i$-$P_i(v, w)$. We also update front$_i$ to be vertex succeeding w in P_i. We now show that the invariant holds in this case as well. The last vertex of Q is w which is on P_i. Also, w occurs on $P(s, u)$ but front$_j$ for any $j \neq i$ occurs after u on P by the choice of u. Hence, the invariant holds in this case as well.

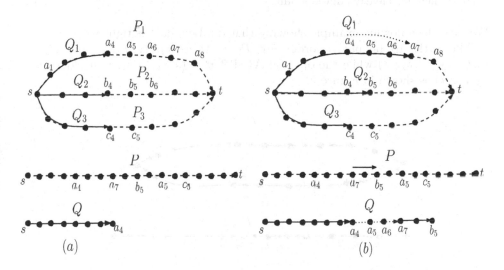

Fig. 1. In (a), we have paths $P_1 = (s, a_1, \dots, t), P_2 = (s, b_1, \dots, t), P_3 = (s, c_1, \dots, t)$, Hamiltonian path P. Q is the current path which respects the ordering of each Q_i where $Q_1 = (s, a_1, \dots, a_4), Q_2 = (s, b_1, \dots, b_4), Q_3 = (s, c_1, \dots, c_4)$. Also front$_1 = a_5$, front$_2 = b_5$, front$_3 = c_5$. Observe that b_5 is the first front vertex in $P(a_4, t)$. Also, a_7 is the last vertex on P_1 which is on $P(a_4, b_5)$. Hence, we extend Q by adding the sub-path $P_1(a_4, a_7)$ and $P(a_7, b_5)$. Q_1 is extended till a_7 and Q_2 till b_5. This is shown in (b).

In every step either one or two paths advance their front vertex (either path i or both path i and path j, using the notation of the above explanation). We iterate till Q ends at t. Clearly, the property that Q visits the vertices of $\cup_i Q_i$ in the order of each Q_i is maintained in each update. See Figure 1 for an example.

We now claim that the total weight of the path Q found is no more than the sum of weights of individual paths, plus k times the weight of the optimal ATSPP solution P. To show this we first argue that the sub-paths of P_i in Q are edge-disjoint for each i. We then show that for any path P_i all jumps out of P_i use disjoint sub-paths of the ATSPP P. Hence, any edge of P can be used at most k times.

The first claim follows from the fact that any subpath of P_i used in Q starts at one vertex before the current front$_i$ and ends at one vertex before the new front$_i$.

Now we prove the second claim. Observe that if we jump out of u and v on P_i and u occurs before v in P_i then the jump at u occurs before the jump at v. Also, v cannot lie on the sub-path of P which is traversed after jumping from P_i at u as otherwise we would jump out at v and not at u. Now, we claim that u lies before v in P and hence u cannot lie on the sub-path of P traversed after jumping from v (which contains nodes occurring after v in P). Indeed, let w be the front vertex of P_j where the jump sub-path starting from u ends. By definition u is the furthest vertex of P_i which precedes w on P. Hence, v lies after w on P and therefore after u. As no two jumps out of P_i have a common vertex, they are clearly edge-disjoint.   □

We give the following example showing that the Lemma 1 is tight when $k = 2$.

The paths to be put in order are $P_1 = (s, a_1, a_2, \ldots, a_{2n}, t)$ and $P_2 = (s, b_1, b_2, \ldots, b_{2n}, t)$ while the optimal ATSPP $P = (s, a_1, b_2, b_1, a_3, a_2, b_4, b_3, a_5, a_4, \ldots, t)$ as shown in Figure 2.

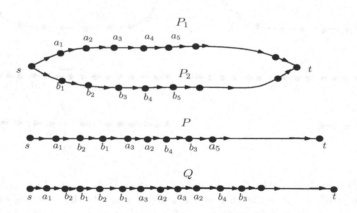

Fig. 2. Tight Example

The weight of the edges (a_{2i+1}, a_{2i}) and (b_{2i}, b_{2i-1}) is one for all $1 \leq i \leq n$. The weight of the edges (a_{2i}, a_{2i+1}) and (b_{2i-1}, b_{2i}) is two for all $1 \leq i \leq n$. The weight of rest of the edges in $P_1 \cup P_2 \cup P$ is zero. Observe that $w(P_1) = 2n$, $w(P_2) = 2n$ and $w(P) = 2n$. The minimum weight spanning walk Q which respects the ordering of both P_1 and P_2 must be $Q = (s, a_1, b_2, b_1, b_2, b_1, a_3, a_2, a_3, a_2, b_4, \ldots, t)$ and $w(Q) = 2n + 2n + 2 \cdot 2n = 8n$ which is exactly $w(P_1) + w(P_2) + 2 \cdot w(P)$.

We do not know whether Lemma 1 is tight when $k > 2$, though there are examples in which any path that spans $\cup_i P_i$ and respects the ordering of each path P_i must weigh at least $\sum_{i=1}^{k} w(P_i) + (k-1) \cdot w(P)$, where P is the minimum ATSPP. Details omitted due to space limitations.

2.3   Algorithm for ATSPP

In this section, we prove Theorem 2. We show that the algorithm *AlgPath* in Figure 3 gives the guarantee as claimed in Theorem 2.

Input: An instance $\mathcal{I} = (G, w, s, t)$ of ATSPP, an α-approximation algorithm *AlgTour* to ATSP, and a parameter $\epsilon > 0$.

1. By trying $O(\frac{\log n}{\epsilon})$ options, obtain a value of g such that $(1 - \frac{\epsilon}{8}) \cdot OPT \leq g \leq OPT$.
2. Remove all edges incident into s and also edges incident out of t. Include the edge (t, s) with the weight as g. Let this modified graph be $\hat{G}$ and the modified weight function $\hat{w}$. Let $KG = (V, E(KG))$ be the complete directed graph on V. Compute $\hat{miw} : E(KG) \rightarrow R_+$, the metric completion of the $\hat{w}$, i.e., $\hat{miw}(u, v)$ is the shortest distance from u to v under the weight function $\hat{w}$.
3. Find the α-approximate solution C given by *AlgTour* on the complete graph KG under the weight function $\hat{miw}$. Let T be the tour obtained in $\hat{G}$ after replacing each edge (u, v) by its corresponding shortest path in $\hat{G}$.
4. Let r be the number of times edge (t, s) is chosen in T. Remove all copies of (t, s) to decompose T into a collection of r (s, t)-paths $\mathcal{P} = \{P_1, \ldots, P_r\}$ which together span V. Shortcut these paths to ensure that each vertex except s and t is in exactly one of them.
5. Return $Q = Weave(G, \mathcal{P}, \epsilon)$.

Output: A $(2 + \epsilon)\alpha$-approximate solution to $\mathcal{I}$.

Fig. 3. Algorithm AlgPath

Figure 4 describes the algorithm *Weave* which given a collection of (s, t) paths $\mathcal{P}$ returns a single (s, t) path Q which respects the order of each path $P_i \in \mathcal{P}$ and is of small weight. This is used as a subroutine in the algorithm *AlgPath*.

Lemma 2. *Given a collection of r (s, t)-paths $\mathcal{P} = \{P_1, \ldots, P_r\}$ and a parameter $\epsilon > 0$, algorithm $Weave(\mathcal{P}, s, t, \epsilon)$ returns a single (s, t)-path spanning all vertices in $\mathcal{P}$ and respecting the order of vertices of weight no more than $\sum_{i=1}^{r} w(P_i) + (1 + \epsilon/8)r \cdot OPT$, where OPT is the weight of the optimal (s, t)-spanning path. The running time of the algorithm is $O(n^{\frac{1}{\epsilon} + O(1)})$.*

Proof. In any iteration, if we replace paths $P_1, \ldots, P_k$ by Q, then Lemma 1 guarantees that such a path exist of weight no more than $w(Q) \leq \sum_{i=1}^{k} w(P_i) +$

Input: A collection $\mathcal{P} = \{P_1, \ldots, P_r\}$ of r (s,t)-paths, and a parameter $\epsilon > 0$.

1. If $r = 1$ return P_1. Otherwise let $k = \min\{\frac{9}{\epsilon}, r\}$.
2. Use dynamic programming to a find a minimum weight path P' that spans the vertices of $(P_1, \ldots, P_k)$ and respects the order of each of the paths $P_1, \ldots P_k$. This can be done by inductively computing the weight of the minimum weight (s,v)-path that spans all vertices of $Q_1, \ldots Q_k$ and ends at v, where for each $1 \leq i \leq k$, Q_i is some prefix path of P_i, and v is the last vertex on one (or more) of the Q_i. For every choice of $Q_1, \ldots Q_k$ and v, a minimum weight corresponding path can be found by examining at most k previously computed weights that correspond to paths in which one of the Q_i is shorter by one vertex.
3. Let $\mathcal{P}' = \mathcal{P} \cup P' \setminus \{P_1, \ldots, P_k\}$.
4. Return $Weave(G, \mathcal{P}', \epsilon)$.

Output: (s,t)-path Q spanning vertices in $\mathcal{P}$, which respects the order of each path $P_i \in \mathcal{P}$ and of weight at most $\sum_{i=1}^{r} w(P_i) + (1 + \epsilon/8)r \cdot OPT$, where OPT is the weight of the optimal (s,t)-spanning path.

Fig. 4. Algorithm Weave

$k \cdot OPT$ which the dynamic program will find. Hence, in each iteration, the number of paths reduce by $k - 1$ and the weight of the new collection of paths increases by $k \cdot OPT$. Hence, the total increase in weight is at most $(l+r-1)OPT$ where l is the number of iterations. But $l \leq \lceil \frac{r}{9/\epsilon-1} \rceil \leq \frac{\epsilon r}{8} + 1$ for $\epsilon < 1$. Hence, the weight of Q is at most $\sum_{i=1}^{r} w(P_i) + (1 + \epsilon/8)r \cdot OPT$.

The running time of the algorithm is a polynomial in n times the number of possible choices of prefix lengths in step 2 of algorithm $Weave(\mathcal{P}, s, t, \epsilon)$. This number is $O(n^{\frac{1}{\epsilon}})$.   □

Now, we prove Theorem 2.

Proof. First, we show that one out of a polynomial number of guesses satisfies the conditions of Step 1. Indeed, the algorithm can first find a lower bound L and upper bound U such that $U \leq nL$(a trivial n-approximation would suffice). We start by setting $g = U$ and running the algorithm. We then decrease g by factor of $(1 - \frac{\epsilon}{8})$ and run it again. We iterate in such a manner till the value of g reaches L. Observe that each guess of g will yield a feasible solution and we can return the best solution obtained. Also, the total number of guesses needed is $\log_{1-\frac{\epsilon}{8}} \frac{L}{U} = O(\frac{\log n}{\epsilon})$. Hence, we assume that we have the guess which satisfies the conditions of Step 1 of the algorithm.

Now, observe that $KG = (V, E(KG))$ with the weight function $\hat{m}w$ satisfies the triangle inequality. Also, the optimal Hamiltonian cycle in KG weighs exactly $OPT + g$ where OPT is the weight of optimal (s,t)-spanning path in G under w. Hence, we must have that weight of Hamiltonian cycle C found by $AlgTour$ is $\hat{m}w(C) \leq \alpha(OPT + g) \leq 2\alpha OPT$ as $g \leq OPT$. If the edge (t,s) is chosen in T r times then removing all copies of (t,s) decomposes T into a collection of r (s,t)-paths $P_1, \ldots, P_r$ which together span V and such that $\sum_{i=1}^{r} w(P_i) \leq 2\alpha \cdot OPT - rg$. Now in Step 5, algorithm $Weave$ returns a single (s,t)-spanning path Q of weight at most $\sum_{i=1}^{r} w(P_i) + (1 + \epsilon/8)r \cdot OPT$ from Lemma 2. Also, $rg \leq \alpha(OPT + g)$ and $g \geq (1 - \epsilon/8) \cdot OPT$ imply that $r \leq (2 + \epsilon/8)\alpha \leq 4\alpha$ for $\epsilon < 1$. Hence, the weight of Q,

$$w(Q) \leq \sum_{i=1}^{r} w(P_i) + (1 + \epsilon/8)r \cdot OPT \leq 2\alpha \cdot OPT - rg + (1 + \epsilon/8)r \cdot OPT$$

$$\leq 2\alpha \cdot OPT + \frac{\epsilon}{4}r \cdot OPT \leq (2 + \epsilon)\alpha \cdot OPT$$

where the last two inequalities follow from the fact that $g \geq (1 - \epsilon/8) \cdot OPT$ and $r \leq 4\alpha$. $\qquad\square$

3   An Improved Approximation Algorithm for ATSP

Kaplan et al [10] show an $\frac{4 \log_2 n}{3 \log_2 3} \simeq .842 \log_2 n$-approximation for the ATSP problem. It is the current best known algorithm as well. In this section, we first show that their analysis is not tight and can be improved to $\frac{\log_2 n}{\log_2 (\sqrt{2}+1)} \simeq .787 \log_2 n$-approximation. Then, we show an improved algorithm which gives a better approximation guarantee of $\frac{2}{3} \log_2 n$.

A subgraph of the input graph is called Eulerian if the indegree of each vertex is equal to its outdegree. A connected Eulerian subgraph has an Eulerian tour that visits every edge exactly once, and moreover, such a tour can be found efficiently. When edge weights of the original graph satisfy the triangle inequality, standard *shortcutting* arguments show that there exists a tour of weight no more than the weight of any Eulerian subgraph. Hence in what follows, we will ensure that we return a connected Eulerian subgraph which has low weight. Also note that for any Eulerian graph the connected components are exactly the strongly connected components.

3.1   Improving the KLSS Analysis

The KLSS algorithm [10] starts from using the following linear program LP-ATSP that enforces sub-tour elimination constraints for subsets of size two.

$$\min \sum_{e \in E} c_e x_e$$

$$\sum_{e \in \delta^+(v)} x_e = \sum_{e \in \delta^-(v)} x_e = 1 \quad \forall v \in V$$

$$x_{(u,v)} + x_{(v,u)} \leq 1 \; \forall u, v \in V$$

$$0 \leq x_e \leq 1 \quad \forall e \in E$$

Here $\delta^+(v)$ is the set of edges going out of v and $\delta^-(v)$ is the set of edges going into v. The above LP-ATSP is a relaxation of the ATSP problem, because for every Hamiltonian cycle, assigning $x_e = 1$ iff edge e belongs to the Hamiltonian cycle is a feasible solution of the LP.

The following is the key lemma used in the KLSS [10] algorithm. The lemma is based on decomposing the optimal solution to LP-ATSP to get the following guarantee.

Lemma 3. *[10] Given an edge-weighted directed graph G, there exists a polynomial time algorithm which using the optimal solution to LP-ATSP finds two cycle covers C_1 and C_2 such that*

1. *C_1 and C_2 do not share any 2-cycle.*
2. *$w(C_1) + w(C_2) \leq 2 \cdot OPT$ where OPT is the weight of the optimal solution to LP-ATSP.*

Their algorithm proceeds as follows

1. Find two cycle covers given by Lemma 3. Choose F to be one of C_1, C_2 and $C_3 = C_1 \cup C_2$ which minimizes the potential function $\frac{w(F)}{\log_2(n_i/c(F))}$, where n_i is the number of nodes in the current iteration and $c(F)$ is the number of components in F.
2. For each connected component pick one representative vertex. Delete the rest of the vertices and iterate till at most one component is left.

Let the number of steps taken by the algorithm be p and let $F_1, \ldots, F_p$ be the edges selected in each iteration. Return the solution $\cup_{i=1}^p F_p$. The following claim is implicit in Kaplan et al.

Claim 1. *[10] If $\frac{w(F_i)}{\log_2(n_i/c(F_i))} \leq \alpha OPT$, then the above algorithm is $\alpha \log_2 n$-approximation.*

Proof. Using the fact that $n_p = 1$, $n_1 = n$ and $n_{i+1} = c(F_i)$, we obtain that the weight of the edges included is

$$\sum_{i=1}^p w(F_i) \leq \sum_{i=1}^p \log_2 \frac{n_i}{c(F_i)} \cdot \alpha \cdot OPT \leq \alpha \cdot OPT \sum_{i=1}^p \log_2 \frac{n_i}{n_{i+1}} = \alpha \cdot \mathsf{OPT} \cdot \log_2 n$$

In their paper, Kaplan et al [10] show that $\alpha = \frac{4}{3 \log_2 3}$ suffices. We show that $\frac{1}{\log_2 \sqrt{2}+1}$ suffices. We need another claim proven in Kaplan et al.

Claim 2. *[10] In any iteration, if C_1 and C_2 are the cycle covers found then $c(C_1) + c(C_2) + c(C_3) \leq n_i$ where n_i is the number of nodes in graph at this iteration.*

Claim 3. *In any iteration i, if F_i is chosen then $\frac{w(F_i)}{\log_2(n_i/c(F_i))} \leq \alpha OPT$ for $\alpha = \frac{1}{\log_2(\sqrt{2}+1)}$.*

Proof. Observe that α is at most the value of the following optimization problem. Here, w_i corresponds to $w(C_i)/OPT$ and c_i corresponds to $c(C_i)/n_i$. These scalings do not affect the value of α.

$$\max z \tag{1}$$

$$z \leq \frac{w_j}{\log_2 \frac{1}{c_j}} \ \forall\, 1 \leq j \leq 3 \tag{2}$$

$$w_1 + w_2 \leq 2 \tag{3}$$

$$w_3 = w_1 + w_2 \tag{4}$$

$$c_1 + c_2 + c_3 \leq 1 \tag{5}$$

$$c_3 \leq c_j \ \forall\, j = 1,2 \tag{6}$$

$$c_j \leq \frac{1}{2} \ \forall\, j = 1,2 \tag{7}$$

$$w_j, c_j \geq 0 \ \forall\, j = 1,2,3 \tag{8}$$

In the above program constraints (2) correspond to z denoting the best potential of the three solution C_1, C_2 or C_3. Constraints (3)-(4) claim that sum of weights of C_1 and C_2 is exactly the weight of C_3 which is most $2OPT$. Variables c_i correspond to $c(C_i)/n_i$ and hence constraint (5) follows from Claim 2. Constraint (6) is valid as $C_3 = C_1 \cup C_2$ and hence will have fewer components. Moreover, as each cycle has length at least two, $c(C_i) \leq n_i/2$.

Let (z^*, w^*, c^*) denote the optimum solution. The objective value is a non-decreasing function of w_3 and hence, without loss of generality, we assume $w_3^* = 2$. We also claim that $w_1^* = w_2^* = 1$ and $c_1^* = c_2^*$. Indeed if that is not the case, then we construct another solution $w_1' = w_2' = 1, w_3' = w_3^*$ and $c_1' = c_2' = \frac{c_1^* + c_2^*}{2}, c_3' = c_3^*$. Let $z' = \min_{j=1,2,3} \frac{w_j'}{\log_2 \frac{1}{c_j'}}$. Observe that (z', w', c') is a feasible solution as it satisfies all constraints $(2) - (8)$. Now, we claim that

$$\frac{w_2'}{\log_2 \frac{1}{c_2'}} = \frac{w_1'}{\log_2 \frac{1}{c_1'}} \geq \min\{\frac{w_1^*}{\log_2 \frac{1}{c_1^*}}, \frac{w_2^*}{\log_2 \frac{1}{c_2^*}}\}$$

Suppose the above relation does not hold. Then using the fact that $w_1' = w_2' = 1$ and $c_1' = c_2' = \frac{c_1^* + c_2^*}{2}$ for each $i = 1,2$ we must have $\frac{1}{\log_2 \frac{2}{c_1^* + c_2^*}} < \frac{w_i^*}{\log_2 \frac{1}{c_i^*}}$.

Cross multiplying and summing over $i = 1,2$, we have

$$\log_2 \frac{1}{c_1^*} + \log_2 \frac{1}{c_2^*} < (w_1^* + w_2^*) \cdot \log_2 \frac{2}{c_1^* + c_2^*}$$

Using the fact $w_1^* + w_2^* = 2$ and the fact that logarithm is an increasing function, we have

$$\log_2 \frac{1}{c_1^* \cdot c_2^*} < 2\log_2 \frac{2}{c_1^* + c_2^*} \implies \frac{1}{c_1^* \cdot c_2^*} < \left(\frac{2}{c_1^* + c_2^*}\right)^2$$

which violates the AM-GM inequality that $\left(\frac{c_1^* + c_2^*}{2}\right)^2 \geq c_1^* \cdot c_2^*$ which holds since $c_1^*, c_2^* \geq 0$.

Hence, we have $z' = \min_{j=1,2,3} \frac{w'_j}{\log_2 \frac{1}{c'_j}} \geq \min\{\min\{\frac{w^*_1}{\log_2 \frac{1}{c^*_1}}, \frac{w^*_2}{\log_2 \frac{1}{c^*_2}}\}, \frac{w'_3}{\log_2 \frac{1}{c'_3}}\} = z^*$. Thus, we assume without loss of generality that $w^*_1 = w^*_2 = 1$ and $c^*_1 = c^*_2$.

Under these condition observe that all three inequalities $z^* \leq \frac{w^*_j}{\log_2 \frac{1}{c^*_j}}$ must hold at equality. Now, solving we obtain that $c^*_1 = c^*_2 = \sqrt{2} - 1$, $c^*_3 = 3 - 2\sqrt{2}$ and $z^* = \frac{1}{\log_2 (\sqrt{2}+1)}$. □

3.2   Modifying the KLSS Algorithm

Here we explain how we can change the algorithm of KLSS to obtain an improved guarantee of $\frac{2}{3}\log_2 n$. The algorithm is very similar. Each time we find the cycle covers C_1 and C_2 as given by Lemma 3. Instead of selecting the best of C_1, C_2 or $C_3 = C_1 \cup C_2$, we decompose C_3 into two Eulerian subgraphs.

The following is the key Lemma used for the decomposition. For every connected component of C_3, we can apply the following lemma.

Lemma 4. *Let C be a connected directed graph with at least three vertices in which every vertex has in-degree 2 and out-degree 2. C is allowed to have parallel edges but no self loops. Then there are either two (vertex disjoint) cycles of length 2 or one cycle of length at least 3 such that removing the edges of these cycles from C leaves C connected.*

Proof. The edges in C can be partitioned into C_1 and C_2 such that each of them induces on C a directed graph with in-degree 1 and out-degree 1. (This was used in Kaplan et al, and the proof of this fact follows easily from the fact that every d-regular bipartite graph is a union of d perfect matchings.) Each of C_1 and C_2 is a collection of cycles that spans all vertices of C. Let c_i be the number of cycles in C_i, for $i \in \{1,2\}$. We now proceed with a case analysis.

Case 1. $c_1 \neq c_2$. Assume in this case without loss of generality that $c_2 > c_1$. One by one, add the cycles of C_2 to C_1. When the process begins, the number of connected components is c_1. When it ends, the number of connected components is 1 (because then we have C). Every cycle of C_2 added in the process either reduces the number of connected components, or leaves it unchanged. The inequality $c_2 \geq (c_1 - 1) + 2$ shows that the addition of at least two of the cycles of C_2 left the number of connected components unchanged. These two cycles can be removed from C while still keeping C connected.

Case 2. $c_1 = c_2$. Let H denote a bipartite graph in which every cycle of C_1 is a left hand side vertex, every cycle of C_2 is a right hand side vertex, and two vertices are connected by an edge if the corresponding cycles share a vertex. Note that H is connected (because C is.) We consider three subcases.

1. H has a vertex of degree at least 3. Hence some cycle (say, cycle C^* of C_2) connects at least three cycles (of C_1). The argument of the case $c_2 > c_1$ can be extended to this case, by making C^* the first cycle of C_2 that is added to C_1. The number of connected components drops by at least 2 in the first iteration, ensuring that at least two other cycles from C_2 do not cause a drop in number of connected components.

2. H has a vertex of degree 1 and no vertex of degree more than 2. Then H is a path (because H is connected). If the path is of length 1, it follows that both C_1 and C_2 are single cycles (of length at least 3) that span the whole of C, and hence either one of them may be removed while keeping C connected. If the path is of length more than 1, then removing the two cycles that correspond to the endpoints of the path keeps C connected (observe that all vertices of the two removed cycles are contained in the set of vertices of their respective neighboring cycles in H).

3. All vertices in H have degree 2. Then H is a cycle. If either C_1 or C_2 contain a cycle of length 3 or more, then this cycle can be removed while keeping H (and hence also C) connected. If all cycles in C_1 and C_2 are of length 2, then it must be the case that C can be decomposed into two anti-parallel cycles (each of length $|C| \geq 3$), and removing any one of them keeps C connected. □

Now, we modify the algorithm in the following manner. Let C_5 be the set of cycles chosen from each component of C_3 without disconnecting each of the components as given by Lemma 4. Observe that C_5 need not be a cycle cover. Let $C_4 = C_3 \setminus C_5$.

Instead of picking the best of C_1, C_2 or C_3 as in Kaplan et al, in each iteration we pick the best of C_4 or C_5 according to the same potential function $w(F)/(\log_2 n_i/c(F))$ where n_i is the number of vertices in the current graph. The rest of the algorithm remains the same. We pick a single representative vertex from each of the connected components of F, delete all vertices and recurse.

Observe that $c(C_4) = c(C_3)$ as the number of components in C_3 and C_4 are equal. Also, $c(C_5) \leq n_i - 2c(C_3)$ as we pick at least 2-cycles of size 2 or a cycle of a size at least 3 from each of the component of C_3.

Claim 4. *In any iteration i, if F_i is chosen then $\dfrac{w(F_i)}{\log_2(n_i/c(F_i))} \leq \alpha OPT$ for $\alpha = 2/3$.*

Proof. Observe that α is the at most the value of the following optimization problem. Here, w_i corresponds to $w(C_i)/OPT$ and c_i corresponds to $c(C_i)/n_i$. These scalings do not affect the value of α.

$$\max z$$
$$z \leq \frac{w_j}{\log_2 \frac{1}{c_j}} \ \forall\, j \in \{4,5\}$$
$$w_4 + w_5 \leq 2$$
$$c_4 = c_3$$
$$c_5 \leq 1 - 2c_3$$
$$w_j, c_j \geq 0 \ \forall\, j = 4, 5$$

At the optimum solution we must have $z = \frac{w_4}{\log_2 \frac{1}{c_4}} = \frac{w_5}{\log_2 \frac{1}{c_5}}$ otherwise we can change w_4 and w_5 so as to make them equal without violating the feasibility and not decreasing the objective function. Also, we must have $w_4 + w_5 = 2$ and $c_5 = 1 - 2c_3$. Using these equalities, we have that $w_4 = \frac{2 \log_2 c_4}{\log_2(c_4(1-2c_4))}$ which gives that the objective function to maximize is $\frac{w_4}{-\log_2 c_4} = \frac{-2}{\log_2(c_4(1-2c_4))}$ which gets maximized when $c_4(1 - 2c_4)$ gets maximized. But $c_4(1 - 2c_4)$ has a maximum value of $1/8$ at $c_4 = 1/4$. This implies that at the optimum solution we have $w_4 = \frac{4}{3}, w_5 = \frac{2}{3}, c_4 = \frac{1}{4}, c_5 = \frac{1}{2}$ and $z = \frac{2}{3}$.   □

Now, proof of Theorem 1 follows from Claim 1 and Claim 4.

References

1. Arora, S.: Polynomial time approximation schemes for Euclidean traveling salesman and other geometric problems. Journal of the ACM 45(5), 753–782 (1998)
2. Berman, P., Karpinski, M.: 8/7-Approximation Algorithm for (1,2)-TSP. In: Proceedings of 17th ACM-SIAM Symposium on Discrete Algorithms, pp. 641–648. ACM Press, New York (2006)
3. Blaser, M.: A New Approximation Algorithm for the Asymmetric TSP with Triangle inequality. In: Proceedings of the Annual ACM-SIAM Symposium on Discrete Algorithms (SODA), pp. 638–645 (2002)
4. Charikar, M., Goemans, M.X., Karloff, H.: On the Integrality Ratio for the Asymmetric Traveling Salesman Problem. Mathematics of Operations Research, 31, 245–252 (2006)
5. Chekuri, C., Pal, M.: An O(logn) Approximation Ratio for the Asymmetric Travelling Salesman Path Problem. In: 9th International Workshop on Approximation Algorithms for Combinatorial Optimization Problems (APPROX), pp. 95–103 (2006)
6. Christofides, N.: Worst-case analysis of a new heuristic for the travelling salesman problem, Report 388, Graduate School of Industrial Administration, CMU (1976)
7. Frieze, A., Galbiati, G., Maffioli, F.: On the worst-case performance of some algorithms for the asymmetric traveling salesman problem. Networks 12, 23–39 (1982)
8. Gutin, G., Punnen, A.P. (eds.): Traveling Salesman problem and its Variations. Springer, Berlin, Germany (2002)
9. Held, M., Karp, R.M.: The Travelling Salesman Problem and Minimum Spanning Trees. Operations Research 18, 1138–1162 (1970)
10. Kaplan, H., Lewenstein, M., Shafrir, N., Sviridenko, M.: Approximation algorithms for asymmetric TSP by decomposing directed regular multigraphs. J. ACM 52(4), 602–626 (2005)
11. Kleinberg, J., Williamson, D.: Unpublished Note (1998)
12. Lam, F., Newman, A.: Travelling Salesman Path Problems, (Manuscript 2005)
13. Papadimitriou, C.H., Vempala, S.: On The Approximability Of The Traveling Salesman Problem. Combinatorica 26(1), 101–120 (2006)
14. Shmoys, D.B., Williamson, D.P.: Analyzing the Held-Karp TSP bound a monotonicity property with application. Information Processing Letters 35(6), 281–285 (1990)
15. Wolsey, L.: Heuristic Analysis, Linear Programming and Branch and Bound. In: Mathematical Programming Studies (1980)

Approximation Algorithms for the Traveling Repairman and Speeding Deliveryman Problems with Unit-Time Windows

Greg N. Frederickson and Barry Wittman

Department of Computer Sciences
Purdue University, West Lafayette, IN 47907
gnf@cs.purdue.edu, bwittman@cs.purdue.edu

Abstract. Constant-factor, polynomial-time approximation algorithms are presented for two variations of the traveling salesman problem with time windows. In the first variation, the traveling repairman problem, the goal is to find a tour that visits the maximum possible number of locations during their time windows. In the second variation, the speeding deliveryman problem, the goal is to find a tour that uses the minimum possible speed to visit all locations during their time windows. For both variations, the time windows are of unit length, and the distance metric is based on a weighted, undirected graph. Algorithms with improved approximation ratios are given for the case when the input is defined on a tree rather than a general graph. A sketch of NP-hardness is also given for the tree metric.

1 Introduction

For decades the traveling salesman problem (TSP) has served as the archetypal hard combinatorial optimization problem that attempts to satisfy requests spread over a metric space [15]. Yet as a model of real-life problems, the TSP is deficient in some respects. In particular, the salesman may not have enough time to visit all desired locations. Furthermore, a visit to any particular location may be of value only if it occurs within a certain limited interval of time.

We consider a fundamental version of such a repairman problem, in which the repairman is presented with a set of *service requests*. Each service request is located at a node in a weighted, undirected graph and is assigned a *time window* during which it is valid. The repairman may start at any time from any location and stop similarly. (This latter assumption is at variance with much of the preceding literature about the repairman problem [1,2]. We choose to frame our problem without specifying initial and final locations because doing so leads to an elegant solution that gives insight into such problems).

We handle two variations of our problem. In the first, each *service event* that the repairman completes will yield a specified *profit*. The goal of the repairman is to choose a subset of requests and then find a *service run* that satisfies those requests so as to maximize the total profit possible. In the second variation,

M. Charikar et al. (Eds.): APPROX and RANDOM 2007, LNCS 4627, pp. 119–133, 2007.
© Springer-Verlag Berlin Heidelberg 2007

all service requests must be satisfied, with the service provider increasing the speed by as little as necessary to make a *service tour*. (To cover a given distance and arrive within a given time, the deliveryman must travel at least as fast as some particular minimum *speed*.) We call this variation the *speeding deliveryman problem*, recognizing that, for example, a pizza delivery driver may need to hurry to deliver the pizza in a timely manner. We appear to be the first to frame this problem in terms of speed-up, a refreshing departure from the standard emphasis on distance traveled or profit achieved.

For both variations, we focus on the case in which all time windows are the same length (i.e., unit-time), and all profits for service events are identical. Additionally, we refer to each service event as being instantaneous, although service times can be absorbed into the structure of the graph in many cases. These restrictions still leave problems that are APX-hard, via a simple reduction from TSP, which has been shown to be APX-hard [16].

Our goal is thus to find polynomial-time approximation algorithms. For the repairman, they produce a service run whose profit is within a constant factor of the profit for an optimal service run. For the deliveryman, they produce a service tour whose maximum speed is within a constant factor of the optimum speed that accommodates all requests. These variations contrast neatly, as the repairman is a maximization problem while the deliveryman is a minimization problem. To the best of our knowledge, we are the first to find approximation algorithms that get within a constant factor for a general metric. In addition, we show how to solve both variations with improved constants in the case that the service network is an edge-weighted tree rather than a general (weighted) graph. The problems are NP-hard for a tree, as we shall show.

Our algorithms make use of several novel ideas: Ideally, we would like to partition requests into subsets, handling each subset in some unified fashion. The greatest challenge is how to play off proximity of location against proximity of time when forming the subsets. Our solutions partition requests by their time windows, so that the requests of any subset in the partition are uniformly available over the entire extent of time under consideration for that subset. We thus partition time into discrete *periods* and *trim* the time window for each request to be the period that was wholly contained in it. Doing so induces at most a linear number of periods which we can consider separately, and either loses the repairman at most a constant fraction of possible profit or increases the necessary speed of the deliveryman by at most a constant factor.

Once we partition requests on the basis of common trimmed windows, we are able, for any given common trimmed window, to solve the variation restricted to trees exactly for the repairman and almost exactly for the deliveryman. We can then stitch together solutions for each different trimmed window, using dynamic programming. For general graphs, we rely on approximation algorithms for any given trimmed window, and again stitch together approximate solutions for each different trimmed window, once again using dynamic programming.

The approximation algorithms for the traveling repairman seem necessarily to be more complex than those for the speeding deliveryman. We adapt a constant approximation algorithm from [1] for the *point-to-point orienteering problem*, in which the goal is to travel from a starting point to an ending point visiting as many intermediate points as possible without exceeding a budget. Even when our time periods are discrete and non-overlapping, we must be careful in our use of dynamic programming as we patch together promising sets into a promising service run. If the repairman is too greedy and uses too much time in one time period, then he or she may end up being too late to handle many service requests in subsequent periods. Thus for each request location, and for each amount of profit that the repairman could arrive with, we need to identify the earliest time at which the repairman could have arrived with at least a constant fraction of that amount of profit in hand. With this more detailed information, we can then effectively use dynamic programming.

Although we seem to be the first to study the speeding deliveryman problem, we are not the first to consider the repairman problem, which is a generalization of a host of repairman, deliveryman and traveling salesman problems such as those in [1,2,6,14,17]. Much work has been done on related problems in a metric space on the line. Assuming unit-time windows, a $\frac{1}{4+\epsilon}$-approximation was given for the repairman on a line in [2]. We improve on this approximation to $\frac{1}{3}$ and in a more general setting, a tree. We are the first to give poly-time constant-ratio algorithms for the unit-time repairman problem on a tree or on a graph.

For general metric spaces and general time windows together, an $\Omega(\frac{1}{(\log n)^2})$-approximation was given in [1], whereas a constant approximation is given in [6], but only when there are a constant number of different time windows. TSP with time windows has been studied in the operations research community, as in [7] and [8], where it was exhaustively solved to optimality.

In Sect. 2, we characterize the effects of contracting the time windows of the service requests. In Sect. 3, we give an approximation algorithm with a bound of $\frac{1}{3}$ for the repairman on a tree. In Sect. 4, we give an approximation algorithm with a bound of $\frac{1}{15}$ for the repairman on a graph. In Sect. 5, we give an approximation algorithm for a deliveryman in a tree with a maximum increase in speed by a factor of $4 + \epsilon$. In Sect. 6, we give an approximation algorithm for a deliveryman in a graph with a maximum increase in speed of a factor of 8. In Sect. 7, we sketch the NP-hardness of the problems on a tree.

2   Trimming Requests

Trimming is a simple and yet powerful technique that can be applied when we deal with unit-time windows. Starting with time 0, we make divisions in time at values which are integer multiples of one half, i.e., 0, .5, 1, and so on. We assume that no request window starts on such a division, because if it did, we could redefine times to be decreased by a negligible amount. Let a *period* be the time interval from one division up to but not including the next division. Because every service request is exactly one unit long in time, half of any request window will be wholly contained within only one period, with the rest divided between the preceding and following periods. We then trim each service request window to coincide with the period wholly contained in it, ignoring those portions of the request window that fall outside of the chosen period.

For the repairman problem, the trimming may well lower the profit of the best service run, but by no more than a constant factor. Let $p(R)$ denote the profit of a service run R.

Theorem 1 (Limited Loss Theorem). *Consider any instance of the repairman problem. Let R^* be an optimal service run with respect to untrimmed requests. There exists a service run R with respect to trimmed requests such that $p(R) \geq \frac{1}{3} p(R^*)$.*

Proof. We use an elegant best-of-three argument. Let the *target interval* of a request be that part of the request window that coincides with the period to which the request is trimmed. Call that part of the request window contained in the previous period its *early interval*, and call that part of the request window contained in the following period its *late interval*.

Now R^* must have at least one third of its service events in either the target intervals, the early intervals, or the late intervals. If at least one third of the service events of R^* occur in target intervals, then we have R follow the same path and schedule as R^* but service only those requests in target intervals.

If at least one third of the service events of R^* occur in early intervals, then we take service run R to be R^* but started .5 units later in time, and with R servicing those requests that were in early intervals of R^* but are now in target intervals of R. Then the number of service events of R will be at least one third of the number of service events for R^*.

Similarly, if at least one third of the service events of R^* occur in late intervals, we take R to be R^* but started .5 units earlier in time, with R servicing those requests that were in late intervals of R^* but are now in target intervals of R.

In each case, it is possible to construct a service run R for trimmed requests that contains at least one third of the service events of an optimal service run for untrimmed requests. Since one of these three cases must always hold, we can always find a desired R.   □

For the deliveryman problem, trimming may well increase the necessary speed of the best service tour, but by no more than a constant factor. Let $s(T)$ denote the minimum speed needed for service tour T to visit all service requests.

Theorem 2 (Small Speedup Theorem). *Consider any instance of the deliveryman problem. Let T^* be an optimal service tour with respect to untrimmed requests. There exists a service tour $\hat{T}$ with respect to trimmed requests such that $s(\hat{T}) \leq 4s(T^*)$.*

Proof. We shall define tour $\hat{T}$ on which the deliveryman travels back and forth along T^* at a speed of $4s(T^*)$. During any two consecutive periods, the deliveryman will make a net advance equal to the advance of T^* over those two periods. Let the starting times of the periods be t_0, t_1, t_2, and so on. Define a function $f(t)$ that for any given time t gives the location of the deliveryman on T^*. Formally, $\hat{T}$ is the sequence of locations given by a function $\hat{f}(t)$ as it sweeps from time 0 to the latest time which is in the domain of $f(t)$. For $t_i \leq t < t_i + 1$, where i is even, let

$$\hat{f}(t) = \begin{cases} f\left(t_i - 1 + 4(t - t_i)\right) & t_i \leq t < t_i + .625 & \text{(forward 5 steps)} \\ f\left(t_i + 1.5 - 4(t - t_i)\right) & t_i + .625 \leq t < t_i + 1 & \text{(backward 3 steps)} \end{cases}$$

Consider a request r serviced at time t in T^*. If $t_i \leq t < t_i + .5$, then the time window of the request will be trimmed to be one of three periods of length .5: $[t_i - .5, t_i)$, $[t_i, t_i + .5)$, or $[t_i + .5, t_i + 1)$. We consider cases when i is even or i is odd separately.

Case 1: i is even
    If the window containing r is trimmed to be $[t_i - .5, t_i)$, service request r at time $t_i + .25((t - t_i) + .5)$. If the window is trimmed to be $[t_i, t_i + .5)$, service request r at time $t_i + .25((t - t_i) + 1)$. If the window is trimmed to be $[t_i + .5, t_i + 1)$, service request r at time $t_i + .25((t - t_i) + 1.5)$.
Case 2: i is odd
    If the window containing r is trimmed to be $[t_i - .5, t_i)$, service request r at time $t_i + .25((t_i - t) + 2)$. If the window is trimmed to be $[t_i, t_i + .5)$, service request r at time $t_i + .25((t_i - t) + 1.5)$. If the window is trimmed to be $[t_i + .5, t_i + 1)$, service request r at time $t_i + .25((t_i - t) + 1)$.   □

3   Repairman Problem for a Tree

A demonstration of the power of trimming is that we can solve the repairman problem on a tree exactly in the case that windows are already trimmed. We first give a dynamic programming algorithm for the repairman problem on a tree when all requests share an identical time window. To find a path from s to t of profit p, we start with the direct path from s to t and then add on low-cost pieces of subtrees that branch off the direct path as necessary to achieve profit p. We do so by contracting the path into a single node r and using dynamic programming to sweep up from the leaves, finding the cheapest paths in the tree for each possible profit.

SWEEP-TREE(node u)

For p from 0 to $\pi(u)$, set $L_u[p] \leftarrow 0$
For each child v of u,
    Call SWEEP-TREE(v), which will generate L_v.
    Add $2d(u,v)$ to each entry in L_v except $L_v[0]$.
    Let $\max_u$ be the largest profit in L_u and $\max_v$ the largest profit in L_v.
    For p from 0 to $\max_u + \max_v$, set $L[p] \leftarrow \infty$.
    For a from 0 to $\max_u$ and b from 0 to $\max_v$,
        $L[a + b] \leftarrow \min\{L[a + b], L_u[a] + L_v[b]\}$.
    $L_u \leftarrow L$

Our recursive subroutine SWEEP-TREE(r) produces a list L_r of the lowest costs at which various profit levels can be achieved by including portions of the tree rooted at r. List L_r is a mapping from profits to costs where $L_r[p]$ is the cost of achieving profit p, if recorded, and ∞ otherwise. Let $\pi(u)$ be the profit gained by visiting u. Note that $\pi(u)$ counts the number of service requests at u.

If we define $\pi(r)$ to be the profit of the direct path, then adding $d(s,t)$ to all the costs in the list L_r yields the costs of the best paths on the full tree starting at s and ending at t for all possible profit levels.

Lemma 1. *For all possible profits, SWEEP-TREE identifies minimum-length paths from s to t in a total of $O(n^2)$ time.*

Proof. Let n and related variables represent the sum of the number of nodes in the subtree rooted at vertex u plus the total profit for the nodes in that subtree. Let n_0 be the amount of n attributable to vertex u, and n_i be the amount of n attributable to the subtree rooted at the ith child of u, for $i = 1, 2, \ldots, k$. Then the time for SWEEP-TREE is defined by

$$T(n) \leq cn_0 + \sum_{i=1}^{k} \left[T(n_i) + cn_i \left(1 + \sum_{j=0}^{i-1} n_j \right) \right]$$

Proof by induction establishes that $T(n) \leq dn^2$ for a suitable constant d.   □

Although SWEEP-TREE might be viewed as being vaguely reminiscent of Sect. 2.6.3 in [5], we note that the presentation in [5] is rather indeterminate and that its claimed running time is not fully polynomial.

Using algorithm SWEEP-TREE, we next give the algorithm REPAIRMAN-TREE for multiple trimmed windows. This algorithm uses dynamic programming to move from period to period, in increasing order by time. As it progresses, it finds service runs of all possible profits from every trimmed request in the current period through some subset of trimmed requests in the current period and arriving at any possible trimmed request in a later period. In this way, for every profit value, we identify the earliest that we can arrive at a request that achieves that profit value. The critical insight is that we may have to leave a certain period very early in order to reach later requests in time. By recording even the low profit service runs and using them as starting points, we never commit ourselves to a service run which appears to be promising in early stages but arrives too late to visit a large number of requests in later stages.

We focus on periods that each contain at least one trimmed request, numbering them from S_1, the period starting at the smallest time value, up to the last period S_m. Let n be the total number of requests. For every period S_i, we arbitrarily number its trimmed requests as s_{ij}. Let R_{ij}^k be the earliest arriving k-profit sequence of service events ending at s_{ij}. Let A_{ij}^k be the arrival time of R_{ij}^k at s_{ij}. For each s_{ij}, we initialize every R_{ij}^1 to be $\{s_{ij}\}$ and every A_{ij}^1 to be 0. For $k > 1$, let R_{ij}^k be initialized to *null*, and let every other A_{ij}^k be initialized to $begin(S_{i+1})$, where $begin(S_i)$ is the first time instant in period S_i.

We use SWEEP-TREE to find a path of shortest length from a given starting request to a given ending request, subject to accumulating a specified profit. Let $time(R)$ be amount of time a path R takes. For each indexed period S_i, from 1 up to m, we process period S_i as described in PROCESS-PERIOD.

PROCESS-PERIOD(S_i)

For each trimmed request s_{ij} in period S_i,
    For each possible profit value p,
        For each subsequent period S_a that contains a trimmed request,
            For each trimmed request s_{ab} in S_a, do the following:
                Let R be the path corresponding to $L_r(p)$ that results from
                    SWEEP-TREE(r) with $s = s_{ij}$ and $t = s_{ab}$, on the set $S_i - \{s_{ij}\}$.
                Let R^- be R with its last leg, ending at s_{ab}, removed.
                For k from 1 to $n - p(R)$,
                    If $A_{ij}^k + time(R^-) < begin(S_{i+1})$, then
                        Let profit q be $k + p(R) - 1$.
                        If $A_{ij}^k + time(R) < A_{ab}^q$, then
                            Set R_{ab}^q to be R_{ij}^k followed by R.
                            Set A_{ab}^q to be $\max\{A_{ij}^k + time(R), begin(S_a)\}$.

After all the periods have been processed, we identify the largest-profit path found, and return that resulting service run $\hat{R}$ as the output of algorithm REPAIRMAN-TREE. A suitably reorganized REPAIRMAN-TREE will take $O(n^4)$ time, since it will make $O(n^2)$ calls to SWEEP-TREE, each of which takes $O(n^2)$ time. Applying the Limited Loss Theorem:

Theorem 3. *The approximation ratio of REPAIRMAN-TREE is at least $\frac{1}{3}$.*

4   Repairman Problem for a Graph

Our approximation algorithm for the repairman on a graph uses approximation algorithms for the optimization problems below. The cost $c(T)$ of path, tour, or tree T is the total cost of its edges under metric d.

Rooted k-Minimum Spanning Tree (k-MST): Given node v and integer k, find a tree of smallest cost that contains at least k nodes, including v.

Rooted k-Traveling Salesman Problem (k-TSP): Given node v and integer k, find a smallest-weight tour containing at least k nodes, including v.

Source-Sink k-Path (k-SSP): Given nodes s and t and integer k, find a path of smallest weight from s to t that contains at least k nodes. (This problem is called min-cost s-t path in [4].)

Besides approximation algorithms for these problems, we also consider two more approximation problems for k-SSP. Let the *excess* of a path P from s to t be $\epsilon(P) = c(P) - d(s,t)$.

Small-Excess k-SSP: Given nodes s and t and integer k, find a path of small excess from s to t containing at least k nodes. (This problem is called min-excess path in [4].)

Reduced-Profit k-SSP: Given nodes s and t and integers k and $\sigma > 1$, find a path from s to t containing at least k/σ nodes and costing no more than an optimal k-SSP.

The final performance bound of our algorithm depends on the approximation ratio of the Reduced-Profit k-SSP algorithm, which in turn depends on several other algorithms. First, we discuss a way to approximate k-SSP using a way to approximate k-MST described in [11]. Second, an algorithm for Small-Excess k-SSP using an approximation for k-SSP is given in [4]. We note that the approach in [4] uses a distance bound, whereas for our problem it seems imperative that we use profit values as a parameter to get an appropriate set of paths. Third, we will describe a technique similar to the one used in [1] to solve the Reduced-Profit k-SSP problem using a σ-approximation to Small-Excess k-SSP. Fourth, we describe how to approximate the repairman problem by nesting the approximation algorithm within a dynamic programming structure. Finally, we coax a better performance bound out of our algorithm by combining two different approximations for k-SSP in a mixed strategy approach [9].

Having outlined our approach in the preceding paragraph, we expand on that brief description:

First – we approximate k-SSP: In [11], Garg used a linear programming formulation based on the Goemans and Williamson technique [12] to find a 2-approximation to the k-MST problem in polynomial time. We use the 2-approximation to k-MST to find a 4-approximation to k-SSP as follows: Let T^* be an optimal solution to k-MST on set V with *two* required nodes s and t. We can force a k-MST to include node t as well by creating a set V' from V with $n-1$ copies of t and then defining $k' = k + n - 1$, where $n = |V|$. We can approximate T^* by using the above 2-approximation to k-MST on V' with parameter k' and then removing all extra copies of t. The resulting tree $\hat{T}$ will have a profit of at least k, contain both s and t, and cost $c(\hat{T}) \le 2c(T^*)$. Path $\hat{P}$ of profit k from s to t can be constructed from $\hat{T}$ by doubling up all its edges except for those on a direct path from s to t in $\hat{T}$. Call this algorithm MST-FOR-SSP.

Lemma 2. *Path $\hat{P}$ produced by MST-FOR-SSP has cost $c(\hat{P}) \le 4c(P^*) - d(s,t)$, where P^* is an optimal solution to k-SSP.*

Proof. Since P^* is a tree of no less cost than T^*, we have
$$c(\hat{P}) \le 2c(\hat{T}) - d(s,t) \le 4c(T^*) - d(s,t) \le 4c(P^*) - d(s,t). \qquad \square$$

Second – we approximate Small-Excess k-SSP: An algorithm from [4] which we shall call SMALL-EXCESS takes an α-approximation to the k-SSP problem and generates a $\left(\frac{3}{2}\alpha - \frac{1}{2}\right)$-approximation to Small-Excess k-SSP. Using MST-FOR-SSP with $\alpha = 4$, this yields an $\frac{11}{2}$-approximation to the Small-Excess k-SSP problem.

Third – we approximate Reduced-Profit k-SSP: We describe approximation algorithm REDUCED-PATH for the Reduced-Profit k-SSP problem which is similar to the one used in [4] for orienteering except that we supply a profit value instead of a distance bound as a parameter. To find a path B, we identify many possible subpaths B_j by choosing all possible pairs of nodes u, v and running SMALL-EXCESS between them with profit parameter k/σ. Of all these possible pairs, we keep the one for which $c(B_j) + d(s, u) + d(v, t)$ is smallest. For that pair, we form path B by appending (s, u) and (v, t) to B_j.

Lemma 3. *Let σ be the smallest whole number no smaller than the approximation bound for SMALL-EXCESS, and let P^* be an optimal k-path from s to t. Then B is a path from s to t such that $c(B) \leq c(P^*)$ and $p(B) \geq \frac{1}{\sigma}p(P^*)$.*

Proof. This proof is a modest generalization of the one in [1] used for orienteering. Split P^* into σ pairwise edge-disjoint subpaths, $P_1^*, \ldots, P_\sigma^*$, such that each subpath P_i^* has a profit of at least k/σ. Note that σ need not divide evenly into k, since we can allow fractional parts of the profit on shared endpoints. Label the starting and ending points of P_i^* as u_{i-1} and u_i, respectively. Thus $s = u_0$ and $t = u_\sigma$. Then

$$\epsilon(P^*) = c(P^*) - d(u_0, u_\sigma) \geq \sum_{i=1}^{\sigma} c(P_i^*) - \sum_{i=1}^{\sigma} d(u_{i-1}, u_i)$$

$$= \sum_{i=1}^{\sigma} [c(P_i^*) - d(u_{i-1}, u_i)] = \sum_{i=1}^{\sigma} \epsilon(P_i^*)$$

If subpath P_j^* has the minimum excess of all σ subpaths, then $\epsilon(P_j^*) \leq \frac{1}{\sigma}\epsilon(P^*)$. Let $\tilde{P}_j$ be the subpath made by running SMALL-EXCESS from u_{j-1} to u_j with profit parameter k/σ. Thus, $\epsilon(\tilde{P}_j) \leq \sigma\epsilon(P_j^*)$. Then, consider a full path $\tilde{P}$ which is made by taking a direct edge from u_0 to u_{j-1}, running from u_{j-1} to u_j along $\tilde{P}_j$, and then finishing by taking another direct edge from u_j to u_σ. We then have

$$c(P^*) = \sum_{i=1}^{\sigma} c(P_i^*) = \sum_{i=1}^{\sigma} [d(u_{i-1}, u_i) + \epsilon(P_i^*)]$$

$$\geq d(u_0, u_{j-1}) + d(u_j, u_\sigma) + \sigma\epsilon(P_j^*) \geq c(\tilde{P})$$

Because we considered all possible pairs to construct B_j,

$$c(B) \leq c(\tilde{P}) \leq c(P^*), \text{ and } p(B) \geq \frac{1}{\sigma} p(P^*). \qquad \square$$

Using the $\frac{11}{2}$-approximation to the Small-Excess k-SSP, we round up to get a value of $\sigma = \lceil\frac{11}{2}\rceil = 6$, and thus a $\frac{1}{6}$-approximation for Reduced-Profit k-SSP.

G.N. Frederickson and B. Wittman

Fourth – we approximate the repairman problem: Our approximation algorithm for the repairman in a graph incorporates the preceding approximation algorithms within the context of a dynamic programming algorithm with the same overall structure as the algorithm for a tree. For each indexed period S_i, from 1 up to m, we process period S_i as in PROCESS-PERIOD. The only difference is that instead of taking R to be the path corresponding to $L_r(p)$ that results from SWEEP-TREE(r), it takes R to be the output of REDUCED-PATH(s_{ij}, s_{ab}, p) on the set $S_i - \{s_{ij}\}$, where s_{ij} is the starting request in the path, s_{ab} is the ending request in the path, and p is the profit of which the path must have a constant fraction. As before, we identify the largest-profit path found and return the resulting service run $\hat{R}$ as the output of algorithm REPAIRMAN-GRAPH.

Theorem 4. *The approximation ratio of REPAIRMAN-GRAPH is at least $\frac{1}{18}$.*

Finally – we incorporate a mixed strategy approach: We can improve the overall performance bound on the approximation to the k-SSP problem by combining the k-MST approximation described above with a k-TSP approximation in a mixed strategy approach [9]. It was claimed in [11] that the same technique used to find a 2-approximation to k-MST can be used to find a 2-approximation to k-TSP. Because details supporting this claim are not available, we assume a constant β-approximation for k-TSP for the analysis below. From results in [10], $\beta \leq 3$.

We now state an algorithm using an approximation for k-TSP. We use a β-approximation to k-TSP to find a $(\beta + 1)$-approximation to k-SSP in a way similar to an algorithm given in [3]. First, we create a new set V' by merging nodes s and t into a single node st. We run an all pairs shortest path algorithm on V' and define metric d' from d using shortest paths. Then, we run the k-TSP approximation on set V' with root st and profit parameter $k' = k - 1$, yielding a tour $\hat{U}'$. If we treat st as the points s and t at the same location, then $\hat{U}'$ is a k-path $\hat{P}'$ from s to t. An optimal tour U'^* is an optimal k-path P'^* on V'.

To obtain a k-path on V with the original metric, we expand the node st back to s and t and delete all edges between s and t. If the degrees of s and t are even, we add back one edge between them. Since the degrees of all other nodes will be even and the degrees of s and t will be odd, we then get a k-path from s to t. We call this algorithm TSP-FOR-SSP. Our algorithm MIXED-REPAIR selects as $\hat{P}$ the smaller of the paths found by MST-FOR-SSP and TSP-FOR-SSP.

Lemma 4. *Path $c(\hat{P})$ from MIXED-REPAIR satisfies $c(\hat{P}) \leq \left(2 + \frac{\beta}{2}\right) c(P^*)$, where P^* is an optimal solution to k-SSP and β is the approximation bound for k-TSP.*

Proof. By Lemma 2, $c(\hat{P}) \leq 4c(P^*) - d(s, t)$.
In the k-TSP formulation, since s and t are at the same location in V',
$$c(\hat{P}') = c(\hat{U}') \leq \beta c(U'^*) = \beta c(P'^*).$$

Since it must be the case that $c(P'^*) \leq c(P^*)$,
$$c(\hat{P}) \leq c(\hat{P}') + d(s, t) \leq \beta c(P'^*) + d(s, t) \leq \beta c(P^*) + d(s, t).$$

When $(4 - \beta)c(P^*) \leq 2d(s,t)$, then the bound from k-MST dominates and

$$c(\hat{P}) \leq 4c(P^*) - d(s,t) \leq 4c(P^*) - \tfrac{1}{2}(4 - \beta)c(P^*) = \left(2 + \tfrac{\beta}{2}\right)c(P^*).$$

When $(4 - \beta)c(P^*) > 2d(s,t)$, then the bound from k-TSP dominates and

$$c(\hat{P}) \leq \beta c(P^*) + d(s,t) < \beta c(P^*) + \tfrac{1}{2}(4 - \beta)c(P^*) = \left(2 + \tfrac{\beta}{2}\right)c(P^*). \quad \square$$

This new performance bound for k-SSP correspondingly affects the bounds for the other approximations. Using our $\left(2 + \frac{\beta}{2}\right)$-approximation to k-SSP yields a $\left(\frac{5}{2} + \frac{3\beta}{4}\right)$-approximation to the Small-Excess k-SSP problem. Noting that a 3-approximation to k-TSP is given in [10], values of β such that $2 \leq \beta \leq 3$ will produce values of 4 or 5 for σ. Thus, the mixed strategy improves the performance bound of the Reduced-Profit k-SSP algorithm to $\frac{1}{4}$ or $\frac{1}{5}$, depending on the value of β. Again, applying the Limited Loss Theorem:

Theorem 5. *The approximation ratio of MIXED-REPAIR is at least $\frac{1}{15}$.*

MIXED-REPAIR uses $O(n^{14} \log n)$ time. It makes $O(n^3)$ calls to REDUCED-PATH, which makes $O(n^2)$ calls to SMALL-EXCESS, which makes $O(n^5)$ calls to the k-MST approximation algorithm. The k-MST algorithm has two procedures $\mathcal{Q}$ and $\mathcal{T}$, each of which is run no more than n times. Procedure $\mathcal{Q}$ runs no more than $O(n)$ Goemans and Williamson linear programs, each of which takes $O(n^2 \log n)$ [12]. With the time for procedure $\mathcal{T}$ subsumed by the time for $\mathcal{Q}$, the time for the k-MST algorithm is $O(n^4 \log n)$.

5   Deliveryman for a Tree

With the Small Speedup Theorem in our arsenal, the speeding deliveryman algorithm on a tree is almost as cleanly conceived as the repairman on a tree. When all requests share the same time window, we can find an optimal solution as follows. For every possible starting request u and ending request v in the period, we identify the direct path between u and v. Remove any leaf and its adjacent edge if the leaf is not u or v or the location of a request, and repeat until every leaf is either u or v or is the location of a request. We then double up every edge in the slimmed down tree that is not the direct path from u to v, and then identify the Euler path from u to v. Since we test all pairs of starting and ending requests, we clearly find the shortest length path and therefore the minimum necessary speed to visit all requests in the tree during a single period. We can find the direct distances between all pairs in the tree and thus the length of the shortest of the Euler paths in $O(n^2)$ time.

To approximate a solution to the problem on a tree over multiple periods, we develop an algorithm to test if a specific speed is fast enough to visit all requests during their periods. We use the idea behind the single-period solution in conjunction with dynamic programming. TEST-SPEED processes every period in order and finds the earliest-arriving paths starting at request u, ending at request v, and visiting all requests in the period for every pair of requests

u and v. It then glues each of these paths to the earliest-arriving paths which visit all requests in previous periods and, for every request v in S_i, keeps the earliest-arriving complete path ending at v. Since it finds an earliest-arriving path for a particular speed, finding a path which visits all requests during their periods implies that the speed is sufficiently fast.

For $u \in S_i$, let the arrival time A_u be the earliest time at which a path visiting all the requests in periods before S_i arrives at request u before visiting any other request in S_i. For $v \in S_i$, let the departure time D_v be the earliest time at which a path visiting all requests in the periods up to and including S_i ends at request v. For $u, v \in S_i$, let $length_{uv}$ be the length of the shortest path from u to v which visits all requests in S_i.

TEST-SPEED(*speed*)

For each request u in S_1, set $A_u \leftarrow 0$.
For i from 1 to m
   For each request v in S_i,
      $D_v \leftarrow \min_{u \in S_i}\{A_u + length_{uv}/speed\}$
      If $D_v >$ last time instant of S_i, then set $D_v \leftarrow \infty$.
   For each request w in S_{i+1},
      $A_w \leftarrow \max\{$ first time instant of $S_{i+1}, \min_{v \in S_i}\{D_v + d(v,w)/speed\}\}$
      If $A_w >$ last time instant of S_{i+1}, then set $A_w \leftarrow \infty$.
If there exists a request $v \in S_m$ such that $D_v < \infty$,
   then return "feasible speed", else return "speed too slow".

In the next section, we will give a technique which allows us to find a service tour $\hat{T}$ on trimmed windows such that $\frac{1}{2}s(\hat{T}) \leq s(T^*) \leq s(\hat{T})$ where T^* is an optimal service tour over trimmed windows. Using TEST-SPEED, our algorithm DELIVERY-TREE binary searches in this range to find a speed within a factor of $1 + \epsilon$ of the optimal speed. We get a total of $O(n^3 \log \frac{1}{\epsilon})$ time for DELIVERY-TREE. Applying the Small Speedup Theorem gives:

Theorem 6. *DELIVERY-TREE finds a service tour of speed at most $4 + \epsilon$ times the optimal speed.*

6   Deliveryman for a Graph

The algorithm for the deliveryman on a graph takes wonderfully direct advantage of the Small Speedup Theorem, using a minimum spanning tree (MST) algorithm as its main workhorse. Given an instance of the problem on trimmed windows, we find an approximately minimum speed service tour in the following way. For each period, we find an MST of the points inside. For every period S_i but the last, we find the shortest path which connects the MST of the points in S_i to the MST of the points in S_{i+1}. Let the point in S_i which is adjacent to that edge be called v_i, and let the point in S_{i+1} be called u_{i+1}.

Within each period S_i, we double up all edges in the MST that are not on the direct path from u_i to v_i and sequence all edges into an Euler path. We then connect these m Euler paths by the edges (v_i, u_{i+1}) for $1 \le i < m$. Thus, we have created a service tour from u_1 to v_m. We then determine the minimum speed at which this service tour can be taken and still visit all requests during their trimmed time windows.

Let $c(u_i, v_j)$ denote the cost of traveling from u_i to v_j along the service tour. For all pairs i and j where $1 \le i \le m$ and $i \le j \le m$, we find the speed needed to cover the tour from u_i to v_j and store these speeds in a set. The value of each speed will be $\frac{2 \cdot c(u_i, v_j)}{j - i + 1}$ as the factor of 2 accounts for the .5 unit-time windows. We search in this list until we find the lowest speed at which we can visit all requests within their periods. As one of these pairs of vertices must define the most constraining speed, we find the minimum speed at which we can travel $\hat{T}$. The total running time for our algorithm DELIVERY-GRAPH will be $O(n^2)$.

Let T^* be the optimal tour over trimmed windows, and let $\hat{T}$ be the tour generated by our algorithm. Let T_i^* be the subtour of tour T^* restricted to requests inside period S_i, and $\hat{T}_i$ be the subtour of tour $\hat{T}$ restricted to S_i. A proof by induction on i establishes:

Lemma 5. *Traveling $\hat{T}$ at a speed no greater than twice the minimum required speed for T^*, for each S_i we never arrive at the first vertex in T_i^* before arriving at the first vertex in $\hat{T}_i$.*

Theorem 7. *DELIVERY-GRAPH finds a service tour of speed at most 8 times the optimal speed.*

7   Sketch of NP-Hardness on a Tree

NP-completeness proofs for many of the time-constrained traveling salesman problems on a line were given in [17], but we know of no proof that the unit-time window repairman problem in [2] is NP-hard. However, we give a sketch of a proof that the repairman problem with unit-length time windows on a tree is NP-hard. We use a reduction from a version of the partition problem restricted to positive values: Given a multiset of n positive integers, decide whether the multiset can be partitioned into two multisets which sum to the same value, i.e. half the sum of all of the integers in the multiset. This problem is NP-complete [13].

The idea behind our reduction is as follows. We assume that the sum of the integers in the multiset is $2K$. First create a central node u in a tree by itself. For each integer in the multiset, create a node and connect it to u with an edge having the cost of the integer. Then create the start and end nodes s and t and connect them to u with edges of cost $6K$. Also create the midpoint node v and connect it to u with an edge of cost K. To complete the input for the repairman problem, we must also associate at least one time window with each node. Associate the time window $[0, 6K]$ with s and $[12K, 18K]$ with t.

Associate the time window $[6K, 12K]$ with the each of the nodes corresponding to an integer and also to the central node u. Finally, associate two time windows, $[3K, 9K]$ and $[9K, 15K]$, with the midpoint node v. Note that the graph is a tree and that all the time windows are exactly $6K$ units long, meeting the unit-length requirement. The optimal tour has just enough time to visit all of the nodes if and only if it starts at the start node, visits a set of nodes whose integers sum to K, visits the midpoint node, visits the remaining nodes which also sum to exactly K, and finally ends at the end node.

While we have sketched a proof that assumes intervals closed on both ends, it is possible to modify the proof for intervals closed on one end and open on the other. A proof that the deliveryman problem on a tree with unit-time windows is NP-hard follows the same form.

References

1. Bansal, N., Blum, A., Chawla, S., Meyerson, A.: Approximation algorithms for deadline-TSP and vehicle routing with time-windows. In: Proc. 36th ACM Symp. on Theory of Computing, pp. 166–174. ACM Press, New York (2004)
2. Bar-Yehuda, R., Even, G., Shahar, S.: On approximating a geometric prize-collecting traveling salesman problem with time windows. J. Algorithms 55(1), 76–92 (2005)
3. Blum, A., Chawla, S., Karger, D.R., Lane, T., Meyerson, A., Minkoff, M.: Approximation algorithms for orienteering and discounted-reward TSP. Preliminary version of [4]
4. Blum, A., Chawla, S., Karger, D.R., Lane, T., Meyerson, A., Minkoff, M.: Approximation algorithms for orienteering and discounted-reward TSP. Proc. 44th IEEE Symp. on Foundations of Computer Science , p. 46 (2003)
5. Chawla, S.: Graph Algorithms for Planning and Partitioning. PhD thesis, Carnegie Mellon University (2005)
6. Chekuri, C., Kumar, A.: Maximum coverage problem with group budget constraints and applications. In: Jansen, K., Khanna, S., Rolim, J.D.P., Ron, D. (eds.) RANDOM 2004 and APPROX 2004. LNCS, vol. 3122, Springer, Heidelberg (2004)
7. Focacci, F., Lodi, A., Milano, M.: Solving TSP with time windows with constraints. In: Proc. 1999 Int. Conf. on Logic Programming, pp. 515–529. MIT Press, Cambridge (1999)
8. Focacci, F., Lodi, A., Milano, M.: Embedding relaxations in global constraints for solving TSP and TSPTW. Annals of Mathematics and Artificial Intelligence 34(4), 291–311 (2002)
9. Frederickson, G.N., Hecht, M.S., Kim, C.E.: Approximation algorithms for some routing problems. SIAM J. Comput. 7(2), 178–193 (1978)
10. Garg, N.: A 3-approximation for the minimum tree spanning k vertices. In: Proc. 37th IEEE Symp. on Foundations of Computer Science, IEEE Computer Society Press, Los Alamitos (1996)
11. Garg, N.: Saving an epsilon: a 2-approximation for the k-MST problem in graphs. In: Proc. 37th ACM Symp. on Theory of Computing, pp. 396–402. ACM Press, New York (2005)
12. Goemans, M.X., Williamson, D.P.: A general approximation technique for constrained forest problems. SIAM J. Comput. 24, 296–317 (1995)

13. Karp, R.M.: Reducibility among combinatorial problems. In: Miller, R.E., Thatcher, J.W. (eds.) Complexity of Computer Computations, pp. 85–103. Plenum Press, New York (1972)
14. Karuno, Y., Nagamochi, H., Ibaraki, T.: Better approximation ratios for the single-vehicle scheduling problems on line-shaped networks. Networks 39(4), 203–209 (2002)
15. Lawler, E.L., Lenstra, J.K., Rinnooy Kan, A.H.G., Schmoys, D.B.: The Traveling Salesman Problem: A Guided Tour of Combinatorial Optimization. John Wiley & Sons, Chichester (1985)
16. Papadimitriou, C.H., Yannakakis, M.: The traveling salesman problem with distances one and two. Math. Oper. Res. 18(1), 1–11 (1993)
17. Tsitsiklis, J.N.: Special cases of traveling salesman and repairman problems with time windows. Networks 22, 263–282 (1992)

Stochastic Steiner Tree with Non-uniform Inflation

Anupam Gupta[1,*], MohammadTaghi Hajiaghayi[2,**], and Amit Kumar[3,***]

[1] Computer Science Department, Carnegie Mellon University,
Pittsburgh PA 15213
[2] Computer Science Department, Carnegie Mellon University,
Pittsburgh PA 15213
[3] Department of Computer Science and Engineering, Indian Institute of Technology,
New Delhi, India – 110016

Abstract. We study the Steiner Tree problem in the model of two-stage stochastic optimization with *non-uniform inflation factors*, and give a poly-logarithmic approximation factor for this problem. In this problem, we are given a graph $G = (V, E)$, with each edge having two costs c_M and c_T (the costs for Monday and Tuesday, respectively). We are also given a probability distribution $\pi :$ $2^V \to [0, 1]$ over subsets of V, and will be given a *client set* S drawn from this distribution on Tuesday. The algorithm has to buy a set of edges E_M on Monday, and after the client set S is revealed on Tuesday, it has to buy a (possibly empty) set of edges $E_T(S)$ so that the edges in $E_M \cup E_T(S)$ connect all the nodes in S. The goal is to minimize the $c_M(E_M) + \mathbf{E}_{S \leftarrow \pi}[\, c_T(\, E_T(S)\,)\,]$.

We give the first poly-logarithmic approximation algorithm for this problem. Our algorithm builds on the recent techniques developed by Chekuri et al. (FOCS 2006) for multi-commodity Cost-Distance. Previously, the problem had been studied for the cases when $c_T = \sigma \times c_M$ for some constant $\sigma \geq 1$ (i.e., the *uniform* case), or for the case when the goal was to find a tree spanning *all the vertices* but Tuesday's costs were drawn from a given distribution $\widehat{\pi}$ (the so-called "stochastic MST case").

We complement our results by showing that our problem is at least as hard as the single-sink Cost-Distance problem (which is known to be $\Omega(\log \log n)$ hard). Moreover, the requirement that Tuesday's costs are fixed seems essential: if we allow Tuesday's costs to dependent on the scenario as in stochastic MST, the problem becomes as hard as Label Cover (which is $\Omega(2^{\log^{1-\epsilon} n})$-hard). As an aside, we also give an LP-rounding algorithm for the multi-commodity Cost-Distance problem, matching the $O(\log^4 n)$ approximation guarantee given by Chekuri et al. (FOCS 2006).

1 Introduction

This paper studies the Steiner tree problem in the framework of two-stage stochastic approximation, which is perhaps best (albeit a bit informally) described as follows. *On*

---

* Supported in part by an NSF CAREER award CCF-0448095, and by an Alfred P. Sloan Fellowship.
** Supported in part by NSF ITR grant CCR-0122581 (The Aladdin Center).
*** Part of this work was done while the author was at Max-Planck-Institut für Informatik, Saarbrücken, Germany.

M. Charikar et al. (Eds.): APPROX and RANDOM 2007, LNCS 4627, pp. 134–148, 2007.

Monday, we are given a graph with two cost functions c_M and c_T on the edges, and a distribution π predicting future demands; we can build some edges E_M at cost c_M. On Tuesday, the actual demand set S arrives (drawn from the distribution π), and we must complete a Steiner tree on the set S, but any edges E_T bought on Tuesday cost c_T. How can we minimize our expected cost

$$c_M(E_M) + \mathbf{E}_{S \leftarrow \pi}[\, c_T(\, E_T(S)\,)\,]\ ?$$

The Stochastic Steiner tree problem has been studied before in the special case when Tuesday's cost function c_T is a scaled-up version of Monday's costs c_M (i.e., there is an constant *inflation factor* $\sigma > 1$ such that $c_T(e) = \sigma \times c_M(e)$); for this case, constant-factor approximations are known [10, 12, 9]. While these results can be generalized in some directions (see Section 1.1 for a detailed discussion), it has been an open question whether we could handle the case when the two costs c_M and c_T are unrelated. (We will refer to this case as the *non-uniform inflation* case, as opposed to the uniform inflation case when the costs c_M and c_T are just scaled versions of each other).

This gap in our understanding was made more apparent by the fact that many other problems such as Facility Location, Vertex Cover and Set Cover, were all shown to admit good approximations in the non-uniform inflation model [22, 24]: in fact, the results for these problems could be obtained even when the edge cost could depend on the day *as well as on the demand set appearing on Tuesday.*

Theorem 1 (Main Theorem). *There is an $O(\log^2(\min(N, \lambda)) \log^4 n \log \log n)$-approximation algorithm for the two-stage stochastic Steiner tree problem with non-uniform inflation costs with N scenarios, on a graph with n nodes. Here $\lambda = \max_{e \in E} c_T$ $(e)/c_M(e)$, i.e., the maximum inflation over all edges.*

This is the first non-trivial approximation algorithm for this problem. Note that the cost of an edge can either increase or decrease on Tuesday; however, we would like to emphasize that our result holds only when Tuesday's costs c_T do not depend on the materialized demand set S. (Read on for a justification of this requirement).

We also show that the two-stage stochastic Steiner tree problem is at least as hard as the single-source cost-distance problem.

Theorem 2 (Hardness). *The two-stage stochastic Steiner tree problem is at least $\Omega(\log \log n)$-hard unless $NP \subseteq DTIME(n^{\log \log \log n})$.*

The hardness result in the above theorem holds even for the special case of Stochastic Steiner tree when the cost of some edges remain the same between days, and the cost of the remaining edges increases on Tuesday by some universal factor.

Finally, we justify the requirement that Tuesday's costs c_T are fixed by showing that the problem becomes very hard without this requirement. Indeed, we can show the following theorem whose proof is deferred to the journal paper.

Theorem 3. *The two-stage stochastic Steiner tree problem when Tuesday's costs are dependent on the materialized demand is at least $\Omega(2^{\log^{1-\varepsilon} n})$ hard for every fixed $\varepsilon > 0$.*

Finally, we also give an LP-rounding algorithm for the multi-commodity Cost-Distance problem, matching the $O(\log^4 n)$ approximation guarantee given by Chekuri et al. [4]; however, we note that the LP we consider is not the standard LP for the problem.

Our Techniques. Our approach will be to reduce our problem to a more general problem which we call Group-Cost-Distance:

Definition 4 (Group-Cost-Distance). *Consider a (multi)graph $G = (V, E)$ with each edge having a buying cost b_e and a renting cost c_e. Given a set of subsets $S_1, S_2, \ldots, S_N \subseteq V$, find for each i a tree T_i that spans S_i, so as to minimize the total cost*

$$\sum_{e \in \cup_i T_i} b_e + \sum_{i=1}^{N} \sum_{e \in T_i} c_e. \qquad (1.1)$$

Defining $\mathbb{F} = \cup_i T_i$ and $x_e = $ number of trees using edge e, we want to minimize $\sum_{e \in \mathbb{F}} (b_e + x_e\, c_e)$.

The problem can also be called *"multicast" cost-distance*, since we are trying to find multicast trees on each group that give the least cost given the concave cost functions on each edge. Note that when each $S_i = \{s_i, t_i\}$, then we get the (Multicommodity) Cost-Distance problem, for which the first poly-logarithmic approximation algorithms were given only recently [4]; in fact, we build on the techniques used to solve that problem to give the approximation algorithm for the Group-Cost-Distance problem.

1.1 Related Work

Stochastic Problems. For the Stochastic Steiner tree problem in the *uniform inflation* case where all the edge-costs increase on Tuesday by the same amount σ, an $O(\log n)$-approximation was given by Immorlica et al. [16], and constant-factor approximations were given by [10, 12, 9]. These results were extended to handle the case when the inflation factors could be random variables, and hence the probability distribution would be over tuples of the form (demand set S, inflation factor σ) [11, 15].

A related result is known for the Stochastic Minimum Spanning Tree problem, where one has to connect *all* the vertices of the graph. In this case, we are given Monday's costs c_M, and the probability distribution is over possible Tuesday costs c_T. For this problem, Dhamdhere et al. [7] gave an $O(\log n + \log N)$ approximation, where N is the number of scenarios. They solve an LP and randomly round the solution; however, their random rounding seems to crucially require that all nodes need to be connected up, and the idea does not seem to extend to the *Steiner* case. (Note that their problem is incomparable to ours: in this paper, we assume that Monday's and Tuesday's costs were deterministic whereas they do not; on the other hand, in our problem, we get a random set of terminals on Tuesday, whereas they have to connect *all* the vertices which makes their task easier).

Approximation algorithms for several other problems have been given in the non-uniform stochastic setting; see [22, 24]. For a general overview of some techniques used in stochastic optimization, see, e.g., [10, 24]. However, nothing has been known for the Stochastic Steiner tree problem with non-uniform inflation costs.

In many instances of the stochastic optimization problem, it is possible that the number of possible scenarios on Tuesday (i.e., the support of the distribution π) is exponentially large. Charikar et al. [2] gave a useful technique by which we could reduce the problem to a a much smaller number of scenarios (polynomial in the problem size and inflation factors) by random sampling. We shall use this tool in our algorithm as well.

Buy-at-Bulk and Cost-Distance Problems. There has been a huge body of work on so-called *buy-at-bulk* problems which model natural economies-of-scale in allocating bandwidth; see, e.g., [3] and the references therein. The (single-source) Cost-Distance problem was defined by Meyerson, Munagala and Plotkin [20]: this is the case of Group-Cost-Distance with a root $r \in V$, and each $S_i = \{t_i, r\}$. They gave a randomized $O(\log k)$-approximation algorithm where $k = |\cup_i S_i|$, which was later derandomized by Chekuri, Khanna and Naor [5]. (An online poly-logarithmic competitive algorithm was given by Meyerson [19].) These results use a randomized pairing technique that keeps the expected demand at each node constant; this idea does not seem to extend to Group-Cost-Distance. The Multicommodity Cost-Distance problem (i.e., with arbitrary source-sink pairs) was studied by Charikar and Karagiozova [3] who gave an $\exp\{\sqrt{\log n \log \log n}\}$-approximation algorithm. Very recently, this was improved to poly-logarithmic approximation ratio by Chekuri, Hajiaghayi, Kortsarz, and Salavatipour [4] (see also [13, 14]). We will draw on several ideas from these results.

Embedding Graph Metrics into Subtrees. Improving a result of Alon et al. [1], Elkin et al. [8] recently showed the following theorem that every graph metric can be approximated by a distribution over its subtrees with a distortion of $O(\log^2 n \log \log n)$.

Theorem 5 (Subtree Embedding Theorem). *Given a graph* $G = (V, E)$, *there exists a probability distribution* $\mathcal{D}_G$ *over spanning trees of* G *such that for every* $x, y \in V(G)$, *the expected distance* $\mathbf{E}_{T \leftarrow \mathcal{D}_G}[d_T(x,y)] \leq \beta_{EEST}\, d_G(x,y)$ *for* $\beta_{EEST} = O(\log^2 n \log \log n)$.

Note that spanning trees T trivially ensure that $d_G \leq d_T$. The parameter β_{EEST} will appear in all of our approximation guarantees.

2   Reduction to Group Cost-Distance

Note that the distribution π may be given as a black-box, and may be very complicated. However, using a theorem of Charikar, Chekuri, and Pál on using sample averages [2, Theorem 2], we can focus our attention on the case when the probability distribution π is the *uniform* distribution over some N sets $S_1, S_2, \ldots, S_N \subseteq V$, and hence the goal is to compute edge sets $E_0 \doteq E_M$, and $E_1, E_2, \ldots, E_N$ (one for each scenario) such that $E_0 \cup E_i$ contains a Steiner tree on S_i. Scaling the objective function by a factor of N, we now want to minimize

$$N \cdot c_M(E_0) + \sum_{i=1}^{N} c_T(E_i) \tag{2.2}$$

Basically, the N sets will just be N independent draws from the distribution π. We set the value $N = \Theta(\lambda^2 \epsilon^{-5} m)$, where λ is a parameter that measures the "relative cost of" information" and can be set to $\max_e c_T(e)/c_M(e)$ for the purposes of this paper, m is the number of edges in G, and ϵ is a suitably small constant. Let ρ_{ST} be the best known approximation ratio for the Steiner tree problem [23]. The following reduction can be inferred from [2, Theorem 3] (see also [25]):

Lemma 1 (Scenario Reduction). *Given an α-approximation algorithm for the above instance of the stochastic Steiner tree problem with N scenarios, run it independently $\Theta(1/\epsilon)$ times and take the best solution. With constant probability, this gives an $O((1 + \epsilon)\alpha)$-approximation to the original stochastic Steiner tree problem on the distribution π.*

Before we go on, note that E_0 and each of the $E_0 \cup E_i$ are acyclic in an optimal solution. We now give the reduction to **Group-Cost-Distance**. Create a new (multi)graph, whose vertex set is still V. For each edge $e \in E$ in the original graph, we add two parallel edges e_1 (with *buying cost* $b_{e_1} = N \cdot c_M(e)$ and *renting cost* $c_{e_1} = 0$) and e_2 (with *buying cost* $b_{e_2} = 0$ and *renting cost* $c_{e_2} = c_T(e)$). The goal is to find a set of N trees $T_1, \ldots, T_N$, with T_i spanning the set S_i, so as to minimize

$$\sum_{e \in \cup_i T_i} b_e + \sum_{i=1}^N c(T_i). \tag{2.3}$$

It is easily verified that the optimal solution to the two objective functions (2.2) and (2.3) are the same when we define the buying and renting costs as described above. Using Lemma 1, we get the following reduction.

Lemma 2. *An α-approximation for the **Group-Cost-Distance** problem implies an $O(\alpha)$-approximation for the non-uniform Stochastic Steiner tree problem.*

(As an aside, if the distribution π consists of N scenarios listed explicitly, we can do an identical reduction to **Group-Cost-Distance**, but now the value of N need not have any relationship to λ.)

3   Observations and Reductions

Recall that a solution to the **Group-Cost-Distance** problem is a collection of trees T_i spanning S_i; their union is $\mathbb{F} = \cup_i T_i$, and x_e is the number of trees that use edge e. Note that if we were just given the set $\mathbb{F} \subseteq E$ of edges, we could use the ρ_{ST}-approximation algorithm for finding the minimum cost Steiner tree to find trees T_i' such that $c(T_i') \leq \rho_{ST} c(T_i)$ for any tree $T_i \subseteq \mathbb{F}$ spanning S_i. Let us define

$$\mathrm{cost}(\mathbb{F}) \doteq b(\mathbb{F}) + \sum_i c(T_i); \tag{3.4}$$

where $T_i \subseteq \mathbb{F}$ is the minimum cost Steiner tree spanning S_i. We use $\mathsf{OPT} = \mathbb{F}^*$ to denote the set of edges used in an optimal solution for the **Group-Cost-Distance** instance, and hence $\mathrm{cost}(\mathsf{OPT})$ is the total optimal cost. Henceforth, we may specify a solution to an instance of the **Group-Cost-Distance** problem by just specifying the set of edges $\mathbb{F} = \cup_i T_i$, where T_i is the tree spanning S_i in this solution.

As an aside, note that $\text{cost}(\mathbb{F})$ is the optimal cost of *any* solution using the edges from $\mathbb{F} \subseteq E$; computing $\text{cost}(\mathbb{F})$ is hard given the set $\mathbb{F}$, but that is not a problem since it will be used only as an accounting tool. Of course, given $\mathbb{F}$, we can build a solution to the Group-Cost-Distance problem of cost within a ρ_{ST} factor of $\text{cost}(\mathbb{F})$.

We will refer to the sets S_i as *demand groups*, and a vertex in one of these groups as a *demand vertex*. For simplicity, we assume that for all i, $|S_i|$ is a power of 2; this can be achieved by replicating some vertices.

3.1 The Pairing Cost-Distance Problem: A Useful Subroutine

A *pairing* of any set A is a perfect matching on the graph $(A, \binom{A}{2})$. The following *tree-pairing lemma* has become an indispensable tool in network design problems (see [21] for a survey):

Lemma 3 ([18]). *Let T' be an arbitrary tree and let $v_1, v_2, \ldots, v_{2q}$ be an even number of vertices in T'. There exists a pairing of the v_i into q pairs so that the unique paths joining the respective pairs are edge-disjoint.*

Let us define another problem, whose input is the same as that for Group-Cost-Distance.

Definition 6 (Pairing Cost-Distance). *Given a graph $G = (V, E)$ with buy and rent costs b_e and c_e on the edges, and a set of demand groups $\{S_i\}_i$, the Pairing Cost-Distance problem seeks to find a pairing $\mathcal{P}_i$ of the nodes in S_i, along with a path connecting each pair of nodes $(x, y) \in \mathcal{P}_i$.*

Let $\mathbb{F}'$ be the set of edges used by these paths, and let x'_e be the number of pairs using the edge $e \in \mathbb{F}'$, then the cost of a solution is $\sum_{e \in \mathbb{F}'}(b_e + x'_e\, c_e)$. As before, given the set $\mathbb{F}'$, we can infer the best pairing that only uses edges in $\mathbb{F}'$ by solving a min-cost matching problem: we let $\text{cost}'(\mathbb{F}')$ denote this cost, and let OPT' be the optimal solution to the Pairing Cost-Distance instance.[1] So, again, we can specify a solution to the Pairing Cost-Distance problem by specifying this set $\mathbb{F}'$. The following lemma relates the costs of the two closely related problems:

Lemma 4. *For any instance, the optimal cost $\text{cost}'(\text{OPT}')$ for Pairing Cost-Distance is at most the optimal cost $\text{cost}(\text{OPT})$ for Group-Cost-Distance.*

Proof. Let $\mathbb{F}$ be the set of edges bought by OPT for the Group-Cost-Distance problem. We construct a solution for the Pairing Cost-Distance problem. Recall that OPT builds a Steiner tree T_i spanning S_i using the edges in $\mathbb{F}$. By Lemma 3, we can pair up the demands in S_i such that the unique paths between the pairs in T_i are pair-wise edge-disjoint. This gives us a solution to Pairing Cost-Distance, which only uses edges in $\mathbb{F}$, and moreover, the number of times an edge is used is at most x_e, ensuring a solution of cost at most $\text{cost}(\text{OPT})$.

---

[1] An important but subtle point: note that x'_e counts the number of paths that pass over an edge. It may happen that all of these paths may connect pairs that belong to the *same* set S_i, and cost might have to pay for this edge only once: regardless, we pay multiple times in cost'.

Lemma 5 (Reducing Group-Cost-Distance to Pairing Cost-Distance). *If there is an algorithm $\mathcal{A}$ for Pairing Cost-Distance that returns a solution $\mathbb{F}'$ with $\text{cost}'(\mathbb{F}') \leq \alpha\,\text{cost}(\text{OPT})$, we get an $O(\alpha \log n)$-approximation for the Group-Cost-Distance problem.*

Note that $\mathcal{A}$ is not a true approximation algorithm for Pairing Cost-Distance, since we compare its performance to the optimal cost for Group-Cost-Distance; hence we will call it an α-*pseudo-approximation algorithm*.

Proof. In each iteration, when we connect up pairs of nodes in S_i, we think of taking the traffic from one of the nodes and moving it to the other node; hence the number of "active" nodes in S_i decreases by a factor of 2. This can only go on for $O(\log n)$ iterations before all the traffic reaches one node in the group, ensuring that the group is connected using these pairing paths. Since we pay at most $\alpha\,\text{cost}(\text{OPT})$ in each iteration, this results in an $O(\alpha \log n)$ approximation for Group-Cost-Distance.

4   An Algorithm for Pairing Cost-Distance

In this section, we give an LP-based algorithm for Pairing Cost-Distance; by Lemma 5 this will imply an algorithm for Group-Cost-Distance, and hence for Stochastic Steiner Tree.

We will prove the following result for Pairing Cost-Distance (PCD):

Theorem 7 (Main Result for Pairing Cost-Distance). *There is an $\alpha = O(\beta_{EEST} \log^2 H \cdot \log n)$ pseudo-approximation algorithm for the Pairing Cost-Distance problem, where $H = \max\{\sum_i |S_i|, n\}$.*

Since $H = O(Nn)$ and we think of $N \geq n$, this gives us an $O(\log^2 N \log^3 n \log \log n)$ pseudo-approximation. Before we present the proof, let us give a high-level sketch. The algorithm for Pairing Cost-Distance follows the general structure of the proofs of Chekuri et al. [4]; the main difference is that the problem in [4] already comes equipped with $\{s, t\}$-pairs that need to be connected, whereas our problem also requires us to figure out which pairs to connect—and this requires a couple of new ingredients.

Loosely, we first show the existence of a "good" *low density* pairing solution—this is a solution that only connects up *some* pairs of nodes in *some* of the sets S_i (instead of pairing up *all* the nodes in *all* the S_i's), but whose "density" (i.e., ratio of cost to pairs-connected) is at most a β_{EEST} factor of the density of OPT. Moreover, the "good" part of this solution will be that all the paths connecting the pairs will pass through a single "junction" node. The existence of this single junction node (which can now be thought of as a *sink*) makes this problem look somewhat like a low-density "pairing" version of single-sink cost-distance. We show how to solve this final subproblem within an $O(\log H \cdot \log n)$ factor of the best possible such solution, which is at most β_{EEST} times OPT's density. Finally, finding these low-density solutions iteratively and using standard set-cover arguments gives us a Pairing Cost-Distance solution with cost $O(\beta_{EEST} \log^2 H \cdot \log n)\,\text{cost}(\text{OPT})$, which gives us the claimed theorem.

4.1 Defining the Density

Consider an instance of the **Pairing Cost-Distance** (PCD) problem in which the current demand sets are $\widehat{S}_i$. Look at a partial PCD solution that finds for each set $\widehat{S}_i$ some set $\mathcal{P}_i$ of $t_i \geq 0$ mutually disjoint pairs $\{x^i_j, y^i_j\}^{t_i}_{j=1}$ along with paths P^i_j connecting these pairs. Let $\mathcal{P} = \cup_i \mathcal{P}_i$ be the (multi)set of all these $t = \sum_i t_i$ paths. We shall use $\mathcal{P}$ to denote both the pairs in it and the paths used to connect them. Denote the cost of this partial solution by $\mathrm{cost}'(\mathcal{P}) = b(\cup_{P \in \mathcal{P}} P) + \sum_{P \in \mathcal{P}} c(P)$. Let $|\mathcal{P}|$ be the number of pairs being connected in the partial solution. The *density* of the partial solution $\mathcal{P}$ is defined as $\frac{\mathrm{cost}'(\mathcal{P})}{|\mathcal{P}|}$. Recall that $H = \max\{\sum_i |S_i|, n\}$ is the total number of terminals in the **Pairing Cost-Distance** instance.

Definition 8 (f-dense **Partial PCD** solution). *Consider an instance $\mathcal{I}$ of the **Pairing Cost-Distance** problem: a **Partial PCD** solution $\mathcal{P}$ is called f-dense if*

$$\frac{\mathrm{cost}'(\mathcal{P})}{|\mathcal{P}|} \leq f \cdot \frac{\mathrm{cost}(\mathrm{OPT})}{H(\mathcal{I})},$$

where $H(\mathcal{I})$ is the total number of terminals in the instance $\mathcal{I}$.

Theorem 9. *Given an algorithm to find f-dense **Partial PCD** solutions, we can find an $O(f \log H)$-pseudo-approximation to **Pairing Cost-Distance**.*

To prove this result, we will use the following theorem which can be proved by standard techniques (see e.g., [17]), and whose proof we omit.

Theorem 10 (**Set Covering Lemma**). *Consider an algorithm working in iterations: in iteration ℓ it finds a subset $\mathcal{P}_\ell$ of paths connecting up $|\mathcal{P}_\ell|$ pairs. Let H_ℓ be the number of terminals remaining before iteration ℓ. If for every ℓ, the solution $\mathcal{P}_\ell$ is an f-dense solution with $\mathrm{cost}'(\mathcal{P}_\ell)/|\mathcal{P}_\ell| \leq f \cdot \frac{\mathrm{cost}(\mathrm{OPT})}{H_\ell}$, then the total cost of the solution output by the algorithm is at most $f \cdot (1 + \ln H) \cdot \mathrm{cost}(\mathrm{OPT})$.*

In the next section, we will show how to find a **Partial PCD** solution which is $f = O(\beta_{EEST} \log H \cdot \log n)$-dense.

4.2 Finding a Low-Density **Partial PCD** Solution

We now show the existence of a partial pairing $\mathcal{P}$ of demand points which is β_{EEST}-dense, and where all the pairs in $\mathcal{P}$ will be routed on paths that pass through a common *junction* point. The theorems of this section are essentially identical to corresponding theorems in [4].

Theorem 11. *Given an instance of **Pairing Cost-Distance** on $G = (V, E)$, there exists a solution $\mathbb{F}'$ to this instance such that (a) the edges in $\mathbb{F}'$ induce a forest, (b) $\mathbb{F}'$ is a subset of OPT' and hence the buying part of $\mathrm{cost}'(\mathbb{F}') \leq b(\mathrm{OPT}')$, and (c) the renting part of $\mathrm{cost}'(\mathbb{F}')$ is at most $O(\beta_{EEST})$ times the renting part of $\mathrm{cost}'(\mathrm{OPT}')$.*

Proof Sketch. The above theorem can be proved by dropping all edges in $E \setminus \mathrm{OPT}'$ and approximating the metric generated by rental costs c_e in each resulting component by a random subtree drawn from the distribution guaranteed by Theorem 5. We chose a subset of OPT$'$, and hence the buying costs cannot be any larger. Since the expected distances increase by at most β_{EEST}, the expected renting cost increases by at most this factor. And since this holds for a random forest, by the probabilistic method, there must exist one such forest with these properties.   □

Definition 12 (Junction Tree). *Consider a solution to the Partial PCD problem with paths $\mathcal{P}$, and which uses the edge set $\mathbb{F}'$. The solution is called a junction tree if the subgraph induced by $\mathbb{F}'$ is a tree and there is a special root vertex r such that all the paths in $\mathcal{P}$ contain the root r.*

As before, the *density* of a solution $\mathcal{P}$ is the ratio of its cost to the number of pairs connected by it. We can now prove the existence of a low-density junction tree for the Partial PCD problem. The proof of this lemma is deferred to the journal paper.

Lemma 6 (Low-Density Existence Lemma). *Given an instance of Pairing Cost-Distance problem, there exists a solution to the Partial PCD solution which is a junction tree and whose density is $2\beta_{EEST} \cdot \frac{\text{cost}'(\text{OPT}')}{H} \leq 2\beta_{EEST} \cdot \frac{\text{cost}(\text{OPT})}{H}$.*

In the following section, we give an $O(\log H \cdot \log n)$-approximation algorithm for finding a junction tree with minimum density. Since we know that there is a "good" junction tree (by Lemma 6), we can combine that algorithm with Lemma 6 to get a Partial PCD solution which is $f = O(\beta_{EEST} \log H \cdot \log n)$-dense.

4.3   Finding a Low-Density Junction Tree

In this section, we give an LP-rounding based algorithm for finding a junction tree with density at most $O(\log H \cdot \log n)$ times that of the min-density junction tree. Our techniques continue to be inspired by [4]; however, in their paper, they were given a fixed pairing by the problem, and had to figure out which ones to connect up in the junction tree. In our problem, we have to both figure out the pairing, and then choose which pairs from this pairing to connect up; we have to develop some new ideas to handle this issue.

The Linear-Programming Relaxation. Recall the problem: we are given sets S_i, and want to find some partial pairings for each of the sets, and then want to route them to some root vertex r so as to minimize the density of the resulting solution. We will assume that we know r (there are only n possibilities), and that the sets S_i are disjoint (by duplicating nodes as necessary).

Our LP relaxation is an extension of that for the Cost-Distance problem given by Chekuri, Khanna, and Naor [5]. The intuition is based on the following: given a junction-tree solution $\mathbb{F}'$, let $\mathcal{P}'$ denote the set of pairs connected via the root r. Now $\mathbb{F}'$ can also be thought of as a solution to the Cost-Distance problem with root r and the terminal set $\cup_{(u,v)\in\mathcal{P}'}\{u,v\}$. Furthermore, the cost $\text{cost}'(E')$ is the same as the optimum of the Cost-Distance problem. (This is the place when the definition of cost' becomes crucial—we can use the fact that the cost measure cost' is paying for the number of *paths* using an edge, not the number of *groups* using it).

Let us write an IP formulation: let $\mathcal{S} = \cup_i S_i$ denote the set of all terminals. For each demand group S_i and each pair of vertices $u, v \in S_i$, the variable z_{uv} indicates whether we match vertices u and v in the junction tree solution or not. To enforce a matching, we ask that $\sum_u z_{uv} = \sum_v z_{uv} \leq 1$. For each $e \in E$, the variable y_e denotes whether the edge e is used; for each path from some vertex u to the root r, we let f_P denote whether P is the path used to connect u to the root. Let $\mathbb{P}_u$ be the set of paths from u to the root r. Clearly, we want $\sum_{P \in \mathbb{P}_u} f_P \leq x_e$ for all $e \in P$. Moreover, $\sum_{P \in \mathbb{P}_u} f_P \geq \sum_{v \in S_i} z_{uv}$ for each $u \in S_i$, since if the node u is paired up to someone, it must be routed to the root. Subject to these constraints (and integrality), we want to minimize

$$\min \frac{\sum_{e \in E} b_e x_e + \sum_{u \in \mathcal{S}} \sum_{P \in \mathbb{P}_u} c(P) f_P}{\sum_{i=1}^{N} \sum_{u,v \in S_i} z_{uv}} \tag{4.5}$$

It is not hard to check that this is indeed an ILP formulation of the min-density junction tree problem rooted at r. As is usual, we relax the integrality constraints, guess the value $M \geq 1$ of the denominator in the optimal solution, and get:

$$\min \quad \sum_{e \in E} b_e x_e + \sum_{u \in \mathcal{S}} \sum_{P \in \mathbb{P}_u} c(P) f_P \tag{LP1}$$

$$\text{s.t.} \quad \sum_{i=1}^{N} \sum_{u,v \subset S_i} z_{uv} = M$$

$$\sum_{P \in \mathbb{P}_u : P \ni e} f_P \leq x_e \qquad \text{for all } u \in \mathcal{S}$$

$$\sum_{P \in \mathbb{P}_u} f_P \geq \sum_{v \in S_i} z_{uv} \qquad \text{for all } u \in S_i, i \in [1..N]$$

$$\sum_{v \in S_i} z_{uv} \leq 1 \qquad \text{for all } u \in S_i, i \in [1..N]$$

$$x_e, f_P, z_{uv} = z_{vu} \geq 0$$

We now show that the integrality gap of the above LP is small.

Theorem 13. *The integrality gap of (LP1) is $O(\log H \cdot \log n)$. Hence there is an $O(\log H \cdot \log n)$-approximation algorithm for finding the minimum density junction tree solution for a given* **Partial PCD** *instance.*

Proof. Consider an optimal fractional solution given by (x^*, f^*, z^*) with value LP^*. We start off with $z = z^*$, and will alter the values in the following proof. Consider each set S_i, and let $w_i = \sum_{u,v \in S_i} z_{uv}^*$ be the total size of the fractional matching within S_i. We find an approximate maximum weight cut in the complete graph on the nodes S_i with edge weights z_{uv}—this gives us a bipartite graph which we denote by B_i; we zero out the z_{uv} values for all edges $(u, v) \in S_i \times S_i$ that do not belong to the cut B_i. How does this affect the LP solution? Since the weight of the max cut we find is at least $w_i/2$, we are left with a solution where $\sum_{i=1}^{N} \sum_{u,v \in S_i} z_{uv} \geq M/2$ (and hence that constraint is *almost* satisfied).

Now consider the edges in the bipartite graph B_i, with edge weights z_{uv}—if this graph has a cycle, by alternatively increasing and decreasing the values of z variables along this even cycle by ϵ in one of the two directions, we can make $z_{u'v'}$ zero for at least one edge (u', v') of this cycle without increasing the objective function. (Note that this operation maintains all the LP constraints.) We then delete the edge (u', v') from B_i, and repeat this operation until B_i is a forest.

Let us now partition the edges of the various such forests $\{B_i\}_{i=1}^N$ into $O(\log H)$ classes based on their current z values. Let $Z_{\max} = \max_{u,v} z_{uv}$, and define $p = 1 + 2\lceil \log H \rceil = O(\log H)$. For each $a \in [0..p]$, define the set C_a to contain all edges (u, v) with $Z_{\max}/2^{a+1} < z_{uv} \le Z_{\max}/2^a$; note that the pairs $(u, v) \notin \cup_{a=1}^p C_a$ have a cumulative z_{uv} value of less than (say) $Z_{\max}/4 \le M/4$. Hence, by an easy averaging argument, there must be a class C_a with $\sum_{(u,v) \in C_a} z_{uv} \ge \Omega(M/\log H)$. Define $Z_a = Z_{\max}/2^a$; hence $|C_a|\, Z_a = \Omega(M/\log H)$.

Since we have restricted our attention to pairs in C_a, we can define $B_{ia} = B_i \cap C_a$, which still remains a forest. For any tree T in this forest, we apply the tree-pairing lemma on the nodes of the tree T, and obtain a matching C'_{ia} on S_i of size $\lceil |V(T)|/2\rceil$. Defining $C'_a = \cup_i C'_{ia}$, we get that $|C'_a|\, Z_a = \Omega(M/\log H)$ as well.

Finally, we create the following instance of the **Cost-Distance** problem. The terminal set contains all the terminals that are matched in C'_a, and the goal is to connect them to the root. Set the values of the variables $\tilde{f}_P = f_P^*/Z_a$ and $\tilde{x}_e = x_e/Z_a$. These settings of variables satisfy the LP defined by [5] for the instance defined above. The integrality gap for this LP is $O(\log n)$ and so we get a solution with cost' at most $O(\log n) \cdot LP^*/Z_a$. However, this connects up $|C'_a| = \Omega(\frac{M}{Z_a \log H})$ pairs, and hence the density is $O(\log H \cdot \log n)\frac{LP^*}{M}$, hence proving the theorem.

5   Reduction from Single-Sink Cost-Distance

Theorem 14. *If there is a polynomial time α-approximation algorithm for the two-stage stochastic Steiner tree problem, then there is a polynomial time α-approximation algorithm for the single-source cost-distance problem.*

The hardness result of Theorem 2 follows by combining the above reduction with a result of Chuzhoy et al. [6] that the single-source cost-distance problem cannot be approximated to better than $\Omega(\log \log n)$ ratio under complexity theory assumptions.

Proof of Theorem 14. Consider an instance of the **Cost-Distance** problem: we are given a graph $G = (V, E)$, a root vertex r, and a set S of terminals. Each edge e has buying cost b_e and rental cost c_e. A solution specifies a set of edges E' which spans the root and all the nodes in S: if the shortest path in (V, E') from $u \in S$ to r is P_u, then the cost of the solution is $b(E') + \sum_{u \in S} c(P_u)$. We take any edge with buying cost b_e and rental cost c_e, and subdivide this edge into two edges, giving the first of these edges a buying cost of b_e and rental cost ∞, and the other edge gets buying cost ∞ and rental cost c_e.

We reduce this instance to the two-stage stochastic Steiner tree problem where the scenarios are explicitly specified. The instance of the stochastic Steiner tree problem has the same graph. There are $|S|$ scenarios (each with probability $1/|S|$), where each scenario has exactly one unique demand from S. For an edge e which can only be bought, we set $c_M(e) = b_e$ and $c_T(e) = \infty$; hence any such edge must necessarily be bought on Monday, if at all. For an e which can only be rented, we set $c_M(e) = c_T(e) = |S| \cdot c_e$; note that there is no advantage to buying such an edge on Monday, since we can buy it on Tuesday for the same cost if needed — in the rest, we will assume that any optimal solution is lazy in this way.

It can now be verified that there is an optimal solution to the Stochastic Steiner Tree problem where the subset F' of edges bought in the first stage are only of the former type, and we have to then buy the "other half" of these first-stage edges to connect to the root in the second stage, hence resulting in isomorphic optimal solutions.   □

6   Summary and Open Problems

In this paper, we gave a poly-logarithmic approximation algorithm for the stochastic Steiner tree problem in the non-uniform inflation model. Several interesting questions remain open. When working in the black-box model, we apply the scenario reduction method of Charikar et al. [2], causing the resulting number of scenarios N to be a polynomial function of the parameter λ, which is bounded by the maximum inflation factor on any edge. Hence our running time now depends on λ, and the approximation ratio depends on $\log \lambda$. Can we get results where these measures depend only on the number of nodes n, and not the number of scenarios N?

In another direction, getting an approximation algorithm with similar guarantees for (a) the stochastic Steiner *Forest* problem, i.e., where each scenario is an instance of the Steiner forest problem, or (b) the k-stage stochastic Steiner tree problem, remain open.

References

1. Alon, N., Karp, R.M., Peleg, D., West, D.: A graph-theoretic game and its application to the k-server problem. SIAM J. Comput. 24(1), 78–100 (1995)
2. Charikar, M., Chekuri, C., Pál, M.: Sampling bounds for stochastic optimization. In: Chekuri, C., Jansen, K., Rolim, J.D.P., Trevisan, L. (eds.) APPROX 2005 and RANDOM 2005. LNCS, vol. 3624, pp. 257–269. Springer, Heidelberg (2005)
3. Charikar, M., Karagiozova, A.: On non-uniform multicommodity buy-at-bulk network design. In: STOC, pp. 176–182. ACM Press, New York (2005)
4. Chekuri, C., Hajiaghayi, M., Kortsarz, G., Salavatipour, M.R.: Approximation algorithms for non-uniform buy-at-bulk network design problems. In: FOCS (2006)
5. Chekuri, C., Khanna, S., Naor, J.S.: A deterministic algorithm for the cost-distance problem. In: SODA, pp. 232–233 (2001)
6. Chuzhoy, J., Gupta, A., Naor, J.S., Sinha, A.: On the approximability of network design problems. In: SODA, pp. 943–951 (2005)
7. Dhamdhere, K., Ravi, R., Singh, M.: On two-stage stochastic minimum spanning trees. In: Jünger, M., Kaibel, V. (eds.) Integer Programming and Combinatorial Optimization. LNCS, vol. 3509, pp. 321–334. Springer, Heidelberg (2005)
8. Elkin, M., Emek, Y., Spielman, D.A., Teng, S.-H.: Lower-stretch spanning trees. In: STOC, pp. 494–503. ACM Press, New York (2005)
9. Gupta, A., Pál, M.: Stochastic Steiner trees without a root. In: Caires, L., Italiano, G.F., Monteiro, L., Palamidessi, C., Yung, M. (eds.) ICALP 2005. LNCS, vol. 3580, pp. 1051–1063. Springer, Heidelberg (2005)
10. Gupta, A., Pál, M., Ravi, R., Sinha, A.: Boosted sampling: Approximation algorithms for stochastic optimization problems. In: STOC, pp. 417–426 (2004)

11. Gupta, A., Pál, M., Ravi, R., Sinha, A.: What about Wednesday? approximation algorithms for multistage stochastic optimization. In: Chekuri, C., Jansen, K., Rolim, J.D.P., Trevisan, L. (eds.) APPROX 2005 and RANDOM 2005. LNCS, vol. 3624, pp. 86–98. Springer, Heidelberg (2005)
12. Gupta, A., Ravi, R., Sinha, A.: An edge in time saves nine: LP rounding approximation algorithms for stochastic network design. In: FOCS, pp. 218–227 (2004)
13. Hajiaghayi, M.T., Kortsarz, G., Salavatipour, M.R.: Approximating buy-at-bulk k-steiner trees. In: Electronic Colloquium on Computational Complexity (ECCC) (2006)
14. Hajiaghayi, M.T., Kortsarz, G., Salavatipour, M.R.: Polylogarithmic approximation algorithm for non-uniform multicommodity buy-at-bulk. In: Electronic Colloquium on Computational Complexity (ECCC) (2006)
15. Hayrapetyan, A., Swamy, C., Tardos, E.: Network design for information networks. In: SODA, pp. 933–942 (2005)
16. Immorlica, N., Karger, D., Minkoff, M., Mirrokni, V.: On the costs and benefits of procrastination: Approximation algorithms for stochastic combinatorial optimization problems. In: SODA, pp. 684–693 (2004)
17. Johnson, D.S.: Approximation algorithms for combinatorial problems. J. Comput. System Sci. 9, 256–278 (1974)
18. Klein, P., Ravi, R.: A nearly best-possible approximation algorithm for node-weighted Steiner trees. J. Algorithms 19(1), 104–115 (1995)
19. Meyerson, A.: Online algorithms for network design. In: SPAA, pp. 275–280 (2004)
20. Meyerson, A., Munagala, K., Plotkin, S.: Cost-distance: Two metric network design. In: FOCS, pp. 624–630 (2000)
21. Ravi, R.: Matching based augmentations for approximating connectivity problems. In: Correa, J.R., Hevia, A., Kiwi, M. (eds.) LATIN 2006. LNCS, vol. 3887, pp. 13–24. Springer, Heidelberg (2006)
22. Ravi, R., Sinha, A.: Hedging uncertainty: Approximation algorithms for stochastic optimization problems. In: IPCO, pp. 101–115 (2004)
23. Robins, G., Zelikovsky, A.: Tighter bounds for graph Steiner tree approximation. SIAM J. Discrete Math. 19(1), 122–134 (2005)
24. Shmoys, D., Swamy, C.: Stochastic optimization is (almost) as easy as deterministic optimization. In: FOCS, pp. 228–237 (2004)
25. Swamy, C., Shmoys, D.B.: Sampling-based approximation algorithms for multi-stage stochastic. In: FOCS, pp. 357–366 (2005)

A   An LP-Based Algorithm for Multicommodity Cost-Distance

In their paper, Chekuri et al. [4] give an approximation algorithm for the Multicommodity Cost-Distance problem using a combination of combinatorial and LP-based techniques. We now give an LP-based algorithm for the Multicommodity Cost-Distance problem with an approximation guarantee of $O(\log^4 n)$, thus matching the result of [4] using a different approach. (Note that this is not the standard LP for the Multicommodity Cost-Distance problem, which we do not currently know how to round).

Take an instance of Multicommodity Cost-Distance: let $S_i = \{s_i, t_i\}$ be each terminal pair we want to connect, and $S = \cup_i S_i$. Consider the following linear program:

$$Z^*_{MCD} = \min \quad \sum_{e \in E} b_e \, z_e(p) + \sum_{u \in S} \sum_{P \in \mathbb{P}_{u*}} c(P) \, f_P \qquad \text{(LP-MCD)}$$

$$\text{s.t.} \qquad \sum_{p \in V} x(u,p) \geq 1 \qquad\qquad \text{for all } u \in S$$

$$\sum_{P \in \mathbb{P}_{u,p}} f_P \geq x(u,p) \qquad\qquad \text{for all } u \in S, p \in V$$

$$\sum_{P \in \mathbb{P}_{u,p}:P \ni e} f_P \leq z_e(p) \qquad\qquad \text{for all } u \in S, p \in V$$

$$x(s_i, p) = x(t_i, p) \qquad\qquad \text{for all } i, p \in V$$

$$x(u,p), f_P, z_e(p) \geq 0$$

To understand this, consider the ILP obtained by adding the constraints that the variables are all $\{0, 1\}$. Each solution assigns each terminal u to one junction node p (specified by $x(u,p)$) such that each s_i-t_i pair is assigned to the same junction (since $x(s_i, p) = x(t_i, p)$). It then sends the unit flow at the terminal using flow f_P to this junction p, ensuring that if an edge e is used, it has been purchased (i.e., $z_e(p) = 1$). The unusual part of this ILP is that the buying costs of the edge are paid not just once, but *once for each junction* that uses this edge. (If all the variables $z_e(p)$ would be replaced by a single variable z_e, we would get back the standard ILP formulation of Multicommodity Cost-Distance).

Lemma 7. *Each integer solution to (LP-MCD) is an solution to the* Multicommodity Cost-Distance *problem with the same cost. Moreover, any solution to the* Multicommodity Cost-Distance *problem with cost* OPT *can be converted into a solution for (LP-MCD) with cost at most* $O(\log n) \times$ OPT.

While we defer the proof of this theorem to the final version of the paper, we note that the former statement is trivial from the discussion above, and the second statement follows from the paper of Chekuri et al. [4, Theorem 6.1]. Hence, it suffices to show how to round (LP-MCD).

A.1 Rounding the LP for Multicommodity Cost-Distance

In this section, we show how to round (LP-MCD) to get an integer solution with cost $O(\log^3 n) \times Z^*_{MCD}$. The basic idea is simple:

- Given a fractional solution, we first construct a (feasible) partial solution with $\hat{x}(u,p) \in \{0, 1\}$ but with fractional f_P and $z_e(p)$ values. The expected cost of this partial solution is $O(\log^2 n) \times Z^*_{MCD}$.
- Since the $\hat{x}$ values are integral in this partial solution, each terminal u sends unit flow to some set of junctions p (chosen by $\hat{x}(u,p)$), such that each pair s_i, t_i sends flow to the same set of junctions. We can choose one of these junctions arbitrarily, and hence each terminal pair is assigned to some junction.
- Finally, note that all the flows to any junction p can be supported by fractional $z_e(p)$ capacities; these capacities are entirely separate from the capacities $z_e(p')$ for any other terminal p' in the graph. Hence the problem decomposes into several single-sink problems (one for each junction), and we can use the rounding algorithm of Chekuri et al. [5] to round the solutions for each of the junctions p *independently* to obtain integer flows f_P and integer capacities $z_e(p)$ while losing only another $O(\log n)$ in the approximation.

The last two steps are not difficult to see, so we will focus on the first rounding step (to make the $x(u,p)$ variables integral). The rounding we use is a fairly natural one, though not the obvious one.

1. Modify the LP solution to obtain a feasible solution where all $x(u,p)$ values lie between $1/n^3$ and 1, and each $x(u,p) = 2^{\eta(u,p)}$ for $\eta(u,p) \in Z_{\geq 0}$. This can be done losing at most $O(1)$ times the LP cost.
2. For each junction p, pick a random threshold $T_p \in_R [1/n^3, 1]$ independently and uniformly at random. Then round up f_P for all $P \in \mathcal{P}_{*p}$, and all $z_e(p)$ by a factor of $1/T$. Note that if $x(u,p) \geq T$, then we can send at least 1 unit of (fractional) flow from u to p using these scaled-up capacities $z_e(p)$. We set $\widehat{x}(u,p) = 1$, and call the terminal u *satisfied* by the junction p. (Note that since $x(s_i, p) = x(t_i, p)$, if s_i is satisfied by u then so is t_i.)
3. We repeat the threshold rounding scheme in the previous step $O(\log n)$ times. We return a solution $\{\widehat{x}, \widehat{f}, \widehat{z}\}$, where $\widehat{x}(u,p)$ is as described above, and $\widehat{f}, \widehat{z}$ are obtained by summing up all the scaled-up values of f and z over all the $O(\log n)$ steps.

We now show that each terminal has been satisfied with high probability, and that the expected cost of each solution produced in the second step above is $O(\log n) \times Z^*_{MCD}$.

Lemma 8. *The expected cost of the second step above is* $O(\log n) \times Z^*_{MCD}$.

Proof. The variable $z_e(p)$ gets scaled up to $z_e(p)/T$, whose expected value is $\int_{T=1/n^3}^{1} z_e(p)/T \, dT = z_e(p) \times O(\log N)$. The same holds for the f_P variables.

Lemma 9. *Each terminal u is satisfied by a single round of threshold rounding with constant probability, and hence will be satisfied in at least one of $O(\log n)$ rounds with probability $1 - 1/\operatorname{poly}(n)$.*

Lemma 10. *The above scheme gives a feasible solution $\{\widehat{x}, \widehat{f}, \widehat{z}\}$ to (LP-MCD) with $\widehat{x} \in \{0,1\}$ and cost at most $O(\log^2 n) \times Z^*_{MCD}$.*

Finally, looking at this solution, decomposing it for each junction p, and using the [5] rounding scheme converts the flows $\widehat{f}$ and the capacities $\widehat{z}$ to integers as well. Combining this with Lemma 7 gives us another proof for the following result.

Theorem 15. *There is an $O(\log^4 n)$ approximation algorithm for the Multicommodity Cost-Distance problem based on LP rounding.*

On the Approximation Resistance of a Random Predicate

Johan Håstad

Royal Institute of Technology, Stockholm, Sweden

Abstract. A predicate is approximation resistant if no probabilistic polynomial time approximation algorithm can do significantly better then the naive algorithm that picks an assignment uniformly at random. Assuming that the Unique Games Conjecture is true we prove that most Boolean predicates are approximation resistant.

1 Introduction

We consider constraint satisfaction problems (CSPs) over the Boolean domain. In our model a problem is defined by a k-ary predicate P and an instance is given by a list of k-tuples of literals. The task is to find an assignment to the variables such that all the k-bit strings resulting from the list of k-tuples of literals under the assignment satisfy the predicate P. In this paper we focus on Max-CSPs which are optimization problems where we try to satisfy as many constraints as possible.

The most famous such problem is probably Max-3-Sat where $k = 3$ and P is simply the disjunction of the three bits. Another problem that (almost) falls into this category is Max-Cut, in which $k = 2$ and P is non-equality. In traditional Max-Cut we do not allow negations among the literals and if we do allow negation the problem becomes Max-E2-Lin-2, linear equations modulo 2 with exactly two variables in each equation.

It is a classical result that most Boolean CSPs are NP-complete. Already in 1978 Schaefer [12] gave a complete characterization giving only 5 classes for which the problem is in P while establishing NP-completeness in the other cases.

Of course if a CSP is NP-complete, the corresponding Max-CSP is NP-hard. The converse is false and several of Schaefer's easy satisfiability problems are in fact NP-hard as optimization problems. We turn to study approximation algorithms. An algorithm is here considered to be a C-approximation if it, on each input, finds an assignment with an objective value that is within a factor C of the optimal solution. We allow randomized approximation algorithms and in such a case it is sufficient that the expected value, over the random choices of the algorithm, of the objective value satisfies the desired bound.

To define what is non-trivial is a matter of taste but hopefully there is some consensus that the following algorithm is trivial: Without looking at the instance pick a random value for each variable. We say that an approximation ratio is non-trivial if it gives a value of C that is better than the value obtained by this

M. Charikar et al. (Eds.): APPROX and RANDOM 2007, LNCS 4627, pp. 149–163, 2007.
© Springer-Verlag Berlin Heidelberg 2007

trivial algorithm. We call a predicate *approximation resistant* if it is NP-hard to achieve a non-trivial approximation ratio.

It is perhaps surprising but many CSPs are approximation resistant and one basic example is Max-3-Sat [6]. The famous algorithm of Goemans and Williamson [1] shows that Max-Cut is not approximation resistant and this result can be extended in great generality and no predicate that depends on two inputs from an arbitrary finite domain can be approximation resistant [7].

Zwick [14] established approximability results for predicates that depend on three Boolean inputs and from this it follows that the only predicates on three inputs that are approximation resistant are those that are implied by parity or its negation. A predicate P is implied by a predicate Q iff whenever $Q(x)$ is true so is $P(x)$ and as an example the negation of parity implies disjunction as if we know that an odd number of variables are true they cannot all be false.

Many scattered results on (families of) wider predicates do exist [4,10] and in particular Hast [5] made an extensive classification of predicates on four inputs. Predicates that can be made equal by permuting the inputs or negating one or more inputs behave the same with respect to approximation resistance and with this notion of equivalence there are 400 different non-constant predicates on 4 Boolean inputs. Hast proved that 79 of these are approximation resistant, established 275 to be non-trivially approximable leaving the status of 46 predicates open. Zwick [13] has obtained numerical evidence suggesting that most of the latter predicates are in fact non-trivially approximable.

The main result of this paper is to give evidence that a random predicate for a large value of k is approximation resistant. The result is only evidence in that it relies on the Unique Games Conjecture (UGC) of Khot [8] but on the other hand we establish that a vast majority of the predicates are approximation resistant under this assumption.

We base our proof on the recent result by Samorodnitsky and Trevisan [11] that establishes that if d is the smallest integer such that $2^d - 1 \geq k$ then there is a predicate of width k that accepts only 2^d of the 2^k possible k-bit strings and which, based on the UGC, is approximation resistant. We extend their proof to establish that any predicate implied by their predicate is approximation resistant.

To establish our main result we proceed to prove that a random predicate is implied by some predicate which is equivalent to the predicate of Samorodnitsky and Trevisan. This is established by a second moment method. A standard random predicate on k bits is constructed by, for each of the 2^k inputs, flipping an unbiased coin to determine whether that input is accepted. It turns out that our results apply to other spaces of random predicates. In fact, if we construct a random predicate by accepting each input with probability k^{-c} for some $c > 0$ we still, with high probability for sufficiently large k, get an approximation resistant predicate. Here c is a number in the range $[1/2, 1]$ that depends on how close k is to the smallest number of the form $2^d - 1$ larger than k.

We make the proof more self contained by reproving one main technical lemma of [11] relating to Gowers uniformity norms and influences of functions. Our proof is similar in spirit to the original proof but significantly shorter and we hence believe it is of independent interest.

Of course the contribution of this paper heavily depends on how one views the Unique Games Conjecture, UGC. At the least one can conclude that it will be difficult to give a non-trivial approximation algorithm for a random predicate. Our results also point to the ever increasing need to settle the UGC.

An outline of the paper is as follows. We start by establishing some notation and giving some definitions in Section 2. We prove the lemmas relating to Gowers uniformity in Section 3 and proceed to establish that any predicate implied by the predicate used by Samorodnitsky and Trevisan is approximation resistant in Section 4. We then present our applications of this theorem by first establishing that a random predicate is approximation resistant in Section 5 and that all very dense predicates are approximation resistant in Section 6. We end with some concluding remarks in Section 7.

2   Preliminaries

We consider functions mapping $\{-1,1\}^n$ into the real numbers and usually into the interval $[-1,1]$. In this paper we use $\{-1,1\}$ as the value set of Boolean variables but still call the values "bits". For $x, x' \in \{-1,1\}^n$ we let $x \cdot x'$ denote the coordinate-wise product. In $\{0,1\}^n$-notation this is the simply the exclusive-or of vectors.

For any $\alpha \subseteq [n]$ we have the character χ_α defined by

$$\chi_\alpha(x) = \prod_{i \in \alpha} x_i$$

and the Fourier expansion is given by

$$f(x) = \sum_{\alpha \subseteq [n]} \hat{f}_\alpha \chi_\alpha(x).$$

We are interested in long codes coding $v \in [L]$. This is a function $\{-1,1\}^L \to \{-1,1\}$ and if A is the long code of v then $A(x) = x_v$. We want our long codes to be folded, which means that they only contain values for inputs with $x_0 = 1$. The value when $x_0 = -1$ is defined to be $-A(-x)$. This ensures that the function is unbiased and that the Fourier coefficient corresponding to the empty set is 0.

For two sets α and β we let $\alpha \Delta \beta$ be the symmetric difference of the two sets.

The influence $\inf_i f$ is the expected variance of f when all variables except x_i are picked randomly and uniformly. It is well known that

$$\inf_i = \sum_{i \in \alpha} \hat{f}_\alpha^2.$$

The following lemma from [10] is useful.

Lemma 2.1. *Let* $(f_j)_{j=1}^k$, $\{-1,1\}^n \to [-1,1]$ *be* k *functions, and*

$$f(x) = \prod_{j=1}^{k} f_j(x).$$

Then, for every $i \in [n]$, $\text{inf}_i(f) \le k \sum_{j=1}^k \text{inf}_i(f_j)$.

The pairwise cross influence of a set of functions $(f_j)_{j=1}^k$ is defined to be the maximal simultaneous influence in any two of the functions or more formally

$$\text{cinf}_i(f_j)_{j=1}^k = \max_{j_1 \ne j_2} \min(\text{inf}_i(f_{j_1}), \text{inf}_i(f_{j_2})).$$

Let P be a predicate on k Boolean inputs. An instance of the problem Max-CSP(P) is given by a list of k-tuples of literals. The task is to find the assignment to the variables that maximizes the number of k-tuples that satisfy P.

An algorithm is a C-approximation if it, for any instance I of this problem, produces an assignment which satisfies at least $C \cdot Opt(I)$ constraints where $Opt(I)$ is the number of constraints satisfied by an optimal solution.

Let $d(P)$ be the fraction of k-bit strings accepted by P. The trivial algorithm that just picks a random assignment satisfies, on the average, a $d(P)$-fraction of the constraints and as an optimal solution cannot satisfy more than all the constraints this yields a (randomized) $d(P)$-approximation algorithm. We have the following definition.

Definition 2.1. *A predicate* P *is* approximation resistant *if, for any* $\epsilon > 0$, *it is NP-hard to approximate Max-CSP(P) within* $d(P) + \epsilon$.

Some predicates have an even stronger property.

Definition 2.2. *A predicate* P *is* hereditary approximation resistant *if any predicate* Q *implied by* P *is approximation resistant.*

3   Gowers Uniformity and Influence

Gowers [2,3] introduced the notion of dimension-d uniformity norm $U^d(f)$ which was used in an essential way by Samorodnitsky and Trevisan [11]. Their result says that if a function does not have an influential variable and is unbiased then the dimension-d uniformity norm is small. More importantly for their application, [11] also proved that if a set of functions has small cross influences and at least one function is unbiased then the corresponding product is small. We slightly extend their result by allowing a small bias of the involved functions. Allowing this extension makes it possible to give a short, direct proof.

We want to emphasize that the results obtained by Samorodnitsky and Trevisan are sufficient for us but we include the results of this section since we believe that our proofs are simpler and that the extension might be interesting on its own and possibly useful in some other context.

Theorem 3.1. *Let $f: \{-1,1\}^n \to [-1,1]$ be a function with $\max_i \inf_i(f) \le \epsilon$ and $|E[f]| \le \delta$, then*

$$\left| E_{x^1,\ldots x^d} \left[\prod_{S \subseteq [d]} f \left(\prod_{i \in S} x^i \right) \right] \right| \le \delta + (2^{d-1} - 1)\sqrt{\epsilon}.$$

Proof. We prove the theorem by induction over d. Clearly it is true for $d = 1$ as the quantity to estimate equals $|f(1^n)E[f]|$.

For the induction step let $g^{x^d}(x) = f(x)f(x \cdot x^d)$. Then, by Lemma 2.1, $\max_i \inf_i g^{x^d} \le 4\epsilon$. Furthermore

$$E_x[g^{x^d}] = 2^{-n} \sum_x f(x)f(x \cdot x^d) = f * f(x^d)$$

and let us for notational simplicity denote this function by $h(x^d)$. As convolution turns into product on the Fourier transform side we have $\hat{h}_\alpha = \hat{f}_\alpha^2$. For any $\alpha \ne \emptyset$ we have $\hat{f}_\alpha^2 \le \max_i \inf_i(f) \le \epsilon$ and hence

$$\|h\|_2^2 = \sum_\alpha \hat{h}_\alpha^2 = \sum_\alpha \hat{f}_\alpha^4 \le \hat{f}_\emptyset^4 + \epsilon \sum_{\alpha \ne \emptyset} \hat{f}_\alpha^2 \le \delta^4 + \epsilon.$$

This implies, using the Cauchy-Schwartz inequality, that

$$E_{x^d}[|E_x[g^{x^d}(x)]|] \le \sqrt{\delta^4 + \epsilon} \le \delta^2 + \sqrt{\epsilon} \le \delta + \sqrt{\epsilon}. \tag{1}$$

Now

$$\left| E_{x^1,\ldots x^d} \left[\prod_{S \subseteq [d]} f \left(\prod_{i \in S} x^i \right) \right] \right| \le E_{x^d} \left| E_{x^1,\ldots x^{d-1}} \left[\prod_{S \subseteq [d-1]} g^{x^d} \left(\prod_{i \in S} x^i \right) \right] \right|,$$

which, by induction, is bounded by

$$E_{x^d} \left[|E_x[g^{x^d}]| + (2^{d-2} - 1)\sqrt{4\epsilon} \right] \le \delta + (2^{d-1} - 1)\sqrt{\epsilon}.$$

Note that by doing some more calculations we can get a better bound as a function of δ by not doing the wasteful replacement of δ^2 by δ in (1). We proceed to allow the functions to be different and require the pairwise cross influence to be small.

Theorem 3.2. *Let $(f_S)_{S \subseteq [d]}$ be a set of functions $\{-1,1\}^n \to [-1,1]$, with $\max_i \operatorname{cinf}_i(f_S) \le \epsilon$ and $\min_{S \ne \emptyset} |E[f_S]| \le \delta$, then*

$$\left| E_{x^1,\ldots x^d} \left[\prod_{S \subseteq [d]} f_S (\prod_{i \in S} x^i) \right] \right| \le \delta + (2^d - 2)\sqrt{\epsilon}.$$

Proof. We use induction over d. The base case $d = 1$ is straightforward and let us do the induction step.

By a change of variables we can assume that $|E[f_{[d]}]| \leq \delta$. Now define a new set of functions by

$$g_S^{x^d}(x) = f_S(x) f_{S \cup \{d\}}(x \cdot x^d),$$

for any $S \subseteq [d-1]$. The cross influence of this set of functions is, by Lemma 2.1, bounded by 4ϵ. Let $h(x^d)$ be the average of $g_{[d-1]}^{x^d}$. Then $h = f_{[d-1]} * f_{[d]}$ and $\hat{h}_\alpha = \hat{f}_{[d-1],\alpha} \hat{f}_{[d],\alpha}$ which yields

$$\|h\|_2^2 = \sum_\alpha \hat{h}_\alpha^2 = \hat{f}_{[d-1],\emptyset}^2 \hat{f}_{[d],\emptyset}^2 + \sum_{\alpha \neq \emptyset} \hat{f}_{[d-1],\alpha}^2 \hat{f}_{[d],\alpha}^2 \leq$$

$$\delta^2 + \sum_{\alpha \neq \emptyset} \min(\hat{f}_{[d-1],\alpha}^2, \hat{f}_{[d],\alpha}^2)(\hat{f}_{[d-1],\alpha}^2 + \hat{f}_{[d],\alpha}^2) \leq \delta^2 + \sum_{\alpha \neq \emptyset} \epsilon(\hat{f}_{[d-1],\alpha}^2 + \hat{f}_{[d],\alpha}^2) \leq \delta^2 + 2\epsilon.$$

Using induction we get

$$\left| E_{x^1,\dots x^d} \left[\prod_{S \subseteq [d]} f_S \left(\prod_{i \in S} x^i \right) \right] \right| \leq E_{x^d} \left| E_{x^1,\dots x^{d-1}} \left[\prod_{S \subseteq [d-1]} g_S \left(\prod_{i \in S} x^i \right) \right] \right| \leq$$

$$E_{x^d} \left[|E[g_{[d-1]}^{x^d}]| + (2^{d-1} - 2)\sqrt{4\epsilon} \right] \leq \delta + 2\sqrt{\epsilon} + (2^d - 4)\sqrt{\epsilon} \leq \delta + (2^d - 2)\sqrt{\epsilon}.$$

4   The ST-Predicate

Assume that $2^{d-1} \leq k \leq 2^d - 1$. For any integer i with $1 \leq i \leq 2^d - 1$ let $\hat{i} \subseteq [d]$ be the set whose characteristic vector is equal to the binary expansion of i. We define $P_{ST}(x)$, a predicate on k-bit strings, to be true if for all triplets i_1, i_2, and i_3 such that $\hat{i}_1 \Delta \hat{i}_2 = \hat{i}_3$ we have $x_{i_1} x_{i_2} = x_{i_3}$. Of course the predicate depends on k but as k (and d) remains fixed we suppress this dependence.

It is not difficult to see that the accepted strings form a linear space of dimension d. In fact the following procedure for picking a random string accepted by P_{ST} is a good way to visualize the predicate. For each i that is a power of two set x_i to a random bit. For other values of i set

$$x_i = \prod_{j \in \hat{i}} x_{2^j}.$$

Now consider Max-CSP(P_{ST}) and the following theorem is from [11].

Theorem 4.1. *Assuming the UGC, for any $\epsilon > 0$, it is NP-hard to approximate Max-CSP(P_{ST}) within $2^{d-k} + \epsilon$.*

Equivalently, the theorem says that P_{ST}, assuming UGC, is approximation resistant, but we need more.

Theorem 4.2. *Assuming UGC, P_{ST} is hereditary approximation resistant.*

It is satisfying to note that for $k = 3$ the predicate P_{ST} is simply parity and hence this instance of the theorem was proved in [6] without using the UGC.

Proof. Let Q be any predicate of arity k implied by P_{ST}. Our proof is very similar to the proof of [11] but we use a slightly different terminology. We assume that the reader is familiar with Probabilistically Checkable Proofs (PCPs) and their relation to inapproximability result for Max-CSPs. Details of the connection can be found in many places, one possible place being [6]. The short summary is that for any $\gamma > 0$ we need to define a PCP where the acceptance condition is given by the predicate Q and such that it is hard to distinguish the case when the maximal acceptance probability is $1 - \gamma$ and the case when the maximal acceptance probability is $d(Q)+\gamma$. It is also needed that the verifier uses $O(\log n)$ random bits when checking proofs of statements of size n. The latter property implies that the proof is of polynomial size.

As in [11] we use a form of the UGC which, using the terminology of [11], is called the k-ary unique games. We have variables $(v_i)_{i=1}^n$ taking values in a finite domain of size L, which we assume to be $[L]$. A constraint is given by a k-tuple, $(v_{i_j})_{j=1}^k$ of variables and k permutations $(\pi_j)_{j=1}^k$. An assignment V strongly satisfies the constraint iff the k elements $\pi_j(V(v_{i_j}))$ are all the same and the assignment weakly satisfies the constraint if these values are not all distinct. The following result, originally by Khot and Regev [9] is stated in [11].

Theorem 4.3. *If the UGC is true then for every k and ϵ there is a $L = L(k, \epsilon)$ such that, given a k-ary unique game problem with alphabet size L, it is NP-hard to distinguish the case in which there is an assignment that strongly satisfies at least a $(1 - \epsilon)$-fraction of the constraints from the case where every assignment weakly satisfies at most a fraction ϵ of the constraints.*

We proceed to construct a PCP based on the k-ary unique game problem. The test is as described in [11] but slightly reformulated.

The written proof is supposed to be coding of an assignment which satisfies a $(1 - \epsilon)$-fraction of the constraints. For each v_i the proof contains the long code A_i of $V(v_i)$. We access these long codes in a folded way as described in the preliminaries. This folding gives rise to negations in the resulting instance of Max-CSP(Q). We let permutations act on vectors by $\pi(x)_j = x_{\pi(j)}$.

As in many PCPs we use noise vectors $\mu \in \{-1, 1\}^L$ which has the property that μ_v is picked randomly and independently and for each $v \in [L]$ it equals 1 with probability $1 - \delta$ and -1 with probability δ, where δ is a parameter to be determined. It is an important parameter of the test and hence we include it explicitly. The verifier of the PCP works as follows.

Q-test(δ)

1. Pick a random k-ary constraint, given by variables $(v_{i_j})_{j=1}^k$, and permutations $(\pi_j)_{j=1}^k$.

2. Pick d independent random unbiased unbiased $x^i \in \{-1,1\}^L$ and k independent noise functions $\mu^j \in \{-1,1\}^L$.
3. Let $y^j = \prod_{i \in \hat{j}} x^i$ and $b_j = A_{i_j}(\pi_j(y^j) \cdot \mu^j)$.
4. Accept if $Q(b) = Q(b_1, b_2, \ldots b_k)$ is true.

We first address completeness.

Lemma 4.1. *For any $\gamma > 0$ there exists $\delta > 0$, $\epsilon > 0$ such that if there is an assignment that strongly satisfies a fraction $1 - \epsilon$ of the constraints in the k-ary unique game problem then the verifier in Q-test(δ) can be made to accept with probability $1 - \gamma$.*

Proof. Assume that each A_j is the correct long code of the value $V(v_j)$ for an assignment V that satisfies at least a $(1 - \epsilon)$-fraction of the constraints. Then assuming that $\mu^j_{V(v_{i_j})} = 1$ and $\pi_j(V(v_{i_j})) = v$ for all j we have

$$b_j = y^j_{\pi_j(V(v_{i_j}))} \cdot \mu^j_{V(v_{i_j})} = y^j_v = \prod_{i \in \hat{j}} x^i_v.$$

Recalling the description of the accepted inputs of P_{ST} it follows that b satisfies P_{ST} and hence also Q. The completeness is hence at least $1 - \epsilon - k\delta$ and choosing ϵ and δ sufficiently small this is at least $1 - \gamma$.

Let us turn to the more challenging task of analyzing the soundness.

Lemma 4.2. *For any $\gamma > 0$, $\delta > 0$ there exist $\epsilon = \epsilon(k, \delta, \gamma) > 0$ such that if the verifier in Q-test(δ) accepts with probability at least $d(Q) + \gamma$ there exists an assignment that weakly satisfies at least a fraction ϵ of the constraints in the k-ary unique game problem.*

Proof. We assume that the verifier accepts with probability $d(Q) + \gamma$ and turn to define a (randomized) assignment that weakly satisfies a fraction of the constraints that only depends on k, δ and γ.

We use the multilinear representation of Q (which is in fact identical to the Fourier transform)

$$Q(b) = \sum_\beta \hat{Q}_\beta \prod_{j \in \beta} b_j.$$

Note that the constant term $\hat{Q}_\emptyset$ is exactly $d(Q)$ and hence if the verifier accepts with probability $d(Q) + \gamma$ there must be some nonempty β such that

$$|E[\prod_{j \in \beta} b_j]| \geq 2^{-k}\gamma, \tag{2}$$

where the expectation is taken over a random constraint of the k-ary unique game and random choices of x^i and μ^j.

Let us first study expectation over the noise vectors and towards this end let

$$B_j(y) = E_\mu[A_j(y \cdot \mu)],$$

which gives $E_{\mu^j}(b_j) = B_{i_j}(\pi_j(y^j))$. It is a standard fact (for a proof see [6]) that

$$\hat{B}_{j,\beta} = (1 - 2\delta)^{|\beta|}\hat{A}_{j,\beta}$$

and hence

$$\sum_{|\beta| \geq t} \hat{B}_{j,\beta}^2 \leq (1 - 2\delta)^{2t} \tag{3}$$

for any t. Now set $\Gamma = 2^{-2(d+k+2)}\gamma^2$ and let $t = O(\delta^{-1}\log\Gamma^{-1})$ be such that

$$(1 - 2\delta)^{2t} \leq \Gamma/2,$$

and define

$$T_j = \{i \mid \inf_i B_j \geq \Gamma\}.$$

As

$$\inf_i B_j = \sum_{i \in \beta} \hat{B}_{j,\beta}^2, \tag{4}$$

by (3) and the definition of t, if $i \in T_j$ then we must have at least a contribution of $\Gamma/2$ from sets of size at most t in (4). Using this it follows that $|T_j| \leq 2t/\Gamma$ for any j.

Consider the probabilistic assignment that for each v_j chooses a random element of T_j. If T_j is empty we choose an arbitrary value for v_j.

By (2) we know that for at least a fraction $2^{-k}\gamma/2$ of the constraints we have

$$\left| E_{x^i,\mu^j}\left[\prod_{j \in \beta} b_j \right] \right| \geq 2^{-k}\gamma/2. \tag{5}$$

Fix any such constraint and define the following family of functions.

For any $j \notin \beta$ or $k < j \leq 2^d - 1$ set $h_{\tilde{j}}$ to be identically one while if $j \in \beta$ we define $h_{\tilde{j}}$ by

$$h_{\tilde{j}}(y) = B_{i_j}(\pi_j(y)).$$

These definitions imply that

$$E_\mu\left[\prod_{j \in \beta} b_i \right] = \prod_{S \subseteq [d]} h_S\left(\prod_{i \in S} x^i \right) \tag{6}$$

and hence we are in a position to apply Theorem 3.2. Note first that, by folding, each h that is non-constant is in fact unbiased and hence, as β is non-empty, the minimum bias of the set of functions is 0.

We now claim that the maximal cross influence of the function set h_S is at least Γ. Indeed suppose that this is not the case. Then, by Theorem 3.2, the expectation of (6), over the choice of vectors x^i, is at most

$$(2^d - 2)\sqrt{\Gamma} < 2^d 2^{-(d+k+2)}\gamma \leq 2^{-k}\gamma/2$$

contradicting (5).

Thus we have $j_1, j_2 \in \beta$ and an i such that $\inf_i h_{\hat{j}_1} \geq \Gamma$ and $\inf_i h_{\hat{j}_2} \geq \Gamma$. Now, by definition, $\inf_i h_{\hat{j}_1}$ is the same as $\inf_{\pi_{j_1}^{-1}(i)}(B_{i_{j_1}})$. We conclude that there is a common element in $\pi_{j_1}(T_{i_{j_1}})$ and $\pi_{j_2}(T_{i_{j_2}})$ and our probabilistic assignment weakly satisfies the constraint with probability at least

$$\frac{1}{|T_{i_{j_1}}|} \cdot \frac{1}{|T_{i_{j_2}}|} \geq \frac{\Gamma^2}{4t^2}.$$

As this happens for at least a fraction $2^{-k}\gamma/2$ of the constraints our probabilistic assignment weakly satisfies, on the average, a fraction at least

$$\frac{2^{-k}\gamma\Gamma^2}{8t^2}$$

of the constraints. Clearly there exists a standard, deterministic assignment that satisfies the same fraction of the constraints. This finishes the proof of Lemma 4.2.

As stated before Lemma 4.1 and Lemma 4.2 together with the fact that the acceptance criteria of Q-test(δ) is given by Q is sufficient to prove Theorem 4.2. Note that the randomness used by the verifier is bounded by $O(\log n)$ and most of the randomness is used to choose a random constraints as all other random choices only require $O(1)$ random bits.

We do not give the details of these standard parts of the proof here. In short, an approximation algorithm for Max-CSP-(Q) can be used to solve the problem established to be hard in Theorem 4.3.

5   Random Predicates

Remember that we allow negation of inputs and permutation of input variables and hence two predicates that can be obtained from each other by such operations are equivalent. Thus Theorem 4.2 does not only apply to P_{ST} but also to any predicate which is equivalent to it.

Consider the following space of random predicates.

Definition 5.1. *Let $Q_{p,k}$ be the probability space of predicates in k variables where each input is accepted with probability p.*

A uniformly random predicate corresponds to a predicate from $Q_{1/2,k}$ but we will consider also smaller values of p. Whenever needed in calculations we assume $p \leq 1/2$.

We want to analyze the probability that a random predicate from $Q_{p,k}$ is implied by a negated and/or permuted variant of P_{ST} and let us just check that it is reasonable to believe that this is the case.

We have $k!$ permutations of the inputs and 2^k possible ways to negate the inputs. Thus the expected number of P_{ST}-equivalent predicates that imply a random predicate from $Q_{p,d}$ is

$$p^{2^d}2^k k!.$$

There is hope if this number is at least one, which, ignoring low order terms, happens as soon as

$$p \geq k^{-k2^{-d}}.$$

This lower bound is between k^{-1} and $k^{-1/2}$ and in particular it is smaller than any constant. In fact this rough estimate turns out to be rather close to the truth and the proof is an application of the second moment method. A problem to be overcome is that some pairs of P_{ST}-equivalent predicate have large intersection of their accepted sets. To address this problem we pick a large subset of the P_{ST}-equivalent predicates with bounded size intersections.

Theorem 5.1. *Assuming UGC and $2^{d-1} \leq k \leq 2^d - 1$ then, there is a value c of the form $c = k2^{-d}(1 - o(1))$, such that, with probability $1 - o(1)$, a random predicate chosen according to $Q(p,k)$ with $p = k^{-c}$ is approximation resistant.*

Proof. In view of Theorem 4.2 we need only prove that a random predicate from $Q_{p,k}$ with high probability is implied by some predicate which can be obtained from P_{ST} by negations and/or permutations of inputs.

Let us denote the set accepted by P_{ST} by L. It is a linear space of dimension d. Negating one or more inputs gives an affine space that is either L or completely disjoint from L. We get 2^{k-d} disjoint affine spaces denoted by $L + \alpha$ where α ranges over a suitable set of cardinality 2^{k-d}. We can also permute the coordinates and this gives a total of $k!2^{k-t}$ sets

$$\pi(L + \alpha)$$

Consider

$$\pi(L + \alpha) \cap \pi'(L + \beta).$$

It is an affine space which is either empty or of dimension of the linear space

$$\pi(L) \cap \pi'(L)$$

The below lemma is useful towards this end. Due to space limitations the proof of the lemma will only appear in the full version.

Lemma 5.1. *Let d_0 and k_0 be sufficiently large constants and let r be a number such that $2^{d-r} \geq d_0$ and assume that $k \geq k_0$. Then, if π and π' are two random permutations we have*

$$Pr[dim(\pi(L) \cap \pi'(L)) \geq r] \leq 2^{(2-r)k}.$$

Set $R = 2^{k(r-2)}$ and let us see how to use Lemma 5.1 to choose R different permutations π_i such that

$$dim(\pi_i(L) \cap \pi_j(L)) \leq r$$

for any $i \neq j$. First pick $2R$ random permutations. The expected number of pairs (i,j), $i < j$, with

$$dim(\pi_i(L) \cap \pi_j(L)) > r$$

is bounded by $2R^22^{(1-r)k} \le R$ and hence there is a choice of $2R$ permutations such that the number of such pairs is bounded by R. Erase one of the two permutations in each such pair and we have the desired set. Let us fix this set $(\pi_i)_{i=1}^R$ once and for all.

Let $X_{i,\alpha}$ be the indicator variable for the event a random predicate from $Q_{k,p}$ is identically one on the set

$$\pi_i(L+\alpha).$$

Set

$$X = \sum_{i,\alpha} X_{i,\alpha}.$$

The probability of the event that the random predicate is not identically one on any $\pi_i(L+\alpha)$ is now exactly $Pr[X=0]$ and we estimate the probability of this event. Clearly

$$E[X] = p^{2^d} 2^{k-d} R. \tag{7}$$

The variance of X equals

$$E\left[\sum_{i_1,i_2,\alpha_1,\alpha_2} (X_{i_1,\alpha_1} - p^{2^d})(X_{i_2,\alpha_2} - p^{2^d})\right]. \tag{8}$$

We have the following lemma.

Lemma 5.2. *We have* $E[(X_{i_1,\alpha_1} - p^{2^d})(X_{i_2,\alpha_2} - p^{2^d})] = 0$ *if* $\pi_{i_1}(L+\alpha_1)$ *and* $\pi_{i_2}(L+\alpha_2)$ *are disjoint while if the size of the intersection is* K *it is bounded by*

$$p^{2^{d+1}-K}.$$

Proof. In fact

$$E[(X_{i_1,\alpha_1} - p^{2^d})(X_{i_2,\alpha_2} - p^{2^d})] = E[X_{i_1,\alpha_1} X_{i_2,\alpha_2}] - p^{2^{d+1}} = p^{2^{d+1}-K} - p^{2^{d+1}}.$$

Let us now estimate (8). Terms with $i_1 = i_2$ are easy as the corresponding sets either have full intersection or are disjoint. These give a contribution that is upper bounded by $E[X]$. Now for $i_1 \ne i_2$ let us fix α_1 and consider

$$\sum_{\alpha_2} E\left[(X_{i_1,\alpha_1} - p^{2^d})(X_{i_2,\alpha_2} - p^{2^d})\right]. \tag{9}$$

It is the case that for some $r' \le r$ we have $2^{d-r'}$ terms with set intersection size $2^{r'}$ while all other intersections are empty leading to the upper estimate

$$2^{d-r'} p^{2^{d+1}-2^{r'}} \le 2^{d-r} p^{2^{d+1}-2^r}$$

(using the assumption $p \le 1/2$) for the sum (9). Summing over all i_1, i_2 and α_1 we get

$$\sigma^2(X) \le E[X] + R^2 2^{k-d} 2^{d-r} p^{2^{d+1}-2^r} = E[X] + R^2 2^{k-r} p^{2^{d+1}-2^r}. \tag{10}$$

We have

$$Pr[X = 0] \le \frac{\sigma^2(X)}{E[X]^2} \le \frac{1}{E[X]} + \frac{R^2 2^{k-r} p^{2^{d+1}-2^r}}{R^2 2^{2(k-d)} p^{2^{d+1}}} \le \frac{1}{E[X]} + 2^{2d-(k+r)} p^{-2^r} \quad (11)$$

We need to choose p and r to make this probability $o(1)$. Set $p = k^{-c}$ for some $c \le 1$. Then provided

$$2^r \log k < (k+r) - 2d - \omega(1)$$

the second term of (11) is small. This is possible to achieve with $r = d - \Theta(\log d)$. Note that this choice also ensures $d - r \in \omega(1)$ as required by Lemma 5.1.

Fixing this value of r the first term of (11) is $o(1)$ provided that

$$p^{2^d} \ge 2^{(2-r)k}$$

which with, $p = k^{-c}$, is equivalent to

$$c \le k2^{-d} \cdot \frac{r-2}{\log k}. \quad (12)$$

As the second factor of the bound in (12) is $(1 - o(1))$ we have proved Theorem 5.1.

Apart from adjustments of the error terms this is the best that can be obtained by the current methods. Namely setting $p = k^{-(k2^{-d}+\epsilon)}$ for $\epsilon > 0$ the probability of a random predicate being implied by some P_{ST}-equivalent predicate goes to 0 as can be seen from calculating the expected value of the number of such predicates.

One can always wonder about reasonable values for p for small values of k. Particularly good values for k are numbers of the form $2^d - 1$ as this gives an unusually sparse predicate P_{ST}. Numerical simulations suggests that a random predicate on 7 bits that accepts M of the 128 inputs has a probability at least $1/2$ of being implied by a P_{ST}-equivalent predicate iff $M \ge 60$. Thus it seems like the asymptotic bound of density essentially k^{-1} is approached slowly.

6   Very Dense Predicates

As P_{ST} only accepts 2^d inputs we can derive approximation resistance of many predicates but let us here give only one immediate application.

Theorem 6.1. Let $2^{d-1} \le k \le 2^d - 1$ and P be any predicate that accepts at least $2^k + 1 - 2^{k-d}$ inputs, then, assuming the UGC, P is approximation resistant.

Proof. We use the same notation as used in the proof of Theorem 5.1.

We need to prove that any such predicate is implied by a P_{ST}-equivalent predicate. This time we need only apply negations and look at $L + \alpha$ for all the 2^{k-d} different representatives α. As P only rejects $2^{k-d} - 1$ different inputs and the sets $L + \alpha$ are disjoint, one such set is included in the accepted inputs of P. The corresponding suitable negated form of P_{ST} hence implies P and Theorem 6.1 follows from Theorem 4.2.

It is an interesting question how dense a non-trivially approximable predicate can be. Let d_k be the maximum value of $d(P)$ for all predicates on k variables which are not approximation resistant. We have $d_2 = d_3 = 3/4$ and Hast [5] proved that $d_4 = \frac{13}{16}$ and, as we can always ignore any input, d_k is an increasing function of k. It is not obvious whether d_k tends to one as k tends to infinity.

Our results show that dense predicates which can be non-trivially approximated need to be extremely structured as they cannot be implied by any P_{ST}-equivalent predicate.

7 Concluding Remarks

The key result in the current paper is to prove that P_{ST} is hereditary approximation resistant. This is another result indicating that the more inputs accepted by the predicate P, the more likely it is to be approximation resistant. One could be tempted to conclude that all approximation resistant predicates are in fact hereditary approximation resistant. We would like to point that this is false and Hast [5] has an example of two predicates P and Q where P is approximation resistant, P implies Q and Q is not approximation resistant.

That a predicate is approximation resistant is almost the ultimate hardness. There is a stronger notion; approximation resistance on satisfiable instances. In such a case no efficient algorithm is able to do significantly better than picking a random assignment even in the case when the instance is satisfiable.

An example of a predicate which is approximation resistant but not approximation resistant on satisfiable instances is Max-E3-Lin-2, linear equations modulo 2 with three variables in each equation. In this case, for a satisfiable instance, it is easy to find an assignment that satisfies all constraints by Gaussian elimination.

In most cases, however, approximation resistant predicates have turned out to be approximation resistant also on satisfiable instances and it would seem reasonable to conjecture that a random predicate is indeed approximation resistant on satisfiable instances. If true it seems hard to prove this fact using the Unique Games Conjecture in that the non-perfect completeness of UGC would tend to produce instances of the CSP which are not satisfiable. There are variants of the unique games conjecture [8] which postulate hardness of label cover problems with perfect completeness but it would be much nicer to take a different route not relying on any conjectures.

Another open problem is of course to establish approximation resistance in absolute terms and not to rely on the UGC or, more ambitiously, to prove the UGC.

Acknowledgment. I am grateful to Per Austrin for useful comments on the current manuscript.

References

1. Goemans, M., Williamson, D.: Improved approximation algorithms for maximum cut and satisfiability problems using semidefinite programming. Journal of the ACM 42, 1115–1145 (1995)
2. Gowers, T.: A new proof of Szemerédi's theorem for progressions of length four. Geometric and Functional Analysis 8, 529–551 (1998)
3. Gowers, T.: A new proof of Szemerédi's theorem. Geometric and Functional Analysis 11, 465–588 (2001)
4. Guruswami, V., Lewin, D., Sudan, M., Trevisan, L.: A tight characterization of NP with 3 query PCPs. In: In: Proceedings of 39th Annual IEEE Symposium on Foundations of Computer Science, 1998, Palo Alto, pp. 8–17. IEEE Computer Society Press, Los Alamitos (1998)
5. Hast, G.: Beating a random assignment. KTH, Stockholm, Ph.D Thesis (2005)
6. Håstad, J.: Some optimal inapproximability results. Journal of ACM 48, 798–859 (2001)
7. Håstad, J.: Every 2-CSP allows nontrivial approximation. In: Proceedings of the 37th Annual ACM Symposium on Theory of Computation, pp. 740–746. ACM Press, New York (2005)
8. Khot, S.: On the power of unique 2-prover 1-round games. In: Proceedings of 34th ACM Symposium on Theory of Computating, pp. 767–775. ACM Press, New York (2002)
9. Khot, S., Regev, O.: Vertex cover might be hard to approximate to within $2 - \varepsilon$. In: Proc. of 18th IEEE Annual Conference on Computational Complexity (CCC), pp. 379–386. IEEE Computer Society Press, Los Alamitos (2003)
10. Samorodnitsky, A., Trevisan, L.: A PCP characterization of NP with optimal amortized query complexity. In: Proceedings of the 32nd Annual ACM Symposium on Theory of Computing, pp. 191–199. ACM Press, New York (2000)
11. Samorodnitsky, A., Trevisan, L.: Gowers uniformity, influence of variables and PCPs. In: Proceedings of the 38th Annual ACM Symposium on Theory of Computing, pp. 11–20. ACM Press, New York (2006)
12. Schaefer, T.: The complexity of satisfiability problems. In: Conference record of the Tenth annual ACM Symposium on Theory of Computing, pp. 216–226. ACM Press, New York (1978)
13. Zwick, U.: Personal Communication
14. Zwick, U.: Approximation algorithms for constraint satisfaction problems involving at most three variables per constraint. In: Proceedings 9th Annual ACM-SIAM Symposium on Discrete Algorithms, pp. 201–210. ACM Press, New York (1998)

Integrality Gaps of Semidefinite Programs for Vertex Cover and Relations to ℓ_1 Embeddability of Negative Type Metrics

Hamed Hatami, Avner Magen, and Evangelos Markakis*

Department of Computer Science,
University of Toronto

Abstract. We study various SDP formulations for VERTEX COVER by adding different constraints to the standard formulation. We rule out approximations better than $2 - O(\sqrt{\log \log n / \log n})$ even when we add the so-called pentagonal inequality constraints to the standard SDP formulation, and thus almost meet the best upper bound known due to Karakostas, of $2 - \Omega(\sqrt{1/\log n})$. We further show the surprising fact that by strengthening the SDP with the (intractable) requirement that the metric interpretation of the solution embeds into ℓ_1 with no distortion, we get an exact relaxation (integrality gap is 1), and on the other hand if the solution is arbitrarily close to being ℓ_1 embeddable, the integrality gap is $2 - o(1)$. Finally, inspired by the above findings, we use ideas from the integrality gap construction of Charikar to provide a family of simple examples for negative type metrics that cannot be embedded into ℓ_1 with distortion better than $8/7 - \epsilon$. To this end we prove a new isoperimetric inequality for the hypercube.

1 Introduction

A vertex cover in a graph $G = (V, E)$ is a set $S \subseteq V$ such that every edge $e \in E$ intersects S in at least one endpoint. Denote by $\mathrm{vc}(G)$ the size of the minimum vertex cover of G. It is well-known that the minimum vertex cover problem has a 2-approximation algorithm, and it is widely believed that for every constant $\epsilon > 0$, there is no $(2 - \epsilon)$-approximation algorithm for this problem. Currently the best known hardness result for this problem, based on the PCP theorem, shows that 1.36-approximation is NP-hard [1]. If we were to assume the Unique Games Conjecture [2] the problem would be essentially settled as $2 - \Omega(1)$ would then be NP-hard [3].

In [4], Goemans and Williamson introduced semidefinite programming as a tool for obtaining approximation algorithms. Since then semidefinite programming has become an important technique, and for many problems the best known approximation algorithms are obtained by solving an SDP relaxation of them.

---

* Currently at the Center for Math and Computer Science (CWI), Amsterdam, the Netherlands.

M. Charikar et al. (Eds.): APPROX and RANDOM 2007, LNCS 4627, pp. 164–179, 2007.

The best known algorithms for VERTEX COVER compete in "how big is the little oh" in the $2 - o(1)$ factor. The best two are in fact based on SDP relaxations: Halperin [5] gives a $(2 - \Omega(\log \log \Delta / \log \Delta))$-approximation where Δ is the maximal degree of the graph while Karakostas obtains a $(2 - \Omega(1/\sqrt{\log n}))$-approximation [6].

The standard way to formulate the VERTEX COVER problem as a quadratic integer program is the following:

$$\text{Min } \sum_{i \in V}(1 + x_0 x_i)/2$$
$$\text{s.t. } (x_i - x_0)(x_j - x_0) = 0 \qquad \forall\, ij \in E$$
$$x_i \in \{-1, 1\} \qquad \forall\, i \in \{0\} \cup V,$$

where the set of the vertices i for which $x_i = x_0$ correspond to the vertex cover. Relaxing this integer program to a semidefinite program, the scalar variable x_i becomes a vector $\mathbf{v}_i$ and we get:

$$\text{Min } \sum_{i \in V}(1 + \mathbf{v}_0 \mathbf{v}_i)/2$$
$$\text{s.t. } (\mathbf{v}_i - \mathbf{v}_0) \cdot (\mathbf{v}_j - \mathbf{v}_0) = 0 \qquad \forall\, ij \in E \qquad (1)$$
$$\|\mathbf{v}_i\| = 1 \qquad \forall\, i \in \{0\} \cup V.$$

Kleinberg and Goemans [7] proved that SDP (1) has integrality gap of $2 - o(1)$: given $\epsilon > 0$ they construct a graph G_ϵ for which $\text{vc}(G_\epsilon)$ is at least $(2 - \epsilon)$ times larger than the solution to SDP (1). They also suggested the following strengthening of SDP (1) and left its integrality gap as an open question:

$$\text{Min } \sum_{i \in V}(1 + \mathbf{v}_0 \mathbf{v}_i)/2$$
$$\text{s.t. } (\mathbf{v}_i - \mathbf{v}_0) \cdot (\mathbf{v}_j - \mathbf{v}_0) = 0 \qquad \forall\, ij \in E$$
$$(\mathbf{v}_i - \mathbf{v}_k) \cdot (\mathbf{v}_j - \mathbf{v}_k) \geq 0 \qquad \forall\, i, j, k \in \{0\} \cup V \qquad (2)$$
$$\|\mathbf{v}_i\| = 1 \qquad \forall\, i \in \{0\} \cup V.$$

Charikar [8] answered this question by showing that the same graph G_ϵ but a different vector solution satisfies SDP (2)[1] and gives rise to an integrality gap of $2 - o(1)$ as before. The following is an equivalent formulation to SDP (2):

$$\text{Min } \sum_{i \in V} 1 - \|\mathbf{v}_0 - \mathbf{v}_i\|^2/4$$
$$\text{s.t. } \|\mathbf{v}_i - \mathbf{v}_0\|^2 + \|\mathbf{v}_j - \mathbf{v}_0\|^2 = \|\mathbf{v}_i - \mathbf{v}_j\|^2 \qquad \forall\, ij \in E$$
$$\|\mathbf{v}_i - \mathbf{v}_k\|^2 + \|\mathbf{v}_j - \mathbf{v}_k\|^2 \geq \|\mathbf{v}_i - \mathbf{v}_j\|^2 \qquad \forall\, i, j, k \in \{0\} \cup V \qquad (3)$$
$$\|\mathbf{v}_i\| = 1 \qquad \forall\, i \in \{0\} \cup V$$

Viewing SDPs as relaxations over ℓ_1: The above reformulation reveals a connection to metric spaces. The second constraint in SDP (3) says that $\|\cdot\|^2$ induces a metric on $\{\mathbf{v}_i : i \in \{0\} \cup V\}$, while the first says that $\mathbf{v}_0$ is on the shortest path between the images of every two neighbours. This suggests a more careful study of the problem from the metric viewpoint which is the purpose of this article. Such connections are also important in the context of

---

[1] To be more precise, Charikar's result was about a slightly weaker formulation than (2) but it is not hard to see that the same construction works for SDP (2) as well.

the SPARSEST CUT problem, where the natural SDP relaxation was analyzed in the breakthrough work of Arora, Rao and Vazirani [9] and it was shown that its integrality gap is at most $O(\sqrt{\log n})$. This later gave rise to some significant progress in the theory of metric spaces [10,11].

Let $f : (X, d) \rightarrow (X', d')$ be an embedding of metric space (X, d) into another metric space (X', d'). The value $\sup_{x,y \in X} \frac{d'(f(x),f(y))}{d(x,y)} \times \sup_{x,y \in X} \frac{d(x,y)}{d'(f(x),f(y))}$ is called the distortion of f. For a metric space (X, d), let $c_1(X, d)$ denote the minimum distortion required to embed (X, d) into ℓ_1. Notice that $c_1(X, d) = 1$ if and only if (X, d) can be embedded isometrically into ℓ_1, namely without changing any of the distances. Consider a vertex cover S and its corresponding solution to SDP (2), i.e., $\mathbf{v}_i = 1$ for every $i \in S \cup \{0\}$ and $\mathbf{v}_i = -1$ for every $i \notin S$. The metric defined by $\| \cdot \|^2$ on this solution (i.e., $d(i, j) = \|\mathbf{v}_i - \mathbf{v}_j\|^2$) is isometrically embeddable into ℓ_1. Thus we can strengthen SDP (2) by allowing any arbitrary list of valid inequalities in ℓ_1 to be added. The triangle inequality is one type of such constraints. The next natural inequality of this sort is the pentagonal inequality: A metric space (X, d) is said to satisfy the pentagonal inequality if for $S, T \subset X$ of sizes 2 and 3 respectively it holds that $\sum_{i \in S, j \in T} d(i, j) \geq \sum_{i,j \in S} d(i, j) + \sum_{i,j \in T} d(i, j)$. Note that this inequality does not apply to every metric, but it does hold for those that are ℓ_1-embeddable. This leads to the following natural strengthening of SDP (3):

$$
\begin{aligned}
&\text{Min} \sum_{i \in V} 1 - \|\mathbf{v}_0 - \mathbf{v}_i\|^2/4 \\
&\text{s.t.} \quad \|\mathbf{v}_i - \mathbf{v}_0\|^2 + \|\mathbf{v}_j - \mathbf{v}_0\|^2 = \|\mathbf{v}_i - \mathbf{v}_j\|^2 && \forall \, ij \in E \\
&\qquad \sum_{i \in S, j \in T} \|\mathbf{v}_i - \mathbf{v}_j\|^2 \geq \sum_{i,j \in S} \|\mathbf{v}_i - \mathbf{v}_j\|^2 + && \forall \, S, T \subseteq \{0\} \cup V, \quad (4) \\
&\qquad \qquad \qquad \qquad \qquad \sum_{i,j \in T} \|\mathbf{v}_i - \mathbf{v}_j\|^2 && |S| = 2, |T| = 3 \\
&\qquad \|\mathbf{v}_i\| = 1 && \forall \, i \in \{0\} \cup V
\end{aligned}
$$

In Theorem 5, we prove that SDP (4) has an integrality gap of $2 - o(1)$. It is important to point out that a-priori there is no reason to believe that local addition of inequalities such as these will be useless; indeed in the case of SPARSEST CUT where triangle inequality is necessary to achieve the $O(\sqrt{\log n})$ bound mentioned above. It is interesting to note that for SPARSEST CUT, it is not known how to show a nonconstant integrality gap against pentagonal (or any other k-gonal) inequalities, although recently a nonconstant integrality gap was shown in [12] and later in [13], in the presence of the triangle inequalities[2].

A recent result by Georgiou, Magen, Pitassi and Tourlakis [14] shows and integrality gap of $2 - o(1)$ for a nonconstant number of rounds of the so-called LS_+ system for VERTEX COVER. It is not known whether this result subsumes Theorem 5 or not, since pentagonal inequalities are not generally implied by any number of rounds of the $LS+$ procedure. We elaborate more on this in the Discussion section.

One can further impose any ℓ_1-constraint not only for the metric defined by $\{\mathbf{v}_i : i \in V \cup \{0\}\}$, but also for the one that comes from $\{\mathbf{v}_i : i \in V \cup \{0\}\} \cup \{-\mathbf{v}_i :$

---

[2] As Khot and Vishnoi note, and leave as an open problem, it is possible that their example satisfies some or all k-gonal inequalities.

$i \in V \cup \{0\}\}$. Triangle inequalities for this extended set result in constraints $\|\mathbf{v}_i - \mathbf{v}_j\|^2 + \|\mathbf{v}_i - \mathbf{v}_k\|^2 + \|\mathbf{v}_j - \mathbf{v}_k\|^2 \leq 2$. The corresponding tighter SDP is used in [6] to get integraility gap of at most $2 - \Omega(\frac{1}{\sqrt{\log n}})$. Karakostas [6] asks whether the integrality gap of this strengthening breaks the "2-o(1) barrier": we answer this negatively in Section 4.3. In fact we show that the above upper bound is almost asymptotically tight, exhibiting integrality gap of $2 - O(\sqrt{\frac{\log \log n}{\log n}})$.

Integrality gap with respect to ℓ_1 embeddability: At the extreme, strengthening the SDP with ℓ_1-valid constraints, would imply the condition that the metric defined by $\|\cdot\|$ on $\{\mathbf{v}_i : i \in \{0\} \cup V\}$, namely $d(i,j) = \|\mathbf{v}_i - \mathbf{v}_j\|^2$ is ℓ_1 embeddable. Doing so leads to the following intractable program:

$$\begin{aligned}
&\text{Min } \sum_{i \in V} 1 - \|\mathbf{v}_0 - \mathbf{v}_i\|^2/4 \\
&\text{s.t. } \|\mathbf{v}_i - \mathbf{v}_0\|^2 + \|\mathbf{v}_j - \mathbf{v}_0\|^2 = \|\mathbf{v}_i - \mathbf{v}_j\|^2 && \forall\, ij \in E \\
&\quad\quad \|\mathbf{v}_i\| = 1 && \forall\, i \in \{0\} \cup V \\
&\quad\quad c_1(\{\mathbf{v}_i : i \in \{0\} \cup V\}, \|\cdot\|^2) = 1
\end{aligned} \quad (5)$$

In [15], it is shown that an SDP formulation of MINIMUM MULTICUT, even with the constraint that the $\|\cdot\|^2$ distance over the variables is isometrically embeddable into ℓ_1, still has a large integrality gap. For the MAX CUT problem, which is more intimately related to our problem, it is easy to see that ℓ_1 embeddability does not prevent integrality gap of $8/9$. It is therefore tempting to believe that there is a large integrality gap for SDP (5) as well. Surprisingly, SDP (5) has no gap at all: we show in Theorem 2, that the value of SDP (5) is exactly the size of the minimum vertex cover. A consequence of this fact is that any feasible solution to SDP (2) that surpasses the minimum vertex cover induces an ℓ_2^2 metric which is not isometrically embeddable into ℓ_1. This includes the integrality gap constructions of Kleinberg and Goemans, and that of Charikar's for SDPs (2) and (3) respectively. The construction of Charikar provides not merely ℓ_2^2 distance function but also a negative type metric, that is an ℓ_2^2 metric that satisfies triangle inequality. See [16] for background and nomenclature.

In contrast to Theorem 2, we show in Theorem 3 that if we relax the embeddability constraint in SDP (5) to $c_1(\{\mathbf{v}_i : i \in \{0\} \cup V\}, \|\cdot\|^2) \leq 1 + \delta$ for any constant $\delta > 0$, then the integrality gap may "jump" to $2 - o(1)$. Compare this with a problem such as SPARSEST CUT in which an addition of such a constraint immediately implies integrality gap at most $1 + \delta$.

Negative type metrics that are not ℓ_1 embeddable: Negative type metrics are metrics which are the squares of Euclidean distances of set of points in Euclidean space. Inspired by Theorem 2, we construct a simple negative type metric space $(X, \|\cdot\|^2)$ that does not embed well into ℓ_1. Specifically, we get $c_1(X) \geq \frac{8}{7} - \epsilon$ for every $\epsilon > 0$. In order to show this we prove a new isoperimetric inequality for the hypercube $Q_n = \{-1, 1\}^n$, which we believe is of independent interest.

Theorem 1. *(Generalized Isoperimetric inequality) For every set $S \subseteq Q_n$,*

$$|E(S, S^c)| \geq |S|(n - \log_2 |S|) + p(S).$$

where $p(S)$ denotes the number of vertices $\mathbf{u} \in S$ such that $-\mathbf{u} \in S$.

Khot and Vishnoi [12] constructed an example of an n-point negative type metric that for every $\delta > 0$ requires distortion at least $(\log\log n)^{1/6-\delta}$ to embed into ℓ_1. Krauthgamer and Rabani [17] showed that in fact Khot and Vishnoi's example requires a distortion of at least $\Omega(\log\log n)$. Later Devanur, Khot, Saket and Vishnoi [13] showed an example with distortion $\Omega(\log\log n)$ even on average when embedded into ℓ_1 (we note that our example is also "bad" on average). Although the above examples require nonconstant distortion to embed into ℓ_1, we believe that Theorem 6 is interesting because of its simplicity (to show triangle inequality holds proves to be extremely technical in [12,13]). Prior to Khot and Vishnoi's result, the best known lower bounds (see [12]) were due to Vempala, 10/9 for a metric obtained by a computer search, and Goemans, 1.024 for a metric based on the Leech Lattice. We mention that by [11] every negative type metric embeds into ℓ_1 with distortion $O(\sqrt{\log n}\log\log n)$.

2   Preliminaries and Notation

A vertex cover of a graph G is a set of vertices that touch all edges. An independent set in G is a set $I \subseteq V$ such that no edge $e \in E$ joins two vertices in I. We denote by $\alpha(G)$ the size of the maximum independent set of G. Vectors are always denoted in bold font (such as $\mathbf{v}$, $\mathbf{w}$, etc.); $\|\mathbf{v}\|$ stands for the Euclidean norm of $\mathbf{v}$, $\mathbf{u} \cdot \mathbf{v}$ for the inner product of $\mathbf{u}$ and $\mathbf{v}$, and $\mathbf{u} \otimes \mathbf{v}$ for their tensor product. Specifically, if $\mathbf{v}, \mathbf{u} \in \mathbb{R}^n$, $\mathbf{u} \otimes \mathbf{v}$ is the vector with coordinates indexed by ordered pairs $(i,j) \in [n]^2$ that assumes value $\mathbf{u}_i\mathbf{v}_j$ on coordinate (i,j). Similarly, the tensor product of more than two vectors is defined. It is easy to see that $(\mathbf{u} \otimes \mathbf{v}).(\mathbf{u}' \otimes \mathbf{v}') = (\mathbf{u} \cdot \mathbf{u}')(\mathbf{v} \cdot \mathbf{v}')$. For two vectors $\mathbf{u} \in \mathbb{R}^n$ and $\mathbf{v} \in \mathbb{R}^m$, denote by $(\mathbf{u}, \mathbf{v}) \in \mathbb{R}^{n+m}$ the vector whose projection to the first n coordinates is $\mathbf{u}$ and to the last m coordinates is $\mathbf{v}$.

Next, we give a few basic definitions and facts about finite metric spaces. A metric space (X, d_X) embeds with distortion at most D into (Y, d_Y) if there exists a mapping $\phi : X \mapsto Y$ so that for all $a, b \in X$ $\gamma \cdot d_X(a,b) \leq d_Y(\phi(a), \phi(b)) \leq \gamma D \cdot d_X(a,b)$, for some $\gamma > 0$. We say that (X, d) is ℓ_1 embeddable if it can be embedded with distortion 1 into $\mathbb{R}^m$ equipped with the ℓ_1 norm. An ℓ_2^2 distance on X is a distance function for which there there are vectors $\mathbf{v}_x \in \mathbb{R}^m$ for every $x \in X$ so that $d(x, y) = \|\mathbf{v}_x - \mathbf{v}_y\|^2$. If, in addition, d satisfies triangle inequality, we say that d is an ℓ_2^2 metric or negative type metric. It is well known [16] that every ℓ_1 embeddable metric is also a negative type metric.

3 ℓ_1 and Integrality Gap of SDPs for Vertex Cover – An "All or Nothing" Phenomenon

It is well known that for SPARSEST CUT there is a tight connection between ℓ_1 embeddability and integrality gap. In fact the integrality gap is bounded above by the least ℓ_1 distortion of the SDP solution. At the other extreme stand problems like MAX CUT and MULTI CUT, where ℓ_1 embeddability does not provide any strong evidence for small integrality gap. In this section we show that VERTEX COVER falls somewhere between these two classes of ℓ_1-integrality gap relationship witnessing a sharp transition in integrality gap in the following sense: while ℓ_1 embeddability implies no integrality gap, allowing a small distortion, say 1.001 does not prevent integrality gap of $2 - o(1)$!

Theorem 2. *For a graph $G = (V, E)$, the answer to the SDP formulated in SDP (5) is the size of the minimum vertex cover of G.*

Proof. Let d be the metric solution of SDP (5). We know that d is the result of an ℓ_2^2 unit representation (i.e., it comes from square norms between unit vectors), and furthermore it is ℓ_1 embeddable. By a well known fact about ℓ_1 embeddable metrics (see, e.g. [16]) we can assume that there exist $\lambda_t > 0$ and $f_t : \{0\} \cup V \to \{-1, 1\}$, $t = 1, \ldots, m$, such that

$$\|\mathbf{v}_i - \mathbf{v}_j\|^2 = \sum_{t=1}^{m} \lambda_t |f_t(i) - f_t(j)|, \qquad (6)$$

for every $i, j \in \{0\} \cup V$. Without loss of generality, we can assume that $f_t(0) = 1$ for every t. For convenience, we switch to talk about INDEPENDENT SET and its relaxation, which is the same as SDP (5) except the objective becomes $\text{Max} \sum_{i \in V} \|\mathbf{v}_0 - \mathbf{v}_i\|^2 / 4$. Obviously, the theorem follows from showing that this is an exact relaxation.

We argue that (i) $I_t = \{i \in V : f_t(i) = -1\}$ is a (nonempty) independent set for every t, and (ii) $\sum \lambda_t = 2$. Assuming these two statements we get

$$\sum_{i \in V} \frac{\|\mathbf{v}_i - \mathbf{v}_0\|^2}{4} = \sum_{i \in V} \frac{\sum_{t=1}^{m} \lambda_t |1 - f_t(i)|}{4} = \sum_{t=1}^{m} \frac{\lambda_t |I_t|}{2} \leq \max_{t \in [m]} |I_t| \leq \alpha(G),$$

and so the relaxation is exact and we are done.

We now prove the two statements. The first is rather straightforward: For $i, j \in I_t$, (6) implies that $d(i, 0) + d(0, j) > d(i, j)$. It follows that ij cannot be an edge else it would violate the first condition of the SDP. (We may assume that I_t is nonempty since otherwise the $f_t(\cdot)$ terms have no contribution in (6).) The second statement is more surprising and uses the fact that the solution is optimal. The falsity of such a statement for the problem of MAX CUT (say) explains the different behaviour of the latter problem with respect to integrality gaps of ℓ_1 embeddable solutions. We now describe the proof.

Let $\mathbf{v}_i' = (\sqrt{\lambda_1/2} f_1(i), \ldots, \sqrt{\lambda_m/2} f_m(i), 0)$. From (6) we conclude that $\|\mathbf{v}_i' - \mathbf{v}_j'\|^2 = \|\mathbf{v}_i - \mathbf{v}_j\|^2$, hence there exists a vector $\mathbf{w} = (w_1, w_2, \ldots, w_{m+1}) \in \mathbb{R}^{m+1}$ and an orthogonal transformation T, such that $\mathbf{v}_i = T(\mathbf{v}_i' + \mathbf{w})$. We know that

$$1 = \|\mathbf{v}_i\|^2 = \|T(\mathbf{v}_i' + \mathbf{w})\|^2 = \|\mathbf{v}_i' + \mathbf{w}\|^2 = w_{m+1}^2 + \sum_{t=1}^{m} (\sqrt{\lambda_t/2} f_t(i) + w_t)^2. \quad (7)$$

Since $\|\mathbf{v}'_i\|^2 = \|\mathbf{v}'_0\|^2 = \sum_{t+1}^m \lambda_t/2$, for every $i \in V \cup \{0\}$, from (7) we get $\mathbf{v}'_0 \cdot \mathbf{w} = \mathbf{v}'_i \cdot \mathbf{w}$. Summing this over all $i \in V$, we have

$$|V|(\mathbf{v}'_0 \cdot \mathbf{w}) = \sum_{i \in V} \mathbf{v}'_i \cdot \mathbf{w} = \sum_{t=1}^m (|V| - 2|I_t|)\sqrt{\lambda_t/2}w_t,$$

or

$$\sum_{t=1}^m |V|\sqrt{\lambda_t/2}w_t = \sum_{t=1}^m (|V| - 2|I_t|)\sqrt{\lambda_t/2}w_t,$$

and therefore

$$\sum_{t=1}^m |I_t|\sqrt{\lambda_t/2}w_t = 0. \tag{8}$$

Now (7) and (8) imply that

$$\max_{t \in [m]} |I_t| \geq \sum_{t=1}^m (\sqrt{\lambda_t/2}f_t(0) + w_t)^2 |I_t| = \sum_{t=1}^m \left(\frac{\lambda_t|I_t|}{2} + w_t^2|I_t|\right) \geq \sum_{t=1}^m \frac{\lambda_t|I_t|}{2}. \tag{9}$$

As we have observed before

$$\sum_{t=1}^m \frac{\lambda_t|I_t|}{2} = \sum_{i \in V} \frac{\|\mathbf{v}_i - \mathbf{v}_0\|^2}{4}$$

which means (as clearly $\sum_{i \in V} \frac{\|\mathbf{v}_i - \mathbf{v}_0\|^2}{4} \geq \alpha(G)$) that the inequalities in (9) must be tight. Now, since $|I_t| > 0$ we get that $\mathbf{w} = \mathbf{0}$ and from (7) we get the second statement, i.e., $\sum \lambda_t = 2$. This concludes the proof. □

Now let δ be an arbitrary positive number, and let us relax the last constraint in SDP (5) to get

$$\begin{aligned}
\textbf{Min}\ & \sum_{i \in V} 1 - \|\mathbf{v}_0 - \mathbf{v}_i\|^2/4 \\
\text{s.t.}\ & \|\mathbf{v}_i - \mathbf{v}_0\|^2 + \|\mathbf{v}_j - \mathbf{v}_0\|^2 = \|\mathbf{v}_i - \mathbf{v}_j\|^2 & \forall\ ij \in E \\
& \|\mathbf{v}_i\| = 1 & \forall\ i \in \{0\} \cup V \\
& c_1(\{\mathbf{v}_i : i \in \{0\} \cup V\}, \|\cdot\|^2) \leq 1 + \delta
\end{aligned}$$

Theorem 3. *For every $\epsilon > 0$, there is a graph G for which $\frac{vc(G)}{sd(G)} \geq 2 - \epsilon$, where $sd(G)$ is the solution to the above SDP.*

The proof appears in the next section after we describe Charikar's construction.

4   Integrality Gap for Stronger Semidefinite Formulations

In this section we discuss the integrality gap for stronger semidefinite formulations of vertex cover. In particular we show that Charikar's construction satisfies both SDPs (11) and (4). We start by describing this construction.

4.1 Charikar's Construction

The graphs used in the construction are the so-called Hamming graphs. These are graphs with vertices $\{-1, 1\}^n$ and two vertices are adjacent if their Hamming distance is exactly an even integer $d = \gamma n$. A result of Frankl and Rödl [18] shows that $vc(G) \geq 2^n - (2 - \delta)^n$, where $\delta > 0$ is a constant depending only on γ. In fact, when one considers the exact dependency of δ in γ it can be shown (see [14]) that as long as $\gamma = \Omega(\sqrt{\log n / n})$ then any vertex cover comprises $1 - O(1/n)$ fraction of the graph. Kleinberg and Goemans [7] showed that by choosing a constant γ and n sufficiently large, this graph gives an integrality gap of $2 - \epsilon$ for SDP (1). Charikar [8] showed that in fact G implies the same result for the SDP formulation in (2) too. To this end he introduced the following solution to SDP (2):

For every $\mathbf{u}_i \in \{-1, 1\}^n$, define $\mathbf{u}'_i = \mathbf{u}_i / \sqrt{n}$, so that $\mathbf{u}'_i \cdot \mathbf{u}'_i = 1$. Let $\lambda = 1 - 2\gamma$, $q(x) = x^{2t} + 2t\lambda^{2t-1}x$ and define $\mathbf{y}_0 = (0, \ldots, 0, 1)$, and

$$
\mathbf{y}_i = \sqrt{\frac{1 - \beta^2}{q(1)}} \left(\underbrace{\mathbf{u}'_i \otimes \ldots \otimes \mathbf{u}'_i}_{2t \text{ times}}, \sqrt{2t\lambda^{2t-1}}\mathbf{u}'_i, 0 \right) + \beta\mathbf{y}_0,
$$

where β will be determined later. Note that $\mathbf{y}_i$ is normalized to satisfy $\|\mathbf{y}_i\| = 1$.

Moreover $\mathbf{y}_i$ is defined so that $\mathbf{y}_i \cdot \mathbf{y}_j$ takes its minimum value when $ij \in E$, i.e., when $\mathbf{u}'_i \cdot \mathbf{u}'_j = -\lambda$. As is shown in [8], for every $\epsilon > 0$ we may set $t = \Omega(\frac{1}{\epsilon}), \beta = \Theta(1/t), \gamma = \frac{1}{4t}$ to get that $(\mathbf{y}_0 - \mathbf{y}_i) \cdot (\mathbf{y}_0 - \mathbf{y}_j) = 0$ for $ij \in E$, while $(\mathbf{y}_0 - \mathbf{y}_i) \cdot (\mathbf{y}_0 - \mathbf{y}_j) \geq 0$ always.

Now we verify that all the triangle inequalities, i.e., the second constraint of SDP (2) are satisfied: First note that since every coordinate takes only two different values for the vectors in $\{\mathbf{y}_i : i \in V\}$, it is easy to see that $c_1(\{\mathbf{y}_i : i \in V\}, \|\cdot\|^2) = 1$. So the triangle inequality holds when $i, j, k \in V$. When $i = 0$ or $j = 0$, the inequality is trivial, and it only remains to verify the case that $k = 0$, i.e., $(\mathbf{y}_0 - \mathbf{y}_i) \cdot (\mathbf{y}_0 - \mathbf{y}_j) \geq 0$, which was already mentioned above. Now $\sum_{i \in V}(1 + \mathbf{y}_0 \cdot \mathbf{y}_i)/2 = \frac{1+\beta}{2} \cdot |V| = (\frac{1}{2} + O(\epsilon)) |V|$. In our application, we prefer to set γ and ϵ to be $\Omega(\sqrt{\frac{\log \log n}{\log n}})$ and since, by the above comment, $vc(G) = (1 - O(1/n))|V|$ the integrality gap we get is

$$
(1 - O(1/n))/(1/2 + O(\epsilon)) = 2 - O(\epsilon) = 2 - O\left(\sqrt{\frac{\log \log |V|}{\log |V|}}\right).
$$

4.2 Proof of Theorem 3

We show that the negative type metric implied by Charikar's solution (after adjusting the parameters appropriately) requires distortion of at most $1 + \delta$. Let $\mathbf{y}_i$ and $\mathbf{u}'_i$ be defined as in Section 4.1. To prove Theorem 3, it is sufficient to prove that $c_1(\{\mathbf{y}_i : i \in \{0\} \cup V\}, \|\cdot\|^2) = 1 + o(1)$. Note that every coordinate of $\mathbf{y}_i$ for all $i \in V$ takes at most two different values. It is easy to see that this implies $c_1(\{\mathbf{y}_i : i \in V\}, \|\cdot\|^2) = 1$. In fact

$$f : \mathbf{y}_i \mapsto \frac{1-\beta^2}{q(1)} \left(\frac{2}{n^t} \underbrace{\mathbf{u}'_i \otimes \ldots \otimes \mathbf{u}'_i}_{2t \text{ times}}, \frac{2}{\sqrt{n}} 2t\lambda^{2t-1} \mathbf{u}'_i \right), \tag{10}$$

is an isometry from $(\{\mathbf{y}_i : i \in V\}, \|\cdot\|^2)$ to ℓ_1. For $i \in V$, we have

$$\|f(\mathbf{y}_i)\|_1 = \frac{1-\beta^2}{q(1)} \left(\frac{2}{n^t} \times \frac{n^{2t}}{n^t} + \frac{2}{\sqrt{n}} 2t\lambda^{2t-1} \frac{1}{\sqrt{n}} + 0 \right) = \frac{1-\beta^2}{q(1)} \times (2 + 4t\lambda^{2t-1})$$

Since $\beta = \Theta(\frac{1}{t})$, recalling that $\lambda = 1 - \frac{1}{2t}$, it is easy to see that for every $i \in V$, $\lim_{t\to\infty} \|f(\mathbf{y}_i)\|_1 = 2$. On the other hand for every $i \in V$

$$\lim_{t\to\infty} \|\mathbf{y}_i - \mathbf{y}_0\|^2 = \lim_{t\to\infty} 2 - 2(\mathbf{y}_i \cdot \mathbf{y}_0) = \lim_{t\to\infty} 2 - 2\beta = 2.$$

So if we extend f to $\{\mathbf{y}_i : i \in V \cup \{0\}\}$ by defining $f(\mathbf{y}_0) = \mathbf{0}$, we obtain a mapping from $(\{\mathbf{y}_i : i \in V \cup \{0\}\}, \|\cdot\|^2)$ to ℓ_1 whose distortion tends to 1 as t goes to infinity.   □

4.3   Karakostas' and Pentagonal SDP Formulations

Karakostas suggests the following SDP relaxation, that is the result of adding to SDP (3) the triangle inequalities applied to the set $\{\mathbf{v}_i : i \in V \cup \{0\}\} \cup \{-\mathbf{v}_i : i \in V \cup \{0\}\}$.

$$
\begin{aligned}
\textbf{Min } &\sum_{i \in V} (1 + \mathbf{v}_0 \mathbf{v}_i)/2 \\
\text{s.t. } &(\mathbf{v}_i - \mathbf{v}_0) \cdot (\mathbf{v}_j - \mathbf{v}_0) = 0 && \forall\, ij \in E \\
&(\mathbf{v}_i - \mathbf{v}_k) \cdot (\mathbf{v}_j - \mathbf{v}_k) \geq 0 && \forall\, i,j,k \in V \\
&(\mathbf{v}_i + \mathbf{v}_k) \cdot (\mathbf{v}_j - \mathbf{v}_k) \geq 0 && \forall\, i,j,k \in V \\
&(\mathbf{v}_i + \mathbf{v}_k) \cdot (\mathbf{v}_j + \mathbf{v}_k) \geq 0 && \forall\, i,j,k \in V \\
&\|\mathbf{v}_i\| = 1 && \forall\, i \in \{0\} \cup V.
\end{aligned}
\tag{11}
$$

Theorem 4. *The integrality gap of SDP (11) is $2 - O(\sqrt{\log\log|V|/\log|V|})$.*

Proof. We show that Charikar's construction satisfies formulation (11). By [8] and from the discussion in Section 4.1, it follows that all edge constraints and triangle inequalities of the original points hold. Hence we need only consider triangle inequalities with at least one nonoriginal point. By homogeneity, we may assume that there is exactly one such point.

Since all coordinates of $\mathbf{y}_i$ for $i > 0$ assume only two values with the same absolute value, it is clear that not only does the metric they induce is ℓ_1 but also taking $\pm\mathbf{y}_i$ for $i > 0$ gives an ℓ_1 metric; in particular all triangle inequalities that involve these vectors are satisfied. In fact, we may fix our attention to triangles in which $\pm\mathbf{y}_0$ is the middle point. This is since

$$(\pm\mathbf{y}_i - \pm\mathbf{y}_j) \cdot (\mathbf{y}_0 - \pm\mathbf{y}_j) = (\pm\mathbf{y}_j - \mathbf{y}_0) \cdot (\mp\mathbf{y}_i - \mathbf{y}_0).$$

Consequently, and using symmetry, we are left with checking the nonnegativity of $(\mathbf{y}_i + \mathbf{y}_0) \cdot (\mathbf{y}_j + \mathbf{y}_0)$ and $(-\mathbf{y}_i - \mathbf{y}_0) \cdot (\mathbf{y}_j - \mathbf{y}_0)$.

$$(\mathbf{y}_i + \mathbf{y}_0) \cdot (\mathbf{y}_j + \mathbf{y}_0) = 1 + \mathbf{y}_0 \cdot (\mathbf{y}_i + \mathbf{y}_j) + \mathbf{y}_i \cdot \mathbf{y}_j \geq 1 + 2\beta + \beta^2 - (1 - \beta^2) = 2\beta(1 + \beta) \geq 0.$$

Finally, $(-\mathbf{y}_i - \mathbf{y}_0) \cdot (\mathbf{y}_j - \mathbf{y}_0) = 1 + \mathbf{y}_0 \cdot (\mathbf{y}_i - \mathbf{y}_j) - \mathbf{y}_i \cdot \mathbf{y}_j = 1 - \mathbf{y}_i \cdot \mathbf{y}_j \geq 0$ as $\mathbf{y}_i, \mathbf{y}_j$ are of norm 1.   □

By now we know that taking all the ℓ_1 constraints leads to an exact relaxation, but not a tractable one. Our goal here is to explore the possibility that stepping towards ℓ_1 embeddability while still maintaining computational feasibility would considerably reduce the integrality gap. A canonical subset of valid inequalities for ℓ_1 metrics is the so-called *Hypermetric inequalities*. Again, taking all these constraints is not feasible, and we instead consider the effect of adding a small number of such constraints. The simplest hypermetric inequalities beside triangle inequalities are the *pentagonal* inequalities. These constraints consider two sets of points of size 2 and 3, and require that the sum of the distances between points in different sets is at least the sum of the distances within sets. Formally, let $S, T \subset X$, $|S| = 2$, $|T| = 3$, then we have the inequality $\sum_{i \in S, j \in T} d(i, j) \geq \sum_{i, j \in S} d(i, j) + \sum_{i, j \in T} d(i, j)$. To appreciate this inequality it is useful to describe where it fails. Consider the graph metric of $K_{2,3}$. Here, the LHS of the inequality is 6 and the RHS is 8, hence $K_{2,3}$ violates the pentagonal inequality. In the following theorem we show that this strengthening past the triangle inequalities fails to reduce the integrality gap significantly.

Theorem 5. *The integrality gap of SDP (4) is $2 - O(\sqrt{\log \log |V| / \log |V|})$.*

Proof. We show that the metric space used in Charikar's construction is feasible. By ignoring $\mathbf{y}_0$ the space defined by $d(i, j) = \|\mathbf{y}_i - \mathbf{y}_j\|^2$ is ℓ_1 embeddable. Therefore, we wish to consider a pentagonal inequality containing $\mathbf{y}_0$ and four other vectors, denoted by $\mathbf{y}_1, \mathbf{y}_2, \mathbf{y}_3, \mathbf{y}_4$. Assume first that the partition of the five points in the inequality puts $\mathbf{y}_0$ together with two other points; then, using the fact that $d(0, 1) = d(0, 2) = d(0, 3) = d(0, 4)$ and triangle inequality we get that such an inequality must hold. It remains to consider a partition of the form $(\{\mathbf{y}_1, \mathbf{y}_2, \mathbf{y}_3\}, \{\mathbf{y}_4, \mathbf{y}_0\})$, and show that:

$$d(1,2) + d(1,3) + d(2,3) + d(0,4) \leq d(1,4) + d(2,4) + d(3,4) + d(0,1) + d(0,2) + d(0,3)$$

Recall that every $\mathbf{y}_i$ is associated with a $\{-1, 1\}$ vector $\mathbf{u}_i$ and with its normalized multiple $\mathbf{u}'_i$. After substituting the distances as functions of the normalized vectors, our goal will then be to show:

$$q(\mathbf{u}'_1 \cdot \mathbf{u}'_2) + q(\mathbf{u}'_1 \cdot \mathbf{u}'_3) + q(\mathbf{u}'_2 \cdot \mathbf{u}'_3) - q(\mathbf{u}'_1 \cdot \mathbf{u}'_4) - q(\mathbf{u}'_2 \cdot \mathbf{u}'_4) - q(\mathbf{u}'_3 \cdot \mathbf{u}'_4) \geq -\frac{2q(1)}{1 + \beta} \quad (12)$$

Let $E = q(\mathbf{u}'_1 \cdot \mathbf{u}'_2) + q(\mathbf{u}'_1 \cdot \mathbf{u}'_3) + q(\mathbf{u}'_2 \cdot \mathbf{u}'_3) - q(\mathbf{u}'_1 \cdot \mathbf{u}'_4) - q(\mathbf{u}'_2 \cdot \mathbf{u}'_4) - q(\mathbf{u}'_3 \cdot \mathbf{u}'_4)$. The rest of the proof analyzes the minima of the function E and ensures that (12) is satisfied at those minima. We first partition the coordinates of the original hypercube into four sets according to the values assumed by $\mathbf{u}_1, \mathbf{u}_2$ and $\mathbf{u}_3$. We may assume that in any coordinate at most one of these get the value 1 (otherwise multiply the values of the coordinate by -1). We get four sets, P_0 for the coordinates in which all three vectors assume value -1, and P_1, P_2, P_3 for the coordinates in which exactly $\mathbf{u}_1, \mathbf{u}_2, \mathbf{u}_3$ respectively assumes value 1.

We now argue about the coordinates of $\mathbf{u}_4$ at a minimum of E.

Proposition 1. *If there is a violating configuration, then there is one in which $\mathbf{u}_4$ is either all 1 or all -1 on each one of P_0, P_1, P_2, P_3*

For P_0, we can in fact say something stronger than we do for P_1, P_2, P_3:

Proposition 2. *If there is a violating configuration, then there is one in which $\mathbf{u}_4$ has all the P_0 coordinates set to -1.*

Proposition 1 is based solely on the (strict) convexity of q. Proposition 2 is more involved and uses more properties of the polynomial q. Due to lack of space we omit the proofs of the propositions from this version.

The above characterizations significantly limit the type of configurations we need to check. The cases that are left are characterized by whether $\mathbf{u}_4$ is 1 or -1 on each of P_1, P_2, P_3. By symmetry all we really need to know is $\xi(\mathbf{u}_4) = |\{i : \mathbf{u}_4 \text{ is } 1 \text{ on } P_i\}|$. If $\xi(\mathbf{u}_4) = 1$ it means that $\mathbf{u}_4$ is the same as one of $\mathbf{u}_1, \mathbf{u}_2$ or $\mathbf{u}_4$ hence the pentagonal inequality reduces to the triangle inequality, which we already know is valid. If $\xi(\mathbf{u}_4) = 3$, it is easy to see that $\mathbf{u}_1'\mathbf{u}_4' = \mathbf{u}_2'\mathbf{u}_3'$, and likewise $\mathbf{u}_2'\mathbf{u}_4' = \mathbf{u}_1'\mathbf{u}_3'$ and $\mathbf{u}_3'\mathbf{u}_4' = \mathbf{u}_1'\mathbf{u}_2'$ hence E is 0 for these cases, which means that (12) is satisfied.

We are left with the cases $\xi(\mathbf{u}_4) \in \{0, 2\}$.

Case 1: $\xi(\mathbf{u}_4) = 0$

Let $x = \frac{2}{n}|P_1|, y = \frac{2}{n}|P_2|, z = \frac{2}{n}|P_3|$. Notice that $x + y + z = \frac{2}{n}(|P_1| + |P_2| + |P_3|) \leq 2$, as these sets are disjoint. Now, think of

$$E = q(1 - (x+y)) + q(1 - (x+z)) + q(1 - (y+z)) - q(1-x) - q(1-y) - q(1-z)$$

as a function from $\mathbb{R}^3$ to $\mathbb{R}$. We will show that E achieves its minimum at points where either x, y or z are zero. Assume that $0 \leq x \leq y \leq z$.

Consider the function $g(\delta) = E(x - \delta, y + \delta, z)$. It is easy to see that $g'(0) = q'(1 - (x + z)) - q'(1 - (y + z)) - q'(1 - x) + q'(1 - y)$. We will prove that $g'(\delta) \leq 0$ for every $\delta \in [0, x]$. This, by the Mean Value Theorem implies that $E(0, x + y, z) \leq E(x, y, z)$, and hence we may assume that $x = 0$. This means that $\mathbf{y}_1 = \mathbf{y}_4$ which reduces to the triangle inequality on $\mathbf{y}_0, \mathbf{y}_2, \mathbf{y}_3$.

Note that in $g'(0)$, the two arguments in the terms with positive sign have the same average as the arguments in the terms with negative sign, namely $\mu = 1 - (x+y+z)/2$. We now have $g'(0) = q'(\mu + b) - q'(\mu + s) - q'(\mu - s) + q'(\mu - b)$, where $b = (x - y + z)/2, s = (-x + y + z)/2$. After calculations:

$$g'(0) = 2t[(\mu + b)^{2t-1} + (\mu - b)^{2t-1} - (\mu + s)^{2t-1} - (\mu - s)^{2t-1}]$$

$$= 4t \sum_{i \text{ even}} \binom{2t - 1}{i} \mu^{2t-1-i}(b^i - s^i)$$

Observe that $\mu \geq 0$. Since $x \leq y$, we get that $s \geq b \geq 0$. This means that $g'(0) \leq 0$. It can be easily checked that the same argument holds if we replace x, y by $x - \delta$ and $y + \delta$. Hence $g'(\delta) \leq 0$ for every $\delta \in [0, x]$, and we are done.

Case 2: $\xi(\mathbf{u}_4) = 2$ The expression for E is now:

$$E = q(1-(x+y))+q(1-(x+z))+q(1-(y+z))-q(1-x)-q(1-y)-q(1-(x+y+z))$$

Although $E(x, y, z)$ is different than in Case 1, the important observation is that if we consider again the function $g(\delta) = E(x - \delta, y + \delta, z)$ then the derivative $g'(\delta)$ is the same as in Case 1 and hence the same analysis shows that $E(0, x + y, z) \leq E(x, y, z)$. But if $x = 0$, then $\mathbf{y}_2$ identifies with $\mathbf{y}_4$ and the inequality reduces to the triangle inequality on $\mathbf{y}_0, \mathbf{y}_1, \mathbf{y}_3$. $\qquad\square$

5 Lower Bound for Embedding Negative Type Metrics into ℓ_1

While, in view of Theorem 3, Charikar's metric does not supply an example that is far from ℓ_1, we may still (partly motivated by Theorem 2) utilize the idea of "tensoring the cube" and then adding some more points in order to achieve negative type metrics that are not ℓ_1 embeddable. Our starting point is an isoperimetric inequality on the cube that generalizes the standard one. Such a setting is also relevant in [12,17] where harmonic analysis tools are used to bound expansion; these tools are unlikely to be applicable to our case where the interest and improvements lie in the constants.

Theorem 1. *(Generalized Isoperimetric inequality) For every set $S \subseteq Q_n$,*

$$|E(S, S^c)| \geq |S|(n - \log_2 |S|) + p(S).$$

where $p(S)$ denotes the number of vertices $\mathbf{u} \in S$ such that $-\mathbf{u} \in S$.

Proof. We use induction on n. Divide Q_n into two sets $V_1 = \{\mathbf{u} : \mathbf{u}_1 = 1\}$ and $V_{-1} = \{\mathbf{u} : \mathbf{u}_1 = -1\}$. Let $S_1 = S \cap V_1$ and $S_{-1} = S \cap V_{-1}$. Now, $E(S, S^c)$ is the disjoint union of $E(S_1, V_1 \setminus S_1)$, $E(S_{-1}, V_{-1} \setminus S_{-1})$, and $E(S_1, V_{-1} \setminus S_{-1}) \cup E(S_{-1}, V_1 \setminus S_1)$. Define the operator $\widehat{}$ on Q_n to be the projection onto the last $n - 1$ coordinates, so for example $\widehat{S_1} = \{\mathbf{u} \in Q_{n-1} : (1, \mathbf{u}) \in S_1\}$. It is easy to observe that $|E(S_1, V_{-1} \setminus S_{-1}) \cup E(S_{-1}, V_1 \setminus S_1)| = |\widehat{S_1} \triangle \widehat{S_{-1}}|$. We argue that

$$p(S) + |S_1| - |S_{-1}| \leq p(\widehat{S_1}) + p(\widehat{S_{-1}}) + |\widehat{S_1} \triangle \widehat{S_{-1}}|. \tag{13}$$

To prove (13), for every $\mathbf{u} \in \{-1, 1\}^{n-1}$, we show that the contribution of $(1, \mathbf{u})$, $(1, -\mathbf{u})$, $(-1, \mathbf{u})$, and $(-1, -\mathbf{u})$ to the right hand side of (13) is at least as large as their contribution to the left hand side: This is trivial if the contribution of these four vectors to $p(S)$ is not more than their contribution to $p(\widehat{S_1})$, and $p(\widehat{S_{-1}})$. We therefore assume that the contribution of the four vectors to $p(S)$, $p(\widehat{S_1})$, and $p(\widehat{S_{-1}})$ are 2, 0, and 0, respectively. Then without loss of generality we may assume that $(1, \mathbf{u}), (-1, -\mathbf{u}) \in S$ and $(1, -\mathbf{u}), (-1, \mathbf{u}) \notin S$, and in this case the contribution to both sides is 2. By induction hypothesis and (13) we get

$$
\begin{aligned}
|E(S, S^c)| &= |E(\widehat{S_1}, Q_{n-1} \setminus \widehat{S_1}| + |E(\widehat{S_{-1}}, Q_{n-1} \setminus \widehat{S_{-1}}| + |\widehat{S_1} \Delta \widehat{S_{-1}}| \\
&\geq |S_1|(n - 1 - \log_2 |S_1|) + p(\widehat{S_1}) + |S_{-1}|(n - 1 - \log_2 |S_{-1}|) + p(\widehat{S_{-1}}) + |\widehat{S_1} \Delta \widehat{S_{-1}}| \\
&\geq |S|n - |S| - (|S_1| \log_2 |S_1| + |S_{-1}| \log_2 |S_{-1}|) + p(\widehat{S_1}) + p(\widehat{S_{-1}}) + |\widehat{S_1} \Delta \widehat{S_{-1}}| \\
&\geq |S|n - (2|S_{-1}| + |S_1| \log_2 |S_1| + |S_{-1}| \log_2 |S_{-1}|) + p(S).
\end{aligned}
$$

Now the lemma follows from the fact that $2|S_{-1}| + |S_1| \log_2 |S_1| + |S_{-1}| \log_2 |S_{-1}| \leq |S| \log_2 |S|$, which can be obtained using easy calculus. $\square$

We call a set $S \subseteq Q_n$ *symmetric* if $-\mathbf{u} \in S$ whenever $\mathbf{u} \in S$. Note that $p(S) = |S|$ for symmetric sets S.

Corollary 1. *For every symmetric set $S \subseteq Q_n$*

$$
|E(S, S^c)| \geq |S|(n - \log_2 |S| + 1).
$$

The corollary above implies the following Poincaré inequality.

Proposition 3. *(Poincaré inequality for the cube and an additional point) Let $f : Q_n \cup \{\mathbf{0}\} \to \mathbb{R}^m$ satisfy that $f(\mathbf{u}) = f(-\mathbf{u})$ for every $\mathbf{u} \in Q_n$, and let $\alpha = \frac{\ln 2}{14 - 8 \ln 2}$. Then the following Poincaré inequality holds.*

$$
\frac{1}{2^n} \cdot \frac{8}{7}(4\alpha + 1/2) \sum_{\mathbf{u}, \mathbf{v} \in Q_n} \|f(\mathbf{u}) - f(\mathbf{v})\|_1 \leq \alpha \sum_{\mathbf{uv} \in E} \|f(\mathbf{u}) - f(\mathbf{v})\|_1 + \frac{1}{2} \sum_{\mathbf{u} \in Q_n} \|f(\mathbf{u}) - f(\mathbf{0})\|_1
$$

Proof. It is enough to prove the above inequality for $f : V \to \{0, 1\}$. We may assume without loss of generality that $f(\mathbf{0}) = 0$. Associating S with $\{\mathbf{u} : f(\mathbf{u}) = 1\}$, the inequality of the proposition reduces to

$$
\frac{1}{2^n} \frac{8}{7}(4\alpha + 1/2)|S||S^c| \leq \alpha|E(S, S^c)| + |S|/2, \tag{14}
$$

where S is a symmetric set, owing to the condition $f(\mathbf{u}) = f(-\mathbf{u})$. From the isoperimetric inequality of Theorem 1 we have that $|E(S, S^c)| \geq |S|(x + 1)$ for $x = n - \log_2 |S|$ and so

$$
\left(\frac{\alpha(x + 1) + 1/2}{1 - 2^{-x}} \right) \frac{1}{2^n} |S||S^c| \leq \alpha|E(S, S^c)| + |S|/2.
$$

It can be shown that $\frac{\alpha(x+1)+1/2}{1-2^{-x}}$ attains its minimum in $[1, \infty)$ at $x = 3$ whence $\frac{\alpha(x+1)+1/2}{1-2^{-x}} \geq \frac{4\alpha+1/2}{7/8}$, and Inequality (14) is proven. $\square$

Theorem 6. *Let $V = \{\tilde{\mathbf{u}} : \mathbf{u} \in Q_n\} \cup \{\mathbf{0}\}$, where $\tilde{\mathbf{u}} = \mathbf{u} \otimes \mathbf{u}$. Then for the semi-metric space $X = (V, \| \cdot \|^2)$ we have $c_1(X) \geq \frac{8}{7} - \epsilon$, for every $\epsilon > 0$ and sufficiently large n.*

Proof. We start with an informal description of the proof. The heart of the argument is showing that the cuts that participate in a supposedly good ℓ_1 embedding of X cannot be balanced on one hand, and cannot be imbalanced on the other. First notice that the average distance in X is almost double that of the distance between $\mathbf{0}$ and any other point (achieving this in a cube structure without violating the triangle inequality was where the tensor operation came in handy). For a cut metric on the points of X, such a relation only occurs for very imbalanced cuts; hence the representation of balanced cuts in a low distortion embedding cannot be large. On the other hand, comparing the (overall) average distance to the average distance between neighbouring points in the cube shows that any good embedding must use cuts with very small edge expansion, and such cuts in the cube must be balanced (the same argument says that one must use the dimension cuts when embedding the hamming cube into ℓ_1 with low distortion). The fact that only *symmetric cuts* participate in the ℓ_1 embedding (or else the distortion becomes infinite due to the tensor operation) enables us to use the stronger isoperimetric inequality which leads to the current lower bound. We now proceed to the proof.

We may view X as a distance function with points in $\mathbf{u} \in Q_n \cup \{\mathbf{0}\}$, and $d(\mathbf{u}, \mathbf{v}) = \|\tilde{\mathbf{u}} - \tilde{\mathbf{v}}\|^2$. We first notice that X is indeed a metric space, i.e., that triangle inequalities are satisfied: notice that $X \setminus \{\mathbf{0}\}$ is a subset of $\{-1, 1\}^{n^2}$. Therefore, the square Euclidean distances is the same (upto a constant) as their ℓ_1 distance. Hence, the only triangle inequality we need to check is $\|\tilde{\mathbf{u}} - \tilde{\mathbf{v}}\|^2 \leq \|\tilde{\mathbf{u}} - \mathbf{0}\|^2 + \|\tilde{\mathbf{v}} - \mathbf{0}\|^2$, which is implied by the fact that $\tilde{\mathbf{u}} \cdot \tilde{\mathbf{v}} = (\mathbf{u} \cdot \mathbf{v})^2$ is always nonnegative.

For every $\mathbf{u}, \mathbf{v} \in Q_n$, we have $d(\mathbf{u}, \mathbf{0}) = \|\tilde{\mathbf{u}}\|^2 = \tilde{\mathbf{u}} \cdot \tilde{\mathbf{u}} = (\mathbf{u} \cdot \mathbf{u})^2 = n^2$, and $d(\mathbf{u}, \mathbf{v}) = \|\tilde{\mathbf{u}} - \tilde{\mathbf{v}}\|^2 = \|\tilde{\mathbf{u}}\|^2 + \|\tilde{\mathbf{v}}\|^2 - 2(\tilde{\mathbf{u}} \cdot \tilde{\mathbf{v}}) = 2n^2 - 2(\mathbf{u} \cdot \mathbf{v})^2$. In particular, if $\mathbf{u}\mathbf{v} \in E$ we have $d(\mathbf{u}, \mathbf{v}) = 2n^2 - 2(n-2)^2 = 8(n-1)$. We next notice that

$$\sum_{\mathbf{u}, \mathbf{v} \in Q_n} d(\mathbf{u}, \mathbf{v}) = 2^{2n} \times 2n^2 - 2\sum_{\mathbf{u}, \mathbf{v}} (\mathbf{u} \cdot \mathbf{v})^2 = 2^{2n} \times 2n^2 - 2\sum_{\mathbf{u}, \mathbf{v}} \left(\sum_i \mathbf{u}_i \mathbf{v}_i\right)^2 = 2^{2n}(2n^2 - 2n),$$

as $\sum_{\mathbf{u}, \mathbf{v}} \mathbf{u}_i \mathbf{v}_i \mathbf{u}_j \mathbf{v}_j$ is 2^{2n} when $i = j$, and 0 otherwise.

Let f be a nonexpanding embedding of X into ℓ_1. Notice that

$$d(\mathbf{u}, -\mathbf{u}) = 2n^2 - 2(\mathbf{u} \cdot \mathbf{v})^2 = 0,$$

and so any embedding with finite distortion must satisfy $f(\mathbf{u}) = f(-\mathbf{u})$. Therefore Inequality (3) can be used and we get that

$$\frac{\alpha \sum_{\mathbf{u}\mathbf{v} \in E} \|f(\tilde{\mathbf{u}}) - f(\tilde{\mathbf{v}})\|_1 + \frac{1}{2} \sum_{\mathbf{u} \in Q_n} \|f(\tilde{\mathbf{u}}) - f(\mathbf{0})\|_1}{\frac{1}{2^n} \sum_{\mathbf{u}, \mathbf{v} \in Q_n} \|f(\tilde{\mathbf{u}}) - f(\tilde{\mathbf{v}})\|_1} \geq \frac{8}{7}(4\alpha + 1/2). \quad (15)$$

On the other hand,

$$\frac{\alpha \sum_{\mathbf{u}\mathbf{v} \in E} d(\mathbf{u}, \mathbf{v}) + \frac{1}{2} \sum_{\mathbf{u} \in Q_n} d(\mathbf{u}, \mathbf{0})}{\frac{1}{2^n} \sum_{\mathbf{u}, \mathbf{v} \in Q_n} d(\mathbf{u}, \mathbf{v})} \frac{8\alpha(n^2 - n) + n^2}{2n^2 - 2n} = 4\alpha + 1/2 + o(1). \quad (16)$$

The discrepancy between (15) and (16) shows that for every $\epsilon > 0$ and for sufficiently large n, the required distortion of V into ℓ_1 is at least $8/7 - \epsilon$.   □

6 Discussion

It is important to understand our results in the context of the Lift and Project system defined by Lovász and Schrijver [19], specifically the one that uses positive semidefinite constraints, called LS_+ (see [20] for relevant discussion). As was mentioned in the introduction, a new result of Georgiou, Magen, Pitassi and Tourlakis [14] shows that after a super-constant number of rounds of LS_+, the integrality gap is still $2 - o(1)$. To relate LS_+ to SDPs one needs to use the conversion $\mathbf{y}_i = 2\mathbf{z}_i - \mathbf{z}_0$, where $\mathbf{y}_i$ is as usual the vectors of the SDP solution and the $\mathbf{z}_i$ are the Cholesky decomposition of the matrix of the lifted variables in the LS_+ system. With this relation in mind, it can be shown that the triangle inequalities with respect to $\mathbf{v}_0$ are implied after as little as one round of LS_+ and so [14] extends Charikar's result on the SDP with these types of triangle inequalities. However, at least for some graphs, triangle inequalities not involving $\mathbf{v}_0$ as well as pentagonal inequalities are *not* implied by any number of rounds of LS_+. To see this, consider the application of LS_+ system to Vertex Cover when the instance is the empty graph. Since for this instance the Linear Program relaxation is tight, lifted inequalities must appear in the first round or not at all. But it is easy to see that even the general triangle inequalities do not appear after one round and thus will never appear. It is important to note that for the graphs used in [14] (which are the same as the ones we use here) we do not know whether the general triangle inequality and whether pentagonal inequalities are implied after a few rounds of LS_+.

Acknowledgment. Special thanks to George Karakostas for very valuable discussions. We also thank the referees for their detailed and insightful comments.

References

1. Dinur, I., Safra, S.: On the hardness of approximating minimum vertex-cover. Annals of Mathematics 162(1), 439–486 (2005)
2. Khot, S.: On the power of unique 2-prover 1-round games. In: Proceedings of the Thirty-Fourth Annual ACM Symposium on Theory of Computing (electronic), New York, pp. 767–775. ACM Press, New York (2002)
3. Khot, S., Regev, O.: Vertex cover might be hard to approximate to within $2 - \epsilon$. In: Proceedings of the 18th IEEE Conference on Computational Complexity, pp. 379–386. IEEE Computer Society Press, Los Alamitos (2003)
4. Goemans, M.X., Williamson, D.P.: Improved approximation algorithms for maximum cut and satisfiability problems using semidefinite programming. J. Assoc. Comput. Mach. 42(6), 1115–1145 (1995)
5. Halperin, E.: Improved approximation algorithms for the vertex cover problem in graphs and hypergraphs. SIAM J. Comput (electronic) 31(5), 1608–1623 (2002)
6. Karakostas, G.: A better approximation ratio for the vertex cover problem. In: Proceedings of the Thirty-Second International Colloquium on Automata, Languages and Programming (2005)
7. Kleinberg, J., Goemans, M.X.: The Lovász theta function and a semidefinite programming relaxation of vertex cover. SIAM J. Discrete Math (electronic) 11(2), 196–204 (1998)

8. Charikar, M.: On semidefinite programming relaxations for graph coloring and vertex cover. In: SODA '02: Proceedings of the thirteenth annual ACM-SIAM symposium on Discrete algorithms, Philadelphia, PA, USA, Society for Industrial and Applied Mathematics, pp. 616–620 (2002)
9. Arora, S., Rao, S., Vazirani, U.: Expander flows, geometric embeddings and graph partitioning. In: Proceedings of the 36th Annual ACM Symposium on Theory of Computing (electronic), New York, pp. 222–231. ACM Press, New York (2004)
10. Chawla, C., Gupta, A., Räcke, H.: Embeddings of negative-type metrics and an improved approximation to generalized sparsest cut. In: SODA '05: Proceedings of the sixteenth annual ACM-SIAM symposium on Discrete algorithms, Vancouer, BC, Canada, pp. 102–111. ACM Press, New York (2005)
11. Arora, S., Lee, J., Naor, A.: Euclidean distortion and the sparsest cut [extended abstract]. In: STOC'05: Proceedings of the 37th Annual ACM Symposium on Theory of Computing, New York, pp. 553–562. ACM Press, New York (2005)
12. Khot, S., Vishnoi, N.: The unique games conjecture, integrality gap for cut problems and embeddability of negative type metrics into ℓ_1. In: Proceedings of The 46-th Annual Symposium on Foundations of Computer Science (2005)
13. Devanur, N., Khot, S., Saket, R., Vishnoi, N.: Integrality gaps for sparsest cut and minimum linear arrangement problems. In: Proceedings of the thirty-eighth annual ACM symposium on Theory of computing, ACM Press, New York (2006)
14. Georgiou, K., Magen, A., Pitassi, T., Tourlakis, I.: Integrality gaps of $2 - o(1)$ for vertex cover sdps in the lovász-schrijver hierarchy. In: ECCCTR: Electronic Colloquium on Computational Complexity, technical reports, TR06-152 (2006)
15. Agarwal, A., Charikar, M., Makarychev, K., Makarychev, Y.: $O(\sqrt{\log n})$ approximation algorithms for min UnCut, min 2CNF deletion, and directed cut problems. In: STOC '05: Proceedings of the thirty-seventh annual ACM symposium on Theory of computing, pp. 573–581. ACM Press, New York, NY, USA (2005)
16. Deza, M., Laurent, M.: Geometry of cuts and metrics, p. 587. Springer-Verlag, Berlin (1997)
17. Krauthgamer, R., Rabani, Y.: Improved lower bounds for embeddings into l_1. In: Proceedings of the ACM-SIAM Symposium on Discrete Algorithms, ACM Press, New York (2006)
18. Frankl, P., Rödl, V.: Forbidden intersections. Trans. Amer. Math. Soc. 300(1), 259–286 (1987)
19. Lovász, L., Schrijver, A.: Cones of matrices and set-functions and 0-1 optimization. SIAM Journal on Optimization 1(2), 166–190 (1991)
20. Arora, S., Alekhnovich, M., Tourlakis, I.: Towards strong nonapproximability results in the lovász-schrijver hierarchy. In: STOC '05: Proceedings of the thirty-seventh annual ACM symposium on Theory of computing, New York, NY, USA, ACM Press (2005)

Optimal Resource Augmentations for Online Knapsack

Kazuo Iwama[1],[*] and Guochuan Zhang[2],[**]

[1] School of Informatics, Kyoto University, Japan
iwama@kuis.kyoto-u.ac.jp
[2] Department of Mathematics, Zhejiang University
Hangzhou, 310027, China
zgc@zju.edu.cn

Abstract. It is known that online knapsack is not competitive. This negative result remains true even if the items are removable. In this paper we consider online removable knapsack with resource augmentation, in which we hold a knapsack of capacity $R \geq 1.0$ and aim at maintaining a feasible packing to maximize the total weight of the items packed. Accepted items can be removed to leave room for newly arriving items. Once an item is rejected/removed it can not be considered again. We evaluate an online algorithm by comparing the resulting packing to an optimal packing that uses a knapsack of capacity one. Optimal online algorithms are derived for both the weighted case (items have arbitrary weights) and the un-weighted case (the weight of an item is equal to its size).

Keywords: On-line algorithms, knapsack, resource augmentation.

1 Introduction

Competitive analysis is a standard approach for evaluating the performance of online algorithms. However, that is not almighty and occasionally gives us only little and/or pessimistic information. For example, a certain type of online problems do not have competitive online algorithms (i.e., the competitive ratio of any online algorithm diverges). Also for some online problem, several algorithms that are very different empirically have almost the same competitive ratio, giving us no useful information, either.

In such a case, there is another popular approach, *resource augmentation* [9], which allows online algorithms to use more resource, by a factor of R, than offline algorithms. In the case of Bin Packing, for example, the online player can use bins of size $R \geq 1.0$ while the offline adversary can use those of size one (and both try to minimize the number of bins to pack given items). Thus the competitive ratio becomes a function in R which is more general than the original one which

[*] Research supported in part by KAKENHI (16092101, 16092215, 16300002).
[**] Research supported in part by NSFC (60573020).

M. Charikar et al. (Eds.): APPROX and RANDOM 2007, LNCS 4627, pp. 180–188, 2007.

is a special case obtained by setting $R = 1.0$. There are several successful results in which we can achieve bounded (and relatively small) competitive ratios even if R is slightly larger than 1.0 [4]. In this paper, we take this approach for the *online removable knapsack problem*.

The problem was introduced in [2]. Items $i_1, i_2, \cdots$ are given sequentially. For each i_j, the online player has to decide if accepting or rejecting it. In order to accept i_j, there must be enough space in the knapsack for the item, and to assure this it is allowed to discard old items in the knapsack. However once an item is rejected or discarded it will never be considered again. It was shown in [2] that one can attain an optimal competitive ratios, $(\sqrt{5}+1)/2$, if items are unweighted (i.e., the weight of each item is the same as its size), but as was shown later by [3], no competitive algorithms exist for the general (i.e., weighted) case.

Our Contribution. Our model in this paper allows the online player to use a knapsack of capacity $R \geq 1$ while the adversary has a knapsack of capacity one. Our results includes: (i) Thanks to the resource augmentation, we can easily prove that the standard greedy algorithm has a bounded competitive ratio, $1/(R - 1)$, for the weighted case. (ii) Interestingly, this simple algorithm is optimal; to show this we use a tricky sequence of sufficiently many items whose weight changes gradually.

Thus, the online player needs twice ($R = 2$) as large capacity as the offline player to attain the same amount of total weight, but can do better for the unweighted case: (iii) For the unweighted input, we can achieve a competitive ratio of $2/(2R-1)$ if $R \geq (1+\sqrt{2})/2$ and $(\sqrt{4R+1}+1)/(2R)$ if $R < (1+\sqrt{2})/2$. (Note that the ratio becomes 1.0 when $R = 1.5$.) This ratio is again optimal. (iv) Resource augmentation also helps for the classical model which does not allow the removal of old items: Without resource augmentation, no competitive algorithm exists even for the unweighted case, but with resource augmentation we can attain a bounded competitive ratio of $1/(R-1)$. Note that the competitive ratio still diverges for the weighted case.

See the following table for the summary. All the ratios in the table are tight. We furthermore have the conjecture that we can design optimal online algorithms for the intermediate case between the weighted and unweighted cases, i.e., for the case that the maximum weight/size ratio of the whole input is known in advance (assuming that the minimum weight/size ratio is one), by a natural extension of the above algorithms.

	Weighted	Unweighted	
With Removal	$1/(R - 1)$, $1 < R \leq 2$	$2/(2R - 1)$ if $(1 + \sqrt{2})/2 \leq R \leq 3/2$,	
		$(\sqrt{4R+1} + 1)/(2R)$ if $1 \leq R < (1 + \sqrt{2})/2$,	
Without Removal	∞	$1/(R - 1)$, $1 < R \leq 2$	

Related Results. There is a rich literature for resource augmentation. For instance, see [1] for bin packing and [8] for machine scheduling. The online knapsack problem first appeared in [5,6] without removal, where the authors introduced stochastic models to bypass the difficulty of the standard competitive analysis. Iwama and Taketomi [2] proposed the removable online knapsack problem and studied its competitive ratio for the unweighted case. Optimal online algorithms for the one-bin case and the multi-bin case (but the gain is counted by the maximum-loaded single bin), with competitive ratios $(\sqrt{5}+1)/2$ and 1.3815, respectively, were derived. Noga and Sarbua[7] considered a similar online knapsack problem with resource augmentation. Their model allows an item to be divided upon its arrival and be packed into the knapsack partially. They gave both a deterministic algorithm and a randomized algorithm, where the deterministic one achieves the best competitive ratio of $2/R$, where $1 \leq R \leq 2$.

Competitive Ratio. Before closing the introductory section we give the standard definition of the competitive ratio with resource augmentation for our problem. Let A be an online algorithm with a knapsack of capacity $R \geq 1$ and let OPT be some optimal (offline) algorithm with a knapsack of capacity one. The competitive ratio (abbreviated as CR hereafter) is then defined as $\sup_L OPT(L)/A(L)$, where $A(L)$ and $OPT(L)$ are the total weight of items accepted by the online algorithm A and the optimal algorithm OPT, respectively, for an instance L.

2   The Weighted Case

If $R = 1.0$ the problem is not competitive, which can be seen as follows: Let (s, w) denote an item whose size and weight are s and w, respectively. Consider the sequence of items

$$(1,1), (\epsilon^2, \epsilon), (\epsilon^2, \epsilon), \cdots,$$

where $\epsilon > 0$ is a small number. Let A be any online algorithm. Then A must take the first item $(1,1)$, otherwise the adversary ends the game immediately. Now the knapsack is already full and A must discard $(1,1)$ to take one of the subsequent small items. If A actually take some (ϵ^2, ϵ), then the adversary stops the game and the competitive ratio is $1/\epsilon$. Otherwise the adversary keeps giving the small items, allowing the offline player to eventually achieve the total weight of $\epsilon \cdot 1/\epsilon^2 = 1/\epsilon$ whereas the online player holds only $(1,1)$. Thus the CR is $1/\epsilon$, too.

Now we assume that $R > 1$ and consider the following simple greedy algorithm. Suppose that the knapsack currently holds items $i_1, i_2, \cdots, i_k$ and the new item is i_0. Then sort those $k+1$ items $i_0, i_1, \cdots, i_k$ by their weight/size ratio in nondecreasing order and take the largest ones as many as possible.

Lemma 1. *The CR of the greedy algorithm is less than $1/(R-1)$.*

Proof. Suppose that n items are given so far and let $(s_1, w_1), (s_2, w_2), \cdots, (s_n, w_n)$ be their sorted (by weight/size ratio) sequence. Then by a simple induction one can see that at any moment the knapsack either includes all those items (obviously the lemma holds) or the first k items in this list such that $s_1 + s_2 + \cdots + s_k \leq R$ and $s_1 + s_2 + \cdots + s_k + s_{k+1} > R$. (Note that the knapsack of the online player may also hold other items which appear after (s_{k+1}, w_{k+1}) in the sorted list.) Thus we have that $s_1 + s_2 + \cdots + s_k > R - 1$ since $s_{k+1} \leq 1$ and hence OPT (=the total weight of the offline knapsack) is less than $(w_1 + w_2 + \cdots + w_k) \cdot (1/(R - 1))$. Since the online player achieves a total weight of $(w_1 + w_2 + \cdots + w_k)$, the lemma holds. $\qquad\square$

Lemma 2. *For any fixed $c > 0$, there does not exist an online algorithm with a CR of $\frac{1}{R-1} - c$ or less.*

Proof. We use a similar sequence of items as were used for the case of $R = 1$. To see the basic idea, let us assume for the moment that $R = 1 + 2/3$. Then our goal is to show that the CR becomes as bad as 1.5 approximately. Suppose that the input sequence looks like

$$(1, 1), (\epsilon, \sigma\epsilon), (\epsilon, \sigma\epsilon), \cdots.$$

The first item must be taken and recall that the online player cannot discard it for ever when $R = 1$. Assume that this is also true for a general $R > 1$. Then the online player could (approximately) make a total weight of

$$ON = \frac{2/3}{\epsilon} \cdot \sigma\epsilon + 1 = \frac{2}{3}\sigma + 1,$$

and the offline player

$$OPT = \frac{1}{\epsilon} \cdot \sigma\epsilon = \sigma.$$

Thus by selecting a sufficiently large σ, the CR would be arbitrarily close to 1.5 and we would be done.

Unfortunately this is too optimistic. Consider the moment when the 2/3 fraction of the knapsack becomes full. Note that the total weight of the small items staying there, $(2/3)\sigma$, is already much more than 1.0 if σ is large. This means the online player *can* discard $(1, 1)$ to take the next (and subsequent) small items without suffering any damage against the adversary.

What if σ is small, say $\sigma = 1.0$? Then the above value $(2/3)\sigma$ is less than 1.0 and the online player cannot discard $(1, 1)$, since if he/she does so, then the game ends and the CR becomes 1.5 at that moment. But as one can see easily, this time, the online player can just keep $(1, 1)$ for ever. ON will be as much as $1 + 2/3$, which is more than enough since $OPT \leq 1.0$.

Thus neither is satisfactory for the adversary. Our solution is to start the sequence with $\sigma = 1.0$ and increase its value *very slowly*. At the beginning, since $\sigma = 1.0$, the online player has to keep $(1, 1)$ as described above, and subsequently he/she will also be forced to do the same if the change of σ is slow. More formally, for any R such that $1.0 < R \leq 2.0$, we use the following sequence of items:

$$(1,1), \ (\epsilon,\epsilon), \ \cdots\cdots, \ (\epsilon,\epsilon),$$
$$(\epsilon,(1+\delta)\epsilon), \ \cdots\cdots, \ (\epsilon,(1+\delta)\epsilon),$$
$$(\epsilon,(1+2\delta)\epsilon), \ \cdots\cdots, \ (\epsilon,(1+2\delta)\epsilon),$$
$$\cdots\cdots\cdots.$$

Note that all the items excepting the first $(1,1)$ have the same size. In the first stage, $1/\epsilon$ small items of weight ϵ come, the same number of items with weight $(1+\delta)\epsilon$ in the second stage and the same number of items with weight $(1+(i-1)\delta)\epsilon$ in the ith stage. Here $\delta > 0$ is a sufficiently small number. Now we can make the following claims:

(i) If the online player discards $(1,1)$ somewhere in the first stage, then the game ends and it is easy to see that the CR is at least $1/(R-1+\epsilon)$.

(ii) Suppose that the online player holds $(1,1)$ at the beginning of the ith stage ($i \geq 2$) and that he/she discards it within that stage. Then the game ends and at that moment, $ON \leq \frac{R-1}{\epsilon}(1+(i-1)\delta)\epsilon+(1+(i-1)\delta)\epsilon = (R-1+\epsilon)(1+(i-1)\delta)$ and $OPT \geq \lceil\frac{1}{\epsilon}\rceil (1+(i-2)\delta)\epsilon$. (Namely the offline player can hold the whole small items of the previous stage.) Thus the CR is at least

$$\frac{\lceil\frac{1}{\epsilon}\rceil \epsilon}{R-1+\epsilon} \cdot \frac{1+(i-2)\delta}{1+(i-1)\delta},$$

which is arbitrarily close to $1/(R-1)$ for any $i \geq 2$ if ϵ and δ are sufficiently small.

(iii) Suppose that the online player keeps holding $(1,1)$. Then at the end of the ith stage ($i \geq 2$), $ON \leq \frac{R-1}{\epsilon}(1+(i-1)\delta)\epsilon+1$ and $OPT \geq \lceil\frac{1}{\epsilon}\rceil (1+(i-1)\delta)\epsilon$. Thus the CR is at least

$$\frac{\lceil\frac{1}{\epsilon}\rceil \epsilon}{R-1} \cdot \frac{1+(i-1)\delta}{1+(i-1)\delta + 1/(R-1)},$$

which approaches to $\lceil\frac{1}{\epsilon}\rceil \epsilon/(R-1)$ arbitrarily for a fixed δ as i grows. So, if ϵ is sufficiently small, the CR is arbitrarily close to $1/(R-1)$.

Thus whatever the online player does, the CR gets as large as the lemma says. □

Theorem 1. *The greedy algorithm is best possible for the weighted case.*

3   The Unweighted Case

In this case we assume that the weight of an item is equal to its size. In other words we are going to maximize the total size of items packed into the knapsack.

Lemma 3. *There is no online algorithm with a knapsack of capacity $R \geq 1$ so that its competitive ratio is less than*

$$\lambda = \begin{cases} 2/(2R-1) & \text{if } (1+\sqrt{2})/2 \leq R \leq 3/2, \\ (\sqrt{4R+1}+1)/(2R) & \text{if } 1 \leq R < (1+\sqrt{2})/2 \end{cases}$$

Proof. Consider the following sequence of four items for any online algorithm A.

$$x + \epsilon, \ R - x, \ 1/2, \ 1/2$$

For the first item, x satisfies $R - 1/2 \le x < 1$ (its exact value is determined later) and $\epsilon > 0$ is a sufficiently small number. Algorithm A has to accept this item, since otherwise the adversary stops the input. The second item is $R - x$. If A decides to accept it, the first item must be removed. In this case the game is over, and we get the first ratio $r_1 = (x + \epsilon)/(R - x)$. (Note that $x + \epsilon \le 1.0$, which can be in the offline knapsack.) If A discards the second item, the 3rd item is of size $1/2$. If A takes it, then A must discard $x + \epsilon$ and we obtain the second ratio $r_2 = (\frac{1}{2} + R - x)/(\frac{1}{2}) = 1 + 2(R - x)$. If A still keeps the first item, the fourth item is again $1/2$. We thus get the third ratio $r_3 = (\frac{1}{2} + \frac{1}{2})/(x + \epsilon)$. In the following we deal with two cases.

If $R \ge (1 + \sqrt{2})/2$, set $x = R - 1/2$. Comparing the three ratios we have

$$r_1 > \frac{x}{R - x} = 2R - 1, \ r_2 = 2, \ r_3 = \frac{2}{2R - 1 + \epsilon} < \frac{2}{2R - 1} \le 2R - 1.$$

Thus

$$\min\{r_1, r_2, r_3\} = r_3 = 2/(2R - 1 + \epsilon) \to 2/(2R - 1), \text{ as } \epsilon \text{ tends to zero.}$$

It follows that the competitive ratio of algorithm A is at least $2/(2R - 1)$.

If $R < (1 + \sqrt{2})/2$, set $x = (\sqrt{4R + 1} - 1)/2 \ge (\sqrt{5} + 1)/2$. Comparing r_2 and r_3 we can see that

$$r_2 - r_3 > 1 + 2(R - x) - 1/x$$
$$\ge 2 + 2R - \sqrt{4R + 1} - (\sqrt{5} + 1)/2$$
$$> 4 - \sqrt{5} - (\sqrt{5} + 1)/2$$
$$> 0$$

Note that $x^2 + x = R$, which implies that $r_1 > r_3$. Thus

$$\min\{r_1, r_2, r_3\} = r_3 \to (\sqrt{4R + 1} + 1)/(2R), \text{ as } \epsilon \text{ tends to zero.}$$

The competitive ratio of algorithm A is at least $(\sqrt{4R + 1} + 1)/(2R)$.   □

To achieve the matching upper bound, we need the following algorithm, denoted by UW, which is much more complicated than the weighted case.

Algorithm UW. The algorithm works in two cases.
(1) $\mathbf{R \ge (1 + \sqrt{2})/2}$. Let B be the set of a big item (of size greater than $1/2$) and S be the set of small items (of size at most $1/2$) currently in the knapsack, respectively. B and S are initially set to be empty. Algorithm UW will output a feasible packing of $B \cup S$.

Case 1.1. If a big item is arriving, compare it with the one in B (if exists), keep the larger in the knapsack and discard the other. If B is updated, S needs updating accordingly: if necessary remove the items of S one by one in any order until the remaining items can be packed into the knapsack (together with the item in B).

Case 1.2. If a small item a is coming, update S: If the knapsack has enough space for a (it can be packed together with the items of S and B), put a to S; otherwise reject a.

(2) $\mathbf{R} < (\mathbf{1} + \sqrt{\mathbf{2}})/\mathbf{2}$. *Case* 2.1. Algorithm UW will end the game if one of the following two trivial cases occurs: (1) the current packing is already of size $\geq \delta_R = (\sqrt{4R+1} - 1)/2$; (2) the coming item has a size of at least δ_R. In the latter case we discard all items and pack this item in the knapsack.

If the above cases never happen, we do packing as follows: Put the first item in the knapsack. *Case* 2.2. As a new item is arriving, pack it into the knapsack if it can fit in. If there is not enough space we deal with two further cases. *Case* 2.3. If the two largest items (among those already in the knapsack and the new one) can be packed together into the knapsack, then pack the items in the non-increasing order of size until some item cannot fit in, and stop; *Case* 2.4. if it is not this case, remove the largest item and pack the remaining ones into the knapsack, which is obviously possible.

Theorem 2. *The competitive ratio of algorithm UW is*

$$\rho_G \leq \begin{cases} 2/(2R-1) & if\,(1+\sqrt{2})/2 \leq R \leq 3/2, \\ (\sqrt{4R+1}+1)/(2R) & if\,1 \leq R < (1+\sqrt{2})/2, \end{cases}$$

and it is thus an optimal online algorithm.

Proof. To avoid the trivial case we assume that some item is discarded during the packing by algorithm UW.

We first deal with the case that $R \geq (1+\sqrt{2})/2$. During the packing of algorithm UW if no small items are removed/discarded, then the total size of the items in the final packing of algorithm UW is not less than the total size of an optimal packing with knapsack of capacity one. To see this note that the packing by algorithm UW accepts all small items and the largest big item, while an optimal packing (with knapsack capacity of one) cannot do better since two big items exceeds 1.0. Thus the competitive ratio is at most one. In the following assume that some small item is removed/discarded by UW. Then the remaining space of the knapsack is less than $1/2$ since the small item cannot fit in the knapsack. Thus the total size of items packed by UW is greater than $R - 1/2$. The competitive ratio $\rho_G \leq 1/(R - 1/2) = 2/(2R - 1)$. Notice that we did not use the condition that $R \geq (1+\sqrt{2})/2$, namely this upper bound holds for the whole range of R.

For $R < (1+\sqrt{2})/2$, we can prove the better upper bound. The bound $(\sqrt{4R+1}+1)/(2R)$ is trivially achieved if the algorithm ends as one of the two cases occurs: (1) the current packing is already of size $\geq \delta_R = (\sqrt{4R+1}-1)/2$; (2) the coming item has a size of at least δ_R. Moreover if the total size of the items is over R but the two largest ones can fit together into the knapsack we hold a packing with remaining space less than $R/3$. The competitive ratio is thus at most $3/(2R)$ that is smaller than $(\sqrt{4R+1}+1)/(2R)$.

So, whenever algorithm UW stops, it already holds enough amount of total weight. Thus we can assume that only Cases 2.2 and 2.4 have happened so far. We can then claim that no small items are removed/rejected during the packing. Note that we discard items in Case 2.4 only, i.e., if the incoming item cannot fit in the knapsack and the sum of the two largest items is larger than R. In this case, we discard the largest item (it is less than δ_R but larger than $1/2$) and thus the remaining items, including all the small ones, must fit in the knapsack. Let y be the largest one (and its size) among all the input items so far and let z be the largest item (and its size) remaining in the final packing. Clearly $y + z > R$ and $y < \delta_R$. It implies that $z > R - \delta_R$. Recall that we do not have any small item lost. Let S be the total size of small items. Thus the optimal value is at most $y + S$ while Algorithm UW holds a packing of size $z + S$. The competitive ratio is at most

$$(y + S)/(z + S) \le y/z < \delta_R/(R - \delta_R) = (\sqrt{4R+1}+1)/(2R). \qquad \square$$

Note that when $R = 1$ it is exactly the bound (golden ratio) shown in [2].

4   The Model Without Removal

If the online player is not allowed to discard old items and without resource augmentation, then the CR diverges even for the unweighted case. This is easily seen by the sequence of only two items, ϵ followed by 1.0. The online player has to take the the first item and not enough space for the second one. With resource augmentation, however, it is enough for him/her to use a simple greedy algorithm. Namely if the current item fits, then it is taken and otherwise discarded. If the item is discarded, at least the $R - 1$ fraction of the knapsack has been filled, which means the CR is at most $1/(R - 1)$. The online player cannot do better; the sequence given by the adversary is a simple extension of the above, namely $(R - 1) + \epsilon$ followed by 1.0.

For the weighted case, resource augmentation is still powerless completely. The adversary gives the sequence of items

$$(1, w_1), (1, w_2), \cdots, (1, w_n),$$

where $w_1 > 0$ and w_i $(i \ge 2)$ is determined to satisfy

$$w_i \ge \sigma(w_1 + w_2 + \cdots + w_{i-1}).$$

Then, even if the online player knows the maximum number n of items, he/she cannot maintain the sequence if $R < n$ or cannot achieve any competitive ratio better than σ. (If $R = n$ then the game is trivial since everything can be held).

References

1. Csirik, J., Woeginger, G.J.: Resource augmentation for online bounded space bin packing. Journal of Algorithms archive Volume 44, 308–320 (2002)
2. Iwama, K., Taketomi, S.: Removable online knapsack problems. In: Widmayer, P., Triguero, F., Morales, R., Hennessy, M., Eidenbenz, S., Conejo, R. (eds.) ICALP 2002. LNCS, vol. 2380, pp. 293–305. Springer, Heidelberg (2002)
3. Iwama, K., Zhang, G.: Removable online knapsack - weighted case. In: Proceedings of the 7th Japan-Korea Workshop on Algorithms and Computation (WAAC), pp. 223–227 (2003)
4. Kalyanasundaram, B., Pruhs, K.: Speed is as powerful as clairvoyance. Journal of the ACM 47, 214–221 (1995)
5. Lueker, G.S.: Average-case analysis of off-line and on-line knapsack problems. In: Proceedings of the Sixth Annual ACM-SIAM SODA, pp.179–188 (1995)
6. Marchetti-Spaccamela, A., Vercellis, C.: Stochastic on-line knapsack problems. Math. Prog. 68, 73–104 (1995)
7. Noga, J., Sarbua, V.: An online partially fractional knapsack problems. In: Proceedings ISPAN 2005, pp. 108–112 (2005)
8. Phillips, C.A., Stein, C., Torng, E., Wein, J.: Optimal time-critical scheduling via resource augmentation. In: Proceedings of STOC1997, pp. 140–149 (1997)
9. Sleator, D., Tarjan, R.: Amortized effciency of list update and paging rules. Comm. ACM 23, 202–208 (1985)

Soft Edge Coloring

Chadi Kari[1], Yoo-Ah Kim[1], Seungjoon Lee[2], and Alexander Russell[1],
and Minho Shin[3]

[1] Department of Computer Science and Engineering,
University of Connecticut, Storrs, CT 06269
{chadi,ykim,acr}@engr.uconn.edu
[2] AT&T Labs - Research, Florham Park, NJ 07932
slee@research.att.com
[3] Department of Computer Science,
University of Maryland, College Park, MD 20742
mhshin@cs.umd.edu

Abstract. We consider the following channel assignment problem arising in wireless networks. We are given a graph $G = (V, E)$, and the number of wireless cards C_v for all v, which limit the number of colors that edges incident to v can use. We also have the total number of channels C_G available in the network. For a pair of edges incident to a vertex, they are said to be *conflicting* if the colors assigned to them are the same. Our goal is to color edges (assign channels) so that the number of conflicts is minimized. We first consider the homogeneous network where $C_v = k$ and $C_G \geq C_v$ for all nodes v. The problem is NP-hard by a reduction from EDGE COLORING and we present two combinatorial algorithms for this case. The first algorithm is a distributed greedy method, which gives a solution with at most $(1 - \frac{1}{k})|E|$ more conflicts than the optimal solution. We also present an algorithm yielding at most $|V|$ more conflicts than the optimal solution. The algorithm generalizes Vizing's algorithm in the sense that it gives the same result as Vizing's algorithm when $k = \Delta + 1$. Moreover, we show that this approximation result is best possible unless $P = NP$. For the case where $C_v = 1$ or k, we show that the problem is NP-hard even when $C_v = 1$ or 2, and $C_G = 2$, and present two algorithms. The first algorithm is completely combinatorial and produces a solution with at most $(2 - \frac{1}{k})OPT + (1 - \frac{1}{k})|E|$ conflicts. We also develop an SDP-based algorithm, producing a solution with at most $1.122OPT + 0.122|E|$ conflicts for $k = 2$, and $(2 - \Theta(\frac{\ln k}{k}))OPT + (1 - \Theta(\frac{\ln k}{k}))|E|$ conflicts in general.

1 Introduction

We consider a channel assignment problem arising in multi-channel wireless networks. In wireless networks nearby nodes *interfere* with each other and cannot simultaneously transmit over the same wireless channel. One way to overcome this limitation is to assign independent channels (that can be used without interference) to nearby links of the network. Consider the example shown in Figure 1. When all links use the same channel, only one pair of nodes may communicate with each other at a time due to conflicts. However, if there are three channels available and each node has two wireless interface

M. Charikar et al. (Eds.): APPROX and RANDOM 2007, LNCS 4627, pp. 189–203, 2007.
© Springer-Verlag Berlin Heidelberg 2007

cards (so it can use two channels), then we may assign a different channel to each link to avoid conflicts among edges in this channel assignment. Channel assignment to utilize multiple channels have recently been studied by many researchers in networking community [1,2,3,4].

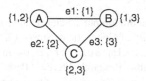

Fig. 1. Each node has two wireless interface cards (thus can use two different channels) and three channels are available in total. We can assign a distinct channel to each link as shown above so that there is no conflict among edges.

We informally define the SOFT EDGE COLORING problem as follows. We are given a graph $G = (V, E)$, and constraints on the number of wireless cards C_v for all v, which limit the number of colors that edges incident to v can use. In addition, we have a constraint on the total number of channels available in the network (denoted as C_G). For a pair of edges incident to a vertex, they are said to be *conflicting* if the colors assigned to them are the same. Our goal is to color edges (assign channels) so that the number of conflicts is minimized while satisfying constraints on the number of colors that can be used.

SOFT EDGE COLORING is a variant of the EDGE COLORING problem. In our problem, coloring need not be proper (two adjacent edges are allowed to use the same color)—the goal is to minimize the number of such conflicts. In addition, each node has its local color constraint, which limits the number of colors that can be used by the edges incident to the node. For example, if a node has two wireless cards ($C_v = 2$), the node can choose two colors and edges incident to the node should use only those two colors.

Our results. We briefly summarize our results. We first consider the homogeneous network where $C_v = k$ and $C_G \geq C_v$ for all nodes v. For an arbitrary k, the problem is NP-hard by a reduction from EDGE COLORING. We present two combinatorial algorithms for this case. The first algorithm is a simple greedy method, which gives a solution with at most $(1 - \frac{1}{k})|E|$ more conflicts than the optimal solution; furthermore, it can be computed in a distributed fashion. We also present an algorithm yielding at most $|V|$ more conflicts than the optimal solution. The algorithm generalizes Vizing's algorithm in the sense that it gives the same result when $k = \Delta + 1$. In fact, our algorithm gives an optimal solution when $d_v \bmod k = k - 1$ for all vertices v. Moreover, we show that this approximation result is best possible unless $P = NP$.

In a heterogeneous network, we consider the case where each node v can have different $C_v = 1$ or k. We show that the problem is NP-hard even when $C_v = 1$ or 2, and $C_G = 2$, and present two algorithms for this case. The first algorithm is completely combinatorial and produces a solution with at most $(2 - \frac{1}{k})OPT + (1 - \frac{1}{k})|E|$ conflicts. We also develop an SDP-based algorithm, producing a solution with at most

$1.122 OPT + 0.122|E|$ conflicts for $k = 2$, and $(2 - \Theta(\frac{\ln k}{k})) OPT + (1 - \Theta(\frac{\ln k}{k}))|E|$ conflicts in general (slightly better than the combinatorial algorithm).

Relationship to MIN K-PARTITION *and* MAX K-CUT. The MIN K-PARTITION problem is the dual of MAX K-CUT problem where we color vertices with k different colors so that the total number of conflicts (monochromatic edges) is minimized. Our problem for the homogeneous network when $C_G = C_v = k$ for all v is an edge coloring version of MIN k-PARTITION problem[1]. Kann *et al.* [5] showed that for $k > 2$ and for every $\epsilon > 0$, there exists a constant α such that the MIN k-PARTITION cannot be approximated within $\alpha|V|^{2-\epsilon}$ unless $P = NP$. In our problem when $C_v = k$ for all v, we have an approximation algorithm with additive term of $|V|$.

For the case when $C_v = 1$ or k, we use a SDP formulation similar to one used for MAX K-CUT and utilize the upperbounds obtained in [6]. To obtain a $(2 - \Theta(\frac{\ln k}{k}))$-approximation, we compare the upperbounds with two lowerbounds — one based on necessary interference at a vertex determined by its degree, and one given by the SDP relaxation.

Other Related Work. Fitzpatrick and Meertens [7] have considered a variant of graph coloring problem (called the SOFT GRAPH COLORING problem) where the objective is to develop a distributed algorithm for coloring vertices so that the number of conflicts is minimized. The algorithm repeatedly recolors vertices to quickly reduce the conflicts to an acceptable level. They have studied experimental performance for regular graphs but no theoretical analysis has been provided. Damaschke [8] presented a distributed soft coloring algorithm for special cases such as paths and grids, and provided the analysis on the number of conflicts as a function of time t. In particular, the conflict density on the path is given as $O(1/t)$ when two colors are used, where the conflict density is the number of conflicts divided by $|E|$.

In the traditional edge coloring problem, the goal is to find the minimum number of colors required to have a proper edge coloring. The problem is NP-hard even for cubic graphs [9]. For a simple graph, a solution using at most $\Delta + 1$ colors can be found by Vizing's theorem [10] where Δ is the maximum degree of a node. For multigraphs, there is an approximation algorithm which uses at most $1.1\chi' + 0.8$ colors where χ' is the optimal number of colors required [11] (the additive term was improved to 0.7 by Caprara *et al.* [12]).

1.1   Problem Definition

We are given a graph $G = (V, E)$ where $v \in V$ is a node in a wireless network and an edge $e = (u, v) \in E$ represents a communication link between u and v. Each node v can use C_v different channels and the total number of channels that can be used in the network is C_G. More formally, let $E(v)$ be the edges incident to v and $c(e)$ be the color assigned to e. Then $|\bigcup_{e \in E(v)} c(e)| \le C_v$ and $|\bigcup_{e \in E} c(e)| \le C_G$.

---

[1] Or it can be considered as MIN k-PARTITION problem when the given graph is a line graph where the line graph of G has a vertex corresponding to each edge of G, and there is an edge between two vertices in the line graph if the corresponding edges are incident on a common vertex in G.

A pair of edges e_1 and e_2 in $E(v)$ are said to be conflicting if the two edges use the same color. Let us define the *conflict number* (CF_e) of an edge $e \in E$ to be the number of other edges that conflict with e. In other words, for an edge $e = (u, v)$, CF_e is the number of edges (other than e itself) in $E(u) \bigcup E(v)$ that use the same channel as e. Our goal is to minimize the total number of conflicts. That is,

$$CF_G = \frac{1}{2} \sum_{e \in E} CF_e. \tag{1}$$

In the remainder of this paper, we mean *channels* by *colors* and use edge coloring and channel assignment, interchangeably. We also use conflicts and interferences interchangeably.

2 Algorithms for Homogeneous Networks

In this section, we consider the case for a homogeneous network where for all nodes v, the number of channels that can be used is the same ($C_v = k$). For an arbitrary k, the problem is NP-hard as the edge coloring problem can be reduced to our problem by setting $k = C_G = \Delta$ where Δ is the maximum degree of nodes.

2.1 Greedy Algorithm

We first present and analyze a greedy algorithm for this problem. The algorithm works as follows: We choose colors from $\{1, \ldots, k\}$ (We only use k colors even when $C_G > k$ and the approximation ratio of our algorithm remains the same regardless of the value of C_G.) For any uncolored edge $e = (u, v)$, we choose a color for edge e that introduces the smallest number of conflicts. More formally, when we assign a color to $e = (u, v)$, we count the number of edges in $E(u) \bigcup E(v)$ that are already colored with c (denoted as $n(c, e)$), and choose color c with the smallest $n(c, e)$.

Theorem 1. *The total number of conflicts by the greedy algorithm in homogeneous networks is at most*

$$CF_G = \frac{1}{2} \sum_{e \in E} CF_e \leq OPT + (1 - \frac{1}{k})|E|. \tag{2}$$

Theorem 1 directly follows from Lemma 2 and 3. The proofs are included in [13].

Lemma 2. *The total number of conflicts when $C_v = k$ for all node v is at least* $\frac{1}{2} \sum_v \frac{d_v^2}{k} - |E|$.

Lemma 3. *The total number of conflicts introduced by Algorithm 1 is at most* $\frac{1}{2} \sum_v \frac{d_v^2}{k} - \frac{|E|}{k}$.

Note that the algorithm can be performed in a distributed manner and each node needs only local information. We can also consider a simple randomized algorithm, in which each edge chooses its color uniformly at random from $\{1, \ldots, k\}$. The algorithm gives the same expected approximation guarantee and it can be easily derandomized using conditional expectations. The following corollary of Lemma 2 will be used to prove approximation factors for heterogenous networks.

Corollary 4. *Given an optimal solution, let $OPT(S)$ ($S \subseteq V$) be the number of conflicts at vertices in S and $|E(S)|$ be $\sum_{v \in S} \frac{d_v}{2}$. Then we have $OPT(S) \geq \frac{1}{2} \sum_{v \in S} \frac{d_v^2}{k} - |E(S)|$.*

2.2 Improved Algorithm

In this section, we give an algorithm with additive approximation factor of $|V|$. Our algorithm is a generalization of Vizing's algorithm in the sense that it gives the same result as Vizing's algorithm when $k = \Delta + 1$ where Δ is the maximum degree of nodes. We first define some notations. For each vertex v, let $m_k = \lfloor \frac{d_v}{k} \rfloor$ and $\alpha_v = d_v - m_v k$. Let $|E_i(v)|$ be the size of the color class of color i at vertex v i.e. the number of edges adjacent to v that have color i.

Definition 5. *A color i is called* strong *on a vertex v if $|E_i(v)| = m_k + 1$. A color i is called* weak *on v if $|E_i(v)| = m_k$. A color i is called* very weak *on v if $|E_i(v)| < m_k$.*

Definition 6. *A vertex v has a* balanced *coloring if the number of strong classes at v is at most $\min(\alpha_v + 1, k - 1)$ and no color class in $E(v)$ is larger than $m_k + 1$. A graph $G = (V, E)$ has a balanced coloring if each vertex $v \in V$ has a balanced coloring.*

In the following we present an algorithm that achieves a balanced coloring for a given graph $G = (V, E)$; we show in Theorem 16 that a balanced coloring implies an additive approximation factor of $|V|$ in terms of number of conflicts. In Algorithm BALANCEDCOLORING(e) described below, we color edge e so that the graph has a balanced coloring (which may require the recoloring of already colored edges to maintain the balanced coloring), assuming that it had a balanced coloring before coloring e. We perform BALANCEDCOLORING for all edges in arbitrary order. The following terms are used in the algorithm description. Let $|S_v|$ denote the number of strong color classes at vertex v.

Definition 7. *For vertex $v \in V$ with $|S_v| < \min(\alpha_v + 1, k - 1)$ or with $|S_v| = k - 1$, i is a* missing color *if i is weak or very weak on v. For vertex $v \in V$ with $|S_v| = \alpha_v + 1$, i is a* missing color *if i is very weak on v*

Definition 8. *An ab-path* between vertices u and v where a and b are colors, is a path connecting u and v and has the following properties:

- *Edges in the path have alternating colors a and b.*
- *Let $e_1 = (u, w_1)$ be the first edge on that path and suppose e_1 is colored a, then u must be missing b and not missing a.*
- *If v is reached by an edge colored b then v must be missing a but not missing b, otherwise if v is reached by an edge colored a then v must be missing b and not missing a.*

Definition 9. *A* flipping *of an ab-path is a recoloring of the edges on the path such that edges previously with color a will be recolored with color b and vice versa.*

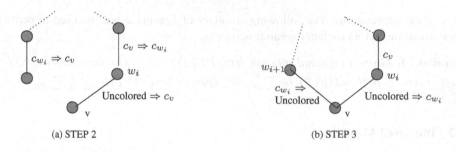

(a) STEP 2   (b) STEP 3

Fig. 2. The figures illustrate how recoloring is performed in BALANCEDCOLORING. The colors beside edges indicate the original color and the color after recoloring.

Algorithm BALANCEDCOLORING($e = (v, w)$)
Let $w_1 = w$. At i-th round ($i = 1, 2, \ldots$), we do the following.
STEP 1: Let $\mathcal{C}_v$ be the set of missing colors on v. If $i = 1$, $\mathcal{C}_{w_1}$ is the set of missing colors on w_1. When $i \geq 2$, $\mathcal{C}_{w_i}$ is the set of missing colors on w_i minus $c_{w_{i-1}}$. ($c_{w_{i-1}}$ is defined in STEP 2 at $(i-1)$-th round). If $\mathcal{C}_v \cap \mathcal{C}_{w_i} \neq \emptyset$, then choose color $a \in \mathcal{C}_v \cap \mathcal{C}_{w_i}$, color edge (v, w_i) with a and terminate.
STEP 2: If $\mathcal{C}_v \cap \mathcal{C}_{w_i} = \emptyset$, choose $c_v \in \mathcal{C}_v$ and $c_{w_i} \in \mathcal{C}_{w_i}$. Find a $c_v c_{w_i}$-path that starts at w_i and does not end at v. If such a path exists, flip this path, color edge (v, w_i) with c_v and terminate.
STEP 3: If all $c_v c_{w_i}$-paths that start at vertex w_i end at v, fix one path and let (v, w_{i+1}) be the last edge on that path. The edge (v, w_{i+1}) must have color c_{w_i}. Uncolor it and color edge (v, w_i) with c_{w_i}. Mark edge (v, w_i) as used and repeat the above steps with edge (v, w_{i+1}) (go to $(i+1)$-th round).

Analysis. In the following, we prove that our algorithm terminates and achieves a balanced coloring. First we prove that we can always find a missing color at each round (Lemma 10 and 11) and at some round $j < d_v$, the algorithm terminates (Lemma 12). Due to the choice of missing colors and ab-path, we can show that our algorithm gives a balanced coloring (Lemma 13 and 14).

Lemma 10. *For the given edge (v, w_1), there is a missing color at v and w_1. That is, $\mathcal{C}_v \neq \emptyset$ and $\mathcal{C}_{w_1} \neq \emptyset$.*

Proof. When $|S_v| < \min(\alpha_v + 1, k - 1)$ or $|S_v| = k - 1$, there must be at least one weak color, which is a missing color. If $|S_v| = \alpha_v + 1$ then we can show that the remaining $k - \alpha_v - 1$ color classes cannot be all weak (i.e. having size m_v). Note that $d_v = m_v k + \alpha_v$, so if there are $\alpha_v + 1$ strong color classes and the remaining color classes have all exactly size m_v then the number of edges at v is strictly larger than d_v, which is not possible. So there must be a very weak class of which size is strictly less than m_v.

For w_i, $i \geq 2$, we need to choose a missing color at w_i other than $c_{w_{i-1}}$. We prove in the following lemma, that there is a missing color other than $c_{w_{i-1}}$.

Lemma 11. *At i-th round ($i \geq 2$), there is a missing color other than $c_{w_{i-1}}$ at w_i.*

Proof. Note first that $c_{w_{i-1}}$ was not a missing color at w_i in $(i-1)$-th round since otherwise we should have stopped at STEP 2. Consider the case that $c_{w_{i-1}}$ was a strong color in $(i-1)$-th round. As it is not possible that all k colors are strong, there must be a weak (or very weak) color c other than $c_{w_{i-1}}$ in $(i-1)$-th round. After uncoloring (u, w_i), the number of strong color classes will be reduced and we now have $|S_v| < \min(\alpha_v + 1, k - 1)$. Then c is missing at w_i in i-th round.

For the case that $c_{w_{i-1}}$ was a weak color in $(i-1)$-th round, $|S_v| = \alpha_v + 1$ in $(i-1)$-th round (otherwise, $c_{w_{i-1}}$ should have been a missing color). After uncoloring (u, w_i), S_v remains the same but $c_{w_{i-1}}$ is a very weak color. We need to show that there is a very weak color other than $c_{w_{i-1}}$. The number of edges that have weak or very weak colors is at most $(d_v - 1) - (\alpha_v + 1)(m_v + 1) = (k - \alpha_v - 1)m_v - 2 = (k - \alpha_v - 3)m_v + 2(m_v - 1)$. Therefore, there must be at least one very weak color other than $c_{w_{i-1}}$.

Lemma 12. *At some round $j < d_v$ there exists a $c_v c_{w_j}$-path starting at w_j and not ending at v.*

Proof. At some round j, if the algorithm does not terminate at Step 1 of some round $k < j$, the colors are going to run out (i.e. the missing color c_{w_j} is the same as color $c_{w_i}, i < j$ and all edges (v, w_i) with color c_{w_i} have been already used by the algorithm). We show that there is no $c_v c_{w_i}$-path connecting v and w_j (thus the algorithm has to terminate at Step 2). Suppose that there exists a $c_v c_{w_i}$-path P connecting v and w_j, v is missing color c_v and not missing color c_{w_i}, so v must be reached on an edge colored c_{w_i}. Let $e = (v, w_i)$ be the edge in P adjacent to v. Since we used in the algorithm all edges with color c_{w_i}, then edge (v, w_i) must have been already used by the algorithm. Now rewind the algorithm to the point where (v, w_i) was uncolored, w_i is missing c_{w_i} and not missing c_v, so if P exists there must be also a $c_v c_{w_i}$-path connecting w_i and w_j. This contradicts that at round $i < j$ we could not find such path.

Lemma 13. *A flipping of an ab-path in a graph with balanced coloring will not violate the balanced coloring.*

Proof. Suppose an ab-path runs from u to v. Suppose u is missing color b and not missing color a. Let $e = (u, w_0)$ be the first edge of that path, so it is colored by a. Flipping the ab-path will recolor e with color b, but since b was missing on u the color class $|E_b(v)|$ will not exceed $m_v + 1$, and also the number of strong classes will not become larger than $\min(\alpha_v + 1, k - 1)$ as we made color a missing on u. The same argument works for v but with possibly b and a interchanged in the argument. For internal vertices on the path nothing changes as the number of edges colored a and b stays the same.

Lemma 14. *Let v be a vertex that has a balanced coloring. Let $e \in E(v)$ be uncolored and let i be a missing color on v. Coloring e with i will not violate a balanced coloring at v.*

Proof. Suppose that at a vertex v, $|S_v| < \min(\alpha_v + 1, k - 1)$. Then the number of strong color classes at v is strictly less than $\alpha_v + 1$ and coloring edge e with i will not violate a balanced coloring at v as $|S_v|$ will not exceed $\min(\alpha_v + 1, k - 1)$ and $E_i(v)$ will not exceed $m_v + 1$ (i is missing on v). Suppose at a vertex v, $|S_v| = \alpha_v + 1$, then we show in Lemma 10 and 11 that there must be a very weak color class. When $|S_v| = k - 1$, the remaining color is very weak as one edge is not colored. Thus coloring e with i will not make $E_i(v)$ a strong color class and the number of strong color classes remains the same. So the balanced coloring at v will not be violated.

Theorem 15. *The above algorithm terminates and achieves a balanced coloring.*

Proof. In Lemma 12 we show that at some round $j, (1 \leq j \leq d_v)$, if we do not terminate at Step 1 of the algorithm then there will be a $c_v c_{w_j}$-path P starting at w_j and not ending at v. Now, if for some $i, (1 \leq i \leq j), C_v \cap C_{w_i} \neq \emptyset$ then vertices v and w_i are missing the same color c_v. By Lemma 14, coloring edge (v, w_i) with color c_v will not violate a balanced coloring at v or at w_i, hence the algorithm terminates at Step 1 with a balanced coloring for G.

If $\forall i, i \leq j, C \cap C_{w_i} = \emptyset$ we show that the algorithm terminates at Step 2 of round j. As mentioned above, at round j there will be a $c_v c_{w_j}$-path P starting at w_j and not ending at v. On this path, the edge adjacent to w_j is colored with c_v since w_j is missing c_{w_j} and not missing c_v. Note that flipping path P will recolor this edge with color c_{w_j} making color c_v missing on w_j. Furthermore, by Lemma 13, flipping P will not violate the balanced coloring at any vertex in P. Thus c_v is now missing at v and w_j and as in Step 1 we can now color edge (v, w_j) with color c_v without violating a balanced coloring at v or w_j. So the algorithm terminates at Step 2 with a balanced coloring.

Theorem 16. *A balanced coloring of a graph achieves a $|V|$ additive approximation factor.*

Proof. We have shown an algorithm that colors the edges of a graph $G = (V, E)$ such that the coloring is balanced at each vertex. Here we show that the algorithm introduces at each vertex $v \in V$ one more conflict than the optimal solution. At each vertex v suppose there is an ordering on the size of the color classes, 1 being a strong class and k being the weakest class. Note that at a vertex v, the number of conflicts is minimized when the number of strong classes is α_v and the remaining colors are weak. As the number of strong classes achieved by our algorithm is at most $\alpha_v + 1$, the first α_v classes introduce the same number of conflicts in both the optimal and our solution.

The $(\alpha_v + 1)^{th}$ color class in a balanced coloring which is strong, exceeds the corresponding color class in OPT (which is necessarily weak) by 1. Then the additional number of conflicts is $\frac{1}{2}(m_v + 1)m_v - \frac{1}{2}m_v(m_v - 1) = m_v$.

Now if there is an additional edge in the $(\alpha_v + 1)^{th}$ color class in a balanced coloring then there must be an additional edge in some color class $i, \alpha_v + 1 < i < k$ in OPT i.e. some color class i is weak in OPT but very weak in our balanced coloring. The number of additional conflicts of OPT in i is $\frac{1}{2}m_v(m_v - 1) - \frac{1}{2}(m_v - 1)(m_v - 2) = m_v - 1$. So, finally the additional number of conflicts introduced by the balanced algorithm is 1 at each vertex. Thus the approximation factor is $|V|$.

Corollary 17. *When $\alpha_v = k - 1$ for all v, the algorithm gives an optimal solution.*

Proof. Note that the balanced coloring gives exactly $k - 1$ strong color classes and one weak color class when $\alpha_v = k - 1$, which is the optimal.

We can show that the approximation ratio given by the algorithm is best possible unless $P = NP$. The proof is in the next theorem.

Theorem 18. *It is NP-hard to approximate the channel assignment problem in homogeneous networks within an additive term of $o(|V|^{1-\epsilon})$, given a constant ϵ.*

Proof. Suppose that we have a simple graph $G = (V, E)$. It is known that finding the edge chromatic number $\chi'(G)$ of G is NP-hard (the edge chromatic number is the minimum number of colors for edge-coloring G) [9]. By the Vizing's theorem [10], the chromatic index of a simple graph G is Δ or $\Delta + 1$ where Δ is the maximum degree of any vertex $v \in V$.

Given a constant ϵ, let $G' = (V', E')$ be the graph which has $|E|^{\frac{1}{\epsilon}-1}$ copies of G. Note that $|E'| = |E|^{\frac{1}{\epsilon}}$. We set $C_G = C_v = \Delta$ If $\chi'(G) = \Delta$ then the optimal solution of the channel assignment problem is 0. Otherwise if $\chi'(G) = \Delta + 1$, then each of component of G' has at least one conflict and therefore, the optimal solution has at least $|E|^{\frac{1}{\epsilon}-1}$ conflicts, which is the same as $|E'|^{1-\epsilon}$. Thus if we have an approximation algorithm with additive term of $o(|E'|^{1-\epsilon})$ for a graph $G' = (V, E')$, we can decide the chromatic index of G, which is NP-hard. Contradiction.

3   Networks Where $C_v = 1$ or k

In this section, we present two algorithms for networks with $C_v = 1$ or k and analyze the approximation factors of the algorithms. The case where $C_v = 1$ or k is interesting since (i) it reflects a realistic setting, in which most of mobile stations are equipped with one wireless card and nodes with multiple wireless cards are placed in strategic places to increase the capacity of networks. (ii) as shown in Theorem 19, the problem is NP-hard even when $C_v = 1$ or 2. (The proof is in [13]).

Theorem 19. *The channel assignment problem to minimize the number of conflicts is NP-hard even when $C_v = 1$ or 2, and $C_G = 2$.*

3.1   Extended Greedy Algorithm

We first present an extended greedy algorithm when $C_v = 1$ or k, and $C_G \geq C_v$. The approximation factor is $2 - \frac{1}{k}$. Even though the algorithm based on SDP (semi-definite programming) gives a slightly better approximation factor (see Section 3.2), the greedy approach gives a simple combinatorial algorithm. The algorithm generalizes the idea of the greedy algorithm for homogeneous networks. In this case, an edge cannot choose its color locally since the color choice of an edge can affect colors for other edges to obey color constraints.

Before describing the algorithm, we define some notations. Let $V_i \subseteq V$ be the set of nodes v with $C_v = i$ (i.e., we have V_1 and V_k). V_1 consists of connected clusters $V_1^1, V_1^2, \ldots V_1^t$, such that nodes $u, v \in V_1$ belong to the same cluster if and only if there is a path composed of nodes in V_1 only. (See Figure 3 for example.) Let E_1^i be a set of edges both of which endpoints are in V_1^i. We also define B_1^i to be a set of edges whose one endpoint is in V_1^i and the other is in V_k. We can think of B_1^i as a set of edges in the boundary of cluster V_1^i. Note that all edges in $E_1^i \bigcup B_1^i$ should have the same color. E_k is a set of edges both of which endpoints are in V_k. E_1 is defined to be $\bigcup_i E_1^i$.

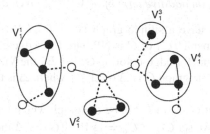

Fig. 3. The figure show an example of clusters V_1^i when $C_v = 1$ or k. Black nodes have only one wireless card and white nodes have k wireless cards. Dotted lines belong to B_1^i.

In the greedy algorithm for homogeneous networks, each edge greedily chooses a color so that the number of interferences it creates (locally) is minimized. Similarly, when $C_v = 1$ or k, edges in the same cluster V_1^i choose a color so that the number of conflicts it creates is minimized. Formally, we choose a color c with minimum value of $\sum_{e=(u,v)\in B_1^i, v\in V_k} n_c(v)$ where $n_c(v)$ is the number of edges $e' \in E(v)$ with color c. Once edges in $E_1^i \bigcup B_1^i$ for all i choose their colors, the remaining edges (edges belonging to E_k) greedily choose their colors.

Any edges (u, v) incident to a vertex in V_1 should use the same color and therefore are conflicting with each other no matter what algorithm we use. Given an optimal solution, consider $OPT(V_1)$ and $OPT(V_k)$ where $OPT(S)$ is the number of conflicts at vertices in $S \subseteq V$. Similarly, we have $CF(V_1)$ and $CF(V_k)$ where $CF(S)$ is the number of conflicts at vertices in $S \subseteq V$ in our solution. Then we have $OPT(V_1) = CF(V_1)$. Therefore, we only need to compare $OPT(V_k)$ and $CF(V_k)$.

Theorem 20. *The number of conflicts created by the extended greedy algorithm at V_k is at most $(2 - 1/k)OPT + (1 - 1/k)|E|$.*

Proof. We will simply show that the number of conflicts created by the extended greedy algorithm at V_k is at most $(2 - 1/k)OPT(V_k) + (1 - 1/k)|E(V_k)|$ where $|E(V_k)|$ is $\sum_{v\in V_k} \frac{d_v}{2}$ as $CF(V_1) = OPT(V_1)$. For each $e \in E \setminus E_1$, $n(e)$ be the number of conflicts at vertices in V_k which are introduced when we assign a channel to e. Then the total number of conflicts at V_k is $\sum n(e)$.

We first consider the number of conflicts created when we assign colors to edges in B_1^i (recall that B_1^i is a set of edges of which endpoints are in V_1^i and V_k). For an edge $e = (u, v)$ where $u \in V_1^i$ and $v \in V_k$, let $d_v(e)$ be the number of edges in $E(v)$ to which a color is assigned before e. Then when we choose a color for $E_1^i \bigcup B_1^i$, the number of conflicts at vertices in V_k with edges not in $E_1^i \bigcup B_1^i$, is at most

$$\frac{\sum_{v\in V_k} \sum_{e\in E(v) \cap B_1^i} d_v(e)}{k}$$

as we choose a color with minimum conflicts. If v has $e_i(v)$ edges in B_1^i, $\frac{1}{2}e_i(v)(e_i(v)-1)$ additional conflicts (between edges in B_1^i) are created. For edges in E_k we use the greedy algorithm presented in Section 2, and therefore, the number of conflicts created when we assign colors in E_k is at most $\frac{\sum_{v \in V_k} \sum_{e \in E(v) \cap E_k} d_v(e)}{k}$.

Summing up all the conflicts,

$$\sum_{e \in E \setminus E_1} n(e) \leq \frac{\sum_{v \in V_k} \sum_{e \in E(v)} d_v(e)}{k} + \frac{\sum_i \sum_{v \in V_k}(e_i^2(v) - e_i(v))}{2}$$

As for each node v, $\sum_{e \in E(v)} d_v(e) \leq \frac{d_v(d_v-1)}{2} - \frac{\sum_i (e_i^2(v) - e_i(v))}{2}$ (colors for edges in B_1^i will be determined at the same time), we have

$$\sum_{e \in E \setminus E_1} n(e) \leq \frac{1}{k} \sum_{v \in V_k}\left(\frac{d_v(d_v-1)}{2} - \frac{\sum_i e_i^2(v) - e_i(v)}{2}\right) + \frac{\sum_i \sum_{v \in V_k}(e_i^2(v) - e_i(v))}{2}$$

$$= \frac{1}{2} \sum_{v \in V_k} \frac{d_v(d_v-1)}{k} + \left(1 - \frac{1}{k}\right)\frac{\sum_i \sum_{v \in V_k}(e_i^2(v) - e_i(v))}{2}$$

$$= \frac{1}{2} \sum_{v \in V_k} \frac{d_v^2}{k} - \frac{1}{2} \sum_{v \in V_k} d_v + \left(1 - \frac{1}{k}\right)\frac{\sum_i \sum_{v \in V_k}(e_i^2(v) - e_i(v))}{2} + \frac{1}{2}\left(1 - \frac{1}{k}\right) \sum_{v \in V_k} d_v$$

$$\leq \left(2 - \frac{1}{k}\right)OPT(V_k) + \left(1 - \frac{1}{k}\right)|E(V_k)|.$$

where $OPT(V_k)$ is the optimal number of conflicts at vertices in V_k and $|E(V_k)|$ be $\sum_{v \in V_k} \frac{d_v}{2}$.

The last inequality comes from the fact that both $\frac{1}{2}\sum_{v \in V_k} \frac{d_v^2}{k} - \frac{1}{2}\sum_{v \in V_k} d_v$ (by Corollary 4) and $\frac{1}{2}\sum_i \sum_{v \in V_k}(e_i^2(v) - e_i(v))$ are lower bounds on the optimal solution.

Note that as in the homogeneous case, we can obtain the same expected approximation guarantee with a randomized algorithm, i.e., choose a color uniformly at random for each cluster V_1^i. Note also that the approximation ratio remains the same for any $C_G \geq k$. In the following section, we obtain a slightly better approximation factor using SDP relaxation when $C_v = 1$ or k and $C_G = k$.

3.2   SDP-Based Algorithm

In this subsection, we assume that k different channels are available in the network and all nodes have 1 or k wireless cards. We formulate the problem using semidefinite programming. Consider the following vector programming (VP), which we can convert to an SDP and obtain an optimal solution in polynomial time. We have an m-dimensional unit vector Y_e for each edge e ($m \leq n$).

$$\textbf{VP:} \quad \min \sum_v \sum_{e_1, e_2 \in E(v)} \frac{1}{k}((k-1)Y_{e_1} \cdot Y_{e_2} + 1) \tag{3}$$

$$|Y_e| = 1 \tag{4}$$

$$Y_{e_1} \cdot Y_{e_2} = 1 \quad \text{if } C_v = 1, \; e_1, e_2 \in E(v) \tag{5}$$

$$Y_{e_1} \cdot Y_{e_2} \geq \frac{-1}{k-1} \quad \text{for } e_1, e_2 \in E(v) \tag{6}$$

We can relate a solution of VP to a channel assignment as follows. Consider k unit length vectors in m-dimensional space such that for any pair of vectors v_i and v_j, the dot product of the vectors is $-\frac{1}{k-1}$. (It has been shown that $-\frac{1}{k-1}$ is the minimum possible value of the maximum of the dot products of k vectors [6,14].) Given an optimal channel assignment of the problem, we can map each channel to a vector v_i. Y_e takes the vector that corresponds to the channel of edge e. If C_v is one, all edges incident to v should have the same color. The objective function is exactly the same as the number of conflicts in the given channel assignment since if $Y_{e_1} = Y_{e_2}$ (e_1 and e_2 have the same color), it contributes one to the objective function, and 0 otherwise. Thus the optimal solution of the VP gives a lower bound of the optimal solution in the channel assignment problem.

The above VP can be converted to a semidefinite programming (SDP) and solved in polynomial time (within any desired precision) [15,16,17,18,19], and given a solution for the SDP, we can find a solution to the corresponding VP, using incomplete Cholesky decomposition [20].

We use the rounding technique used for MAXCUT by Goeman and Williamson [21] when $k = 2$ and show that the expected number of interferences in the solution is at most $1.122OPT + 0.122|E|$. When $k > 2$, we obtain the approximation guarantee of $(2 - \frac{1}{k} - \frac{2(1+\epsilon)\ln k}{k} + O(\frac{k}{(k-1)^2})$ with additive term of $(1 - \frac{2(1+\epsilon)\ln k}{k-1}(1 - \frac{k}{(k-1)^2}))|E|$, where $\epsilon(k) \sim \frac{\ln \ln k}{(\ln k)^{\frac{1}{2}}}$.

When $k = 2$: We select a random unit vector r, and assign channel one to all edges with $Y_e \cdot r \geq 0$ and channel two to all other edges.

Lemma 21. *[21] For $-1 \leq t \leq 1$, $\frac{\arccos t}{\pi} \geq \frac{\alpha}{2}(1 - t)$, where $\alpha > .87856$.*

Theorem 22. *The expected number of total conflicts by our algorithm is at most $1.122OPT + 0.122|E|$.*

Proof. See [13].

When $k > 2$: We use the rounding algorithm for MAX k-CUT when $k > 2$ [6]. Given an optimal solution for VP, we obtain a coloring as follows. We first select k random vectors, denoted as $R = \{r_1, r_2, \cdots, r_k\}$. Each random vector $r_i = (r_{i,1}, r_{i,2}, \ldots r_{i,n})$ is selected by choosing each component $r_{i,j}$ independently at random from a standard normal distribution $N(0, 1)$. For each edge e, assign e to vector r_i if r_i is the closest vector to Y_e (i.e., the vector with the maximum value of $Y_e \cdot r_i$). Ties are broken arbitrarily.

Let $\beta_{ij} = Y_{e_i} \cdot Y_{e_j}$. Let P include all pairs of edges in $E(v)$ for any $v \in V_k$. For a pair $(i, j) \in P$, (i, j) is included in pP (positive pairs)$\subseteq P$ if $\beta_{ij} \geq 0$ and (i, j) is included in nP (negative pairs) $\subseteq P$ if $\beta_{ij} < 0$. We utilized the following two lemmas from [6].

Lemma 23. *[6] For $(i, j) \in nP$, $E[X_{ij}] = \frac{1}{k} + 2(1 + \epsilon)\frac{\ln k}{k}\beta_{ij} + O(\beta_{ij}^2)$ where $\epsilon(k) \sim \frac{\ln \ln k}{(\ln k)^{\frac{1}{2}}}$ [22]*

Lemma 24. *For $(i, j) \in pP$, $E[X_{ij}] \leq \frac{1}{k}((k-1)\beta ij + 1)$*

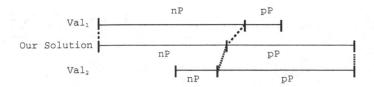

Fig. 4. The expected number of conflicts in nP is bounded by $Val_1(nP)$ and conflicts in pP is bounded by $Val_2(pP)$

Proof. As $E[X_{ij}] = \frac{1}{k}((k-1)\beta_{ij}+1)$ when $\beta_{ij} = 0$ and 1, and $E[X_{ij}]$ is a convex function in $[0,1]$ [6], we have the lemma.

Note the if we simply compare the lowerbound obtained by SDP and the upperbound given in Lemma 23 and 24, we cannot obtain a constant factor approximation. However, by carefully combining the lowerbound in Corollary 4, we can obtain a slightly better approximation factor than the greedy algorithm. We define $Val_1(S)$ to be $\frac{1}{k}|S|$ for any set $S \subseteq P$. In addition, let $Val_2(S)$ be $\frac{1}{k}\sum_{(i,j)\in S}((k-1)\beta_{ij}+1)$. That is, $Val_2(S)$ is a lowerbound obtained by SDP relaxation and $Val_1(S)$ is a lower bound based on the fact that all edges incident to a vertex can interfere with each other. As shown in Figure 4, simply combining two lowerbounds gives a 2-approximation. To prove a better bound than greedy, we first prove the following lemmas.

Lemma 25. *The number of conflicts by the algorithm is at most* $Val_2(P) + \delta(\frac{k-1}{2(1+\epsilon)\ln k}$
$-1) + \sum_{(i,j)\in nP} O(\beta_{ij}^2)$ *where* $\delta = -\frac{2(1+\epsilon)\ln k}{k}\sum_{(i,j)\in nP}\beta_{ij}$.

Proof

$$\sum_{(i,j)\in P} E[X_{ij}] = \sum_{(i,j)\in pP} E[X_{ij}] + \sum_{(i,j)\in nP} E[X_{ij}]$$

$$\leq \sum_{(i,j)\in pP} \frac{1}{k}((k-1)\beta ij + 1) + \sum_{(i,j)\in nP}(\frac{1}{k} + 2(1+\epsilon)\frac{\ln k}{k}\beta_{ij} + O(\beta_{ij}^2))$$

$$= Val_2(P) + \sum_{(i,j)\in nP}(\frac{1}{k} + 2(1+\epsilon)\frac{\ln k}{k}\beta_{ij}) - \sum_{(i,j)\in nP}\frac{1}{k}((k-1)\beta ij + 1) + \sum_{(i,j)\in nP} O(\beta_{ij}^2)$$

$$\leq Val_2(P) + \delta(\frac{k-1}{2(1+\epsilon)\ln k} - 1) + \sum_{(i,j)\in nP} O(\beta_{ij}^2).$$

$$(7)$$

Lemma 26. *The number of conflicts by the algorithm is at most* $Val_1(P) - \delta + Val_2(P)$
$(1 - \frac{1}{k}) + \sum_{(i,j)\in nP} O(\beta_{ij}^2)$ *where* $\delta = -\frac{2(1+\epsilon)\ln k}{k}\sum_{(i,j)\in nP}\beta_{ij}$.

Proof. By Lemma 23 the number of conflicts of pair of edges in nP is at most $Val_1(nP)$
$-\delta + \sum_{(i,j)\in nP} O(\beta_{ij}^2)$ and by Lemma 24 the number of conflicts of pair of edges in pP is at most $Val_1(pP) + \frac{k-1}{k}\sum_{(i,j)\in pP}\beta_{ij}$ so we have:

$$\sum_{(i,j)\in P} E[X_{ij}] \le Val_1(nP) - \delta + Val_1(pP) + \frac{k-1}{k} \sum_{(i,j)\in pP} \beta ij + \sum_{(i,j)\in nP} O(\beta_{ij}^2)$$

$$= Val_1(P) - \delta + \frac{k-1}{k} \sum_{(i,j)\in pP} \beta ij + \sum_{(i,j)\in nP} O(\beta_{ij}^2)$$

$$\le Val_1(P) - \delta + Val_2(P)(1 - \frac{1}{k}) + \sum_{(i,j)\in nP} O(\beta_{ij}^2) \qquad (8)$$

Theorem 27. *The number of conflicts created by the algorithm is at most* $((2 - \frac{1}{k} - \frac{2(1+\epsilon)\ln k}{k} + O(\frac{k}{(k-1)^2}))OPT + (1 - \frac{2(1+\epsilon)\ln k}{k-1}(1 - \frac{k}{(k-1)^2}))|E|,$ *where* $\epsilon(k) \sim \frac{\ln\ln k}{(\ln k)^{\frac{1}{2}}}.$

Proof. By Lemma 25 and 26 the number of conflicts is upper bounded by $\min(Val_2(P) + \delta(\frac{k-1}{2(1+\epsilon)\ln k} - 1), Val_1(P) - \delta + Val_2(P)(1 - \frac{1}{k})) + \sum_{(i,j)\in nP} O(\beta_{ij}^2)$, which is maximized when $\delta = \frac{2(1+\epsilon)\ln k}{k-1}(Val_1(P) - \frac{1}{k}Val_2(P))$. Let $f(k) = \frac{2(1+\epsilon)\ln k}{k-1}$. Then the maximum number of conflicts is

$$(1 - f(k))Val_1(P) + (1 - \frac{1}{k} + \frac{f(k)}{k})Val_2(P) + \sum_{nP} O(\beta_{ij}^2).$$

Note that $Val_1(P) \le OPT(V_k) + |E(V_k)|$ and $Val_2(P) \le OPT(V_k)$. Therefore, the total conflict at V_k is at most

$$(2 - \frac{1}{k} - \frac{(k-1)f(k)}{k})OPT(V_k) + (1 - f(k)|E(V_k)| + \sum_{nP} O(\beta_{ij}^2).$$

Since $\beta_{ij}^2 \le \frac{1}{(k-1)^2}$, we have the theorem.

4   Discussion

Note that in all of our algorithms the total number of different colors used in the network is only $\max C_v$ rather than C_G. Although the number of conflicts may be reduced using more colors than $\max C_v$ (see for example Figure 1), it is not easy to make sure that each edge has at least one channel which are available at both endpoints in that case. In fact, it may be possible that the size of the set of common channels is small, which may result in creating more conflicts. One possible solution is to further improve the solution by recoloring edges with additional colors after obtaining the solution by the algorithm. It will be an interesting future work to analyze how much we can improve the performance by such recoloring.

Acknowledgements. The second author would like to thank Nikhil Bansal for useful discussions.

References

1. Raniwala, A., Chiueh, T.: Architecture and algorithms for an IEEE 802.11-based multi-channel wireless mesh network. In: Proceedings of Infocom (March 2005)
2. Kyasanur, P., Vaidya, N.H.: Routing and interface assignment in multi-channel multi-interface wireless networks. In: Proceedings of WCNC (March 2005)
3. Alicherry, M., Bhatia, R., Li, L.E.: Joint channel assignment and routing for throughput optimization in mu lti-radio wireless mesh networks. In: Proceedings of ACM MobiCom, ACM Press, New York (2005)
4. Shin, M., Lee, S., Kim, Y.A.: Distributed channel assignment for multi-radio wireless networks. In: The Third IEEE International Conference on Mobile Ad-hoc and Sensor Systems (MASS), IEEE Computer Society Press, Los Alamitos (2006)
5. Kann, V., Khanna, S., Lagergren, J., Panconesi, A.: On the hardness of max k-cut and its dual. In: Proc. 5th Israel Symposium on Theory and Computing Systems (ISTCS), pp. 61–67 (1996)
6. Frieze, A., Jerrum, M.: Improved approximation algorithms for MAX k-CUT and MAX BISECTION. In: Balas, E., Clausen, J. (eds.) Integer Programming and Combinatorial Optimization. LNCS, vol. 920, pp. 1–13. Springer, Heidelberg (1995)
7. Fitzpatrick, S., Meertens, L.: An experimental assessment of a stochastic, anytime, decentralized, soft colourer for sparse graphs. In: Steinhöfel, K. (ed.) SAGA 2001. LNCS, vol. 2264, Springer, Heidelberg (2001)
8. Damaschke, P.: Distributed soft path coloring. In: Alt, H., Habib, M. (eds.) STACS 2003. LNCS, vol. 2607, pp. 523–534. Springer, Heidelberg (2003)
9. Holyer, I.: The np-completeness of edge-coloring. SIAM J. Computing 10(4), 718–720 (1981)
10. Vizing, V.G.: On an estimate of the chromatic class of a p-graph (russian). Diskret. Analiz. 3, 25–30 (1964)
11. Nishizeki, T., Kashiwagi, K.: On the 1.1 edge-coloring of multigraphs. SIAM J. Disc. Math. 3(3) 391–410 (1990)
12. Caprara, A., Rizzi, R.: Improving a family of approximation algorithms to edge color multigraphs. Information Processing Letters 68(1), 11–15 (1998)
13. Kari, C., Kim, Y.A., Lee, S., Russell, A., Shin, M.: Soft edge coloring. (2007) Full version available at http://www.engr.uconn.edu/~ykim/soft.pdf
14. Karger, D., Motwani, R., Sudan, M.: Approximate graph coloring by semidefinite programming. In: Proc. 35th IEEE Symposium on Foundations of Computer Science, pp. 2–13. IEEE Computer Society Press, Los Alamitos (1994)
15. Alizadeh, F.: Interior point methods in semidefinite programming with applications to combinatorial optimization. SIAM Journal on Optimization 5(1), 13–51 (1995)
16. Grotschel, M., Lovasz, L., Schrijver, A.: The ellipsoid method and its consequences in combinatorial optimization. Combinatorica 1, 169–197 (1981)
17. Grotschel, M., Lovasz, L., Schrijver, A.: Geometric algorithms and combinatorial optimization. Springer, Heidelberg (1987)
18. Nesterov, V., Nemirovskii, A.: Self-concordant functions and polynomial time methods in convex programming. Central Economical and Mathematical Institute, U.S.S.R. Academy of Science, Moscow (1990)
19. Nesterov, V., Nemirovskii, A.: Interior-point polynomial algorithms in convex programming. SIAM (1994)
20. Golub, G.H., Loan, C.F.V.: Matrix Computations. The Johns Hopkins University Press, Baltimore, MD (1983)
21. Goemans, M.X., Williamson, D.P.: Improved approximation algorithms for maximum cut and satisfiability problems using semidefinite programming. Journal of the ACM 42, 1115–1145 (1995)
22. Galambos, J.: The asymptotic theory of extreme order statistics. Wiley, New York (1978)

Approximation Algorithms for the Max-Min Allocation Problem

Subhash Khot* and Ashok Kumar Ponnuswami**

College of Computing
Georgia Institute of Technology
Atlanta GA 30309, USA
{khot,pashok}@cc.gatech.edu

Abstract. The Max-Min allocation problem is to distribute indivisible goods to people so as to maximize the minimum utility of the people. We show a $(2k-1)$-approximation algorithm for Max-Min when there are k people with subadditive utility functions. We also give a k/α-approximation algorithm (for $\alpha \leq k/2$) if the utility functions are additive and the utility of an item for a person is restricted to 0, 1 or U for some $U > 1$. The running time of this algorithm depends exponentially on the parameter α. Both the algorithms are combinatorial, simple and easy to analyze.

1   Introduction

The problem of allocating goods to people has received extensive attention in economics. The properties expected of the allocation may depend on the kind of goods available, the motives of the distributor or of the people interested in the goods, and finally, the utility that a person can derive from a bundle of goods. In this paper, we consider the Max-Min allocation problem where:

- There are m indivisible items.
- There are k people interested in the items. Our goal (or, the distributor's goal) is to maximize the minimum utility of the people. Each person is assumed to answer questions about their utilities truthfully.
- The utility functions of the people are subadditive. If the utility functions are additive, we call the problem Max-Min$^+$.

Variants of Max-Min have received extensive attention in literature. Mjelde [8] studies the problem of allocating resources to various activities with the goal of maximizing the minimum return from the activities. Mjelde considers the case where the resources are infinitely divisible and the return from an activity is a concave function of the weighted resources invested in it. Younis et al. [11]

---

* Research supported in part by Sloan Foundation Fellowship, Microsoft New Faculty Fellowship and NSF CAREER award CCF-0643626.
** Research supported in part by Subhash Khot's Sloan Foundation Fellowship and Microsoft New Faculty Fellowship.

M. Charikar et al. (Eds.): APPROX and RANDOM 2007, LNCS 4627, pp. 204–217, 2007.
© Springer-Verlag Berlin Heidelberg 2007

consider the problem of assigning workloads to sensors so as to maximize the time before the battery of some sensor runs out. Ramirez-Marquez et al. [9] look at the problem of designing series-parallel systems with the goal of maximizing the minimum reliability of the subsystems. The "on-line" version of Max-Min$^+$ has been studied by [2,5,10].

We are interested in algorithms that can compute the value of the maximum attainable minimum utility. It can be shown that computing the optimum value exactly is NP-hard. Bezáková and Dani [4] showed that approximating this quantity within a factor of $2 - \epsilon$ is also NP-hard. They also gave a $(m - k + 1)$-approximation algorithm for Max-Min$^+$. For the case when the utility functions are "maximal", they gave an exact algorithm. Golovin [6] gave a k-approximation algorithm for Max-Min$^+$ by rounding the solution to a natural LP for the problem. Golovin also gave a l-approximation algorithm if we can ignore $1/l$ fraction of the people for any l and a $(m-k+1)$-approximation algorithm if the utility functions are submodular. For a special case of Max-Min$^+$ called "Big Goods/Small Goods", where the small goods have utility 0 or 1 to a person and the big goods have utility 0 or x ($x > 1$), Golovin gave a $O(\sqrt{k})$-approximation.

Bansal and Sviridenko [3] define a *configuration* LP for Max-Min$^+$ and show that it has a gap of $\Omega(\sqrt{k})$. For the special case when the utility of the jth item to a person is either 0 or u_j, they use the LP to approximate the optimum to within $O(\log \log k / \log \log \log k)$. Asadpour and Saberi [1] recently showed that the gap of the configuration LP for Max-Min$^+$ demonstrated by [3] is essentially tight. They give a randomized rounding procedure that produces an $O(\sqrt{k} \log^3 k)$ approximate solution. To construct a separation oracle for the configuration LP, one needs to solve the single person "minimum knapsack" problem. This can be done efficiently if the utility functions are additive, but not when they are subadditive. Golovin's [6] approach also does not immediately yield a result when the utility functions are subadditive because it depends on fractionally allocating items of a certain value to every person.

1.1 Our Results

Our first main result is to give a combinatorial $(2k-1)$-approximation algorithm for Max-Min.

Theorem 1. *There exists an algorithm that when given a Max-Min instance over k people (with subadditive utility functions) and m items finds a $(2k - 1)$-approximate allocation in time polynomial in k and m.*

Starting with an estimate a on the optimum, the algorithm finds the set of people who *must* be given a "big" item (that is, an item of utility $\geq b = a/(2k - 1)$, the target) in any solution of value a and gives them one big item each. For the rest of the people, the algorithm finds a bundle of "small" items that has utility $\geq b$ for them. To accomplish this, the algorithm starts by giving small items of value b to some people. Whenever the algorithm runs out of items even though there are people who are yet to get any items, the algorithm tries to reallocate the bundles to people who can derive greater utility from them and free some items.

We point out the limitations of the algorithm and suggest how they can be overcome to construct a better combinatorial algorithm. Using these ideas, we get a better approximation for the special case of Max-Min$^+$ where the utility of an item to a person is either 0, 1 or U for some $U > 1$. The $(0, 1, U)$-Max-Min$^+$ problem seems to be interesting because it still captures the fact that an item may have varying degrees of utility to different people, something that makes the problem very hard to tackle. It is also worth noting that the LP gap examples for the *configuration LP* shown by Bansal and Sviridenko [3] are of this restricted form. Note that this model is more general than the model of "Big Goods/Small Goods" studied by Golovin [6] since here, an item can be big for one person and small for another. We generalize the flow problem defined by Golovin [6] for Big Goods/Small Goods to $(0, 1, U)$-Max-Min$^+$ and obtain a a k/α-approximation algorithm.

Theorem 2. *There exists an algorithm that when given a $(0, 1, U)$-Max-Min$^+$ instance (over k people and m items) and a number $\alpha \leq k/2$ as input finds an allocation of value $\geq \alpha \cdot OPT/k$ in time $m^{O(1)}k^{O(\alpha)}$, where OPT is the optimum for the Max-Min$^+$ instance.*

For example, this gives a non-trivial running time of $m^{O(1)}2^{\tilde{O}(k^{2/3})}$ if we are interested in a $k^{1/3}$-approximation.

In the next section, we define the problem formally. In Section 3 we describe the $(2k - 1)$-approximation algorithm for Max-Min. Finally, in Section 4 we describe our algorithm for $(0, 1, U)$-Max-Min$^+$.

2   Preliminaries

Definition 1. *The function $u : 2^T \to \mathbb{R}$ is said to be monotone if $u(S_1 \cup S_2) \geq u(S_1)$ for all disjoint subsets S_1 and S_2 of T. The function u is said to be a utility function (or a valuation) if u is monotone and $u(\emptyset) = 0$. With a little abuse of notation, for $t \in T$ we use $u(t)$ to denote $u(\{t\})$. A function $u : 2^T \to \mathbb{R}$ is said to be a subadditive if $u(S_1 \cup S_2) \leq u(S_1) + u(S_2)$ for all disjoint subsets S_1 and S_2 of T. It is submodular if $u(S_1 \cup S_2) + u(S_1 \cap S_2) \leq u(S_1) + u(S_2)$ for all subsets S_1 and S_2 of T. A function u is additive if $u(S_1 \cup S_2) = u(S_1) + u(S_2)$ for all disjoint subsets S_1 and S_2 of T.*

We refer the reader to Lehmann et al. [7] for a discussion about the class of subadditive functions and its various subclasses.

Let $P = \{p_1, p_2, \ldots, p_k\}$ be the set of people. Let $T = \{t_1, t_2, \ldots, t_m\}$ be the set of indivisible items. Let u_i be the utility function of person p_i. An allocation is a partition $T_1, T_2, \ldots, T_k$ of the set T into k sets, one for each person. The value of this allocation is $\min_{i \in [k]} u_i(T_i)$, where $[k]$ denotes the set $\{1, 2, \ldots, k\}$. The Max-Min problem is to find an allocation that maximizes this quantity.

The optimum of the Max-Min problem is the value of such an allocation. Unless otherwise stated, we assume that the utility functions under consideration are subadditive. We also assume that we are given valuation oracles for the utility functions (see [7]). That is, we can query the utility of a subset S of items to any person p_i and get back the answer $u_i(S)$ in constant time.

Max-Min$^+$ is the special case of Max-Min where all the utility functions are additive. We will consider a version of the Max-Min$^+$ problem where the utility $u_i(t_j)$ of an item t_j for a person p_i can either be 0, 1, or U for some $U > 1$ (U can be a part of the input). We call this version the $(0, 1, U)$-Max-Min$^+$ problem.

3   A $(2k - 1)$-Approximation Algorithm for Max-Min

We first give an algorithm that finds a $(2k - 1)$-approximate solution when the value of the optimum is known. We will use this to construct an algorithm that can find an approximate solution without knowing the optimum.

3.1   Finding an Approximate Solution When the Optimum Is Known

Let the optimum of the problem be given to be $a = (2k - 1)b$. We are expected to find an allocation of value $\geq b$. We first give a general outline of Algorithm 1.

The algorithm first checks if for every person, the net worth of her "small" items is at least a.

Definition 2. *An item t_j is said to be small for p_i if $u_i(t_j) < b$.*

Note that an item that is small for one person may be "big" for another. Therefore, we must always talk of an item being big or small with reference to some person. If the net value of all small items for a person p_i is less than a, then in any solution of value a, p_i must get a big item. But once we give a person a big item, we need not give her any other item in a solution of value b.

Using the above observation as the basis, the algorithm constructs a set $Q_0 \subseteq P$ of people who need to be given at least one big item each in any solution of value a. For the purpose of finding a allocations of value b, we can safely restrict our attention to only those allocations where all people in Q_0 get big items. The algorithm allocates everyone in Q_0 one big item each. It then tries to satisfy one person from $P \backslash Q_0$ at a time. For these people, the algorithm completely ignores any contribution from big items.

Definition 3. *Given a utility function $u : 2^T \to \mathbb{R}$, define $u^b : 2^T \to \mathbb{R}$ to be $u^b(S) = u(\{t_j : t_j \in S \text{ and } u(t_j) < b\})$*

Informally, u^b is a modified utility function that ignores big items. Note that if u is subadditive, so is u^b. Using u_i^b as the utility function for people in $P \backslash Q_0$ only makes the problem harder to solve (Ignoring big items is convenient when analyzing how the small items of a person are distributed). The algorithm may allocate items to the first i people only to discover that the leftover set of items do not have much utility for p_{i+1}. The algorithm is forced to reallocate the items to make p_i happy. A person is *happy* if she has been allocated a subset of items of value at least b to her. For the purpose of reallocating goods, the algorithm uses a *reallocation graph.*

Definition 4. *Given* $(Q_0, T_0, (T_i)_{i \in [k]})$, *where* $Q_0 \subseteq P$ *and* $(T_0, T_1, \ldots, T_k)$ *is a partition of* T, *define the reallocation graph as follows:*

- *There is one vertex for every person in* $P \setminus Q_0$ *and a vertex for* T_0.
- *There is an edge from* p_{i_1} *to* p_{i_2} *if* $u^b_{i_1}(T_{i_2}) \geq 2b$.
- *There is an edge from* p_{i_1} *to* T_0 *if* $u^b_{i_1}(T_0) \geq b$.

Here, T_i is the set of items given to person p_i (T_0 is the set of unallocated items). In a sense, the reallocation graph shows where a person can find b worth of small items as a bundle, either as a bundle of value $2b$ from some other person or from the set of free items. We exclude the people in Q_0 from the graph because the algorithm will never reallocate items given to them. Based on the reallocation graph, the algorithm reallocates items using one of the two rules (we will show that at least one of the two rules is always applicable):

- *(Rule 1:)* If there is a simple cycle (a cycle on which every vertex appears only once), then the number of unallocated items $|T_0|$ can be increased without changing the number of happy people (see lines 8-13). To accomplish this, give the ith person on the cycle the items that the $(i+1)^{th}$ person currently has. We will show that each person can return at least one item back to T_0 and still stay happy.
- *(Rule 2:)* If there is a simple path from an person p_i with no items to T_0, then p_i can be made happy without affecting any other person's happiness (see lines 16-19). To accomplish this, give the lth person on the path items that the $(l+1)^{th}$ person currently has. Give the last person on the path items from T_0.

During each application of the above rules, we ensure that a person keeps only a minimal set of items that makes her happy. This gives a bound on the utility a person derives from the bundle given to her.

Lemma 1. *At any time in the algorithm, for any* $p_i \in P \setminus Q_0$, *either* T_i *is empty or* $u^b_i(T_i) \in [b, 2b)$.

Proof. Note that if $t_j \in S$, then $u^b_i(S) - u^b_i(S \setminus t_j) \leq u^b_i(t_j) < b$ (the first inequality follows from subadditivity of u^b_i). This implies that if $u^b_i(S) \geq 2b$, it can not be a minimal set satisfying $u^b_i(S) \geq b$. □

Theorem 3. *For any value of the parameter a such that $0 < a < \infty$, Algorithm 1 either:*

- *Fails at Line 4, in which case the optimum is less than a, or*
- *Returns an allocation of value $\geq b = a/(2k-1)$ in time polynomial in k and m.*

Algorithm 1. An algorithm to approximate Max-Min when the optimum is specified

---

Input: $k, m, (u_i)_{i \in [k]}, a = (2k-1)b$, where $0 < a < \infty$
Output: An allocation $(T_1, \ldots, T_k)$ of value $\geq b$

1 Set $S_i \leftarrow \{t_j \in T : u_i(t_j) < b\}$, the set of small items for person i. Compute $a_i = u_i(S_i)$, the net worth of small items to p_i.
2 Let Q_0 be the set of people p_i satisfying $a_i < a$.
3 Set $T_0 \leftarrow T$. Set $T_i \leftarrow \emptyset$ for all $i \in [k]$.
4 Give all people in Q_0 an arbitrary big item each (Update the T_is accordingly). If it is not possible, stop.
5 **while** *there is an unhappy person* **do**
6   **if** *there is a cycle in the current reallocation graph* **then**
    /*Implements Rule 1.   */
7   Let $p_{i_1}, p_{i_2}, \ldots, p_{i_L}$ be a simple cycle (with the edges being from p_{i_l} to $p_{i_{l+1}}$).
8   Set $T' \leftarrow T_{i_1}$.
9   Take away the items T_{i_1} currently with p_{i_1}.
10   **for** $l = 1$ **to** $L - 1$ **do**
11   Take away the items $T_{i_{l+1}}$ currently with p_{i_l}. Give a minimal set of these items of value $\geq b$ to p_{i_l} and return the rest to T_0.
12   **end**
13   Give a minimal subset of T' of value $\geq b$ to p_{i_L}. Return the rest of the items in T' to T_0.
14   **else if** *there is a path from an unhappy person p_i to T_0 in the reallocation graph* **then**
    /*Implements Rule 2.   */
15   Let $p_i = p_{i_1}, p_{i_2}, \ldots, p_{i_L}, T_0$ be a simple path from p_i to T_0.
16   **for** $l = 1$ **to** $L - 1$ **do**
17   Take away the items $T_{i_{l+1}}$ currently with $p_{i_{l+1}}$. Give a minimal set of these items of value $\geq b$ to p_{i_l} and return the rest to T_0.
18   **end**
19   Give a minimal set of items of value $\geq b$ from T_0 to p_{i_L}.
20 **end**
21 Return the allocation $(T_1, T_2, \ldots, T_{k-1}, T_k \cup T_0)$ as the answer.

Proof. If $p_i \in Q_0$, then p_i must get a big item in any allocation of value $\geq a$. This is because $u_i(S) \leq u_i(S_i) < a \ \forall S \subseteq S_i$ (by monotonicity of u_i). Therefore, if these people can not all be given *some* big item simultaneously, there can not be an allocation of value a. Therefore, if the algorithm fails, it does so because the optimum is less than a.

Now assume the algorithm does not fail at Line 4. If the reallocation graph has a cycle, every person on the cycle is first given a set of items of value at least $2b$ (since an edge from p_{i_1} to p_{i_2} means that the utility of the items currently with p_{i_2} for p_{i_1} is $\geq 2b$). The algorithm then takes one or more items away from these people and returns to T_0 (since from Lemma 1, a set of value $\geq 2b$ can't be a minimal set of value b). Since every time we reallocate items for people on some cycle, we return an item back to $T_0 \subseteq T$, in at most m applications of Rule 1, we would have eliminated all cycles.

Assume now that the graph G has no cycle. Assume a person p_i is unhappy. The set of items T is currently partitioned into sets $T_0, T_1, T_2, \ldots, T_k$, of which at most k are non-empty (since T_i is empty). Also if $i' \in P \setminus Q_0$, $u_{i'}^b(T) \geq a$. Suppose $i' \in P \setminus Q_0$ does not have an edge to T_0. Then at least one of the $k - 1$ non-empty sets in $T_1, T_2, \ldots, T_k$ is worth at least $(a - b)/(k - 1) = 2b$ to her (by the subadditivity of $u_{i'}^b$ and also since $u_{i'}^b(T_0) < b$). But this set can not be a set of items given to a person in Q_0 (since people in Q_0 only get one item and a single item is worth less than b under $u_{i'}^b$). Hence every person in $P \setminus Q_0$ has an outgoing edge in G, either to T_0 or to some other person. But if T_0 is not reachable from p_i, there must be a cycle in the graph, a contradiction.   □

We can now perform a binary search on the possible values of the optimum and get an approximation factor of $O(k)$. But using some properties of Algorithm 1, we next obtain a $(2k - 1)$-approximation by doing a binary search over a set of size $O(km)$.

3.2   Finding a $(2k - 1)$-Approximation to Max-Min When the Optimum Is Not Known

Using Algorithm 1 as a subroutine, Algorithm 2 tries to find an allocation that is a $(2k - 1)$-approximation for Max-Min. The main idea is as follows. Suppose we increase the input parameter a to Algorithm 1 starting from a very small number. Suppose for some $a = a_0$, Algorithm 1 successfully returns an allocation, but fails when called with $a = a_0 + \epsilon$ for infinitesimally small $\epsilon > 0$. Then one of the two events must happen at $a = a_0 + \epsilon$

- An item that is big for some person at $a = a_0$ is no longer big at $a = a_0 + \epsilon$.
- A new person p_i enters Q_0 because $a_i = a_0$, where $a_i = u_i(\{t_j \in T : u_i(t_j) < a_0/(2k - 1)\})$.

Therefore, it is enough to check the behavior of Algorithm 1 at these $mk = O(m^2)$ values of a. It is sufficient to do a binary search on this list of $O(m^2)$ values of a to find a number for which the algorithm succeeds, but fails for the next. We will now prove Theorem 1.

Proof. We will show that Algorithm 2 satisfies the conditions of the theorem. Assume that that the optimum of the input instance is greater than zero. If Algorithm 1 is called with $a = (2k - 1)b_1$, all items of non-zero value are big for a person since $b = \min_{i,j} u_i(t_j)$. Therefore, Algorithm 1 just attempts to allocate one item of non-zero value to every person. Hence, this call to Algorithm 1 succeeds since the optimum is non-zero.

Suppose a call to Algorithm 1 with parameter $a = a'$ succeeds for some $a' \in A$, but fails with parameter $a = a''$ for the next larger value of $a'' \in A$. From the analysis of Algorithm 1, we know that the optimum is less than a''. But suppose the optimum OPT satisfies $a' < OPT < a''$. Algorithm 1 would

Algorithm 2. An algorithm to approximate Max-Min

Input: $k, m, (u_i)_{i \in [k]}$
Output: An allocation $(T_1, \ldots, T_m)$ of value within a factor $2k - 1$ of the
optimum

1 Check if there is no allocation of non-zero value. If this is the case, then return
any allocation as the answer.
2 Let $B = \{u_i(t_j) : u_i(t_j) > 0\}$, the set of non-zero values of singleton sets to the
people. Let the elements of the set be $b_1, b_2, \ldots$ in increasing order.
3 Let A denote the set of values of a we will try. Set $A \leftarrow \{(2k - 1)b : b \in B\}$
4 **for** $b_l \in B$ **do**
5   Let $S_i = \{t_j : u_i(t_j) < b_l\}$, that is, the set of items that will be small for p_i
with respect to $b_l + \epsilon$. Let $a_i = u_i(S_i)$.
6   Add all elements of $\{u_i(S_i)\}_{i \in [k]}$ between b_l and b_{l+1} to A (If b_l is the
largest number in B, use $b_{l+1} = \infty$).
7 **end**
8 By a binary search, find an $a \in A$ for which Algorithm 1 succeeds, but fails for
the next larger value in A.
9 Return the allocation returned by Algorithm 1 for this value of a as the answer.

---

have succeeded if called with parameter $a = OPT$. Let $b_0 = OPT/(2k - 1)$
and $b'' = a''/(2k - 1)$. Note that the set of big items for a person is the same
with respect to any $b \in (a'/(2k - 1), b'']$ (since no $u_i(t_j)$ lies in the interval
$(a'/(2k - 1), b'')$). This also implies that a_i, the net worth of all small items
to person p_i, is the same when calculated with respect to $b = b_0$ or $b = b''$.
But we know that no $a_i/(2k - 1)$ lies in the interval $[b_0, b'')$ either. Therefore,
if Algorithm 1 succeeds when called with parameter OPT, it will also succeed
with parameter a''. Therefore, our assumption that $a' < OPT$ must be wrong. If
$OPT \le a'$, then the allocation found using Algorithm 1 with $a = a'$ must have
value $\ge a'/(2k - 1) \ge OPT/(2k - 1)$.

If the call to Algorithm 1 with the largest $a \in A$ succeeds, then we must have
found a solution of value $u_i(T)$ for some i, which is obviously optimum. $\square$

3.3   Limitations of Algorithm 1

We first give an example to show that the analysis is the best possible for Algorithm 1. Consider the problem over k people with additive utility functions.
There are two groups of items. There are $k - 1$ items in the first group such that
the jth item is of value $2k - 1$ to p_j and value zero to everyone else. There are
$2k - 1$ items that are of unit value to everyone. The optimum is to give all the unit
valued items to p_k and the $k - 1$ big items to the first $k - 1$ people. But when called
with $a = 2k - 1$, Algorithm 1 treats every item as big. The algorithm then just
matches every person to one item and returns a $(2k - 1)$-approximate solution.

Suppose we want to use the same idea to find a k-approximation (or even just
a $(2 - \epsilon)k$-approximation). Since the optimum in the above example is $2k - 1$, and
we should find a solution of value 2 (since our target $b = \frac{2k-1}{(2-\epsilon)k} > 1$). Since the

unit valued items are now small for everyone, and there are $2k - 1$ small items, we will conclude no one needs to get big items. But then the best allocation that can be found has value one.

Intuitively, one limitation of the algorithm is that once it decides who gets big items, it does not re-evaluate its decision when it gets stuck. In the above example, it is clear that many of the people share the same set of small items. Hence most of them must be given big items to find a significantly better solution. This limits the algorithm to a k-approximation because a person could lose her small items to $k - 1$ other people, who all might have got big items in a better solution. An extra factor of 2 comes in because even if we shoot for a solution of value b, we might have to give some people close to $2b$ worth of small items. One can also come up with examples where the right choice of big items is critical. In the next section we present an algorithm that takes care of these issues, albeit for a restricted version of Max-Min.

4   A (k/α)-Approximation Algorithm for $(0, 1, U)$-Max-Min$^+$

We now give a (k/α)-approximation for $(0, 1, U)$-Max-Min$^+$ that runs in time $m^{O(1)}k^{O(\alpha)}$. We call the items of utility U to a person big for that person. Suppose the value of the optimum is OPT. Then there is a solution of value $OPT/2$ in which for a person p_i, p_i either gets only unit valued items or only gets big items. We will restrict our attention to such allocations only, where all the items a person gets have same utility to her. We call such allocations *uniform*. We will find a $\beta = k/(2\alpha)$-approximation over such allocations.

We first give an algorithm that finds an approximate solution when the value of the optimum over uniform allocations is provided as extra information. Since the optimum over uniform allocations is in the set $\{1, 2, \ldots, m\} \cup \{U, 2U, \ldots, Um\}$, we can just call the algorithm for each of these $O(m)$ values.

Algorithm 3 maintains a set of people Q such that there exists a uniform allocation of value a in which all the people in Q get a big items. The goal is to allocate $n_U = \lceil \frac{a}{\beta U} \rceil$ big items to everyone in Q and $n_1 = \lceil \frac{a}{\beta} \rceil$ unit valued items to everyone else. The algorithm must somehow figure out which big items to give to a person in Q. It should also be able to check if there are enough small items for people not in Q. For this purpose we define a *allocation-flow* problem.

Definition 5. *Given $Q \subseteq P$, define the network flow problem N_Q as follows.*

- *The set of vertices is $\{s, t\} \cup P \cup T$, where s and t are the source and sink respectively.*
- *There is an edge from s to every person $p_i \in Q$ with capacity n_U.*
- *There is an edge from s to every person $p_i \notin Q$ with capacity n_1.*
- *There is an edge from $p_i \in Q$ to item t_j with infinite capacity if $u_i(t_j) = U$.*
- *There is an edge from $p_i \notin Q$ to item t_j with infinite capacity if $u_i(t_j) = 1$.*
- *There is an edge from every item to the sink t with capacity 1.*

We call N_Q the allocation-flow problem.

If there is a flow that saturates all edges out of s in N_Q, it defines an allocation of value $\lceil a/\beta \rceil$ in a natural way. Otherwise, the algorithm tries to grow Q based on the following lemma.

Lemma 2. *Consider a uniform allocation of value a, in which everyone in Q gets big items. Suppose $\{s\}$ is not a minimum cut in N_Q. Suppose $\{s\} \cup R \cup T'$ be a minimum cut for some $R \subseteq P$ and $T' \subseteq T$. Then in this allocation, all but $l = \lfloor \frac{n_1 \cdot |R \setminus Q| - 1}{a} \rfloor$ people from R get big items too.*

Proof. First observe that there are no edges from R to $T \setminus T'$ (since all such edges have infinite capacity). Therefore, all items that are big for people in $Q \cap R$ are in T'. At the same time, all unit-valued items for people in $R \setminus Q$ are also in T'. Also, the value of the cut $\{s\}$ is $n_1 \cdot |P \setminus Q| + n_U \cdot |Q|$. The value of the cut $\{s\} \cup R \cup T'$ is $n_1 \cdot |(P \setminus R) \setminus Q| + n_U \cdot |(P \setminus R) \cap Q| + |T'|$. Hence the number of items in T' is upper bounded by $n_1 \cdot |R \setminus Q| + n_U \cdot |R \cap Q| - 1$, since otherwise $\{s\}$ is also a minimum cut. If we give n_U big items to every person in $Q \cap R$, it leaves at most $n_1 \cdot |R \setminus Q| - 1$ items in T'. But then we cannot give a unit-valued items to more than $l = \lfloor \frac{n_1 \cdot |R \setminus Q| - 1}{a} \rfloor$ people from R. Therefore, the other people in R must get big items. $\qquad\square$

Algorithm 3. An algorithm to approximate $(0, 1, U)$-Max-Min$^+$ given the optimum over uniform allocations.

Input: $(k, m, (u_i)_{i \in [k]}, a)$ where $a \geq \beta \geq 4$ is the optimum over uniform allocations
Output: An allocation of value $\geq a/\beta$.
1   Set $Q \leftarrow \emptyset$, $Q_0 \leftarrow \emptyset$, $c = 0$ /*The Q_cs record how Q grows with time. */
2   **while** $\{s\}$ *is not a minimum cut* **do**
3   Set $c \leftarrow c + 1$
4   Find a minimum cut in N_Q. Let R_c denote the set of people on the same side as s in this cut.
5   **if** R_c *contains only people from Q* **then stop**.
6   Non-deterministically pick some $l_c = \left\lfloor \dfrac{n_1 \cdot \|R_c \setminus Q\| - 1}{a} \right\rfloor$ people from $R_c \setminus Q$. Add everyone except these l_c people to Q.
7   Set $Q_c \leftarrow Q$.
8   **end**
9   Find an integral flow in N_Q saturating the edges leaving s.
10 Give item t_j to person p_i if there is a flow of value 1 from t_j to p_i.

Theorem 4. *Let $k/(2a) = \beta \geq 4$ and $a \geq \beta$. If there is a uniform allocation of value a, then one of the non-deterministic branches of the algorithm succeeds in finding an allocation of value a/β. Also, Algorithm 3 can be converted to a deterministic algorithm that runs in time $m^{O(1)} k^{O(\alpha)}$.*

Proof. Initially, $Q_0 = \emptyset$ and hence trivially, there is a uniform allocation of value a in which everyone in Q_0 gets big item. If the algorithm stops at Line 5, then there is no solution of value a in which everyone in Q_{c-1} gets big items. By Lemma 2, assuming everyone in Q_{c-1} gets big items in some uniform allocation, at least one of the non-deterministic branches at Line 6 generates a set Q_c such that everyone in Q_c also gets big items in this allocation. Therefore, every branch proceeds as if it computed a valid set Q_c.

If $l_c = 0$, everyone in R_c gets added to Q_c and no non-determinism is used. In this case, the size of Q goes up by at least 1 since $R_c \setminus Q_{c-1} \neq \emptyset$. We will now show that every time the algorithm uses non-determinism, the size of Q goes up by $O(1/\alpha)$ fraction of the people. If $l_c \geq 1$, then

$$|R_c \setminus Q_{c-1}| \geq l_c a/n_1 = \frac{l_c a}{\lceil a/\beta \rceil} \geq \frac{l_c a}{2a/\beta} = \frac{l_c k}{4\alpha}$$

The second inequality follows from our assumption $a \geq \beta$. Therefore, every time we use non-determinism, the size of Q goes up by $l_c k/(4\alpha) - l_c = l_c(k/(4\alpha) - 1) \geq l_c k/(8\alpha)$. But this means that $\sum_c l_c k/(8\alpha) \leq k$, which implies $\sum_c l_c \leq 8\alpha$.

At any time at Line 6, the number of non-deterministic branches generated is at most k^{l_c}. Construct a "non-determinism tree" where a node, representing a branch of computation, has a child for every non-deterministic branch it creates. If we label each node by the l_c it generated, its fan-out is bounded by k^{l_c}. Also, the sum of labels from the root to any leaf is at most 8α. Therefore the number of nodes in the non-determinism tree is at most $k^{O(\alpha)}$. $\qquad\square$

If the optimum over uniform allocations is $\leq \beta$, a β-approximate uniform allocation is just any allocation of value ≥ 1. This is trivial to find since $U > 1$ and hence we just need to give one item of non-zero value to everyone. If $k/8 \leq \alpha \leq k/2$, we can solve the problem by brute force (just try all 2^k possible values for Q). The running time in this case is $2^{O(\alpha)} m^{O(1)}$. This completes the proof of Theorem 2.

5   Conclusion

We showed a $(2k-1)$-approximation algorithm for the Max-Min allocation problem. An improvement in running time can be obtained if we are willing to restrict ourselves to submodular utility functions. Algorithm 1 can be made greedy in this case. The detailed algorithm and analysis can be found in Appendix A.

We also showed an approximation algorithm for $(0, 1, U)$-Max-Min$^+$ where there is a trade-off between the approximation factor and running time. It would be interesting to see if this trade-off can be improved. The more interesting problem would be to generalize the algorithm to Max-Min$^+$. Even if there are only three non-zero values for items, it is not clear how the algorithm can be generalized. In this case, when $\{s\}$ is not a minimum cut in the allocation-flow problem, we have no good bound on the number of people who must be given larger items.

Another challenging open problem is to improve the factor $2 - \epsilon$ hardness of approximation shown by Bezáková and Dani [4].

Acknowledgments

We would like to thank Amin Saberi for suggesting the $(0, 1, \infty)$ version of the problem. Working on this special case provided us with valuable ideas to develop our combinatorial algorithm for Max-Min. We would also like to thank Nikhil Devanur for suggesting the greedy approach for submodular utility functions.

References

1. Asadpour, A., Saberi, A.: An approximation algorithm for max-min fair allocation of indivisible goods. In: STOC '07, (to appear)
2. Azar, Y., Epstein, L.: On-line machine covering. J. Sched. 1(2), 67–77 (1998)
3. Bansal, N., Sviridenko, M.: The santa claus problem. In: STOC '06: Proceedings of the 38th Annual ACM Symposium on Theory of Computing, pp. 31–40. ACM Press, New York, NY, USA (2006)
4. Bezáková, I., Dani, V.: Allocating indivisible goods. SIGecom Exchanges 5(3), 11–18 (2005)
5. Epstein, L.: Tight bounds for bandwidth allocation on two links. Discrete Appl. Math. 148(2), 181–188 (2005)
6. Golovin, D.: Max-min fair allocation of indivisible goods. Technical Report CMU-CS-05-144, Carnegie Mellon University (2005)
7. Lehmann, B., Lehmann, D., Nisan, N.: Combinatorial auctions with decreasing marginal utilities. In: EC '01: Proceedings of the Third ACM Conference on Electronic Commerce, New York, NY, USA, 2001 pp. 18–28. ACM Press, New York (2001)
8. Mjelde, K.M.: Max-min resource allocation. BIT Numerical Mathematics 23(4), 529–537 (1983)
9. Ramirez-Marquez, J.E., Coit, D.W., Konak, A.: Redundancy allocation for series-parallel systems using a max-min approach. IIE Transactions, 891–898 (2004)
10. Tan, Z., He, Y., Epstein, L.: Optimal on-line algorithms for the uniform machine scheduling problem with ordinal data. Inf. Comput. 196(1), 57–70 (2005)
11. Younis, M., Akkaya, K., Kunjithapatham, A.: Optimization of task allocation in a cluster-based sensor network. In: ISCC '03: Proceedings of the Eighth IEEE International Symposium on Computers and Communications, Washington, DC, USA, 2003 p. 329. IEEE Computer Society Press, Los Alamitos (2003)

A   A Greedy $(2k - 1)$-Approximation Algorithm for Submodular Max-Min

We will describe a greedy algorithm that finds a $(2k-1)$-approximation when the optimum is given as an additional input. We can then use this as a subroutine for Algorithm 2 since submodular functions are also subadditive and since failing conditions of Algorithms 1 and 4 are the same. As in the subadditive case, we find a set of people who must get a big item in any solution of value a. Then we pick one unallocated item at a time and let the unhappy people bid for it (Recall a person is unhappy if she has less than b worth of items). A person bids the *incremental utility* that she would get from the item. The incremental utility of an item t_j when a person already has a subset S of the items is defined as $u(t_j|S) = u(t_j \cup S) - u(S)$, where u is the utility function[1]. The item is given to someone who has the highest bid. It is critical that we let only the currently unhappy people to participate in the bidding since otherwise the items may end up with just a few people.

Algorithm 4. An algorithm to to approximate Max-Min with submodular utility functions when the optimum is specified

> **Input:** $k, m, (u_i)_{i \in [k]}, a = (2k-1)b$, where $0 < a < \infty$
> **Output:** An allocation $(T_1, \ldots, T_m)$ of value $\geq b$
>
> 1 Set $S_i \leftarrow \{t_j \in T : u_i(t_j) < b\}$, the set of small items for person i. Compute $a_i = u_i(S_i)$, the net worth of small items to p_i.
> 2 Let Q_0 be the set of people p_i satisfying $a_i < a$.
> 3 Set $T_0 \leftarrow T$. Set $T_i \leftarrow \emptyset$ for all $i \in [k]$.
> 4 Give all people in Q_0 a big item each (Update the T_is accordingly). If it is not possible, stop.
> 5 Set $H \leftarrow Q_0$, the set of happy people.
> 6 **while** $H \neq P$ **do**
> 7   Pick an item t_j from T_0. Set $T_0 \leftarrow T_0 \setminus t_j$.
> 8   Calculate the bid of person p_i as $B_{ij} = u_i^b(t_j|T_i)$. /*Refer to Definition
>   3.   */
> 9   Let $p_i' \in P \setminus H$ be an unhappy person with the highest bid.
> 10   Set $T_{i'} \leftarrow T_{i'} \cup t_j$.
> 11   **if** $u_{i'}^b(T_{i'}) \geq b$ **then** add $p_{i'}$ to H. /*$p_{i'}$ is also happy now.   */
> 12 **end**
> 13 Return the allocation $(T_1, T_2, \ldots, T_{k-1}, T_k \cup T_0)$ as answer.

Lemma 3. *At any moment, a person $p_i \notin Q_0$ is either unhappy or $u_i^b(T_i) < 2b$.*

Proof. If u_i is submodular, then so is u_i^b. For any item $t_j \notin T_i$, $u_i^b(T_i \cup t_j) \leq u_i^b(T_i) + u_i^b(t_j) < u_i^b(T_i) + b$ from submodularity of u_i^b. Since $u_i^b(T_i)$ only increases in steps less than b and a person can not bid once $u_i^b(T_i) \geq b$, $u_i^b(T_i) < 2b$. □

Lemma 4. *At Line 7 in Algorithm 4, T_0 is never empty.*

Proof. Suppose T_0 is empty at Line 7 when p_1 is still unhappy. At this moment, $(T_1, T_2, \ldots, T_k)$ is a partition of T. At most $|Q_0|$ items that are small for p_1 were given away to people in Q_0. For any item with $t_j \in T_i$ with some other person p_i at this moment, p_1 bid some number B_{1j} that was at most the bid B_{ij} of

---

[1] As shown in Lehmann et al. [7], an equivalent definition of submodularity of u is that $\forall t_j \in T, S_1, S_2 \subseteq T : u(t_j|S_1) \geq u(t_j|S_1 \cup S_2)$.

p_i. If $T_1' \subseteq T_1$ was the set of items with p_1 when t_j was picked from T_0, then $B_{ij} \geq B_{1j} = u_1^b(t_j|T_1') \geq u_1^b(t_j|T_1) \geq u_1^b(t_j|T_1 \cup S)$ for any $S \subseteq T$. The last two inequalities follow from submodularity of u_1^b.

Relabel the items so that the first $|Q_0|$ items are with people in Q_0 and the last m' items are with people in $P \setminus Q_0 \setminus p_1$, the $m - |Q_0| - m'$ items in the middle being with p_1. Then,

$$u_1(S_1) = u_1^b(T) = \sum_{j \in [m]} u_1^b(t_j|\{t_1, t_2, \ldots, t_{j-1}\})$$

$$\leq \sum_{j \leq |Q_0|} u_1^b(t_j) + u_1^b(T_1) + \sum_{j > m - m'} u_1^b(t_j|\{t_1, t_2, \ldots, t_{j-1}\})$$

$$< b|Q_0| + b + \sum_{j > m - m'} \text{winning bid for } t_j$$

$$= b(|Q_0| + 1) + \sum_{p_i \in P \setminus Q_0 \setminus p_1} \left(\sum_{t_j \in T_i} B_{ij} \right)$$

$$= b(|Q_0| + 1) + \sum_{p_i \in P \setminus Q_0 \setminus p_1} u_i^b(T_i)$$

$$< b(|Q_0| + 1) + 2b(k - |Q_0| - 1) \leq (2k - 1)b,$$

which contradicts $p_1 \notin Q_0$. The second inequality follows from the analysis in the previous paragraph.   □

Hardness of Embedding Metric Spaces of Equal Size

Subhash Khot* and Rishi Saket**

Georgia Institute of Technology
{khot,saket}@cc.gatech.edu

Abstract. We study the problem embedding an n-point metric space into another n-point metric space while minimizing distortion. We show that there is no polynomial time algorithm to approximate the minimum distortion within a factor of $\Omega((\log n)^{1/4-\delta})$ for any constant $\delta > 0$, unless NP $\subseteq$ DTIME($n^{\text{poly}(\log n)}$). We give a simple reduction from the METRIC LABELING problem which was shown to be inapproximable by Chuzhoy and Naor [10].

1   Introduction

Given an *embedding* $f : X \mapsto Y$ from a finite metric space (X, d_X) into another metric space (Y, d_Y), define

$$\text{Expansion}(f) := \max_{i,j \in X, i \neq j} \frac{d_Y(f(i), f(j))}{d_X(i, j)},$$

$$\text{Contraction}(f) := \max_{i,j \in X, i \neq j} \frac{d_X(i, j)}{d_Y(f(i), f(j))}.$$

The *distortion* of f is the product of Expansion(f) and Contraction(f).

The problem of embedding one metric space into another has been well studied in Mathematics, especially in the context of bi-lipschitz embeddings. Metric embeddings have also played an increasing role in Computer Science; see Indyk's survey [12] for an overview of their algorithmic applications. Much of the research in this area has focussed on embedding finite metrics into a *useful* target space such as ℓ_2, ℓ_1 or distributions over tree metrics. For example, Bourgain's Theorem [8] shows that every n-point metric space embeds into ℓ_2 with distortion $O(\log n)$. Fakcharoenphol, Rao, and Talwar [11], improving on the work of Bartal [7], show that every n-point metric embeds into a distribution over tree metrics with distortion $O(\log n)$. Recently, the class of *negative type* metrics has received much attention, as they arise naturally as a solution of a SDP-relaxation for the SPARSEST CUT problem. Arora, Lee, and Naor [3], building on techniques from Arora, Rao, and Vazirani [4], show that every n-point negative type metric embeds into ℓ_2 with distortion $O(\sqrt{\log n} \log \log n)$.

---

* Research supported in part by Sloan Foundation Fellowship, Microsoft New Faculty Fellowship and NSF CAREER award CCF-0643626.

** Research supported in part by Subhash Khot's Sloan Foundation Fellowship and Microsoft New Faculty Fellowship.

M. Charikar et al. (Eds.): APPROX and RANDOM 2007, LNCS 4627, pp. 218–227, 2007.

Given two spaces (X, d_X) and (Y, d_Y), let $c_Y(X)$ denote the minimum distortion needed to embed X into Y. A natural computational problem is to determine $c_Y(X)$, exactly or approximately (call it the MIN-DISTORTION problem). Of course, the complexity of the problem depends on the nature of the two spaces. It is known that $c_{\ell_2}(X)$ can be computed in polynomial time for any n-point metric X, using a straightfoward SDP-relaxation. It is however NP-hard to determine whether $c_{\ell_1}(X) = 1$. When the target space is the line metric, the problem is hard, even to approximate within a factor n^β for some constant $0 < \beta < 1$, as shown by Badoiu, Chuzhoy, Indyk, and Sidiropoulos [6].

One variant of the problem is when X and Y are of equal size and explicitly given (call it the MIN-DISTORTION$^=$ problem). Though a very natural problem, it has not been studied until recently, and not much is known about its complexity. The problem was formulated by Kenyon, Rabani, and Sinclair [14] motivated by applications to shape matching and object recognition. Note that when both are shortest path metrics on n-vertex graphs, determining whether $c_Y(X) = 1$ is same as determining whether the two graphs are isomorphic. On the positive side, Kenyon *et al.* give a dynamic programming based algorithm when the metrics are one-dimensional and the distortion is at most $2 + \sqrt{5}$. In this case, the bijection f between the two point sets must obey certain properties regarding how the points are relatively ordered, which enables efficient computation of all such possible embeddings in a recursive manner. On the negative side, Papadimitriou and Safra [15] showed that $c_Y(X)$ (when $|X| = |Y|$) is hard to approximate within a factor of 3, and the result holds even when both point sets are in $\mathbb{R}^3$. They give a clever reduction from GRAPH 3-COLORING.

In this paper, we prove the following inapproximability result.

Theorem 1. *(Inapproximability of* MIN-DISTORTION$^=$*) For any constant $\delta > 0$, there is no polynomial time algorithm to approximate the distortion required to embed an n-point metric space into another n-point metric space within a factor of $(\log n)^{1/4-\delta}$, unless* NP $\subseteq$ DTIME$(n^{\mathrm{poly}(\log n)})$.

2 Preliminaries

In this section we formally state the problem MIN-DISTORTION$^=$, and some of the tools we require for the construction in the next section. We will be concerned only with finite metric spaces i.e. metrics on finite sets of points.

Definition 1. *The problem of* MIN-DISTORTION$^=$ *is, given two n point metric spaces (X, d_X) and (Y, d_Y), computing an embedding f of X into Y with the minimum distortion.*

A related problem is METRIC LABELING which was introduced by Kleinberg and Tardos [13].

Definition 2. *The problem of* METRIC LABELLING *is the following. Given a weighted graph $G(V, E, \{w_e\}_{e \in E})$ and a metric space (X, d), along with a cost function $c : V \times X \mapsto \mathbb{R}$, the goal is to find a mapping h of V to X such that the following quantity is minimized,*

$$\sum_{(u,v)\in E(G)} w(u,v) \cdot d(h(u), h(v)) + \sum_{u\in V} c(u, f(u)).$$

The problem is essentially of finding a 'labeling' of the vertices of the graph G with the points in the metric space X, so as to minimize the sum of the connection costs (cost of labelling vertices in G with points in X) and weighted 'stretch' of the edges of G. A special case of METRIC LABELING is the $(0, \infty)$-EXTENSION problem in which, essentially, the connection costs are 0 or ∞. It is formally defined as follows.

Definition 3. *The problem of* $(0, \infty)$-EXTENSION *is the following. Given a weighted graph* $G(V, E, \{w_e\}_{e\in E})$ *and a metric space* (X, d), *along with a subset of allowed labels* $s(u) \subseteq X$ *for every vertex* $u \in V(G)$, *the goal is to find a mapping* h *of* V *to* X, *satisfying* $h(u) \in s(u)$ *for all* $u \in V(G)$, *such that the following quantity is minimized,*

$$\sum_{(u,v)\in E(G)} w(u,v) \cdot d(h(u), h(v)).$$

It has been shown that METRIC LABELING is equivalent to its special case of $(0, \infty)$-EXTENSION [9]. For METRIC LABELING Kleinberg and Tardos [13] obtained an $O(\log n \log\log n)$ approximation, where n is the size of the metric X, which was improved to $O(\log n)$ [11]. Chuzhoy and Naor [10] give a hardness of approximation factor of $(\log n)^{1/2-\delta}$ for METRIC LABELING. The instance that they construct is the $(0, \infty)$-EXTENSION version of METRIC LABELING, and moreover G is unweighted.

Definition 4. *A* 3-SAT(5) *formula* ϕ *is a* 3-CNF *formula in which every variable appears in exactly* 5 *clauses.*

The reduction in [10] starts with an instance of the MAX-3-SAT(5) problem in which, given a 3-SAT(5) formula ϕ, the goal is to find an assignment to the variables that satisfies the maximum number of clauses. The following is the well known PCP theorem [2] [5].

Theorem 2. *(PCP theorem): There exists a positive constant* ε *such that, given an instance* ϕ *of* MAX-3-SAT(5) *there is no polynomial time algorithm to decide whether there is an assignment to the variables of* ϕ *that satisfies all the clauses (Yes instance) or that no assignment satisfies more than* $1 - \varepsilon$ *fraction of the clauses, unless* P = NP.

Overview of reduction. The construction of [10] yields a $(0, \infty)$-EXTENSION version with graph G and target metric H, where both G and H are unweighted graphs, with the metric on H being the shortest path metric. They show that in the Yes case, there is an embedding of $V(G)$ into $V(H)$ such that the end points of an edge of G are mapped onto end points of some edge of H, so that the stretch of every edge in G is 1. However, in the No case in any such embedding, for at least a constant fraction of edges of G, their end points are mapped onto pairs of vertices in H which are $\Omega(k)$ distance apart, where k is a parameter used in the construction so that $k = \theta(\log |V(H)|)^{1/2-\delta}$. This yields a hardness of approximation factor of $(\log n)^{1/2-\delta}$ for METRIC LABELING for any $\delta > 0$.

The main idea behind our reduction is to start with the instance constructed in [10] and view the graph G also as a metric, and the mapping h as an embedding of G into H. This makes sense since their instance is a $(0, \infty)$-EXTENSION version and moreover G is unweighted, in which case the quantity to be optimized is just the average stretch of every edge in G by the embedding h.

The reduction proceeds in three steps where the first step constructs the instance of [10]. The second involves adding additional points in the two metrics in order to equalize the number of points in both the metrics and third step enforces the constraints of allowed labels, by adding some more points in the two metrics. The entire construction and analysis is presented in the following section.

3   Reduction from MAX-3-SAT(5)

Our reduction from MAX-3-SAT(5) is based on the construction in [10] for the hardness of METRIC LABELING. We modify the instance of METRIC LABELING to obtain a 'gap' instance of MIN-DISTORTION$^=$. In this section we describe the entire construction and analysis of the hardness factor.

3.1   Construction

We describe the k-prover protocol used in [10]. Let ϕ be a MAX-3-SAT(5) formula, and let n be the number of clauses in ϕ. Let $P_1, P_2, \ldots, P_k$ be k provers (the parameter k will be set to $\theta(poly(\log n))$). The protocol is as follows,

- For each (i, j), $1 \le i < j \le k$, the verifier chooses a clause C_{ij} from ϕ uniformly at random, and randomly selects x_{ij} a distinguished variable from C_{ij}. P_i is sent C_{ij} and returns an assignment to the variables of C_{ij}. P_j is sent x_{ij} and returns as assignment to the variable x_{ij}. Every other prover is sent both C_{ij} and x_{ij} and returns assignments to all the variables of the clause. Hence, the verifier sends $\binom{k}{2}$ coordinates to each prover.
- The verifier accepts if the answers of all the provers are consistent and satisfy all the clauses of the query.

We will denote the set of random strings used by the verifier as R. For $r \in R$, let $q_i(r)$ be the query sent to prover P_i when r is the random string chosen by the verifier. Let $Q_i = \cup_r q_i(r)$ be the set of all possible queries of to P_i. For $q \in Q_i$, let $\mathcal{A}_i(q)$ be the set of all answers to q that satisfy all the clauses in q. Let P_i and P_j $(1 \le i < j \le k)$ be any two provers, and $q_i \in Q_i$ and $q_j \in Q_j$ be two queries such that for some $r \in R$, $q_i = q_i(r)$ and $q_j = q_j(r)$. Let $A_i \in \mathcal{A}_i(q_i)$ and $A_j \in \mathcal{A}_j(q_j)$ be respective answers to these queries. Then, A_i and A_j are called *weakly consistent* if the assignments to C_{ij} in A_i and x_{ij} in A_j are consistent and satisfy the clause C_{ij}. They are called *strongly consistent* if they are also consistent in all the other coordinates and also satisfy all the other clauses. The following theorem is due to Chuzhoy and Naor [10].

Theorem 3. *There is a constant $0 < \varepsilon < 1$ such that if ϕ is a Yes instance, then there is a strategy of the k provers such that the verifier accepts, and if ϕ is a No instance, then for any pair of provers P_i and P_j, the probability that their answer is weakly consistent is at most $1 - \frac{\varepsilon}{3}$.*

We now construct our instance of MIN-DISTORTION$^=$ starting from a MAX-3-SAT(5) formula ϕ. The reduction proceeds in three stages. At each stage we obtain two metric spaces, with the first stage yielding exactly the METRIC LABELING instance of [10] and at the end of the third stage we obtain two metric spaces of equal size which constitute the desired instance of MIN-DISTORTION$^=$.

STEP I. In this step we construct two graphs G_1 and H_1 which are same as the graphs constructed in the METRIC LABELING instance of [10]. The construction of the graph G_1 is as follows.

- For every prover P_i and every $q \in Q_i$, there is a vertex $v(i, q)$.
- For every random string $r \in R$, there is a vertex $v(r)$.
- There is an edge of length 1 between $v(r)$ and $v(i, q)$ if $q = q_i(r)$.

Call the vertices of G_1 as 'query' vertices. The graph H_1 is constructed in the following manner.

- For every i ($1 \leq i \leq k$), every $q \in Q_i$, and answer $A_i \in \mathcal{A}_i(q)$, there is a vertex $v(i, q, A_i)$.
- For every $r \in R$ and every pairwise strongly consistent answers $A_1, A_2, \ldots, A_k$ to the queries $q_1(r), q_2(r), \ldots, q_k(r)$ respectively (where $A_i \in \mathcal{A}_i(q_i(r))$ for $1 \leq i \leq k$), there is a vertex $v(r, A_1, A_2, \ldots, A_k)$.
- There is an edge of length 1 between $v(i, q, A_i)$ and $v(r, A_1, A_2, \ldots, A_k)$ if $q = q_i(r)$ and the ith coordinate of the tuple $(A_1, A_2, \ldots, A_k)$ is A_i.

The vertices of H_1 will be referred to as 'label' vertices. Figure 1 shows the local structure of G_1 and H_1. It can be seen that for every 'query' vertex u in G_1, there is a set of 'label' vertices $s(u)$ in H_1, such that $\{s(u)\}_{u \in V(G_1)}$ is a partition of $V(H_1)$. For the sake of convenience, we modify our notation to let $V(G_1) = \{u_1, u_2, \ldots, u_N\}$,

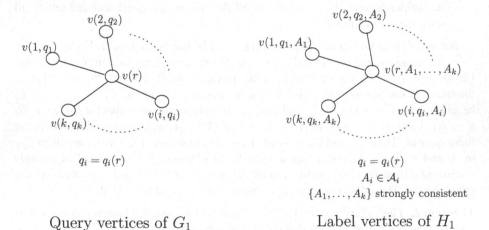

$$q_i = q_i(r)$$

$$q_i = q_i(r)$$
$$A_i \in \mathcal{A}_i$$
$$\{A_1, \ldots, A_k\} \text{ strongly consistent}$$

Query vertices of G_1 Label vertices of H_1

Fig. 1. Vertices of G_1 and H_1. All edges are of length 1.

where $N = |V(G_1)|$ and $V(H_1) = \cup_{i=1}^{N} s(u_i)$, and let $\ell_1^i, \ell_2^i, \ldots, \ell_{m_i}^i$ be the elements of $s(u_i)$, where $m_i = |s(u_i)|$. Also note that there is an edge between the sets $s(u_i)$ and $s(u_j)$ in H_1 only if there is an edge between u_i and u_j in G_1. The graphs G_1 and H_1 with the partition $\{s(u_i)\}_{u_i \in V(G_1)}$ constitute the instance of METRIC LABELING of [10]. They show that in the Yes case, there is a labeling of every $u_i \in V(G_1)$ with a label from $s(u_i)$ such that each edge in G_1 is mapped to an edge in H_1, while in the No case in every such labeling, a constant fraction of edges in G_1 are mapped to pairs of vertices $\Omega(k)$ distance apart in H_1. Figure 2 illustrates this structure of G_1 and H_1. Let d_{G_1} and d_{H_1} denote the shortest path metric on G_1 and H_1 respectively. Note that eventually we want two metric spaces of equal cardinality, in the next step of the construction we achieve that goal.

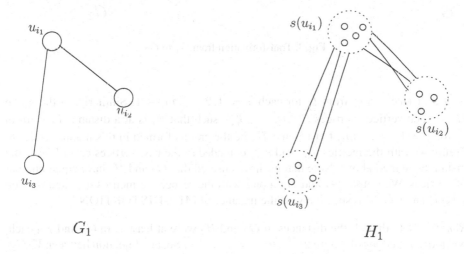

$$G_1 \qquad\qquad\qquad\qquad H_1$$

Fig. 2. Structure of G_1 and H_1

STEP II. We first modify G_1 as follows. For every vertex u_i ($1 \leq i \leq N$) in G_1, we add $m_i - 1$ vertices $t_1^i, t_2^i, \ldots t_{m_i-1}^i$ and add an edge from each t_j^i to u_i of length $\sqrt{k}$. Let G_2 be the new graph created in this process and d_{G_2} be the shortest path metric on G_2. The transformation is shown in Figure 3.

We truncate d_{H_1} from below to $\sqrt{k}$ and from above to $10k$, i.e. if the distance between a particular pair of points is less than $\sqrt{k}$ then it is set to $\sqrt{k}$ and if it is greater than $10k$ then it is set to $10k$, otherwise it remains the same. We also truncate d_{G_1} from above to $10k$, but not from below. Let H_2 be the new resultant graph and d_{H_2} be this new metric on $V(H_2)$. Observe that G_2 and H_2 have the same number of vertices.

STEP III. We now have two graphs G_2 and H_2 with equal number of vertices. However, we desire that any 'good' embedding of vertices of G_2 into H_2 map the set $\{u_i, t_1^i, \ldots, t_{m_i-1}^i\}$ onto the set $s(u_i)$. For this we add new vertices to both the graphs in the following manner. Let $\eta_i = 2^{-iN}$. For every i ($1 \leq i \leq N$), we add m_i vertices $\alpha_1^i, \alpha_2^i, \ldots, \alpha_{m_i}^i$ to G_2, where $\alpha_{m_i}^i$ is at a distance η_i from u_i and similarly α_j^i

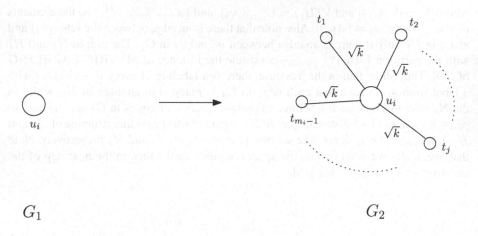

$$G_1 \qquad\qquad\qquad\qquad\qquad\qquad\qquad G_2$$

Fig. 3. Transformation from G_1 to G_2

is at a distance of η_i from t_j^i for each $j = 1, 2, \ldots, m_i - 1$. Similarly, in the graph H, we add vertices vertices $\beta_1^i, \beta_2^i, \ldots, \beta_{m_i}^i$, such that β_j^i is at a distance of η_i from ℓ_j^i for $j = 1, 2, \ldots, m_i$. Let G_3 and H_3 be the graphs formed in this manner (refer to Figure 4), with the metrics d_{G_3} and d_{H_3} extended to the new vertices according to the distances defined above. Note that we have ensured that G_3 and H_3 have equal number of vertices. We output G_3 and H_3 along with the respective metrics d_{G_3} and d_{H_3} on $V(G_3)$ and $V(H_3)$ respectively, as the instance of MIN-DISTORTION$^=$.

Remark. Note that all the distances in G_2 and H_2 were at least 1. In G_3 and H_3 each, we added m_i edges of length 2^{-iN}, for $i = 1, \ldots, N$. So, any bijection between $V(G_3)$ and $V(H_3)$ must map edges of length 2^{-iN} in G_3 to edges of length 2^{-iN} in H_3 for $i = 1, \ldots, N$, otherwise at least one edge will be stretched or contracted by a factor of at least 2^N. Therefore, this ensures that in any embedding of G_3 into H_3, for any given i ($1 \leq i \leq N$) the set $\{\alpha_1^i, \ldots, \alpha_{m_i}^i\}$ is mapped onto the set $\{\beta_1^i, \ldots, \beta_{m_i}^i\}$ and the set $\{u_i, t_1^i, \ldots, t_{m_i-1}^i\}$ is mapped onto the set $\{\ell_1^i, \ldots, \ell_{m_i}^i\}$, otherwise the distortion is at least 2^N.

3.2   Analysis

Yes instance. Suppose that the MAX-3-SAT(5) formula had a satisfying assignment, say σ to the variables. We will construct an embedding f of $V(G_3)$ into $V(H_3)$ such that no edge in G_3 is contracted, whereas any edge in G_3 is stretched by at most a factor of $O(\sqrt{k})$, which leads to an embedding with distortion at most $O(\sqrt{k})$. For any i ($1 \leq i \leq N$), consider the vertex u_i. We recall that u_i was a 'query' vertex, and $s(u_i)$ is the set of 'label' vertices corresponding to u_i. We set $f(u_i)$ to be a label vertex in $s(u_i)$ given by the satisfying assignment σ. The map f takes $t_1^i, t_2^i, \ldots, t_{m_i-1}^i$ arbitrarily to the remaining $m_i - 1$ vertices in $s(u_i)$. The vertex α_j^i ($1 \leq j \leq m_i - 1$) is

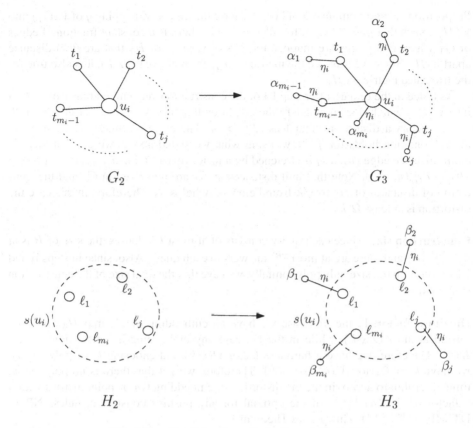

Fig. 4. Transformation from G_2 to G_3 and H_2 to H_3

mapped to $\beta^i_{j'}$ such that $\beta^i_{j'}$ at distance η_i to $f(t^i_j)$. Similarly, the vertex $\alpha^i_{m_i}$ is mapped to $\beta^i_{j''}$ which is at a distance of η_i from $f(u_i)$.

Consider any two vertices u_i and $u_{i'}$ in in G_1 that are adjacent. Since σ is a satisfying assignment, we have that the vertices $f(u_i)$ and $f(u_{i'})$ are also adjacent in H_1. And since adjacent vertices remain adjacent in G_3 and adjacent vertices in H_1 have distance $\sqrt{k}$ in H_3, therefore the stretch of the edge (u_i, u_j) in G_3 is $\sqrt{k}$. The edge (u_i, t^i_j) $(1 \leq j \leq m_i - 1)$ has length $\sqrt{k}$ and since the diameter of the metric on H_3 (and G_3) is $O(k)$, the stretch of the edge is at most $O(\sqrt{k})$. Also, clearly the distances of the edges $(t^i_1, \alpha^i_1), \ldots, (t^i_{m_i-1}, \alpha^i_{m_i-1}), (u_i, \alpha_{m_i})$ are not stretched or contracted. One key fact that we have utilized is that there is an edge (in H_1) between $s(u_i)$ and $s(u_{i'})$ only if u_i and $u_{i'}$ are adjacent in G_1. This ensures that there is no edge (in G_1) contracts, which guarantees that no distance in G_3 contracts. Therefore, there is an embedding of the vertices of G_3 into H_3 with distortion at most $O(\sqrt{k})$.

No instance. Suppose that the MAX-3-SAT(5) formula has no satisfying assignment that satisfies more that $1 - \varepsilon$ fraction of the clauses. In this case, as a consequence of

Proposition 4.4 and Lemma 4.5 of [10], we have that in every mapping g of $V(G_1)$ into $V(H_1)$, such that $g(u) \in s(u)$, for all $u \in V(G)$, there is a constant fraction of edges of G_1 whose end points are mapped to pairs of vertices of H_1 that are $\theta(k)$ distance apart in H_1. Since we truncate the metric on d_{H_1} from above to $\Omega(k)$, it is also true for the truncated metric on H_1.

As noted in the remark after step III of the construction, any embedding f of $V(G_3)$ into $V(H_3)$ which does not satisfy the condition that $f(u_i) \in s(u_i)$ for all $i = 1, 2, \ldots, N$ incurs a distortion of at least $2^N > k$. Therefore we may assume that the above condition holds for f. Now, using what we stated above, we get that there is a unit distance edge $(u_i, u_{i'})$ is stretched by a factor of $\Omega(k)$, i.e. $d_{H_3}(f(u_i), f(u_{i'})) \geq \Omega(k) \cdot d_{G_1}(u_i, u_{i'})$. Note that unit distances in G_1 are preserved in G_3 and the truncation of distances of H_1 to $\sqrt{k}$ from below only helps us. Therefore, in this case the distortion is at least $\Omega(k)$.

Construction size. Since each query consists of at most k^2 clauses, the size of R is at most $3 \cdot n^{k^2}$ and there are at most 7^{k^2} answers to each query. Also, since in steps II and III we blow up the size only polynomially, we have that the total size of the construction is at most $n^{O(k^2)}$.

Hardness factor. In the Yes case we have an embedding of G_3 into H_3 with distortion at most $O(\sqrt{k})$, while in the No case any embedding has distortion at least $\Omega(k)$. Therefore, we have a hardness factor $\Omega(\sqrt{k})$ and choosing $k = \text{poly}(\log n)$, we have $k = \Omega((\log |V(G_3)|)^{1/2-\delta})$. Therefore, we get that there is no polynomial time algorithm to approximate the distortion of embedding two n-point metrics within a factor of $(\log(n))^{1/4-\delta}$ of the optimal for any positive constant δ, unless NP $\subseteq$ DTIME$(n^{\text{poly}(\log n)})$. This proves Theorem 1.

4   Conclusion

In this paper we have shown a hardness factor of $(\log n)^{1/4-\delta}$ for approximating the distortion required to embed two general n point metrics. An interesting question is whether such a superconstant lower bound also holds for constant dimensional metrics. Papadimitriou and Safra [15] prove a 3 factor hardness of approximating the distortion for 3-dimensional metrics. Extending such a constant factor hardness to 2-dimensional, or 1-dimensional metrics is an important question. On the algorithmic side the only positive result is for the case when the metrics are 1-dimensional and the optimum distortion is at most $2 + \sqrt{5}$ [14], and extending it to higher dimensions remains an open question. For general n point metrics, no non-trivial upper bound is known, and it would be interesting if a reasonable upper bound can be derived for this problem.

References

1. Arora, S.: Polynomial time approximation schemes for Euclidean Traveling Salesman and other Geometric problems. Journal of the ACM 45(5), 753–782 (1998)
2. Arora, S., Lund, C., Motawani, R., Sudan, M., Szegedy, M.: Proof verification and the hardness of approximation problems. Journal of the ACM 45(3), 501–555 (1998)

3. Arora, S., Lee, J.R., Naor, A.: Euclidean distortion and the sparsest cut. In: Proc. STOC, pp. 553–562 (2005)
4. Arora, S., Rao, S., Vazirani, U.V.: Expander flows, geometric embeddings and graph partitioning. In: Proc. STOC, pp. 222–231 (2004)
5. Arora, S., Safra, S.: Probabilistic checking of proofs: A new characterization of NP. Journal of the ACM 45(1), 70–122 (1998)
6. Badoiu, M., Chuzhoy, J., Indyk, P., Sidiropoulos, A.: Low-distortion embeddings of general metrics into the line. In: Proc. STOC, pp. 225–233 (2005)
7. Bartal, Y.: On approximating arbitrary metrices by tree metrics. In: Proc. STOC, pp. 161–168 (1998)
8. Bourgain, J.: On Lipschitz embeddings of finite metrics in Hilbert space. Israel Journal of Mathematics 52, 46–52 (1985)
9. Chekuri, C., Khanna, S., Naor, J., Zosin, L.: Approximation algorithms for the metric labeling problem via a new linear programming formulation. In: Proc. SODA, pp. 109–118 (2001)
10. Chuzhoy, J., Naor, J.: The hardness of Metric Labeling. In: Proc. FOCS, pp. 108–114 (2004)
11. Fakcharoenphol, J., Rao, S., Talwar, K.: A tight bound on approximating arbitrary metrics by tree metrics. J. Comput. Syst. Sci. 69(3), 485–497 (2004)
12. Indyk, P.: Algorithmic applications of low-distortion embeddings. In: Proc. FOCS, pp. 10–33 (2001)
13. Kleinberg, J., Tardos, E.: Approximation algorithms for classification problems with pairwise relationships: metric labeling and Markov random fields. Journal of the ACM 49, 616–630 (2002)
14. Kenyon, C., Rabani, Y., Sinclair, A.: Low distortion maps between point sets. In: Proc. STOC, pp. 272–280 (2004)
15. Papadimitriou, C.H., Safra, S.: The complexity of low-distortion embeddings between point sets. In: Proc. SODA, pp. 112–118 (2005)

Coarse Differentiation and Multi-flows
in Planar Graphs

James R. Lee* and Prasad Raghavendra

University of Washington
{prasad,james}@cs.washington.edu

Abstract. We show that the multi-commodity max-flow/min-cut gap
for series-parallel graphs can be as bad as 2. This improves the largest
known gap for planar graphs from $\frac{3}{2}$ to 2. Our approach uses a technique
from geometric group theory called *coarse differentiation* in order to
lower bound the distortion for embedding a particular family of shortest-
path metrics into L_1.

1 Introduction

In the past 15 years, low-distortion metric embeddings—following the initial
work of Linial, London, and Rabinovich [25]—have become an integral part of
theoretical computer science, and the geometry of metric spaces seems to lie
at the heart of some of the most important open problems in approximation
algorithms (for some recent examples see, e.g. [4,3,1,16,13] and the discussions
therein). For background on the field of metric embeddings and their applications
in computer science, we refer to Matoušek's book [27, Ch. 15], the surveys [20,24],
and the compendium of open problems [26].

One of the central connection between embeddings and theoretical CS lies in
the correspondence between low-distortion L_1 embeddings, on the one hand, and
the *Sparsest Cut problem* (see, e.g. [25,5,4,3]) and *concurrent multi-commodity
flows* (see, e.g. [18,12]) on the other. This relationship allows us to bring sophisticated geometric and analytic techniques to bear on classical problems in
graph partitioning and in the theory of network flows. In the present paper, we
show how techniques developed initially in geometric group theory can be used
to shed new light on the connections between sparse cuts and multi-commodity
flows in planar graphs.

Multi-commodity Flows and Sparse Cuts. Let $G = (V, E)$ be a graph
(network), with a *capacity* $C(e) \geq 0$ associated to every edge $e \in E$. Assume that
we are given k pairs of vertices $(s_1, t_1), ..., (s_k, t_k) \in V \times V$ and $D_1, ..., D_k \geq 1$.
We think of the s_i as *sources*, the t_i as *targets*, and the value D_i as the *demand*
of the *terminal pair* (s_i, t_i) for some *commodity* κ_i.

In the *MaxFlow* problem the objective is to maximize the *fraction* λ of the
demand that can be shipped simultaneously for all the commodities, subject to

---

* Research supported by NSF CAREER award CCF-0644037.

M. Charikar et al. (Eds.): APPROX and RANDOM 2007, LNCS 4627, pp. 228–241, 2007.
© Springer-Verlag Berlin Heidelberg 2007

the capacity constraints. Denote this maximum by λ^*. A trivial upper bound on λ^* is the *sparsest cut ratio*. Given any subset $S \subseteq V$, we write

$$\Phi(S) = \frac{\sum_{uv \in E} C(uv) \cdot |\mathbf{1}_S(u) - \mathbf{1}_S(v)|}{\sum_{i=1}^k D_i \cdot |\mathbf{1}_S(s_i) - \mathbf{1}_S(t_i)|},$$

where $\mathbf{1}_S$ is the characteristic function of S. The value $\Phi^* = \min_{S \subseteq V} \Phi(S)$ is the minimum over all cuts (partitions) of V, of the ratio between the total capacity crossing the cut and the total demand crossing the cut. In the case of a single commodity (i.e. $k = 1$) the classical MaxFlow-MinCut theorem states that $\lambda^* = \Phi^*$, but in general this is no longer the case. It is known [25,5] that $\Phi^* = O(\log k)\lambda^*$. This result is perhaps the first striking application of metric embeddings in combinatorial optimization (specifically, it uses Bourgain's embedding theorem [7]).

Indeed, the connection between L_1 embeddings and multi-commodity flow/cut gaps can be made quite precise. For a graph G, let $c_1(G)$ represent the largest distortion necessary to embed any shortest-path metric on G into L_1 (i.e. the maximum over all possible assignments of non-negative lengths to the edges of G). Then $c_1(G)$ gives an upper bound on the ratio between the sparsest cut ratio and the maximum flow for any multi-commodity flow instance on G (i.e. with any choices of capacities and demands) [25,5]. Furthermore, this connection is tight in the sense that there is always a multi-commodity flow instance on G that achieves a gap of $c_1(G)$ [18].

Despite significant progress [29,18,12,8], some fundamental questions are still left unanswered. As a prime example, consider the well-known *planar embedding conjecture* (see, e.g., [18,24,26,12]):

> *There exists a constant C such that every planar graph metric embeds into L_1 with distortion at most C.*

In initiating a systematic study of L_1 embeddings [18] for minor-closed families, Gupta, Newman, Rabinovich, and Sinclair put forth the following vast generalization of this conjecture (we refer to [14] for the relevant graph theory).

Conjecture 1 (Minor-closed embedding conjecture). If $\mathcal{F}$ is any non-trivial minor-closed family, then $\sup_{G \in \mathcal{F}} c_1(G) < \infty$.

Lower Bounds on the Multi-commodity Max-Flow/Min-Cut Ratio in Planar Graphs. While techniques for proving upper bounds on the L_1-distortion required to embed such families has steadily improved, progress on lower bounds has been significantly slower, and recent breakthroughs in lower bounds for L_1 embeddings of discrete metric spaces that rely on discrete Fourier analysis [22,21] do not apply to excluded-minor families.

The best previous lower bound on $c_1(G)$ when G is a planar graph occured for $G = K_{2,n}$, i.e. the complete $2 \times n$ bipartite graph. By a straightforward generalization of the lower bound in [29], it is possible to show that $c_1(K_{2,n}) \to \frac{3}{2}$ as $n \to \infty$ (see also [2] for a particularly simple proof of this fact in the dual setting).

We show that, in fact there is an infinite family of series-parallel (and hence, planar) graphs $\{G_n\}$ such that $\lim_{n\to\infty} c_1(G_n) = 2$.

1.1 Results and Techniques

Our lower bound approach is based on exhibiting *local rigidity* for pieces of metric spaces under low-distortion embeddings into L_1. This circle of ideas, and the relationship to theory of metric differentiation are long-studied phenomena in geometric analysis (see e.g. [19,30,6,10]). More recently, they have been applied to the study of L_1 embeddings [11] based on local rigidity results for sets of finite perimeter in the Heisenberg group [17]; see [23] for the relevance to integrality gaps for the Sparsest Cut problem.

Our basic approach is simple; we know that $c_1(K_{2,n}) \leq \frac{3}{2}$ for every $n \geq 1$. But consider $s, t \in V(K_{2,n})$ which constitute the partition of size 2. Say that a cut $S \subseteq V(K_{2,n})$ is *monotone with respect to s and t* if every simple s-t path in $K_{2,n}$ has at most one edge crossing the cut $(S, \bar{S})$. It is not difficult to show that if an L_1 embedding is composed entirely of cuts which are monotone with respect to s and t, then that embedding must have distortion at least $2 - \frac{2}{n}$.

Consider now the recursively defined family of graphs $K_{2,n}^{\oslash k}$, where $K_{2,n}^{\oslash 1} = K_{2,n}$ and $K_{2,n}^{\oslash k}$ arises by replacing every edge of $K_{2,n}^{\oslash k-1}$ with a copy of $K_{2,n}$. The family $\{K_{2,2}^{\oslash k}\}_{k \geq 1}$ are the well-known *diamond graphs* of [28,18]. We show that in any low-distortion embedding of $K_{2,n}^{\oslash k}$ into L_1, for $k \geq 1$ large enough, it is possible to find a (metric) copy of $K_{2,n}$ for which the induced embedding is composed almost entirely of monotone cuts. The claimed distortion bound follows, i.e. $\lim_{n,k\to\infty} c_1(K_{2,n}^{\oslash k}) = 2$. In Section 5, we exhibit embeddings which show that for every fixed n, $\lim_{k\to\infty} c_1(K_{2,n}^{\oslash k}) < 2$ and also for every fixed k, $\lim_{n\to\infty} c_1(K_{2,n}^{\oslash k}) < 2$, thus it is necessary for the lower bound to be asymptotic in both parameters. (E.g. for the diamond graphs, $\lim_{k\to\infty} c_1(K_{2,2}^{\oslash k}) \leq \frac{4}{3}$.)

The ability to find these monotone copies of $K_{2,n}$ inside a low-distortion L_1 embedding of $K_{2,n}^{\oslash k}$ arises from two sources. The first is the *coarse differentiation* technique of Eskin, Fischer, and Whyte [15] which gives a discrete approach to finding local regularity in distorted paths; this is carried out in Section 3. The second aspect is the relationship between regularity and monotonicity for L_1 embeddings which is expounded upon in Section 3.2.

2   Preliminaries

For a graph G, we will use $V(G), E(G)$ to denote the sets of vertices and edges. Further, let d_G denote the shortest path metric on the graph G. For an integer n, let $K_{2,n}$ denote the complete bipartite graph with 2 and n vertices on either side.

An s-t graph G, is a graph two of whose vertices are labelled s and t. For an s-t graph G, we will use $s(G), t(G)$ to denote the vertices in G labelled s and t respectively. Throughout this article, the graphs $K_{2,n}$ are considered s-t graphs in the natural way (the two vertices forming a partition are labeled s and t).

We define the operation $\oslash$ as follows.

Definition 1. *Given two s-t graphs G and H, define $H \oslash G$ to be the graph obtained by replacing each edge $e = (u, v) \in E(H)$ by a copy of G. Formally,*

- $V(H \oslash G) = V(H) \cup E(H) \times (V(G) - \{s(G), t(G)\})$
- *For every edge $e = (u, v) \in E(H)$, there are $|E(G)|$ edges*

$$E(H) = \{(e_{v_1}, e_{v_2})|(v_1, v_2) \in E(G - s - t)\} \cup \{(u, e_w)|(s, w) \in E(G)\}$$
$$\cup \{(e_w, v)|(w, v) \in E(G)\}$$

- $s(H \oslash G) = s(H), t(H \oslash G) = t(H)$.

Note that, there are two ways to substitute an edge with an *s-t* graph. Thus, in the above definition we are assuming some arbitrary orientation for each edge. However for our purposes, the graphs $K_{2,n}$ are symmetric and thus the distinction does not arise. For a *s-t* graph G and integers k, define $G^{\oslash k}$ recursively as $G^{\oslash 1} = G$ and $G^{\oslash k} = G^{\oslash k-1} \oslash G$.

Definition 2. *For two graphs G, H, a subset of vertices $X \subseteq V(H)$ is said to be a copy of G if there exists a bijection $f : V(G) \to X$ such that $d_H(f(u), f(v)) = C \cdot d_G(u, v)$ for some constant C, i.e. there is a distortion 1 map from G to X.*

Now we make the following two simple observations about *copies* in the graph $G \oslash H$.

Observation 1. *The graph $H \oslash G$ contains $|E(H)|$ copies of the graph G, one copy corresponding to each edge in H.*

Observation 2. *The subset of vertices $V(H) \subseteq V(H \oslash G)$ form a copy of H. In particular, $d_{H \oslash G}(u, v) = d_G(s, t) \cdot d_H(u, v)$*

For a graph G, we can write the graph $G^{\oslash N}$ as $G^{\oslash N} = G^{\oslash k-1} \oslash G \oslash G^{N-k}$. By observation 1, there are $|E(G^{\oslash k-1})| = |E(G)|^{k-1}$ copies of G in $G^{\oslash k-1} \oslash G$. Now using observation 2, we obtain $|E(G)|^{k-1}$ copies of G in $G^{\oslash N}$. We refer to these as the *level k copies* of G, and their vertices as *level k vertices*.

In the case of $K_{2,n}^{\oslash N}$, we will use a compact notation to refer to the *copies* of $K_{2,n}$. We will refer to the n vertices other than s, t in $K_{2,n}$ by $M = \{m_i\}_{i=1}^n$. For two *level k* vertices x, y in $K_{2,n}^N$, we will use $K_{2,n}^{(x,y)}$ to denote the *copy* of $K_{2,n}$ between x and y, i.e x and y are the vertices labelled s and t in the *copy*. Note that such a copy does not exist between all pairs of *level k* vertices. Further, for any copy $K_{2,n}^{(x,y)}$ of $K_{2,n}$, we will use the notation $M(K_{2,n}^{(x,y)}) = \{m_i(K_{2,n}^{(x,y)})\}$ to refer to the corresponding vertices in the copy.

Embeddings and Distortion. If $(X, d_X), (Y, d_Y)$ are metric spaces, and $f : X \to Y$, then we write $\|f\|_{\text{Lip}} = \sup_{x \neq y \in X} \frac{d_Y(f(x), f(y))}{d_X(x, y)}$. If f is injective, then the *distortion* of f is $\|f\|_{\text{Lip}} \cdot \|f^{-1}\|_{\text{Lip}}$. If $d_Y(f(x), f(y)) \leq d(x, y)$ for every $x, y \in X$, we say that f is *non-expansive*.

For a metric space X, we use $c_1(X)$ to denote the least distortion required to embed X into L_1. The value $c_1(G)$ for a graph G refers to the minimum L_1 distortion of the shortest path metric on G.

3   Coarse Differentiation

The lower bounds results in this paper rely on the technique of coarse differentiation from geometric group theory. In this section, we will derive the basic lemma of coarse differentiation that we will use later.

Given a non-expansive map $f : Y \to X$ between two metric spaces (X, d) and (Y, d'), the objective of the coarse differentiation argument is to obtain, some subset of Y on which the function f appears fairly regular(differentiable). To make this precise, we present the following definition:

Definition 3. *A path* $\mathcal{P} = \{x_1, x_2, \ldots, x_k\}$ *in a metric space* (X, d) *is said to* ϵ-*efficient if*

$$\sum_{i=1}^{k-1} d(x_i, x_{i+1}) \geq (1 + \epsilon) d(x_1, x_k)$$

Note that the quantity on the left is always at least $d(x_1, x_k)$ by triangle inequality. Thus, the above inequality implies that the triangle inequality is close to being tight. i.e the path $\{x_1, x_2, \ldots, x_k\}$ is "nearly straight".

Definition 4. *A function* $f : Y \to X$ *is said to be* ϵ-*efficient with respect to a path* $\mathcal{P} = \{y_1, y_2, \ldots, y_k\} \in Y$ *if the path* $f(\mathcal{P}) = \{f(y_1), f(y_2), \ldots f(y_k)\}$ *is* ϵ-*efficient in* X.

For the sake of exposition, we first present the coarse differentiation argument on a function from $[0, 1]$. We will later derive the more general result used in the lower bound argument.

Let $f : [0, 1] \to X$ be a non expansive map in to a metric space (X, d). Let $M \in \mathbb{N}$ be given, and for each $k \in \mathbb{N}$, let $L_k = \{jM^{-k}\}_{j=0}^{M^k} \subseteq [0, 1]$ be the set of *level-k points*, and let $S_k = \{(jM^{-k}, (j+1)M^{-k}) : j \in \{1, \ldots, M^k - 1\}\}$ be the set of *level-k pairs*.

For an interval $I = [a, b]$, $f|_I$ denotes the restriction of f to the interval I. Now we say that $f|_I$ is ε-*efficient at granularity* M if

$$\sum_{j=0}^{M-1} d\left(f\left(a + \frac{(b-a)j}{M} \right), f\left(a + \frac{(b-a)(j+1)}{M} \right) \right) \leq (1 + \varepsilon)\, d(f(a), f(b)).$$

Further, we say that a function f is (ε, δ)-*inefficient at level* k if

$$\left| \{(a, b) \in S_k : f|_{[a,b]} \text{ is not } \varepsilon\text{-efficient at granularity } M \} \right| \geq \delta M^k.$$

In other words, the probability that a randomly chosen level k restriction $f|_{[a,b]}$ is not ε-efficient is at least δ. Otherwise, we say that f is (ε, δ)-*efficient at level* k. The main theorem of this section follows.

Theorem 3 (Coarse differentiation). *If a non-expansive map* $f : [0, 1] \to X$ *is* (ε, δ)-*inefficient at an* α-*fraction of levels* $k = 1, 2, \ldots, N$, *then* $\mathsf{dist}(f|_{L_{N+1}}) \geq \frac{1}{2}\varepsilon\alpha\delta N$.

Proof. Let $D = \text{dist}(f|_{L_{N+1}})$, and let $1 \leq k_1 < \cdots < k_h \leq N$ be the $h \geq \lfloor \alpha N \rfloor$ levels at which f is (ε, δ)-inefficient.

Let us consider the first level k_1. Let $S'_{k_1} \subseteq S_{k_1}$ be a subset of size $|S'_{k_1}| \geq \lfloor \delta |S_{k_1}| \rfloor$ for which

$$(a, b) \in S'_{k_1} \implies f|_{[a,b]} \text{ is not } \varepsilon\text{-efficient at granularity } M$$

For any such $(a, b) \in S'_{k_1}$, we know that

$$\sum_{j=0}^{M-1} d\left(f\left(a + jM^{-k_1-1}\right), f\left(a + (j+1)M^{-k_1-1}\right)\right) > (1 + \varepsilon)d(f(a), f(b))$$

$$\geq d(f(a), f(b)) + \varepsilon \frac{M^{-k_1}}{D}.$$

by the definition of (not being) ε-efficient, and the fact that $d(f(a), f(b)) \geq |a - b|/D$. For all segments $(a, b) \in S_{k_1} - S'_{k_1}$, the triangle inequality yields

$$\sum_{j=0}^{M-1} d\left(f\left(a + jM^{-k_1-1}\right), f\left(a + (j+1)M^{-k_1-1}\right)\right) \geq d(f(a), f(b))$$

By summing the above inequalities over all the segments in S_{k_1}, we get

$$\sum_{(u,v)\in S_{k_1+1}} d\left(f(u), f(v)\right) \geq \sum_{(a,b)\in S_{k_1}} d(f(a), f(b)) + \frac{\varepsilon\delta}{2D},$$

where the extra factor 2 in the denominator on the RHS just comes from removing the floor from $|S'_{k_1}| \geq \lfloor \delta |S_{k_1}| \rfloor$. Similarly, for each of the levels $k_2, \ldots, k_h$, we will pick up an excess term of $\varepsilon\delta/(2D)$. We conclude that

$$1 \geq \sum_{(u,v)\in S_{N+1}} d\left(f(u), f(v)\right) \geq \frac{\varepsilon\delta h}{2D},$$

where the LHS comes from the fact that f is non-expansive. Simplifying achieves the desired conclusion.

3.1 Family of Maps

Let $\mathcal{F}$ denote a family of non-expansive maps from $L_k = \{jM^{-k}\}_{j=0}^{M^k} \subset [0, 1]$ to a metric space (X, d). For $r < k$, we say that the family $\mathcal{F}$ is (ϵ, δ)-inefficient at level r if

$$\left|\{(a, b) \in S_r, f \in \mathcal{F} : f|_{[a,b]} \text{ is not } \varepsilon\text{-efficient at granularity } M\}\right| \geq \delta M^r |\mathcal{F}|.$$

Theorem 4. *If a family $\mathcal{F}$ of non-expansive maps from $L_{N+1} = \{jM^{-N-1}\}_{j=0}^{M^{N+1}}$ to (X, d) is (ε, δ)-inefficient at an α-fraction of levels $k = 1, 2, \ldots, N$, then for some $f \in \mathcal{F}$, $\text{dist}(f|_{\mathcal{P}}) \geq \frac{1}{2}\varepsilon\alpha\delta N$.*

Proof. Define the metric d' on $X^{|\mathcal{F}|}$ by

$$d'(x,y) = \frac{1}{|\mathcal{F}|} \sum_{i=1}^{|\mathcal{F}|} d(x_i, y_i)$$

The family $\mathcal{F} = \{f_1, f_2, \ldots, f_{|\mathcal{F}|}\}$ defines a non-expansive map from $F : L_k \to X^{|\mathcal{F}|}$ by $F(x) = (f_1(x), f_2(x), \ldots, f_{|\mathcal{F}|}(x))$. Observe that the proof of Theorem 3 holds for non-expansive maps defined only on $L_{N+1} \subset [0,1]$. The result follows by applying Theorem 3 to the function F.

3.2   Efficient L_1-Valued Maps and Monotone Cuts

Any map $f : X \to L_1$ is a probability distribution over cuts of X. In this section, we will study how these cuts act on paths in X with respect to which f is ϵ-efficient. Towards this, we make the following definition:

Definition 5. *A path $\mathcal{P} = \{x_1, x_2, \ldots, x_k\}$ is said to be monotone with respect to a cut $(S, \overline{S})$, if the cut separates at most one of the edges of the path $\mathcal{P}$. Formally, $S \cap \mathcal{P} = \{x_1, x_2, \ldots, x_i\}$ for some $1 \leq i \leq k$.*

Lemma 1. *If a function $f : X \to L_1$ is ϵ-efficient with respect to a path $\mathcal{P} = \{x_1, x_2, \ldots, x_k\}$ then $\mathcal{P}$ is monotone with respect to at least $1 - \epsilon$ fraction of the cuts(under the probability distribution on cuts induced by f).*

Proof. If the path $\mathcal{P}$ is not monotone with respect to a cut $(S, \overline{S})$, then

$$\sum_{i=1}^{k-1} |1_S(x_i) - 1_S(x_{i+1})| \geq 2|1_S(x_1) - 1_S(x_k)|$$

Pick a random cut $(S, \overline{S})$ with the probability distribution induced by f. Let $\mathcal{E}$ denote the event that the cut is not monotone with respect to $\mathcal{P}$. For the purpose of contradiction, suppose the path $\mathcal{P}$ is not monotone with respect to ϵ-fraction of the cuts, i.e $\Pr[\mathcal{E}] > \epsilon$. Then we have,

$$\sum_{i=1}^{k-1} \|f(x_i) - f(x_{i+1})\|_1 = \mathbb{E}\Big[\sum_{i=1}^{k-1} |1_S(x_i) - 1_S(x_{i+1})| \Big| \mathcal{E}\Big] \Pr[\mathcal{E}]$$

$$+ \mathbb{E}\Big[\sum_{i=1}^{k-1} |1_S(x_i) - 1_S(x_{i+1})| \Big| \overline{\mathcal{E}}\Big] \Pr[\overline{\mathcal{E}}]$$

$$\geq 2\mathbb{E}\Big[|1_S(x_1) - 1_S(x_k)|\Big] \Pr[\mathcal{E}]$$

$$+ \mathbb{E}\Big[|1_S(x_1) - 1_S(x_k)|\Big] (1 - \Pr[\mathcal{E}])$$

$$\geq (1 + \Pr[\mathcal{E}]) \mathbb{E}\Big[|1_S(x_1) - 1_S(x_k)|\Big]$$

$$> (1 + \epsilon)\|f(x_1) - f(x_k)\|_1$$

This is a contradiction, since f is ϵ-efficient with respect to $\mathcal{P}$.

4 L_1 Distortion Lower Bound

Our lower bound examples are the recursively defined family of graphs - $\{K_{2,n}^{\otimes k}\}_{k=1}^{\infty}$. In the case $n = 2$, the graphs $K_{2,n}^{\otimes k}$ are the same as diamond graphs [28,18].

Lemma 2. *Given an s-t graph G and $\epsilon, D > 0$, there exists an integer N such that the following holds : For any non-expansive map $f : G^{\otimes N} \to X$ with $\mathrm{dist}(f) \le D$, there exists a copy G' of G in $G^{\otimes N}$ such that f is ϵ-efficient on all s-t shortest paths in G'.*

Proof. Let M denote the length of the shortest s-t path in G. Let $\mathcal{P}(G)$ denote the family of s-t shortest paths in G. Fix $\delta = \frac{1}{|\mathcal{P}(G)|}$, $\alpha = \frac{1}{2}$ and $N = \lceil \frac{8D}{\epsilon\delta} \rceil$.

Let $\mathcal{P}$ denote the family of all s-t shortest paths in $G^{\otimes N}$. Each path in $\mathcal{P}$ is of length M^N. For each path $P \in \mathcal{P}$, define $f_P : L_N = \{jM^{-N}\}_{j=0}^{M^N} \to X$ by $f_P(jM^{-N}) = f(v_j)$ where v_j is the jth vertex along the path P. Let $\mathcal{F}$ denote the family of maps $\{f_P | P \in \mathcal{P}\}$. From the choice of parameters, observe that $\frac{1}{2}\epsilon\alpha\delta N > D = \mathrm{dist}(f)$. Applying Theorem 4 on the family $\mathcal{F}$, $\mathcal{F}$ is (ϵ, δ)-*efficient* at an $\alpha = \frac{1}{2}$-fraction of levels $j = 1, 2, \dots N$. Specifically, there exists a level $k < N$ such that $\mathcal{F}$ is (ϵ, δ)-*efficient* at level k.

Let $\mathcal{G}_k$ denote the set of all level k copies of G in $G^{\otimes N}$. If for at least one of the level k *copies*, all the s-t shortest paths are ϵ-efficient then the proof is complete.

On the contrary, suppose each level k *copy* has an s-t path that is not ϵ-efficient. Then in each level k *copy* at least a $\delta = \frac{1}{|\mathcal{P}(G)|}$ fraction of the s-t paths are ϵ-inefficient. Thus for a random choice of a path $P \in \mathcal{P}$ and $0 \le j < M^k$ the subpath $\{v_{jM^{N-k}}, v_{(j+1)M^{N-k}}, \dots v_{(j+M)M^{N-k}}\}$ of P is ϵ-inefficient with probability at least δ. Equivalently, for a random choice of function $f_P \in \mathcal{F}$ and a level k pair $[a, b]$, f_P is ϵ-inefficient on $[a, b]$ with probability at least δ. This is a contradiction, since the family $\mathcal{F} = \{f_P | P \in \mathcal{P}\}$ is (ϵ, δ)-efficient at level k.

Lemma 3. *For $\epsilon < \frac{1}{2}$ and any function $f : K_{2,n} \to L_1$ that is ϵ/n-efficient with respect to each of the paths from $\{s, m_i, t\}, 1 \le i \le n$, the distortion $\mathrm{dist}(f) \ge 2 - \frac{2}{n} - 16\epsilon$.*

Proof. Without loss of generality, we assume that in the embedding f, the probability of trivial cuts $(V(K_{2,n}), \phi)$ is 0. Suppose for the function f, there is a nonzero probability of trivial cuts $(V(K_{2,n}), \phi)$, then we can create a function $\tilde{f}$ with the same distortion as f, by just discarding the trivial cuts.

Using lemma 1, each of the paths $\{s, m_i, t\}$ is monotone with respect to all but ϵ/n fraction of the cuts. By a union bound, we conclude that for all but ϵ-fraction of the cuts all the paths $\{s, m_i, t\}$ are monotone.

Consider a cut $(S, \overline{S})$ that is monotone with respect to all the paths $\{s, m_i, t\}$. Let us refer to these cuts as *good* cuts. For a *good* cut $(S, \overline{S})$, we will now show that

$$\sum_{i,j \in n} |1_S(m_i) - 1_S(m_j)| \le \frac{n^2}{4}|1_S(s) - 1_S(t)|$$

Without loss of generality, let us assume $s \in S$. There are two cases:

Case I: $(t \in S)$ For each i, at most one of the edges $(s, m_i), (m_i, t)$ are cut by $(S, \overline{S})$. From this, we conclude that each vertex m_i is also in S. Hence $\overline{S} = \phi$, or in other words the cut $(S, \overline{S})$ is trivial. By assumption about f, this case does not arise.

Case II: $(t \notin S)$ Suppose $|S| = r + 1$, then r of the vertices $\{m_i\}_{i=1}^{n}$ are in S. Thus we can write

$$|1_S(s) - 1_S(t)| = 1 \qquad \sum_{i,j \in n} |1_S(m_i) - 1_S(m_j)| = r(n - r)$$

In particular, the above equations imply

$$\sum_{i,j \in n} |1_S(m_i) - 1_S(m_j)| \leq \frac{n^2}{4}|1_S(s) - 1_S(t)|$$

For a random cut, let $\mathcal{E}$ denote the event that it is *good*. As already observed, the probability $\Pr[\overline{\mathcal{E}}]$ is less than ϵ.

$$\sum_{i,j \in [n]} \|f(m_i) - f(m_j)\|_1 = \mathbb{E}\Big[\sum_{i,j \in [n]} |1_S(m_i) - 1_S(m_j)| \Big| \mathcal{E}\Big] \Pr[\mathcal{E}]$$

$$+ \mathbb{E}\Big[\sum_{i,j \in [n]} |1_S(m_i) - 1_S(m_j)| \Big| \overline{\mathcal{E}}\Big] \Pr[\overline{\mathcal{E}}]$$

$$\leq \frac{n^2}{4} \mathbb{E}\Big[|1_S(s) - 1_S(t)| \Big| \mathcal{E}\Big] \Pr[\mathcal{E}] + \epsilon n^2 \qquad (1)$$

From the two cases above, clearly for every non trivial *good* cut $(S, \overline{S})$ we have $|1_S(s) - 1_S(t)| = 1$. Consequently we have

$$\|f(s) - f(t)\|_1 \geq \mathbb{E}\Big[|1_S(s) - 1_S(t)| \Big| \mathcal{E}\Big] \Pr[\mathcal{E}] \geq (1 - \epsilon) > 1/2$$

Using this in equation 1 we get,

$$\sum_{i,j \in [n]} \|f(m_i) - f(m_j)\|_1 \leq \frac{n^2}{4} \|f(s) - f(t)\|_1 + 2\epsilon n^2 \|f(s) - f(t)\|_1$$

$$\leq \frac{(1 + 8\epsilon)n^2}{4} \|f(s) - f(t)\|_1 \qquad (2)$$

In the metric d on $K_{2,n}$, $d(m_i, m_j) = d(s, t) = 2$. Hence we have

$$\sum_{i,j \in [n]} d(m_i, m_j) = \binom{n}{2} d(s, t)$$

Along with the inequality 2, this implies

$$\mathrm{dist}(f) \geq \frac{\binom{n}{2}}{\frac{(1+8\epsilon)n^2}{4}} = \frac{1}{1 + 8\epsilon}\left(2 - \frac{2}{n}\right) \geq \left(2 - \frac{2}{n}\right) - 16\epsilon$$

Theorem 5. *For any* $n \geq 2$, $\lim_{k \to \infty} c_1(K_{2,n}^{\oslash k}) \geq 2 - \frac{2}{n}$.

Proof. For any $\epsilon' > 0$, let N be the integer obtained by applying Lemma 2 with $\epsilon = \epsilon'/n$, $D = 2$ and $G = K_{2,n}$. We will show that for any map $f : K_{2,n}^{\oslash N} \to L_1$, $\mathrm{dist}(f) \geq 2 - \frac{2}{n} - 16\epsilon'$. Without loss of generality, we can assume that f is non-expansive, otherwise just rescale the map f. Suppose $\mathrm{dist}(f) \leq 2$, then from Lemma 2 there exists a *copy* of $K_{2,n}$ in which all the s-t paths are $\frac{\epsilon'}{n}$ efficient. Using Lemma 3, on this *copy* of $K_{2,n}$ we get $\mathrm{dist}(f) \geq 2 - \frac{2}{n} - 16\epsilon'$.

Hence for any $\epsilon' > 0$, there exists large enough N such that $c_1(K_{2,n}^{\oslash N}) \geq 2 - \frac{2}{n} - 16\epsilon'$. The required result follows, since $c_1(K_{2,n}^{\oslash k})$ is monotonically increasing with k.

5   Embeddings of $K_{2,n}^{\oslash k}$

In this section, we show that for every fixed $n, k \in \mathbb{N}$, we have $c_1(K_{2,n}^{\oslash k}) < 2$. In particular, the lower bound of Theorem 5 has to be asymptotic in nature. More precisely, we will show that in order to achieve $c_1(K_{2,n}^{\oslash k}) = 2 - o(1)$, we must take both $k, n \to \infty$.

A Next-Embedding Operator. Let T be a random variable ranging over subsets of $V(K_{2,n}^{\oslash k})$, and let S be a random variable ranging over subsets of $V(K_{2,n})$. We define a random subset $P_S(T) \subseteq V(K_{2,n}^{\oslash k+1})$ as follows. One moves from $K_{2,n}^{\oslash k}$ to $K_{2,n}^{\oslash k+1}$ by replacing every edge $(x, y) \in E(K_{2,n}^{\oslash k})$ with a copy of $K_{2,n}$ which we will call $K_{2,n}^{(x,y)}$. For every edge $(x, y) \in K_{2,n}^{\oslash k}$, let $S^{(x,y)}$ be an independent copy of the cut S (which ranges over subsets of $V(K_{2,n})$). We form the cut $P_S(T) \subseteq V(K_{2,n}^{\oslash k+1})$ as follows. If $(x, y) \in E(K_{2,n}^{\oslash k})$, then for $v \in V(K_{2,n}^{(x,y)})$, we put

$$1_{P_S(T)}(v) = \begin{cases} 1_{P_S(T)}\left(s(K_{2,n}^{(x,y)})\right) & \text{if } 1_{S^{(x,y)}}(v) = 1_{S^{(x,y)}}\left(s(K_{2,n}^{(x,y)})\right) \\ 1_{P_S(T)}\left(t(K_{2,n}^{(x,y)})\right) & \text{otherwise} \end{cases}$$

We note that, strictly speaking, the operator P_S depends on n and k, but we allow these to be implicit parameters.

5.1   Embeddings for Small n

Our first embedding gives an optimal upper bound on $\lim_{k \to \infty} c_1(K_{2,n}^{\oslash k})$ for every $n \geq 1$. Consider the graph $K_{2,n}$ with vertex set $V = \{s, t\} \cup M$. An embedding in the style of [18] would define a random subset $S \subseteq V$ by selecting $M' \subseteq M$ to contain each vertex from M independently with probability $\frac{1}{2}$, and then setting $S = \{s\} \cup M'$. The resulting embedding has distortion 2 since, for every pair $x, y \in M$, we have $\Pr[1_S(x) \neq 1_S(y)] = \frac{1}{2}$. To do slightly better, we choose a uniformly random subset $M' \subseteq M$ of size $\lfloor \frac{n}{2} \rfloor$ and set $S = \{s\} \cup M'$ or $S = \{s\} \cup (M \setminus M')$ each with probability half. In this case, we have

$$\Pr[1_S(x) \neq 1_S(y)] = \frac{\lfloor \frac{n}{2} \rfloor \cdot \lfloor \frac{n+1}{2} \rfloor}{\binom{n}{2}} > \frac{1}{2},$$

resulting in a distortion slightly better than 2. A recursive application of these ideas results in $\lim_{k \to \infty} c_1(K_{2,n}^{\otimes k}) < 2$ for every $n \geq 1$, though the calculation is complicated by the fact that the worst distortion is incurred for a pair $\{x, y\}$ with $x \in M(H)$ and $y \in M(G)$ where H is a copy of $K_{2,n}^{\otimes k_1}$ and G is a copy of $K_{2,n}^{\otimes k_2}$, and the relationship between k_1 and k_2 depends on n. (For instance, $c_1(K_{2,2}) = 1$ while $\lim_{k \to \infty}(K_{2,2}^{\otimes k}) = \frac{4}{3}$.)

Theorem 6. *For any $n, k \in \mathbb{N}$, we have $c_1(K_{2,n}^{\otimes k}) \leq 2 - \frac{2}{2^{\lceil \frac{n}{2} \rceil + 1}}$.*

Proof. For simplicity, we prove the bound for $K_{2,2n}^{\otimes k}$. A similar analysis holds for $K_{2,2n+1}^{\otimes k}$. We define a random cut $S_k \subseteq V(K_{2,2n}^{\otimes k})$ inductively. For $k = 1$, choose a uniformly random partition $M(K_{2,2n}^{\otimes 1}) = M_s \cup M_t$ with $|M_s| = |M_t| = n$, and let $S_1 = \{s(K_{2,2n}^{\otimes 1})\} \cup \{M_s\}$. The key fact which causes the distortion to be less than 2 is the following: For any $x, y \in M(K_{2,2n}^{\otimes 1})$, we have

$$\Pr[1_{S_1}(x) \neq 1_{S_1}(y)] = \frac{n^2}{\binom{2n}{2}} = \frac{n}{2n - 1} > \frac{1}{2}. \tag{3}$$

This follows because there are $\binom{2n}{2}$ pairs $\{x, y\} \in M(K_{2,2n}^{\otimes 1})$ and n^2 are separated by S_1.

Assume now that we have a random subset $S_k \subseteq V(K_{2,2n}^{\otimes k})$. We set $S_{k+1} = P_{S_1}(S_k)$ where P_{S_1} is the operator defined above, which maps random subsets of $V(K_{2,2n}^{\otimes k})$ to random subsets of $V(K_{2,2n}^{\otimes k+1})$. In other words $S_k = P_{S_1}^{k-1}(S_1)$.

Let $s_0 = s(K_{2,2n}^{\otimes k})$ and $t_0 = t(K_{2,2n}^{\otimes k})$. It is easy to see that the cut $S = S_k$ defined above is always monotone with respect to every s_0-t_0 shortest path in $K_{2,2n}^{\otimes k}$, thus every such path has exactly one edge cut by S_k, and furthermore the cut edge is uniformly chosen from along the path, i.e. $\Pr[1_S(x) \neq 1_S(y)] = 2^{-k}$ for every $(x, y) \in E(K_{2,2n}^{\otimes k})$. In particular, it follows that if $u, v \in V(K_{2,2n}^{\otimes k})$ lie along the same simple s_0-t_0, then $\Pr[1_S(u) \neq 1_S(v)] = 2^{-k}d(u, v)$.

Now consider any $u, v \in V(K_{2,2n}^{\otimes k})$. Fix some shortest path P from u to v. By symmetry, we may assume that P goes left (toward s_0) and then right (toward t_0). Let s be the left-most point of P. In this case, $s = s(H)$ for some subgraph H which is a copy of $K_{2,2n}^{\otimes k'}$ with $k' \leq k$, and such that $u, v \in V(H)$; we let $t = t(H)$. We also have $d(u, v) = d(u, s) + d(s, v)$. Let $M = M(H)$, and fix $x, y \in M$ which lie along the s-u-t and s-v-t shortest-paths, respectively. Without loss of generality, we may assume that $d(s, v) \leq d(s, y)$. We need to consider two cases (see Figure 5.1).

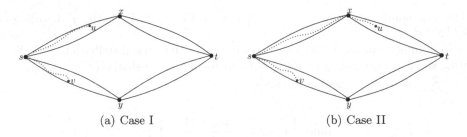

(a) Case I   (b) Case II

Fig. 1. The two cases of Theorem 6

Case I: $d(u,s) \leq d(x,s)$.
For any pair $a,b \in V(K_{2,2n}^{\otimes k})$, we let $\mathcal{E}_{a,b}$ be the event $\{1_S(a) \neq 1_S(b)\}$. In this case, we have $\Pr[\mathcal{E}_{u,v}] = \Pr[\mathcal{E}_{s,t}] \cdot \Pr[\mathcal{E}_{u,v} \mid \mathcal{E}_{s,t}]$. Since s,t clearly lie on a shortest s_0-t_0 path, we have $\Pr[\mathcal{E}_{s,t}] = 2^{-k}d(s,t)$. For any event $\mathcal{E}$, we let $\mu[\mathcal{E}] = \Pr[\mathcal{E} \mid \mathcal{E}_{s,t}]$. Now we calculate using (3),

$$\mu[\mathcal{E}_{u,v}] \geq \mu[\mathcal{E}_{x,y}] \left(\mu[\mathcal{E}_{x,s} \mid \mathcal{E}_{x,y}] \mu[\mathcal{E}_{u,s} \mid \mathcal{E}_{x,s}, \mathcal{E}_{x,y}] + \mu[\mathcal{E}_{x,t} \mid \mathcal{E}_{x,y}] \mu[\mathcal{E}_{v,s} \mid \mathcal{E}_{x,t}, \mathcal{E}_{x,y}] \right)$$

$$= \frac{n}{2n-1} \left(\frac{1}{2} \cdot \frac{d(u,s)}{d(x,s)} + \frac{1}{2} \cdot \frac{d(v,s)}{d(y,s)} \right)$$

$$= \frac{n}{2n-1} \frac{d(u,v)}{d(s,t)}.$$

Hence in this case, $\Pr[1_S(u) \neq 1_S(v)] \geq \frac{n}{2n-1} \cdot 2^{-k}d(u,v)$.

Case II: $d(u,s) \geq d(x,s)$.
Here, we need to be more careful about bounding $\mu[\mathcal{E}_{u,v}]$. It will be helpful to introduce the notation $a \mapsto b$ to represent the event $\{1_S(a) = 1_S(b)\}$. We have,

$$\mu[\mathcal{E}_{u,v}] = \mu[x \mapsto t, y \mapsto s] + \mu[x \mapsto t, y \mapsto t, v \mapsto s] + \mu[x \mapsto s, y \mapsto s, u \mapsto t]$$

$$+ \mu[x \mapsto s, y \mapsto t, u \mapsto t, v \mapsto s] + \mu[x \mapsto s, y \mapsto t, u \mapsto s, v \mapsto t]$$

$$= \frac{1}{2} \frac{n}{2n-1} + \frac{n-1}{2n-1} \frac{d(v,y) + d(u,x)}{d(s,t)} + \frac{1}{2} \frac{n}{2n-1} \left(\frac{d(u,x)d(v,y) + d(u,t)d(v,s)}{d(x,t)d(y,s)} \right)$$

If we set $A = \frac{d(v,s)}{d(s,t)}$ and $B = \frac{d(u,x)}{d(s,t)}$, then $\frac{d(u,v)}{d(s,t)} = \frac{1}{2} + A + B$ and simplifying the expression above, we have

$$\mu[\mathcal{E}_{u,v}] = \frac{1}{2} + B + \frac{A}{2n-1} - \frac{4n}{2n-1}AB$$

Since the shortest path from u to v goes through s by assumption, we must have $A + B \leq \frac{1}{2}$. Thus we are interested in the minimum of $\mu[\mathcal{E}_{u,v}]/(\frac{1}{2}+A+B)$ subject to the constraint $A + B \leq \frac{1}{2}$. It is easy to see that the minimum is achieved at $A + B = \frac{1}{2}$, thus setting $B = \frac{1}{2} - A$, we are left to find

$$\min_{0 \leq A \leq \frac{1}{2}} \left\{ 1 - 2A + \frac{4nA^2}{2n-1} \right\} = \frac{2n+1}{4n}.$$

(The minimum occurs at $A = \frac{1}{2} - \frac{1}{4n}$.) So in this case, $\Pr[\mathbf{1}_S(u) \neq \mathbf{1}_S(v)] \geq \frac{2n+1}{4n}2^{-k}d(u,v)$.

Combining the above two cases, we conclude that the distribution $S = S_k$ induces an L_1 embedding of $K_{2,2n}^{\oslash k}$ with distortion at most $\max\{\frac{2n-1}{n}, \frac{4n}{2n+1}\} = 2 - \frac{2}{2n+1}$. A similar calculation yields

$$c_1(K_{2,2n+1}^{\oslash k}) \leq \left(\min_{0 \leq A \leq \frac{1}{2}} \left\{ 1 - 2A + \frac{4(n+1)A^2}{2n+1} \right\} \right)^{-1} = 2 - \frac{2}{2n+3}.$$

Embeddings for small k are deferred to the full version of the paper.

References

1. Agarwal, A., Charikar, M., Makarychev, K., Makarychev, Y.: $O(\sqrt{\log n})$ approximation algorithms for Min UnCut, Min 2CNF Deletion, and directed cut problems. In: 37th Annual ACM Symposium on Theory of Computing, ACM, New York (2005)
2. Andoni, A., Deza, M., Gupta, A., Indyk, P., Raskhodnikova, S.: Lower bounds for embedding of edit distance into normed spaces. In: Proceedings of the 14th annual ACM-SIAM Symposium on Discrete Algorithms, ACM Press, New York (2003)
3. Arora, S., Lee, J.R., Naor, A.: Euclidean distortion and the Sparsest Cut. In: 37th Annual Symposium on the Theory of Computing, pp. 553–562, Journal of the AMS (to appear 2005)
4. Arora, S., Rao, S., Vazirani, U.: Expander flows, geometric embeddings, and graph partitionings. In: 36th Annual Symposium on the Theory of Computing, pp. 222–231. J. ACM, New York (to appear 2004)
5. Aumann, Y., Rabani, Y.: An $O(\log k)$ approximate min-cut max-flow theorem and approximation algorithm. SIAM J. Comput (electronic) 27(1), 291–301 (1998)
6. Benyamini, Y., Lindenstrauss, J.: Geometric nonlinear functional analysis. vol. 1(48) of American Mathematical Society Colloquium Publications. American Mathematical Society, Providence, RI (2000)
7. Bourgain, J.: On Lipschitz embedding of finite metric spaces in Hilbert space. Israel J. Math. 52(1-2), 46–52 (1985)
8. Brinkman, B., Karagiozova, A., Lee, J.R.: Vertex cuts, random walks, and dimension reduction in series-parallel graphs. STOC (to appear)
9. Chakrabarti, A., Lee, J.R., Vincent, J.: Parity and contraction in planar graph embeddings (2007)
10. Cheeger, J.: Differentiability of Lipschitz functions on metric measure spaces. Geom. Funct. Anal. 9(3), 428–517 (1999)
11. Cheeger,J., Kleiner, B.: Differentiating maps into L^1 and the geometry of BV functions. Preprint, 2006.
12. Chekuri, C., Gupta, A., Newman, I., Rabinovich, Y., Sinclair, A.: Embedding k-outerplanar graphs into l_1. SIAM J. Discrete Math. 20(1), 119–136 (2006)
13. Chlamtac, E., Makayrchev, K., Makarychev, Y.: How to play unique game using embeddings. In: 47th Annual Syposium on Foundations of Computer Science (2006)
14. Diestel, R.: *Graph theory*. Graduate Texts in Mathematics, vol. 173. Springer, Berlin (2005)

15. Eskin, A., Fisher, D., Whyte, K.: Quasi-isometries and rigidity of solvable groups. Preprint (2006)
16. Feige, U., Hajiaghayi, M.T., Lee, J.R.: Improved approximation algorithms for minimum-weight vertex separators. In: 37th Annual ACM Symposium on Theory of Computing. ACM, 2005., SIAM J. Comput. New York (to appear)
17. Franchi, B., Serapioni, R., Serra Cassano, F.: Rectifiability and perimeter in the Heisenberg group. Math. Ann. 321(3), 479–531 (2001)
18. Gupta, A., Newman, I., Rabinovich, Y., Sinclair, A.: Cuts, trees and l_1-embeddings of graphs. Combinatorica 24(2), 233–269 (2004)
19. Heinonen, J.: Lectures on analysis on metric spaces. Universitext. Springer, New York (2001)
20. Indyk, P.: Algorithmic applications of low-distortion geometric embeddings. In: 42nd Annual Symposium on Foundations of Computer Science, pp. 10–33. IEEE Computer Society Press, Los Alamitos (2001)
21. Khot, S., Naor, A.: Nonembeddability theorems via Fourier analysis. Math. Ann. 334(4), 821–852 (2006)
22. Khot, S., Vishnoi, N.: The unique games conjecture, integrality gap for cut problems and embeddability of negative type metrics into ℓ_1. In: 46th Annual Symposium on Foundations of Computer Science, pp. 53–62. IEEE Computer Society Press, Los Alamitos (2005)
23. Lee, J.R., Naor, A.: l_p metrics on the Heisenberg group and the Goemans-Linial conjecture. In: 47th Annual Symposium on Foundations of Computer Science, IEEE Computer Society Press, Los Alamitos, CA (2006)
24. Linial, N.: Finite metric-spaces—combinatorics, geometry and algorithms. In: Proceedings of the International Congress of Mathematicians, vol. III, pp. 573–586. Higher Ed. Press, Beijing (2002)
25. Linial, N., London, E., Rabinovich, Y.: The geometry of graphs and some of its algorithmic applications. Combinatorica 15(2), 215–245 (1995)
26. Matoušek, J.: Open problems on low-distortion embeddings of finite metric spaces. Online, http://kam.mff.cuni.cz/~matousek/metrop.ps
27. Matoušek, J.: Lectures on discrete geometry. Graduate Texts in Mathematics, vol. 212. Springer-Verlag, New York (2002)
28. Newman, I., Rabinovich, Y.: A lower bound on the distortion of embedding planar metrics into Euclidean space. Discrete Comput. Geom. 29(1), 77–81 (2003)
29. Okamura, H., Seymour, P.D.: Multicommodity flows in planar graphs. J. Combin. Theory Ser. B 31(1), 75–81 (1981)
30. Pansu, P.: Métriques de Carnot-Carathéodory et quasiisométries des espaces symétriques de rang un. Ann. of Math (2) 129(1), 1–60 (1989)

Maximum Gradient Embeddings
and Monotone Clustering*
(Extended Abstract)

Manor Mendel[1] and Assaf Naor[2]

[1] The Open University of Israel
[2] Courant Institute

Abstract. Let (X, d_X) be an n-point metric space. We show that there exists a distribution $\mathscr{D}$ over non-contractive embeddings into trees $f : X \to T$ such that for every $x \in X$,

$$\mathbb{E}_{\mathscr{D}} \left[\max_{y \in X \setminus \{x\}} \frac{d_T(f(x), f(y))}{d_X(x, y)} \right] \leq C (\log n)^2,$$

where C is a universal constant. Conversely we show that the above quadratic dependence on $\log n$ cannot be improved in general. Such embeddings, which we call *maximum gradient embeddings*, yield a framework for the design of approximation algorithms for a wide range of clustering problems with monotone costs, including fault-tolerant versions of k-median and facility location.

1 Introduction

We introduce a new notion of embedding, called *maximum gradient embeddings*, which is "just right" for approximating a wide range of clustering problems. We then provide optimal maximum gradient embeddings of general finite metric spaces, and use them to design approximation algorithms for several natural clustering problems. Rather than being encyclopedic, the main emphasis of the present paper is that these embeddings yield a generic approach to many problems.

Due to their special structure, it is natural to try to embed metric spaces into trees. This is especially important for algorithmic purposes, as many hard problems are tractable on trees. Unfortunately, this is too much to hope for in the bi-Lipschitz category: As shown by Rabinovich and Raz [22] the n-cycle incurs distortion $\Omega(n)$ in any embedding into a tree. However, one can relax this idea and look for a *random* embedding into a tree which is faithful on average. Such an approach has been developed in recent years by mathematicians and computer scientists. In the mathematical literature this is referred to as embeddings into products of trees, and it is an invaluable tool in the study of negatively curved spaces (see for example [7, 10, 20]).

---

* Full version appears in [19].

M. Charikar et al. (Eds.): APPROX and RANDOM 2007, LNCS 4627, pp. 242–256, 2007.
© Springer-Verlag Berlin Heidelberg 2007

Probabilistic embeddings into *dominating* trees became an important algorithmic paradigm due to the work of Bartal [3, 4] (see also [1, 11] for the related problem of embedding graphs into distributions over spanning trees). This work led to the design of many approximation algorithms for a wide range of NP hard problems. In some cases the best known approximation factors are due to the "probabilistic tree" approach, while in other cases improved algorithms have been subsequently found after the original application of probabilistic embeddings was discovered. But, in both cases it is clear that the strength of Bartal's approach is that it is generic: For a certain type of problem one can quickly get a polylogarithmic approximation using probabilistic embedding into trees, and then proceed to analyze certain particular cases if one desires to find better approximation guarantees. However, probabilistic embeddings into trees do not always work. In [5] Bartal and Mendel introduced the weaker notion of multi-embeddings, and used it to design improved algorithms for special classes of metric spaces. Here we *strengthen* this notion to maximum gradient embeddings, and use it to design approximation algorithms for harder problems to which regular probabilistic embeddings do not apply.

Let (X, d_X) and (Y, d_Y) be metric spaces, and fix a mapping $f : X \to Y$. The mapping f is called *non-contractive* if for every $x, y \in X$, $d_Y(f(x), f(y)) \geq d_X(x, y)$. The *maximum gradient* of f at a point $x \in X$ is defined as

$$|\nabla f(x)|_\infty = \sup_{y \in X \setminus \{x\}} \frac{d_Y(f(x), f(y))}{d_X(x, y)}. \tag{1}$$

Thus the *Lipschitz constant* of f is given by $\|f\|_{\mathrm{Lip}} = \sup_{x \in X} |\nabla f(x)|_\infty$.

In what follows when we refer to a tree metric we mean the shortest-path metric on a graph-theoretical tree with weighted edges. Recall that (U, d_U) is an ultrametric if for every $u, v, w \in U$ we have $d_U(u, v) \leq \max\{d_U(u, w), d_U(w, v)\}$. It is well known that ultrametrics embed isometrically into tree metrics. The following result is due to Fakcharoenphol, Rao and Talwar [12], and is a slight improvement over an earlier theorem of Bartal [4]. For every n-point metric space (X, d_X) there is a distribution $\mathscr{D}$ over non-contractive embeddings into ultrametrics $f : X \to U$ such that

$$\max_{\substack{x, y \in X \\ x \neq y}} \mathbb{E}_{\mathscr{D}} \frac{d_U(f(x), f(y))}{d_X(x, y)} = O(\log n). \tag{2}$$

The logarithmic upper bound in (2) cannot be improved in general.

Inequality (2) is extremely useful for optimization problems whose objective function is linear in the distances, since by linearity of expectation it reduces such tasks to trees, with only a logarithmic loss in the approximation guarantee. When it comes to non-linear problems, the use of (2) is very limited. We will show that this issue can be addressed using the following theorem, which is our main result.

Theorem 1. *Let (X, d_X) be an n-point metric space. Then there exists a distribution $\mathscr{D}$ over non-contractive embeddings into ultrametrics $f : X \to U$ (thus* *both the ultrametric* (U, d_U) *and the mapping* f *are random) such that for every* $x \in X$, $\mathbb{E}_{\mathscr{D}}|\nabla f(x)|_\infty \leq C(\log n)^2$, *where* C *is a universal constant.*

On the other hand, there exists a universal constant $c > 0$ *and arbitrarily large n-point metric spaces* Y_n *such that for any distribution over non-contractive embeddings into trees* $f : Y_n \to T$ *there is necessarily some* $x \in Y_n$ *for which* $\mathbb{E}_{\mathscr{D}}|\nabla f(x)|_\infty \geq c(\log n)^2$.

We call embeddings as in Theorem 1, i.e. embeddings with small expected maximum gradient, *maximum gradient embeddings into distributions over trees* (in what follows we will only deal with distributions over trees, so we will drop the last part of this title when referring to the embedding, without creating any ambiguity). The proof of the upper bound in Theorem 1 is a modification of an argument of Fakcharoenphol, Rao and Talwar [12], which is based on ideas from [3, 8]. Alternative proofs of the main technical step of the proof of the upper bound in Theorem 1 can be also deduced from the results of [18] or an argument in the proof of Lemma 2.1 in [13].

The heart of this paper is the lower bound in Theorem 1. The metrics Y_n in Theorem 1 are the diamond graphs of Newman and Rabinovich [14], which have been previously used as counter-examples in several embedding problems— see [6, 14, 17, 21].

1.1 A Framework for Clustering Problems with Monotone Costs

We now turn to some algorithmic applications of Theorem 1. The general reduction in Theorem 2 below should also be viewed as an explanation why maximum gradient embeddings are so natural— they are precisely the notion of embedding which allows such reductions to go through. . In the full version of this paper we also analyze in detail two concrete optimization problems which belong to this framework.

A very general setting of the clustering problem is as follows. Let X be an n-point set, and denote by $\mathrm{MET}(X)$ the set of all metrics on X. A *possible clustering solution* consists of sets of the form $\{(x_1, C_1), \ldots, (x_k, C_k)\}$ where $x_1, \ldots, x_k \in X$ and $C_1, \ldots, C_k \subseteq X$. We think of $C_1, \ldots, C_k$ as the clusters, and x_i as the "center" of C_i. In this general framework we do not require that the clusters cover X, or that they are pairwise disjoint, or that they contain their centers. Thus the space of possible clustering solutions is $\mathcal{P} = 2^{X \times 2^X}$ (though the exact structure of $\mathcal{P}$ does not play a significant role in the proof of Theorem 2). Assume that for every point $x \in X$, every metric $d \in \mathrm{MET}(X)$, and every possible clustering solution $P \in \mathcal{P}$, we are given $\Gamma(x, d, P) \in [0, \infty]$, which we think of as a measure of the dissatisfaction of x with respect to P and d. Our goal is to minimize the average dissatisfaction of the points of X. Formally, given a measure of dissatisfaction (which we also call in what follows a *clustering cost function*) $\Gamma : X \times \mathrm{MET}(X) \times \mathcal{P} \to [0, \infty]$, we wish to compute for a given metric $d \in \mathrm{MET}(X)$ the value

$$\mathrm{Opt}_\Gamma(X, d) \stackrel{\text{def}}{=} \min \left\{ \sum_{x \in X} \Gamma(x, d, P) : P \in \mathcal{P} \right\}$$

(Since we are mainly concerned with the algorithmic aspect of this problem, we assume from now on that Γ can be computed efficiently.)

We make two natural assumptions on the cost function Γ. First of all, we will assume that it scales homogeneously with respect to the metric, i.e. for every $\lambda > 0$, $x \in X$, $d \in \mathrm{MET}(X)$ and $P \in \mathcal{P}$ we have $\Gamma(x, \lambda d, P) = \lambda \Gamma(x, d, P)$. Secondly we will assume that Γ is monotone with respecting to the metric, i.e. if $d, \overline{d} \in \mathrm{MET}(X)$ and $x \in X$ satisfy $d(x,y) \le \overline{d}(x,y)$ for every $y \in X$ then $\Gamma(x, d, P) \le \Gamma(x, \overline{d}, P)$. In other words, if all the points in X are further with respect to $\overline{d}$ from x then they are with respect to d, then x is more dissatisfied. This is a very natural assumption to make, as most clustering problems look for clusters which are small in various (metric) senses. We call clustering problems with Γ satisfying these assumptions *monotone clustering problems*. A large part of the clustering problems that have been considered in the literature fall into this framework.

The following theorem is a simple application of Theorem 1. It shows that it is enough to solve monotone clustering problems on ultrametrics, with only a polylogarithmic loss in the approximation factor.

Theorem 2 (Reduction to ultrametrics). *Let X be an n-point set and fix a homogeneous monotone clustering cost function $\Gamma : X \times \mathrm{MET}(X) \times \mathcal{P} \to [0, \infty]$. Assume that there is a randomized polynomial time algorithm which approximates $\mathrm{Opt}_\Gamma(X, \rho)$ to within a factor $\alpha(n)$ on any ultrametric $\rho \in \mathrm{MET}(X)$. Then there is a polynomial time randomized algorithm which approximates $\mathrm{Opt}_\Gamma(X, d)$ on any metric $d \in \mathrm{MET}(X)$ to within a factor of $O\left(\alpha(n)(\log n)^2\right)$.*

Proof. Let (X, d) be an n-point metric space and let $\mathscr{D}$ be the distribution over random ultrametrics ρ on X from Theorem 1 (which is computable in polynomial time, as follows directly from our proof of Theorem 1). In other words, $\rho(x, y) \ge d(x, y)$ for all $x, y \in X$ and

$$\max_{x \in X} \mathbb{E}_{\mathscr{D}} \max_{y \in X \setminus \{x\}} \frac{\rho(x, y)}{d(x, y)} \le C(\log n)^2.$$

Let $P \in \mathcal{P}$ be a clustering solution for which $\mathrm{Opt}_\Gamma(X, d) = \sum_{x \in X} \Gamma(x, d, P)$. Using the monotonicity and homogeneity of Γ we see that

$$\mathbb{E}_{\mathscr{D}} \mathrm{Opt}_\Gamma(X, \rho) \le \mathbb{E}_{\mathscr{D}} \sum_{x \in X} \Gamma(x, \rho, P) \qquad (P \text{ is sub-optimal in } \rho)$$

$$\le \mathbb{E}_D \sum_{x \in X} \Gamma\left(x, \left[\max_{y \in X \setminus \{x\}} \frac{\rho(x, y)}{d(x, y)}\right] \cdot d, P\right) \qquad (\text{Monotonicity of } \Gamma)$$

$$= \mathbb{E}_{\mathscr{D}} \sum_{x \in X} \left[\max_{y \in X \setminus \{x\}} \frac{\rho(x, y)}{d(x, y)}\right] \cdot \Gamma(x, d, P) \qquad (\text{Homogeneity of } \Gamma)$$

$$= \sum_{x \in X} \left(\mathbb{E}_{\mathscr{D}} \left[\max_{y \in X \setminus \{x\}} \frac{\rho(x, y)}{d(x, y)}\right]\right) \cdot \Gamma(x, d, P)$$

$$\le C(\log n)^2 \cdot \mathrm{Opt}_\Gamma(X, d) \qquad (\text{Theorem 1}).$$

Hence, with probability at least $\frac{1}{2}$, $\mathrm{Opt}_\Gamma(X, \rho) \leq 2C(\log n)^2 \cdot \mathrm{Opt}_\Gamma(X, d)$. For such ρ compute a clustering solution $Q \in \mathcal{P}$ satisfying

$$\sum_{x \in X} \Gamma(x, \rho, Q) \leq \alpha(n)\mathrm{Opt}_\Gamma(X, \rho) \leq 2C\alpha(n)(\log n)^2 \cdot \mathrm{Opt}_\Gamma(X, d).$$

Since $\rho \geq d$ it remains to use the monotonicity of Γ once more to deduce that

$$\sum_{x \in X} \Gamma(x, \rho, Q) \geq \sum_{x \in X} \Gamma(x, d, Q) \geq \mathrm{Opt}_\Gamma(X, d).$$

Thus Q is a $O\left(\alpha(n)(\log n)^2\right)$ approximate solution to the clustering problem on X with cost Γ.   □

Due to Theorem 2 we see that the main difficulty in monotone clustering problems lies in the design of good approximation algorithms for them on ultrametrics. This is a generic reduction, and in many particular cases it might be possible use a case-specific analysis to improve the $O\left((\log n)^2\right)$ loss in the approximation factor. However, as a general reduction paradigm for clustering problems, Theorem 2 makes it clear why maximum gradient embeddings are so natural.

We next demonstrate the applicability of the monotone clustering framework to a concrete example called *fault-tolerant k-median*. In the full version of the paper, we analyze another clustering problem, called $\Sigma \ell_p$ clustering.

Fault-tolerant k-median. Fix $k \in \mathbb{N}$. The k-median problem is as follows. Given an n-point metric space (X, d_X), find $x_1, \ldots, x_k \in X$ that minimize the objective function

$$\sum_{x \in X} \min_{j \in \{x_1, \ldots, x_k\}} d_X(x, x_j). \tag{3}$$

This very natural and well studied problem can be easily cast as monotone clustering problem by defining $\Gamma(x, d, \{(x_1, C_1), \ldots, (x_m, C_m)\})$ to be ∞ if $m \neq k$, and otherwise $\Gamma(x, d, \{(x_1, C_1), \ldots, (x_m, C_m)\}) = \min_{j \in \{x_1, \ldots, x_k\}} d(x, x_j)$.

The linear structure of (3) makes it a prime example of a problem which can be approximated using Bartal's probabilistic embeddings. Indeed, the first non-trivial approximation algorithm for k-median clustering was obtained by Bartal in [4]. Since then this problem has been investigated extensively: The first constant factor approximation for it was obtained in [9] using LP rounding, and the first combinatorial (primal-dual) constant-factor algorithm was obtained in [15]. In [2] an analysis of a natural local search heuristic yields the best known approximation factor for k-median clustering.

Here we study the following fault-tolerant version of the k-median problem. Let (X, d) be an n-point metric space and fix $k \in \mathbb{N}$. Assume that for every $x \in X$ we are given an integer $j(x) \in X$ (which we call the fault-tolerant parameter of x). Given $x_1, \ldots, x_k$ and $x \in X$ let $x_j^*(x; d)$ be the j-th closest point to x in $\{x_1, \ldots, x_k\}$. In other words, $\{x_j^*(x; d)\}_{j=1}^k$ is a re-ordering of $\{x_j\}_{j=1}^k$ such that $d(x, x_1^*(x; d)) \leq \cdots \leq d(x, x_k^*(x; d))$. Our goal is to minimize the objective function

$$\sum_{x \in X} d\left(x, x_{j(x)}^*(x; d)\right). \tag{4}$$

To understand (4) assume for the sake of simplicity that $j(x) = j$ for all $x \in X$. If $\{x_j\}_{j=1}^k$ minimizes (4) and $j-1$ of them are corrupted (due to possible noise), then the optimum value of (4) does not change. In this sense the clustering problem in (4) is fault-tolerant. In other words, the optimum solution of (4) is insensitive to (controlled) noise. Observe that for $j = 1$ we return to the k-median clustering problem.

We remark that another fault-tolerant version of k-median clustering was introduced in [16]. In this problem we connect each point x in the metric space X to $j(x)$ centers, but the objective function is the sum over $x \in X$ of the sum of the distances from x to all the $j(x)$ centers. Once again, the linearity of the objective function seems to make the problem easier, and in [23] a constant factor approximation is achieved (this immediately implies that our version of fault-tolerant k-median clustering, i.e. the minimization of (4), has a $O\left(\max_{x \in X} j(x)\right)$ approximation algorithm). In particular, the LP that was previously used for k-median clustering naturally generalizes to this setting. This is not the case for our fault-tolerant version in (4). Moreover, the local search techniques for k-median clustering (see for example [2]) do not seem to be easily generalizable to the case $j > 1$, and in any case seem to require $n^{\Omega(j)}$ time, which is not polynomial even for moderate values of j.

Arguing as above in the case of k-median clustering we see that the fault-tolerant k-median clustering problem in (4) is a monotone clustering problem. In the full version of this paper we show that it can be solved exactly in polynomial time on ultrametrics. Thus, in combination with Theorem 2, we obtain a $O\left((\log n)^2\right)$ approximation algorithm for the minimization of (4) on general metrics.

2   Proof of Theorem 1

We begin by sketching the proof of the upper bound in Theorem 1. The full version of this paper has a complete self-contained proof of it.

By the arguments appearing in [13, 18], for every N-point metric spacemetric space (X, d_X) there exist a distribution $\mathscr{D}$ of non-contractive embeddings $f : X \to U$ such that for every $x \in X$, and $t \geq 1$,

$$\Pr_{\mathcal{D}}\left[\exists y \in X, \frac{d_U(f(x), f(y))}{d_X(x, y)} \geq t\right] \leq \frac{128 \log_2 n}{t}.$$

By using a known trick one can modify the construction of $\mathscr{D}$ to also satisfy for every $x \in X$,

$$\Pr_{\mathscr{D}}\left[\exists y \in X, \frac{d_U(f(x), f(y))}{d_X(x, y)} \geq 4n\right] = 0.$$

Hence for every $x \in X$,

$$
\begin{aligned}
\mathbb{E}_{\mathscr{D}} |\nabla f|_\infty &= \mathbb{E}_{\mathscr{D}} \sup_{y \neq x} \frac{d_U(f(x), f(y))}{d_X(x, y)} \\
&\leq \sum_{i=0}^{\infty} 2^{i+1} \cdot \Pr\left[\exists y \in X, \frac{d_U(f(x), f(y))}{d_X(x, y)} \geq 2^i\right] \\
&\leq \sum_{i=0}^{2+\log_2 n} 2^i \cdot \frac{128 \log_2 n}{2^i} = O\left((\log n)^2\right).
\end{aligned}
$$

We next prove a matching lower bound. As mentioned in the introduction, the metrics Y_n in Theorem 1 are the diamond graphs of Newman and Rabinovich [14], which will be defined presently. Before passing to this more complicated lower bound, we will analyze the simpler example of cycles.

Let C_n, $n > 3$, be the unweighted path on n-vertices. We will identify C_n with the group $\mathbb{Z}_n$ of integers modulo n. We first observe that in this special case the upper bound in Theorem 1 can be improved to $O(\log n)$. This is achieved by using Karp's embedding of the cycle into spanning paths— we simply choose an edge of C_n uniformly at random and delete it. Let $f : C_n \to \mathbb{Z}$ be the randomized embedding thus obtained, which is clearly non-contractive.

Karp noted that it is easy to see that as a probabilistic embedding into trees f has distortion at most 2. We will now show that as a maximum gradient embedding, f has distortion $\Theta(\log n)$. Indeed, fix $x \in C_n$, and denote the deleted edge by $\{a, a+1\}$. Assume that $d_{C_n}(x, a) = t \leq n/2 - 1$. Then the distance from $a + 1$ to x changed from $t + 1$ in C_n to $n - t - 1$ in the path. It is also easy to see that this is where the maximum gradient is attained. Thus

$$
\mathbb{E}|\nabla f(x)|_\infty \approx \frac{2}{n} \sum_{0 \leq t \leq n/2} \frac{n - t - 1}{t + 1} = \Theta(\log n).
$$

We will now show that any maximum gradient embedding of C_n into a distribution over trees incurs distortion $\Omega(\log n)$. For this purpose we will use the following lemma from [22].

Lemma 1. *For any tree metric T, and any non-contractive embedding $g : C_n \to T$, there exists an edge $(x, x+1)$ of C_n such that $d_T(g(x), g(x+1)) \geq \frac{n}{3} - 1$.*

Now, let $\mathscr{D}$ be a distribution over non-contractive embeddings of C_n into trees $f : C_n \to T$. By Lemma 1 we know that there exists $x \in C_n$ such that $d_T(f(x), f(x+1)) \geq \frac{n-3}{3}$. Thus for every $y \in C_n$ we have that

$$
\max\{d_T(f(y), f(x)), d_T(f(y), f(x+1))\} \geq \frac{n-3}{6}.
$$

On the other hand $\max\{d_{C_n}(y, x), d_{C_n}(y, x+1)\} \leq d_{C_n}(x, y) + 1$. It follows that,

$$
|\nabla f(y)|_\infty \geq \frac{n-3}{6 d_{C_n}(x, y) + 6}.
$$

Summing this inequality over $y \in C_n$ we see that

$$\sum_{y \in C_n} |\nabla f(y)|_\infty \geq \sum_{0 \leq k \leq n/2} \frac{n-3}{6k+6} = \Omega(n \log n).$$

Thus

$$\max_{y \in C_n} \mathbb{E}_{\mathscr{D}} |\nabla f(y)|_\infty \geq \frac{1}{n} \sum_{y \in C_n} \mathbb{E}_{\mathscr{D}} |\nabla f(y)|_\infty = \Omega(\log n),$$

as required.

We now pass to the proof of the lower bound in Theorem 1. We start by describing the diamond graphs $\{G_k\}_{k=1}^\infty$, and a special labelling of them that we will use throughout the ensuing arguments. The first diamond graph G_1 is a cycle of length 4, and G_{k+1} is obtained from G_k by replacing each edge by a quadrilateral. Thus G_k has 4^k edges and $\frac{2 \cdot 4^k + 4}{3}$ vertices. As we have done before, the required lower bound on maximum gradient embeddings of G_k into trees will be proved if we show that for every tree T and every non-contractive embedding $f : G_k \to T$ we have

$$\frac{1}{4^k} \sum_{e \in E(G_k)} \sum_{x \in e} |\nabla f(x)|_\infty = \Omega\left(k^2\right). \tag{5}$$

Note that the inequality (5) is different from the inequality that we proved in the case of the cycle in that the weighting on the vertices of G_k that it induces is not uniform— high degree vertices get more weight in the average in the left-hand side of (5).

We will prove (5) by induction on k. In order to facilitate such an induction, we will first strengthen the inductive hypothesis. To this end we need to introduce a useful labelling of G_k. For $1 \leq i \leq k$ the graph G_k contains 4^{k-i} canonical copies of G_i, which we index by elements of $\{1,2,3,4\}^{k-i}$, and denote $\left\{G_{[\alpha]}^{(k)}\right\}_{\alpha \in \{1,2,3,4\}^{k-i}}$. These graphs are defined as follows. For $k = 2$ they are shown in Figure 1.

For $k = 3$ these canonical subgraphs are shown in Figure 2.

Formally, we set $G_{[\emptyset]}^{(k)} = G_k$, and assume inductively that the canonical subgraphs of G_{k-1} have been defined. Let H_1, H_2, H_3, H_4 be the top-right, top-left, bottom-right and bottom-left copies of G_{k-1} in G_k, respectively. For $\alpha \in \{1,2,3,4\}^{k-1-i}$ and $j \in \{1,2,3,4\}$ we denote the copy of G_i in H_j corresponding to $G_{[\alpha]}^{(k-1)}$ by $G_{[j\alpha]}^{(k)}$.

For every $1 \leq i \leq k$ and $\alpha \in \{1,2,3,4\}^{k-i}$ let $T_{[\alpha]}^{(k)}, B_{[\alpha]}^{(k)}, L_{[\alpha]}^{(k)}, R_{[\alpha]}^{(k)}$ be the topmost, bottom-most, left-most, and right-most vertices of $G_{[\alpha]}^{(k)}$, respectively. Fixing an embedding $f : G_k \to T$, we will construct inductively a set of simple cycles $\mathscr{C}_{[\alpha]}$ in $G_{[\alpha]}^{(k)}$ and for each $C \in \mathscr{C}_{[\alpha]}$ an edge $\varepsilon_C \in E\left(\mathscr{C}_{[\alpha]}\right)$, with the following properties.

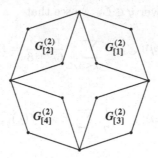

Fig. 1. The graph G_2 and the labelling of the canonical copies of G_1 contained in it

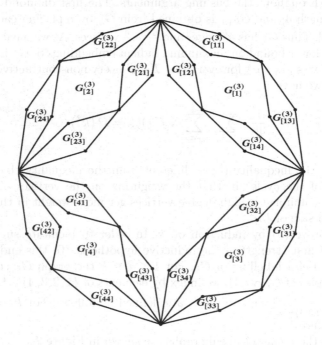

Fig. 2. The graph G_3 and the induced labeling of canonical copies of G_1 and G_2

1. The cycles in $\mathscr{C}_{[\alpha]}$ are edge-disjoint, and they all pass through the vertices $T_{[\alpha]}^{(k)}, B_{[\alpha]}^{(k)}, L_{[\alpha]}^{(k)}, R_{[\alpha]}^{(k)}$. There are 2^{i-1} cycles in $\mathscr{C}_{[\alpha]}$, and each of them contains 2^{i+1} edges. Thus in particular the cycles in $\mathscr{C}_{[\alpha]}$ form a disjoint cover of the edges in $G_{[\alpha]}^{(k)}$.

2. If $C \in \mathscr{C}_{[\alpha]}$ and $\varepsilon_C = \{x, y\}$ then $d_T(f(x), f(y)) \geq \frac{2^{i+1}}{3} - 1$.

3. Denote $E_{[\alpha]} = \{\varepsilon_C : C \in \mathscr{C}_{[\alpha]}\}$ and $\Delta_i = \bigcup_{\alpha \in \{1,2,3,4\}^{k-i}} E_{[\alpha]}$. The edges in Δ_i will be called the *designated edges* of level i. For $\alpha \in \{1,2,3,4\}^{k-i}$, $C \in \mathscr{C}_{[\alpha]}$ and $j < i$ let $\Delta_j(C) = \Delta_j \cap E(C)$ be the designated edges of level j on C. Then we require that each of the two paths $T^{(k)}_{[\alpha]} - L^{(k)}_{[\alpha]} - B^{(k)}_{[\alpha]}$ and $T^{(k)}_{[\alpha]} - R^{(k)}_{[\alpha]} - B^{(k)}_{[\alpha]}$ in C contain exactly 2^{i-j-1} edges from $\Delta_j(C)$.

The construction is done by induction on i. For $i = 1$ and $\alpha \in \{1,2,3,4\}^{k-1}$ we let $\mathscr{C}_{[\alpha]}$ contain only the 4-cycle $G^{(k)}_{[\alpha]}$ itself. Moreover by Lemma 1 there is and edge $\varepsilon_{G^{(k)}_{[\alpha]}} \in E\left(G^{(k)}_{[\alpha]}\right)$ such that if $\varepsilon_{G^{(k)}_{[\alpha]}} = \{x,y\}$ then $d_T(f(x), f(y)) \geq \frac{1}{3}$. This completes the construction for $i = 1$. Assuming we have completed the construction for $i - 1$ we construct the cycles at level i as follows. Fix arbitrary cycles $C_1 \in \mathscr{C}_{[1\alpha]}$, $C_2 \in \mathscr{C}_{[2\alpha]}$, $C_3 \in \mathscr{C}_{[3\alpha]}$, $C_4 \in \mathscr{C}_{[4\alpha]}$. We will use these four cycles to construct two cycles in $\mathscr{C}_{[\alpha]}$. The first one consists of the $T^{(k)}_{[\alpha]} - R^{(k)}_{[\alpha]}$ path in C_1 which contains the edge ε_{C_1}, the $R^{(k)}_{[\alpha]} - B^{(k)}_{[\alpha]}$ path in C_3 which does not contain the edge ε_{C_3}, the $B^{(k)}_{[\alpha]} - L^{(k)}_{[\alpha]}$ path in C_4 which contains the edge ε_{C_4}, and the $L^{(k)}_{[\alpha]} - T^{(k)}_{[\alpha]}$ path in C_2 which does not contain the edge c_{C_2}. The remaining edges in $E(C_1) \cup E(C_2) \cup E(C_3) \cup E(C_4)$ constitute the second cycle that we extract from C_1, C_2, C_3, C_4. Continuing in this manner by choosing cycles from $\mathscr{C}_{[1\alpha]} \setminus \{C_1\}$, $\mathscr{C}_{[2\alpha]} \setminus \{C_2\}$, $\mathscr{C}_{[3\alpha]} \setminus \{C_3\}$, $\mathscr{C}_{[4\alpha]} \setminus \{C_4\}$ and repeating this procedure, and then continuing until we exhaust the cycles in $\mathscr{C}_{[1\alpha]} \cup \mathscr{C}_{[2\alpha]} \cup \mathscr{C}_{[3\alpha]} \cup \mathscr{C}_{[4\alpha]}$, we obtain the set of cycles $\mathscr{C}_{[\alpha]}$. For every $C \in \mathscr{C}_{[alpha]}$ we then apply Lemma 1 to obtain an edge ε_C with the required property.

For each edge $e \in E(G_k)$ let $\alpha \in \{1,2,3,4\}^{k-i}$ be the unique multi-index such that $e \in E\left(G^{(k)}_{[\alpha]}\right)$. We denote by $C_i(e)$ the unique cycle in $\mathscr{C}_{[\alpha]}$ containing e. We will also denote $\widehat{e}_i(e) = \varepsilon_{C_i(e)}$. Finally we let $a_i(e) \in e$ and $b_i(e) \in \widehat{e}_i(e)$ be vertices such that

$$d_T(f(a_i(e)), f(b_i(e))) = \max_{\substack{a \in e \\ b \in \widehat{e}_i(e)}} d_T(f(a), f(b)).$$

Note that by the definition of $\widehat{e}_i(e)$ and the triangle inequality we are assured that

$$d_T(f(a_i(e)), f(b_i(e))) \geq \frac{1}{2}\left(\frac{2^{i+1}}{3} - 1\right) \geq \frac{2^i}{12}. \tag{6}$$

Recall that we plan to prove (5) by induction on k. Having done all of the above preparation, we are now in position to strengthen (5) so as to make the inductive argument easier. Given two edges $e, h \in G_k$ we write $e \frown_i h$ if both e, h are on the same canonical copy of G_i in G_k, $C_i(e) = C_i(h) = C$, and furthermore e and h on the same side of C. In other words, $e \frown_i h$ if there is $\alpha \in \{1,2,3,4\}^{k-i}$ and $C \in \mathscr{C}_{[\alpha]}$ such that if we partition the edges of C into two disjoint $T^{(k)}_{[\alpha]} - B^{(k)}_{[\alpha]}$ paths, then e and h are on the same path.

Let $m \in \mathbb{N}$ be a universal constant that will be specified later. For every integer $\ell \leq k/m$ and any $\alpha \in \{1,2,3,4\}^{k-m\ell}$ define

$$L_\ell(\alpha) = \frac{1}{4^{m\ell}} \sum_{e \in E\left(G_{[\alpha]}^{(k)}\right)} \max_{\substack{i \in \{1,\dots,\ell\} \\ e \frown_{im} \widehat{e}_{im}(e)}} \frac{d_T(f(a_{im}(e)), f(b_{im}(e))) \wedge 2^{im}}{d_{G_k}(e, \widehat{e}_{im}(e)) + 1}.$$

We also write $L_\ell = \min_{\alpha \in \{1,2,3,4\}^{k-m\ell}} L_\ell(\alpha)$. We will prove that $L_\ell \geq L_{\ell-1} + c\ell$, where $c > 0$ is a universal constant. This will imply that for $\ell = \lfloor k/m \rfloor$ we have $L_\ell = \Omega(k^2)$ (since m is a universal constant). By simple arithmetic (5) follows. Observe that for every $\alpha \in \{1,2,3,4\}^{k-m\ell}$ we have

$$L_\ell(\alpha) = 4^{-m} \sum_{\beta \in \{1,2,3,4\}^m} 4^{-m(\ell-1)} \sum_{e \in E\left(G_{[\beta\alpha]}^{(k)}\right)} \max_{\substack{i \in \{1,\dots,\ell\} \\ e \frown_{im} \widehat{e}_{im}(e)}} \frac{d_T(f(a_{im}(e)), f(b_{im}(e))) \wedge 2^{im}}{d_{G_k}(e, \widehat{e}_{im}(e)) + 1}$$

$$= 4^{-m} \sum_{\beta \in \{1,2,3,4\}^m} 4^{-m(\ell-1)} \sum_{e \in E\left(G_{[\beta\alpha]}^{(k)}\right)} \max_{\substack{i \in \{1,\dots,\ell-1\} \\ e \frown_{im} \widehat{e}_{im}(e)}} \frac{d_T(f(a_{im}(e)), f(b_{im}(e))) \wedge 2^{im}}{d_{G_k}(e, \widehat{e}_{im}(e)) + 1}$$

$$+ \frac{1}{4^{m\ell}} \sum_{e \in E\left(G_{[\alpha]}^{(k)}\right)} \max \left\{ 0, \frac{d_T(f(a_{\ell m}(e)), f(b_{\ell m}(e))) \wedge 2^{\ell m}}{d_{G_k}(e, \widehat{e}_{\ell m}(e)) + 1} \cdot \mathbf{1}_{\{e \frown_{\ell m} \widehat{e}_{\ell m}(e)\}} \right.$$

$$\left. - \max_{\substack{i \in \{1,\dots,\ell-1\} \\ e \frown_{im} \widehat{e}_{im}(e)}} \frac{d_T(f(a_{im}(e)), f(b_{im}(e))) \wedge 2^{im}}{d_{G_k}(e, \widehat{e}_{im}(e)) + 1} \right\}$$

$$= \frac{1}{4^m} \sum_{\beta \in \{1,2,3,4\}^m} L_{\ell-1}(\beta\alpha)$$

$$+ \frac{1}{4^{m\ell}} \sum_{e \in E\left(G_{[\alpha]}^{(k)}\right)} \max \left\{ 0, \frac{d_T(f(a_{\ell m}(e)), f(b_{\ell m}(e))) \wedge 2^{\ell m}}{d_{G_k}(e, \widehat{e}_{\ell m}(e)) + 1} \cdot \mathbf{1}_{\{e \frown_{\ell m} \widehat{e}_{\ell m}(e)\}} \right.$$

$$\left. - \max_{\substack{i \in \{1,\dots,\ell-1\} \\ e \frown_{im} \widehat{e}_{im}(e)}} \frac{d_T(f(a_{im}(e)), f(b_{im}(e))) \wedge 2^{im}}{d_{G_k}(e, \widehat{e}_{im}(e)) + 1} \right\}$$

$$\geq L_{\ell-1}$$

$$+ \frac{1}{4^{m\ell}} \sum_{e \in E\left(G_{[\alpha]}^{(k)}\right)} \max \left\{ 0, \frac{d_T(f(a_{\ell m}(e)), f(b_{\ell m}(e))) \wedge 2^{\ell m}}{d_{G_k}(e, \widehat{e}_{\ell m}(e)) + 1} \cdot \mathbf{1}_{\{e \frown_{\ell m} \widehat{e}_{\ell m}(e)\}} \right.$$

$$\left. - \max_{\substack{i \in \{1,\dots,\ell-1\} \\ e \frown_{im} \widehat{e}_{im}(e)}} \frac{d_T(f(a_{im}(e)), f(b_{im}(e))) \wedge 2^{im}}{d_{G_k}(e, \widehat{e}_{im}(e)) + 1} \right\}.$$

Thus it is enough to show that

$$A \overset{\text{def}}{=} 4^{-m\ell} \sum_{e \in E\left(G_{[\alpha]}^{(k)}\right)} \max\left\{0, \frac{d_T(f(a_{\ell m}(e)), f(b_{\ell m}(e))) \wedge 2^{\ell m}}{d_{G_k}(e, \widehat{e}_{\ell m}(e)) + 1} \cdot \mathbf{1}_{\{e \frown_{\ell m} \widehat{e}_{\ell m}(e)\}}\right.$$

$$\left. - \max_{\substack{i \in \{1,\ldots,\ell-1\} \\ e \frown_{im} \widehat{e}_{im}(e)}} \frac{d_T(f(a_{im}(e)), f(b_{im}(e))) \wedge 2^{im}}{d_{G_k}(e, \widehat{e}_{im}(e)) + 1}\right\} = \Omega(\ell). \quad (7)$$

To prove (7), denote for $C \in \mathscr{C}_{[\alpha]}$

$$S_C = \left\{e \in E(C): \ \varepsilon_C \frown_{\ell m} e \ \text{ and }\right.$$

$$\max_{\substack{i \in \{1,\ldots,\ell-1\} \\ e \frown_{im} \widehat{e}_{im}(e)}} \frac{d_T(f(a_{im}(e)), f(b_{im}(e))) \wedge 2^{im}}{d_{G_k}(e, \widehat{e}_{im}(e)) + 1}$$

$$\left. \geq \frac{1}{2} \cdot \frac{d_T(f(a_{\ell m}(e)), f(b_{\ell m}(e))) \wedge 2^{\ell m}}{d_{G_k}(e, \widehat{e}_{\ell m}(e)) + 1}\right\}.$$

Then using (6) we see that

$$A \geq \frac{1}{2 \cdot 4^{m\ell}} \sum_{C \in \mathscr{C}_{[\alpha]}} \sum_{\substack{e \in E(C)\backslash S_C \\ \varepsilon_C \frown_{\ell m} e}} \frac{d_T(f(a_{\ell m}(e)), f(b_{\ell m}(e))) \wedge 2^{\ell m}}{d_{G_k}(e, \widehat{e}_{\ell m}(e)) + 1}$$

$$\geq \frac{1}{2 \cdot 4^{m\ell}} \sum_{C \in \mathscr{C}_{[\alpha]}} \sum_{\substack{e \in E(C) \\ \varepsilon_C \frown_{\ell m} e}} \frac{d_T(f(a_{\ell m}(e)), f(b_{\ell m}(e))) \wedge 2^{\ell m}}{d_{G_k}(e, \widehat{e}_{\ell m}(e)) + 1}$$

$$\frac{1}{2 \cdot 4^{m\ell}} \sum_{C \in \mathscr{C}_{[\alpha]}} \sum_{e \in S_C} \frac{d_T(f(a_{\ell m}(e)), f(b_{\ell m}(e))) \wedge 2^{\ell m}}{d_{G_k}(e, \widehat{e}_{\ell m}(e)) + 1}$$

$$\geq \frac{1}{2 \cdot 4^{m\ell}} \sum_{C \in \mathscr{C}_{[\alpha]}} \sum_{i=1}^{2^{m\ell-1}} \frac{2^{m\ell}}{12i} - \frac{1}{2 \cdot 4^{m\ell}} \sum_{C \in \mathscr{C}_{[\alpha]}} \sum_{e \in S_C} \frac{d_T(f(a_{\ell m}(e)), f(b_{\ell m}(e))) \wedge 2^{\ell m}}{d_{G_k}(e, \widehat{e}_{\ell m}(e)) + 1}$$

$$= \Omega\left(\frac{1}{4^{m\ell}} \cdot |\mathscr{C}_{[\alpha]}| \cdot 2^{m\ell} \cdot m\ell\right)$$

$$- \frac{1}{2 \cdot 4^{m\ell}} \sum_{C \in \mathscr{C}_{[\alpha]}} \sum_{e \in S_C} \frac{d_T(f(a_{\ell m}(e)), f(b_{\ell m}(e))) \wedge 2^{\ell m}}{d_{G_k}(e, \widehat{e}_{\ell m}(e)) + 1}$$

$$= \Omega(m\ell) - \frac{1}{2 \cdot 4^{m\ell}} \sum_{C \in \mathscr{C}_{[\alpha]}} \sum_{e \in S_C} \frac{d_T(f(a_{\ell m}(e)), f(b_{\ell m}(e))) \wedge 2^{\ell m}}{d_{G_k}(e, \widehat{e}_{\ell m}(e)) + 1}. \quad (8)$$

To estimate the negative term in (8) fix $C \in \mathscr{C}_{[\alpha]}$. For every edge $e \in S_C$ (which implies in particular that $\widehat{e}_{\ell m}(e) = \varepsilon_C$) we fix an integer $i < \ell$ such that $e \frown_{im} \widehat{e}_{im}(e)$ and

$$\frac{2^{im}}{d_{G_k}(e, \widehat{e}_{im}(e)) + 1} \geq \frac{d_T(f(a_{im}(e)), f(b_{im}(e))) \wedge 2^{im}}{d_{G_k}(e, \widehat{e}_{im}(e)) + 1}$$

$$\geq \frac{1}{2} \cdot \frac{d_T(f(a_{\ell m}(e)), f(b_{\ell m}(e))) \wedge 2^{\ell m}}{d_{G_k}(e, \widehat{e}_{\ell m}(e)) + 1}$$

$$\geq \frac{1}{12} \cdot \frac{2^{\ell m}}{d_{G_k}(e, \varepsilon_C) + 1},$$

or

$$d_{G_k}(e, \widehat{e}_{im}(e)) + 1 \leq 2^{(i-\ell)m+4} [d_{G_k}(e, \varepsilon_C) + 1]. \tag{9}$$

We shall call the edge $\widehat{e}_{im}(e)$ the designated edge that inserted e into S_C. For a designated edge $\varepsilon \in E(C)$ of level im (i.e. $\varepsilon \in \Delta_{im}(C)$) we shall denote by $\mathscr{E}_C(\varepsilon)$ the set of edges of C which ε inserted to S_C. Denoting $D_\varepsilon = d_{G_k}(\varepsilon, \varepsilon_C) + 1$ we see that (9) implies that for $e \in \mathscr{E}_C(\varepsilon)$ we have

$$\left| D_\varepsilon - [d_{G_k}(e, \varepsilon_C) + 1] \right| \leq 2^{(i-\ell)m+4} [d_{G_k}(e, \varepsilon_C) + 1]. \tag{10}$$

Assuming that $m \geq 5$ we are assured that $2^{(i-\ell)m+4} \leq \frac{1}{2}$. Thus (10) implies that

$$\frac{D_\varepsilon}{1 + 2^{(i-\ell)m+4}} \leq d_{G_k}(e, \varepsilon_C) + 1 \leq \frac{D_\varepsilon}{1 - 2^{(i-\ell)m+4}}.$$

Hence

$$\sum_{e \in S_C} \frac{d_T(f(a_{\ell m}(e)), f(b_{\ell m}(e))) \wedge 2^{\ell m}}{d_{G_k}(e, \widehat{e}_{\ell m}(e)) + 1}$$

$$\leq \sum_{i=1}^{\ell-1} \sum_{\varepsilon \in \Delta_{im}(C)} \sum_{e \in \mathscr{E}_C(\varepsilon)} \frac{2^{\ell m}}{d_{G_k}(e, \varepsilon_C) + 1}$$

$$\leq 2 \sum_{i=1}^{\ell-1} \sum_{\varepsilon \in \Delta_{im}(C)} \sum_{\substack{j \in \mathbb{N} \\ \frac{D_\varepsilon}{1+2^{(i-\ell)m+4}} \leq j \leq \frac{D_\varepsilon}{1-2^{(i-\ell)m+4}}}} \frac{2^{\ell m}}{j}$$

$$= O(1) \cdot 2^{\ell m} \sum_{i=1}^{\ell-1} |\Delta_{im}(C)| \cdot \log \left(\frac{1 + 2^{(i-\ell)m+4}}{1 - 2^{(i-\ell)m+4}} \right)$$

$$= O(1) \cdot 2^{\ell m} \ell \cdot 2^{(\ell-i)m} \cdot 2^{(i-\ell)m} = O(1) \cdot 2^{\ell m} \ell.$$

Thus, using (8) we see that

$$A = \Omega(m\ell) - O(1) \cdot \frac{1}{4^{\ell m}} \cdot |\mathscr{C}_{[\alpha]}| 2^{m\ell} \ell = \Omega(m\ell) - O(1)\ell = \Omega(\ell),$$

provided that m is a large enough absolute constant. This completes the proof of the lower bound in Theorem 1. $\qquad \square$

References

[1] Alon, N., Karp, R., Peleg, D., West, D.: A graph-theoretic game and its application to the k-server problem. SIAM J. Comput. 24(1) (1995)

[2] Arya, V., Garg, N., Khandekar, R., Meyerson, A., Munagala, K., Pandit, V.: Local search heuristics for k-median and facility location problems. SIAM J. Comput. 33(3), 544–562 (2004)

[3] Bartal, Y.: Probabilistic approximation of metric spaces and its algorithmic applications. In: 37th Annual Symposium on Foundations of Computer Science, pp. 184–193. IEEE Comput. Soc. Press, Los Alamitos (1996)

[4] Bartal, Y.: On approximating arbitrary metrics by tree metrics. In: Proceedings of the 30th Annual ACM Symposium on Theory of Computing, pp. 161–168. ACM, New York (1998)

[5] Bartal, Y., Mendel, M.: Multi-embedding of metric spaces. SIAM J. Comput. 34(1), 248–259 (2004)

[6] Brinkman, B., Charikar, M.: On the impossibility of dimension reduction in l1. J. ACM 52(5), 766–788 (2005)

[7] Buyalo, S., Schroeder, V.: Embedding of hyperbolic spaces in the product of trees. Geom. Dedicata 113, 75–93 (2005)

[8] Calinescu, G., Karloff, H., Rabani, Y.: Approximation algorithms for the 0- extension problem. SIAM J. Comput. 34(2), 358–372 (2004/2005)

[9] Charikar, M., Guha, S., Tardos, É., Shmoys, D.B.: A constant-factor approximation algorithm for the k-median problem. J. Comput. System Sci. 65(1), 129–149 (2002)

[10] Dranishnikov, A.N.: On hypersphericity of manifolds with nite asymptotic dimenion. Trans. Amer. Math. Soc. 355(1), 155–167 (2003)

[11] Elkin, M., Emek, Y., Spielman, D.A., Teng, S.-H.: Lower-stretch spanning trees. In: Proceedings of the 37th Annual ACM Symposium on Theory of Computing, pp. 494–503. ACM, New York (2005)

[12] Fakcharoenphol, J., Rao, S., Talwar, K.: A tight bound on approximating arbitrary metrics by tree metrics. In: Proceedings of the 35th annual ACM Symposium on Theory of Computing, pp. 448–455 (2003)

[13] Gupta, A., Hajiaghayi, M., Räcke, H.: Oblivious Network Design. In: Proceedings of the Seventeenth Annual ACM-SIAM Symposium on Discrete Algorithms (Miami, FL, 2006). pp. 970–979. ACM, New York (2006)

[14] Gupta, A., Newman, I., Rabinovich, Y., Sinclair, A.: Cuts, trees and l1- embeddings of graphs. Combinatorica 24(2), 233–269 (2004)

[15] Jain, K., Vazirani, V.: Approximation algorithms for metric facility location and k-median problems using the primal-dual schema and Lagrangian relaxation. J. ACM 48(2), 274–296 (2001)

[16] Jain, K., Vazirani, V.V.: An approximation algorithm for the fault tolerant metric facility location problem. Algorithmica (Approximation algorithms) 38(3), 433–439 (2004)

[17] Lee, J.R., Naor, A.: Embedding the diamond graph in Lp and dimension reduction in L1, Geom. Funct. Anal. 14(4), 745–747 (2004)

[18] Mendel, M., Naor, A.: Ramsey partitions and proximity data structures. J. European Math. Soc 9(2), 253–275 (2007), http://arxiv.org/cs/0511084

[19] Mendel, M., Naor, A.: Maximum gradient embedding and monotone clustering, available at: http://arxiv.org/cs.DS/0606109

[20] Naor, A., Peres, Y., Schramm, O., Sheffield, S.: Markov chains in smooth Banach spaces and Gromov hyperbolic metric spaces, available at: http://arxiv.org/math/0410422
[21] Newman, I., Rabinovich, Y.: A lower bound on the distortion of embedding planar metrics into Euclidean space. Discrete Comput. Geom. 29(1), 77–81 (2003)
[22] Rabinovich, Y., Raz, R.: Lower bounds on the distortion of embedding nite metric spaces in graphs. Discrete Comput. Geom. 19(1), 79–94 (1998)
[23] Swamy, C., Shmoys, D.B.: Fault-tolerant facility location. In: Proceedings of the Fourteenth Annual ACM-SIAM Symposium on Discrete Algorithms, pp. 735–736. ACM, New York (2003)

Poly-logarithmic Approximation Algorithms for Directed Vehicle Routing Problems

Viswanath Nagarajan and R. Ravi*

Tepper School of Business, Carnegie Mellon University, Pittsburgh PA 15213
{viswa,ravi}@cmu.edu

Abstract. This paper studies vehicle routing problems on asymmetric metrics. Our starting point is the *directed k-TSP* problem: given an asymmetric metric (V, d), a root $r \in V$ and a target $k \leq |V|$, compute the minimum length tour that contains r and at least k other vertices. We present a polynomial time $O(\log^2 n \cdot \log k)$-approximation algorithm for this problem. We use this algorithm for directed k-TSP to obtain an $O(\log^2 n)$-approximation algorithm for the *directed orienteering* problem. This answers positively, the question of poly-logarithmic approximability of directed orienteering, an open problem from Blum et al. [2]. The previously best known results were quasi-polynomial time algorithms with approximation guarantees of $O(\log^2 k)$ for directed k-TSP, and $O(\log n)$ for directed orienteering (Chekuri & Pal [4]). Using the algorithm for directed orienteering within the framework of Blum et al. [2] and Bansal et al. [1], we also obtain poly-logarithmic approximation algorithms for the directed versions of *discounted-reward TSP* and the *vehicle routing problem with time-windows*.

1 Introduction

Vehicle routing problems (VRPs) form a large set of variants of the basic Traveling Salesman Problem, that are also encountered in practice. Some of the problems in this class are the capacitated VRP [10], the distance constrained VRP [15], the Dial-a-Ride problem [19], and the orienteering problem [9]. Many different objectives are encountered in VRPs: for example, minimizing cost of a tour (capacitated VRP & the Dial-a-Ride problem), minimizing number of vehicles (distance constrained VRP), and maximizing profit (the orienteering problem).

The Operations Research literature contains several papers dealing with exact or heuristic approaches for VRPs (ex. [18,20,13,14,5]). The techniques used in these papers include dynamic programming, local search, simulated annealing, genetic algorithms, branch and bound, and cutting plane algorithms. There has also been some interesting work in approximation algorithms for VRPs. The undirected orienteering problem involves finding a bounded length path starting

---

* Authors supported in part by NSF grants CCF-0430751 and ITR grant CCR-0122581 (The ALADDIN project).

M. Charikar et al. (Eds.): APPROX and RANDOM 2007, LNCS 4627, pp. 257–270, 2007.

at a fixed vertex that covers the maximum number of vertices; Blum et al. [2] obtained the first constant factor approximation algorithm for this problem, which was improved to a factor of 3 in Bansal et al. [1]. Blum et al. also obtained a constant factor approximation algorithm for *discounted-reward TSP*, a problem motivated by robot navigation. Bansal et al. [1] used orienteering as a subroutine to also obtain poly-logarithmic approximation algorithms for some generalizations of orienteering, namely *deadline TSP* and *vehicle routing problem with time windows*.

Most of the work on VRPs focuses on symmetric metric spaces. In asymmetric metrics, the best known approximation guarantee for even the basic traveling salesman problem is $O(\log n)$ [7]; improving this bound is an important open question. Obtaining good approximation algorithms for the directed orienteering problem was stated as an open problem in Blum et al. [2]. Recently, Chekuri & Pal [4] obtained a general approximation algorithm for a class of VRPs on asymmetric metrics, that runs in *quasi-polynomial time*. In particular, their result implies an $O(\log n)$-approximation algorithm for the orienteering problem on directed graphs (in quasi-polynomial time). We are not aware of any non-trivial polynomial-time approximation algorithms for this problem. In this paper, we study polynomial time approximation algorithms for some vehicle routing problems on asymmetric metrics.

1.1   Problem Definition

All the problems that we consider are defined over an asymmetric metric space (V, d) on $|V| = n$ vertices. In the *directed k-TSP* problem, we are given a root $r \in V$ & a target $k \le n$, and the goal is to compute a minimum length tour that contains r and at least k other vertices. Any tour containing the root r is called an r-tour. Directed k-TSP is a generalization of the asymmetric traveling salesman problem (ATSP). A related problem is the *minimum ratio ATSP* problem, which involves finding a tour containing the root r that minimizes the ratio of the length of the tour to the number of vertices in it. If the requirement that the tour contain the root is dropped, the ratio problem becomes the minimum mean weight cycle problem, which is solvable in polynomial time. However, the rooted version which we are interested in is NP-complete.

In the *orienteering* problem, we are given a metric space, a specified origin s and a length bound D, and the goal is to find a path of length at most D, that starts at s and visits the maximum number of vertices. We actually consider a more general version of this problem, which is the directed version of point-to-point orienteering [1]. In the *directed orienteering* problem, we are given specified origin s & destination t vertices, and a length bound D, and the goal is to compute a path from s to t of length at most D, that visits the maximum number of vertices. The orienteering problem can also be extended to the setting where there is some profit at each vertex, and the goal is to maximize total profit.

Many problems we deal with in this paper have the following form, where $\mathcal{S}$ is a feasible set, $C : \mathcal{S} \to \mathbb{R}^+$ is a cost function, $N : \mathcal{S} \to \mathbb{N}$ is a coverage function, and k is the target:

$$\min\{C(x) : x \in \mathcal{S}, N(x) \geq k\}$$

For example, in the k-TSP problem, $\mathcal{S}$ is the set of all tours containing r, and for any $x \in \mathcal{S}$, $C(x)$ is the length of tour x and $N(x)$ is the number of vertices (other than r) covered in the tour. For any problem of the above form, a polynomial time algorithm $\mathcal{A}$ is said to be an (α, β) bi-criteria approximation if on each problem instance, $\mathcal{A}$ obtains a solution $y \in \mathcal{S}$ satisfying $C(y) \leq \alpha \cdot OPT$ and $N(y) \geq \frac{k}{\beta}$, where $OPT = \min\{C(x) : x \in \mathcal{S}, N(x) \geq k\}$ is the optimal value of this instance.

1.2   Results and Paper Outline

We present a polynomial time $O(\log^2 n \cdot \log k)$-approximation algorithm for the directed k-TSP problem. This is based on an $O(\log^2 n)$-approximation algorithm for the minimum ratio ATSP problem. To the best of our knowledge, this problem has not been studied earlier. An important ingredient in this algorithm is a splitting-off theorem on directed Eulerian graphs due to Frank [6] and Jackson [12]. This algorithm is described in Section 2. Our proof also implies a $\lceil \log n \rceil$ upper bound on the integrality gap of a natural LP relaxation for ATSP.

We then use the approximation algorithm for minimum ratio ATSP, to obtain a bi-criteria approximation algorithm for the *directed k-path* problem (Section 3). We also observe that the reductions in Blum et al. [2] & Bansal et al. [1] (in undirected metrics) from the k-path problem to the orienteering problem, can be easily adapted to the directed case. Together with the approximation algorithm for the directed k-path problem, we obtain an $O(\log^2 n)$ approximation guarantee for directed orienteering. This answers in the affirmative, the question of poly-logarithmic approximability of directed orienteering [2].

Finally, we note that the techniques used for discounted-reward TSP [2], and vehicle routing with time-windows [1], also work in the directed setting. Since these algorithms use the orienteering (or minimum excess) problem in a black-box fashion, our results imply approximation algorithms with guarantees: $O(\log^2 n)$ for discounted-reward TSP, and $O(\log^4 n)$ for VRP with time-windows.

In independent work, Chekuri et al. [3] also obtain many of the results reported in this paper, although via different techniques. Their approximation guarantees are slightly better than what we obtain for these prblems. They obtain an $O(\log^3 k)$-approximation algorithm for the directed k-TSP problem, and an $O(\log^2 OPT)$-approximation algorithm for directed orienteering (where $OPT \leq n$ is the optimal value of the orienteering instance).

2   The Directed k-TSP Problem

The *directed k-TSP* problem is a generalization of the asymmetric traveling salesman problem (ATSP), for which the best known approximation guarantee is $O(\log n)$. In this section, we obtain an $O(\log^2 n \cdot \log k)$-approximation algorithm for this problem. We first obtain an $O(\log^2 n)$-approximation algorithm for minimum ratio ATSP (Theorem 5), and then show how this implies the result for directed k-TSP (Theorem 6). Our algorithm for minimum ratio ATSP is based on the integrality gap (Theorem 2) of a suitable LP relaxation for ATSP, which we study next.

2.1   A Linear Relaxation for ATSP

In this section, we consider the following LP relaxation for ATSP, where (V, d) is the input metric.

$$\min \sum_e d_e \cdot z_e$$
$$s.t.$$
$$(ALP) \quad \begin{array}{ll} z(\delta^+(v)) = z(\delta^-(v)) & \forall v \in V \\ z(\delta^+(S)) \geq 1 & \forall \phi \neq S \neq V \\ z_e \geq 0 & \forall \text{ arcs } e \end{array}$$

This relaxation was also studied in Vempala & Yannakakis [21], where the authors proved a structural property about basic solutions to (ALP). We show that the integrality gap of (ALP) is at most $\lceil \log n \rceil$. The following stronger LP relaxation (with additional degree = 1 constraints) was shown to have an integrality gap of at most $\lceil \log n \rceil$ in Williamson [22].

$$\min \sum_e d_e \cdot z_e$$
$$s.t.$$
$$(ALP') \quad \begin{array}{ll} z(\delta^+(v)) = 1 & \forall v \in V \\ z(\delta^-(v)) = 1 & \forall v \in V \\ z(\delta^+(S)) \geq 1 & \forall \phi \neq S \neq V \\ z_e \geq 0 & \forall \text{ arcs } e \end{array}$$

However, we are not aware of any previously known upper bound on the integrality gap of (ALP). We first give an independent proof that upper bounds the integrality gap of (ALP) (Theorem 2), and then show that for any asymmetric metric (V, d), the optimal values of (ALP) and (ALP') coincide (Theorem 3). Our proof makes use of the following directed splitting-off theorem due to Mader [16].

Theorem 1 (Mader [16]). *Let $D = (U + x, A)$ be a directed graph such that indegree equal to outdegree at x, and the directed connectivity between any pair of vertices in U is at least k. Then for every arc $(x, v) \in A$ there exists an arc $(u, x) \in A$ so that after replacing the two arcs (u, x) & (x, v) by an arc (u, v), the directed connectivity between every pair of vertices in U remains at least k.*

This operation of replacing two arcs (u, x) & (x, v) by the single arc (u, v) is called *splitting-off*.

Theorem 2. *The integrality gap of (ALP) is at most $\lceil \log n \rceil$.*

Proof: This proof has the same outline as the proof for the stronger LP relaxation (ALP') in Williamson [22]. We use the $\lceil \log n \rceil$ approximation algorithm for ATSP due to Frieze et al. [7], which works by computing cycle covers repeatedly (in at most $\lceil \log n \rceil$ iterations). In this algorithm, if $U \subseteq V$ is the set of representative vertices in some iteration, the cost incurred in this iteration equals the minimum cycle cover on U. Let $ALP(U)$ denote the LP relaxation ALP restricted to a subset U of the original vertices, and $opt(ALP(U))$ its optimal value. Then we have:

Claim 1. *For any subset $U \subseteq V$, the minimum cycle cover on U has cost at most $opt(ALP(U))$.*

Proof: Consider the following linear relaxation for cycle cover.

$$\min \ \sum_e d_e \cdot x_e$$
$$s.t.$$
$$x(\delta^+(v)) - x(\delta^-(v)) = 0 \qquad \forall v \in U$$
$$(CLP) \qquad x(\delta^+(v)) \geq 1 \qquad \forall v \in U$$
$$x_e \geq 0 \qquad \forall \text{ arcs } e$$

These constraints are equivalent to a circulation problem on network N which contains two vertices v_{in} & v_{out} for each vertex $v \in U$. The arcs in N are: $\{(u_{out}, v_{in}) : \forall\, u, v \in U,\ u \neq v\}$, and $\{(v_{in}, v_{out}) : \forall v \in U\}$. The cost of each (u_{out}, v_{in}) arc is $d(u, v)$, and each (v_{in}, v_{out}) arc costs 0. It is easy to see that the minimum cost circulation on N that places at least one unit of flow on each arc in $\{(v_{in}, v_{out}) : \forall v \in U\}$ is exactly the optimal solution to (CLP). But the linear program for minimum cost circulation is integral (network matrices are totally unimodular, c.f. [17]), and so is (CLP).

Any integral solution to (CLP) defines an Eulerian subgraph H with each vertex in U having degree at least 1. Each connected component C of H is Eulerian and can be shortcut to get a cycle on the vertices of C. Since triangle inequality holds, the cost of each such cycle is at most that of the original component. So this gives a cycle cover of U of cost at most $opt(CLP(U))$, the optimal value of (CLP). But the linear program $ALP(U)$ is more constrained than $CLP(U)$; so the minimum cycle cover on U costs at most $opt(ALP(U))$. ∎

We now establish the *monotonicity property* of ALP, namely:

$$opt(ALP(U)) \leq opt(ALP(V)) \qquad \forall U \subseteq V$$

Consider any subset $U \subseteq V$, vertex $v \in U$, and $U' = U - v$; we will show that $opt(ALP(U')) \leq opt(ALP(U))$. Let z be any fractional solution to $ALP(U)$ so that $L \cdot z$ is integral for some large enough $L \in \mathbb{N}$. Define a multigraph H on vertex set U with $L \cdot z_{w_1, w_2}$ arcs going from w_1 to w_2 (for all $w_1, w_2 \in U$). From the feasibility of z in $ALP(U)$, H is Eulerian and has arc-connectivity at least L. Now applying Theorem 1 repeatedly on vertex $v \in U$ (until its degree is zero), we obtain a multigraph H' on $U' = U - v$ such that the arc-connectivity of H' is still at least L. Further, due to the triangle inequality, the total cost of H' is at most that of H. Finally, scaling down H' by L we obtain a fractional solution to $ALP(U')$ of cost at most $d \cdot z$. Thus, $opt(ALP(U')) \leq opt(ALP(U))$, and using this inductively we have monotonicity for ALP.

This suffices to prove the theorem, as the cost incurred in each iteration of the Frieze et al. [7] algorithm can be bounded by $opt(ALP(V))$, and there are at most $\lceil \log n \rceil$ iterations.   ∎

We note that in order to prove the monotonicity property for the linear program (ALP'), Williamson [22] used the equivalence of (ALP') and the Held-Karp bound [11], and showed that the Held-Karp lower bound is monotone. Using splitting-off, we obtain a more direct proof of monotonicity. In fact, we can prove a stronger statement than Theorem 2, which relates the optimal values of (ALP) and (ALP'). It was shown in [22] that the optimal value of (ALP') equals the Held-Karp lower bound [11]; so the next theorem shows that for any ATSP instance, the values of the Help-Karp bound, (ALP') and (ALP) are all equal. A similar result for the symmetric case was proved in Goemans & Bertsimas [8], which was also based on splitting-off (for undirected graphs).

Theorem 3. *The optimal values of (ALP) and (ALP') are equal.*

Proof: Clearly the optimal value of (ALP') is at most that of (ALP). We will show that any fractional solution z to (ALP) can be modified to a fractional solution z' to (ALP'), such that $\sum_e d_e \cdot z'_e \leq \sum_e d_e \cdot z_e$, which would prove the theorem. As in the proof of Theorem 2, let $L \in \mathbb{N}$ be large enough so that $L \cdot z$ is integral, and let H denote a multi di-graph with $L \cdot z_{u,v}$ arcs from u to v, for all $u, v \in V$. From the feasibility of z in (ALP), we know that H is Eulerian and has arc-connectivity at least L.

If some $v \in V$ has degree strictly greater than L, we reduce its degree by one as follows. Let v' be any vertex in $V \setminus v$, and $\mathcal{P}_{v,v'}$ denote a minimal set of arcs that constitutes exactly L arc-disjoint paths from v to v'. Due to minimality, the number of arcs in $\mathcal{P}_{v,v'}$ incident to v is exactly L and they are all arcs leaving v. Since the degree of v is at least $L + 1$, there is an arc $(v, w) \in H \setminus \mathcal{P}_{v,v'}$. Applying Theorem 1 to arc (v, w), we obtain arc $(u, v) \in H \setminus \mathcal{P}_{v,v'}$ such that the arc-connectivity of vertices $V \setminus v$ in $H' = (H \setminus \{(u,v), (v,w)\}) \cup (u,w)$ remains at least L. Further, by the choice of (v, w), $\mathcal{P}_{v,v'} \subseteq H'$; so the arc-connectivity from v to v' in H' is at least L. Since H' is Eulerian, it now follows that the arc-connectivity of vertices V in H' is also at least L. Thus we obtain a multigraph H' from H which maintains connectivity and decreases the degree of vertex v by 1. Repeating this procedure for all vertices in V having degree greater than L, we obtain (an Eulerian) multigraph G having arc-connectivity L such that the degree of each vertex equals L.

Note that in the degree reducing procedure above, the only operation we used was splitting-off. Since d satisfies triangle inequality, the total cost of arcs in G (under length d) is at most that of H. Finally, scaling down G by L, we obtain the claimed fractional solution z' to (ALP').   ∎

2.2   Minimum Ratio ATSP

We now describe the $O(\log^2 n)$-approximation algorithm for minimum ratio ATSP, which uses Theorem 2. In addition, we require the following strengthening of Mader's splitting-off Theorem, in the case of Eulerian digraphs.

Theorem 4 (Frank [6] (Theorem 4.3) and Jackson [12]). *Let $D = (U + x, A)$ be a directed Eulerian graph. For each arc $f = (x, v) \in A$ there exists an arc $e = (u, x) \in A$ so that after replacing arcs e & f by arc (u, v), the directed connectivity between every pair of vertices in U is preserved.*

Theorem 5. *There is a polynomial time $O(\log^2 n)$-approximation algorithm for the minimum ratio ATSP problem.*

Proof: The approximation algorithm for minimum ratio ATSP is based on the following LP relaxation for this problem.

$$\min \ \sum_e d_e \cdot x_e$$

s.t.

$$x(\delta^+(v)) = x(\delta^-(v)) \qquad \forall v \in V$$
$$x(\delta^+(S)) \geq y_v \qquad \forall S \subseteq V - \{r\} \ \forall v \in S$$
$$(RLP) \quad \sum_{v \neq r} y_v \geq 1$$
$$x_e \geq 0 \qquad \forall \text{ arcs } e$$
$$0 \leq y_v \leq 1 \qquad \forall v \in V - \{r\}$$

To see that this is indeed a relaxation, consider the optimal integral r-tour C^* that covers l vertices (excluding r). We construct a solution to (RLP) by setting $y_v = \frac{1}{l}$ for all vertices $v \in C^*$, and $x_e = \frac{1}{l}$ for all arcs $e \in C^*$. It is easy to see that this solution is feasible and has cost $\frac{d(C^*)}{l}$ which is the optimal ratio. The linear program (RLP) can be solved in polynomial time using the Ellipsoid algorithm. The algorithm is as follows:

1. Let (x, y) denote an optimal solution to (RLP).
2. Discard all vertices $v \in V \setminus r$ with $y_v \leq \frac{1}{2n}$; all remaining vertices have y-values in the interval $[\frac{1}{2n}, 1]$.
3. Define $g = \lceil \log_2 n + 1 \rceil$ groups of vertices where group G_i (for $i = 1, \cdots, g$) consists of all vertices v having $y_v \in (\frac{1}{2^i}, \frac{1}{2^{i-1}}]$.
4. Run the Frieze et al. [7] algorithm on each of $G_i \cup \{r\}$ and output the r-tour with the smallest ratio.

Note that the total y-value of vertices remaining after step 2 is at least $1/2$. Consider any group G_i; let $L_i \in \mathbb{N}$ be large enough so that $L_i \cdot 2^i \cdot x$ is integral. Define a multigraph H_i with $L_i \cdot 2^i \cdot x_{u,v}$ arcs from u to v for all $u, v \in V$. Below, for a directed graph D and vertices $u, v \in D$ the directed arc-connectivity from u to v is denoted $\lambda(u, v; D)$. From the feasibility of x in RLP, it is clear that H_i is Eulerian. Further, for all $v \in G_i$, $\lambda(r, v; H_i) = \lambda(v, r; H_i) \geq L_i \cdot 2^i \cdot y_v \geq L_i$. Now we split-off vertices in $V \setminus (G_i \cup \{r\})$ one by one, using Theorem 4, which preserves the arc-connectivity of $G_i \cup \{r\}$. This results in an Eulerian multigraph H_i' on vertices $G_i \cup r$ satisfying $\lambda(r, v; H_i'), \lambda(v, r; H_i') \geq L_i$ for all $v \in G_i$. Further, due to triangle inequality the total weight of arcs in H_i' is at most that in H_i. Now, scaling down H_i' by L_i, we obtain a fractional solution z^i to $ALP(G_i \cup \{r\})$ of cost $d \cdot z^i \leq 2^i (d \cdot x)$. Now Theorem 2 implies that there exists an r-tour on G_i of cost at most $\beta = \lceil \log n \rceil$ times $d \cdot z^i$. In fact, the Frieze et al. [7] algorithm applied on $G_i + r$ produces such a tour. We now claim that one of the r-tours found in step 4 (over all $i = 1, \cdots g$) has a small ratio:

$$\min_{i=1}^{g} \frac{\beta(d \cdot z^i)}{|G_i|} \leq \min_{i=1}^{g} \frac{2^i \beta(d \cdot x)}{|G_i|} \leq \frac{\beta \sum_{i=1}^{g} d \cdot x}{\sum_{i=1}^{g} |G_i|/2^i} \leq 4g\beta \cdot (d \cdot x)$$

The last inequality follows from the fact that after step 2, $\frac{1}{2} \leq \sum_{v \neq r} y_v \leq \sum_{i=1}^{g} \frac{1}{2^{i-1}} |G_i| = 2 \sum_{i=1}^{g} \frac{|G_i|}{2^i}$, since there is a total y-weight of at least $1/2$ even after step 2. Thus we have a $4g\beta = O(\log^2 n)$ approximation algorithm for minimum ratio ATSP.   ∎

We note that for this proof of Theorem 5 to work, just a bound on the integrality gap of (ALP') [22] is insufficient. The Eulerian multigraph H_i' that gives rise to the fractional ATSP solution z^i on $G_i \cup \{r\}$ may not have degree L_i at all vertices; so z^i may be infeasible for $ALP'(G_i \cup \{r\})$. This is the reason we need to consider the LP relaxation (ALP).

2.3   Directed k-TSP

We now describe how minimum ratio ATSP can be used to obtain an approximation algorithm for the directed k-TSP problem.

Theorem 6. *There is a polynomial time $O(\log^2 n \cdot \log k)$ approximation algorithm for the directed k-TSP problem.*

Proof: We use the $\alpha = O(\log^2 n)$-approximation algorithm for the related minimum ratio ATSP problem. Let OPT denote the optimal value of the directed k-TSP instance. By performing binary search, we may assume that we know the value of OPT within a factor 2. We only consider vertices $v \in V$ satisfying $d(r, v), d(v, r) \leq OPT$; this does not affect the optimal solution. Then we invoke the minimum ratio ATSP algorithm repeatedly (each time restricted to the currently uncovered vertices) until the total number of covered vertices $t \geq \frac{k}{2}$. Note that for every instance of the ratio problem that we solve, there is a feasible solution of ratio $\leq \frac{2 \cdot OPT}{k}$ (namely, the optimal k-TSP tour covering at least $k/2$ residual vertices). Thus we obtain an r-tour on $t \geq \frac{k}{2}$ vertices having ratio $\leq \frac{2\alpha \cdot OPT}{k}$; so the length of this r-tour is at most $\frac{2\alpha t \cdot OPT}{k}$. Now, we split this r-tour into $l = \lceil \frac{2t}{k} \rceil$ di-paths, each containing at least $\frac{t}{l} \geq \frac{k}{4}$ vertices (this can be done in a greedy fashion). By averaging, the minimum length di-path in this collection has length at most $\frac{2\alpha t OPT/k}{l} \leq \alpha \cdot OPT$. Joining the first and last vertices in this di-path to r, we obtain an r-tour containing at least $\frac{k}{4}$ vertices, of length at most $(\alpha + 2) \cdot OPT$. So we get an $(O(\alpha), 4)$ bi-criteria approximation for directed k-TSP. This algorithm can now be used as follows: until k vertices are covered, repeat: if k' denotes the number of vertices covered so far, run the bi-criteria approximation algorithm with a target of $k-k'$, restricted to currently uncovered vertices. A standard set cover based analysis implies that this is an $O(\alpha \cdot \log k)$-approximation algorithm for directed k-TSP. ∎

3   The Directed Orienteering Problem

In this section, we consider the orienteering problem in asymmetric metrics. As mentioned before, this is in fact the directed counterpart of the point-to-point orienteering problem [1]. We adapt the framework of Blum et al. [2] (for undirected orienteering) to the directed case. The algorithm for directed orienteering is based on the following sequence of reductions: directed k-path to minimum ratio ATSP (Theorem 7), directed minimum excess to directed k-path (Theorem 8), and directed orienteering to directed minimum excess (Theorem 9). The last two reductions are identical to the corresponding reductions for undirected orienteering in Blum et al. [2] and Bansal et al. [1]. The *directed k-path* problem is the following: given an asymmetric metric (V, d), origin (s) & destination (t) vertices, and a target k, find an s-t di-path of minimum length that visits at least k other vertices. We prove the following bi-criteria approximation guarantee for this problem.

Theorem 7. *A ρ-approximation algorithm for minimum ratio ATSP implies a $(3, 4\rho)$ bi-criteria approximation algorithm for the directed k-path problem.*

Proof: We assume (by performing a binary search) that we know the optimal value OPT of the directed k-path instance within a constant factor, and let G denote the directed graph corresponding to metric (V, d) (which has an arc of length $d(u, v)$ from u to v for every pair of vertices $u, v \in V$). We modify graph G to obtain graph H as follows: **(a)** discard all vertices v such that $d(s, v) > OPT$ or $d(v, t) > OPT$; and **(b)** add an extra arc from t to s of length OPT. In the rest of this proof, we refer to the shortest path metric induced by H as (V, l). Note that each tour in metric l corresponds to a tour in graph H (using shortest paths in H for each metric arc); below, any tour in metric l will refer to the corresponding tour in graph H. Since there is an s-t path of length OPT (in metric d) covering k vertices, appending the (t, s) arc, we have an s-tour σ^* of length at most $2 \cdot OPT$ (in metric l) covering $k + 1$ vertices.

Now, we run the minimum ratio ATSP algorithm with root s in metric l repeatedly until either **(1)** $\frac{k}{2}$ vertices are covered and the extra (t, s) arc is never used in the current tour in graph H; or **(2)** the extra (t, s) arc is used for the first time in the current tour in H. Let σ be the s-tour obtained (in graph H) at the end of this iteration, and h the number of vertices covered. Note that each s-tour added in a single call to minimum ratio ATSP, may use the extra (t, s) arc at most once (by an averaging argument). So in case (1), the (t, s) arc is absent in σ, and in case (2), the (t, s) arc is used exactly once & it is the last arc in σ. Note also that during each call of minimum ratio ATSP, there is a feasible solution of ratio $\frac{2OPT}{k}$ (σ^* restricted to the remaining vertices); so the ratio of the s-tour σ, $\frac{l(\sigma)}{h} \leq \rho \cdot \frac{2OPT}{k}$. From σ we now obtain a feasible s-t path τ in metric d as follows. In case (1), add a direct (s,t) arc: $\tau = \sigma \cdot (s,t)$; in case (2), remove the only copy of the extra (t,s) arc (occurring at the end of σ): $\tau = \sigma \setminus \{(t,s)\}$. In either case, s-t path τ contains h vertices and has length $d(\tau) \leq \frac{2\rho h}{k} OPT + OPT$. Note that in case (1), $h \geq \frac{k}{2}$; and in case (2), since the extra (t,s) arc is used, $\frac{OPT}{h} \leq \frac{l(\sigma)}{h} \leq 2\rho \frac{OPT}{k}$, so $h \geq \frac{k}{2\rho}$. Hence $d(\tau) \leq \frac{4\rho h}{k} OPT$. We now greedily split τ into maximal paths, each of which has length at most OPT; the number of subpaths obtained is at most $\frac{d(\tau)}{OPT} \leq \frac{4\rho h}{k}$. So one of these paths contains at least $h/(\frac{4\rho h}{k}) = \frac{k}{4\rho}$ vertices. Adding direct arcs from s to the first vertex on this path and from the last vertex on this path to t, we obtain an s-t path of length at most $3 \cdot OPT$ containing at least $\frac{k}{4\rho}$ vertices.   ∎

As in Blum et al. [2], we define the *excess* of an s-t di-path as the difference of the path length and the shortest path distance from s to t. The *directed min-excess* problem is then: given an asymmetric metric (V,d), origin (s) & destination (t) vertices, and a target k, find an s-t di-path of minimum excess that visits at least k other vertices. The next theorem reduces the directed minimum excess problem to the k-path problem, for which we just obtained an approximation algorithm.

Theorem 8. *An (α, β) bi-criteria approximation algorithm for the directed k-path problem implies a $(2\alpha - 1, \beta)$ bi-criteria approximation algorithm for the directed minimum excess problem.*

Proof: This algorithm is essentially identical to the one in Blum et al. [2] for undirected metrics; we outline the analysis here for the sake of completeness. We first show that the optimal s-t path π can be divided into segments that have a certain structure. Then we show how to approximate solutions of this structure using a dynamic program.

Breaking ties arbitrarily, we assume there is an ordering $s = v_1, \cdots, v_n$ on the vertex set V such that $0 = d(s, v_1) < d(s, v_2) < d(s, v_3) < \cdots < d(s, v_n)$. For any real number $l \geq 0$, define $f(l)$ to be the number of arcs (x,y) in the optimal path π with $d(s,x) < l \leq d(s,y)$. For values of $l \geq d(s,t)$ such that $f(l) = 1$, we redefine $f(l) = 2.$[1] Clearly $f(l) \geq 1$ for all $0 \leq l \leq d(s,t)$. We break the real line into two types of intervals: *type 1 intervals* are maximal intervals where $f(l) = 1$, and *type 2 intervals* are maximal intervals where $f(l) \geq 2$. Let the interval boundaries be labeled $0 = b_1 < b_2 < \cdots < b_m$; note that $b_{m-1} < d(s,t) \leq b_m$. Then the i-th interval is (b_i, b_{i+1}), and the interval types

---

[1] This modification is done for ease in defining the intervals as types 1 & 2. Without the modification, values of $l \geq d(s,t)$ with $f(l) = 1$ would be placed in a type 1 interval; but the subpath of π covering vertices in such an interval is *not* monotone. By ensuring that they are placed in a type 2 interval, we allow these vertices to be covered using the minimum excess algorithm. Also, the entire length of the subpath in such an interval contributes to the excess of path π; so our algorithm is allowed to approximate the *length* of this subpath.

alternate between 1 & 2. Let $V_i = \{v \in V : b_i < d(s,v) \leq b_{i+1}\}$ be the vertices in the i-th interval; since one of any two consecutive intervals is of type 1, the optimal path π is monotone across vertex sets $V_1, \cdots, V_{m-1}$. In other words, the vertices V_i form a contiguous subpath $S_i = \pi \cap V_i$ on π. To simplify what follows, we place vertices at interval boundaries, so every arc of π can be assumed to be in some V_i (as mentioned in Blum et al. [2], the analysis holds even without this assumption). So the length of π can be written as $d(\pi) = \sum_i l_i$ where $l_i = d(S_i)$ is the length of π in V_i. Similar to Lemma 3.1 in Blum et al. [2], we have the following.

Claim 2. *If V_i is a type 1 interval, then $l_i \geq b_{i+1} - b_i$. If V_i is a type 2 interval not containing t, then $l_i \geq 2(b_{i+1} - b_i)$. If V_i is a type 2 interval containing t, then $l_i \geq 2(d(s,t) - b_i)$.*

If ϵ denotes the optimal value of the min-excess problem, we have $d(s,t) + \epsilon = d(\pi) = \sum_i l_i$. Now from Claim 2, $d(s,t) - (d(s,t) - b_{m-1}) + \sum_{i=1}^{m-2}(b_{i+1} - b_i) \leq \sum_{i:type1} l_i + \frac{1}{2}\sum_{i:type2} l_i$. In other words, $\sum_{i:type1} l_i + \frac{1}{2}\sum_{i:type2} l_i \geq d(s,t) = d(\pi) - \epsilon = \sum_i l_i - \epsilon$. Thus $\sum_{i:type2} l_i \leq 2 \cdot \epsilon$.

Based on this structure of the optimal solution, we write a dynamic program that stitches together intervals & paths within them to obtain a single s-t di-path. The subproblems in this dynamic program are defined by tuples (a,b,u,v,p,i) where $a,b \in V$ define the interval boundaries (a is closest to s & b is farthest from s), $u,v \in V$ are the start & end vertices of the path within this interval, p is the number of vertices to be covered in this interval, and $i \in \{1,2\}$ is the type of the interval.

Solving type 1 intervals: This can be done exactly in polynomial time by a simple dynamic program.

Solving type 2 intervals: This uses the directed k-path algorithm assumed in the theorem, to find an α-approximately minimum length u-v di-path covering at least p/β vertices.

Combining intervals: The dynamic program for combining intervals will consider the segmentation of the optimal path π in one of its guesses. The length of the solution resulting from this is at most $\sum_{i:type\ 1} l_i + \sum_{i:type\ 2} \alpha \cdot l_i = \sum_i l_i + (\alpha - 1)\sum_{i:type\ 2} l_i \leq d(\pi) + 2(\alpha - 1)\epsilon = d(s,t) + (2\alpha - 1)\epsilon$; so the excess of this path is at most $(2\alpha - 1)\epsilon$. Further, the number of vertices covered by any feasible solution generated by this dynamic program is at least $\frac{k}{\beta}$. Thus we get the desired bi-criteria approximation guarantee. ∎

We now give the final reduction used in approximating directed orienteering.

Theorem 9. *An (α, β) bi-criteria approximation algorithm for the directed minimum excess problem implies an $\lceil \alpha \rceil \cdot \beta$ approximation algorithm for the directed orienteering problem.*

Proof: This algorithm is identical to Theorem 1 in Bansal et al. [1]. Again, we outline the proof only for the sake of completeness. Consider splitting the optimal s-t orienteering path π into $l = \lceil \alpha \rceil$ pieces, each having at least k/l vertices (where k is the optimal value of the orienteering problem). Let the boundary vertices of these pieces be $s = u_0, u_1, \cdots, u_l = t$ in the order in which they appear on π. For vertices u_i & u_j (with $i < j$), $\pi(u_i, u_j)$ denotes the subpath in π from u_i to u_j, and we define $\epsilon'(u_i, u_j) = d(\pi(u_i, u_j)) - d(u_i, u_j)$ to be the excess along path π. It was proved in Bansal et al. [1] (and it also holds in the directed case) that ϵ' is sub-additive: for any $0 \le h < i < j \le l$, $\epsilon'(u_h, u_i) + \epsilon'(u_i, u_j) \le \epsilon'(u_h, u_j)$. Among the l pieces of π, consider the one $\pi(u_i, u_{i+1})$ with minimum value of $\epsilon'(u_i, u_{i+1})$. Consider the s-t path $\sigma = s, u_i, \pi(u_i, u_{i+1}), u_{i+1}, t$; note that $d(\pi) - d(\sigma) = \epsilon'(u_0, u_i) + \epsilon'(u_{i+1}, u_l) \ge (l - 1) \cdot \epsilon'(u_i, u_{i+1})$, since the excess function ϵ' is sub-additive and the i-th piece has minimum ϵ'. The algorithm guesses the optimal value k, vertices u_i and u_{i+1} and runs the min-excess algorithm with source u_i, destination u_{i+1}, and target of k/l vertices. For the correct guess, $\pi(u_i, u_{i+1})$ is a feasible solution, and the min-excess approximation algorithm finds a $u_i - u_{i+1}$ path covering at least $\frac{k}{\beta l}$ vertices, of length at most $d(u_i, u_{i+1}) + \alpha \cdot \epsilon'(u_i, u_{i+1}) = d(\pi(u_i, u_{i+1})) + (\alpha - 1)\epsilon'(u_i, u_{i+1})$. So the resulting s-t di-path (obtained by appending arcs (s, u_i) and (u_{i+1}, t)) covers at least $\frac{k}{\beta \cdot l}$ vertices, and has length at most $d(s, u_i) + d(\pi(u_i, u_{i+1})) + d(u_{i+1}, t) + (\alpha - 1)\epsilon'(u_i, u_{i+1}) = d(\sigma) + (\alpha - 1)\epsilon'(u_i, u_{i+1}) \le d(\sigma) + (l - 1)\epsilon'(u_i, u_{i+1}) \le d(\pi) \le D$ the length bound. ∎

We now obtain our main result, that relates the directed orienteering problem and minimum ratio ATSP.

Corollary 1. *A ρ-approximation algorithm for the minimum ratio ATSP problem implies an $O(\rho)$-approximation algorithm for the directed orienteering problem. Conversely, a ρ-approximation algorithm for directed orienteering implies an $O(\rho)$-approximation algorithm for minimum ratio ATSP.*

Proof: The first direction follows directly from Theorems 7,8 & 9. For the other direction, we are given a ρ-approximation algorithm for directed orienteering. Let D denote the length of some minimum ratio tour σ^*, t the last vertex visited by σ^* (before returning to the root r), and h the number of vertices it covers; so the optimal ratio is $\frac{D}{h}$. The algorithm for minimum ratio ATSP first guesses a value D' such that $D' \le D \le 2 \cdot D'$, and the last vertex t. Note that we can guess powers of 2 for the value of D', which gives $O(\log_2(n \cdot d_{max}))$ possibilities for D' (where d_{max} is the length of the longest arc). Also, the number of possibilities for t is at most n; so the algorithm only makes a polynomial number of guesses. The algorithm then runs the directed orienteering algorithm with r & t as the start/end vertices and a length bound of $2D' - d(t, r) \ge D - d(t, r)$. Note that removing the last (t, r) arc from σ^* gives a feasible solution to this orienteering instance that covers h vertices. Hence the ρ-approximation algorithm is guaranteed to find an r-t di-path covering at least $\frac{h}{\rho}$ vertices, having length at most $2D' - d(t, r)$. Now, adding the (t, r) arc to this path gives an r-tour of ratio at most $2D'/(\frac{h}{\rho}) \le 2\rho \frac{D}{h}$. ∎

Corollary 1 and Theorem 5 imply an $O(\log^2 n)$-approximation algorithm for the directed orienteering problem. Further, any improvement in the approximation guarantee of minimum ratio ATSP implies a corresponding improvement for directed orienteering.

3.1   Extensions

Discounted reward TSP: In this problem [2], we are given a metric space with rewards on vertices, and a discount factor $\gamma < 1$; the goal is to find a path that maximizes the total discounted reward (where the reward for a vertex visited at time t is discounted by a factor γ^t). The approximation algorithm for the undirected version of this problem (Blum et al. [2]) uses the minimum excess problem as a subroutine within a dynamic program. It can be verified directly that this reduction also works in the directed case, and so the $(O(1), O(\log^2 n))$ bi-criteria approximation for directed minimum excess implies an $O(\log^2 n)$-approximation algorithm for directed discounted reward TSP.

Vehicle routing problem with time windows: In this VRP, we are given a metric space with a specified depot vertex and all other vertices having a time window (that specifies a release time and a deadline), and the goal is to find a path starting at the depot that maximizes the number of vertices visited in their time window. Note that orienteering is a special case when all vertices have the same time window. Bansal et al. [1] use the point-to-point orienteering problem as a subroutine, and show that an α-approximation algorithm for orienteering implies an $O(\alpha \cdot \log^2 n)$-approximation for vehicle routing with time-windows. In fact, all the steps used in these reductions can be adapted to the case of directed metrics as well. So there is an $O(\log^4 n)$-approximation algorithm for VRP with time-windows on asymmetric metrics. A special case of the VRP with time-windows occurs when each vertex has the same release time, and only the deadline is vertex dependent; this problem is *deadline TSP*. The results of Bansal et al. [1] for this problem, along with the directed orienteering algorithm imply an $O(\log^3 n)$-approximation algorithm for directed deadline TSP.

References

1. Bansal, N., Blum, A., Chawla, S., Meyerson, A.: Approximation Algorithms for Deadline-TSP and Vehicle Routing with Time Windows. In: Proceedings of the 36th Annual ACM Symposium on Theory of Computing, pp. 166–174 (2004)
2. Blum, A., Chawla, S., Karger, D.R., Lane, T., Meyerson, A., Minkoff, M.: Approximation Algorithms for Orienteering and Discounted-Reward TSP. In: Proceedings of the 44th Annual IEEE Symposium on Foundations of Computer Science, pp. 46–55 (2003)
3. Chekuri, C., Korula, N., Pal, M.: Improved Algorithms for Orienteering and Related Problems. Manuscript (2007)
4. Chekuri, C., Pal, M.: A recursive greedy algorithm for walks in directed graphs. In: Proceedings of the 46th Annual IEEE Symposium on Foundations of Computer Science, pp. 245–253 (2005)
5. Desrochers, M., Desrosiers, J., Solomon, M.: A New Optimization Algorithm for the Vehicle Routing Problem with Time Windows. Operation Research 40, 342–354 (1992)

6. Frank, A.: On Connectivity properties of Eulerian digraphs. Annals of Discrete Mathematics, 41 (1989)
7. Frieze, A., Galbiati, G., Maffioli, F.: On the worst-case performance of some algorithms for the asymmetric travelling salesman problem. Networks 12, 23–39 (1982)
8. Goemans, M.X., Bertsimas, D.J.: On the parsimonious property of connectivity problems. In: Proceedings of the 1st annual ACM-SIAM symposium on Discrete algorithms, pp. 388–396 (1990)
9. Golden, B.L., Levy, L., Vohra, R.: The Orienteering Problem. Naval Research Logistics 34, 307–318 (1987)
10. Haimovich, M., Rinnooy, A.H.G.: Bounds and heuristics for capacitated routing problems. Mathematics of Operations Research 10, 527–542 (1985)
11. Held, M., Karp, R.M.: The travelling salesman problem and minimum spanning trees. Operations Research 18, 1138–1162 (1970)
12. Jackson, B.: Some remarks on arc-connectivity, vertex splitting, and orientation in digraphs. Journal of Graph Theory 12(3), 429–436 (1988)
13. Kantor, M., Rosenwein, M.: The Orienteering Problem with Time Windows. Journal of the Operational Research Society 43, 629–635 (1992)
14. Kohen, A., Kan, A.R., Trienekens, H.: Vehicle Routing with Time Windows. Operations Research 36, 266–273 (1987)
15. Li, C.-L., Simchi-Levi, D., Desrochers, M.: On the distance constrained vehicle routing problem. Operations Research 40, 790–799 (1992)
16. Mader, W.: Construction of all n-fold edge-connected digraphs (German). European Journal of Combinatorics 3, 63–67 (1982)
17. Nemhauser, G.L., Wolsey, L.A.: Integer and Combinatorial Optimization (1999)
18. Savelsbergh, M.: Local Search for Routing Problems with Time Windows. Annals of Operations Research 4, 285–305 (1985)
19. Savelsbergh, M.W.P., Sol, M.: The general pickup and delivery problem. Transportation Science 29, 17–29 (1995)
20. Tan, K.C., Lee, L.H., Zhu, K.Q., Ou, K.: Heuristic Methods for Vehicle Routing Problems with Time Windows. Artificial Intelligence in Engineering, pp. 281–295 (2001)
21. Vempala, S., Yannakakis, M.: A convex relaxation for the asymmetric tsp. In: Proceedings of the 10th annual ACM-SIAM symposium on Discrete algorithms, pp. 975–976 (1999)
22. Williamson, D.: Analysis of the held-karp heuristic for the traveling salesman problem. Master's thesis, MIT Computer Science (1990)

Encouraging Cooperation in Sharing Supermodular Costs

(Extended Abstract)

Andreas S. Schulz[1] and Nelson A. Uhan[2]

[1] Sloan School of Management, Massachusetts Institute of Technology
77 Massachusetts Avenue, E53-361, Cambridge, MA 02139
schulz@mit.edu
[2] Operations Research Center, Massachusetts Institute of Technology
77 Massachusetts Avenue, E40-130, Cambridge, MA 02139
uhan@mit.edu

Abstract. We study the computational complexity and algorithmic aspects of computing the least core value of supermodular cost cooperative games, and uncover some structural properties of the least core of these games. We provide motivation for studying these games by showing that a particular class of optimization problems has supermodular optimal costs. This class includes a variety of problems in combinatorial optimization, especially in machine scheduling. We show that computing the least core value of supermodular cost cooperative games is NP-hard, and design approximation algorithms based on oracles that approximately determine maximally violated constraints. We apply our results to schedule planning games, or cooperative games where the costs arise from the minimum sum of weighted completion times on a single machine. By improving upon some of the results for general supermodular cost cooperative games, we are able to give an explicit formula for an element of the least core of schedule planning games, and design a fully polynomial time approximation scheme for computing the least core value of these games.

1 Introduction

Consider a situation where a set of agents agree to share the cost of their joint actions, and need to determine how to distribute the costs amongst themselves in a fair manner. For example, a set of agents may agree to process their jobs together on a machine, and share the cost of optimally scheduling their jobs. This kind of situation can be modeled naturally as a *cooperative game*. A cooperative game is a pair (N, v) where $N = \{1, \ldots, n\}$ represents a set of agents, and $v(S)$ represents the cost to agents in S.

In this work, we are concerned with cooperative games (N, v) where v is nonnegative, *supermodular*, and $v(\emptyset) = 0$. We call such games *supermodular cost cooperative games*. A set function $v : 2^N \mapsto \mathbb{R}$ is supermodular if

$$v(S \cup \{i\}) - v(S) \leq v(T \cup \{i\}) - v(T) \qquad \text{for all } S \subseteq T \subseteq N \setminus \{i\}. \tag{1}$$

M. Charikar et al. (Eds.): APPROX and RANDOM 2007, LNCS 4627, pp. 271–285, 2007.

In words, supermodularity captures the notion of increasing marginal costs. Our primary motivation behind studying these games is that many problems from combinatorial optimization—especially in machine scheduling—have optimal costs that are supermodular. Cooperative games whose costs are determined by combinatorial optimization problems have been considered previously: these include assignment games [28], minimum-cost spanning tree games [13], traveling salesman games [21], facility location games [10], scheduling-related games [2,17,19], and many others.

The central concern in cooperative game theory is the fair allocation of costs amongst agents. The prominent solution concept for cooperative games is the *core* [8]. The core of a cooperative game (N, v) consists of all cost allocations x that distribute $v(N)$—the cost incurred when all agents cooperate—in a way such that no subset of agents has incentive to forsake the rest of the agents and act on its own behalf. Formally,

$$\text{core}(N, v) = \left\{ x \in \mathbb{R}^N : x(N) = v(N), \ x(S) \leq v(S) \text{ for all } S \subseteq N \right\}.$$

(For notational convenience, for any vector x we define $x(S) = \sum_{i \in S} x_i$ for any $S \subseteq N$.) It is well known that cooperative games with submodular[1] costs always have nonempty cores [26]. This result is quite intuitive. As a coalition grows, the cost of adding a particular agent to the coalition decreases, making the idea of sharing costs more appealing. On the other hand, a supermodular cost cooperative game (N, v) has an empty core (as long as v is not modular[2]). Similar intuition still holds: the cost of adding a particular agent to a coalition increases as the coalition grows, diminishing the appeal of sharing costs.

When a cooperative game has an empty core, one might wonder if it is possible to allocate costs so that no subset of agents has incentive to deviate, and a fraction α of the total cost $v(N)$ can be recovered. This notion is captured in the α-core solution concept. Formally, for any $\alpha \in (0, 1]$,

$$\alpha\text{-core}(N, v) = \left\{ x \in \mathbb{R}^N : x(N) \geq \alpha v(N), \ x(S) \leq v(S) \text{ for all } S \subseteq N \right\}.$$

The α-core has been studied for a variety of games [3,4,15,20]. Unfortunately, in supermodular cost cooperative games, the largest fraction α one could hope to recover under a fair allocation is $\sum_{i \in N} v(\{i\})/v(N)$, which may be arbitrarily small.

Since the prospect for cooperation in sharing supermodular costs is bleak, we are led to ask: how much do we need to penalize a coalition for defecting in order to encourage cooperation amongst all agents? This notion is captured in the *least core value* of a cooperative game. The *least core* of a cooperative game (N, v) is the set of cost allocations x that are optimal solutions to the *least core optimization problem*

$$
\begin{aligned}
z^* = \ \text{minimize} \quad & z \\
\text{subject to} \quad & x(N) = v(N) \\
& x(S) \leq v(S) + z \quad \text{for all } S \subseteq N, \ S \neq \emptyset, N.
\end{aligned}
\tag{2}
$$

---

[1] A set function $v : 2^N \mapsto \mathbb{R}$ is *submodular* if $-v$ is supermodular.

[2] A set function is *modular* if it is submodular and supermodular.

The optimal value z^* of (2) is the least core value[3] of the game (N, v). By computing the least core value, we gain insight into the value a coalition of agents places on the ability to act on their own. The least core solution concept was introduced by Shapley and Shubik [27], and later named by Maschler, Peleg and Shapley [18]. Computing an element in the least core has been studied in several contexts. Faigle, Kern and Paulusma [5] showed that computing an element in the least core of minimum-cost spanning tree games is NP-hard. Kern and Paulusma [16] presented a polynomial description of the least core optimization problem for cardinality matching games. Properties of the least core value, on the other hand, seem to have been largely ignored.

In this work, we consider various theoretical aspects of computing the least core value of supermodular cost cooperative games, and its applications to sharing optimal scheduling costs. In Section 2, we motivate the interest in supermodular cost cooperative games by providing a class of optimization problems whose optimal costs are supermodular. This class of optimization problems includes a variety of classical scheduling problems and other combinatorial optimization problems. Then, in Section 3, we show that finding the least core value of supermodular cost cooperative games is NP-hard, and design approximation algorithms based on oracles that approximately determine maximally violated constraints. Finally, in Section 4, we apply our results to *schedule planning games*, or cooperative games where the costs arise from the minimum sum of weighted completion times on a single machine. By improving on some of the results for general supermodular cost cooperative games, we are able to give an explicit formula for an element of the least core of schedule planning games, and design a fully polynomial time approximation scheme for computing the least core value of these games.

Due to space limitations, we have omitted several proofs in this extended abstract. We refer the reader to the full version of the paper for those proofs.

2   A Class of Optimization Problems with Supermodular Optimal Costs

We begin by providing some motivation for looking at cooperative games with supermodular costs. The problem of minimizing a linear function over a *supermodular polyhedron*—a polyhedron of the form $\{x \in \mathbb{R}^N : x(S) \geq v(S) \text{ for all } S \subseteq N\}$ where $v : 2^N \mapsto \mathbb{R}$ is supermodular—arises in many areas of combinatorial optimization, especially in scheduling. For example, Queyranne [22] showed that the convex hull of feasible completion time vectors on a single machine is a supermodular polyhedron. Queyranne and Schulz [23] showed that the convex hull of feasible completion time vectors for unit jobs on parallel machines with nonstationary speeds is a supermodular polyhedron. The scheduling problem they considered includes various classical scheduling problems as special cases. Goemans et al. [9] showed that the convex hull of mean busy time vectors of all

---

[3] Adding the inequalities $x_i \leq v(\{i\}) + z$ for all $i \in N$ and using the equality $x(N) = v(N)$, we can bound z^* below by $(v(N) - \sum_{i \in N} v(\{i\}))/|N|$. So as long as costs are finite, the least core value is well defined.

preemptive schedules of jobs with release dates on a single machine is a super-modular polyhedron.

In this section, we show that the optimal cost of minimizing a nonnegative linear function over a supermodular polyhedron is a supermodular function. As a result, by studying supermodular cost cooperative games, we are able to gain insight into the sharing of optimal costs for a wide class of combinatorial optimization problems.

Theorem 1. *Let N be a finite set, and let $u : 2^N \mapsto \mathbb{R}$ be a supermodular function such that $u(\emptyset) = 0$. If $d_j \geq 0$ for all $j \in N$, then the function $v : 2^N \mapsto \mathbb{R}$ defined by*

$$v(S) = \min \left\{ \sum_{j \in S} d_j x_j : x(A) \geq u(A) \text{ for all } A \subseteq S \right\} \quad \text{for all } S \subseteq N$$

is supermodular on N.

Using identical techniques, we can also show that maximizing a nonnegative linear function over a submodular polyhedron—a polyhedron of the form $\{x \in \mathbb{R}^N : x(S) \leq v(S) \text{ for all } S \subseteq N\}$ where $v : 2^N \mapsto \mathbb{R}$ is submodular—has submodular optimal values. Some combinatorial optimization problems that can be formulated as maximizing a linear function over a submodular polyhedron include the maximum weighted forest problem, and more generally, finding the maximum weight basis of a matroid.

As mentioned above, by the work of Queyranne [22], Queyranne and Schulz [23], and Goemans et al. [9], we immediately have the following corollary of Theorem 1.

Corollary 1. *If for all $S \subseteq N$, $v(S)$ is the objective value of optimally scheduling jobs in S for the problem[4] (a) $1 \mid\mid \sum w_j C_j$, (b) $Q \mid p_j = 1 \mid \sum w_j C_j$, (c) $P \mid p_j = 1, r_j \text{ integral} \mid \sum w_j C_j$, (d) $P \mid\mid \sum C_j$, or (e) $1 \mid r_j, pmtn \mid \sum w_j M_j$, then v is supermodular.*

Unfortunately, Corollary 1(d) does not extend to the case with arbitrary weights and processing times. In addition, one can show that the scheduling problems $1 \mid r_j \mid \sum C_j$ and $1 \mid prec \mid \sum C_j$ do not have supermodular optimal costs.

3 Complexity and Approximation

Now that we have some notion of what kind of combinatorial optimization problems have supermodular optimal costs, we turn our attention to the computational complexity and approximability of computing the least core value of supermodular cost cooperative games. Note that an arbitrary supermodular function v may not be compactly encoded. Therefore, for the remainder of this section we assume that values of v are given by an oracle. In addition, for the remainder of the paper, we assume that there are at least two agents ($n \geq 2$).

---

[4] We describe these problems using the notation of Graham et al. [12].

3.1   Computational Complexity

Theorem 2. *Computing the least core value of supermodular cost cooperative games is strongly NP-hard.*

Proof. We show that any instance of the strongly NP-hard maximum cut problem on an undirected graph [7] can be reduced to an instance of computing the least core value of a supermodular cost cooperative game. Consider an arbitrary undirected graph $G = (N, E)$. Let $\kappa : 2^N \mapsto \mathbb{R}$ be the *cut function* of G; that is, $\kappa(S) = |\{\{i, j\} \in E : i \in S, j \in N \setminus S\}|$. Also, let the function $\eta : 2^N \mapsto \mathbb{R}$ be defined as $\eta(S) = |\{\{i, j\} \in E : i \in S, j \in S\}|$. Clearly, η is nonnegative. Using the increasing marginal cost characterization of supermodularity (1), it is straightforward to see that η is supermodular. Using counting arguments, it is also straightforward to show that $\eta(S) + \eta(N \setminus S) + \kappa(S) = \eta(N)$ for any $S \subseteq N$.

Now consider the supermodular cost cooperative game (N, v), where $v(S) = 2\eta(S)$ for all $S \subseteq N$. For each agent $i \in N$, we define the cost allocation $x_i = \deg(i)$, where $\deg(i)$ denotes the degree of node i in G. In addition, let $z = \max_{S \subseteq N, S \neq \emptyset, N} \kappa(S)$. First, we show that (x, z) is feasible in (2). Note that $x(N) = \sum_{i \in N} \deg(i) = v(N)$. In addition, we have that for any $S \subseteq N$, $S \neq \emptyset, N$,

$$z \geq \kappa(S) = (2\eta(S) + \kappa(S)) - 2\eta(S) = x(S) - v(S).$$

Suppose (x^*, z^*) is optimal in (2). Adding the inequalities $x^*(S) \leq v(S) + z^*$ and $x^*(N \setminus S) \leq v(N \setminus S) + z^*$ for any $S \subseteq N$, $S \neq \emptyset, N$, and using the equality $x^*(N) = v(N)$, we have that

$$2z^* \geq v(N) - v(S) - v(N \setminus S) \quad \text{for all } S \subseteq N, S \neq \emptyset, N.$$

Therefore, $z^* \geq \kappa(S)$ for all $S \subseteq N, S \neq \emptyset, N$. Since (x, z) is feasible in (2), it follows that $z^* = z = \max_{S \subseteq N, S \neq \emptyset, N} \kappa(S)$. In other words, finding the least core value of (N, v) is equivalent to finding the value of a maximum cut on $G = (N, E)$. $\square$

In our proof of the above theorem, we show that for any instance of the maximum cut problem on an undirected graph, there exists a supermodular cost cooperative game whose least core value is exactly equal to the value of the maximum cut. Since the maximum cut problem is not approximable within a factor of 1.0624 [14], we immediately obtain the following inapproximability result:

Corollary 2. *There is no ρ-approximation algorithm[5] for computing the least core value of supermodular cost cooperative games, where $\rho < 1.0624$, unless P=NP.*

---

[5] A ρ-approximation algorithm ($\rho \geq 1$) is an algorithm that always finds a solution whose objective value is within a factor ρ of the optimal value, and whose running time is polynomial in the input size.

3.2   Approximation by Fixing a Cost Allocation

The above negative results indicate that it is rather unlikely that we will be able to compute the least core value of supermodular cost cooperative games exactly in polynomial time. This leads us to design methods with polynomial running time that approximate the least core value of these games.

Suppose (N, v) is a cooperative game, with v supermodular. As a first attempt at approximation, we fix a cost allocation x such that $x(N) = v(N)$, and then try to determine the minimum value of z such that (x, z) is feasible in the least core optimization problem (2). Since we are looking for the smallest value z such that $z \geq x(S) - v(S)$ for all $S \subseteq N$, $S \neq \emptyset, N$, we can determine z by solving the maximization problem

$$z = \max_{S \subseteq N, S \neq \emptyset, N} \{x(S) - v(S)\}.$$

This motivates defining for any cooperative game (N, v) and cost allocation x such that $x(N) = v(N)$, the function $f^x : 2^N \mapsto \mathbb{R}$ as $f^x(S) = x(S) - v(S)$, and the following problem:

---

x-maximally violated constraint problem for (N, v) (x-MVC).
For a cost allocation x such that $x(N) = v(N)$, find a subset S^* such that

$$f^x(S^*) = \max_{S \subseteq N, S \neq \emptyset, N} f^x(S) = \max_{S \subseteq N, S \neq \emptyset, N} \{x(S) - v(S)\}.$$

Note that the x-maximally violated constraint problem for a supermodular cost cooperative game is an instance of submodular function maximization. For any value of z, if $z \geq f^x(S^*)$, then (x, z) is feasible in (2). If $z < f^x(S^*)$, then $x(S^*) \leq v(S^*) + z$ is a constraint that is maximally violated by (x, z). Intuitively, we want to find a value z that is as close to $f^x(S^*)$ as possible, but *larger* than $f^x(S^*)$, since (x, z) is feasible if and only if $z \geq f^x(S^*)$.

How should we fix x? Consider the *base polytope* of the set function v : $2^N \mapsto \mathbb{R}$,

$$B(v) = \{x \in \mathbb{R}^N : x(N) = v(N), \ x(S) \geq v(S) \text{ for all } S \subseteq N\}.$$

For arbitrary set functions v, computing an element of the base polytope $B(v)$ of v may indeed be hard. Fortunately, when v is supermodular, the vertices of $B(v)$ are polynomial-time computable, and even have explicit formulas (see [6] for details). It turns out that for any cost allocation x in $B(v)$, we can show that the optimal value of the x-maximally violated constraint problem is always within a factor of 2 of the least core value of (N, v).

Theorem 3. *Suppose (N, v) is a supermodular cost cooperative game, and x is a cost allocation in the base polytope $B(v)$ of v. Let $f^x(S^*)$ be the optimal value of the x-maximally violated constraint problem for (N, v), and let z^* be the least core value of (N, v). Then $f^x(S^*) \leq 2z^*$.*

Proof. Let (x^*, z^*) be an optimal solution to (2). As in the proof of Theorem 2, we have that

$$2z^* \geq v(N) - v(S) - v(N \setminus S) \quad \text{for all } S \subseteq N, \ S \neq \emptyset, N.$$

Since $x \in B(v)$, we can deduce that for any $S \subseteq N$, $S \neq \emptyset, N$,

$$2z^* \geq v(N) - v(S) - v(N \setminus S) = x(S) - v(S) + x(N \setminus S) - v(N \setminus S) \geq f^x(S).$$

Since the above lower bound on $2z^*$ holds for any $S \subseteq N$, $S \neq \emptyset, N$, it follows that $2z^* \geq f^x(S^*)$. $\qquad\qquad\qquad\qquad\qquad\qquad\qquad\qquad\qquad\qquad\qquad\qquad\qquad$ □

In some sense, Theorem 3 tells us that any cost allocation x in the base polytope $B(v)$ of v is "almost" an element of the least core of (N, v). We use this observation, in conjunction with a polynomial-time-computable cost allocation x in $B(v)$ and a ρ-approximation algorithm for the x-maximally violated constraint problem for (N, v), to approximate the least core value of (N, v).

Theorem 4. *Suppose (N, v) is a supermodular cost cooperative game, and x is a polynomial-time-computable cost allocation in the base polytope $B(v)$ of v. If there exists a ρ-approximation algorithm for the x-maximally violated constraint problem for (N, v), then there exists a 2ρ-approximation algorithm for computing the least core value of (N, v).*

Proof. Consider the following algorithm for approximating the least core value of (N, v):

Input: supermodular cost cooperative game (N, v)
Output: an approximation z to the least core value of (N, v)

1. Compute a cost allocation x in the base polytope $B(v)$ of v.
2. Run the ρ-approximation algorithm for the x-maximally violated constraint problem for (N, v). Let $\bar{S}$ be the output of this approximation algorithm.
3. Output $z = \rho f^x(\bar{S})$.

Since $x \in B(v)$, we have that $x(N) = v(N)$. Since $\bar{S}$ is output from a ρ-approximation algorithm for the x-maximally violated constraint problem for (N, v), it follows that $z = \rho f^x(\bar{S}) \geq f^x(S^*) \geq x(S) - v(S)$ for all $S \subseteq N$, $S \neq \emptyset, N$. So (x, z) is a feasible solution. By Theorem 3, it follows that $z = \rho f^x(\bar{S}) \leq \rho f^x(S^*) \leq 2\rho z^*$. $\qquad\qquad\qquad\qquad\qquad$ □

3.3 Approximation Without Fixing a Cost Allocation

Until now, we have considered approximating the least core value of a supermodular cost cooperative game (N, v) by fixing a cost allocation x and then finding z such that (x, z) is feasible in the least core optimization problem (2). Suppose that, instead of fixing a cost allocation in advance, we compute a cost allocation along with an approximation to the least core value. Let us assume that we have a ρ-approximation algorithm for the x-maximally violated constraint problem for (N, v), for *every* x such that $x(N) = v(N)$. Note that since v is supermodular and $v(\emptyset) = 0$, for any x such that $x(N) = v(N)$, we have $\max_{S \subseteq N, S \neq \emptyset, N} f^x(S) \geq 0$. This ensures that the notion of a ρ-approximation algorithm for the x-maximally violated constraint problem is sensible. By using the ellipsoid method and binary search, we can establish one of the main results of this work:

Theorem 5. *Suppose (N, v) is a supermodular cost cooperative game, and there exists a ρ-approximation algorithm for the x-maximally violated constraint problem for (N, v), for every cost allocation x such that $x(N) = v(N)$. Then there exists a ρ-approximation algorithm for computing the least core value of (N, v).*

4   A Special Case from Single-Machine Scheduling

In this section, we study a particular supermodular cost cooperative game. Consider a situation where agents each have a job that needs to be processed on a machine, and any coalition of agents can potentially open their own machine. Suppose each agent $i \in N$ has a job whose processing time is $p_i > 0$ and weight is $w_i \geq 0$. Jobs are independent, and are scheduled non-preemptively on a single machine, which can process at most one job at a time. A *schedule planning game* is a cooperative game (N, v) where $v(S)$ is the minimum sum of weighted completion times of jobs in S. The least core value of schedule planning games has a natural interpretation: it is the amount we need to charge any coalition for opening a new machine in order to achieve cooperation.

Various cooperative games that arise from scheduling situations have been studied previously. In sequencing games [2], agents—each with a job that needs to be processed—start with a feasible solution on a fixed number of machines, and the profit assigned to a coalition of agents is the maximal cost savings the coalition can achieve by rearranging themselves. Schedule planning games have received somewhat limited attention in the past; several authors have developed axiomatic characterizations of various cost sharing rules for these games [17,19].

From Corollary 1, it follows that schedule planning games are indeed supermodular cost cooperative games, and the results from Section 3 apply. We will apply the results of Section 3.2, in which approximation is based on fixing a cost allocation, to finding the least core value of schedule planning games. Before doing so, however, we establish some useful and interesting properties of the least core of schedule planning games. These properties will in turn help us determine the computational complexity of this special case, and choose a specific cost allocation $\bar{x}$ in the base polytope $B(v)$ of v with especially nice features. In particular, we will be able to design approximation algorithms for the $\bar{x}$-maximally violated constraint problem for schedule planning games, as well as show a stronger translation in approximability between the $\bar{x}$-maximally violated constraint problem and the approximability of the least core value of these games.

4.1 Key Properties of the Least Core of Schedule Planning Games

The structure of the cost function for schedule planning games allows us to explicitly express an element of the least core of schedule planning games and recast the least core optimization problem as the maximization of a set function defined solely in terms of the cost function v.

Smith [29] showed that scheduling jobs in nonincreasing order of w_j/p_j minimizes the sum of weighted completion times on one machine. To simplify the analysis, for the remainder of this section we assume without loss of generality that $w_1/p_1 \geq \cdots \geq w_n/p_n$.

We consider the cost allocation $\bar{x}$ defined as follows:

$$\bar{x}_i = \frac{1}{2}\big(v(S^i) - v(S^{i-1})\big) + \frac{1}{2}\big(v(N \setminus S^{i-1}) - v(N \setminus S^i)\big) \tag{3}$$

$$= \frac{1}{2}w_i \sum_{j=1}^{i} p_j + \frac{1}{2}p_i \sum_{j=i}^{n} w_j \tag{4}$$

for $i = 1, \ldots, n$, where $S^i = \{1, \ldots, i\}$ and $S^0 = \emptyset$. It is straightforward to show that $\bar{x} \in B(v)$; in fact, it is a convex combination of two vertices of $B(v)$. Since $\bar{x} \in B(v)$, we have that $\bar{x}(S) \geq v(S)$. As it turns out, for schedule planning games, we are able to show a more precise relationship between the cost allocation $\bar{x}(S)$ of a coalition S and its cost $v(S)$.

Lemma 1. *Suppose (N, v) is a schedule planning game. Then, the cost allocation $\bar{x}$ as defined in (4) satisfies $\bar{x}(S) - v(S) = \frac{1}{2}(v(N) - v(S) - v(N \setminus S))$ for all $S \subseteq N$.*

With this lemma in hand, we can show the following key properties of the least core of schedule planning games.

Theorem 6. *Suppose (N, v) is a schedule planning game.*

1. *The cost allocation $\bar{x}$ as defined in (4) is an element of the least core of (N, v).*
2. *The least core value of (N, v) is*

$$z^* = \frac{1}{2} \max_{S \subseteq N, S \neq \emptyset, N} \big\{v(N) - v(S) - v(N \setminus S)\big\}.$$

Proof. Let $h = \max_{S \subseteq N, S \neq \emptyset, N}\{v(N) - v(S) - v(N \setminus S)\}$. First, we show that $(\bar{x}, h/2)$ is a feasible solution to (2). By Lemma 1, we have that $\bar{x}(N) = v(N)$, and for any $S \subseteq N$, $S \neq \emptyset, N$,

$$\frac{h}{2} \geq \frac{1}{2}\big(v(N) - v(S) - v(N \setminus S)\big) = \bar{x}(S) - v(S).$$

Suppose (x^*, z^*) is an optimal solution in (2). As in the proof of Theorem 2, we obtain the following lower bound on $2z^*$:

$$2z^* \geq v(N) - v(S) - v(N \setminus S) \quad \text{for all } S \subseteq N, S \neq \emptyset, N.$$

Therefore, $z^* \geq h/2$. It follows that $\bar{x}$ is an element of the least core of (N, v), and the least core value of (N, v) is $h/2$.   □

In addition to being an element of the least core, it happens that the cost allocation $\bar{x}$ as defined in (4) is the Shapley value of schedule planning games [19]. This is quite special: we can show that when v is nonnegative and supermodular, the Shapley value is not necessarily an element of the least core of (N, v).

One might wonder if the cost allocation $\bar{x}$ as defined in (3) is an element of the least core for general supermodular cost cooperative games. We can show that when v is nonnegative and supermodular, the cost allocation $\bar{x}$ as defined in (3) is not necessarily a least core element of (N, v), for any ordering of N. We can also show that when (N, v) is a schedule planning game, not every cost allocation in $B(v)$ is necessarily an element of the least core of (N, v).

4.2   Computational Complexity

Although computing the least core value of supermodular cost cooperative games is strongly NP-hard, it is still unclear if this remains the case for schedule planning games. In the previous subsection, we showed that we can efficiently compute an element of the least core of schedule planning games. In fact, we have an explicit formula for a least core element. Computing the least core value of schedule planning games, however, remains NP-hard.

Theorem 7. *Computing the least core value of scheduling planning games is NP-hard, even when $w_j = p_j$ for all $j \in N$.*

Proof. By Theorem 6, the least core value of schedule planning games is

$$z^* = \frac{1}{2}v(N) - \frac{1}{2}\min_{S \subseteq N, S \neq \emptyset, N} \{v(S) + v(N \setminus S)\}.$$

Note that the minimization problem above is equivalent to the problem of minimizing the sum of weighted completion times of jobs in N, with weight w_j and processing time p_j for each job $j \in N$, on two identical parallel machines. Bruno et al. [1] showed that this two-machine problem is NP-hard, even when $w_j = p_j$ for all jobs $j \in N$. $\square$

The above result is in stark contrast to the underlying problem defining the costs in schedule planning games—minimizing the sum of weighted completion times on a single machine—for which *any* order is optimal when each job has its weight equal to its processing time.

4.3   Tighter Bounds on Approximation Based on Fixing a Cost Allocation

In Section 3.2, we showed that for any supermodular cost cooperative game (N, v) and a cost allocation x in the base polytope $B(v)$ of v, a ρ-approximation algorithm for the x-maximally violated constraint problem implies a 2ρ-approximation algorithm for computing the least core value of (N, v). It is reasonable to believe, though, that for schedule planning games, we may be in a position to do better, since the cost allocation $\bar{x}$ as defined in (4) is in fact an element of the least core. This is indeed the case: from Lemma 1 and Theorem 6, it follows that

$$z^* = \max_{S \subseteq N, S \neq \emptyset, N} \left\{ \bar{x}(S) - v(S) \right\} = \max_{S \subseteq N, S \neq \emptyset, N} f^{\bar{x}}(S). \tag{5}$$

This is exactly the $\bar{x}$-maximally violated constraint problem for schedule planning games! Therefore, we obtain the following strengthening of Theorem 4.

Theorem 8. *Suppose there exists a ρ-approximation algorithm for the $\bar{x}$-maximally violated constraint problem for schedule planning games, where the cost allocation $\bar{x}$ is as defined in (4). Then there exists a ρ-approximation algorithm for computing the least core value of schedule planning games.*

4.4 Approximately Solving $\bar{x}$-MVC for Schedule Planning Games

The proofs from some of the previous subsections give us some insight into how to design oracles that approximately solve $\bar{x}$-MVC for schedule planning games. By carefully looking at the proof of Lemma 1, we can show that x-MVC for schedule planning games is actually a special case of finding a maximum weighted cut in a complete undirected graph. In the proof of Theorem 7, we see that $\bar{x}$-MVC for schedule planning games is actually equivalent (with respect to optimization) to the scheduling problem $P2 \, || \, \sum w_j C_j$, implying that we might be able to use or modify existing algorithms to approximately solve $\bar{x}$-MVC.

A maximum-cut based approximate oracle. From the proof of Lemma 1, we can show that

$$f^{\bar{x}}(S^*) = \frac{1}{2} \max_{S \subseteq N, S \neq \emptyset, N} \left\{ \sum_{j \in S} \sum_{i \in N \setminus S} \mu_{ij} \right\}$$

where

$$\mu_{ij} = \begin{cases} w_j p_i & \text{if } i < j \\ w_i p_j & \text{if } i > j \end{cases} \quad \text{for all } i \neq j.$$

Observe that $\mu_{ij} = \mu_{ji}$. So $f^{\bar{x}}(S^*)$ is proportional to the value of a maximum cut on a complete undirected graph with node set N and capacity μ_{ij} for arc $\{i, j\}$. Therefore, if we have a ρ-approximation algorithm for the maximum cut problem, then by Theorem 8, we have a ρ-approximation algorithm for finding the least core value of schedule planning games. For example, using the approximation algorithm of Goemans and Williamson [11] based on a semidefinite relaxation of the maximum cut problem yields a 1.1382-approximation algorithm for finding the least core value of schedule planning games. However, using the algorithm of Goemans and Williamson does not exploit the special structure of this particular maximum cut problem, since their method applies for maximum cut problems on general undirected graphs.

A fully polynomial-time approximation scheme based on two-machine scheduling. In this subsection, we provide a fully polynomial time approximation scheme[6] (FPTAS) for the $\bar{x}$-maximally violated constraint problem for schedule planning games. By Theorem 6, we know that $\bar{x}$-MVC is in fact

$$\max_{S \subseteq N, S \neq \emptyset, N} f^{\bar{x}}(S) = \frac{1}{2} \max_{S \subseteq N, S \neq \emptyset, N} \left\{ v(N) - v(S) - v(N \setminus S) \right\}.$$

For simplicity of exposition, we consider maximizing $g(S) = 2f^{\bar{x}}(S)$ for the remainder of this subsection.

As mentioned earlier, maximizing $g(S) = v(N) - v(S) - v(N \setminus S)$ is equivalent to minimizing $v(S) + v(N \setminus S)$, which is the scheduling problem $P2 \mid\mid \sum w_j C_j$. $P2 \mid\mid \sum w_j C_j$ is NP-complete [1], and has an FPTAS [24]. Although the two problems are equivalent from the optimization perspective, because of the constant $v(N)$, it is not immediately obvious if they are equivalent in terms of approximability. We present a dynamic program that solves $\bar{x}$-MVC exactly for schedule planning games in psuedopolynomial time, and then convert this dynamic program into an FPTAS. This development is inspired by the FPTAS for $P2 \mid\mid \sum w_j C_j$ in [24]. The analysis is similar to the analysis of the FPTAS for $P2 \mid\mid C_{max}$ in [25].

We think of determining the maximizer S^* by scheduling the jobs in N on two machines: the jobs scheduled on machine 1 will form S^*, and the jobs scheduled on machine 2 will form $N \setminus S^*$. As usual, we consider the jobs in order of nonincreasing weight-to-processing-time ratios (i.e. $1, \ldots, n$). We can partition the jobs into S^* and $N \setminus S^*$ sequentially using the following dynamic program. The state space E is partitioned into n disjoint sets, $E_1, \ldots, E_n$. A schedule σ for jobs $\{1, \ldots, k\}$ on two machines corresponds to a state $(a, b, c) \in E_k$. The first coordinate a is the sum of processing times of all jobs scheduled by σ on machine 1. The second coordinate b is the sum of processing times of all jobs scheduled by σ on machine 2. The third coordinate c is the running objective value: $v(\{1, \ldots, k\})$ *minus* the sum of weighted completion times on two machines for σ.

Suppose jobs $1, \ldots, k-1$ have already been scheduled, and job k is under consideration. If job k is scheduled on machine 1, then the running objective value increases by $w_k(a + b + p_k) - w_k(a + p_k) = w_k b$. If job k is scheduled on machine 2, then the running objective value increases by $w_k(a + b + p_k) - w_k(b + p_k) = w_k a$. This suggests the following dynamic programming algorithm.

Algorithm 1 (Exact dynamic program)

> **Input:** weights w_i, processing times p_i for all $i \in N$
> **Output:** the optimal value of $\bar{x}$-MVC for schedule planning games, $f^{\bar{x}}(S^*)$
>
> $E_1 = \{(p_1, 0, 0), (0, p_1, 0)\}$

---

[6] A *fully polynomial time approximation scheme* is an algorithm that finds a solution whose objective function value is within a factor $(1 + \epsilon)$ of the optimal value for any $\epsilon > 0$, and whose running time is polynomial in the input size and $1/\epsilon$.

For $k = 2, \ldots, n$
  For every vector $(a, b, c) \in E_{k-1}$
    Put $(a + p_k, b, c + w_k b)$ and $(a, b + p_k, c + w_k a)$ in E_k
  Find $(a, b, c) \in E_n$ with maximum c value, c^*
  Return $f^{\bar{x}}(S^*) = \frac{1}{2}g(S^*) = \frac{1}{2}c^*$

Let $P = \sum_{i=1}^{n} p_i$ and $W = \sum_{i=1}^{n} w_i$. Each state corresponds to a point in $\{(a, b, c) \in \mathbb{Z}^3 : 0 \leq a \leq P, 0 \leq b \leq P, 0 \leq c \leq WP\}$. Therefore, the running time of this dynamic program is $O(nWP^3)$.

Let $\delta = (1 + \frac{\epsilon}{2n})^{-1}$ for some $0 < \epsilon < 1$. Note that $\delta \in (0, 1)$. In addition, define $L = \lceil \log_{1/\delta} P \rceil$ and $M = \lceil \log_{1/\delta} WP \rceil$. Observe that we can bound L and M from above as follows:

$$L = \left\lceil \frac{\log P}{\log 1/\delta} \right\rceil \leq \left\lceil \left(1 + \frac{2n}{\epsilon}\right) \log P \right\rceil, \quad M = \left\lceil \frac{\log WP}{\log 1/\delta} \right\rceil \leq \left\lceil \left(1 + \frac{2n}{\epsilon}\right) \log WP \right\rceil$$

(The inequalities follow since $\log z \geq (z - 1)/z$ for any $z \in (0, 1]$.) Consider the grid formed by the points $(\delta^{-r}, \delta^{-s}, \delta^{-t})$, $r = 1, \ldots, L$, $s = 1, \ldots, L$, $t = 1, \ldots, M$. We divide each of the state sets E_k, $k = 1, \ldots, n$, into the boxes formed by the grid:

$$B(r, s, t) = \{(a, b, c) \in \mathbb{R}^3 : a \in [\delta^{-r+1}, \delta^{-r}], \ b \in [\delta^{-s+1}, \delta^{-s}], \ c \in [\delta^{-t+1}, \delta^{-t}]\}$$
$$r = 1, \ldots, L, \ s = 1, \ldots, L, \ t = 1, \ldots, M.$$

Observe that if (a_1, b_1, c_1) and (a_2, b_2, c_2) are in the same box,

$$\delta a_1 \leq a_2 \leq \frac{a_1}{\delta}, \qquad \delta b_1 \leq b_2 \leq \frac{b_1}{\delta}, \qquad \delta c_1 \leq c_2 \leq \frac{c_1}{\delta} . \tag{6}$$

We simplify the state sets E_k by using *a single point in each box* as a representative for all vectors in the same box. We denote these simplified state sets by E_k^{δ}. The "trimmed" dynamic program is as follows.

Algorithm 2 (Dynamic program with "trimmed" state space)

Input: weights w_i, processing times p_i for all $i \in N$
Output: an approximation to the optimal value of $\bar{x}$-MVC for schedule planning games, $f^{\bar{x}}(\bar{S})$

Pick $\epsilon \in (0, 1)$, calculate δ
$E_1^{\delta} = \{(p_1, 0, 0), (0, p_1, 0)\}$
For $k = 2, \ldots, n$
  For every vector $(a, b, c) \in E_{k-1}^{\delta}$
    Put corresponding representatives of $(a + p_k, b, c + w_k b)$ and
      $(a, b + p_k, c + w_k a)$ in E_k^{δ}
  Find $(a, b, c) \in E_n^{\delta}$ with maximum c value, $\bar{c}$
  Return $f^{\bar{x}}(\bar{S}) = \frac{1}{2}g(\bar{S}) = \frac{1}{2}\bar{c}$

The key property of the "trimmed" state space used in Algorithm 2 is that every element in the original state space has an element in the "trimmed" state space that is relatively close. More specifically,

Lemma 2. *For every $(a, b, c) \in E_k$, there exists a vector $(a', b', c') \in E_k^\delta$ such that $a' \geq \delta^k a$, $b' \geq \delta^k b$, and $c' \geq \delta^k c$.*

Using this lemma, we can analyze the performance and running time of the trimmed dynamic programming algorithm.

Theorem 9. *Algorithm 2 is a fully polynomial time approximation scheme for the $\bar{x}$-maximally violated constraint problem for schedule planning games.*

Combining Theorem 8 and Theorem 9 gives us the following result.

Theorem 10. *There exists a fully polynomial time approximation scheme for computing the least core value of schedule planning games.*

Acknowledgments

This research was supported by the National Science Foundation (DMI-0426686).

References

1. Bruno, J., Coffman, E.G., Sethi, R.: Scheduling independent tasks to reduce mean finishing time. Communications of the ACM 17, 382–387 (1974)
2. Curiel, I., Pederzoli, G., Tijs, S.: Sequencing games. European Journal of Operational Research 40, 344–351 (1989)
3. Faigle, U., Fekete, S.P., Hochstättler, W., Kern, W.: On approximately fair cost allocation for Euclidean TSP games. OR Spektrum 20, 29–37 (1998)
4. Faigle, U., Kern, W.: Approximate core allocation for binpacking games. SIAM Journal on Discrete Mathematics 11, 387–399 (1998)
5. Faigle, U., Kern, W., Paulusma, D.: Note on the computational complexity of least core concepts for min-cost spanning tree games. Mathematical Methods of Operations Research 52, 23–38 (2000)
6. Fujishige, S.: Submodular Functions and Optimization, 2nd edn. Annals of Discrete Mathematics, vol. 58. Elsevier, Amsterdam (2005)
7. Garey, M.R., Johnson, D.S., Stockmeyer, L.: Some simplified NP-complete graph problems. Theoretical Computer Science 1, 237–267 (1976)
8. Gillies, D.B.: Solutions to general non-zero-sum games. In: Tucker, A.W., Luce, R.D. (eds.) Contributions to the Theory of Games, Volume IV. Annals of Mathematics Studies, vol. 40, pp. 47–85. Princeton University Press, Princeton, NJ (1959)
9. Goemans, M.X., Queyranne, M., Schulz, A.S., Skutella, M., Wang, Y.: Single machine scheduling with release dates. SIAM Journal on Discrete Mathematics 15, 165–192 (2002)
10. Goemans, M.X., Skutella, M.: Cooperative facility location games. Journal of Algorithms 50, 194–214 (2004)
11. Goemans, M.X., Williamson, D.P.: Improved approximation algorithms for maximum cut and satisfiability problems using semidefinite programming. Journal of the ACM 42, 1115–1145 (1995)

12. Graham, R.L., Lawler, E.L., Lenstra, J.K., Rinnooy Kan, A.H.G.: Optimization and approximation in deterministic sequencing and scheduling: a survey. Annals of Discrete Mathematics 5, 287–326 (1979)
13. Granot, D., Huberman, G.: Minimum cost spanning tree games. Mathematical Programming 21, 1–18 (1981)
14. Håstad, J.: Some optimal inapproximability results. In: Proceedings of the 29th ACM Symposium on Theory of Computing, 1997, pp. 1–10. ACM Press, New York (1997)
15. Immorlica, N., Mahdian, M., Mirrokni, V.: Limitations of cross-monotonic cost sharing schemes. In: Proceedings of the 16th ACM-SIAM Symposium on Discrete Algorithms, pp. 602–611. ACM Press, New York (2005)
16. Kern, W., Paulusma, D.: Matching games: the least core and the nucleolus. Mathematics of Operations Research 28, 294–308 (2003)
17. Maniquet, F.: A characterization of the Shapley value in queueing problems. Journal of Economic Theory 109, 90–103 (2003)
18. Maschler, M., Peleg, B., Shapley, L.S.: Geometric properties of the kernel, nucleolus, and related solution concepts. Mathematics of Operations Research 4, 303–338 (1979)
19. Mishra, D., Rangarajan, B.: Cost sharing in a job scheduling problem using the Shapley value. In: Proceedings of the 6th ACM Conference on Electronic Commerce, 2005, pp. 232–239. ACM Press, New York (2005)
20. Pál, M., Tardos, É.: Group strategyproof mechanisms via primal-dual algorithms. In: Proceedings of the 44th Annual IEEE Symposium on Foundations of Computer Science, 2003, pp. 584–593. IEEE Computer Society Press, Los Alamitos (2003)
21. Potters, J., Curiel, I., Tijs, S.: Traveling salesman games. Mathematical Programming 53, 199–211 (1991)
22. Queyranne, M.: Structure of a simple scheduling polyhedron. Mathematical Programming 58, 263–285 (1993)
23. Queyranne, M., Schulz, A.S.: Scheduling unit jobs with compatible release dates on parallel machines with nonstationary speeds. In: Balas, E., Clausen, J. (eds.) Proceedings of the 4th International Conference on Integer Programming and Combinatorial Optimization. LNCS, vol. 920, pp. 307–320. Springer, Heidelberg (1995)
24. Sahni, S.: Algorithms for scheduling independent tasks. Journal of the ACM 23, 116–127 (1976)
25. Schuurman, P., Woeginger, G.J.: Approximation schemes - a tutorial. In: Möhring, R.H., Potts, C.N., Schulz, A.S., Woeginger, G.J., Wolsey, L.A. (eds.) Preliminary version of a chapter for "Lectures on Scheduling"
26. Shapley, L.S.: Cores of convex games. International Journal of Game Theory 1, 11–26 (1971)
27. Shapley, L.S., Shubik, M.: Quasi-cores in a monetary economy with nonconvex preferences. Econometrica 34, 805–827 (1966)
28. Shapley, L.S., Shubik, M.: The assignment game I: the core. International Journal of Game Theory 1, 111–130 (1971)
29. Smith, W.E.: Various optimizers for single-stage production. Naval Research Logistics Quarterly 3, 59–66 (1956)

Almost Exact Matchings

Raphael Yuster

Department of Mathematics, University of Haifa, Haifa, Israel
raphy@math.haifa.ac.il

Abstract. In the exact matching problem we are given a graph G, some of whose edges are colored red, and a positive integer k. The goal is to determine if G has a perfect matching, exactly k edges of which are red. More generally if the matching number of G is $m = m(G)$, the goal is to find a matching with m edges, exactly k edges of which are red, or determine that no such matching exists. This problem is one of the few remaining problems that have efficient randomized algorithms (in fact, this problem is in RNC), but for which no polynomial time deterministic algorithm is known.

Our first result shows that, in a sense, this problem is as close to being in P as one can get. We give a polynomial time deterministic algorithm that either correctly decides that no maximum matching has exactly k red edges, or exhibits a matching with $m(G) - 1$ edges having exactly k red edges. Hence, the additive error is one.

We also present an efficient algorithm for the exact matching problem in families of graphs for which this problem is known to be tractable. We show how to count the number of exact perfect matchings in $K_{3,3}$-minor free graphs (these include all planar graphs as well as many others) in $O(n^{3.19})$ worst case time. Our algorithm can also count the number of perfect matchings in $K_{3,3}$-minor free graphs in $O(n^{2.19})$ time.

1 Introduction

The *exact matching problem*, which is a generalization of the maximum matching problem, is defined as follows. Given a graph G with some edges colored red, and an integer k, determine if G has a maximum matching that consists of *exactly* k red edges. A special case is to determine whether there is a perfect matching with exactly k red edges. This problem was first introduced by Papadimitriou and Yannakakis in [14].

The exact matching problem is one of the few remaining natural problems that are not known to be in P, but for which there exists a polynomial time randomized algorithm, namely it is in the complexity class RP. In fact, following Karp, Upfal, and Wigderson [7] who proved that a maximum matching can be found in RNC, it has been shown by Mulmuley, Vazirani, and Vazirani that the exact matching problem is in RNC [13].

Our first result is a deterministic polynomial time algorithm that solves the exact matching problem with an additive error of 1. More formally, let $m(G)$ denote the cardinality of a maximum matching of G.

M. Charikar et al. (Eds.): APPROX and RANDOM 2007, LNCS 4627, pp. 286–295, 2007.
© Springer-Verlag Berlin Heidelberg 2007

Theorem 1. *There is a polynomial time algorithm that given a graph G with some edges colored red, and an integer k, either correctly asserts that no matching of size $m(G)$ contains exactly k red edges, or exhibits a matching of size at least $m(G) - 1$ with exactly k red edges.*

Thus, in a sense, exact matching is an example of a problem that is as close as one can get to P, without showing membership in P. As far a we know, the exact matching problem is now the only example of a natural problem in RP (and in RNC) with such an additive approximation error of 1. The proof of Theorem 1 is given is Section 2.

A large class of graphs for which the exact matching problem can be solved in polynomial time is the class of $K_{3,3}$-minor free graphs (this class includes all planar graphs and many others). This follows, with certain additional effort, from a result of Little [11] showing that $K_{3,3}$-minor free graphs have a Pfaffian orientation (see Section 3 for a definition). In fact, Vazirani has given in [16] an NC algorithm for deciding whether a $K_{3,3}$-minor free graph has a perfect exact matching. Although these results yield polynomial time algorithms, their sequential running times are far from optimal. Our next result is an efficient deterministic algorithm for computing the number of perfect matchings, and perfect exact matchings, in $K_{3,3}$-minor free graphs. In the following theorem, $\omega < 2.376$ denotes the exponent of fast matrix multiplication [2].

Theorem 2. *Given a $K_{3,3}$-minor free n-vertex graph G with some edges colored red, and an integer k, there is an algorithm whose running time is $\tilde{O}(n^{2+\omega/2}) < O(n^{3.19})$, that computes the number of perfect matchings with exactly k red edges. If $k = 0$ the running time is only $\tilde{O}(n^{1+\omega/2}) < O(n^{2.19})$.*

The algorithm is based upon several recent (and also not so recent) results concerning the computation of determinants of adjacency matrices of powers of fixed minor free graphs. Note that the case $k = 0$ implies that the number of perfect matchings in $K_{3,3}$-minor free graphs can be computed in $O(n^{2.19})$ time.

2   Proof of Theorem 1

For convenience, we shall assume that the non-red edges of G are blue. A maximum matching M is called *red-maximum* if every other maximum matching contains at most as many red edges as M.

Finding a red-maximum matching can be done in polynomial time via general weighted matching algorithms, such as the algorithm of Gabow and Tarjan [3]. Assign to each blue edge the weight m and to each red edge the weight $m + 1$, where $m = m(G)$. Notice that every maximum weighted matching must contain m edges. Indeed, if not, then its weight is at most $(m+1)(m-1) = m^2 - 1$, while every maximum matching in the unweighted graph has weight at least m^2 in the weighted graph. Now, since the weight of red edges is larger that the weight of blue edges, every maximum weighted matching maximizes the number of red edges in it.

Let, therefore, M_R and M_B be red-maximum and blue-maximum matchings, respectively. If M_R contains less than k red edges, we are done. We correctly assert that no maximum matching contains k red edges. Similarly, if M_B contains less than $m - k$ blue edges, we are done.

Consider the union of M_R and M_B. It is a spanning subgraph U of G (considered as a multigraph with edge multiplicity 2) having maximum degree 2. Each component of U is either a path (possibly a singleton vertex which is not matched in neither M_R nor M_B) or an even cycle. Cycles of length 2 in U are multiple edges having the same color (they are formed by edges that appear in both M_R and M_B).

We first claim that U has no odd length paths. Suppose that U has t odd length paths. Denote the sum of the lengths of the even cycles and the even paths by x, and the sum of the lengths of the odd paths by y. This means that U has $2m = x + y$ edges and hence $m = (x + y)/2$. On the other hand, in each even cycle or even path we can find a set of independent edges whose size is half the length of the cycle (resp. path), and hence $x/2$ independent edges in all even cycles and even paths. We can also find, in each odd path of length z, a set of $(z + 1)/2$ independent edges. Overall we can find in G a set of $x/2 + y/2 + t/2$ independent edges, but this is more than m if $t > 0$. Since G has no matching of size greater than m, we must have $t = 0$.

If S is an even cycle or even path, we say that S is of type (x, y, z) for $x \leq y \leq z$ if the length of S is $2z$, and one of the maximum matchings with z edges (that does not contain both end-edges in case S is an even path) in S has x red edges and the complimentary maximum matching has y red edges. See Figure 1 for an example. We let $min(S) = x$, $max(S) = y$ and $length(S) = 2z$. We enumerate the connected components of U by $S_1, \ldots, S_t$, and let $x_i = min(S_i)$, $y_i = max(S_i)$ and $length(S_i) = 2z_i$ for $i = 1, \ldots, t$.

Since the number of edges of U is $2m$ (an edge forming a cycle of length 2 is counted twice) we have that $\sum_{i=1}^{t} z_i = m$. Also notice that in each component S_i, one of the matchings is a subset of M_R and the complimentary matching is a subset of M_B, and hence

$$\sum_{i=1}^{t} x_i \leq k \leq \sum_{i=1}^{t} y_i.$$

Our goal is to find $m - 1$ independent edges, exactly k of which are red.

For $i = 0, \ldots, t$, let $f_i = \sum_{j=1}^{i} y_j + \sum_{j=i+1}^{t} x_j$. In particular, notice that $f_0 \leq k$ and $f_t \geq k$, and that f is monotone non-decreasing.

If some $f_i = k$ then we are done. From the first i components $S_1, \ldots, S_i$ we will take the maximum matching with $max(S_j) = y_j$ red edges, and from the remaining components we will take the maximum matching with $min(S_j) = x_j$ red edges. Altogether we obtain a matching of size m consisting of exactly k red edges.

Otherwise, let i be the unique index for which $f_{i-1} < k$ and $f_i > k$. This means that there is a unique integer p_i so that $x_i < p_i < y_i$ and so that

$$y_1 + \cdots + y_{i-1} + p_i + x_{i+1} + \ldots + x_t = k.$$

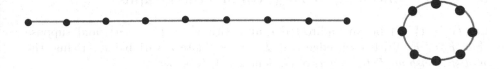

Fig. 1. An even path and an even cycle of type $(2, 3, 4)$

From each S_j for $j = 1, \ldots, i-1$ we take the maximum matching with $max(S_j) = y_j$ red edges. From each S_j for $j = i+1, \ldots, t$ we take the maximum matching with $min(S_j) = x_j$ red edges. It remains to show that in S_i we can select $z_i - 1$ independent edges, exactly p_i of which are red.

More generally, we show that if S is an even cycle or even path of type (x, y, z) and $x < p < y$, we can find in S a set of $z - 1$ independent edges, exactly p of which are red. In fact, if S is an even path we can actually find a matching of size z with exactly p red edges. Indeed, suppose the vertices of S are $v_0, \ldots, v_{2z}$, and assume, without loss of generality, that the matching (v_{2i}, v_{2i+1}) for $i = 0, \ldots, z-1$ is the one with x red edges and the complementary matching is the one with y red edges. For $q = 0, \ldots, z$, consider the matching P_q obtained by taking the edges (v_{2i-2}, v_{2i-1}) for $i = 1, \ldots, q$ and taking the the edges (v_{2i-1}, v_{2i}) for $i = q+1, \ldots, z$. Clearly, P_0 has y red edges and P_z has x red edges. The crucial point to observe is that the difference in the number of red edges between P_q and P_{q+1} is at most one, since they differ in at most one edge. Thus, there must be a q for which P_q has exactly p red edges. The proof for even cycles is similar. Since $x \neq y$, the cycle is not completely red nor completely blue. Thus, there is a vertex incident with a red and a blue edge. Deleting this vertex, we obtain a path of even length of type $(x - 1, y, z - 1)$ or of type $(x, y - 1, z - 1)$. If $p = y - 1$ we are done. Otherwise we can use the proof for the path case, this time, however, we only find a matching with $z - 1$ edges, exactly p of which are red. □

Two additional heuristics can be added to the algorithm of Theorem 1. In the proof we order the connected components $S_1, \ldots, S_t$ of U arbitrarily. Every ordering yields different values for the f_i's. There may be a specific order for which $f_i = k$ for some i, in which case we can actually find a maximum matching with k red edges. We can determine, in polynomial time, if such an order exists, as this is just a subset-sum problem on the set of values $max(S_i) - min(S_i)$ for $i = 1, \ldots, t$ where we wish to find a subset sum of value $k - \sum_{i=1}^{t} min(S_i)$. This subset sum problem can be easily solved using dynamic programming. Another thing to notice is that, in case the algorithm finds a matching with $m - 1$ edges, k of which are red, then the blue subgraph induced by the vertices not incident with these k red edges already contains a matching with $m - 1 - k$ blue edges, and, since $m = m(G)$ it cannot contain a matching with more than $m - k$ blue edges. We therefore need to find just one additional augmenting path in this subgraph, using any maximum matching algorithm (we may fail, however, since it may be the case that this blue subgraph has maximum matching size $m - 1 - k$).

3   Exact Matching in $K_{3,3}$-Minor Free Graphs

Let $G = (V, E)$ be an undirected graph with $V = \{1, \ldots, n\}$, and suppose that $F \subset E$. With each edge $e \in E$ we associate a variable x_e. Define the F-distinguishing Tutte matrix of G, denoted $A_F(G)$, by:

$$a_{ij} = \begin{cases} +x_{ij}, & \text{if } ij \in E \setminus F \text{ and } i < j; \\ -x_{ji}, & \text{if } ij \in E \setminus F \text{ and } i > j; \\ +yx_{ij}, & \text{if } ij \in F \text{ and } i < j; \\ -yx_{ji}, & \text{if } ij \in F \text{ and } i > j; \\ 0, & \text{otherwise.} \end{cases}$$

Notice that if $F = \emptyset$ then the indeterminate y is not needed and $A_\emptyset(G)$ is just the usual Tutte matrix of G. Notice also that $A_F(G)$ is skew-symmetric, and hence its determinant $det(A_F(G))$ is always a *square* of a polynomial in the matrix entries. This polynomial (determined uniquely up to a sign) is the *Pfaffian* of $A_F(G)$, denoted $Pf(A_F(G))$.

For $M \subset E$, let $x(M) = \prod_{e \in M} x_e$. Tutte proved [15] that there is a bijection between the terms in $Pf(A_\emptyset(G))$ and the perfect matchings of G. Namely each term in $Pf(A_\emptyset(G))$ (no matter what its sign is) equals some $x(M)$ where M is a perfect matching. In particular, $Pf(A_\emptyset(G)) \neq 0$ if and only if G has a perfect matching. Tutte's argument immediately generalizes to show that there is a bijection between the perfect matchings of G containing exactly k edges from F, and the terms of the form $y^k x(M)$ in $Pf(A_F(G))$. Although Tutte's result is an important combinatorial insight, it is not computationally attractive, as we cannot compute the determinant of a symbolic matrix (with $|E| + 1$ symbols, in fact) efficiently.

The theory of Pfaffian orientations was introduced by Kasteleyn [8] to solve some enumeration problems arising from statistical physics. These orientations can be used in order to replace the variables x_e in the Tutte matrix with $+1$ and -1 so that each term in the Pfaffian of the resulting matrix is positive, if its sign is positive, and negative, if its sign is negative. Thus, perfect matchings can be efficiently counted in *Pfaffian orientable* graphs. Kasteleyn [8] proved that all planar graphs are Pfaffian orientable. This result was extended by Little [11] who proved that all $K_{3,3}$-minor free graphs (these include all planar graphs, by Kuratowski's and Wagner's Theorems) are Pfaffian orientable.

Let $G = (V, E)$ be an undirected graph, C an even cycle in G, and $\mathbf{G}$ an orientation of G. We say that C is *oddly oriented by* $\mathbf{G}$ if, when traversing C in either direction, the number of co-oriented edges (i.e., edges whose orientation in $\mathbf{G}$ and in the traversal is the same) is odd.

Definition 1. *An orientation $\mathbf{G}$ of G is* Pfaffian *if the the following condition holds: for any two perfect matchings M, M' in G, every cycle in $M \cup M'$ is oddly oriented by $\mathbf{G}$.*

Note that all cycles in the union of two perfect matchings are even. Not all graphs have a Pfaffian orientation. For example, $K_{3,3}$ does not have one.

Given a Pfaffian orientation $\boldsymbol{G}$ of a Pfaffian orientable graph G, replace each variable x_{ij} where $i < j$ with $+1$ if $(i,j) \in E(\boldsymbol{G})$ and with -1 if $(j,i) \in E(\boldsymbol{G})$. Denote the resulting F-distinguishing matrix by $A_F(\boldsymbol{G})$. Kasteleyn proved the following result [8]:

Lemma 1. *For any Pfaffian orientation $\boldsymbol{G}$ of G, $|Pf(A_\emptyset(\boldsymbol{G}))|$ is the number of perfect matchings of G (or, stated otherwise, $\det(A_\emptyset(\boldsymbol{G}))$ is the square of the number of perfect matchings.*

Kasteleyn's result immediately generalizes to the F-distinguishing Tutte matrix.

Corollary 1. *For any Pfaffian orientation $\boldsymbol{G}$ of G, the coefficient of y^k in $Pf(A_F(\boldsymbol{G}))$ is the number of perfect matchings with exactly k edges from F.*

In order to prove Theorem 2 we need to show that, given an input graph G, we can first efficiently find a Pfaffian orientation of G (or else determine that G is not $K_{3,3}$-minor free). Once we do that, we can use Corollary 1 to compute the determinant of $A_F(\boldsymbol{G})$, where F is the set of red edges. Although this can be done in polynomial time (the matrix has only one symbol, y), this will not be so efficient (the fastest deterministic algorithm for this problem runs in $O(n^{\omega+1})$ time even if $k = 0$ and the matrix is symbol-free). Thus, we need to use a different approach.

We say that a graph $G - (V, E)$ has a (k, α)-*separation*, if V can be partitioned into three parts, A, B, C so that $A \cap B = \emptyset$, $|A \cup C| \le \alpha|V|$, $|B \cup C| \le \alpha|V|$, $|C| \le k$, and if $uv \in E$ and $u \in A$ then $v \notin B$. We say that A and B are separated by C, that C is a separator, and that the partition (A, B, C) *exhibits* a (k, α)-separation.

By the seminal result of Lipton and Tarjan [10], n-vertex planar graphs have an $(O(n^{1/2}), 2n/3)$-separation. In fact, they also show how to compute such a separation in linear time. Subsequently, Alon, Seymour, and Thomas [1] extended the result of Lipton and Tarjan to H-minor free graphs. The running time of their algorithm is $O(n^{1.5})$ for every fixed H. Both algorithms do not assume that the input graph satisfies the conditions. Namely, if the algorithms fail to obtain the desired separator, they conclude that the graph is non-planar (in the Lipton-Tarjan algorithm) or contains an H-minor (in the Alon-Seymour-Thomas algorithm).

When the existence of an $(f(n), \alpha)$-separation can be proved for each n-vertex graph belonging to a hereditary family (closed under taking subgraphs), one can recursively continue separating each of the separated parts A and B until the separated pieces are small enough. This obviously yields a *separator tree*. Notice that planarity, as well as being H-minor free, is a hereditary property. More formally, we say that a graph $G = (V, E)$ with n vertices has an $(f(n), \alpha)$-*separator tree* if there exists a full rooted binary tree T so that the following holds:

(i) Each $t \in V(T)$ is associated with some $V_t \subset V$.
(ii) The root of T is associated with V.
(iii) If $t_1, t_2 \in V(T)$ are the two children of $t \in V(T)$ then $V_{t_1} \subset V_t$ and $V_{t_2} \subset V_t$. Furthermore, if $A = V_{t_1}$, $B = V_{t_2}$ and $C = V_t \setminus (V_{t_1} \cup V_{t_2})$ then (A, B, C) exhibits an $(f(|V_t|), \alpha)$-separation of $G[V_t]$ (the subgraph induced by G_t).
(iv) If t is a leaf then $|V_t| = O(1)$.

By using divide and conquer, the result of Alon, Seymour, and Thomas mentioned above can be stated as follows.

Lemma 2. *For a fixed graph H, an H-minor free graph with n vertices has an $(O(n^{1/2}), 2/3)$-separator tree and such a tree can be found in $\tilde{O}(n^{1.5})$ time.*

Let A be an $n \times n$ matrix. The *representing graph* of A, denoted $G(A)$, is defined by the vertex set $\{1, \ldots, n\}$ where, for $i \neq j$ we have an edge ij if and only if $a_{ij} \neq 0$ or $a_{ji} \neq 0$.

Generalizing the nested dissection method of George [4], Lipton, Rose, and Tarjan [9] and Gilbert and Tarjan [5] proved the following.

Lemma 3. *Let B be a symmetric positive definite $n \times n$ matrix. If, for some positive constant $\alpha < 1$, and for some constant $\beta \geq 1/2$, $G(B)$ has bounded degree and an $(O(n^{1/2}), \alpha)$-separator tree, and such a tree is given, then Gaussian elimination on B can be performed with $O(n^{\omega/2})$ arithmetic operations. The resulting LU factorization of B is given by matrices L and D, $B = LDL^T$, where L is unit lower-triangular and has $\tilde{O}(n)$ non-zero entries, and D is diagonal.*

The requirement that B should be positive definite is needed in the algorithm of Lemma 3 only in order to guarantee that no zero diagonal entries are encountered, and hence no row or column pivoting is needed during the elimination process. This was also observed in [12]. We can easily modify Lemma 3 as follows.

Lemma 4. *Let A be an $n \times n$ integer matrix where each entry has absolute value at most N, and each row and column of A contain only a bounded number of nonzero entries. Let $B = AA^T$. If, for some positive constant $\alpha < 1$, and for some constant $\beta \geq 1/2$, $G(B)$ has an $(O(n^{1/2}), \alpha)$-separator tree, and such a tree is given, then $\det(B)$ can be computed in $\tilde{O}(n^{\omega/2+1} \log N)$ time.*

Proof: Clearly, B has only has a bounded number of non-zero entries in each row or column, and hence $G(B)$ has bounded degree. Notice also that $B = AA^T$ and hence is a symmetric positive semi-definite matrix, and, in fact, if A is non-singular then B is positive definite. Thus we can apply Lemma 3 to B with the additional observation that if we encounter a zero on the diagonal during the elimination process we conclude that $\det(B) = 0$. Notice that Lemma 3 immediately yields the determinant, as this is just the product of the diagonal entries of D. The number of arithmetic operations in Lemma 3 is $O(n^{\omega/2})$, but this is not the actual time complexity. Notice that each element of B has absolute value at most $\Theta(N)$, and therefore, when performing the Gaussian elimination, each rational number encountered has its numerator and denominator no larger in absolute value than $n!\Theta(N)^n$ (in fact, much smaller), and hence the number of bits of the numerator and denominator is $\tilde{O}(n \log N)$. Consequently, each arithmetic operation requires $\tilde{O}(n \log N)$ time. Thus, the bit complexity of the algorithm is $\tilde{O}(n^{\omega/2+1} \log N)$.   □

The requirement that $G(A)$ have bounded degree in Lemma 4 is very limiting. Our input graphs are $K_{3,3}$-minor free, but may have vertices with very high degree. There is a general technique that transforms every graph G to another graph G' so that the latter has maximum degree at most r, where $r \geq 3$, and so that the number of perfect exact matchings of G and G' is the same. Furthermore, there is an easy translation of maximum exact matchings in G to maximum exact matchings in G' and vice versa.

Suppose G has a vertex u of degree at least $r + 1$. Pick two neighbors of u, say v, w. Add two new vertices u' and u'', add the edges $uu', u'u'', u''v, u''w$ and delete the original edges uv, uw. Clearly, this vertex-splitting operation does not change the number of perfect matchings (and increases the size of the maximum matching by 1). Another thing to notice is that if we do not color the new edges uu' and $u'u''$ and let the color of $u''v$ be the same as the color of uv and let the color of $u''w$ be the same as the color of uw then also the number of perfect matchings with exactly k red edges does not change. Finally, another pleasing property is that if G has a Pfaffian orientation before the splitting, it also has one after the splitting; just orient uu' and $u'u''$ in the same direction as a directed path of length 2, orient $u''v$ the same as uv, and orient $u''w$ the same as uw. By repeatedly performing vertex splitting until there are no vertices with degree greater than r, we obtain a desired *vertex split* graph G'. Clearly, if G has n vertices and $O(n)$ edges (as do, say, all fixed-minor-free graphs), then G' has $O(n)$ vertices and $O(n)$ edges as well.

Unfortunately, vertex splitting does not preserve H-minor freeness. We could have that G is H-minor free, but G', its vertex splitted counterpart, contains an H-minor. Luckily, however, a result of [17] (Lemma 2.1 there) shows that G' still has an $(O(n^{1/2}), \alpha)$-separator tree, and one can still use the algorithm of Lemma 2 to produce it, in the same running time. We restate Lemma 2.1 from [17] for the special case of $K_{3,3}$ which is what we need here:

Lemma 5. *Given a $K_{3,3}$-minor free graph G with n vertices, there is a vertex-split graph G' of G of bounded maximum degree so that G' has an $(O(n^{1/2}), \alpha)$-separator tree where $\alpha < 1$ is an absolute constant. Furthermore, such a separator tree for G' can be constructed in $O(n^{1.5})$ time.*

Proof of Theorem 2: Given the input graph G (assumed to be $K_{3,3}$-minor free) with F being its set of red edges, we first apply Lemma 5 and obtain a vertex split graph G' of G and a separator tree T for G'. The time to produce G' and T is $\tilde{O}(n^{1.5})$ (notice that if the input graph is not $K_{3,3}$-minor free the algorithm of Lemma 5 may still succeed, but if it fails we are certain that our input graph is not $K_{3,3}$-minor free, so we halt). Let n' denote the number of vertices of G' and notice that $n' = O(n)$.

Next, we find a Pfaffian orientation for G. This can be done in linear time using the algorithm of Vazirani [16]. We note that Vazirani's algorithm for constructing a Pfaffian orientation of a $K_{3,3}$-minor free graph is an NC algorithm, and is a bit wasteful w.r.t. its sequential work; its main bottleneck is the need to compute triconnected components, and these can be done in linear time using the result of Hopcroft and Tarjan [6]. (Notice that we could have settled for an even slower algorithm, as our overall claimed running time in Theorem 2 is more than quadratic. Also notice that if G is not $K_{3,3}$-minor free then the algorithm may fail to produce a Pfaffian orientation and we halt.).

From the Pfaffian orientation $\boldsymbol{G}$ of G we directly construct, in linear time, a Pfaffian orientation $\boldsymbol{G'}$ of G', as shown in the above description of vertex splitting. Let, therefore, $A' = A_F(\boldsymbol{G'})$ be the resulting F-distinguishing matrix. Notice that if $F = \emptyset$ then A' is just a matrix with entries in $\{-1,0,1\}$. Otherwise, we shall replace each occurrence of the indeterminate y with $N = (n')^{n'}$ and denote the resulting matrix by A. If $F = \emptyset$ then we simply define $A = A'$ and $N = 1$. In either case, A is an integer matrix with absolute value of each entry at most N, and with each row and column having a bounded number of nonzero entries.

Clearly, T is a separator tree for $G(A)$, the representing graph of A. Now, let $B = AA^T$. In order to apply Lemma 4 we need to show that $G(B)$ also has a separator tree, and we must construct such a tree. Notice, however, that the graph $G(B)$ is just the square of the graph $G(A)$ (whose underlying graph is G'). It was observed in [12] that if we take any separator S of G' (that is, S corresponds to a vertex of T), and replace it with the *thick* separator S' consisting of S and all of its neighbors, we obtain a separator for the square of G'. Since G' has bounded degree, we have that $|S'| = \Theta(|S|)$ and hence we can immediately construct from T an $(O(n^{1/2}, \alpha))$-separator tree for $G(B)$. We can now apply Lemma 4 and compute $det(B)$ in $\tilde{O}(n^{\omega/2+1} \log N)$ time, which is $\tilde{O}(n^{\omega/2+1})$ if $F = \emptyset$ and $\tilde{O}(n^{\omega/2+2})$ if $F \neq \emptyset$.

It remains to show how the number of perfect exact matchings can be retrieved from $det(B)$. First notice that $det(B) = det(AA^T) = det(A)^2$ and hence we have $det(A)$. Next notice that A is skew-symmetric and hence we also have $Pf(A) = \sqrt{det(A)}$. Now, if $F = \emptyset$ then, by Lemma 1, $Pf(A)$ is just the number of perfect matchings of G', which, in turn, is the same as the number of perfect matchings of G. If on the other hand, $F \neq \emptyset$, then our choice of replacing y with $(n')^{n'}$ enables us to construct $Pf(A')$ from $Pf(A)$ by considering $Pf(A)$ as a number written in base $(n')^{n'}$ and noticing that there is no "carry" since the number of perfect matchings is less than $n! < (n')^{n'}$. Thus, by Lemma 1 the coefficient of y^k in $Pf(A')$ (or, in turn, the $k + 1$'th least significant digit of $Pf(A)$ written in base $(n')^{n'}$) is the number of perfect matchings of G' with exactly k red edges. This is identical to the number of perfect matchings of G with exactly k red edges.   □

References

1. Alon, N., Seymour, P.D., Thomas, R.: A separator theorem for nonplanar graphs. J. Amer. Math. Soc. 3(4), 801–808 (1990)
2. Coppersmith, D., Winograd, S.: Matrix multiplication via arithmetic progressions. J. Symbol. Comput. 9, 251–280 (1990)
3. Gabow, H.N., Tarjan, R.E.: Faster scaling algorithms for general graph matching problems. Journal of the ACM 38(4), 815–853 (1991)
4. George, A.: Nested dissection of a regular finite element mesh. SIAM J. Numer. Analysis 10, 345–363 (1973)

5. Gilbert, J.R., Tarjan, R.E.: The analysis of a nested dissection algorithm. Numerische Mathematik 50(4), 377–404 (1987)
6. Hopcroft, J.E., Tarjan, R.E.: Dividing a Graph into Triconnected Components. SIAM J. Comput. 2(3), 135–158 (1973)
7. Karp, R., Upfal, E., Wigderson, A.: Constructing a perfect matching is in Random NC. Combinatorica 6, 35–48 (1986)
8. Kasteleyn, P.W.: Graph theory and crystal physics. In: Graph Theory and Theoretical Physics, pp. 43–110. Academic Press, London (1967)
9. Lipton, R.J., Rose, D.J., Tarjan, R.E.: Generalized nested dissection. SIAM J. Numer. Analysis 16(2), 346–358 (1979)
10. Lipton, R.J., Tarjan, R.E.: A separator theorem for planar graphs. SIAM J. Applied Math. 36(2), 177–189 (1979)
11. Little, C.H.C.: An extension of Kasteleyn's method of enumerating the 1-factors of planar graphs. In: Holton, D. (ed.) Combinatorial Mathematics. Proc. 2nd Australian Conference. Lecture Notes in Mathematics, vol. 403, pp. 63–72. Springer, Heidelberg (1974)
12. Mucha, M., Sankowski, P.: Maximum matchings in planar graphs via Gaussian elimination. In: Albers, S., Radzik, T. (eds.) ESA 2004. LNCS, vol. 3221, pp. 532–543. Springer, Heidelberg (2004)
13. Mulmuley, K., Vazirani, U.V., Vazirani, V.V.: Matching is as easy as matrix inversion. Combinatorica 7(1), 105–113 (1987)
14. Papadimitriou, C.H., Yannakakis, M.: The complexity of restricted spanning tree problems. Journal of the ACM 29(2), 285–309 (1982)
15. Tutte, W.T.: The factorization of linear graphs. J. London Math. Soc. 22, 107–111 (1947)
16. Vazirani, V.V.: NC algorithms for computing the number of perfect matchings in $K_{3,3}$-free graphs and related problems. Inf. Comput. 80(2), 152–164 (1989)
17. Yuster, R., Zwick, U.: Maximum matching in graphs with an excluded minor. In: Proc. of 18th SODA, ACM/SIAM, pp. 108–117 (2007)

On Approximating the Average Distance
Between Points

Kfir Barhum, Oded Goldreich, and Adi Shraibman

Weizmann Institute of Science, Rehovot, Israel

Abstract. We consider the problem of approximating the average distance between pairs of points in a high-dimensional Euclidean space, and more generally in any metric space. We consider two algorithmic approaches:

1. Referring only to Euclidean Spaces, we randomly reduce the high-dimensional problem to a one-dimensional problem, which can be solved in time that is almost-linear in the number of points. The resulting algorithm is somewhat better than a related algorithm that can be obtained by using the known randomized embedding of Euclidean Spaces into ℓ_1-metric.
2. An alternative approach consists of selecting a random sample of pairs of points and outputting the average distance between these pairs. It turns out that, for any metric space, it suffices to use a sample of size that is linear in the number of points. Our analysis of this method is somewhat simpler and better than the known analysis of Indyk (STOC, 1999). We also study the existence of corresponding deterministic algorithms, presenting both positive and negative results. In particular, in the Euclidean case, this approach outperforms the first approach.

In general, the second approach seems superior to the first approach.

Introduction

As observed by Feige [3], natural objects give rise to functions for which approximating the average value of a function is easier than approximating the average value of a general function with a corresponding domain and range. For example, the average degree of a connected n-vertex graph can be approximated upto some constant factor (i.e., 2) based on $\sqrt{n}$ samples, whereas the average value of a general function from $[n]$ to $[n-1]$ cannot be approximated to within any constant factor based on $o(n)$ samples. Indeed, the discrepancy is due to the restrictions imposed on functions that represent quantities that correspond to the type of object considered (i.e., degrees of a graph).

Goldreich and Ron initiated a general study of approximating average parameters of graphs [4]. In particular, they considered the problems of approximating the average degree of a vertex in a graph as well as approximating the average distance between pairs of vertices. They considered both queries to the quantity

M. Charikar et al. (Eds.): APPROX and RANDOM 2007, LNCS 4627, pp. 296–310, 2007.

of interest (e.g., the degree of a vertex) and natural queries to the corresponding object (e.g., neighborhood queries in a graph). (Barhum [1, Chap. 2] extended their average-degree approximation algorithm to k-uniform hypergraphs).

In the present paper, we consider the problem of approximating the average distance between points in a (high-dimensional) Euclidean space, and more generally for points in any metric space. Although this study may be viewed as an imitation of [4], the specific context (i.e., geometry rather than graph theory) is different and indeed different techniques are employed.

Our aim is beating the obvious algorithm that computes the exact value of the aforementioned average (by considering all pairs of points). But, unlike in the graph theoretic setting (cf. [4]), we cannot hope for approximation algorithms that run in time that is sub-linear in the number of points (because a single "exceptional" point may dominate the value of the average of all pairwise distances). Thus, we seek approximation algorithms that run in time that is almost linear in the number of points. We consider two algorithmic approaches.

1. *Manipulating the object itself.* This algorithmic approach (presented in Section 1) applies only to the case of Euclidean Spaces. The algorithm operates by randomly reducing the high-dimensional problem to a one-dimensional problem. This approach is closely related to the known randomized embedding of Euclidean Spaces into ℓ_1-metric; see discussion in Section 1.6.
2. *Sampling and averaging.* The straightforward approach (presented in Section 2) consists of selecting a random sample of pairs of points and outputting the average distance between these pairs. Our analysis of this method is somewhat simpler and better than the known analysis of Indyk [5]. We also study the existence of corresponding deterministic algorithms, presenting both positive and negative results.

It turns out that the second algorithmic approach is superior to the first approach. Furthermore, we believe that Sections 2.2 and 2.3 may be of independent interest. In particular, they yield a simple proof to the fact that the graph metric of every constant-degree expander cannot be embedded in a Euclidean space without incurring logarithmic distortion (cf. [7, Prop. 4.2]). We note that, in general, Sections 2.2 and 2.3 touch on themes that are implicit in [7].

1   Euclidean Spaces and the Random Projection Algorithm

In this section, we present an almost linear time algorithm for approximating the sum of the distances between points in a high-dimensional Euclidean space; that is, given $P_1, ..., P_n \in \mathsf{R}^d$, the algorithm outputs an approximation of $\sum_{i,j \in [n]} \|P_i - P_j\|$. The algorithm is based on randomly reducing the high-dimensional case to the one-dimensional case, where the problem is easily solvable. Specifically, the d-dimensional algorithm repeatedly selects a uniformly distributed direction, projects all points to the corresponding line, and computes the sum of the corresponding distances (on this line). Each such experiment yields an *expected* value

298   K. Barhum, O. Goldreich, and A. Shraibman

that is a $\rho(d)$ fraction of the sum that we seek, where $\rho(d)$ denotes the expected length of the projection of a uniformly distributed unit vector on a fixed direction. Furthermore, as we shall see, $O(\epsilon^{-2})$ repetitions suffice for obtaining a $1 \pm \epsilon$ factor approximation (with error probability at most $1/3$).

1.1 The One-Dimensional Case

Our starting point is the fact that an almost linear-time algorithm that computes the exact value (of the sum of all pairwise distances) is known in the one-dimensional case. This algorithm proceeds by first sorting the input points $p_1, ..., p_n \in \mathbb{R}$ such that $p_1 \leq p_2 \leq \cdots \leq p_n$, then computing $\sum_{j=1}^{n} |p_1 - p_j|$ (in a straightforward manner), and finally for $i = 1, ..., n-1$ computing in constant-time the value $\sum_{j=1}^{n} |p_{i+1} - p_j|$ based on $\sum_{j=1}^{n} |p_i - p_j|$. Specifically, we use the fact that

$$
\sum_{j=1}^{n} |p_{i+1} - p_j| = \sum_{j=1}^{i} (p_{i+1} - p_j) + \sum_{j=i+1}^{n} (p_j - p_{i+1})
$$

$$
= (i - (n-i)) \cdot (p_{i+1} - p_i) + \sum_{j=1}^{i} (p_i - p_j) + \sum_{j=i+1}^{n} (p_j - p_i)
$$

$$
= (2i - n) \cdot (p_{i+1} - p_i) + \sum_{j=1}^{n} |p_i - p_j|.
$$

1.2 A Simple Deterministic Approximation for the d-Dimensional Case

Combining the foregoing algorithm with the basic inequalities regarding norms (i.e., the relation of Norm2 to Norm1), we immediately obtain a (deterministic) $\sqrt{d}$-factor approximation algorithm for the d-dimensional case. Specifically, consider the points $P_1, ..., P_n \in \mathbb{R}^d$, where $P_i = (p_{i,1}, ..., p_{i,d})$, and let $\|P_i - P_j\|$ denote the Euclidean (i.e., Norm2) distance between P_i and P_j. Then it holds that

$$
\frac{1}{\sqrt{d}} \cdot \sum_{i,j \in [n]} \sum_{k=1}^{d} |p_{i,k} - p_{j,k}| \leq \sum_{i,j \in [n]} \|P_i - P_j\| \leq \sum_{i,j \in [n]} \sum_{k=1}^{d} |p_{i,k} - p_{j,k}|. \quad (1)
$$

Thus, $\sum_{i,j \in [n]} \|P_i - P_j\|$ can be approximated by $\sum_{i,j \in [n]} \sum_{k=1}^{d} |p_{i,k} - p_{j,k}|$, which is merely the sum of d one-dimensional problems (i.e., $\sum_{k=1}^{d} \sum_{i,j \in [n]} |p_{i,k} - p_{j,k}|$).

1.3 The Main Algorithm

However, we seek a better approximation than the $\sqrt{d}$-approximation just described. Indeed, the main contribution of this section is an almost linear-time (randomized) approximation scheme for the value of $\sum_{i,j \in [n]} \|P_i - P_j\|$. The key conceptual observation is that the rough bounds provided by Eq. (1) reflect (extremely different) worst-case situations, whereas "on the average" there is a tight relation between the Norm2 and the Norm1 values. Recall that while the Norm2 value is invariant of the system of coordinates, Norm1 is defined based on such a system and is very dependent on it. This suggests that, rather than computing the Norm1 value according to an arbitrary system of coordinates (which leaves some slackness w.r.t the Norm2 value that we seek), we should compute the Norm1 value according to a random system of coordinates (i.e., a system that is selected uniformly at random).

To see what will happen when we use a random system of coordinates (i.e., orthonormal basis of R^d), we need some notation. Let $\langle u, v \rangle$ denote the inner-product of the (d-dimensional) vectors u and v. Then, the Norm1 value of the vector v according to the system of coordinates (i.e., orthonormal basis) $b_1, ..., b_d$ equals $\sum_{k=1}^{d} |\langle v, b_k \rangle|$. The key technical observation is that, for an orthonormal basis $b_1, ..., b_d$ that is chosen uniformly at random, it holds that

$$\mathbf{E}_{b_1,...,b_d} \left[\sum_{k=1}^{d} |\langle v, b_k \rangle| \right] = d \cdot \mathbf{E}_{b_1} \left[|\langle v, b_1 \rangle| \right] = d \cdot \|v\| \cdot \mathbf{E}_{b_1} \left[|\langle \overline{v}, b_1 \rangle| \right], \quad (2)$$

where $\overline{v} = v/\|v\|$ is a unit vector in the direction of v. Furthermore, for any unit vector $u \in \mathsf{R}^d$, the value $\mathbf{E}_{b_1} [|\langle u, b_1 \rangle|]$ is independent of the specific vector u, while b_1 is merely a uniformly distributed unit vector (in R^d). Thus, letting r denote a uniformly distributed unit vector, we define $\rho(d) \stackrel{\text{def}}{=} \mathbf{E}_r [|\langle u, r \rangle|]$ and observe that

$$\mathbf{E}_{b_1} [|\langle v, b_1 \rangle|] = \|v\| \cdot \rho(d). \quad (3)$$

Moreover, a closed form expression for $\rho(d)$, which is linearly related to $1/\sqrt{d}$, is well-known (see Appendix).

Turning back to Eq. (3), we have $\|v\| = \mathbf{E}_r [|\langle v, r \rangle|] / \rho(d)$, where r is a random unit vector. It follows that

$$\sum_{i,j \in [n]} \|P_i - P_j\| = \frac{1}{\rho(d)} \cdot \mathbf{E}_r \left[\sum_{i,j \in [n]} |\langle P_i - P_j, r \rangle| \right] \quad (4)$$

Noting that $|\langle P_i - P_j, r \rangle| = |\langle P_i, r \rangle - \langle P_j, r \rangle|$, this completes the randomized reduction of the d-dimensional case to the one-directional case; that is, the reduction selects a random unit vector r, and computes $\sum_{i,j \in [n]} |\langle P_i, r \rangle - \langle P_j, r \rangle|$. Note that we have obtained an *unbiased estimator*[1] for $\sum_{i,j \in [n]} \|P_i - P_j\|$. Furthermore, as shown in the Appendix, this estimator is strongly concentrated around its expected value; in particular, the square root of the variance of this estimator is linearly related to its expectation. We thus obtain:

Theorem 1. *There exists a randomized algorithm that, given an approximation parameter $\epsilon > 0$ and points $P_1, ..., P_n \in \mathsf{R}^d$, runs for $\tilde{O}(\epsilon^{-2} \cdot |P_1, ..., P_n|)$-time and*

---

[1] A random variable X (e.g., the output of a randomized algorithm) is called an unbiased estimator of a value v if $\mathbf{E}[X] = v$.

with probability at least 2/3 outputs a value in the interval $[(1-\epsilon)\cdot A, (1+\epsilon)\cdot A]$, *where* $A = \sum_{i,j\in[n]} \|P_i - P_j\|/n^2$.

Let us spell-out the algorithm asserted in Theorem 1 and complete its analysis. This algorithm consists of repeating the following procedure $O(\epsilon^{-2})$ times:

1. Uniformly select a unit vector $r \in \mathsf{R}^d$.
2. For $i = 1, ..., n$, compute the projection $p_i = \langle P_i, r\rangle$.
3. Compute $\frac{1}{\rho(d)\cdot n^2} \sum_{i,j\in[n]} |p_i - p_j|$, by invoking the procedure described in Section 1.1 and using the value $\rho(d)$ computed as in the Appendix.

The algorithm outputs the average of the values obtained in the various iterations. Step 2 can be implemented using $n \cdot d$ real (addition and multiplication) operations, whereas the complexity of Step 3 is dominated by sorting n real values.

The issues addressed next include the exact implementation of a single iteration (i.e., approximating real-value computations), and providing an analysis of a single iteration (thus proving that $O(\epsilon^{-2})$ iterations suffice). Let us start with the latter.

1.4 Probabilistic Analysis of a Single Iteration

Let us denote by X the random value computed by a single iteration, and let $Z = (\rho(d) \cdot n^2) \cdot X$. Recall that $Z = \sum_{i,j\in[n]} |\langle P_i, r\rangle - \langle P_j, r\rangle|$, which equals $\sum_{i,j\in[n]} |\langle P_i - P_j, r\rangle|$, where r is a uniformly distributed unit vector. Note that

$$\mathbf{E}[Z] = \mathbf{E}_r\left[\sum_{i,j\in[n]} |\langle P_i - P_j, r\rangle|\right]$$

$$= \rho(d) \cdot \sum_{i,j\in[n]} \|P_i - P_j\|,$$

where the second equality is due to Eq. (4). This establishes the claim that *each iteration provides an unbiased estimator of* $\sum_{i,j\in[n]} \|P_i - P_j\|/n^2$. As usual, the usefulness of a single iteration is determined by the variance of the estimator. A simple upper-bound on the variance of Z may be obtained as follows

$$\mathbf{V}[Z] = \mathbf{V}_r\left[\sum_{i,j\in[n]} |\langle P_i - P_j, r\rangle|\right]$$

$$\leq \mathbf{E}_r\left[\left(\sum_{i,j\in[n]} |\langle P_i - P_j, r\rangle|\right)^2\right]$$

$$\leq \left(\sum_{i,j\in[n]} \|P_i - P_j\|\right)^2$$

where the second inequality uses the fact that $|\langle P_i - P_j, r \rangle| \leq \|P_i - P_j\|$ holds (for any unit vector r). This implies that $\mathbf{V}[Z] \leq \rho(d)^{-2} \cdot \mathbf{E}[Z]^2 = O(d \cdot \mathbf{E}[Z]^2)$. In the Appendix, we will show that it actually holds that $\mathbf{V}[Z] = O(\mathbf{E}[Z]^2)$.

Applying Chebyshev's Inequality, it follows that the average value of t iterations (of the procedure) yields an $(1 \pm \epsilon)$-factor approximation with probability at least $1 - \frac{\mathbf{V}[Z]}{(\epsilon \cdot \mathbf{E}[Z])^2 \cdot t}$. Thus, for $\mathbf{V}[Z] = O(\mathbf{E}[Z]^2)$, setting $t = O(\epsilon^{-2})$ will do.

1.5 Implementation Details (i.e., The Required Precision)

By inspecting the various operations of our algorithm, one may verify that it suffices to conduct all calculations with $O(\log(1/\epsilon))$ bits of precision (see [1] for details); that is, such implementation also yields a $(1 \pm \epsilon)$-factor approximation of the desired value. In particular, this holds with respect to the selection of $r \in \mathsf{R}^d$, which is the only randomization that occurs in a single iteration. It follows that each iteration can be implemented using $m \stackrel{\text{def}}{=} O(d \cdot \log(1/\epsilon))$ coin tosses (i.e., $O(\log(1/\epsilon))$ bits of precision per each coordinate of r).

Note that the foregoing implementation of a single iteration yields a random value having an expectation of $(1 \pm \epsilon) \cdot \sum_{i,j \in [n]} \|P_i - P_j\|/n^2$. A full-derandomization of this implementation yields a deterministic $(1 \pm \epsilon)$-approximation algorithm of running-time $2^m \cdot \tilde{O}(|P_1, ..., P_n|)$. However, this is inferior to the result presented in Section 2.3.

1.6 Reflection

In retrospect, the foregoing algorithm is an incarnation of the "embedding paradigm" (i.e., the fact that the Euclidean metric can be embedded with little distortion in the ℓ_1-metric). Specifically, our algorithm may be described as a two-step process in which the Euclidean problem is first randomly reduced (by a random projection) to a problem regarding the ℓ_1-metric in $d' = O(\epsilon^{-2})$ dimensions, and next the latter problem is reduced to d' one-dimensional problems. As shown above, with probability at least $2/3$, the sum of the distances in the Euclidean problem is approximated upto a factor of $1 \pm \epsilon$ by the sum of the distances in the d'-dimensional ℓ_1-metric.

It is well-known (cf. [6]) that a random projection of n points in Euclidean space into a ℓ_1-metric in $d'' = O(\epsilon^{-2} \log n)$ dimensions preserves each of the $\binom{n}{2}$ distances upto a factor of $1 \pm \epsilon$. This yields a reduction of the Euclidean problem to $d'' \gg d'$ one-dimensional problems. Thus, we gained a factor of $d''/d' = \Theta(\log n)$ by taking advantage of the fact that, for our application, we do not require a good (i.e., small distortion) embedding of all pairwise distances, but rather a good (i.e., small distortion) embedding of the average pairwise distances.

2 General Metric and the Sampling Algorithm

The straightforward algorithm for approximating the average pairwise distances consists of selecting a random sample of m pairs of points and outputting the average distance between these pairs. This algorithm works for any metric space. The question is how large should its sample be; that is, how should m relate to the number of points, denoted n. Indeed, m should be proportional to $\mathbf{V}[Z]/\mathbf{E}[Z]^2$, where Z represents the result of a single "distance measurement" (i.e., the distance between a uniformly selected pair of points). Specifically, to obtain an $(1 \pm \epsilon)$-approximation of the average of all pairwise distances, it suffices to take $m = O(\mathbf{V}[Z]/(\epsilon \cdot \mathbf{E}[Z])^2)$. Thus, we first upper-bound the ratio $\mathbf{V}[Z]/\mathbf{E}[Z]^2$, showing that it is at most linear in the number of points (see Section 2.1). We later consider the question of derandomization (see Sections 2.2 and 2.3).

2.1   The Approximation Provided by a Random Sample

We consider an arbitrary metric $(\delta_{i,j})_{i,j\in[n]}$ over n points, where $\delta_{i,j}$ denote the distance between the ith and jth point. Actually, we shall only use the fact that the metric is non-negative and symmetric (i.e., for every $i, j \in [n]$ it holds that $\delta_{i,j} = \delta_{j,i} \geq 0$) and satisfies the triangle inequality (i.e., for every $i, j, k \in [n]$ it holds that $\delta_{i,k} \leq \delta_{i,j} + \delta_{j,k}$). Recall that Z is a random variable representing the distance between a uniformly selected pair of points; that is, $Z = \delta_{i,j}$, where $(i, j) \in [n] \times [n]$ is uniformly distributed.

Proposition 2. *For Z as above, it holds that* $\mathbf{V}[Z] = O(n \cdot \mathbf{E}[Z]^2)$.

Proof. By an averaging argument, it follows that there exists a point c (which may be viewed as a "center") such that

$$\frac{1}{n} \cdot \sum_{j\in[n]} \delta_{c,j} \leq \frac{1}{n^2} \cdot \sum_{i,j\in[n]} \delta_{i,j} . \tag{5}$$

Using such a (center) point c, we upper-bound $\mathbf{E}[Z^2]$ as follows:

$$\mathbf{E}[Z^2] = \frac{1}{n^2} \cdot \sum_{i,j\in[n]} \delta_{i,j}^2$$

$$\leq \frac{1}{n^2} \cdot \sum_{i,j\in[n]} (\delta_{i,c} + \delta_{c,j})^2$$

$$\leq \frac{1}{n^2} \cdot \sum_{i,j\in[n]} (2\delta_{i,c}^2 + 2\delta_{c,j}^2)$$

$$= \frac{4n}{n^2} \cdot \sum_{j\in[n]} \delta_{c,j}^2$$

where the first inequality is due to the triangle inequality and the last equality uses the symmetry property. Thus, we have

$$\mathbf{E}[Z^2] \leq 4n \cdot \sum_{j \in [n]} \left(\frac{\delta_{c,j}}{n} \right)^2$$

$$\leq 4n \cdot \left(\sum_{j \in [n]} \frac{\delta_{c,j}}{n} \right)^2$$

$$\leq 4n \cdot \left(\sum_{i,j \in [n]} \frac{\delta_{i,j}}{n^2} \right)^2$$

where the last inequality is due to Eq. (5). Thus, we obtain $\mathbf{E}[Z^2] \leq 4n \cdot \mathbf{E}[Z]^2$, and the proposition follows (because $\mathbf{V}[Z] \leq \mathbf{E}[Z^2]$).   ∎

Tightness of the bound. To see that Proposition 2 is tight, consider the metric $(\delta_{i,j})_{i,j \in [n]}$ such that $\delta_{i,j} = 1$ if either $i = v \neq j$ or $j = v \neq i$ and $\delta_{i,j} = 0$ otherwise. (Note that this metric can be embedded on the line with v at the origin (i.e. location 0) and all other points co-located at 1.) In this case $\mathbf{E}[Z] = 2(n-1)/n^2 < 2/n$ while $\mathbf{E}[Z^2] = \mathbf{E}[Z]$, which means that $\frac{\mathbf{V}[Z]}{\mathbf{E}[Z]^2} = \frac{1}{\mathbf{E}[Z]} - 1 > \frac{n}{2} - 1$.

Conclusion. Our analysis implies that it suffices to select a random sample of size $m = O(n/\epsilon^2)$. This improves over the bound $m = O(n/\epsilon^{7/2})$ established by Indyk [5, Sec. 8].

2.2   On the Limits of Derandomization

A "direct" derandomization of the sampling-based algorithm requires trying all pairs of points, which foils our aim of obtaining an approximation algorithm that runs faster than the obvious "exact" algorithm. Still, one may ask whether a better derandomization exists. We stress that such a derandomization should work well for all possible metric spaces. The question is how well can a *fixed* sample of pairs (over $[n]$) approximate the average distance between all the pairs (of points) in *any* metric space (over n points). The corresponding notion is formulated as follows.

Definition 3. (universal approximator): *A multi-set of pairs $S \subseteq [n] \times [n]$ is called a* universal (L, U)-approximator *if for every metric $(\delta_{i,j})_{i,j \in [n]}$ it holds that*

$$L(n) \cdot n^{-2} \cdot \sum_{i,j \in [n]} \delta_{i,j} \leq |S|^{-1} \cdot \sum_{(i,j) \in S} \delta_{i,j} \leq U(n) \cdot n^{-2} \cdot \sum_{i,j \in [n]} \delta_{i,j}. \quad (6)$$

In such a case, we also say that S is a universal U/L-approximator.

Needless to say, $[n] \times [n]$ itself is a universal 1-approximator, but we seek universal approximators of almost linear (in n) size. We shall show an explicit construction (of almost linear size) that provides a logarithmic-factor approximation, and prove that this is the best possible.

We note that universal approximators can be represented as n-vertex directed graphs (possibly with parallel and anti-parallel edges). In some cases, we shall present universal approximators as undirected graphs, while actually meaning the corresponding directed graph obtained by replacing each undirected edge with a pair of anti-parallel directed edges.

A construction. For an integer parameter k, we shall consider the generalized k-dimensional hypercube having n vertices, which are viewed as k-ary sequences over $[n^{1/k}]$ such that two vertices are connected by an edges if and only if (as k-long sequences) they differ in one position. That is, the vertices $\langle \sigma_1, ..., \sigma_k \rangle \in [n^{1/k}]^k$ and $\langle \tau_1, ..., \tau_k \rangle$ are connected if and only if $|\{i \in [k] : \sigma_i \neq \tau_i\}| = 1$. In addition, we add k self-loops to each vertex, where each such edge corresponds to some $i \in [k]$. Thus, the degree of each vertex in this n-vertex graph equals $k \cdot n^{1/k}$. We shall show that this graph constitutes a universal $O(k)$-approximator.

Theorem 4. *The generalized k-dimensional n-vertex hypercube is a universal $(1/k, 2)$-approximator.*

In particular, the binary hypercube (i.e., $k = \log_2 n$) on n vertices constitutes a universal $O(\log n)$-approximator.

Proof. For every two vertices $u, v \in [n]$, we consider a canonical path of length k between u and v. This path, denoted $P_{u,v}$, corresponds to the sequence of vertices $w^{(0)}, ..., w^{(k)}$ such that $w^{(i)} = \langle \sigma_1, ..., \sigma_{k-i}, \tau_{k-i+1}, ..., \tau_k \rangle$, where $u = \langle \sigma_1, ..., \sigma_k \rangle$ and $v = \langle \tau_1, ..., \tau_k \rangle$. (Here is where we use the self-loops.) Below, we shall view these paths as sequences of edges (i.e., $P_{u,v}$ is viewed as the k-long sequence $(w^{(0)}, w^{(1)}), ..., (w^{(k-1)}, w^{(k)})$). An important property of these canonical paths is that each edge appears on the same number of paths.

Letting E denoted the directed pairs of vertices that are connected by an edge, and using the triangle inequality and the said property of canonical paths, we note that

$$n^{-2} \cdot \sum_{u,v \in [n]} \delta_{u,v} \leq n^{-2} \cdot \sum_{u,v \in [n]} \sum_{(w,w') \in P_{u,v}} \delta_{w,w'}$$

$$= n^{-2} \cdot \sum_{(w,w') \in E} |\{(u, v) \in [n] \times [n] : (w, w') \in P_{u,v}\}| \cdot \delta_{w,w'}$$

$$= n^{-2} \cdot \sum_{(w,w') \in E} \frac{k \cdot n^2}{|E|} \cdot \delta_{w,w'}$$

which equals $k \cdot |E|^{-1} \cdot \sum_{(w,w') \in E} \delta_{w,w'}$. On the other hand, letting $\Gamma(u) = \{v : (u, v) \in E\}$, and using the triangle inequality and the regularity of the graph, we note that

$$|E|^{-1} \cdot \sum_{(u,v) \in E} \delta_{u,v} \leq |E|^{-1} \cdot \sum_{(u,v) \in E} n^{-1} \cdot \sum_{w \in [n]} (\delta_{u,w} + \delta_{w,v})$$

$$= |E|^{-1} n^{-1} \cdot \sum_{u,w \in [n]} \sum_{v \in \Gamma(u)} (\delta_{u,w} + \delta_{w,v})$$

$$= |E|^{-1}n^{-1} \cdot \left(\sum_{u,w\in[n]} |\Gamma(u)| \cdot \delta_{u,w} + \sum_{v,w\in[n]} |\Gamma^{-1}(v)| \cdot \delta_{u,w} \right)$$

$$= |E|^{-1}n^{-1} \cdot 2 \cdot \frac{|E|}{n} \cdot \sum_{u,w\in[n]} \delta_{u,w}$$

which equals $2 \cdot n^{-2} \cdot \sum_{u,w\in[n]} \delta_{u,w}$.   ∎

Comment: The second part of the proof of Theorem 4 only uses the fact that the hypercube is a regular graph. Thus, any regular graph is a universal $(0,2)$-approximator. Relaxing the regularity hypothesis, we note that every n-vertex graph in which the maximum degree is at most t times the average-degree is a universal $(0,2t)$-approximator. On the other hand, the first part of the proof uses the fact that all vertex-pairs (in the hypercube) can be connected by paths such that no edge is used in more than $k \cdot n^2/|E|$ of the paths. The argument generalizes to arbitrary connected graphs in which no edge is used in more than $K \cdot \frac{n^2}{|E|} \ (\leq n^2)$ of the paths, implying that such a graph is a universal $(K^{-1}, n+2)$-approximator.

A Lower-Bound. We now show that the construction provided in Theorem 4 is optimal. Indeed, our focus is on the case $k < \log_2 n$ (and actually, even $k = o(\log n)$).

Theorem 5. *A universal k-approximator for n points must have $\frac{n^{(k+1)/k}}{4k}$ edges.*

Proof. Let $G = ([n], E)$ be (a directed graph representing) a universal k-approximator. We first note that no vertex can have (out)degree exceeding $2k \cdot (|E|/n)$ (even when not counting self-loops). The reason being that if vertex v has a larger degree, denoted d_v, then we reach contradiction by considering the metric $(\delta_{i,j})_{i,j\in[n]}$ such that $\delta_{i,j} = 1$ if either $i = v \neq j$ or $j = v \neq i$ and $\delta_{i,j} = 0$ otherwise. (Note that this metric can be embedded on the line with v at the origin (i.e. location 0) and all other points co-located at 1.) In this case $n^{-2} \cdot \sum_{i,j\in[n]} \delta_{i,j} < 2/n$, whereas $|E|^{-1} \cdot \sum_{(i,j)\in E} \delta_{i,j} = |E|^{-1} \cdot d_v$ (which is greater than $2k/n$).

We now consider the metric induced by the graph G itself; that is, $\delta_{i,j}$ equals the distance between vertices i and j in the graph G. Clearly, $|E|^{-1} \cdot \sum_{(i,j)\in E} \delta_{i,j} = 1$, but (as we shall see) the average distance between pairs of vertices is much larger. Specifically, letting $d \leq 2k \cdot (|E|/n)$ denote the maximum degree of a vertex in G, we have

$$n^{-2} \cdot \sum_{u,v\in[n]} \delta_{u,v} \geq \min_{u\in[n]} \left\{ n^{-1} \cdot \sum_{v\in[n]} \delta_{u,v} \right\}$$

$$\geq n^{-1} \cdot \sum_{i=0}^{t} d^i \cdot i$$

where t is the smallest integer such that $\sum_{i=0}^{t} d^i \geq n$, which implies $t > \log_d((1 - d^{-1}) \cdot n) \approx \frac{\ln n - d^{-1}}{\ln d}$. Thus, $n^{-2} \cdot \sum_{u,v \in [n]} \delta_{u,v}$ is lower-bounded by $\frac{d^t}{n} \cdot t \geq (1 - d^{-1}) \cdot t > (1 - 2d^{-1}) \cdot \frac{\ln n}{\ln d}$ which must be at most at most k (because otherwise G cannot be a universal k-approximator). Using $(1 - 2d^{-1}) \cdot \frac{\ln n}{\ln d} \leq k$ it follows that $d' \overset{\text{def}}{=} d^{1/(1-2d^{-1})} \geq n^{1/k}$, which (using $d' < 2d$) implies $2d > n^{1/k}$. Finally, using $d \leq 2k \cdot (|E|/n)$, we get $|E| \geq \frac{dn}{2k} > \frac{n^{(k+1)/k}}{4k}$. ∎

Comment: The proof actually applies to any universal (k^{-1}, k)-approximator.

2.3   On the Limits of Derandomization, Revisited

Note that while the first part of the proof of Theorem 5 (i.e., bounding the maximum degree in terms of the average-degree) uses an Euclidean metric, the main part of the proof refers to a graph metric (which may not have a Euclidean embedding). Thus, Theorem 5 does not rule out the existence of sparse graphs that provide good approximations for points in a Euclidean space.

Definition 6. (universal approximator, restricted): *A multi-set of pairs $S \subseteq [n] \times [n]$ is called a (L, U)-approximator for the class $\mathcal{M}$ if Eq. (6) holds for any n-point metric of the class $\mathcal{M}$.*

Needless to say, any universal (L, U)-approximator is (L, U)-approximator for the Euclidean metric, but the converse does not necessarily hold. Indeed, we shall see that approximators for the Euclidean metric can have much fewer edges than universal approximators (for any metric).

Theorem 7. *For every constant $\epsilon > 0$, there exists a efficiently constructible $(1 + \epsilon)$-approximator for the Euclidean metric that has $O(n/\epsilon^2)$ edges. Furthermore, such approximators can be constructed in linear-time.*

Theorem 7 follows by reducing the general case of (high-dimensional) Euclidean metric to the line-metric (i.e., one-dimensional Euclidean metric) and presenting an approximator for the latter metric.

Proposition 8. *Suppose that $S \subseteq [n] \times [n]$ is an f-approximator for the line-metric. Then S constitutes an f-approximator for the Euclidean metric.*

Proof. Considering a d-dimensional Euclidean space with points $P_1, ..., P_n \in \mathbb{R}^d$, we let r denote a uniformly distributed unit vector in $\mathbb{R}^d$. By Eq. (3), for every vector $v \in \mathbb{R}^d$ it holds that $\mathbf{E}_r[|\langle v, r \rangle|] = \rho(d) \cdot \|v\|$. Thus, for every $i, j \in [n]$ it holds that $\|P_i - P_j\| = \rho(d)^{-1} \cdot \mathbf{E}_r[|\langle P_i, r \rangle - \langle P_j, r \rangle|]$ and so

$$\sum_{(i,j) \in S} \|P_i - P_j\| = \rho(d)^{-1} \cdot \mathbf{E}_r \left[\sum_{(i,j) \in S} |\langle P_i, r \rangle - \langle P_i, r \rangle| \right]$$

$$\sum_{i,j \in [n]} \|P_i - P_j\| = \rho(d)^{-1} \cdot \mathbf{E}_r \left[\sum_{i,j \in [n]} |\langle P_i, r \rangle - \langle P_i, r \rangle| \right]$$

The proposition follows by applying the hypothesis to (each value of r in) the r.h.s of each of the foregoing equalities. ∎

Strong expanders. Good approximators for the line-metric are provided by the following notion of graph expansion. We say that the (undirected) graph $G = ([n], E)$ is a $(1 - \epsilon)$-**strong expander** if for every $S \subset [n]$ it holds that

$$\frac{|E(S, [n] \setminus S)|}{|E|} = (1 \pm \epsilon) \cdot \frac{|S| \cdot (n - |S|)}{n^2/2} \tag{7}$$

where $E(V_1, V_2) \overset{\text{def}}{=} \{\{u, v\} \in E : u \in V_1 \wedge v \in V_2\}$. As we shall see (in Proposition 10), sufficiently good expanders (i.e., having relative eigenvalue bound $\epsilon/2$) are strong expanders (i.e., $(1 - \epsilon)$-strong expander). We first establish the connection between strong expanders and good approximators for the line-metric.

Proposition 9. *Suppose that the graph $G = ([n], E)$ is a $(1-\epsilon)$-strong expander. Then it yields a $(1 + \epsilon)$-approximator for the line-metric.*

The sufficient condition regarding G is also necessary (e.g., for any cut $(S, [n] \setminus S)$, consider the points $p_1, ..., p_n \in R$ such that $p_i = 0$ if $i \in S$ and $p_i = 1$ otherwise).

Proof. For any sequence of points $p_1, ..., p_n \in R$, consider the "sorting permutation" $\pi : [n] \rightarrow [n]$ such that for every $i \in [n - 1]$ it holds that $p_{\pi(i)} \leq p_{\pi(i+1)}$. By counting the contribution of each "line segment" $[p_{\pi(i)}, p_{\pi(i+1)}]$ to $\sum_{i,j \in [n]} |p_i - p_j|$, we get

$$\sum_{i,j \in [n]} |p_i - p_j| = \sum_{i=1}^{n-1} 2i \cdot (n - i) \cdot (p_{\pi(i+1)} - p_{\pi(i)}) \tag{8}$$

Similarly, for $S_i = \{\pi(1), ..., \pi(i)\}$, we have

$$\sum_{i,j:\{i,j\} \in E} |p_i - p_j| = \sum_{i=1}^{n-1} 2|E(S_i, [n] \setminus S_i)| \cdot (p_{\pi(i+1)} - p_{\pi(i)}) \tag{9}$$

Using the proposition's hypothesis, we have for every $i \in [n - 1]$,

$$\frac{2 \cdot |E(S_i, [n] \setminus S_i)|}{2 \cdot |E|} = (1 \pm \epsilon) \cdot \frac{2 \cdot i \cdot (n - i)}{n^2} \tag{10}$$

and the proposition follows by combining Eq. (8)–(10). ∎

Proposition 10. *Suppose that the graph $G = ([n], E)$ is a d-regular graph with a second eigenvalue bound $\lambda < d$. Then G is a $(1 - (2\lambda/d))$-strong expander.*

Thus any family of constructible $O(\epsilon^{-2})$-regular Ramanujan graphs (e.g., [8]) yields a constructible family of $(1 - \epsilon)$-strong expanders. Furthermore, such graphs can be constructed in almost linear time (i.e., each edge can be determined by a constant number of arithmetic operations in a field of size smaller than n).

Proof. The claim follows from the Expander Mixing Lemma, which refers to any two sets $A, B \subseteq [n]$, and asserts that

$$\left| \frac{|E(A,B)|}{d \cdot n} - \rho(A) \cdot \rho(B) \right| \leq \frac{\lambda}{d} \cdot \sqrt{\rho(A) \cdot \rho(B)} \qquad (11)$$

where $\rho(S) = |S|/n$ for every set $S \subseteq [n]$. Applying the lemma to the special case of $A = B$, we infer that $|E(A,A)|$ resides in the interval $[\rho(A) \cdot d \cdot |A| \pm \lambda \cdot |A|]$. Assuming, without loss of generality, that $|A| \leq n/2$, we conclude that $|E(A, [n] \setminus A)|/(d \cdot n)$ resides in $[\rho(A) \cdot (1 - \rho(A)) \pm (\lambda/d) \cdot \rho(A)]$, which is a sub-interval of $[(1 \pm (2\lambda/d)) \cdot \rho(A) \cdot (1 - \rho(A))]$ (because $1 - \rho(A) \geq 1/2$). ∎

Conclusion. The foregoing three propositions imply the *existence of* (efficiently constructible) $O(\epsilon^{-2})$-*regular graphs that are* $(1 - \epsilon)$-*approximators for the Euclidean metric.* It is easy to see that the argument extends to the ℓ_1-metric. We note that combining the foregoing fact with the proof of Theorem 5 it follows that no constant-degree expander graph can be embedded in a Euclidean space (resp., ℓ_1-metric) without incurring logarithmic distortion. Details follow.

Recall that a metric (e.g., a graph metric) on n points, denoted $(\delta_{i,j})_{i,j \in [n]}$ is said to be embedded in a Euclidean space with distortion ρ if the distance between points i and j in the embedding is at least $\delta_{i,j}$ and at most $\rho \cdot \delta_{i,j}$. Thus, if a $(1-\epsilon)$-strong expander can be embedded in a Euclidean space with distortion ρ, then this graph constitutes a $(1+\epsilon) \cdot \rho$-approximator of the graph metric induced by itself. But then the proof of Theorem 5 implies that for $k = (1 + \epsilon) \cdot \rho$ this graph must have at least $n^{(k+1)/k}/4k$ edges. Actually, if the graph in question is regular, then we may skip the first part of the said proof and use $d = 2|E|/n$ (rather than $d = 2k|E|/n$), thus obtaining degree lower-bound of $n^{1/k}/2$. Either way, if the graph has constant degree then it must hold that $k = \Omega(\log n)$ (and, for any fixed $\epsilon > 0$, it follows that $\rho = \Omega(\log n)$).

On the other hand, using the fact (cf. [2]) that every metric on n points can be embedded in the ℓ_2-metric with distortion at most $O(\log n)$, it follows that constant-degree expander graphs yield universal log-factor approximators (for any metric on n points). This improves over the special case of Theorem 4 that refers to graphs of logarithmic degree.

Acknowledgments

We are grateful to Bernard Chazelle, Dan Halperin, Robi Krauthgamer, Nati Linial, David Peleg, Gideon Schechtman, and Avi Wigderson for helpful discussions. We are also grateful to the anonymous reviewers of RANDOM'07 for their comments. This research was partially supported by the Israel Science Foundation (grant No. 460/05).

References

1. Barhum, K.: Approximating Averages of Geometrical and Combinatorial Quantities. M.Sc. Thesis, Weizmann Institute of Science,(February 2007) Available from http://www.wisdom.weizmann.ac.il/~oded/msc-kfir.html
2. Bourgain, J.: On Lipschitz Embedding of Finite Metric Spaces in Hilbert Space. Israel J. Math. 52, 46–52 (1985)
3. Feige, U.: On Sums of Independent Random Variables with Unbounded Variance, and Estimating the Average Degree in a Graph. In: Proc. of the 36th STOC, 2004, pp. 594–603 (2004)
4. Goldreich, O., Ron, D.: Approximating Average Parameters of Graphs. In: Díaz, J., Jansen, K., Rolim, J.D.P., Zwick, U. (eds.) APPROX 2006 and RANDOM 2006. LNCS, vol. 4110, pp. 363–374. Springer, Heidelberg (2006)
5. Indyk, P.: Sublinear Time Algorithms for Metric Space Problems. In: Proc. of the 31st STOC, pp. 428–434 (1999)
6. Johnson, W.B., Lindenstrauss, J.: Extensions of Lipschitz Mappings into a Hilbert Space. In:Conf. in Modern Analysis and Probability, pp. 189–206 (1984)
7. Linial, N., London, E., Rabinovich, Y.: The Geometry of Graphs and Some of its Algorithmic Applications. Combinatorica 15(2), 215–245 (1995)
8. Lubotzky, A., Phillips, R., Sarnak, P.: Ramanujan Graphs. Combinatorica 8, 261–277 (1988)

Appendix: On $\rho(d)$ and the Related Variance $\sigma^2(d)$

Recall that $\rho(d) \overset{\text{def}}{=} \mathbf{E}_r\left[|\langle u, r\rangle|\right]$, where u is an arbitrary unit vector (in R^d) and r is a uniformly distributed unit vector (in R^d). Analogously, we define the corresponding variance $\sigma^2(d) \overset{\text{def}}{=} \mathbf{V}_r\left[|\langle u, r\rangle|\right]$. Note that both $\rho(d)$ and $\sigma^2(d)$ are actually independent of the specific vector u.

Theorem 11. (folklore): $\rho(d) = \frac{1}{(d-1)\cdot A_{d-2}}$, where $A_0 = \pi/2$, $A_1 = 1$, and $A_k = \frac{k-1}{k} \cdot A_{k-2}$.

In particular, $\rho(2) = 2/\pi \approx 0.63661977$ and $\rho(3) = 1/2$. In general, $\rho(d) = \Theta(1/\sqrt{d})$. A proof of Theorem 11 can be found in [1, Sec. 3.6]. Using similar techniques (see [1, Sec. 3.7]), one may obtain

Theorem 12. (probably also folklore): $\sigma^2(d) = O(1/d)$. Furthermore, for any two unit vectors $u_1, u_2 \in \mathsf{R}^d$ and for a uniformly distributed unit vector $r \in \mathsf{R}^d$, it holds that $\mathbf{E}_r\left[|\langle u_1, r\rangle| \cdot |\langle u_2, r\rangle|\right] = O(1/d)$.

In fact, the furthermore clause follows from $\sigma^2(d) = O(1/d)$ (and $\rho(d)^2 = O(1/d)$) by using the Cauchy-Schwartz Inequality.[2]

---

[2] That is, $\mathbf{E}_r[|\langle u_1, r\rangle| \cdot |\langle u_2, r\rangle|] \leq \sqrt{\mathbf{E}_r[|\langle u_1, r\rangle|^2] \cdot \mathbf{E}_r[|\langle u_2, r\rangle|^2]} = \mathbf{E}_r[|\langle u_1, r\rangle|^2] = \sigma^2(d) + \rho(d)^2$.

Improved bound for the variance of Z. Recalling that $Z = \sum_{i,j\in[n]} |\langle P_i - P_j, r\rangle|$ and using Theorem 12, we have

$$\mathbf{V}[Z] \leq \mathbf{E}[Z^2] = \mathbf{E}_r \left[\sum_{i_1,j_1,i_2,j_2\in[n]} |\langle P_{i_1} - P_{j_1}, r\rangle| \cdot |\langle P_{i_2} - P_{j_2}, r\rangle| \right]$$

$$= O(1/d) \cdot \sum_{i_1,j_1,i_2,j_2\in[n]} \|P_{i_1} - P_{j_1}\| \cdot \|P_{i_2} - P_{j_2}\|$$

$$= O(1/d) \cdot \left(\sum_{i,j\in[n]} \|P_i - P_j\| \right)^2.$$

Recalling that $\mathbf{E}[Z] = \rho(d) \cdot \sum_{i,j\in[n]} \|P_i - P_j\|$, it follows that $\mathbf{V}[Z] = O(1/d) \cdot (\mathbf{E}[Z]/\rho(d))^2$. Using $\rho(d) = \Omega(1/\sqrt{d})$, we conclude that $\mathbf{V}[Z] = O(\mathbf{E}[Z]^2)$.

On Locally Decodable Codes, Self-correctable Codes, and t-Private PIR⋆

Omer Barkol, Yuval Ishai, and Enav Weinreb

Dept. of Computer Science, Technion, Haifa, Israel
{omerb,yuvali,weinreb}@cs.technion.ac.il

Abstract. A k-query *locally decodable code* (LDC) allows to probabilistically decode any bit of an encoded message by probing only k bits of its corrupted encoding. A stronger and desirable property is that of *self-correction*, allowing to efficiently recover not only bits of the message but also arbitrary bits of its encoding. In contrast to the initial constructions of LDCs, the recent and most efficient constructions are not known to be self-correctable. The existence of self-correctable codes of comparable efficiency remains open.

A closely related problem with a very different motivation is that of *private information retrieval* (PIR). A k-server PIR protocol allows a user to retrieve the i-th bit of a database, which is replicated among k servers, without revealing information about i to any *individual* server. A natural generalization is *t-private PIR*, which keeps i hidden from any t colluding servers. In contrast to the initial PIR protocols, it is not known how to generalize the recent and most efficient protocols to yield t-private protocols of comparable efficiency.

In this work we study both of the above questions, showing that they are in fact related. We start by presenting a general transformation of any 1-private PIR protocol (equivalently, LDC) into a t-private protocol with a similar amount of communication per server. Combined with the recent result of Yekhanin (STOC 2007), this yields a significant improvement over previous t-private PIR protocols. A major weakness of our transformation is that the number of servers in the resulting t-private protocols grows exponentially with t. We show that if the underlying LDC satisfies the stronger *self-correction* property, then there is a similar transformation in which the number of servers grows only *linearly* with t, which is the best one can hope for. Finally, we study the question of closing the current gap between the complexity of the best known LDC and that of self-correctable codes, and relate this question to a conjecture of Hamada concerning the algebraic rank of combinatorial designs.

1 Introduction

In this work we study two natural questions concerning locally decodable codes and the related notion of private information retrieval:

- **Local decoding vs. self-correction.** Can state-of-the-art locally decodable codes be enhanced to satisfy a stronger self-correction property at a minor additional cost?

⋆ Research supported by grant 1310/06 from the Israel Science Foundation, grant 2004361 from the U.S.-Israel Binational Science Foundation, and the Technion VPR fund. Part of this research was done while visiting IPAM.

M. Charikar et al. (Eds.): APPROX and RANDOM 2007, LNCS 4627, pp. 311–325, 2007.

- **1-private vs. t-private PIR.** Can state-of-the-art PIR protocols be strengthened to provide privacy against t colluding servers at a minor additional cost?

We obtain new insights on these questions and show that, somewhat surprisingly, they are related to each other. We elaborate on the two questions below.

1.1   Local Decoding vs. Self-correction

A *locally decodable code* (LDC) is an error-correcting code which supports sublinear-time decoding. More concretely, a k-query LDC allows to *probabilistically* decode any symbol of an encoded message by probing only k symbols of its corrupted encoding. A stronger and desirable property is that of *self-correction*. In a k-query *self-correctable code* (SCC) the probabilistic decoder is required to recover an arbitrary symbol of the *encoding* of the message rather than a symbol of the message itself. Self-correction is stronger than local decoding in the sense that any linear SCC can be used as an LDC via a systematic encoding[1] of the message. In the context of information storage, the additional self-correction property allows the encoding to efficiently "repair" itself in the presence of mobile adversarial corruptions. LDCs were used in several complexity theoretic contexts, such as worst-case to average-case reductions [26,7,6,30] and randomness extraction [27]. SCCs originate from program checkers [12,26] and were later applied in the context of probabilistically checkable proofs [16,5,2,3]. We refer the reader to [32] for a survey of applications of LDCs and SCCs in complexity theory.

The complexity of LDCs with a fixed number of queries k was first explicitly studied by Katz and Trevisan [23], and has since been the subject of a large body of work (see [32,17] for surveys). Until this day, there is a nearly exponential gap between the best known upper bounds on the length of k-query LDCs [10,36] and the corresponding lower bounds [23,24,33,34], for any constant $k \geq 3$. The initial constructions of LDCs were based on Reed-Muller codes and allowed to encode an n-bit message by a k-query LDC of length $\exp(n^{1/(k-1)})$ [7,5,14,8].[2] These LDCs also enjoy the stronger self-correction property, a fact that blurred the distinction between the two notions. (Indeed, some definitions of LDC actually required self-correction.) In contrast to these initial constructions, the recent and most efficient constructions [10,36] are not known to be self-correctable. In particular, the recent breakthrough result of Yekhanin [36] (see also [28]) gives a 3-query LDC of a sub-exponential length (more precisely, of length $\exp(n^{1/\log\log n})$ for infinitely many n), assuming the existence of infinitely many Mersenne primes, or of length $\exp(n^{10^{-7}})$ using the largest known Mersenne prime. In contrast, the best known k-query SCC is still of length $\exp(n^{1/(k-1)})$, as in the initial constructions. The existence of better self-correctable codes remains open.

1.2   1-Private vs. t-Private PIR

A *private information retrieval* (PIR) [14] protocol allows to query a remote database while hiding the identity of the items being retrieved even from the servers holding the

---

[1] An encoding $E : \mathbb{F}^n \to \mathbb{F}^m$ is *systematic* if for every $i \in [n]$ there is $j \in [m]$ such that $E(x)_j = x_i$ for every $x \in \mathbb{F}^n$. Any linear code has a systematic encoding.

[2] Here and in the following, $\exp(f(n))$ stands for $2^{O(f(n))}$.

database. More concretely, an (information-theoretic)[3] k-server PIR protocol allows a user to retrieve the i-th bit of a database $x \in \{0,1\}^n$, which is replicated among k servers, without revealing *any* information about i to any *individual* server. A trivial way to solve the PIR problem is simply by communicating x to the user. If there is only a single server holding the database, this trivial solution is optimal. However, Chor et al. [14] demonstrated that replicating the database among $k \geq 2$ servers gives rise to nontrivial protocols whose communication complexity is sublinear in the database size n. Most of the subsequent research in this area (see [17] for a survey) focused on the goal of minimizing the communication complexity of k-server PIR as a function of n.

Katz and Trevisan [23] showed that there is a close relation between PIR and LDC. In particular, a k-server PIR protocol in which the user sends $c(n)$ bits to each server and receives a single bit in return corresponds to a k-query binary LDC of length $\exp(c(n))$. Indeed, Yekhanin's LDC [36] gives rise to a 3-server PIR protocol with $n^{o(1)}$ bits of communication (assuming the existence of infinitely many Mersenne primes).

A natural generalization of PIR is t-*private PIR*, in which the user's selection i should be kept hidden from any t colluding servers. The initial PIR protocols from [14,1] were generalized to yield t-private protocols of comparable efficiency [21,8,9,35]. (Such a generalization inherently requires to increase the number of servers proportionally to t, since a t-private k-server protocol implies a 1-private $\lfloor k/t \rfloor$-server protocol of the same complexity.) In contrast, no similar generalization was known for the recent PIR protocols of Beimel et al. [10] and Yekhanin [36]. Even in the case $t = 2$, the best 2-private k-server protocol prior to the current work required $\Omega(n^{1/(k-1)})$ bits of communication.

1.3 Our Results

From PIR to t-private PIR. We present a general transformation of any linear k-server PIR protocol, or alternatively a k-query LDC, into a t-private k^t-server protocol with a similar amount of communication per server. Combined with Yekhanin's protocol [36] and assuming the infinitude of Mersenne primes, this yields a t-private protocol with a constant number of servers and $n^{o(1)}$ bits of communication, for any constant t. (In fact, for small values of t the improvement over the best previous t-private protocols [9,35] holds unconditionally.) See Corollaries 1 and 2 for these results.

Better t-private PIR via SCC. A major weakness of the above transformation is that the number of servers in the resulting t-private protocols grows exponentially with t. We show that if the underlying LDC satisfies the stronger *self-correction* property, then there is a similar transformation in which the number of servers grows only *linearly* with t. As noted, this is the best one could hope for. See Corollary 3 for this result.

On LDC vs. SCC. Finally, we study the question of closing the current complexity gap between LDC and SCC. This question is further motivated by the above application to

---

[3] In this work we are not concerned with the alternative *computational* model for PIR [13,25], in which the user's privacy should only hold against computationally bounded servers. Information-theoretic PIR protocols have some additional advantages over their computational counterparts beyond the type of privacy they guarantee; see [10] for discussion.

t-private PIR. While we do not resolve this question, we provide some indication for its difficulty by relating it to a conjecture of Hamada [18] (see also [19,4,11]) concerning the algebraic rank of combinatorial designs. Roughly speaking, Hamada's conjecture asserts that block designs which originate from Reed-Muller codes minimize the rank of their incidence matrix. As noted above, the best known SCCs are based on Reed-Muller codes. We show that a refutation of Hamada's conjecture would lead to better SCCs. This implies that proving the known SCCs to be optimal is at least as hard as proving Hamada's conjecture. We also show a weak converse of this statement, namely that a natural variant of Hamada's conjecture implies the optimality of SCCs of a certain natural form. This provides some evidence for the difficulty of improving the current Reed-Muller based constructions of SCCs.

Organization. In Section 2, we start with some preliminaries. Then, in Section 3, we describe two methods for boosting the privacy of PIR protocols: one that applies to general PIR protocols (Section 3.1) and a more efficient one that applies to PIR protocols with the additional self-correcting property (Section 3.2). In Section 4, we formally define SCCs and prove their equivalence to the above mentioned version of PIR protocols. In Section 5, we discuss a connection between SCCs and block designs, and finally, in Section 6, we mention some open problems.

2   Preliminaries

A k-server PIR protocol involves k servers $S_1, \ldots, S_k$ (throughout this paper k is constant), each holding the same string $x \in \{0,1\}^n$, and a user $\mathcal{U}$ who wants to retrieve the bit x_i without revealing the value of i. The following definition of a PIR protocol is similar to standard definitions from the literature (e.g., [9]). It will be convenient for our purposes to make the domains from which the user's queries and the servers' answers are taken explicit in the specification of the protocol.

Definition 1   (PIR). *A k-server PIR protocol $\mathcal{P} = (\mathcal{R}, \mathcal{D}_Q, \mathcal{D}_A, \mathcal{Q}, \mathcal{A}, \mathcal{M})$ consists of a probability distribution $\mathcal{R}$, a query domain $\mathcal{D}_Q$, an answer domain $\mathcal{D}_A$, and three types of algorithms: a query algorithm $\mathcal{Q}$, an answering algorithm $\mathcal{A}$, and a reconstruction algorithm $\mathcal{M}$. (The domains $\mathcal{D}_Q$ and $\mathcal{D}_A$ will sometimes be omitted from the specification of $\mathcal{P}$.) At the beginning of the protocol, the user picks a random string r from the distribution $\mathcal{R}$. It then computes a k-tuple of queries $\mathcal{Q}(i,r) = (q_1, \ldots, q_k) \in \mathcal{D}_Q{}^k$ and sends to each server S_h the corresponding query q_h. Each server S_h responds with an answer $a_h = \mathcal{A}(h, x, q_h)$, where $a_h \in \mathcal{D}_A$. (In the case of errorless PIR we can assume, w.l.o.g., that the servers are deterministic.) Finally, the user recovers x_i by applying the reconstruction algorithm $\mathcal{M}(i, r, a_1, \ldots, a_k)$. A k-server protocol as above is a t-private ϵ-correct PIR protocol, if it satisfies the following requirements:*

Correctness. *The user errs on computing the correct value with probability at most ϵ. Formally, for every $i \in [n]$ and every database $x \in \{0,1\}^n$,*

$$Pr_{r \in \mathcal{R}} \left[\mathcal{M}(i, r, \mathcal{A}(1, x, \mathcal{Q}(i,r)_1), \ldots, \mathcal{A}(k, x, \mathcal{Q}(i,r)_k)) = x_i \right] \geq 1 - \epsilon.$$

By default, we assume PIR protocols to be perfectly correct, *namely 0-correct. A statistical PIR protocol is a $\frac{1}{3}$-correct protocol. (The error of such a protocol can be driven down to $2^{-\sigma}$ using $O(\sigma)$ repetitions).*

t-**Privacy.** *Each collusion of up to t servers has no information about the bit that the user tries to retrieve. Formally, for every two indices $i_1, i_2 \in [n]$ and for every $T \subseteq [k]$ of size $|T| \leq t$, the distributions $\mathcal{Q}_T(i_1, r)_{r \in \mathcal{R}}$ and $\mathcal{Q}_T(i_2, r)_{r \in \mathcal{R}}$ are identical.*

The *communication complexity* of a PIR protocol is the total number of bits communicated between the user and the k servers, maximized over all choices of x, i and r. We refer to $\log |\mathcal{D}_Q|$ as the *query complexity* of the protocol, and to $\log |\mathcal{D}_A|$ as its *answer complexity*. A PIR protocol is called *binary* if $\mathcal{D}_A = \{0, 1\}$. In a *linear* PIR protocol over a finite field $\mathbb{F}$, the database x is viewed as a vector in $\mathbb{F}^n$ and for any fixed server h and query q_h the answering algorithm $\mathcal{A}(h, x, q_h)$ computes a linear function of x. All PIR protocols from the literature are linear.

Definition 2 (Locally Decodable Code (LDC)). *Let $0 \leq \epsilon, \delta < \frac{1}{2}$. For integers k, n, and m, we say that a code $C : \Sigma^n \to \Gamma^m$ is a (k, ϵ, δ)-LDC if there exists a probabilistic decoding algorithm D, such that for every $x \in \Sigma^n$, every $i \in [n]$, and every $y \in \Gamma^m$ within Hamming distance at most δm from $C(x)$, it holds that: (i) $Pr[D^y(i) = x_i] \geq 1 - \epsilon$. (ii) D probes at most k symbols of y. We assume D to be non-adaptive, in the sense that its choice which symbols to probe is done independently of y. By default, we consider binary LDCs with $\Sigma = \Gamma = \{0, 1\}$.*

We will typically consider a family of binary LDCs $C_n : \{0, 1\}^n \to \{0, 1\}^{m(n)}$ for which the parameters ϵ and δ are constants that do not depend on n. In such a case we will refer to the family as a k-query LDC and denote it by C. The main open question in the area is to find the best asymptotic complexity of the code length $m(n)$ in a k-query LDC. Katz and Trevisan [23] showed a close relation between this question and the complexity of PIR:

Theorem 1 (implicit in [23]). *(i) If there is a k-query LDC $C : \{0, 1\}^n \to \{0, 1\}^m$ then there is a k-server statistical PIR protocol with query complexity $O(\log m)$ and answer complexity $O(1)$. (ii) If there is a statistical PIR protocol with query complexity $q(n)$ and answer complexity $a(n)$, then there is a k-query LDC $C : \{0, 1\}^n \to (\{0, 1\}^{a(n)})^m$ where $m = O(k2^{q(n)})$.*

Theorem 1 was proved using the notion of smooth codes, which serve as an intermediate notion between LDC and PIR, and is essentially equivalent to both.

3   Boosting the Privacy of PIR Protocols

In this section we describe two methods for boosting the privacy threshold t of PIR protocols at the price of an increase in the number of servers. In Section 3.1, we show this for arbitrary PIR protocols, but with an exponential (in t) increase in the number of servers. In Section 3.2, we define a stronger notion of *self-retrievable* PIR protocols (SRPIR) and show that they can be upgraded to t-private PIR using much fewer servers. SRPIR protocols are closely related to SCCs (see Section 4.1), analogously to the relation between PIR and LDCs. Thus, better SCCs would imply better t-private PIR protocols.

3.1   From PIR to t-Private PIR

In this section we describe a method to transform an arbitrary 1-private k-server PIR protocol into a t-private k^t-server PIR protocol. This is done via a general composition of any t_1-private k_1-server PIR protocol with any t_2-private k_2-server PIR protocol which yields a $(t_1 + t_2)$-private $k_1 k_2$-server PIR protocol. Similar composition techniques were used, for different purposes, in [15,22].

We start with the general composition claim. For simplicity of exposition we first consider the binary case. Let $\mathcal{P}_1 = (\mathcal{R}_1, \mathcal{Q}_1, \mathcal{A}_1, \mathcal{M}_1)$ and $\mathcal{P}_2 = (\mathcal{R}_2, \mathcal{Q}_2, \mathcal{A}_2, \mathcal{M}_2)$ be binary t_1-private k_1-server and t_2-private k_2-server PIR protocols, respectively. We construct a $(t_1 + t_2)$-private $k_1 k_2$-server PIR protocol. Partition the $k_1 k_2$ servers into k_1 sets of size k_2: $\mathcal{T}_1, \ldots, \mathcal{T}_{k_1}$ where $\mathcal{T}_v = \{S_{1,v}, S_{2,v}, \ldots, S_{k_2,v}\}$. It will be convenient to view the database as a string indexed by $Z_n = \{0, 1, \ldots, n-1\}$, that is $x = x_0 x_1 \cdots x_{n-1}$, instead of $x = x_1 x_2 \cdots x_n$. To retrieve the ith bit in the new PIR protocol, the user $\mathcal{U}$ chooses two random numbers $i_1, i_2 \in [n]$ such that $i_1 + i_2 \equiv i \pmod{n}$. Next, the user executes the query algorithm $\mathcal{Q}_1$ on i_1, to generate a set of queries $q_1^1, \ldots, q_{k_1}^1$. The query q_h^1 is then sent to every server in the set $\mathcal{T}_h$.

Denote by $x \ll \ell$ the database that results from an ℓ-bit cyclic shift of the database x. Each server in the set $\mathcal{T}_h$ prepares an array of n answers to the query q_h^1 by executing the answering algorithm $\mathcal{A}_1$, on the following n databases: $x \ll 0, x \ll 1, \ldots, x \ll n-1$. After this stage, each server in the set $\mathcal{T}_h$ holds the following binary string:

$$y^h \overset{\text{def}}{=} \mathcal{A}_1(h, x \ll 0, q_h^1), \ldots, \mathcal{A}_1(h, x \ll n-1, q_h^1). \tag{1}$$

In order to reconstruct the bit x_i, the user $\mathcal{U}$ is interested in the answers of the servers on the queries $q_1^1, \ldots q_{k_1}^1$ on the shifted database $x \ll i_2$. That is, the user needs the bits $y_{i_2}^1, \ldots, y_{i_2}^k$. To retrieve the bit $y_{i_2}^h$, the user and the k_2 servers of the set $\mathcal{T}_h$ execute $\mathcal{P}_2$ where the user's input is i_2 and the servers' input is y^h. The random choices of $\mathcal{U}$ are done independently for every set $\mathcal{T}_h$. Note that for every $h \in [k_1]$ it holds that $y_{i_2}^h = \mathcal{A}_1(h, x \ll i_2, q_h^1)$. Using the reconstruction algorithm $\mathcal{M}_1$ on i_1 and $\mathcal{A}_1(1, x \ll i_2, q_1^1), \ldots, \mathcal{A}_1(k_1, x \ll i_2, q_{k_1}^1)$, the user reconstructs the i_1 bit of $x \ll i_2$, which is equal to $x_{(i_1 + i_2 \bmod n)} = x_i$.

The correctness of the new PIR protocol easily follows from that of the original protocols. If the query complexities of the original protocols are $q_1(n)$ and $q_2(n)$ bits per server, then the query complexity of the new protocol is $q_1(n) + q_2(n)$ bits per server. The following claim captures the enhanced privacy of the new protocol. The proof is deferred to the full version of the paper.

Claim 1. *The above PIR protocol is $(t_1 + t_2)$-private.*

In the non-binary case where the protocols $\mathcal{P}_1$ and $\mathcal{P}_2$ have answer complexities $a_1(n)$ and $a_2(n)$, respectively, we change the protocol as follows: At the end of the first stage the servers of each set $\mathcal{T}_h$ are holding the string y^h, as in (1). In this case y^h is of length $n \cdot a_1(n)$, and in order to retrieve the $a_1(n)$ size string $\mathcal{A}_1(h, x \ll i_2, q_h^1)$, the user and the k_2 servers of the set $\mathcal{T}_h$ perform $a_1(n)$ *executions* of $\mathcal{P}_2$ where in the ℓth execution, $0 \le \ell < a_1(n)$, the user's input is $i_2 a_1(n) + \ell$ and the servers' input is y^h. Note that these $a_1(n)$ execution need not be independent, as it is known that the $a_1(n)$ retrieved bits form a sequence in the string y^h. Hence, only one query of $\mathcal{P}_2$ must be sent. Therefore, the overall query complexity remains $q_1(n) + q_2(n)$. The answer complexity, however, changes to $a_1(n)a_2(n)$.

Theorem 2. *Given two PIR protocols that are t_i-private k_i-server PIR protocol with query complexity $q_i(n)$ and answer complexity $a_i(n)$, for $i \in \{1,2\}$, one can construct a $(t_1 + t_2)$-private $k_1 k_2$-server PIR protocol with query complexity $q_1(n) + q_2(n)$ and answer complexity $a_1(n)a_2(n)$.*

The following theorem is proved by composing a PIR protocol with itself t times, applying an induction on t using Theorem 2 for the inductive step.

Theorem 3 (Boosting PIR privacy). *If there is a 1-private k-server (statistical) PIR protocol with query complexity $q(n)$ and answer complexity $a(n)$, then there exists a t-private k^t-server (statistical) PIR protocol with query complexity $O(tq(n))$ and answer complexity $O(a(n)^t)$.*

Theorems 1 and 3 imply the following.

Corollary 1 (From LDC to t-private PIR). *If there is a k-query LDC $C : \{0,1\}^n \to \{0,1\}^m$ then, for every constant t, there is a t-private k^t-server statistical PIR protocol with query complexity $O(\log m)$ and answer complexity $O(1)$.*

Combining this with Yekhanin's LDC [36] we get the following corollary.

Corollary 2 (Improved t-private PIR). *For every constant t, there exists a t-private 3^t-server PIR protocol with communication complexity $O(n^{10^{-7}})$. Furthermore, if there are infinitely many Mersenne primes then, for every constant t, there is a t-private 3^t-server PIR protocol whose communication complexity is $n^{O(1/\log\log n)}$ for infinitely many values of n.*

3.2 Better t-Private PIR from Self-retrievable PIR

In this section, we introduce the notion of *self-retrievable PIR* (SRPIR) protocols. Such protocols allow not only to privately retrieve each entry of the database, but also to privately retrieve the answer to any legal query. This means that if the user wants to know what a server h would answer, given the database x and a query q, he can do this using the protocol without revealing any information about q to the servers.

Definition 3 (SRPIR). *A tuple $\mathcal{P}' = (\mathcal{R}, \mathcal{D}_Q, \mathcal{D}_A, \mathcal{Q}^0, \mathcal{Q}^1, \mathcal{A}, \mathcal{M}^0, \mathcal{M}^1)$ is a self-retrievable PIR protocol if $\mathcal{P} = (\mathcal{R}, \mathcal{D}_Q, \mathcal{D}_A, \mathcal{Q}^0, \mathcal{A}, \mathcal{M}^0)$ is a PIR protocol, $\mathcal{Q}^1$ is an additional query algorithm and $\mathcal{M}^1$ is a query reconstruction algorithm with the following properties: If $\mathcal{U}$ wants to retrieve x_i it acts according to $\mathcal{P}$. If $\mathcal{U}$ wants to retrieve the answer of server h for a query $q \in \mathcal{D}_Q$, it computes queries $\mathcal{Q}^1((h,q),r) =$*

$(q_1, \ldots, q_k) \in \mathcal{D}_Q{}^k$, and the protocol proceeds as before. Finally, the user computes the answer by executing one of the reconstruction algorithms $\mathcal{M}^0(i, r, a_1, \ldots, a_k)$ or $\mathcal{M}^1((h, q), r, a_1, \ldots, a_k)$. A k-server protocol as above is a t-private SRPIR, if it satisfies the correctness and the t-privacy requirements from Definition 1, along with the following further requirements:

Correctness. For every $q \in \mathcal{D}_Q$, h, r and x,

$$\mathcal{M}^1((h, q), r, \mathcal{A}(1, x, \mathcal{Q}^1((h, q), r)_1), \ldots, \mathcal{A}(k, x, \mathcal{Q}^1((h, q), r)_k)) = \mathcal{A}(h, x, q).$$

This can be relaxed to ϵ-correctness similarly to Definition 1.

t-Privacy. Each collusion of up to t servers has no information about q and h. Formally, for every two possible queries $(h_1, q_1), (h_2, q_2) \in [k] \times \mathcal{D}_Q$, and for every $\mathcal{T} \subseteq [k]$, of size $|\mathcal{T}| \leq t$, the distributions $\mathcal{Q}^1{}_\mathcal{T}((h_1, q_1), r)_{r \in \mathcal{R}}$ and $\mathcal{Q}^1{}_\mathcal{T}((h_2, q_2), r)_{r \in \mathcal{R}}$ are identical.

In Section 4.1, we show that SRPIR protocols are essentially equivalent to SCCs. We now describe a method to convert a 1-private k-server SRPIR protocol into a t-private k'-server PIR protocol, where $k' = t(k - 1) + 1$. Our protocol relies on a recursive construction of a formula with T_2^k gates that computes a single $T_{t+1}^{k'}$ gate. (Here T_t^k is the t-out-of-k threshold function $T_t^k : \{0, 1\}^k \to \{0, 1\}$ defined as $T_t^k(z) = 1$ if and only if $\sum_{h=1}^k z_h \geq t$.) The construction of the formula is a straightforward generalization of a construction from [20] for the special cases $k = 3$ and $k = 4$.

Lemma 1. There exists a formula with T_2^k threshold gates that computes a $T_{t+1}^{k'}$ threshold gate. In particular, there is such a formula of size $O(\binom{k'}{t})$ and depth $O(t \log k')$.

We proceed to describe the construction of the t-private PIR protocol. We start with a general overview. Let $\mathcal{P} = (\mathcal{R}, \mathcal{D}_Q, \mathcal{D}_A, \mathcal{Q}^0, \mathcal{Q}^1, \mathcal{A}, \mathcal{M}^0, \mathcal{M}^1)$ be a 1-private k-server SRPIR protocol (See Definition 3). Denote by Φ the formula promised by Lemma 1. That is, Φ computes the function $T_{t+1}^{k'}$ using T_2^k gates. Denote by $z_1, \ldots, z_{k'}$ the input variables of Φ. The queries of the new PIR protocol are computed in a recursive manner according to the structure of the formula Φ. We associate each gate of Φ with a query from $\mathcal{D}_Q$, going from the output gate to the leaves. Eventually, each input wire of the formula, labeled by an input z_h, is associated with a query from $\mathcal{D}_Q$. For each $h' \in [k']$, the query sent to $S_{h'}$ in the new PIR protocol is a list of all $\mathcal{P}$-queries associated with input wires labeled by $z_{h'}$. Details follow.

Let $i \in [n]$ be the input to the user. Denote by g_{out} the output T_2^k gate of Φ and by $g_1, \ldots, g_k$ the gates whose output wires are the inputs wires to g_{out}. The user generates a random string $r_{\text{out}} \in_R \mathcal{R}$ and executes $\mathcal{Q}^0(i, r_{\text{out}})$ to get a list of queries $q_1, \ldots q_k$. Note that given the answers to these k queries the user can reconstruct the value of x_i. For each $h \in [k]$, the gate g_h is associated with the pair $(h, q_h) \in [k] \times \mathcal{D}_Q$.

Next, the user proceeds to associate a query with all the gates in the subformulae rooted in the gates $g_1, \ldots g_k$. Fix $h \in [k]$. Denote by $g_{h,1}, g_{h,2}, \ldots, g_{h,k}$ the k gates whose output wires are the input wires of g_h. The user chooses a random string (independently from all previously chosen strings) $r_h \in_R \mathcal{R}$ and executes $\mathcal{Q}^1((h, q_h), r_h)$

to get k queries $q_{h,1}, \ldots, q_{h,k}$. As before, for each $\ell \in [k]$, the gate $g_{h,\ell}$ is associated with the pair $(\ell, q_{h,\ell})$. Note that the answers of $\mathcal{A}$ on the queries $q_{h,1}, \ldots, q_{h,k}$, imply the answer of $\mathcal{A}$ on the query q_h. The user proceeds recursively to associate a pair $(h, q) \in [k] \times \mathcal{D}_Q$ with every gate of the formula Φ. Finally, every input wire of Φ is associated with a pair from $[k] \times \mathcal{D}_Q$. For each $h' \in [k']$, the user sends the server $S_{h'}$ all the queries associated with input wires that are labeled by $z_{h'}$.

The answering algorithm is simple: Given a set of queries, the server $S_{h'}$ simply answers each of the queries using $\mathcal{A}$, the answering algorithm of the original SRPIR protocol. Finally, in the reconstruction algorithm, the user proceeds through the formula Φ, going from the input wires towards the output gate. Given the answers to the queries associated with the input wires to a given gate g, the user executes the reconstruction algorithm of the original SRPIR protocol to compute the answer on the query associated with g. Finally, the user has the k answers of the queries associated with the input wires of the output gate, from which he can compute the value of x_i.

The correctness of the above PIR protocol is straightforward. The t-privacy of the protocol relies on the fact that a set $\mathcal{T}$ of at most t servers corresponds to an input that is rejected by the formula Φ (this is the input that one gets by setting 1 to $z_{h'}$ if $h' \in \mathcal{T}$ and is 0 otherwise). The proof is then implied by the privacy of the original SRPIR and the independence of the randomness used at each gate of Φ when generating the query of the new PIR protocol. The proof is deferred to the full version of the paper.

Theorem 4 (Boosting privacy for SRPIR). *If there exists a 1-private k-server (statistical) SRPIR protocol with query complexity $q(n)$ and answer complexity $a(n)$, then for every constant t, there exists a t-private $(t(k - 1) + 1)$-server (statistical) PIR protocol with query complexity $O(q(n))$ and answer complexity $O(a(n))$. More precisely, the query complexity is $O(\binom{kt}{t} q(n))$ and the answer complexity is $O(\binom{kt}{t} a(n))$.*

4 On Self-correctable Codes

A self-correctable code (SCC) is an error-correcting codes allowing each symbol of the encoded message to be retrieved (or "self-corrected") by reading only a few other symbols of the encoded message. Katz and Trevisan [23] showed LDCs to be essentially equivalent to PIR protocols. In Section 4.1 we show that SCCs are essentially equivalent to SRPIR protocols. Together with the results of the previous section, this shows that better SCCs would result in better t-private PIR protocols. In Section 4.2, we describe the best known constructions of SCCs. Our definition of SCCs is similar to the definition of smooth codes [23], except that it refers to retrieving a symbol of the *encoding of the message* rather than a symbol of the message itself.

Definition 4 (Self-correctable Code (SCC)). *Let $0 \leq \epsilon < \frac{1}{2}$. We say that $C : \{0, 1\}^n \rightarrow \{0, 1\}^m$ is a (k, k', ϵ)-SCC, if there exists a probabilistic non-adaptive correction algorithm M (the corrector), such that for every $x \in \{0, 1\}^n$ and for every two indices $j, j' \in [m]$, the following hold: (i) C is injective. (ii) $Pr[M^{C(x)}(j) = C(x)_j] \geq 1 - \epsilon$ (iii) M probes at most k symbols from $C(x)$. (iv) $\Pr[M^{(\cdot)}(j) \text{ probes symbol } j'] \leq \frac{k'}{m}$.*

We refer to a $(k, k, 0)$-SCC simply as a k-*query SCC*. We define an *exact* (k, ϵ)-SCC to be one in which the correction algorithm makes exactly k probes and each index is probed with probability $\frac{k}{m}$. Similarly to the case of smooth codes, one can get an exact SCC out of any SCC at the expense of increasing ϵ.

Lemma 2. *If* $C : \{0, 1\}^n \to \{0, 1\}^m$ *is a* (k, k', ϵ)-*SCC, then it is also an exact* (k, ϵ')-*SCC for* $\epsilon' = \frac{1}{2} - \frac{(\frac{1}{2} - \epsilon)^2}{2k'}$.

4.1 SCC and SRPIR

We prove a strong relation between SCCs and SRPIR protocols. This relation is analogous to the relation between LDCs and PIR protocols. The following straightforward generalizations of the claims of [23] are proven in the full version of the paper.

Claim 2 (From SCC to SRPIR). *Let* $C : \{0, 1\}^n \to \{0, 1\}^{m(n)}$ *be a systematic exact* (k, ϵ)-*SCC. Then, there exists a* k-*server,* ϵ-*correct binary SRPIR protocol with query complexity* $\log m$.

Combining Lemma 2 and Claim 2 with Theorem 4, we get the following corollary which is analogous to Corollary 1:

Corollary 3 (From SCC to t-private PIR). *Let* $C : \{0, 1\}^n \to \{0, 1\}^{m(n)}$ *be a systematic* (k, k', ϵ)-*SCC for some constants* k, k', *and* ϵ. *Then, there exists a* t-*private* $(t(k - 1) + 1)$-*server statistical SRPIR protocol with query complexity* $O(\log m)$ *and answer complexity* $O(1)$.

Claim 3 (From SRPIR to SCC). *If there exists a* k-*server,* ϵ-*correct binary SRPIR protocol with queries of size* $\log m$ *for databases in* $\{0, 1\}^n$. *Then there exists a* $(k, 2k, \epsilon)$-*SCC,* $C : \{0, 1\}^n \to \{0, 1\}^{2km}$.

4.2 Best Known SCCs

The best known SCCs are based on Reed-Muller (RM) codes. RM codes over a field $\mathbb{F}$ of size $|\mathbb{F}| > k$ give rise to a natural self-correction procedure [7], yielding (non-binary) k-query SCCs of length $\exp(n^{1/(k-1)})$. Binary SCCs of a similar length can be derived from the PIR protocol of [8, Theorem 2], which is based on RM codes over extensions of $\mathbb{F}_2$. For k of the form $k = 2^r - 1$, this PIR protocol is in fact an SRPIR. (We discuss this in detail in the full version of the paper.) Thus, by Claim 3, it implies a binary SCC with the above parameters.

5 SCCs and Block Designs

In this section, we show a connection between SCCs and an open question from the theory of combinatorial designs. We start by defining a design.

Definition 5 (Design). *Let* m, k *and* λ *be positive integers and* $\mathcal{B}$ *be a family of subsets of* $[m]$. *The pair* $\mathcal{D} = \langle [m], \mathcal{B} \rangle$ *is a* 2-(m, k, λ) *design, if all the sets (called blocks) in* $\mathcal{B}$ *are of size* k *and every* 2 *distinct points in* $[m]$ *appear together in exactly* λ *blocks.*

We now define an algebraic measure of designs, called p-rank.

Definition 6 (p-rank **of a Design**). *Let $\mathcal{D}$ be a* 2-(m, k, λ) *design, and let $\mathbb{F}$ be a finite field of characteristic p, where p is a prime. Consider the incidence matrix of $\mathcal{D}$ denoted $A \stackrel{def}{=} A_{\mathcal{D}}$: The columns of A are labeled by the blocks of $\mathcal{B}$ and the rows of A are labeled by the points of $[m]$. The entry $A[i, B]$, for $B \in \mathcal{B}$ and $i \in [m]$ equals 1 if $i \in B$ and 0 otherwise. The p-rank of a design $\mathcal{D}$ is defined to be $\mathrm{rank}_{\mathbb{F}}(A) = \mathrm{rank}_{\mathbb{F}_p}(A)$.*

The next claim shows that a construction of a design with certain parameters and a low p-rank yields a construction of a good SCC. Roughly speaking, the SCC is generated by a matrix dual to the incidence matrix of the design. To correct the jth bit in a codeword, for $j \in [m]$, the corrector uses the blocks of the design containing j. A complete proof appears in the full version of the paper.

Claim 4 (From design to SCC). *Let $\mathcal{D} = \langle [m], \mathcal{B} \rangle$ be a* 2-(m, k, λ) *design (with $k \geq 2$), A be the incidence matrix of $\mathcal{D}$, and $s = \mathrm{rank}_{\mathbb{F}}(A)$, where $\mathbb{F}$ is a finite field. Let $C : \mathbb{F}^{m-s} \rightarrow \mathbb{F}^m$ be a code dual to the code generated by A. Then C is a $(k-1)$-query SCC of dimension $m - s$.*

The designs with the lowest known p-rank are geometric designs.

Example 1 (Affine Geometric Designs). *Let $q = p^r$ be a prime power and m be a positive integer. The affine geometric design $\mathrm{AG}(m, q)$ is defined as follows. The points of the design are the vectors of the linear space $\mathbb{F}_q^m$. The blocks are the 1-dimensional affine subspaces of $\mathbb{F}_q^m$, also called lines. First note that $\mathrm{AG}(m, q)$ is a* 2-$(q^m, q, 1)$ *design, as every line is of size q and every two points are together in exactly one line.*

We now analyze the p-rank of the design $\mathrm{AG}(m, q)$. We use the properties of Reed-Muller codes over a field $\mathbb{F}_q$ on m variables and degree $q-2$. Recall that the coordinates of this code are labeled with vectors from $\mathbb{F}_q^m$ and thus it is of length q^m. The code is generated by the monomials on m variables $x_1, \ldots, x_m$ of total degree at most $q - 2$. Thus, its dimension is $\binom{m+q-2}{q-2}$. It is known that the incidence vectors of the blocks of $\mathrm{AG}(m, q)$ span exactly the dual code of this Reed-Muller code. Hence, the p-rank of $\mathrm{AG}(m, q)$ is $q^m - \binom{m+q-2}{q-2}$. Moreover, applying Claim 4 on this design for $p = 2$ gives the binary SCC implicit in [8], with the number of servers being $k = q - 1$ (See Section 4.2).

Hamada [18] (see also [19,4,11]) conjectured that the p-rank of every affine geometric design is minimal among all designs with the same parameters, but proving his conjecture is an open problem in design theory. In view of Claim 4, proving strong lower bounds on SCCs, that is, proving that the currently known SCCs are optimal, requires proving the following important special case of Hamada's conjecture.[4]

Conjecture 1 (Hamada [18]). *Consider the design $AG(m, q)$, where $q = p^r$. Then $AG(m, q)$ has the smallest p-rank from all the designs with the same parameters.*

---

[4] The original conjecture was stronger, asserting that every design with similar parameters is isomorphic to $AG(m, q)$. The weaker conjecture brought herein appears in, e.g., [4].

5.1 A Generalization of Hamada's Conjecture

Tonchev [31] defined a generalization of the p-rank measure, called the *p-dimension* of a design. In this section, we show that a generalization of Hamada's conjecture to this measure may serve as an indication that LDCs are more efficient than SCCs, or that improving over current SCC constructions might be difficult. We define a framework for the construction of SCCs and LDCs based on threshold secret sharing schemes. Both the best known SCC [8] and the best known LDC [36] fall into this framework.

Definition 7 (p-dimension of a Design). *Let D be a 2-(m, k, λ) design, and $\mathbb{F} = \mathbb{F}_p$, where p is a prime power. Consider a matrix A with columns labeled by the blocks of B, rows labeled by the points of $[m]$, and entries from $\mathbb{F}$. We say that A matches the design D if for every block $B \in B$ and point $i \in [m]$, the following two conditions holds: (i) If $i \notin B$, then $A[i, B] = 0$ (ii) if $i \in B$, then $A[i, B] \neq 0$. The p-dimension of D is defined to be minimum of $\text{rank}_{\mathbb{F}}(A)$ over all matrices A that match D.*

The known SCCs [8] and all the known LDCs (e.g. [8,10,36,28]) are constructed according to the following paradigm: The codeword indices are viewed as vectors from a vector space $\mathbb{K}^\ell$, where $\mathbb{K}$ is a finite field.[5] The set of the codeword locations probed to decode/correct a symbol is chosen according to a 2-out-of-k threshold secret sharing scheme. The scheme is applied independently on each of the ℓ coordinates of the vector representing the symbol to be decoded/corrected. We concentrate on the following two well-known secret sharing schemes.

Definition 8 (Extended Shamir Secret Sharing Scheme). *Let $\mathbb{K}$ be a finite field satisfying $|\mathbb{K}| \geq k + 1$ and let $\alpha_1, \ldots, \alpha_k \in \mathbb{K}$ be k distinct non-zero field elements. To share a secret $s \in \mathbb{K}^\ell$, the dealer chooses a random vector $a \in \mathbb{K}^\ell$, and gives the server h the share $z^h \in \mathbb{K}^\ell$, where for every $j \in [\ell]$ it holds that $z_j^h = s_j + \alpha_h a_j$.*

Definition 9 (Leading-Coefficient Secret Sharing Scheme). *Let $\mathbb{K}$ be a finite field satisfying $|\mathbb{K}| \geq k$ and let $\alpha_1, \ldots, \alpha_k \in \mathbb{K}$ be k distinct field elements. To share a secret $s \in \mathbb{K}^\ell$, the dealer chooses a random vector $a \in \mathbb{K}^\ell$, and gives the server h the share $z^h \in \mathbb{K}^\ell$, where for every $j \in [\ell]$ it holds that $z_j^h = a_j + \alpha_h s_j$.*

We turn to describe a subclass of SCCs which we call *threshold-based* SCCs.

Definition 10 (Threshold-Based SCC). *Let C be a linear k-SCC $C : \mathbb{F}^n \to \mathbb{F}^m$, where $m = \tilde{p}^\ell$ for some prime $\tilde{p}$ and positive integer ℓ. We identify the set $[m]$ of code indices with $\mathbb{K}^\ell$, where $\mathbb{K} = \mathbb{F}_{\tilde{p}}$, and view it as a domain of secrets. For an index $s \in \mathbb{K}^\ell$, consider the family Z_s of subsets of $\mathbb{K}^\ell$ of size at most k used by the corrector M to correct the index labeled by s in a codeword. The code C with corrector M is called a threshold-based SCC (e.g., Shamir-based SCC or leading-coefficient-based SCC), if the families $\{Z_s\}_{s \in \mathbb{K}^\ell}$ correspond to a threshold secret sharing scheme. That is, the sets of indices, probed by the corrector to correct the codeword index labeled by s, are exactly the sets of possible secret shares given to k servers in the secret sharing applied on the secret s.*

---

[5] We note that we use here a field $\mathbb{K}$, since the field $\mathbb{F}$ continues to serve as the field of the SCC, while $\mathbb{K}$ serves as the field of the vector space that represents the *indices* of the code.

The next claim shows how to transform a Shamir-based SCC into a design. Each block of the design corresponds to a secret and a set of shares used by the corrector of the SCC.

Claim 5 (From Shamir-based SCC to Design). *Let* $m = \tilde{p}^\ell$ *for some prime* $\tilde{p}$. *If there exists Shamir-based* k-*SCC* $C : \mathbb{F}^n \to \mathbb{F}^m$, *where* $\mathbb{F}$ *is a field of characteristic* p, *then there exists a* 2-$(m, k + 1, \lambda)$ *design of* p-*dimension* $m - n$, *for some integer* λ.

Leading coefficient secret sharing schemes yield the following non-perfect version of combinatorial designs.

Definition 11 (Almost Design). *Let* $\mathcal{B}$ *be a family of subsets of* $[m]$. *The pair* $\mathcal{D} = \langle [m], \mathcal{B} \rangle$ *is a* 2-(m, k, λ) ϵ-*almost design, where* $0 \leq \epsilon < 1$, *if all the sets in* $\mathcal{B}$ *are of size* k, *every* 2 *distinct points in* $[m]$ *appear together in at most* λ *blocks, and for every* $i \in [m]$, *it holds that for at least* $(1 - \epsilon)m$ *values* $j \in [m]$ *the set* $\{B \in \mathcal{B} : i, j \in B\}$ *is of size exactly* λ.

The definition of p-dimension generalizes naturally to almost designs. In the full version of the paper we show that almost designs with low p-dimension imply SCCs with good parameters (similarly to Claim 4) and that leading-coefficient-based SCCs imply almost designs with low p-dimension (similarly to Claim 5).

Next, we propose a generalization of Hamada's conjecture, which relaxes the original conjecture in considering p-dimension rather than p-rank.

Conjecture 2 (Generalized Hamada). *Consider the design* $AG(m, q)$ *from Example 1, where* q *is a prime power. Then, the* q-*dimension of the design* $AG(m, q)$ *is not larger than the* q-*dimension of any design with the same parameters.*

Moreover, we conjecture that any *almost design* with the same parameters will not have asymptotically smaller q-dimension. Furthermore, letting q' be a fixed prime power coprime to q, we anticipate that the q'-dimension of any design of the above parameters will not be asymptotically smaller than the q-dimension of $AG(m, q)$.

Threshold-based LDCs can be defined in a similar way to threshold-based SCCs, with the only difference that here the domain of secrets is only the message indices $[n]$ and not the codewords indices $[m]$. In particular, the best known LDC of [36] is leading-coefficient-based. However, a threshold-based LDC does not imply a result on the dimension of designs, as the sets used to correct a message symbol do not correspond to a design. This may serve as an explanation to the fact that if Conjecture 2 is true, then in a framework of Shamir-based or leading-coefficient-based codes, there is a strong separation between LDCs and SCCs. For instance, for $k = 3$ there is a leading-coefficient-based LDC of complexity $\exp(n^{-10^7})$ while for SCCs, assuming the conjectures above, the complexity is $\exp(n^{1/2})$.

6 Open Questions

Our results leave open the following two interesting questions:

- Is there a true gap between the complexity of LDCs and SCCs? Are there better SCCs than those based on Reed-Muller codes? For instance, is there a 3-query

binary SCC of length $2^{o(n^{1/2})}$? Our results of Section 5 may be viewed as providing evidence that making progress on this front would be difficult.
- Is there a better general transformation from any k-server PIR to t-private PIR? Our general transformations from Section 3 either require k^t servers or require the underlying PIR to have the stronger "self-retrieval" property.

Acknowledgments. We wish to thank Amos Beimel, Tuvi Etzion, Eyal Kushilevitz, and Ronny Roth for helpful discussions.

References

1. Ambainis, A.: Upper bound on the communication complexity of private information retrieval. In: Degano, P., Gorrieri, R., Marchetti-Spaccamela, A. (eds.) ICALP 1997. LNCS, vol. 1256, pp. 401–407. Springer, Heidelberg (1997)
2. Arora, S., Safra, S.: Probabilistic checking of proofs: a new characterization of NP. J. of the ACM 45(1), 70–122 (1998)
3. Arora, S., Lund, C., Motwani, R., Sudan, M., Szegedy, M.: Proof Verification and the Hardness of Approximation Problems. J. of the ACM 45(3), 501–555 (1998)
4. Assmus, E.F., Key, J.D.: Designs and codes: An update. Designs, Codes and Cryptography 9(1), 7–27 (1996)
5. Babai, L., Fortnow, L., Levin, L.A., Szegedy, M.: Checking Computations in Polylogarithmic Time. In: Proc. STOC '91, pp. 21–31 (1991)
6. Babai, L., Fortnow, L., Nisan, N., Wigderson, A.: BPP Has Subexponential Time Simulations Unless EXPTIME has Publishable Proofs. Computational Complexity 3, 307–318 (1993)
7. Beaver, D., Feigenbaum, J.: Hiding instances in multioracle queries. In: Proc. of STACS '90, pp. 37–48 (1990)
8. Beimel, A., Ishai, Y.: Information-theoretic private information retrieval: A unified construction. In: Orejas, F., Spirakis, P.G., van Leeuwen, J. (eds.) ICALP 2001. LNCS, vol. 2076, pp. 912–926. Springer, Heidelberg (2001)
9. Beimel, A., Ishai, Y., Kushilevitz, E.: General constructions for information-theoretic private information retrieval. J. of Computer and Systems Sciences 71(2), 213–247 (2005)
10. Beimel, A., Ishai, Y., Kushilevitz, E., Raymond, J.F.: Breaking the $O(n^{\frac{1}{2k-1}})$ barrier for information-theoretic private information retrieval. In: FOCS. Proc. of the 43rd Annual IEEE Symposium on Foundations of Computer Science 2002, pp. 261–270. IEEE Computer Society Press, Los Alamitos (2002)
11. Beth, T., Jungnickel, D., Lenz, H.: Design Theory, 2nd edn. vol. 1. Cambridge University Press, Cambridge (1999)
12. Blum, M., Kannan, S.: Designing programs that check their work. J. of the ACM 42(1), 269–291 (1995)
13. Chor, B., Gilboa, N.: Computationally private information retrieval. In: Proc. of STOC '97, pp. 304–313 (1997)
14. Chor, B., Goldreich, O., Kushilevitz, E., Sudan, M.: Private information retrieval. In: Proc. of the 36th Annual IEEE Symposium on Foundations of Computer Science (FOCS '95), pp. 41–50. IEEE Computer Society Press, Los Alamitos (1995)
15. Di-Crescenzo, G., Ishai, Y., Ostrovsky, R.: Universal service-providers for private information retrieval. J. of Cryptology 14(1), 37–74 (2001)
16. Feige, U., Goldwasser, S., Lovasz, L., Safra, S., Szegedy, M.: Interactive Proofs and the Hardness of Approximating Cliques. J. of the ACM 43(2), 268–292 (1996)

17. Gasarch, W.: A survey on private information retrieval. Bulletin of the European Association for Theoretical Computer Science 82, 72–107 (2004)
18. Hamada, N.: On the p-rank of the incidence matrix of a balanced or partially balanced incomplete block design and its application to error-correcting codes. Hiroshima Math J. 3, 153–226 (1973)
19. Hamada, N.: The geometric structure and the p-rank of an affine triple system derived from a nonassociative moufang loop with the maximum associative center. J. Comb. Theory, Ser. A 30(3), 285–297 (1981)
20. Hirt, M., Maurer, U.M.: Player simulation and general adversary structures in perfect multiparty computation. J. of Cryptology 13(1), 31–60 (2000)
21. Ishai, Y., Kushilevitz, E.: Improved upper bounds on information-theoretic private information retrieval. In: Proc. STOC '99, pp. 79–88 (1999)
22. Ishai, Y., Kushilevitz, E.: On the Hardness of Information-Theoretic Multiparty Computation. In: Cachin, C., Camenisch, J.L. (eds.) EUROCRYPT 2004. LNCS, vol. 3027, pp. 439–455. Springer, Heidelberg (2004)
23. Katz, J., Trevisan, L.: On the efficiency of local decoding procedures for error-correcting codes. In: STOC '00, pp. 80–86 (2000)
24. Kerenidis, I., de Wolf, R.: Exponential lower bound for 2-query locally decodable codes. J. of Computer and Systems Sciences , 395–420 (2004)
25. Kushilevitz, E., Ostrovsky, R.: Replication is not needed: Single database, computationally-private information retrieval. In: FOCS '97, pp. 364–373 (1997)
26. Lipton, R.: Efficient checking of computations. In: STACS '90, pp. 207–215 (1990)
27. Lu, C.-J., Reingold, O., Vadhan, S.P., Wigderson, A.: Extractors: optimal up to constant factors. In: Proc.STOC '03, pp. 602–611 (2003)
28. Raghavendra, P.: A note on Yekhanin's locally decodable codes. In: Electronic Colloquium on Computational Complexity (ECCC) (2007)
29. Shamir, A.: How to share a secret. Communications of the ACM 22, 612–613 (1979)
30. Sudan, M., Trevisan, L., Vadhan, S.P.: Pseudorandom Generators without the XOR Lemma. J. of Computer and Systems Sciences 62(2), 236–266 (2001)
31. Tonchev, V.D.: Linear perfert codes and a characterization of the classical designs. Designs, Codes and Cryptography 17, 121–128 (1999)
32. Trevisan, L.: Some applications of coding theory in computational complexity. Quaderni di Matematica 13, 347–424 (2004)
33. Wehner, S., de Wolf, R.: Improved lower bounds for locally decodable codes and private information retrieval. In: Caires, L., Italiano, G.F., Monteiro, L., Palamidessi, C., Yung, M. (eds.) ICALP 2005. LNCS, vol. 3580, pp. 1424–1436. Springer, Heidelberg (2005)
34. Woodruff, D.: New lower bounds for general locally decodable codes. In: ECCC, Report No. 6 (2007)
35. Woodruff, D., Yekhanin, S.: A geometric approach to information-theoretic private information retrieval. In: proc. of CCC '05, pp. 275–284 (2005)
36. Yekhanin, S.: Towards 3-Query Locally Decodable Codes of Subexponential Length. In: Proc. STOC 2007 (2007)

A Sequential Algorithm for Generating Random Graphs

Mohsen Bayati[1], Jeong Han Kim[2], and Amin Saberi[1]

[1] Stanford University
{bayati,saberi}@stanford.edu
[2] Yonsei University
jehkim@yonsei.ac.kr

Abstract. We present the fastest FPRAS for counting and randomly generating simple graphs with a given degree sequence in a certain range. For degree sequence $(d_i)_{i=1}^{n}$ with maximum degree $d_{\max} = O(m^{1/4-\tau})$, our algorithm generates almost uniform random graph with that degree sequence in time $O(m\, d_{\max})$ where $m = \frac{1}{2}\sum_i d_i$ is the number of edges in the graph and τ is any positive constant. The fastest known FPRAS for this problem [22] has running time of $O(m^3 n^2)$. Our method also gives an independent proof of McKay's estimate [33] for the number of such graphs.

Our approach is based on *sequential importance sampling* (SIS) technique that has been recently successful for counting graphs [15,11,10]. Unfortunately validity of the SIS method is only known through simulations and our work together with [10] are the first results that analyze the performance of this method.

Moreover, we show that for $d = O(n^{1/2-\tau})$, our algorithm can generate an asymptotically uniform d-regular graph. Our results are improving the previous bound of $d = O(n^{1/3-\tau})$ due to Kim and Vu [30] for regular graphs.

1 Introduction

The focus of this paper is on generating random simple graphs (graphs with no multiple edge or self loop) with a given degree sequence. Random graph generation has been studied extensively as an interesting theoretical problem (see [42,11] for detailed surveys). It has also become an important tool in a variety of real world applications including detecting motifs in biological networks [37] and simulating networking protocols on the Internet topology [41,19,32,14,2]. The most general and well studied approach for this problem is the Markov chain Monte Carlo (MCMC) method [22,23,24,25,16,18,26,8]. However, current MCMC-based algorithms have large running times, which make them unusable for real-world networks that have tens of thousands of nodes (for example, see [20]). This has constrained practitioners to use simple heuristics that are non-rigorous and have often led to wrong conclusions [36,37]. Our main contribution in this paper is to provide a much faster fully polynomial randomized approximation scheme (FPRAS) for generating random graphs; this we can do in almost

M. Charikar et al. (Eds.): APPROX and RANDOM 2007, LNCS 4627, pp. 326–340, 2007.

linear time. An FPRAS provides an arbitrary close approximaiton in time that depends only polynomially on the input size and the desired error. (For precise definitions of this, see Section 2).

Recently, sequential importance sampling (SIS) has been suggested as a more suitable approach for designing fast algorithms for this and other similar problems [15,11,31,4]. Chen et al. [15] used the SIS method to generate bipartite graphs with a given degree sequence. Later Blitzstein and Diaconis [11] used a similar approach to generate general graphs. Almost all existing work on SIS method are justified only through simulations and for some special cases counter examples have been proposed [9]. However the simplicity of these algorithms and their great performance in several instances, suggest further study of the SIS method is necessary.

Our Result. Let $d_1, \ldots, d_n$ be non-negative integers given for the degree sequence and let $\sum_{i=1}^{n} d_i = 2m$. Our algorithm is as follows: start with an empty graph and sequentially add edges between pairs of non-adjacent vertices. In every step of the procedure, the probability that an edge is added between two distinct vertices i and j is proportional to $\hat{d}_i \hat{d}_j (1 - d_i d_j / 4m)$ where $\hat{d}_i$ and $\hat{d}_j$ denote the remaining degrees of vertices i and j. We will show that our algorithm produces an asymptotically uniform sample with running time of $O(m\, d_{\max})$ when maximum degree is of $O(m^{1/4-\tau})$ and τ is any positive constant. Then we use a simple SIS method to obtain an FPRAS for any $\epsilon, \delta > 0$ with running time $O(m\, d_{\max} \epsilon^{-2} \log(1/\delta))$ for generating graphs with $d_{\max} = O(m^{1/4-\tau})$. Moreover, we show that for $d = O(n^{1/2-\tau})$, our algorithm can generate an asymptotically uniform d-regular graph. Our results are improving the bounds of Kim and Vu [30] and Steger and Wormald [40] for regular graphs.

Related Work. McKay and Wormald [33,35] give asymptotic estimates for number of graphs within the range $d_{\max} = O(m^{1/3-\tau})$. But, the error terms in their estimates are larger than what is needed to apply Jerrum, Valiant and Vazirani's [21] reduction to achieve asymptotic sampling. Jerrum and Sinclair [22] however, use a random walk on the self-reducibility tree and give an FPRAS for sampling graphs with maximum degree of $o(m^{1/4})$. The running time of their algorithm is $O(m^3 n^2 \epsilon^{-2} \log(1/\delta))$ [39]. A different random walk studied by [23,24,8] gives an FPRAS for random generation for all degree sequences for bipartite graphs and almost all degree sequences for general graphs. However the running time of these algorithms is at least $O(n^4 m^3 d_{\max} \log^5(n^2/\epsilon) \epsilon^{-2} \log(1/\delta))$.

For the weaker problem of generating asymptotically uniform samples (not an FPRAS) the best algorithm was given by McKay and Wormald's switching technique on configuration model [34]. Their algorithm works for graphs with $d_{\max}^3 = O(m^2/\sum_i d_i^2)$ and $d_{\max}^3 = o(m + \sum_i d_i^2)$ with average running time of $O(m + (\sum_i d_i^2)^2)$. This leads to $O(n^2 d^4)$ average running time for d-regular graphs with $d = o(n^{1/3})$. Very recently and independently from our work, Blanchet [10] have used McKay's estimate and SIS technique to obtain an FPRAS with running time $O(m^2)$ for sampling bipartite graphs with given degrees when $d_{\max} = o(m^{1/4})$. His work is based on defining an appropriate Lyapunov function as well as using Mckay's estimates.

Our Technical Contribution. Our algorithm and its analysis are based on beautiful works of Steger and Wormald [40] and Kim and Vu [29]. The technical contributions of our work beyond their analysis are the followings:

1. In both [40,29] the output distribution of proposed algorithms are asymptotically uniform. Here we use SIS technique to obtain an FPRAS.
2. Both [40,29] use McKay's estimate [33] in their analysis. In this paper we give a combinatorial argument to show the failure probability of the algorithm is small and attain a new proof for McKay's estimate.
3. We exploit combinatorial structure and a simple martingale tail inequality to show concentration results in the range $d = O(n^{1/2-\tau})$ for regular graphs that previous polynomial inequalities [28] do not work.

Other Applications and Extensions. Our algorithm and its analysis provide more insight into the modern random graph models such as the configuration model or random graphs with a given *expected* degree sequence [17]. In these models, the probability of having an edge between vertices i and j of the graph is proportional to $d_i d_j$. However, one can use our analysis or McKay's formula [33] to see that in a random simple graph this probability is proportional to $d_i d_j (1 - d_i d_j / 2m)$. We expect that, by adding the correction term and using the concentration result of this paper, it is possible to achieve a better approximation for which sandwiching theorems similar to [30] can be applied.

We have also used similar ideas to generate random graphs with large girth [5]. These graphs are used to design high performance Low-Density Parity-Check (LDPC) codes. One of the methods for constructing these codes is to generate a random bipartite graph with a given optimized degree sequence [3]. Two common heuristics for finding good codes are as follows: (i) generate many random copies and pick the one with the highest girth; (ii) grow progressively the graph while avoiding short loops. While the former implies an exponential slowing down in the target girth, the latter induces a systematic and uncontrolled bias in the graph distribution. Using the same principles, we can remove the bias and achieve a more efficient algorithm that sequentially adds the edges avoiding small cycles.

Organization of the Paper. The rest of the paper has the following structure. The algorithm and the main results are stated in Section 2. In Section 3 we explain the intuition behind the weighted configuration model and our algorithm. It also includes the FPRAS using SIS approach. Section 4 is dedicated to analysis and proofs. We have included the main proofs in this section but due to space limitations some key ingredients such as combinatorial proof for probability of failure and also proofs for controlling the variance of SIS estimator are given in the extended version of this paper [6].

2   Our Algorithm

Suppose we are given a sequence of n nonnegative integers $d_1, d_2, \ldots d_n$ with $\sum_{i=1}^n d_i = 2m$ and a set of vertices $V = \{v_1, v_2, \ldots, v_n\}$. Assume the sequence of given degrees $d_1, \ldots, d_n$ is *graphical*. That is there exists at least one simple graph with those degrees. We propose the following procedure for sampling (counting) an element (number of elements) of set $\mathcal{L}(\bar{d})$ of all labeled simple graphs G with $V(G) = V$ and degree sequence $\bar{d} = (d_1, d_2, \cdots, d_n)$. Throughout this paper $m = \sum_{i=1}^{n} d_i/2$ is number of edges in the graph, $d_{\max} = \max_{i=1}^{n} \{d_i\}$ and for regular graphs d refers to degrees; i.e. $d_i = d$ for all $i = 1, \ldots, n$.

Procedure A

(1) Let E be a set of edges, $\hat{d} = (\hat{d}_1, \ldots, \hat{d}_n)$ be an n-tuple of integers and P be a number. Initialize them by $E = \emptyset$, $\hat{d} = \bar{d}$, and $P = 1$.
(2) Choose two vertices $v_i, v_j \in V$ with probability proportion to $\hat{d}_i \hat{d}_j (1 - \frac{d_i d_j}{4m})$ among all pairs i, j with $i \neq j$ and $(v_i, v_j) \notin E$. Denote this probability by p_{ij} and multiply P by p_{ij}. Add (v_i, v_j) to E and reduce each of $\hat{d}_i$, $\hat{d}_j$ by 1.
(3) Repeat step (2) until no more edge can be added to E.
(4) If $|E| < m$ report *failure* and output $N = 0$, otherwise output $G = (V, E)$ and $N = (m! \, P)^{-1}$.

Note that for regular graphs the factors $1 - d_i d_j / 4m$ are redundant and Procedure A is the same as Steger-Wormald's [40] algorithm. Next two theorems characterize output distribution of Procedure A.

Theorem 1. *Let $\tau > 0$: For any degree sequence $\bar{d}$ with maximum degree of $O(m^{1/4-\tau})$, Procedure A terminates successfully in one round with probability $1 - o(1)$ and any graph G with degree sequence $\bar{d}$ is generated with probability within $1 \pm o(1)$ factor of uniform. Expected running time of Procedure A is at most $O(m \, d_{\max})$.*

For regular graphs a similar result can be shown in a larger range of degrees. Let $\mathcal{L}(n, d)$ denotes the set of all d regular graphs with n vertices:

Theorem 2. *Let $\tau > 0$: If $d = O(n^{1/2-\tau})$, then for all graphs $G \in \mathcal{L}(n, d)$ except a set of size $o(|\mathcal{L}(n, d)|)$ Procedure A generates G with probability within $1 \pm o(1)$ factor of uniform. In other words as $n \to \infty$ output distribution of Procedure A converges to uniform in total variation metric.*

The above results show that output distribution of Procedure A is asymptotically uniform only as $n \to \infty$. But for finite values of n we use SIS to obtain an FPRAS for calculating $|\mathcal{L}(\bar{d})|$ and randomly generating its elements.

Definition 1. *An algorithm for approximately counting (randomly generating) graphs with degree sequence $\bar{d}$ is called an FPRAS if for any $\epsilon, \delta > 0$, it runs in time polynomial in $m, 1/\epsilon, \log(1/\delta)$ and with probability at least $1 - \delta$ the output of the algorithm is number of graphs (a graph with uniform probability) up to a multiplicative error of $1 \pm \epsilon$. For convenience we define a real valued random variable X to be an (ϵ, δ)-estimate for a number y if $\mathbb{P}\{(1-\epsilon)y \leq X \leq (1+\epsilon)y\} \geq 1 - \delta$.*

The following theorem summarizes our main result.

Theorem 3. *Let* $\tau > 0$*: For any degree sequence* $\bar{d}$ *with maximum degree of* $O(m^{1/4-\tau})$ *and any* $\epsilon, \delta > 0$ *we give an FPRAS for counting and generating graphs with degree sequence* $\bar{d}$ *with running time* $O(m\, d_{\max} \epsilon^{-2} \log(1/\delta))$.

3   Definitions and the Main Idea

Before explaining our approach let us quickly review the configuration model [12,7]. Let $W = \cup_{i=1}^{n} W_i$ be a set of $2m = \sum_{i=1}^{n} d_i$ labeled mini-vertices with $|W_i| = d_i$. If one picks two distinct mini-vertices uniformly at random and pairs them together then repeats that for remaining mini-vertices, after m steps a perfect matching $\mathcal{M}$ on the vertices of W is generated. Such matching is also called a *configuration* on W. One can see that number of all distinct configurations is equal to $(1/m!) \prod_{r=0}^{m-1} \binom{2m-2r}{2}$. The same equation shows that this sequential procedure generates all configurations with equal probability. Given a configuration $\mathcal{M}$, combining all mini-vertices of each W_i to form a vertex v_i, a graph $G_{\mathcal{M}}$ is generated whose degree sequence is $\bar{d}$.

Note that the graph $G_{\mathcal{M}}$ might have self edge loops or multiple edges. In fact McKay and Wormald's estimate [35] show that this happens with very high probability except when $d_{\max} = O(\log^{1/2} m)$. In order to fix this problem, at any step one [40] can only look at pairs of mini-vertices that lead to simple graphs (denote these by *suitable pairs*) and pick one uniformly at random. For d-regular graphs when $d = O(n^{1/28-\tau})$ Steger-Wormald [40] have shown this approach asymptotically samples regular graphs with uniform distribution and Kim-Vu [29] have extended that result to $d = O(n^{1/3-\tau})$.

Weighted Configuration Model. Unfortunately for general degree sequences some graphs can have probabilities that are far from uniform. In this paper we will show that for non-regular degree sequences suitable pairs should be picked non-uniformly. In fact Procedure A is a weighted configuration model where at any step a suitable pair $(u, v) \in W_i \times W_j$ is picked with probability proportional to $1 - d_i d_j / 4m$.

Here is a rough intuition behind Procedure A. Define the *execution tree* T of the configuration model as follows. Consider a rooted tree where root (the vertex in level zero) corresponds to the empty matching in the beginning of the model and level r vertices correspond to all partial matchings that can be constructed after r steps. There is an edge in T between a partial matching $\mathcal{M}_r$ from level r to a partial matching $\mathcal{M}_{r+1}$ from level $r + 1$ if $\mathcal{M}_r \subset \mathcal{M}_{r+1}$. Any path from the root to a leaf of T corresponds to one possible way of generating a random configuration.

Let us denote those partial matchings $\mathcal{M}_r$ whose corresponding partial graph $G_{\mathcal{M}_r}$ is simple by "valid" matchings. Denote the number of invalid children of $\mathcal{M}_r$ by $\Delta(\mathcal{M}_r)$. Our goal is to sample valid leaves of the tree T uniformly at random. Steger-Wormald's improvement to configuration model is to restrict the algorithm at step r to the valid children of $\mathcal{M}_r$ and picking one uniformly at random. This approach leads to almost uniform generation for regular graphs

[40,29] since the number of valid children for all partial matchings at level r of the T, is almost equal. However it is crucial to note that for non-regular degree sequences if the $(r+1)^{th}$ edge matches two elements belonging to vertices with larger degrees, the number of valid children for $\mathcal{M}_{r+1}$ will be smaller. Thus there will be a bias towards graphs that have more of such edges.

In order to find a rough estimate of the bias fix a graph G with degree sequence $\bar{d}$. Let $\mathrm{M}(G)$ be set of all leaves $\mathcal{M}$ of the tree T that lead to graph G; i.e. Configurations $\mathcal{M}$ with $G_{\mathcal{M}} = G$. It is easy to see that $|\mathrm{M}(G)| = m! \prod_{i=1}^{n} d_i!$. Moreover for exactly $(1-q_r)|\mathrm{M}(G)|$ of these leaves a fixed edge (i,j) of G appears in the first r edges of the path leading to them; i.e. $(i,j) \in \mathcal{M}_r$. Here $q_r = (m-r)/m$. Furthermore we can show that for a typical such leaf after step r, number of unmatched mini-vertices in each W_i is roughly $d_i q_r$. Thus expected number of un-suitable pairs (u,v) is about $\sum_{i \sim_G j} d_i d_j q_r^2 (1 - q_r)$. Similarly expected number of unsuitable pairs corresponding to self edge loops is approximately $\sum_{i=1}^{n} \binom{d_i q_r}{2} \approx 2m q_r^2 \lambda(\bar{d})$ where $\lambda(\bar{d}) = \sum_{i=1}^{n} \binom{d_i}{2} / (\sum_{i=1}^{n} d_i)$. Therefore defining $\gamma_G = \sum_{i \sim_G j} d_i d_j / 4m$ and using $\binom{2m-2r}{2} \approx 2m^2 q_r^2$ we can write

$$\mathbb{P}(G) \approx m! \prod_{i=1}^{n} d_i! \prod_{r=0}^{m-1} \frac{1}{2m^2 q_r^2 - 2m q_r^2 \lambda(\bar{d}) - 4m(1 - q_r)q_r^2 \gamma_G}$$

$$\approx e^{\lambda(\bar{d})+\gamma_G} m! \prod_{i=1}^{n} d_i! \prod_{r=0}^{m-1} \frac{1}{\binom{2m-2r}{2}} \propto e^{\gamma_G}$$

Hence adding the edge (i,j) roughly creates an $\exp(d_i d_j / 4m)$ bias. To cancel that effect we need to reduce probability of picking (i,j) by $\exp(-d_i d_j / 4m) \approx 1 - d_i d_j / 4m$. We will rigorously prove the above argument in Section 4.

Sequential Importance Sampling (SIS). For finite values of m output distribution of Procedure A can be slightly nonuniform but sequential nature of Procedure A gives us extra information that can be exploited to obtain very close approximations of any target distribution, including uniform. In particular the random variable N that is output of Procedure A is an unbiased estimator for $|\mathcal{L}(\bar{d})|$. More specifically:

$$\mathbb{E}_A(N) = \mathbb{E}_A(N1_{\{A \text{ succeeds}\}}) = \sum_{G \in \mathcal{L}(\bar{d})} \sum_{\pi_G} [m! \, \mathbb{P}_A(\pi_G)]^{-1} \mathbb{P}_A(\pi_G) = |\mathcal{L}(\bar{d})|$$

where π_G denotes one of the $m!$ ways of generating a graph G by Procedure A. Therefore we suggest the following algorithm for approximating $|\mathcal{L}(\bar{d})|$.

Algorithm: **CountGraphs**(ϵ, δ)

(1) For $k = k(\epsilon, \delta)$ run Procedure A exactly k times and denote the corresponding values of the random variable N by $N_1, \ldots, N_k$.
(2) Output $X = \frac{N_1 + \cdots + N_k}{k}$ as estimator for $|\mathcal{L}(\bar{d})|$.

We will show in the next section that if variance of the random variable N is small enough then Algorithm CountGraphs gives the desired approximation when $k(\epsilon, \delta) = O(\epsilon^{-2} \log(1/\delta))$. In order to generate elements of $\mathcal{L}(\bar{d})$ with probabilities very close to uniform we will use the estimator X and apply SIS to a different random variable. For that the algorithm is as follows:

Algorithm: **GenerateGraph**(ϵ, δ)

(1) Let X be an (ϵ, δ)-estimate for $|\mathcal{L}(\bar{d})|$ given by Algorithm CountGraphs.
(2) Repeat Procedure A to obtain one successful outcome G.
(3) Find an (ϵ, δ)-estimate, P_G, for $\mathbb{P}_A(G)$ using Procedure B.
(4) Report G with probability[1] $\frac{\frac{1}{X}}{5P_G}$ and stop. Otherwise go back to step (2).

The crucial step of Algorithm GenerateGraph is step (3) that is finding (ϵ, δ)-estimate for $\mathbb{P}_A(G)$. For that we use Procedure B which is exactly similar to Procedure A except for steps (2), (4).

Procedure B(G, ϵ, δ)

(1) Let E be a set of edges, $\hat{d} = (\hat{d}_1, \ldots, \hat{d}_n)$ be an n-tuple of integers and P be a number. Initialize them by $E = \emptyset$, $\hat{d} = \bar{d}$, and $P = 1$.
(2) Choose an edge $e = (v_i, v_j)$ of G that is not in E uniformly at random. Assume that (v_i, v_j) is chosen with probability proportion to $\hat{d}_i \hat{d}_j (1 - \frac{d_i d_j}{4m})$ among all pairs i, j with $i \neq j$ and $(v_i, v_j) \notin E$. Denote this probability by p_{ij} and multiply P by p_{ij}. Add (v_i, v_j) to E and reduce each of $\hat{d}_i$, $\hat{d}_j$ by 1.
(3) Repeat step (2) until no more edge can be added to E.
(4) Repeat steps (1) to (3) exactly $\ell = \ell(\epsilon, \delta)$ times and let $P_1, \ldots, P_\ell$ be corresponding values for P. Output $P_G = m! \frac{P_1 + \cdots + P_\ell}{\ell}$ as estimator for $\mathbb{P}_A(G)$.

Note that random variable P at the end of step (3) is exactly $\mathbb{P}_A(\pi_G)$ for a random permutation of the edges of G. It is easy to see that $\mathbb{E}_B(P) = \mathbb{P}_A(G)/m!$. Therefore P_G is an unbiased estimator for $\mathbb{P}_A(G)$. Later we will show that it is an (ϵ, δ)-estimator as well by controlling the variance of random variable P.

4   Analysis

Let us fix a simple graph G with degree sequence $\bar{d}$. Denote the set of all edge-wise different perfect matchings on mini-vertices of W that lead to graph G by R(G). Any two elements of R(G) can be obtained from one another by permuting labels of the mini-vertices in any W_i. We will find probability of generating a fixed element $\mathcal{M}$ in R(G). There are $m!$ different orders for picking edges of $\mathcal{M}$ sequentially and different orderings could have different probabilities. Denote the set of these orderings by S$(\mathcal{M})$. Thus

---

[1] In section 4 we will show that $\frac{1}{X} < 5P_G$ for large enough n (independent of $\epsilon, \delta > 0$).

$$\mathbb{P}_A(G) = \sum_{\mathcal{M} \in R(G)} \sum_{\mathcal{N} \in S(\mathcal{M})} \mathbb{P}_A(\mathcal{N}) = \prod_{i=1}^{n} d_i! \sum_{\mathcal{N} \in S(\mathcal{M})} \mathbb{P}_A(\mathcal{N}).$$

For any ordering $\mathcal{N} = \{e_1, \ldots, e_m\} \in S(\mathcal{M})$ and $0 \le r \le m-1$ denote probability of picking the edge e_{r+1} by $\mathbb{P}(e_{r+1})$. Hence $\mathbb{P}_A(\mathcal{N}) = \prod_{r=0}^{m-1} \mathbb{P}(e_{r+1})$ and each term $\mathbb{P}(e_{r+1})$ equals to

$$\mathbb{P}\left(e_{r+1} = (i,j)\right) = \frac{(1 - d_i d_j / 4m)}{\sum_{(u,v) \in E_r} d_u^{(r)} d_v^{(r)} (1 - d_u d_v / 4m)}$$

where $d_i^{(r)}$ denotes the residual degree of vertex i after r steps which shows number of unmatched mini-vertices of W_i at step r. The set E_r consists of all possible edges after picking $e_1, \ldots, e_r$. Denominator of the above fraction for $\mathbb{P}(e_{r+1})$ can be written as $\binom{2m-2r}{2} - \Psi(\mathcal{N}_r)$ where $\Psi(\mathcal{N}_r) = \Delta(\mathcal{N}_r) + \sum_{(i,j) \in E_r} d_i^{(r)} d_j^{(r)} d_i d_j / 4m$. This is because $\sum_{(u,v) \in E_r} d_u^{(r)} d_v^{(r)}$ counts number of suitable pairs in step r and is equal to $\binom{2m-2r}{2} - \Delta(\mathcal{N}_r)$. The quantity $\Psi(\mathcal{N}_r)$ can be also viewed as sum of the weights of unsuitable pairs. Now using $1 - x = e^{-x + O(x^2)}$ for $0 \le x \le 1$ we can write $\mathbb{P}_A(G)$ in the range $d_{\max} = O(m^{1/4 - \tau})$ as following

$$\mathbb{P}_A(G) = \prod_{i=1}^{n} d_i! \prod_{i \sim_G j} \left(1 - \frac{d_i d_j}{4m}\right) \sum_{\mathcal{N} \in S(\mathcal{M})} \prod_{r=0}^{m-1} \frac{1}{\binom{2m-2r}{2} - \Psi(\mathcal{N}_r)}$$

$$= \prod_{i=1}^{n} d_i! \, e^{-\gamma_G + o(1)} \sum_{\mathcal{N} \in S(\mathcal{M})} \prod_{r=0}^{m-1} \frac{1}{\binom{2m-2r}{2} - \Psi(\mathcal{N}_r)}$$

The next step is to show that with very high probability $\Psi(\mathcal{N}_r)$ is close to a number $\psi_r(G)$ independent of the ordering $\mathcal{N}$. More specifically for $\psi_r(G) = (2m - 2r)^2 \left(\frac{\lambda(\bar{d})}{2m} + \frac{r \sum_{i \sim_G j} (d_i - 1)(d_j - 1)}{4m^3} + \frac{(\sum_{i=1}^{n} d_i^2)^2}{32m^3} \right)$ the following is true

$$\sum_{\mathcal{N} \in S(\mathcal{M})} \prod_{r=0}^{m-1} \frac{1}{\binom{2m-2r}{2} - \Psi(\mathcal{N}_r)} = (1 + o(1)) \, m! \prod_{r=0}^{m-1} \frac{1}{\binom{2m-2r}{2} - \psi_r(G)} \quad (1)$$

The proof of this concentration uses Kim-Vu's polynomial [28] and is quite technical. It generalizes Kim-Vu's [29] calculations to general degree sequences. To the interest of space we omit this cumbersome analysis but it is given in in Section 5 of extended version of this paper [6]. But in Section 4.1 we show concentration for regular graphs in a larger region using a new technique.

Next step is to show the following equation for $d_{\max} = O(m^{1/4 - \tau})$.

$$\prod_{r=0}^{m-1} \frac{1}{\binom{2m-2r}{2} - \psi_r(G)} = \prod_{r=0}^{m-1} \frac{1}{\binom{2m-2r}{2}} e^{\lambda(\bar{d}) + \lambda^2(\bar{d}) + \gamma_G + o(1)}. \quad (2)$$

Proof of equation (2) is algebraic and is given in Section 5.2 of [6]. Following lemma summarizes the above analysis.

Lemma 1. *For $d_{\max} = O(m^{1/4-\tau})$ Procedure A generates all graphs with degree sequence $\bar{d}$ with asymptotically equal probability. More specifically*

$$\sum_{\mathcal{N} \in S(\mathcal{M})} \mathbb{P}(\mathcal{N}) = \frac{m!}{\prod_{r=0}^{m} \binom{2m-2r}{2}} e^{\lambda(\bar{d}) + \lambda^2(\bar{d}) + o(1)}$$

Proof (of Theorem 1). In order to show that $\mathbb{P}_A(G)$ is uniform we need an important piece. Lemma 1 shows that $\mathbb{P}_A(G)$ is independent of G but this probability might be still far from uniform. In other words we need to show that Procedure A always succeed with probability $1 - o(1)$. We will show this in section 6 of [6] by proving the following lemma.

Lemma 2. *For $d_{\max} = O(m^{1/4-\tau})$ the probability of failure in one trial of Procedure A is $o(1)$.*

Therefore all graphs G are generated asymptotically with uniform probability. Note that this will also give an independent proof of McKay's formula [33] for number of graphs.

Finally we are left with analysis of the running time which is given by the following lemma.

Lemma 3. *The Procedure A can be implemented so that the expected running time is $O(m\, d_{\max})$ for $d_{\max} = O(m^{1/4-\tau})$.*

Proof of Lemma 3 is given in Section 7 of [6]. This completes proof of Theorem 1. ∎

Proof (of Theorem 3). First we will prove the following result about output X of Algorithm GraphCount.

Lemma 4. *For any $\epsilon, \delta > 0$ there exist $k(\epsilon, \delta) = O(\epsilon^{-2} \log(1/\delta))$ where X is an (ϵ, δ)-estimator for $|\mathcal{L}(\bar{d})|$.*

Proof. By definition

$$\mathbb{P}\left((1-\epsilon)|\mathcal{L}(\bar{d})| < X < (1+\epsilon)|\mathcal{L}(\bar{d})|\right) = \mathbb{P}\left(-\frac{\epsilon \mathbb{E}_A(N)}{\sqrt{\frac{\mathrm{Var}_A(N)}{k}}} < \frac{X - \mathbb{E}_A(X)}{\sqrt{\frac{\mathrm{Var}_A(N)}{k}}} < \frac{\epsilon \mathbb{E}_A(N)}{\sqrt{\frac{\mathrm{Var}_A(N)}{k}}}\right)$$

On the other hand by central limit theorem $\lim_{k \to \infty} \frac{X - \mathbb{E}_A(X)}{\sqrt{\mathrm{Var}_A(N)/k}} \stackrel{d}{=} Z \sim N(0,1)$.

Therefore $\frac{\epsilon \mathbb{E}_A(N)}{\sqrt{\mathrm{Var}_A(N)/k}} > z_\delta$ guarantees that X is (ϵ, δ)-estimator $|\mathcal{L}(\bar{d})|$ where $\mathbb{P}(|Z| > z_\delta) = \delta$. This condition is equivalent to: $k > z_\delta^2 \epsilon^{-2} \mathrm{Var}_A(N)/\mathbb{E}_A(N)^2$. Moreover for tail of normal distribution we have $\mathbb{P}(Z > x) \approx x^{-1} e^{-x^2/2} (2\pi)^{-1}$ as $x \to +\infty$ which gives $z_\delta^2 = o(\log(1/\delta))$. This means if we show $\mathrm{Var}_A(N)/\mathbb{E}_A(N)^2 < \infty$ then $k = O(\log(1/\delta)\epsilon^{-2})$. In fact we will prove a stronger statement

$$\frac{\mathrm{Var}_A(N)}{\mathbb{E}_A(N)^2} = o(1) \tag{3}$$

Proof of equation (3) is given in Section 8.1 of [6].

Last step is to find running time of Algorithm CountGraph. By Theorem 1 each trial of Procedure A takes time $O(m\,d_{\max})$. Therefore running time of Algorithm CountGraph is exactly k times $O(m\,d_{\max})$ which is $O(m\,d_{\max}\epsilon^{-2}\log(1/\delta))$. This finishes proof of Lemma 4. ∎

Now we are left with analysis of Algorithm GenerateGraph which is given by following lemma.

Lemma 5. *For any $\epsilon, \delta > 0$ there exist $\ell(\epsilon, \delta)$ for Procedure B such that Algorithm GenerateGraph uses $O(m\,d_{\max}\epsilon^{-2}\log(1/\delta))$ operations and its output distribution is an (ϵ, δ)-estimator for uniform distribution.*

Similar calculations as in proof of Lemma 4 show that provided $\mathrm{Var}_B(P)/\mathbb{E}_B(P)^2 < \infty$ then for any graph G there exist $\ell(\epsilon, \delta) = O(\epsilon^{-2}\log(1/\delta))$ such that estimator P_G is an (ϵ, δ)-estimator for $\mathbb{P}_A(G)$. Similarly we will show the following stronger result in section 8.2 of [6].

$$\frac{\mathrm{Var}_B(P)}{\mathbb{E}_B(P)^2} = o(1) \tag{4}$$

Now we are ready to prove Lemma 5.

Proof (Proof of Lemma 5). From Theorem 1 there exist integer $n_0 > 0$ independent of ϵ, δ such that for $n > n_0$ the following is true $0.5|\mathcal{L}(\bar{d})| \leq \mathbb{P}_A(G) \leq 1.5|\mathcal{L}(\bar{d})|$. Thus using Lemma 4 for any $\epsilon', \delta' > 0$ and $k = k(\epsilon', \delta')$ and also $\ell = k(\epsilon', \delta')$, with probability $1 - 2\delta'$ we have:

$$0.5\frac{1-\epsilon'}{1+\epsilon'} \leq \frac{\frac{1}{X}}{P_G} \leq 1.5\frac{1+\epsilon'}{1-\epsilon'} \tag{5}$$

Note that for $\delta' < \delta/2$ and $\epsilon' < \min(0.25, \epsilon/2)$ the upper bound in equation (5) is strictly less than 5 which means $5P_G X > 1$ and this validates step (4) of Algorithm GenerateGraph. Moreover the lower bound in equation (5) is at least $1/8$ which shows that expected number of repetitions in Algorithm GenerateGraph is at most 8. On the other hand probability of generating a graph G by one trial of Algorithm GenerateGraph is equal to $\mathbb{P}_A(G)/(5XP_G)$. This means probability of generating an arbitrary graph G in one round of Algorithm GenerateGraph is an (ϵ, δ)-estimator for $\frac{1}{5X}$. Therefore final output distribution of Algorithm GenerateGraph is an (ϵ, δ)-estimator for uniform distribution.

Expected running time of Algorithm GenerateGraph is at most $8k$ times expected running time of Procedure A plus expected running time of Algorithm GraphCount which is $O(m\,d_{\max}\epsilon^{-2}\log(1/\delta))$. This finishes proof of Lemma 5 and therefore proof of Theorem 3. ∎

4.1 Concentration Inequality for Random Regular Graphs

Recall that $\mathcal{L}(n, d)$ denotes the set of all simple d-regular graphs with $m = nd/2$ edges. Same as before let G be a fixed element of $\mathcal{L}(n, d)$ and $\mathcal{M}$ be a fixed matching on W with $G_\mathcal{M} = G$. The main goal is to show that for $d = o(n^{1/2-\tau})$ probability of generating G is at least $(1 - o(1))$ of uniform; i.e. $\mathbb{P}_A(G) = (d!)^n \sum_{\mathcal{N} \in S(\mathcal{M})} \mathbb{P}(\mathcal{N}) \geq (1 - o(1)) / |\mathcal{L}(n, d)|$. Our proof builds upon the steps in Kim and Vu [30]. The following summarizes the analysis of Kim and Vu [30] for $d = O(n^{1/3-\tau})$. Let $m_1 = \frac{m}{d^2 \omega}$ where ω goes to infinity very slowly; e.g. $O(\log^\delta n)$ for small $\delta > 0$ then:

$$|\mathcal{L}(n, d)|(d!)^n \sum_{\mathcal{N} \in S(\mathcal{M})} \mathbb{P}(\mathcal{N}) \overset{(a)}{=} \frac{1 - o(1)}{m!} \sum_{\mathcal{N} \in S(\mathcal{M})} \prod_{r=0}^{m-1} \frac{\binom{2m-2r}{2} - \mu_r}{\binom{2m-2r}{2} - \Delta(\mathcal{N}_r)}$$

$$\overset{(b)}{\geq} \frac{1 - o(1)}{m!} \sum_{\mathcal{N} \in S(\mathcal{M})} \prod_{r=0}^{m_1} \left(1 + \frac{\Delta(\mathcal{N}_r) - \mu_r}{\binom{2m-2r}{2} - \Delta(\mathcal{N}_r)}\right)$$

$$\overset{(c)}{\geq} (1 - o(1)) \prod_{r=0}^{m_1} \left(1 - 3\frac{T_r^1 + T_r^2}{(2m - 2r)^2}\right)$$

$$\overset{(d)}{\geq} (1 - o(1)) \exp\left(-3e \sum_{r=0}^{m_1} \frac{T_r^1 + T_r^2}{(2m - 2r)^2}\right) \tag{6}$$

Here we explain these steps in more details. First define $\mu_r = \mu_r^1 + \mu_r^2$ where $\mu_r^1 = (2m - 2r)^2 (d - 1)/4m$ and $\mu_r^2 = (2m - 2r)^2 (d - 1)^2 r / 4m^2$. Step (a) follows from equation (3.5) of [30] and McKay-Wormald's estimate [35] for regular graphs. Also algebraic calculations in page 10 of [30] justify (b).

The main step is (c) which uses large deviations. For simplicity write Δ_r instead of $\Delta(\mathcal{N}_r)$ and let $\Delta_r = \Delta_r^1 + \Delta_r^2$ where Δ_r^1 and Δ_r^2 show the number of unsuitable pairs in step r corresponding to self edge loops and double edges respectively. For $p_r = r/m$, $q_r = 1 - p_r$ Kim and Vu [30] used their polynomial inequality [28] to derive bounds T_r^1, T_2^r and to show with very high probability $|\Delta_r^1 - \mu_r^1| < T_r^1$ and $|\Delta_r^2 - \mu_r^2| < T_r^2$. More precisely for some constants c_1, c_2

$$T_r^1 = c_1 \log^2 n \sqrt{nd^2 q_r^2 (2dq_r + 1)}, \qquad T_r^2 = c_2 \log^3 n \sqrt{nd^3 q_r^2 (d^2 q_r + 1)}$$

Now it can be shown that $\mu_r^i = o\left((2m - 2r)^2\right)$ and $T_r^i = o\left((2m - 2r)^2\right)$, $i = 1, 2$ which validates the step (c). The step (d) is straightforward using $1 - x \geq e^{-ex}$ for $1 \geq x \geq 0$.

Kim and Vu show that for $d = O(n^{1/3-\tau})$ the exponent in equation (6) is of $o(1)$. Using similar calculations as equation (3.13) in [30] it can be shown that in the larger region $d = O(n^{1/2-\tau})$ for $m_2 = (m \log^3 n)/d$:

$$\sum_{r=0}^{m_1} \frac{T_r^1}{(2m - 2r)^2} = o(1), \qquad \sum_{r=m_2}^{m_1} \frac{T_r^2}{(2m - 2r)^2} = o(1)$$

But unfortunately the remaining $\sum_{r=0}^{m_2} \frac{T_r^2}{(2m-2r)^2}$ is of order $\Omega(d^3/n)$. In fact it turns out the random variable Δ_r^2 has large variance in this range.

Let us explain the main difficulty for moving from $d = O(n^{1/3-\tau})$ to $d = O(n^{1/2-\tau})$. Note that Δ_r^2 is defined on a random subgraphs $G_{\mathcal{N}_r}$ of the graph G

which has exactly r edges. Both [40] and [29,30] have approximated the $G_{\mathcal{N}_r}$ with G_{p_r} in which each edge of G appears independently with probability $p_r = r/m$. Our analysis shows that when $d = O(n^{1/2-\tau})$ this approximation causes the variance of Δ_r^2 to blow up.

In order to fix this problem we modify Δ_r^2 before moving to G_{p_r}. It can be shown via simple algebraic calculations that: $\Delta_r^2 - \mu_r^2 = X_r + Y_r$ where $X_r = sum_{u \sim_{G_{\mathcal{N}_r}} v} [d_u^{(r)} - q_r(d-1)][d_v^{(r)} - q_r(d-1)]$ and $Y_r = q_r(d-1) \sum_u [(d_u^{(r)} - q_r d)^2 - dp_r q_r]$. This modification is critical since the equality $\Delta_r^2 - \mu_r^2 = X_r - Y_r$ does not hold in G_{p_r}.

Next task is to find a new bound $\hat{T}_r^2$ such that $|X_r - Y_r| < \hat{T}_r^2$ with very high probability and $\sum_{r=0}^{m_2} \frac{\hat{T}_r^2}{(2m-2r)^2} = o(1)$. It is easy to see that in G_{p_r} both X_r and Y_r have zero expected value. At this time we will move to G_{p_r} and show that X_r and Y_r are concentrated around zero. In the following we will show the concentration of X_r in details. For Y_r it can be done in exact same way.

Consider the edge exposure martingale (page 94 of [1]) for the edges that are picked up to step r. That is for any $0 \le \ell \le r$ define $Z_\ell^r = \mathbb{E}(X_r \mid e_1, \dots, e_\ell)$. For simplicity of notation let us drop the index r from $Z_\ell^r, d_u^{(r)}, p_r$ and q_r.

Next step is to bound the martingale difference $|Z_i - Z_{i-1}|$. Note that for $e_i = (u, v)$:

$$|Z_i - Z_{i-1}| = \left| (d_u - (d-1)q)(d_v - (d-1)q) + \sum_{u' \sim_{G_p} u} (d_{u'} - (d-1)q) + \sum_{v' \sim_{G_p} v} (d_{v'} - (d-1)q) \right|$$

(7)

Bounding the above difference should be done carefully since the standard worst case bounds are weak for our purpose. Note that the following observation is crucial. For a typical ordering $\mathcal{N}$ the residual degree of the vertices are roughly $dq \pm \sqrt{dq}$. We will make this more precise. For any vertex $u \in G$ consider the following $L_u = \{|d_u - dq| \le c \log^{1/2} n (dq)^{1/2}\}$ where $c > 0$ is a large constant. Then the following lemma holds:

Lemma 6. For all $0 \le r \le m_2$ the following is true: $\mathbb{P}(L_u^c) = o(\frac{1}{m^4})$.

Proof of Lemma 6 uses generalization of the Chernoff inequality from [1] and is given in Section 9 of [6].

To finish bounding the martingale difference we look at other terms in equation (7). For any vertex u consider the event

$$K_u = \left\{ |\sum_{u' \sim_{G_p} u} (d_{u'} - (d-1)q)| \le c \left((dq)^{3/2} + qd + dq^{1/2} \right) \log n \right\}$$

where $c > 0$ is a large constant. We will use the following lemma to bound the martingale difference.

Lemma 7. For all $0 \le r \le m_2$ the followings hold $\mathbb{P}(K_u^c) = o(\frac{1}{m^4})$.

Proof of Lemma 7 is given in Section 9 of [6].

Now we are ready to bound the martingale difference. Let $L = \bigcap_{r=0}^{m_2} \bigcap_{u=1}^{n} (L_u \cap K_u)$. Using Lemmas 6, 7 and the union bound $\mathbb{P}(L^c) = o(1/m^2)$. Hence for the martingale difference we have $|Z_i - Z_{i-1}| 1_L \leq O(dq + dq^{1/2} + (dq)^{3/2}) \log n$.

Next we state and use the following variation of Azuma's inequality.

Proposition 1 (Kim [27]). *Consider a martingale $\{Y_i\}_{i=0}^{n}$ adaptive to a filtration $\{\mathcal{B}_i\}_{i=0}^{n}$. If for all k there are $A_{k-1} \in \mathcal{B}_{k-1}$ such that $\mathbf{E}[e^{\omega Y_k} | \mathcal{B}_{k-1}] \leq C_k$ for all $k = 1, 2, \cdots, n$ with $C_k \geq 1$ for all k, then*

$$\mathbf{P}(Y - \mathbf{E}[Y] \geq \lambda) \leq e^{-\lambda \omega} \prod_{k=1}^{n} C_k + \mathbf{P}(\cup_{k=0}^{n-1} A_k)$$

Proof (of Theorem 2). Applying the above proposition for a large enough constant $c' > 0$ gives:

$$\mathbb{P}\left(|X_r| > c' \sqrt{6r \log^3 n \left(dq + d(q)^{1/2} + (dq)^{3/2}\right)^2}\right) \leq e^{-3\log n} + \mathbb{P}(L^c) = o\left(\frac{1}{m^2}\right)$$

The same equation as above holds for Y_r since the martingale difference for Y_r is of $O(|dq_r(d_u - q_r d)|) = O((dq)^{3/2} \log^{1/2} n))$ using Lemma 6.

Therefore defining $\hat{T}_r^2 = c'(dq + d(q)^{1/2} + (dq)^{3/2})\sqrt{6r \log^3 n}$ we only need to show the following is $o(1)$.

$$\sum_{r=0}^{m_2} \frac{(dq + d(q)^{1/2} + (dq)^{3/2})\sqrt{6r \log^3 n}}{(2m - 2r)^2}$$

But using $ndq = 2m - 2r$:

$$\sum_{r=0}^{m_2} \frac{(dq + dq^{1/2} + (dq)^{3/2})\sqrt{6r \log^3 n}}{n^2 d^2 q^2}$$

$$= \sum_{r=0}^{m_2} O\left(\frac{d^{1/2}\log^{1.5} n}{n^{1/2}(2m - 2r)} + \frac{d\log^{1.5} n}{(2m - 2r)^{3/2}} + \frac{d^{1/2}\log^{1.5} n}{n(2m - 2r)^{1/2}}\right)$$

$$= O\left(\frac{d^{1/2}\log(nd)}{n^{1/2}} + \frac{d}{(n\log^3 n)^{1/2}} + \frac{d}{n^{1/2}}\right)\log^{1.5} n$$

which is $o(1)$ for $d = O(n^{1/2-\tau})$.   ∎

Acknowledgement

We would like to thank Joe Blitzstein, Persi Diaconis, Adam Guetz, Milena Mihail, Alistair Sinclair, Eric Vigoda and Ying Wang for insightful discussions and useful comments on earlier version of this paper.

References

1. Alon, N., Spencer, J.: The Probabilistic Method. Wiley, NY (1992)
2. Alderson, D., Doyle, J., Willinger, W.: Toward and Optimization-Driven Framework for Designing and Generating Realistic Internet Topologies. HotNets (2002)
3. Amraoui, A., Montanari, A., Urbanke, R.: How to Find Good Finite-Length Codes: From Art Towards Science (preprint, 2006), arxiv.org/pdf/cs.IT/0607064
4. Bassetti, F., Diaconis, P.: Examples Comparing Importance Sampling and the Metropolis Algorithm (2005)
5. Bayati, M., Montanari, A., Saberi, A.: (work in progress, 2007)
6. Bayati, M., Kim, J.H., Saberi, A.: A Sequential Algorithm for Generating Random Graphs (2007), Extended Version, available at http://arxiv.org/abs/cs/0702124
7. Bender, E.A., Canfield, E.R.: The asymptotic number of labeled graphs with given degree sequence. J. Combinatorial Theory Ser. A 24(3), 296–307 (1978)
8. Bezáková, I., Bhatnagar, N., Vigoda, E.: Sampling Binary Contingency tables with a Greedy Start. In: SODA (2006)
9. Bezáková, I., Sinclair, A., Štefankovič, D., Vigoda, E.: Negative Examples for Sequential Importance Sampling of Binary Contingency Tables (preprint, 2006)
10. Blanchet, J.: Efficient Importance Sampling for Counting (preprint, 2006)
11. Blitzstein, J., Diaconis, P.: A sequential importance sampling algorithm for generating random graphs with prescribed degrees (submitted)
12. Bollobás, B.: A probabilistic proof of an asymptotoic forumula for the number of labelled regular graphs. European J. Combin. 1(4), 311–316 (1980)
13. Britton, T., Deijfen, M., Martin-Löf, A.: Generating simple random graphs with prescribed degree distribution (preprint)
14. Bu, T., Towsley, D.: On Distinguishing between Internet Power Law Topology Generator. In: INFOCOM (2002)
15. Chen, Y., Diaconis, P., Holmes, S., Liu, J.S.: Sequential Monte Carlo methods for statistical analysis of tables. Journal of the American Statistical Association 100, 109–120 (2005)
16. Cooper, C., Dyer, M., Greenhill, C.: Sampling regular graphs and peer-to-peer network. Combinatorics, Probability and Computing (to appear)
17. Chung, F., Lu, L.: Conneted components in random graphs with given expected degree sequence. Ann. Comb. 6(2), 125–145 (2002)
18. Diaconis, P., Gangolli, A.: Rectangular arrays with fixed margins. In: Discrete probability and algorithms (Minneapolis, MN, 1993). IMA Vol. Math. Appl., vol. 72, pp. 15–41. Springer, Heidelberg (1995)
19. Faloutsos, M., Faloutsos, P., Faloutsos, C.: On Power-law Relationships of the Internet Topology. SIGCOM (1999)
20. Gkantsidis, C., Mihail, M., Zegura, E.: The Markov Chain Simulation Method for Generating Connected Power Law Random Graphs. Alenex (2003)
21. Jerrum, M., Valiant, L., Vazirani, V.: Random generation of combinatorial structures from a uniform distribution, Theoret. Comput. Sci. 43, 169-188. 73, 1, 91-100 (1986)
22. Jerrum, M., Sinclair, A.: Approximate counting, uniform generation and rapidly mixing Markov chains. Inform. and Comput. 82(1), 93–133 (1989)
23. Jerrum, M., Sinclair, A.: Fast uniform generation of regular graphs. Theoret. Comput. Sci. 73(1), 91–100 (1990)
24. Jerrum, M., Sinclair, A., McKay, B.: When is a graphical sequence stable? In: Random graphs Vol 2 (Poznań 1989), pp. 101–115. Wiley-Intersci. Publ. Wiley, New York (1992)

25. Jerrum, M., Sinclair, A., Vigoda, E.: A Polynomial-Time Approximation Algorithm for the Permanent of a Matrix with Non-Negative Entries. Journal of the ACM 51(4), 671–697 (2004)
26. Kannan, R., Tetali, P., Vempala, S.: Simple Markov chain algorithms for generating bipartite graphs and tournaments, (1992) Random Structures and Algorithms 14, 293-308 (1999)
27. Kim, J.H.: On Brooks' Theorem for Sparse Graphs. Combi. Prob. & Comp. 4, 97–132 (1995)
28. Kim, J.H., Vu, V.H.: Concentration of multivariate polynomials and its applications. Combinatorica 20(3), 417–434 (2000)
29. Kim, J.H., Vu, V.H.: Generating Random Regular Graphs, STOC, 213-222 (2003)
30. Kim, J.H., Vu, V.: Sandwiching random graphs. Advances in Mathematics 188, 444–469 (2004)
31. Knuth, D.: Mathematics and computer science: coping with finiteness. Science 194(4271), 1235–1242 (1976)
32. Medina, A., Matta, I., Byers, J.: On the origin of power laws in Internet topologies. ACM Computer Communication Review 30(2), 18–28 (2000)
33. McKay, B.: Asymptotics for symmetric 0-1 matrices with prescribed row sums. Ars Combinatorica 19A, 15–25 (1985)
34. McKay, B., Wormald, N.C.: Uniform generation of random regular graphs of moderate degree. J. Algorithms 11(1), 52–67 (1990b)
35. McKay, B., Wormald, N.C.: Asymptotic enumeration by degree sequence of graphs with degrees $o(n^{1/2})$. Combinatorica 11(4), 369–382 (1991)
36. Milo, R., Kashtan, N., Itzkovitz, S., Newman, M., Alon, U.: On the uniform generation of random graphs with prescribed degree sequences (2004), http://arxiv.org/PS_cache/cond-mat/pdf/0312/0312028.pdf
37. Milo, R., ShenOrr, S., Itzkovitz, S., Kashtan, N., Chklovskii, D., Alon, U.: Network motifs: Simple building blocks of complex networks. Science 298, 824–827 (2002)
38. Molloy, M., Reed, B.: A critical point for random graphs with a given degree sequence. Random Structures and Algorithms 6, 2–3, 161–179 (1995)
39. Sinclair, A.: Personal communication (2006)
40. Steger, A., Wormald, N.C.: Generating random regular graphs quickly (English Summary) Random graphs and combinatorial structures (Oberwolfach, 1997). Combin. Probab. Comput. 8(4), 377–396 (1997)
41. Tangmunarunkit, H., Govindan, R., Jamin, S., Shenker, S., Willinger, W.: Network Topology Generators: Degree based vs. Structural. ACM SIGCOM (2002)
42. Wormald, N.C.: Models of random regular graphs. In: Surveys in combinatorics (Canterbury). London Math. Soc. Lecture Notes Ser., vol. 265, pp. 239–298. Cambridge Univ. Press, Cambridge (1999)
43. Vu, V.H.: Concentration of non-Lipschitz functions and applications, Probabilistic methods in combinatorial optimization. Random Structures Algorithms 20(3), 267–316 (2002)

Local Limit Theorems for the Giant Component of Random Hypergraphs

Michael Behrisch[1,*], Amin Coja-Oghlan[2,**], and Mihyun Kang[1,***]

[1] Humboldt-Universität zu Berlin, Institut für Informatik,
Unter den Linden 6, 10099 Berlin, Germany
[2] Carnegie Mellon University, Department of Mathematical Sciences,
Pittsburgh, PA 15213, USA
{behrisch,coja,kang}@informatik.hu-berlin.de

Abstract. Let $H_d(n, p)$ signify a random d-uniform hypergraph with n vertices in which each of the $\binom{n}{d}$ possible edges is present with probability $p = p(n)$ independently, and let $H_d(n, m)$ denote a uniformly distributed d-uniform hypergraph with n vertices and m edges. We establish a *local limit theorem* for the number of vertices and edges in the largest component of $H_d(n, p)$ in the regime $(d - 1)^{-1} + \varepsilon < \binom{n-1}{d-1}p = O(1)$, thereby determining the joint distribution of these parameters precisely. As an application, we derive an asymptotic formula for the probability that $H_d(n, m)$ is connected, thus obtaining a formula for the asymptotic number of connected hypergraphs with a given number of vertices and edges. While most prior work on this subject relies on techniques from enumerative combinatorics, we present a new, purely probabilistic approach.

1 Introduction

Notation. A *d-uniform hypergraph* H is a set $V(H)$ of *vertices* together with a set $E(H)$ of *edges* $e \subset V(H)$, each connecting $|e| = d$ vertices. The *order* of H is the number $|V(H)|$ of vertices of H, and the *size* of H is the number $|E(H)|$ of edges. Thus, a 2-uniform hypergraph is just a graph. Moreover, we say that $v \in V(H)$ is *reachable* from $w \in V(H)$ if either $v = w$ or there exist edges $e_1, \ldots, e_k \in E(H)$ such that $v \in e_1$, $w \in e_k$, and $e_i \cap e_{i+1} \neq \emptyset$ for $1 \leq i < k$. Clearly, reachability is an equivalence relation. The equivalence classes are the *components* of H.

We let $\mathcal{N}(H)$ signify the maximum order of a component of H. Furthermore, throughout the paper the vertex set $V(H)$ will consist of integers. Therefore, the subsets of $V(H)$ can be ordered lexicographically, and we call the lexicographically first component of H that has order $\mathcal{N}(H)$ the *largest component* of H. In addition, we denote by $\mathcal{M}(H)$ the size of the largest component.

---

* Supported by the DFG research center MATHEON in Berlin.
** Supported by the Deutsche Forschungsgemeinschaft (DFG COJ 646).
*** Supported by the Deutsche Forschungsgemeinschaft (DFG Pr 296).

M. Charikar et al. (Eds.): APPROX and RANDOM 2007, LNCS 4627, pp. 341–352, 2007.

We deal with two models of random d-uniform hypergraphs: $H_d(n,p)$ and $H_d(n,m)$. The random hypergraph $H_d(n,p)$ has the vertex set $V = \{1,\dots,n\}$, and each of the $\binom{n}{d}$ possible edges is present with probability p independently of all others. Moreover, $H_d(n,m)$ is a uniformly distributed d-uniform hypergraph with vertex set $V = \{1,\dots,n\}$ and with exactly m edges. In the case $d = 2$, the notation $G(n,p) = H_2(n,p)$, $G(n,m) = H_2(n,m)$ is common. We say that the random hypergraph $H_d(n,p)$ (resp. $H_d(n,m)$) enjoys a certain property $\mathcal{P}$ *with high probability* (w.h.p. for short) if the probability that $\mathcal{P}$ holds in $H_d(n,p)$ (resp. $H_d(n,m)$) tends to 1 as $n \to \infty$.

The Giant Component. Since the pioneering work of Erdős and Rényi [11,12] on the "giant component" of the random graph $G(n,m)$, the component structure of random discrete objects (e.g., graphs, hypergraphs, digraphs) has been among the main subjects of probabilistic combinatorics. One reason for this is the connection to statistical physics and percolation (as "mean field models"); another reason is the impact of these considerations on computer science (e.g., due to relations to computational problems such as MAX CUT or MAX 2-SAT [10]).

In [12] Erdős and Rényi investigated the component structure of *sparse* random graphs with $O(n)$ edges. The main result is that the order $\mathcal{N}(G(n,m))$ of the largest component undergoes a *phase transition* as $2m/n \sim 1$. Let us state a more general version due to Schmidt-Pruzan and Shamir [25], which covers d-uniform hypergraphs: let either $H = H_d(n,m)$ and $c = dm/n$, or $H = H_d(n,p)$ and $c = \binom{n-1}{d-1}p$; we refer to c as the *average degree* of H.

(1) If $c < (d-1)^{-1} - \varepsilon$ for an arbitrarily small but fixed $\varepsilon > 0$, then $\mathcal{N}(H) = O(\ln n)$ w.h.p.
(2) If $c > (d-1)^{-1} + \varepsilon$, then H has a unique component of order $\Omega(n)$ w.h.p., which is called the *giant component*. More precisely, as shown in [9],

$$\mathcal{N}(H) = (1 - \rho)n + o(n) \text{ w.h.p.,} \tag{1}$$

where ρ is the unique solution to the transcendental equation

$$\rho = \exp(c(\rho^{d-1} - 1)) \tag{2}$$

that lies strictly between 0 and 1. Furthermore, the second largest component has order $O(\ln n)$.

In the present paper we analyze the distribution of $\mathcal{N}(H)$ and $\mathcal{M}(H)$ for a random hypergraph $H = H_d(n,m)$ or $H = H_d(n,p)$ more precisely. We first establish a *local limit theorem* for $\mathcal{N}(H_d(n,p))$, thus determining the asymptotic distribution of $\mathcal{N}(H_d(n,p))$ precisely. Moreover, we obtain a local limit theorem for the *joint* distribution of the order and size of the giant component of $H_d(n,p)$. Furthermore, from these local limit theorems we infer a formula for the asymptotic probability that $H_d(n,m)$ is connected or, equivalently, an asymptotic formula for the number of connected hypergraphs of given order and size. Thus, we solve a (highly non-trivial) *enumerative* problem via probabilistic techniques.

While in the case of *graphs* (i.e., $d = 2$) these results are either known or can be derived from prior work (cf. the related work section below), all our results are new for d-uniform hypergraphs with $d > 2$. Furthermore, our proof relies on new probabilistic techniques, which, in contrast to most prior work, do not rely on involved enumerative techniques. In effect, our techniques are fairly generic and may apply to further problems of a related nature.

This extended abstract is a condensed version of our two recent preprints [4,5], which contain the complete proofs of all of the results discussed in the present paper.

2   Main Results

While (1) determines $\mathcal{N}(H_d(n, p))$ up to fluctuations of order $o(n)$, the following *local limit theorem* yields the exact limiting distribution of this random variable.

Theorem 1. *Let $d \geq 2$ be a fixed integer. For any two compact intervals $\mathcal{I} \subset \mathbf{R}$, $\mathcal{J} \subset ((d-1)^{-1}, \infty)$ and for any $\delta > 0$ there is $n_0 > 0$ such that the following holds. Let $p = p(n)$ be a sequence such that $c = c(n) = \binom{n-1}{d-1}p \in \mathcal{J}$ for all n, let $0 < \rho = \rho(n) < 1$ be the unique solution to (2), and let*

$$\sigma_{\mathcal{N}}^2 = \sigma_{\mathcal{N}}(n)^2 = \frac{\rho\left[1 - \rho + c(d-1)(\rho - \rho^{d-1})\right]n}{(1 - c(d-1)\rho^{d-1})^2}. \tag{3}$$

If $n \geq n_0$ and if ν is an integer such that $\sigma_{\mathcal{N}}^{-1}(\nu - (1 - \rho)n) \in \mathcal{I}$, then

$$\left| 1 - \sqrt{2\pi}\sigma_{\mathcal{N}} \exp\left[\frac{(\nu - (1-\rho)n)^2}{2\sigma_{\mathcal{N}}^2} \right] \mathrm{P}\left[\mathcal{N}(H_d(n, p)) = \nu \right] \right| \leq \delta.$$

Theorem 1 shows that for any integer ν such that $|\nu - (1 - \rho)n| = O(\sigma_{\mathcal{N}})$ we have

$$\mathrm{P}\left[\mathcal{N}(H_d(n, p)) = \nu \right] \sim \frac{1}{\sqrt{2\pi}\sigma_{\mathcal{N}}} \exp\left[-\frac{(\nu - (1-\rho)n)^2}{2\sigma_{\mathcal{N}}^2} \right]. \tag{4}$$

This implies that $\frac{\mathcal{N}(H_d(n,p)) - (1-\rho)n}{\sigma_{\mathcal{N}}}$ converges *in distribution* to the standard normal distribution. However, Theorem 1 is actually *much* stronger than this latter statement, because (4) implies that convergence to the normal distribution even holds on the level of the event that $\mathcal{N}(H_d(n, p))$ hits a *specific value* ν. In fact, observe that the expression on the r.h.s. of (4) is $\Theta(n^{-\frac{1}{2}})$. We sketch the proof of Theorem 1 in Section 4.

Building upon Theorem 1, we can further establish a local limit theorem for the *joint distribution* of $\mathcal{N}, \mathcal{M}(H_d(n, p))$.

Theorem 2. *Let $d \geq 2$ be a fixed integer. For any two compact sets $\mathcal{I} \subset \mathbf{R}^2$, $\mathcal{J} \subset ((d-1)^{-1}, \infty)$, and for any $\delta > 0$ there exists $n_0 > 0$ such that the following holds. Let $p = p(n)$ be a sequence such that $c = c(n) = \binom{n-1}{d-1}p \in \mathcal{J}$ for all n and* let $0 < \rho = \rho(n) < 1$ be the unique solution to (2). Further, let $\sigma_{\mathcal{N}}$ be as in (3) and set

$$\sigma_{\mathcal{M}}^2 = c^2 \rho^d \frac{2 + c(d-1)(\rho^{2d-2} - 2\rho^{d-1} + \rho^d) - \rho^{d-1} - \rho^d}{(1 - c(d-1)\rho^{d-1})^2} n + (1 - \rho^d)\frac{cn}{d},$$

$$\sigma_{\mathcal{N}\mathcal{M}} = c\rho \frac{1 - \rho^d - c(d-1)\rho^{d-1}(1-\rho)}{(1 - c(d-1)\rho^{d-1})^2} n.$$

Suppose that $n \geq n_0$ and that ν, μ are integers such that $x = \nu - (1-\rho)n$ and $y = \mu - (1-\rho^d)\binom{n}{d}p$ satisfy $n^{-\frac{1}{2}}\binom{x}{y} \in \mathcal{I}$. Then letting

$$P(x,y)^{-1} = 2\pi \sqrt{\sigma_{\mathcal{N}}^2 \sigma_{\mathcal{M}}^2 - \sigma_{\mathcal{N}\mathcal{M}}^2}$$

$$\times \exp\left[\frac{\sigma_{\mathcal{N}}^2 \sigma_{\mathcal{M}}^2}{2(\sigma_{\mathcal{N}}^2 \sigma_{\mathcal{M}}^2 - \sigma_{\mathcal{N}\mathcal{M}}^2)}\left(\frac{x^2}{\sigma_{\mathcal{N}}^2} - \frac{2\sigma_{\mathcal{N}\mathcal{M}}xy}{\sigma_{\mathcal{N}}^2 \sigma_{\mathcal{M}}^2} + \frac{y^2}{\sigma_{\mathcal{M}}^2}\right)\right],$$

we have

$$(1-\delta)P(x,y) \leq \mathrm{P}\left[\mathcal{N}(H_d(n,p)) = \nu \wedge \mathcal{M}(H_d(n,p)) = \mu\right] \leq (1+\delta)P(x,y).$$

Theorem 2 implies that (centered and scaled version of) $\mathcal{N}, \mathcal{M}(H_d(n,p))$ converge to a bivariate normal distribution. However, the precise statement is actually much stronger, because it yields the asymptotic probability that $\mathcal{N}$ and $\mathcal{M}$ hit any two specific values ν, μ, provided that ν and μ satisfy $x = \nu - (1-\rho)n = O(\sqrt{n})$ and $y = \mu - (1-\rho^d)\binom{n}{d}p = O(\sqrt{n})$. The proof of Theorem 2 can be found in Section 5.

As an application of the local limit theorem for $H_d(n,p)$ (Theorem 2), we obtain the following formula for the asymptotic probability that a random hypergraph $H_d(\nu,\mu)$ is connected. Here we use the notation $H_d(\nu,\mu)$ instead of $H_d(n,m)$, because it will be more convenient in the course of proving the theorem, cf. Section 6.

Theorem 3. Let $d \geq 2$ be a fixed integer. For any compact set $\mathcal{J} \subset (d(d-1)^{-1}, \infty)$, and for any $\delta > 0$ there exists $\nu_0 > 0$ such that the following holds. Let $\mu = \mu(\nu)$ be a sequence of integers such that $\zeta = \zeta(\nu) = d\mu/\nu \in \mathcal{J}$ for all ν. Then there exists a unique number $0 < r = r(\nu) < 1$ such that $r = \exp\left(-\zeta \cdot \frac{(1-r)(1-r^{d-1})}{1-r^d}\right)$. Let $\Phi(\zeta) = r^{\frac{r}{1-r}}(1-r)^{1-\zeta}(1-r^d)^{\frac{\zeta}{d}}$. Furthermore, let

$$R_2(\nu,\mu) = \frac{1 + r - \zeta r}{\sqrt{(1+r)^2 - 2\zeta r}} \exp\left(\frac{2\zeta r + \zeta^2 r}{2(1+r)}\right) \cdot \Phi(\zeta)^\nu,$$

$$R_d(\nu,\mu) = \frac{1 - r^d - (1-r)\zeta(d-1)r^{d-1}}{\sqrt{(1 - r^d + \zeta(d-1)(r - r^{d-1}))(1 - r^d) - \zeta dr(1 - r^{d-1})^2}}$$

$$\times \exp\left(\frac{\zeta(d-1)(r - 2r^d + r^{d-1})}{2(1 - r^d)}\right) \cdot \Phi(\zeta)^\nu \quad \text{if } d > 2.$$

Finally, let $c_d(\nu,\mu)$ signify the probability that $H_d(\nu,\mu)$ is connected. Then for all $\nu > \nu_0$ we have $(1-\delta)R_d(\nu,\mu) < c_d(\nu,\mu) < (1+\delta)R_d(\nu,\mu)$.

3 Related Work

Random Graphs. Stepanov [27] was the first to obtain a local limit theorem for $\mathcal{N}(G(n,p))$, thereby proving the $d = 2$ case of Theorem 1. However, his approach does not yield the joint distribution of $\mathcal{N}(G(n,p))$ and $\mathcal{M}(G(n,p))$. Moreover, Pittel [21] proved that $\mathcal{N}(G(n,p))$ and $\mathcal{N}(G(n,m))$ (suitably centered and scaled) are asymptotically normal. The arguments in both [21,27] are of enumerative and analytic nature.

Bender, Canfield, and McKay [6] computed the asymptotic probability that $G(n,m)$ is connected for *any* ratio m/n. Although they employ a probabilistic result from Łuczak [18] to simplify their arguments, their proof is based on enumerative methods. Łuczak and Łuczak [19] used the result from [6] to establish a local limit theorem for the order of the largest component for the cluster-scaled model of random graphs (a generalization of $G(n,p)$). Furthermore, Pittel and Wormald [22,23] applied enumerative arguments to derive an improved version of the main result of [6] and to obtain (among other results) a local limit theorem for the joint distribution of $\mathcal{N}, \mathcal{M}(G(n,p))$. In summary, in [6,7,19,22,23] enumerative methods were used to infer the distributions of $\mathcal{N}, \mathcal{M}(G(n,p))$. By contrast, in the present paper we use the converse approach: employing purely probabilistic methods, we first determine the distributions of $\mathcal{N}, \mathcal{M}(G(n,p))$, and from this we derive the number of connected graphs with given order and size.

Furthermore, a few authors have applied probabilistic arguments to problems related to the present work. For instance, Barraez, Boucheron, and Fernandez de la Vega [2] exploited the analogy between the component structure of $G(n,p)$ and branching processes to show that $\mathcal{N}(G(n,p))$ is asymptotically normal. However, their techniques do not yield the asymptotic probability that $\mathcal{N}(G(n,p))$ hits a specific value, i.e., a *local* limit theorem. Finally, van der Hofstad and Spencer [13] used a novel perspective on the branching process argument to rederive the formula of Bender, Canfield, and McKay [6] for the number of connected graphs.

Random Hypergraphs. In contrast to the case of graphs ($d = 2$), little is known about the phase transition and the connectivity probability of random d-uniform hypergraphs with $d > 2$.

Karoński and Łuczak [16] derived an asymptotic formula for the number of connected d-uniform hypergraphs of order n and size $m = \frac{n}{d-1} + o(\ln n/\ln \ln n)$ via combinatorial techniques. Since the minimum number of edges necessary for connectivity is $\frac{n-1}{d-1}$, this formula addresses *sparsely* connected hypergraphs. Building upon this result, the same authors investigated the phase transition in $H_d(n,m)$ and $H_d(n,p)$ [17]. They obtained local limit theorems for the joint distribution of $\mathcal{N}, \mathcal{M}(H_d(n,m))$ and $\mathcal{N}, \mathcal{M}(H_d(n,p))$ in the *early supercritical phase*, i.e., their results apply to $m = \binom{n}{d}p = \frac{n}{d(d-1)} + o(n^{2/3}(\ln n/\ln \ln n)^{1/3})$. Furthermore, Andriamampianina and Ravelomanana [1] extended the result from [16] to the regime $m = \frac{n}{d-1} + o(n^{1/3})$ via enumerative techniques. In addition, relying on [1], Ravelomanana and Rijamamy [24] extended [17] to $m = \binom{n}{d}p = \frac{n}{d(d-1)} + o(n^{7/9})$. All of these results either deal with *very sparsely* connected hypergraphs (i.e., $m = \frac{n}{d-1} + o(n)$), or with the *early* supercritical phase (i.e., $m = \binom{n}{d}p = \frac{n}{d(d-1)} + o(n)$). By comparison, in the present paper we deal with connected hypergraphs with $m = \frac{n}{d-1} + \Omega(n)$ edges and the component structure of random hypergraphs $H_d(n, m)$ or $H_d(n, p)$ with $m = \binom{n}{d}p = \frac{n}{d(d-1)} + \Omega(n)$, thus complementing the aforementioned work.

The regime of m and p that we deal with in the present work was previously studied by Coja-Oghlan, Moore, and Sanwalani [9] using probabilistic arguments. Setting up an analogy between a certain branching process and the component structure of $H_d(n, p)$, they computed the expected order and size of the largest component of $H_d(n, p)$ along with the variance of $\mathcal{N}(H_d(n, p))$. Furthermore, they computed the probability that $H_d(n, m)$ is connected *up to a constant factor*. Hence, Theorem 3 improves on this result, as the new result yields *tight* asymptotics for the connectivity probability. Nonetheless, in the present work we use some of the ideas from [9] (e.g., exposing the edges of $H_d(n, p)$ in several rounds).

4   The Local Limit Theorem for $\mathcal{N}(H_d(n, p))$

Throughout this section, we assume that the conditions stated in Theorem 1 are satisfied. That is, we assume that $\binom{n-1}{d-1}p > (d-1)^{-1}$, and that $|\nu - (1 - \rho)n| = O(\sqrt{n})$, where $0 < \rho < 1$ is the solution to (2).

In order to compute the probability that $\mathcal{N}(H_d(n, p)) = \nu$, we expose the edges of the random hypergraph $H_d(n, p)$ in several rounds. To this end, let $\varepsilon > 0$ be small enough but fixed such that $(1 - \varepsilon)\binom{n-1}{d-1}p > (d - 1)^{-1} + \varepsilon$. Let $p_1 = (1 - \varepsilon)p$, and let p_2 be the solution to the equation $p_1 + p_2 - p_1 p_2 = p$; thus, $p_2 \sim \varepsilon p$. We expose the edges of $H_d(n, p)$ in four rounds as follows.

R1. Let H_1 be a random hypergraph obtained by including each of the $\binom{n}{d}$ possible edges with probability p_1 independently. Let G signify its largest component.

R2. Let H_2 be the hypergraph obtained from H_1 by adding each edge $e \notin H_1$ that lies completely outside of G (i.e., $e \subset V \setminus G$) with probability p_2 independently.

R3. Obtain H_3 by adding each possible edge $e \notin H_1$ that contains vertices of both G and $V \setminus G$ with probability p_2 independently.

R4. Finally, obtain H_4 by including each possible edge $e \notin H_1$ such that $e \subset G$ with probability p_2 independently.

Note that for each possible edge $e \subset V$ the probability that e is present in H_4 equals $p_1 + (1 - p_1)p_2 = p$; hence, $H_4 = H_d(n, p)$.

As $\binom{n-1}{d-1}p_1 > (d - 1)^{-1} + \varepsilon$ by our choice of ε, the main result of Schmidt-Pruzan and Shamir [25] entails that H_1 has exactly one component of linear size $\Omega(n)$ (the "giant component") w.h.p. Moreover, since w.h.p. also the final hypergraph H_4 has exactly one component of linear size (again by [25]), the set G is contained in the largest component of H_4 w.h.p. Hence, $\mathcal{N}(H_3) = \mathcal{N}(H_4)$

(because all the edges added in R4 lie completely inside of the largest component of H_3). Therefore, in order to analyze the distribution of $\mathcal{N}(H_4)$, we shall study

1. the distribution of $|G| = \mathcal{N}(H_1)$,
2. the number $\mathcal{S} = \mathcal{N}(H_3) - |G|$ of vertices that get attached to G by R2–R3.

Regarding the first problem, we have the following result.

Proposition 4. *Let* $c_1 = \binom{n-1}{d-1}p_1$ *and* $c_3 = \binom{n-1}{d-1}p$. *Moreover, let* $0 < \rho_3 < \rho_1 < 1$ *signify the solutions to the transcendental equations* $\rho_j = \exp\left[c_j(\rho_j^{d-1} - 1)\right]$ *and let*

$$\mu_j = (1 - \rho_j)n, \quad \sigma_j^2 = \frac{\rho_j\left[1 - \rho_j + c_j(d-1)(\rho_j - \rho_j^{d-1})\right]n}{(1 - c_j(d-1)\rho_j^{d-1})^2} \quad (j = 1,3).$$

Then $(\mathcal{N}(H_j) - \mu_j)/\sigma_j$ *converges in distribution to the standard normal distribution for* $j = 1,3$.

The proof of Proposition 4 is based on *Stein's method*. Instead of investigating $\mathcal{N}(H_d(n,p))$ directly, we actually study the number $n - \mathcal{N}(H_d(n,p))$ of vertices of $H_d(n,p)$ that belong to components of order $O(\ln n)$. To analyze this quantity, we extend an argument from Barbour, Karoński, and Ruciński [3], who showed that the number of components of a *given* size $k = O(1)$ in the random *graph* $G(n,p)$ is asymptotically normal. By comparison, to establish Proposition 4, we need to take into account small components outside the largest component of *any* size $1 \leq k = O(\ln n)$ in the random *hypergraphs* $H_d(n,p_1)$ and $H_d(n,p)$ (details omitted).

Furthermore, the following proposition shows that $\mathcal{S}$ *given that* $|G| = n_1$ satisfies a local limit theorem.

Proposition 5. *There are numbers* $\mu_\mathcal{S} = \Theta(n)$, $\lambda_\mathcal{S} = \Theta(1)$, *and* $\sigma_\mathcal{S} = \Theta(\sqrt{n})$ *such that for all integers* n_1 *satisfying* $|n_1 - \mu_1| \leq n^{0.6}$ *the following is true. If* s *is an integer such that* $|\mu_\mathcal{S} + \lambda_\mathcal{S}(n_1 - \mu_1) - s| \leq O(\sqrt{n})$, *then*

$$P\left[\mathcal{S} = s \big| |G| = n_1\right] \sim \frac{1}{\sqrt{2\pi}\sigma_\mathcal{S}} \exp\left(-\frac{(\mu_\mathcal{S} + \lambda_\mathcal{S}(n_1 - \mu_1) - s)^2}{2\sigma_\mathcal{S}^2}\right). \quad (5)$$

Proof (sketch). Let us assume for simplicity that $d = 2$. A similar application of Stein's method as in the proof of Proposition 4 shows that given $|G| = n_1$ the random variable $(\mathcal{S} - \mu_\mathcal{S} - \lambda_\mathcal{S}(n_1 - \mu_1))\sigma_\mathcal{S}^{-1}$ is asymptotically normal. To establish the *local* formula (5), we decompose $\mathcal{S} = \mathcal{S}_{\text{big}} + \mathcal{S}_{\text{iso}}$, where $\mathcal{S}_{\text{iso}}$ is the number of isolated vertices in $V \setminus G$ that get attached to G in step R3. Since $d = 2$, $\mathcal{S}_{\text{iso}}$ is *binomially distributed* with mean $\Omega(n)$. Consequently, $\mathcal{S}_{\text{iso}}$ has a local limit theorem. This implies that for any two numbers s_1, s_2 such that $|\mu_\mathcal{S} + \lambda_\mathcal{S}(n_1 - \mu_1) - s_i| \leq O(\sqrt{n})$ and $|s_1 - s_2| = o(\sqrt{n})$ we have $P\left[\mathcal{S} = s_1 \big| |G| = n_1\right] \sim P\left[\mathcal{S} = s_2 \big| |G| = n_1\right]$. In combination with the fact that $\mathcal{S}$ given $|G| = n_1$ is asymptotically normal, this implies the assertion. (Note that the proof does *not* require an explicit analysis of $\mathcal{S}_{\text{big}}$.) $\qquad\square$

Since $\mathcal{N}(H_3) = |G| + \mathcal{S}$, Propositions 4 and 5 yield $\mu_3 = \mu_1 + \mu_S + o(\sqrt{n})$. Therefore, combining Propositions 4 and 5, we conclude that for any integer ν such that $z = (\nu - \mu_3)/\sigma_3 = O(1)$ we have

$$P\left[\mathcal{N}(H_3) = \nu\right] \sim \sum_{s+n_1=\nu} P\left[\mathcal{S} = s \middle| |G| = n_1\right] P\left[|G| = n_1\right]$$

$$\sim \frac{1}{2\pi\sigma_S} \int_{-\infty}^{\infty} \exp\left[-\frac{x^2}{2} - \frac{1}{2}\left(x \cdot (1 + \lambda_S)\frac{\sigma_1}{\sigma_S} - z \cdot \frac{\sigma_3}{\sigma_S}\right)^2\right] dx.$$

Integrating the right hand side, we obtain the desired formula for $\mathcal{N}(H_3) = \mathcal{N}(H_4) = \mathcal{N}(H_d(n, p))$, thus completing the proof of Theorem 1.

5   The Bivariate Local Limit Theorem

In this section we keep the assumption and the notation of Theorem 2. That is, we assume that $\binom{n-1}{d-1}p > (d-1)^{-1}$, and that ν, μ are integers such that $x = \nu - (1-\rho)n, y = \mu - (1-\rho^d)\binom{n}{d}p = O(\sqrt{n})$, where $0 < \rho < 1$ is the solution to (2). Moreover, we let $\sigma^2 = \binom{n}{d}p$.

In this section we derive the bivariate local limit theorem stated in Theorem 2 from the univariate local limit theorem (Theorem 1) via Fourier analytic arguments. The first step is to observe that the local limit theorem for $\mathcal{N}(H_d(n, p))$ (Theorem 1) implies a bivariate local limit theorem for the joint distribution of $\mathcal{N}(H_d(n, p))$ and the number $\bar{\mathcal{M}}(H_d(n, p))$ of edges *outside* the largest component. Indeed, *given* that $\mathcal{N}(H_d(n, p)) = \nu \sim (1 - \rho)n$, the random variable $\bar{\mathcal{M}}(H_d(n, p))$ is asymptotically binomial $\mathrm{Bin}(\binom{n-\nu}{d}, p)$; this simply follows from the fact that w.h.p. $H_d(n, p)$ has a *unique* component of order $\Omega(n)$ (cf. [25]). Consequently, setting $\bar{\mu} = \binom{n}{d}p - \mu$, we get

$$P\left[\mathcal{N}(H_d(n, p)) = \nu \wedge \bar{\mathcal{M}}(H_d(n, p)) = \bar{\mu}\right]$$

$$\sim P\left[\mathcal{N}(H_d(n, p)) = \nu\right] \cdot P\left[\mathrm{Bin}\left[\binom{n-\nu}{d}, p\right] = \bar{\mu}\right]. \quad (6)$$

As Theorem 1 (and the well-known local limit theorem for the binomial distribution – e.g., [8, Chapter 1]) yields an explicit formula for the r.h.s. of (6), we can thus infer an explicit formula for $P\left[\mathcal{N}(H_d(n, p)) = \nu \wedge \bar{\mathcal{M}}(H_d(n, p)) = \bar{\mu}\right]$. However, this does *not* yield a result on the joint distribution of $\mathcal{N}(H_d(n, p))$ and $\mathcal{M}(H_d(n, p))$, because the *total* number of edges in $H_d(n, p)$ is random.

To get around this problem, we make a detour to the $H_d(n, m)$ model, in which the total number of edges is fixed (namely, m). Hence, $\bar{\mathcal{M}}(H_d(n, m)) = m - \mathcal{M}(H_d(n, m))$. Moreover, given that the total number of edges in $H_d(n, p)$ equals m, $H_d(n, p)$ is distributed as $H_d(n, m)$. In effect,

$$P\left[\mathcal{N}(H_d(n, p)) = \nu \wedge \bar{\mathcal{M}}(H_d(n, p)) = \bar{\mu}\right] \quad (7)$$

$$= \sum_{m=0}^{\binom{n}{d}} P\left[\mathrm{Bin}\left(\binom{n}{d}, p\right) = m\right] \cdot P\left[\mathcal{N}(H_d(n, m)) = \nu \wedge \bar{\mathcal{M}}(H_d(n, m)) = \bar{\mu}\right].$$

Now, the crucial idea is to "solve" (7) for

$$P\left[\mathcal{N}(H_d(n,m)) = \nu \wedge \bar{\mathcal{M}}(H_d(n,m)) = \bar{\mu}\right]$$

via Fourier inversion. To this end, recall that (6) yields an explicit expression for the l.h.s. of (7). Moreover, the local limit theorem for the binomial distribution provides an explicit formula for the first factor on the r.h.s. of (7). Furthermore, the terms $P\left[\mathcal{N}(H_d(n,m)) = \nu \wedge \bar{\mathcal{M}}(H_d(n,m)) = \bar{\mu}\right]$ are *independent of* p, while equation (7) holds *for all* p. In other words, (7) is a (grossly over-determined) system of equations that we would like to solve for the quantities $P\left[\mathcal{N}(H_d(n,m)) = \nu \wedge \bar{\mathcal{M}}(H_d(n,m)) = \bar{\mu}\right]$.

In order to put this to work, let $p_z = p + z\sigma\binom{n}{d}^{-1}$, $m_z = \lceil\binom{n}{d}p_z\rceil = \lceil\binom{n}{d}p + z\sigma\rceil$, and set $z^* = \ln^2 n$. Moreover, consider the two functions defined as follows: for $z \in [-z^*, z^*]$,

$$f(z) = f_{n,\nu,\mu}(z) = nP\left[\mathcal{N}(H_d(n,p_z)) = \nu \wedge \bar{\mathcal{M}}(H_d(n,p_z)) = \bar{\mu}\right],$$
$$g(z) = g_{n,\nu,\mu}(z) = nP\left[\mathcal{N}(H_d(n,m_z)) = \nu \wedge \bar{\mathcal{M}}(H_d(n,m_z)) = \bar{\mu}\right]$$

and for $z \in \mathbf{R} \setminus [-z^*, z^*]$, $f(z) = g(z) = 0$. Thus,

$$P\left[\mathcal{N}(H_d(n,m)) = \nu \wedge \bar{\mathcal{M}}(H_d(n,m)) = \bar{\mu}\right] = g(0)/n, \qquad (8)$$

i.e., our goal is to derive an explicit expression for g. Furthermore, we can restate (7) in terms of the functions f, g as follows.

Lemma 6. *Let* $\phi(z) = \frac{1}{\sqrt{2\pi}}\exp(-\frac{z^2}{2})$. *Then* $\|f - g*\phi\|_2 = o(1)$, *where* $g*\phi(z) = \int_{\mathbf{R}} g(z - \xi)\phi(\xi)d\xi$ *signifies the convolution of* g *and* ϕ.

Proof (sketch). This essentially follows by approximating the binomial distribution in (7) by a normal distribution, thereby replacing the sum by an integral. $\square$

As a next step, we derive the following explicit formula for f.

Lemma 7. *Set* $\lambda = \frac{d\sigma(\rho^d - \rho)}{\sigma_{\mathcal{N}}(1 - c(d-1)\rho^{d-1})}$ *and let*

$$F(z) = \frac{n}{2\pi\rho^{d/2}\sigma\sigma_{\mathcal{N}}} \qquad (9)$$

$$\times \exp\left[-\frac{1}{2}\left((x\sigma_{\mathcal{N}}^{-1} - z\lambda)^2 + \rho^d(y\rho^{-d}\sigma^{-1} - c\rho^{-1}\sigma^{-1}x + z)^2\right)\right].$$

Then $\|f - F\|_\infty = o(1)$.

Lemma 7 is a direct consequence of (6), Theorem 1, and the local limit theorem for the binomial distribution.

Further, it is easy (though somewhat tedious) to compute a function h such that $F = h * \phi$.

Lemma 8. *Let λ be as Lemma 7, and define*

$$\varsigma = \lambda^2 + \rho^d, \quad \kappa = -\left[\frac{\lambda}{\sigma_{\mathcal{N}}} + \frac{c\rho^{d-1}}{\sigma}\right]x + \frac{y}{\sigma}, \quad \theta = \frac{x^2}{\sigma_{\mathcal{N}}^2} + \frac{(c\rho^{d-1}x - y)^2}{\rho^d\sigma^2}, \quad and$$

$$h(z) = \frac{n}{2\pi\rho^{d/2}\sqrt{1 - \varsigma\sigma_{\mathcal{N}}\sigma}} \exp\left[-\frac{\varsigma\theta - \kappa^2}{2\varsigma} - \frac{(\varsigma z + \kappa)^2}{2(\varsigma - \varsigma^2)}\right]. \tag{10}$$

Then $F = h * \phi$, and consequently $\|f - h * \phi\|_2 = o(1)$.

Of course, if it *were* true that $f = F = g * \phi$ and $F = h * \phi$, then we could infer immediately that $g = h$; for the Fourier transforms would satisfy $\hat{g}\phi = \hat{h}\phi$, whence $\hat{g} = \hat{h}$, and thus $g = h$ by Fourier inversion. However, Lemmas 6 and 8 only yield that $\|f * \phi - g * \phi\|_2 = o(1)$. Nonetheless, using the Fourier transform in a slightly more sophisticated way, we can derive that $\|g - h\|_\infty = o(1)$. Due to (8) and because $\mathcal{M}(H_d(n,m)) + \bar{\mathcal{M}}(H_d(n,m)) = m$, this implies that

$$P\left[\mathcal{N}(H_d(n,m)) = \nu \wedge \mathcal{M}(H_d(n,m)) = \mu\right] \sim h(0)/n. \tag{11}$$

In other words, we have derived the following *local limit theorem* for the $H_d(n,m)$ model.

Theorem 9. *Let*

$$\tau_{\mathcal{N}}^2 = \rho\frac{1 - (c+1)\rho - c(d-1)\rho^{d-1} + 2cd\rho^d - cd\rho^{2d-1}}{(1 - c(d-1)\rho^{d-1})^2}n,$$

$$\tau_{\mathcal{M}}^2 = c\rho^d\frac{1 - c(d-2)\rho^{d-1} - (c^2d - cd + 1)\rho^d - c^2(d-1)\rho^{2d-2}}{d(1 - c(d-1)\rho^{d-1})^2}n$$

$$+ c\rho^d\frac{2c(cd-1)\rho^{2d-1} - c^2\rho^{3d-2}}{d(1 - c(d-1)\rho^{d-1})^2}n,$$

$$\tau_{\mathcal{N}\mathcal{M}} = c\rho^d\frac{1 - c\rho - c(d-1)\rho^{d-1} + (c + cd - 1)\rho^d - c\rho^{2d-1}}{(1 - c(d-1)\rho^{d-1})^2}n,$$

$$Q(x,y)^{-1} = 2\pi\sqrt{\tau_{\mathcal{N}}^2\tau_{\mathcal{M}}^2 - \tau_{\mathcal{N}\mathcal{M}}^2}$$

$$\times \exp\left[\frac{\tau_{\mathcal{N}}^2\tau_{\mathcal{M}}^2}{2(\tau_{\mathcal{N}}^2\tau_{\mathcal{M}}^2 - \tau_{\mathcal{N}\mathcal{M}}^2)}\left(\frac{x^2}{\tau_{\mathcal{N}}^2} - \frac{2\tau_{\mathcal{N}\mathcal{M}}xy}{\tau_{\mathcal{N}}^2\tau_{\mathcal{M}}^2} + \frac{y^2}{\tau_{\mathcal{M}}^2}\right)\right].$$

Then $P\left[\mathcal{N}(H_d(n,m)) = \nu \wedge \mathcal{M}(H_d(n,m)) = \mu\right] \sim Q(x,y)$.

Theorem 9 just follows from (11) by bringing the expression for h from Lemma 8 into the standard form of a bivariate normal distribution.

Finally, it is easy to derive Theorem 2 from Theorem 9: the $H_d(n,p)$ model is related to the $H_d(n,m)$ model by the formula

$$P\left[\mathcal{N}(H_d(n,p)) = \nu \wedge \mathcal{M}(H_d(n,p)) = \mu\right] = \tag{12}$$

$$= \sum_{m=0}^{\binom{n}{d}} P\left[\text{Bin}\left(\binom{n}{d}, p\right) = m\right] \cdot P\left[\mathcal{N}(H_d(n,m)) = \nu \wedge \mathcal{M}(H_d(n,m)) = \mu\right].$$

Due to Theorem 9 we can compute the sum on the r.h.s. explicitly, thereby obtaining the expression stated in Theorem 2.

6   The Number of Connected Hypergraphs

In this section we keep the assumptions of Theorem 3. Thus, suppose that ν, μ are integers such that $\mu/\nu > (d-1)^{-1}$. We shall derive a formula for the probability $c_d(\nu, \mu)$ that $H_d(\nu, \mu)$ is connected, or, equivalently, for the number $C_d(\nu, \mu)$ of connected hypergraphs of order ν and size μ from the local limit theorem Theorem 2. To this end, we set up a relation between $C_d(\nu, \mu)$ and the probability that in a somewhat larger random hypergraph $H_d(n, p)$ the event $\mathcal{N}(H_d(n, p)) = \nu$ and $\mathcal{M}(H_d(n, p)) = \mu$ occurs. This idea is originally due to Łuczak [18], and has been investigated in detail in [9].

More precisely, given ν and μ, there exist $\nu < n = O(\nu)$ and $0 < p < 1$, $\binom{n-1}{d-1}p > (d-1)^{-1}$, such that

$$\mathrm{E}(\mathcal{N}(H_d(n, p))) = \nu + o(\sqrt{\nu}) \text{ and } \mathrm{E}(\mathcal{M}(H_d(n, p))) = \mu + o(\sqrt{\mu})$$

(cf. [9, Section 3.1]). Since $\binom{n-1}{d-1}p > (d-1)^{-1}$, $H_d(n, p)$ has a *unique* "giant component" of order $\Omega(n)$ w.h.p. Moreover, given that the giant component has order ν and size μ, the giant component clearly is a *uniformly* distributed connected hypergraph with these parameters. Therefore, we obtain the expression

$$\mathrm{P}\left[\mathcal{N}(H_d(n, p)) = \nu \wedge \mathcal{M}(H_d(n, p)) = \mu\right] \sim \binom{n}{\nu} C_d(\nu, \mu) p^\mu (1 - p)^{\binom{n}{d} - \binom{n-\nu}{d} - \mu}.$$

Here $\binom{n}{\nu}$ is the number of ways to choose the vertex set G of the giant component, $C_d(\nu, \mu)$ is the number of ways to choose the connected hypergraph induced on G, p^μ signifies the probability that all of the edges of the chosen connected hypergraph are present, and the $(1 - p)$-factor accounts for the probability that G is not connected to $V \setminus G$.

Finally, since the l.h.s. is known due to Theorem 2, we can solve for $C_d(\nu, \mu)$. Simplifying the resulting formula yields the expression stated in Theorem 3.

References

1. Andriamampianina, T., Ravelomanana, V.: Enumeration of connected uniform hypergraphs. In: Proceedings of FPSAC (2005)
2. Barraez, D., Boucheron, S., Fernandez de la Vega, W.: On the fluctuations of the giant component. Combinatorics, Probability and Computing 9, 287–304 (2000)
3. Barbour, A.D., Karonski, M., Rucinski, A.: A central limit theorem for decomposable random variables with applications to random graphs. J. Combin. Theory Ser. B 47, 125–145 (1989)
4. Behrisch, M., Coja-Oghlan, A., Kang, M.: The order of the giant component of random hypergraphs (preprint, 2007), available at http://arxiv.org/abs/0706.0496
5. Behrisch, M., Coja-Oghlan, A., Kang, M.: Local limit theorems and the number of connected hypergraphs (preprint, 2007), available at http://arxiv.org/abs/0706.0497
6. Bender, E.A., Canfield, E.R., McKay, B.D.: The asymptotic number of labeled connected graphs with a given number of vertices and edges. Random Structures and Algorithms 1, 127–169 (1990)

7. Bender, E.A., Canfield, E.R., McKay, B.D.: Asymptotic properties of labeled connected graphs. Random Structures and Algorithms 3, 183–202 (1992)
8. Bollobás, B.: Random graphs, 2nd edn. Cambridge University Press, Cambridge (2001)
9. Coja-Oghlan, A., Moore, C., Sanwalani, V.: Counting connected graphs and hypergraphs via the probabilistic method. In: Random Structures and Algorithms (to appear)
10. Coppersmith, D., Gamarnik, D., Hajiaghayi, M., Sorkin, G.B.: Random MAX SAT, random MAX CUT, and their phase transitions. Random Structures and Algorithms 24, 502–545 (2004)
11. Erdős, P., Rényi, A.: On random graphs I. Publicationes Mathematicae Debrecen 5, 290–297 (1959)
12. Erdős, P., Rényi, A.: On the evolution of random graphs. Publ. Math. Inst. Hung. Acad. Sci. 5, 17–61 (1960)
13. van der Hofstad, R., Spencer, J.: Counting connected graphs asymptotically. European Journal on Combinatorics 27, 1294–1320 (2006)
14. Janson, S.: The minimal spanning tree in a complete graph and a functional limit theorem for trees in a random graph. Random Structures and Algorithms 7, 337–355 (1995)
15. Janson, S., Łuczak, T., Ruciński, A.: Random Graphs. Wiley, Chichester (2000)
16. Karoński, M., Łuczak, T.: The number of connected sparsely edged uniform hypergraphs. Discrete Math. 171, 153–168 (1997)
17. Karoński, M., Łuczak, T.: The phase transition in a random hypergraph. J. Comput. Appl. Math. 142, 125–135 (2002)
18. Łuczak, T.: On the number of sparse connected graphs. Random Structures and Algorithms 1, 171–173 (1990)
19. Luczak, M., Łuczak, T.: The phase transition in the cluster-scaled model of a random graph. Random Structures and Algorithms 28, 215–246 (2006)
20. O'Connell, N.: Some large deviation results for sparse random graphs. Prob. Th. Relat. Fields 110, 277–285 (1998)
21. Pittel, B.: On tree census and the giant component in sparse random graphs. Random Structures and Algorithms 1, 311–342 (1990)
22. Pittel, B., Wormald, N.C.: Asymptotic enumeration of sparse graphs with a minimum degree constraint. J. Combinatorial Theory, Series A 101, 249–263 (2003)
23. Pittel, B., Wormald, N.C.: Counting connected graphs inside out. J. Combin. Theory, Series B 93, 127–172 (2005)
24. Ravelomanana, V., Rijamamy, A.L.: Creation and growth of components in a random hypergraph process. In: Chen, D.Z., Lee, D.T. (eds.) COCOON 2006. LNCS, vol. 4112, pp. 350–359. Springer, Heidelberg (2006)
25. Schmidt-Pruzan, J., Shamir, E.: Component structure in the evolution of random hypergraphs. Combinatorica 5, 81–94 (1985)
26. Stein, C.: A bound for the error in the normal approximation to the distribution of a sum of dependent variables. In: Proc. 6th Berkeley Symposium on Mathematical Statistics and Probability, pp. 583–602 (1970)
27. Stepanov, V.E.: On the probability of connectedness of a random graph $g_m(t)$. Theory Prob. Appl. 15, 55–67 (1970)

Derandomization of Euclidean Random Walks

Ilia Binder[1],[*] and Mark Braverman[2],[**]

[1] Dept. of Mathematics, University of Toronto
[2] Dept. of Computer Science, University of Toronto

Abstract. We consider the problem of derandomizing random walks in the Euclidean space $\mathbb{R}^k$. We show that for $k = 2$, and in some cases in higher dimensions, such walks can be simulated in Logspace using only poly-logarithmically many truly random bits.

As a corollary, we show that the Dirichlet Problem can be deterministically simulated in space $O(\log n \sqrt{\log \log n})$, where $1/n$ is the desired precision of the simulation.

1 Introduction

1.1 Space-Bounded Derandomization and the Dirichlet Problem on Graphs

We are interested in derandomizing some problems that can be solved by a *probabilistic* log-space Turing Machine. There are many ways to view a probabilistic log-space computation. By definition, a probabilistic space bounded TM has a special state where it can request a random bit. It has to use its working tape if it wants to store the bits it has seen so far. If the machine M uses $S = S(n)$ cells on its working tape and queries at most $R = R(n)$ random bits, it is said to have a space/randomness complexity of (S, R). We require the machine to terminate in $2^{O(S)}$ steps, thus making sure that for all possible random bits the machine halts (cf. discussion in [1]).

Denote the set of valid configurations of M by C_M, $|C_M| = 2^{O(S)}$. Denote the set of accepting configurations (i.e. configurations where M has terminated in the accepting state) by C_{acc}, and the set of rejecting configurations by C_{rej}. Denote the initial configuration by c_{init}. We can view the evaluation path of the machine M on input x as a random process $M(x)_t$ where $M(x)_t \in C_M$ is the configuration of the computation of M on x at time t. Let $T(x) = 2^{O(S)}$ be the time at which the computation terminates. $T(x)$ is a random variable and furthermore, by definition, $M(x)_{T(x)} \in C_{acc} \cup C_{rej}$. The "result" of the computation of M on an input x is its acceptance probability, $p_{acc}(x) := P[M(x)_{T(x)} \in C_{acc}]$. *Derandomizing* the machine M involves giving an algorithm for computing p_{acc} within some error ε, with ε usually being 2^{-S}.

One way to present the computation of the probabilistic log-space machine M is by considering configurations graph G of the machine. The nodes of G are

---

[*] Partially supported by NSERC Discovery grant 5810-2004-298433.
[**] Partially supported by NSERC CGS Scholarship.

M. Charikar et al. (Eds.): APPROX and RANDOM 2007, LNCS 4627, pp. 353–365, 2007.
© Springer-Verlag Berlin Heidelberg 2007

C_M. If a configuration $U \in C_M$ corresponds to a random bit querying state, then it is connected to two configurations, V_0 and V_1, one corresponding to the configuration when the requested random bit is 0, and the other when it is 1. If U corresponds to any other configuration, then it is connected to the unique next state V of U, unless $U \in C_{acc} \cup C_{rej}$, in which case it is connected to itself. There is a natural correspondence between runs of the machine and random walks on G originating at c_{init} and terminating on the set $C_{acc} \cup C_{rej}$. The probability that the random walk terminates on C_{acc} is exactly $p_{acc}(x)$. We can formulate this problem in slightly more general terms. The name, which arises from connection to the classical Dirichlet Problem, will be explained later.

Definition 1. *The Directed Dirichlet Problem* **DirDP** *is defined as follows.*

Input: *A directed graph G, a vertex v_0, a set B of vertexes, a parameter ε and a function $f : B \to [0,1]$.*

Output: *Assuming all vertexes of $V - B$ have out-degree of at least 1 and the random walk on the graph originating at v_0 hits the set B in time $Poly(|V|, 1/\varepsilon)$ with probability at least $1 - \varepsilon/2$, the output should be the expected value of $f(b)$, where b is the first vertex in B that a random walk originating at v_0 hits, computed with precision ε. We denote this value by $\Phi(v_0)$.*

Here the "random walk" takes all the edges from a vertex v with equal probability. Note that we do not need to worry about the representation of the function f, because it is not hard to see that in order to estimate $\Phi(v_0)$ with precision ε we only need to know f with precision $\Theta(\varepsilon)$.

From the discussion above, it follows that derandomizing **DirDP** is as hard as derandomizing space-bounded machines.

Theorem 1. *The following problems are (deterministic) space-$O(S)$ reducible to each other:*

- *Given a probabilistic machine M running in space $S = S(n)$ and randomness $R = 2^{O(S)}$, and an input x, $|x| = n$, compute $p_{acc}(x)$ within an error of $\varepsilon = 2^{-S}$.*
- *Solve the* **DirDP** *problem on a graph of size 2^S within an error of $\varepsilon = 2^{-S}$.*

An equivalent view on the general Dirichlet problem on graphs is a global one. Suppose that instead of considering only *one* starting point v_0, we consider *all* possible starting points. Assuming that for any initial v the random walk originating at v eventually hits B with probability 1, we see that the function $\Phi(v)$ satisfies the following equations:

$$\begin{cases} \Phi(v) = \frac{1}{deg(v)} \cdot \sum_{(v,u) \in E} \Phi(u) & \text{for } v \notin B \\ \Phi(v) = f(v) & \text{for } v \in B \end{cases} \tag{1}$$

It can be shown that under the conditions above, the equation (1) has a unique solution. Note that the first condition can be restated as $(\Delta G)\Phi = 0$, where ΔG is the Laplacian matrix of the graph G.

Attempts to solve **DirDP**, which is at least as hard as derandomizing the class **BPSPACE**(S), can now be restricted to different classes of graphs. One restriction would be to consider the undirected graphs, to obtain the corresponding **UndirDP** problem. To our knowledge, even in this case no results better than the general space $O(S^{3/2})$ derandomization [2] are known. In this paper we consider an important special case of the problem, where the underlying graph has a geometric Euclidean structure.

1.2   The Classical Dirichlet Problem and Its Derandomization

First, we describe the classical Dirichlet problem on $\mathbb{R}^k$. Given a bounded domain Ω and a continuous function on the boundary of Ω, $f : \partial\Omega \to [0,1]$, the goal is to find a function $\Phi : \Omega \to \mathbb{R}$ that satisfies:

$$\begin{cases} \Delta\Phi(x) \equiv \sum_{i=1}^{k} \frac{\partial^2\Phi(x)}{\partial x_i^2} = 0 & \text{for } x \in \text{Interior}(\Omega) \\ \Phi(x) = f(x) & \text{for } x \in \partial\Omega \end{cases} \qquad (2)$$

This classical problem, dating back to the 1840s has numerous applications in Science and Engineering (see for example [3]).

Equation (2) can be viewed as a continuous version of equation (1) because the condition $\Delta\Phi(x) = 0$ on the interior of Ω can be shown to be equivalent to the following condition. For a point x denote by $B(x,\varepsilon)$ the (open) ball of radius ε around x. Then for any $x \in \text{Interior}(\Omega)$, and for any ε such that $B(x,\varepsilon) \subset \text{Interior}(\Omega)$, $\Phi(x)$ is equal to the average value of $\Phi(x)$ on $B(x,\varepsilon)$. Thus, just as in equation (1), we have that for any x, $\Phi(x)$ is equal to the average value of Φ on its "neighbors".

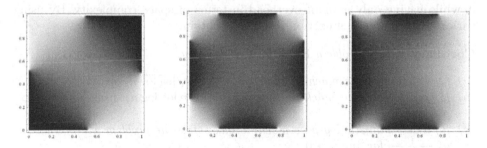

Fig. 1. Examples of solutions to the two-dimensional Dirichlet problem where Ω is a square domain and the boundary condition f is either 1 (black) or 0 (white); the color inside Ω represents the value of Φ. It is equal to the probability that a Brownian motion originating at a point will hit a black segment on the boundary.

As in the graph case above, Brownian motion, the continuous version of a random walk, can be used to solve the Dirichlet problem. For any $x \in \text{Interior}(\Omega)$, denote by B_t the Brownian motion process originating at x: $B_0 = x$. Let the random variable T be the first time B_t hits the boundary $\partial\Omega$. Then the solution to (2) is the expected value of f at B_T:

$$\Phi(x) = E[f(B_T)].$$

Solutions of a Dirichlet problem are illustrated on Fig. 1.

In order to approximately solve the Dirichlet problem in practice, one would need to discretize it first. This is possible under some mild conditions on the continuous problem, that will be discussed in Section 2. We define a discrete *grid* version of the continuous Dirichlet problem in $\mathbb{R}^k$.

Definition 2. *Consider the subdivision of the unit cube in $\mathbb{R}^k$ by a grid with step $1/n$. Let Ω be a subset of the unit cube formed by a collection of small cubes in the grid. The boundary of Ω is comprised of a finite collection C of $k - 1$-dimensional squares. Let the boundary condition $f : \partial\Omega \to [0,1]$ be given within a precision of n^{-3}. f is continuous and linear on each of the squares of the boundary. In other words, it is specified by the values it takes on each piece $s \in C$ of the boundary.*

The discrete Euclidean Dirichlet problem *is, given a grid point x inside Ω, to compute the solution $\Phi(x)$ within an error of $1/n$. We call this problems* **EucDP**.

The most interesting case of **EucDP** is for $\mathbb{R}^2$, because of its connections to the Riemann Mapping Problem and to conformal geometry (see, for example, [4]). We almost completely derandomize the problem in this case.

Theorem 2. *The* **EucDP** *over $\mathbb{R}^2$ is solvable by a randomized TM in space $S = O(\log n)$ using $R = O(\log^4 n)$ random bits.*

The randomness complexity of the machine in Theorem 2 is very low compared to the $O(n)$ complexity of the naïve solution. This allows us to further derandomize it while paying only a small overhead in terms of space complexity. By using various known derandomization results, we obtain:

Corollary 1. *The problem* **EucDP** *is solvable by*

(a) *[2] a deterministic machine that uses $O(\log n \sqrt{\log\log n})$ space;*
(b) *[5] a deterministic machine that uses $O(\log n \log\log n)$ space and runs in poly-time;*
(c) *[6] a deterministic logspace machine that solves the problem within an error of $\frac{1}{2^{\log^{1-\gamma} n}}$ for any $\gamma > 0$.*

As an application, one obtains a derandomized space-efficient algorithm for computing conformal maps. For a simply-connected planar domain $\Omega \subsetneq \mathbb{C}$ with $w \in \Omega$, *Riemann Uniformization Theorem* states that there is unique conformal map ψ of Ω onto the unit disk $\mathbb{D}$ with $\psi(w) = 0$, $\psi'(0) > 0$. Conformal maps and Uniformization are used extensively in many areas, such as solving Partial Differential Equations [7], Hydrodynamics [8], Electrostatics [9], and in computer tomography such as brain mapping [10].

Theorem 3. *There is an algorithm A of complexity described in Corollary 1 that computes the uniformizing map in the following sense.*

Let Ω be a bounded simply-connected domain, and $w_0 \in \Omega$. Suppose that for some n, $\partial\Omega$ is given to A with precision $\frac{1}{n}$ by $O(n^2)$ pixels. Then A computes the absolute value of the uniformizing map $\psi : (\Omega, w) \to (\mathbb{D}, 0)$ within an error of $O(1/n)$. Furthermore, the algorithm computes the value of $\psi(w)$ with precision $1/n$ as long as $|\psi(w)| < 1 - 1/n$.

The reduction of Theorem 3 to Corollary 1 is given in [11].

The rest of the paper is organized as follows. In Section 2, we discuss the discretization of the continuous Dirichlet problem. In Section 3, we describe our algorithm and prove the main lemma, Lemma 3. The lemma implies Theorem 2 and, using the methods of [11], Theorem 3. In Section 4, we discuss the generalization of the Theorem 2 to higher dimensions.

2   Discretizing the Euclidean Dirichlet Problem

In this section we discuss the discretization of continuous Dirichlet problem. We start with a technical result about stability of the Dirichlet problem. Informally, it states that a slight change in boundary and boundary data induces an insignificant change in the solution.

Lemma 1. *Let γ_1, γ_2 be two closed Jordan curves, $\mathrm{dist}(\gamma_1, \gamma_2) < 1/n^3$. Let $f_1(t)$, $f_2(t)$ be two continuous functions on γ_1, γ_2 respectively with the following continuity property: if $x \in \gamma_1$, $y \in \gamma_2$, and $|x - y| < 1/n^2$, then $|f_1(x) - f_2(y)| < 1/n$. Let Φ_1 and Φ_2 be the solutions of the corresponding Dirichlet problems. Let z be a point inside both γ_1 and γ_2 that is at least $1/n$-away from both curves. Then $|\Phi_1(z) - \Phi_2(z)| < 2/n$.*

Here $\mathrm{dist}(\gamma_1, \gamma_2) < 1/n^3$ if no point in γ_1 is more than $1/n^3$ away from γ_2 and vice versa. The lemma follows from the standard estimates in Geometric Function Theory (see, for example, [12]).

Lemma 1 allows us to approximate the solution of the Dirichlet problem on a Jordan curve γ by approximating it using a $1/n^3$-grid curve, and by approximating the boundary data by a continuous piecewise-linear function on the grid-curve, a process illustrated on Fig. 2. Thus the continuous Dirichlet problem, at least for domains bounded by finitely-many Jordan curves, can be solved with an arbitrarily high precision using **EucDP**.

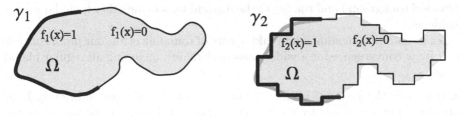

Fig. 2. Discretization of the continuous Dirichlet Problem

3   Derandomizing the Dirichlet Problem

Let $\Omega \subset \mathbb{R}^2$ be as in the definition of **EucDP** (Definition 2). Let $x \in \Omega$ with $R(x) = \text{dist}(x, \partial\Omega)$. We define a stochastic jump point process x_t inductively. At each iteration the process jumps one half of the distance to the boundary at random direction. More precisely, x_t is defined, using complex numbers to parameterize $\mathbb{R}^2$, by

$$x_0 = x, \quad x_t = x_{t-1} + \frac{1}{2}e^{2\pi i\theta_t}R(x_{t-1}).$$

Here $\theta_1, \theta_2, \ldots$ is a sequence of independent random variables, uniformly distributed on the interval $[0, 1]$.

In the limit, this process can be used to solve Dirichlet problem. Specifically, the following Kakutani's theorem is classical.

Lemma 2 (Kakutani's theorem). *Let f be a continuous function on $\partial\Omega$, Φ be the solution of the corresponding Dirichlet problem. With probability 1, $\lim_{t\to\infty} x_t = x_\infty$ exists, $x_\infty \in \partial\Omega$, and $\mathbf{E}[f(x_\infty)] = \Phi(x)$.*

For the proof of the lemma, see for example [13], Appendix G. Let us remark that the theorem is also true for domains in $\mathbb{R}^k$.

Our algorithm is based on a discretized version of Lemma 2 and the following observation. The stochastic process $y_t = \log \text{dist}(x_t, \partial\Omega)$ is a supermartingale with $\Theta(1)$-jump at every step. Note that y_t is bounded from above by a constant, since Ω is bounded. Hence y_t is a random walk on $(-\infty, C)$ with constant-magnitude steps and a downwards drift. Thus it is expected to hit the point $-T$ in time $t = O(T^2)$. In particular, y_t will drop below $-\log n$ in time $\log^3 n$ with high probability. At that point, x_t will be $1/n$-close to $\partial\Omega$.

In the discretized world, we need to take care of the rounding errors, which may affect both the expected solution and the rate of convergence of the algorithm.

Historic Remarks. Our algorithm is a variant of a celebrated "Walk on Spheres" algorithm, first proposed by M. Muller in [14]. The exponential rate of convergence of the process x_t to the boundary was established for convex domains by M. Motoo in [15] and was later generalized for a wider class of planar and 3-dimensional domains by G. A. Mikhailov in [16]. See also [17] for additional historical background and the use of the method for solving other boundary value problems.

Let us note that in addition to taking care of rounding error, our proof controls the rate of convergence for a wider class of domains, including all regular planar domains.

To discretize the process, we fix a square grid of size n^{-c} for sufficiently large constant c (we can take any $c > 6$). The process will only run on points on the grid. Note that storing the coordinates of a point on the grid requires $O(\log n)$ space.

For a point $x \in \mathbb{R}^2$, let the snapping $S(x)$ be one of the points on the grid closest to x. Note that $|x - S(x)| < n^{-c}$. Let $R_d(x)$ be the distance from x to the (discretized) boundary of Ω computed with precision of n^{-c} (and requiring only $O(\log n)$ bits to store). Note that it is easy to compute in space $O(\log n)$, since the discretized boundary is just a polygon with $O(n^{2c})$ vertexes.

Let us now discretize the process x_t. Let X_t be the stochastic process on the grid defined by

$$X_0 = S(x), \quad X_t = S\left(X_{t-1} + \frac{1}{2}e^{2\pi i \theta_t} R_d(X_{t-1})\right).$$

Here $\theta_1, \theta_2, \ldots$ is a sequence of independent random variables taking values $0, 1/n^{2c}, 2/n^{2c}, \ldots, (n^{2c} - 1)/n^{2c}$ with equal probabilities. Note that computing one iteration of X_t requires $O(\log n)$ random bits. The process X_t is illustrated on Fig. 3.

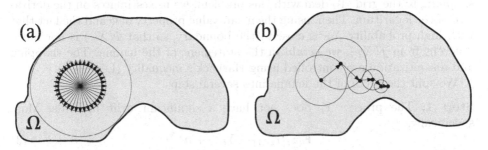

Fig. 3. Illustration of the process X_t. (a) one step of the process; (b) a sample path of five steps of the process.

Now let us define the stopping time T by $T = \min\left(\min\{t : R_d(X_t) < n^{-3}\}, B \log^3 n\right)$, for a sufficiently large constant B. In other words, we stop the process once we are n^{-3}-close to the boundary but never after time $t = B \log^3 n$.

By the definition of T and our discussion about the complexity of computing each jump, X_T is computable in space $O(\log n)$ using $O(\log^4 n)$ random bits. We claim that the values $f(X_T)$ computed with precision n^{-2} can be used to approximate the solution of the continuous Dirichlet problem.

Note that the boundary data f from **EucDP** satisfies the following condition

If x, $y \in \partial\Omega$ with $|x - y| < n^{-3}$, then
$$|f(x) - f(y)| < An^{-2} \text{ for some constant } A. \quad (*)$$

If f is a continuous function satisfying $(*)$, then we can extend it to a grid point $x \in \Omega$ with $R_d(x) < n^{-3}$ by assigning $f(x)$ to be equal to $f(y)$, where y is a closest to x point of $\partial\Omega$. By $(*)$, it is well defined up to an error of $O(n^{-2})$. For other points of the grid we take $f(x) = 0$.

Lemma 3. *Let f be a continuous function on $\partial\Omega$ satisfying* (*). *Let Φ be the solution of the Dirichlet problem with boundary values f. Let $R_d(x) > 1/n$. Then*

$$|\mathbf{E}[f(X_T)] - \Phi(x)| = O(n^{-2}).$$

Lemma 3 implies Theorem 2 by the discussion above.

To simplify the notations, we normalize Ω so that $\text{Diameter}(\Omega) \leq 1$. Define the one-dimensional stochastic process $Y_t = \log R_d(X_t)$.

Strategy of the proof of Lemma 3. First, we establish that with high probability X_T is n^{-3}-close to the boundary. We use the mean-value property of the logarithm in $\mathbb{R}^2$ to show that the expected value of Y_t is non-increasing with time. We also show that the variance of each step of Y_t is bounded from below by some constant. Note that Y_t is always in the interval $[-4\log n, 0]$, and this implies that Y_t will hit the lower boundary with probability at least $1/2$ in $O(\log^2 n)$ steps, and with probability at least $1 - n^{-2}$ in $O(\log^3 n)$ steps.

We also have to take care of the effects of discretization of the jumps and snapping to the grid. To deal with this problem, we use estimates on the derivative of the logarithm. Then, using the mean-value property of Φ and the fact that with high probability X_T is close to the boundary, so that $\Phi(X_T)$ is not much different from $f(X_T)$, we establish the statement of the lemma. The snapping and discretization are controlled using Harnack's inequality (Lemma 4).

We split the proof of the lemma into several steps.

Step 1: The process Y_t does not have a significant drift upwards. More specifically,

$$\mathbf{E}[Y_t|Y_{t-1}] < Y_{t-1} + n^{-3} \tag{3}$$

Proof. Let z be the point near $\partial\Omega$ realizing $R_d(X_t)$. Note also, that for any two numbers $a_1, a_2 > n^{-3}$ we have $|\log a_1 - \log a_2| < n^3|a_1 - a_2|$. Using the fact that the function $g(x) = \log|z - x|$ is harmonic on $\mathbb{R}^2$, we obtain

$$Y_{t-1} = g(X_{t-1}) = \int_0^1 g\left(X_{t-1} + \frac{R_d(X_{t-1})}{2}e^{2\pi i\theta}\right) d\theta >$$

$$\frac{1}{n^{2c}}\sum_{j=0}^{n^{2c}-1} g\left(X_{t-1} + \frac{R_d(X_{t-1})}{2}e^{2\pi i\frac{j}{n^{2c}}}\right) -$$

$$\pi R_d(X_{t-1})n^{-2c} \cdot (2/R_d(X_{t-1})) \tag{4}$$

To obtain the last inequality, we approximate the integral using n^{2c} equally spaced points, the distance between adjacent points is $\pi R_d(X_{t-1})n^{-2c}$, and the derivative of g is bounded by $2/R_d(X_{t-1})$. We continue the chain of inequalities (4) by noting that the snapping operator S only changes the value of $|x - z|$ by at most n^{-c}, and hence the value of $g(x)$ is changed by at most n^{3-c},

$$Y_{t-1} > \frac{1}{n^{2c}}\sum_{j=0}^{n^{2c}-1} g\left(S\left(X_{t-1} + \frac{R_d(X_{t-1})}{2}e^{2\pi i\frac{j}{n^{2c}}}\right)\right) - 2\pi n^{-2c} - n^{3-c} >$$

$$\mathbf{E}[Y_t|Y_{t-1}] - n^{-3}. \tag{5}$$

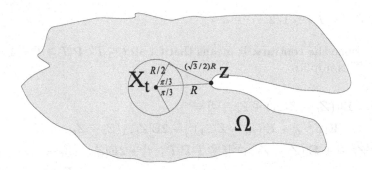

Fig. 4. with probability at least $1/3$, the jump brings X_t closer to z by a factor of at least $\frac{\sqrt{3}}{2}$

Step 2: For some absolute constant $\beta > 0$, $\mathbf{E}[(Y_t - Y_{t-1})^2 | t \leq T] > 2\beta > 0$

Proof. Let z be the point near $\partial \Omega$ realizing $R_d(X_t)$. With probability at least $1/3$, $Y_t \leq \log|z - X_t| < \log \frac{\sqrt{3}}{2} + \log(|z - X_{t-1}| + n^{-c}) < \log \frac{\sqrt{3}}{2} + \log(1 + n^{3-c}) + \log|z - X_{t-1}| < -0.1 + Y_{t-1}$, provided X_{t-1} is a away from the boundary (see Fig. 4) Thus

$$\mathbf{E}[(Y_t - Y_{t-1})^2 | t \leq T] > 1/3(-0.1)^2 > 0.002 > 0,$$

hence we can take $\beta = 0.001$.

Define yet another process $Z_t = Y_t - n^{-3}t$, when $t \leq T$, and $Z_t = Z_T$ for $t > T$. Note that by Step 1 the process $Z_t = Y_t - n^{-3}t$ is a supermartingale. By Doob-Meyer decomposition (see [18]), we can write $Z_t = M_t + I_t$, where M is a martingale and I_t is a decreasing adapted process.

We have $Z_t \leq Y_t \leq 0$, since the diameter of Ω is at most 1. Also, by the definition of the stopping time T, $Z_t \geq -3\log n - n^{-3}\log^3 n > -4\log n$. Observe that, like Y_t, Z_t has definite quadratic variation at each jump. More specifically,

$$\mathbf{E}[(Z_t - Z_{t-1})^2 | t \leq T] =$$
$$\mathbf{E}[(Y_t - Y_{t-1})^2 | t \leq T] + 2n^{-3}\mathbf{E}[(Y_t - Y_{t-1})|t \leq T] + n^{-6} > \beta, \quad (6)$$

provided n is large enough.

Step 3: $\mathbf{E}[Z_{t-1}(Z_t - Z_{t-1})] \geq 0$.

Proof

$$\mathbf{E}[Z_{t-1}(Z_t - Z_{t-1})] = \mathbf{E}[Z_{t-1}\mathbf{E}[Z_t - Z_{t-1}|Z_{t-1}]] =$$
$$\mathbf{E}[Z_{t-1}\mathbf{E}[M_t - M_{t-1}|Z_{t-1}]] + \mathbf{E}[Z_{t-1}\mathbf{E}[I_t - I_{t-1}|Z_{t-1}]] \geq 0, \quad (7)$$

since $\mathbf{E}[M_t - M_{t-1}|Z_{t-1}] = 0$, $I_t - I_{t-1} \leq 0$, $Z_{t-1} \leq 0$.

Let $T' = C\log^2 n$, where C is a large constant to be defined later.

Step 4: $\mathbf{P}[T < T'] > 1/2$.

Proof. Assume the contrary. It means that for all $t \leq T'$, $\mathbf{P}[T \geq t] = 1 - \mathbf{P}[T < t] \geq 1/2$. This implies

$$
\begin{aligned}
\mathbf{E}[Z_t^2] &= \mathbf{E}[((Z_t - Z_{t-1}) + Z_{t-1})^2] = \\
&\quad \mathbf{E}[Z_{t-1}^2] + \mathbf{E}[(Z_t - Z_{t-1})^2] + 2\mathbf{E}[Z_{t-1}(Z_t - Z_{t-1})] = \\
\mathbf{E}[Z_{t-1}^2] &+ \mathbf{E}[(Z_t - Z_{t-1})^2 | t \leq T]\mathbf{P}[T \geq t] + 2\mathbf{E}[Z_{t-1}(Z_t - Z_{t-1})] \geq \\
&\qquad\qquad\qquad\qquad\qquad\qquad\qquad\qquad \mathbf{E}[Z_{t-1}^2] + \beta/2. \quad (8)
\end{aligned}
$$

The last inequality follows from the equations (6) and (7). It implies that $\mathbf{E}[Z_{T'}^2] \geq \frac{\beta}{2} T' = C\frac{\beta}{2} \log^2 n$. But $0 \geq Z_{T'} > -4\log n$, so $\mathbf{E}[Z_{T'}^2] < 16 \log^2 n$. Thus, if we choose $C = 32/\beta$, we get a contradiction.

Step 5: $\mathbf{P}[T < B\log^3 n] > 1 - n^{-2}$. In other words, with high probability X_T stops near the boundary before the time expires.

Proof. By repeating the process from Step 4 independently $2\log n$ times, we get that after $2C\log^3 n$ steps,

$$
\mathbf{P}[T < 2C\log^3 n] > 1 - n^{-2}.
$$

It means that with probability at least $1 - n^{-2}$, for some $t < 2C\log^3 n$, X_t is n^{-3}-close to $\partial\Omega$. Thus, if we take $B = 2C$ in the definition of the stopping time T, we arrive at the desired conclusion.

In the next step, we use Harnack's inequality (see, for example, [19]) to estimate the distortion of the map Φ:

Lemma 4 (Harnack's inequality). *Let $\Omega \subset \mathbb{R}^k$, x, $y \in \Omega$, $\operatorname{dist}(x, \partial\Omega) = R$, $|x - y| = r$. Let Ψ be a positive harmonic function in Ω. Then*

$$
\left(\frac{R - r}{R + r}\right)^k \leq \frac{\Psi(x)}{\Psi(y)} \leq \left(\frac{R + r}{R - r}\right)^k
$$

Let us define two random processes we use to estimate $|\mathbf{E}[f(x_T)] - \Phi(x)|$:

$$
\Phi_t^+ = \Phi(X_t) + n^{3-c}t, \qquad \Phi_t^- = \Phi(X_t) - n^{3-c}t
$$

Step 6: Φ_t^+ is a submartingale and Φ_t^- is a supermartingale

Proof. The proof is very similar to the proof of Step 1. Harnack's Inequality and $0 < \Phi < 1$ imply that

$$\Phi(X_{t-1}) = \int_0^1 \Phi\left(X_{t-1} + \frac{R_d(X_{t-1})}{2}e^{2\pi i\theta}\right)d\theta >$$

$$\frac{1}{n^{2c}}\sum_{j=0}^{n^{2c}-1}\Phi\left(X_{t-1} + \frac{R_d(X_{t-1})}{2}e^{2\pi i\frac{j}{n^{2c}}}\right)\left(1 - \pi R_d(X_{t-1})n^{-2c}\cdot(2/R_d(X_{t-1}))\right) >$$

$$\frac{1}{n^{2c}}\sum_{j=0}^{n^{2c}-1}\Phi\left(S\left(X_{t-1} + \frac{R_d(X_{t-1})}{2}e^{2\pi i\frac{j}{n^{2c}}}\right)\right)\left(1 - 2\pi n^{-2c} - n^{3-c}\right) >$$

$$\frac{1}{n^{2c}}\sum_{j=0}^{n^{2c}-1}\Phi\left(S\left(X_{t-1} + \frac{R_d(X_{t-1})}{2}e^{2\pi i\frac{j}{n^{2c}}}\right)\right) - 2\pi n^{-2c} - n^{3-c} >$$

$$\mathbf{E}[\Phi(X_t)|\Phi(X_{t-1})] - n^{-3}. \quad (9)$$

It implies the first statement. The proof of the second statement is the same.

Now we are in position to prove Lemma 3.

Proof (Proof of Lemma 3). By the Submartingale Theorem (see [20]), $\Phi(x) = \Phi_0^+ \le \mathbf{E}[\Phi_T^+]$. But $\Phi_T^+ = \Phi(X_T) + n^{3-c}T \le \Phi(X_T) + n^{3-c}\log^3 n$. Thus

$$\Phi(x) \le \mathbf{E}[\Phi(X_T)] + n^{3-c}\log^3 n \le \mathbf{E}[f(X_T)] + n^{3-c}\log^3 n +$$

$$\mathbf{E}[\Phi(X_T)|T = B\log^3 n]\mathbf{P}[T = B\log^3 n] +$$

$$\max_{T < B\log^3 n}(\Phi(X_T) - f(X_T)) \le \mathbf{E}[f(X_T)] + 3Ln^{-2}$$

for sufficiently large L and c. Here we use the fact that the modulus of continuity of Φ is no greater than the modulus of continuity of f (see [13]). We also use the Step 4 to obtain an estimate on $\mathbf{P}[T = B\log^3 n]$. Using the same reasoning for Φ_t^- we get

$$\Psi(x) \ge \mathbf{E}[f(X_T)] - 3Ln^{-2}.$$

Finally, by combining the two inequalities we obtain the statement of the lemma.

4 Higher Dimensions: Planar-Like Domains

The proposed algorithm for solving the planar Dirichlet problem does not work for general higher dimensional domains. The only obstacle to literally repeating our proof in higher dimensions comes from the absence of the superharmonicity of the function $\log \text{dist}(x, \partial\Omega)$ (the mean-value property used in equation (4)). In other words, it is no longer true that the value of $\log \text{dist}(X_t, \partial\Omega)$ is non-increasing on average. It turns out that the difficulty leads to an example of a domain for which our method does not converge in logarithmic number of steps, and will thus require a large number of random bits.

Let us describe an example of such a domain in $\mathbb{R}^3$. The domain Ω will be the unit cube with cubes of size $1/n$ removed around every nonzero point of a $2/\sqrt{n}$ grid. Standard results about the random walk in $\mathbb{R}^3$ (see, for example, [20]) imply that a random walk started at the center of the cube hits its surface before it hits the removed cubes with probability greater than $1/2$. Thus if the Dirichlet boundary condition f is changed on the surface of the unit cube, $\Phi(center)$ will change significantly. On the other hand, since for any point $x \in \Omega$, $\text{dist}(x, \partial\Omega) < 1/\sqrt{n}$, it will take $\Omega(\sqrt{n})$ steps for our process to reach the boundary of the unit cube. So our algorithm will require $\Omega(\sqrt{n}\log n)$ random bits.

On the other hand, for a large class of domains, which we call planar-like, our algorithm still works.

Definition 3. *A domain $\Omega \subset \mathbb{R}^k$ is called planar-like if for every point $y \notin \Omega$ there is a $k-2$-dimensional plane containing y and not intersecting Ω.*

Note that since $2 - 2 = 0$-dimensional planes are points, every planar domain is automatically planar-like. We can also observe that any convex domain is planar-like, since in this case for every $y \notin \Omega$ there is a $k-1$-dimensional plane containing y and not intersecting Ω.

It is very easy to see that the function $\log \text{dist}(x, \partial\Omega)$ is superharmonic for planar-like domains (because logarithm of the distance to a $k-2$-dimensional plane is harmonic away from the plane). This allows us to slightly modify the proof of Lemma 3 (only Step 6 requires a minimal change in the constants in Harnack's principle) to obtain the following Theorem.

Theorem 4. *The EucDP over $\mathbb{R}^k$ is solvable by a randomized TM in space $S = O(\log n)$ using $R = O(log^4 n)$ random bits for planar-like domains.*

We defer the details of the proof to the full version of the paper.

Acknowledgments

We would like to thank Elchanan Mossel for his valuable advice on stochastic processes. The authors are also grateful to Stephen Cook for the numerous enlightening discussions.

References

1. Saks, M.: Randomization and derandomization in space-bounded computation. In: Proceedings of 11th Annual IEEE Conference on Computational Complexity (CCC), pp. 128–149. IEEE Computer Society Press, Los Alamitos (1996)
2. Saks, M., Zhou, S.: $BP_H SPACE(S) \subseteq DSPACE(S^{3/2})$. Journal of Computer and System Sciences 58(2), 376–403 (1999)
3. Courant, R., Hilbert, D.: Methods of Mathematical Physics, vol. 1. Wiley, New York (1989)
4. Ahlfors, L.: Complex Analysis. McGraw-Hill, New York (1953)
5. Nisan, N.: $RL \subset SC$. Computational Complexity 4(1), 1–11 (1994)
6. Nisan, N., Zuckerman, D.: Randomness is linear in space. Journal of Computer and System Sciences 52(1), 43–52 (1996)

7. Fuchs, B.A., Shabat, B.V.: Functions of a Complex Variable and Some of Their Applications, vol. 1. Pergamon Press, London (1964)
8. Lavrentiev, M.A., Shabat, B.V.: Problems of Hydrodynamics and Their Mathematical Models, Nauka, Moscow (1973)
9. Feynman, R.P., Leighton, R.B., Sands, M.: The Feynman Lectures on Physics, vol. 2. Addison-Wesley, Reading (1989)
10. Gu, X., Wang, Y., Chan, T.F., Thompson, P.M., Yau, S.T.: Genus zero surface conformal mapping and its application to brain surface mapping. IEEE Transactions on Medical Imaging 23(8), 949–958 (2004)
11. Binder, I., Braverman, M., Yampolsky, M.: On computational complexity of riemann mapping. Arkiv för Matematik (to appear, 2007)
12. Pommerenke, C.: Boundary Behaviour of Conformal Maps. Springer, Heidelberg (1992)
13. Garnett, J.B., Marshall, D.E.: Harmonic Measure. Cambridge Univ. Press, Cambridge (2004)
14. Muller, M.E.: Some continuous Monte Carlo methods for the Dirichlet problem. Ann. Math. Statist. 27, 569–589 (1956)
15. Motoo, M.: Some evaluations for continuous Monte Carlo method by using Brownian hitting process. Ann. Inst. Statist. Math. Tokyo 11, 49–54 (1959)
16. Mihaĭlov, G.A.: Estimation of the difficulty of simulating the process of "random walk on spheres" for some types of regions. Zh. Vychisl. Mat. i Mat. Fiz. 19(2), 510–515, 558–559 (1979)
17. Elepov, B.S., Kronberg, A.A., Mihaĭlov, G.A., Sabel'fel'd, K.K.: Reshenie kraevykh zadach metodom Monte-Karlo. "Nauka" Sibirsk. Otdel., Novosibirsk (1980)
18. Karatzas, I., Shreve, S.E.: Brownian Motion and Stochastic Calculus, 2nd edn. Springer, Heidelberg (1991)
19. Stein, E.M.: Harmonic analysis: real-variable methods, orthogonality, and oscillatory integrals. Princeton University Press, Princeton, NJ (1993)
20. Durett, R.: Probability, Theory and Examples. Wadsworth and Brooks (1991)

High Entropy Random Selection Protocols

(Extended Abstract)

Harry Buhrman[1], Matthias Christandl[2], Michal Koucký[3], Zvi Lotker[4],
Boaz Patt-Shamir[5], and Nikolai Vereshchagin[6]

[1] CWI, Amsterdam
buhrman@cwi.nl
[2] Cambridge University
mc380@cam.ac.uk
[3] Institute of Mathematics of Czech Academy of Sciences
mkoucky@math.cas.cz
[4] Ben Gurion University
zvilo@cse.bgu.ac.il
[5] Tel Aviv University
boaz@eng.tau.ac.il
[6] Lomonosov Moscow State University
ver@mech.math.msu.su

Abstract. We study the two party problem of randomly selecting a string among all the strings of length n. We want the protocol to have the property that the output distribution has high *entropy*, even when one of the two parties is dishonest and deviates from the protocol. We develop protocols that achieve high, close to n, entropy.

In the literature the randomness guarantee is usually expressed as being close to the uniform distribution or in terms of resiliency. The notion of entropy is not directly comparable to that of resiliency, but we establish a connection between the two that allows us to compare our protocols with the existing ones.

We construct an explicit protocol that yields entropy $n - O(1)$ and has $4 \log^* n$ rounds, improving over the protocol of Goldreich et al. [3] that also achieves this entropy but needs $O(n)$ rounds. Both these protocols need $O(n^2)$ bits of communication.

Next we reduce the communication in our protocols. We show the existence, non-explicitly, of a protocol that has 6 rounds, $2n + 8 \log n$ bits of communication and yields entropy $n - O(\log n)$ and min-entropy $n/2 - O(\log n)$. Our protocol achieves the same entropy bound as the recent, also non-explicit, protocol of Gradwohl et al. [4], however achieves much higher min-entropy: $n/2 - O(\log n)$ versus $O(\log n)$.

Finally we exhibit very simple explicit protocols. We connect the security parameter of these geometric protocols with the well studied Kakeya problem motivated by harmonic analysis and analytical number theory. We are only able to prove that these protocols have entropy $3n/4$ but still $n/2 - O(\log n)$ min-entropy. Therefore they do not perform as well with respect to the explicit constructions of Gradwohl et al. [4] entropy-wise, but still have much better min-entropy. We conjecture that these simple protocols achieve $n - o(n)$ entropy. Our geometric construction and its relation to the Kakeya problem follows a new and different approach to the random selection problem than any of the previously known protocols.

M. Charikar et al. (Eds.): APPROX and RANDOM 2007, LNCS 4627, pp. 366–379, 2007.

1 Introduction

We study the following communication problem. Alice and Bob want to select a random string. They are not at the same location so they do not see what the other player does. They communicate messages according to some protocol and in the end they output a string of n bits which is a function of the messages communicated. This string should be as random as possible, in our case we measure the amount of randomness by the entropy of the probability distribution that is generated by this protocol.

The messages they communicate may depend on random experiments the players perform and on messages sent so far. The outcome of an experiment is known only to the party which performs it so the other party cannot verify the outcome of such an experiment or whether the experiment was carried out at all. One or both the parties may deviate from the protocol and try to influence the selected string (*cheat*). We are interested in the situation when a party honestly follows the protocol and wants to have some guarantee that the selected string is indeed as random as possible. The measure of randomness we use is the *entropy* of probability distribution that is the outcome of the protocol.

In this paper we present protocols for this problem. In particular we show a protocol that achieves entropy $n - O(1)$ if at least one party is honest and that uses $4\log^* n$ rounds and communicates $n^2 + O(n\log n)$ bits. The round complexity of our protocol is optimal up-to a constant factor; the optimality follows from a result of Sanghvi and Vadhan [8]. We further consider the question of reducing the communication complexity of our protocols. We show non-constructively that there are protocols with linear communication complexity that achieve entropy $n - \log n$ in just 3 rounds, and in 6 rounds achieves in addition min-entropy $n/2 - O(\log n)$ which is close to the optimal bound of $n/2$, that follows from Goldreich et al. [3] and from a bound on quantum coin-flipping due to Kitaev (see [2]). We propose several explicit and very simple protocols that have entropy $3n/4$ and we conjecture that they have entropy $n - o(n)$. Our proofs establish a connection between the security guarantee of our protocols and the well studied problem of Kakeya over finite fields motivated by Harmonic analysis and analytic number theory (see [5,6] for background information on Kakeya Problem). Although these constructive protocols do not achieve the same parameters as the best known constructive protocols (see next section), our (geometric) protocols are quite different in nature and much simpler to implement and still yield much higher min-entropy.

1.1 Previous Work

There is a large body of previous work which considers the problem of random string selection, and related problems such as a leader selection and fault-tolerant computation. We refer the reader to [8] for an overview of the literature. In this paper we assume that both parties have unlimited computational power, i.e., so called *full information model*. Several different measures for the randomness guarantee of the protocol are used in the literature. The most widely used is the (μ, ϵ)-resilience and the statistical distance from the uniform distribution. Informally a protocol is (μ, ϵ)-resilient if for every set $S \subset \{0,1\}^n$ with density μ (cardinality $\mu 2^n$), the output of the protocol is in S with probability at most ϵ. In this paper we study however another very natural randomness guarantee, namely the entropy of the resulting output distribution. There is a certain relationship between the entropy and resilience, but these parameters are not interchangeable.

In [3], Goldreich et al. constructs a protocol that is $(\mu, \sqrt{\mu})$-resilient for all $\mu > 0$. This protocol runs in $O(n)$ rounds and communicates $O(n^2)$ bits. We show that their security guarantee also implies entropy $n - O(1)$. Hence, our first protocol, that uses $4 \log^* n$ is an improvement in the number of rounds with respect to the entropy measure over that protocol.

Sanghvi and Vadhan [8] give a protocol for every constant $\delta > 0$ that is $(\mu, \sqrt{\mu + \delta})$-resilient and that has constant statistical distance from the uniform distribution. This type of resilience essentially guarantees security only for sets of constant density. Indeed, their protocol allows the cheating party to bias the output distribution so that a particular string has a constant probability of being the output. Hence, their protocol only guarantees constant *min-entropy* and entropy $(1 - \epsilon)n$ for $\epsilon > 0$. Sanghvi and Vadhan also show a lower bound $\Omega(\log^* n)$ on the number of rounds of any random selection protocol that achieves constant statistical distance from the uniform distribution. We show that entropy $n - O(1)$ implies being close to uniform distribution so the lower bound translates to our protocols.

Recently, Gradwohl et al. [4], who also considered protocols with more than 2 players, constructed for each μ a $O(\log^* n)$-round protocol that is $(\mu, O(\sqrt{\mu}))$-resilient and that uses linear communication. Our results are not completely comparable with those of [4]; the protocols of [4] only achieve entropy $n - O(\log n)$ whereas the entropy $n - O(1)$ of our protocol implies only $(\mu, O(1/\log(1/\mu)))$-resilience for all $\mu > 0$. Their $(1/n^2, O(1/n))$-resilient protocol, non-explicit matches our non-explicit protocol from Section 4.1 in terms of entropy but our protocol can be extended to also achieve high $(n/2 - O(\log n))$ min-entropy at the cost of additional 3 rounds.

This extensibility comes from the fact that all our protocols are asymmetric. When Bob is honest (and Alice dishonest) the min-entropy of the output is guaranteed to be as high as $n - O(\log n)$, which implies, by the aforementioned result of Kitaev [2] that the min-entropy is only $O(\log n)$ when Bob is dishonest (and Alice honest). The protocols of Gradwohl et al. in general do not have this feature. Whenever their protocols achieve high $(n - O(\log n))$ entropy the min-entropy is only $O(\log n)$.

Finally our explicit geometric protocol only obtains $3n/4$ entropy and thus performs worse than the explicit protocol from [4], that achieves for $\mu = 1/\log n$ entropy $n - o(n)$. Our explicit protocols though still have min-entropy $n/2 - O(\log n)$ outperforming [4], that only gets min-entropy $O(\log n)$.

The paper is organized as follows. In the next section we review the notion of entropy and of other measures of randomness, and we establish some relationships among them. Section 3 contains our protocol that achieves entropy $n - O(1)$. In Section 4 we address the problem of reducing the communication complexity of our protocols. Due to space limitations we omit almost all the proofs from this extended abstract.

2 Preliminaries

Let $\mathbf{Y}$ be a random variable with a finite range S. The *entropy of* $\mathbf{Y}$ is defined by:

$$H(\mathbf{Y}) = -\sum_{s \in S} \Pr[\mathbf{Y} = s] \cdot \log \Pr[\mathbf{Y} = s].$$

If for some $s \in S$, $\Pr[\mathbf{Y} = s] = 0$ then the corresponding term in the sum is considered to be zero. All logarithms are based two.

Let $\mathbf{X}, \mathbf{Y}$ be (possibly dependent) jointly distributed random variable with ranges T, S, respectively. The *entropy of* $\mathbf{Y}$ *conditional to* $\mathbf{X}$ is defined by:

$$H(\mathbf{Y}|\mathbf{X}) = \sum_{t \in T} \Pr[\mathbf{X} = t] H(\mathbf{Y}|\mathbf{X} = t),$$

where $\mathbf{Y}|\mathbf{X} = t$ stands for the random variable whose range is S and which takes outcome $s \in S$ with probability $\Pr[\mathbf{Y} = s|\mathbf{X} = t]$.

The following are basic facts about the entropy:

$$H(f(\mathbf{Y})) \leq H(\mathbf{Y}) \text{ for any function } f, \tag{1}$$
$$H(\mathbf{Y}) < \log |S|, \tag{2}$$
$$H(\mathbf{Y}|\mathbf{X}) \leq H(\mathbf{Y}), \tag{3}$$
$$H(\langle \mathbf{X}, \mathbf{Y} \rangle) = H(\mathbf{Y}|\mathbf{X}) + H(\mathbf{X}), \tag{4}$$
$$H(\mathbf{X}) \leq H(\langle \mathbf{Y}, \mathbf{X} \rangle) \text{ (follows from (4))}, \tag{5}$$
$$H(\langle \mathbf{Y}, \mathbf{X} \rangle) \leq H(\mathbf{Y}) + H(\mathbf{X}) \text{ (follows from (3) and (4))}. \tag{6}$$

Here $\langle \mathbf{Y}, \mathbf{X} \rangle$ stands for the random variable with range $S \times T$, which takes the outcome $\langle s, t \rangle$ with probability $\Pr[\mathbf{X} = t, \mathbf{Y} = s]$. We will abbreviate $H(\langle \mathbf{Y}, \mathbf{X} \rangle)$ as $H(\mathbf{Y}, \mathbf{X})$ in the sequel.

The following corollaries of these facts are used in the sequel

1. Let Y_i be random variables with the same range S and let $\mathbf{Y}$ be obtained by picking an index $i \in \{1, \ldots, n\}$ uniformly at random and then drawing a random sample according to $\mathbf{Y}_i$. Then $H(\mathbf{Y}) \geq \frac{1}{n} \sum_{i=1}^{n} H(\mathbf{Y}_i)$. (Indeed, let $\mathbf{X}$ stand for the random variable uniformly distributed in $\{1, \ldots, n\}$. Then $H(\mathbf{Y}) \geq H(\mathbf{Y}|\mathbf{X}) = \frac{1}{n} \sum_{i=1}^{n} H(\mathbf{Y}_i)$.)
2. Let $\ell \geq 1$ be an integer and $f : S \to T$ be a function from a set S to a set T. Let $\mathbf{Y}$ be a random variable with range S. If $\forall t \in T, |f^{-1}(t)| \leq \ell$ then $H(f(\mathbf{Y})) \geq H(\mathbf{Y}) - \log \ell$. (Indeed, let $\mathbf{X}$ be the index of $\mathbf{Y}$ in $f^{-1}(\mathbf{Y})$. Then $H(\mathbf{Y}) = H(f(\mathbf{Y}), \mathbf{X}) \leq H(f(\mathbf{Y})) + H(\mathbf{X}) \leq H(f(\mathbf{Y})) + \log \ell$.)

The *min-entropy* of a random variable $\mathbf{X}$ with a finite range S is

$$H_\infty(\mathbf{X}) = \min\{-\log \Pr[\mathbf{X} = s] : s \in S\}.$$

The *statistical distance* between random variables $\mathbf{X}, \mathbf{Y}$ with the same finite range S is defined as the maximum

$$|\Pr[\mathbf{X} \in A] - \Pr[\mathbf{Y} \in A]|$$

over all subsets A of S. It is easy to see that the maximum is attained for A consisting of all s with $\Pr[\mathbf{X} = s] > \Pr[\mathbf{Y} = s]$ (as well as for its complement). For every integer $n \geq 1$, we denote by $\mathbf{U}_n$ the uniform probability distribution of strings $\{0,1\}^n$.

In order to apply a lower-bound from [8] to show that our main protocol needs $\Omega(\log^* n)$ rounds we establish a relation between entropy and constant statistical distance.

Lemma 1. *For every real c there is a real $q < 1$ such that the following holds. If $\mathbf{X}$ is a random variable with range $\{0,1\}^n$ and $H(\mathbf{X}) \geq n - c$ then the statistical distance of $\mathbf{X}$ and $\mathbf{U}_n$ is at most q.*

Definition. Let r, n be natural numbers. A deterministic strategy of a player (Alice or Bob) is a function that maps each tuple $\langle x_1, \ldots, x_i \rangle$ of binary strings where $i < r$ to a binary string (the current message of the player provided $\langle x_1, \ldots, x_i \rangle$ is the sequence of previous messages). A randomized strategy of a player (Alice or Bob) is a probability distribution over deterministic strategies.

A *protocol running in r rounds* is a function f that maps each r-tuple $\langle x_1, \ldots, x_r \rangle$ of binary strings to a binary string of length n (the first string x_1 is considered as Alice's message, the second string x_2 as Bob's message and so on) and a pair $\langle \mathbf{S}_A, \mathbf{S}_B \rangle$ of randomized strategies.

If S_A, S_B are deterministic strategies of Alice and Bob then the outcome of the protocol for S_A, S_B is defined as $f(x_1, \ldots, x_r)$ where $x_1, \ldots, x_r$ are defined recursively: $x_{2i+1} = S_A(\langle x_1, \ldots, x_{2i} \rangle)$ and $x_{2i+2} = S_B(\langle x_1, \ldots, x_{2i+1} \rangle)$.

If $\mathbf{S}_A, \mathbf{S}_B$ are randomized strategies of Alice and Bob then the outcome of the protocol is a random variable generated as follows: select independently Alice's and Bob's strategies S_A, S_B with respect to probability distributions $\mathbf{S}_A$ and $\mathbf{S}_B$, respectively, and output the result of the protocol for S_A, S_B.

We say that Alice follows the protocol (is *honest*) if she uses the strategy $\mathbf{S}_A$. We say that Alice deviates from the protocol (*cheats*) if she uses any other randomized strategy. Similarly for Bob.

We say that a protocol P for random string selection is (k, l)-good if the following properties hold:

1. If both Alice and Bob follow the protocol then the outcome is a fully random string of length n.
2. If Alice follows the protocol and Bob deviates from it then the outcome has entropy at least k.
3. If Bob follows the protocol and Alice deviates from it then the outcome has entropy at least l.

(End of Definition.)

Throughout the paper we use the following easy observation that holds for every protocol:

Lemma 2. *Assume that Alice's strategy $\mathbf{S}_A$ guarantees that the entropy of the outcome is at least α for all deterministic strategies of Bob. Then the same guarantee holds for all randomized strategies of Bob as well. A similar statement is true for min-entropy in place of entropy.*

In [8], Sanghvi and Vadhan establish that any protocol for random selection that guarantees a constant statistical distance of the output from the uniform distribution requires at least $\Omega(\log^* n)$ rounds. Hence we obtain the following corollary to the previous lemma.

Corollary 1. *If P is a protocol that is $(n - O(1), n - O(1))$-good then P has at least $\Omega(\log^* n)$ rounds.*

For $\mu, \epsilon > 0$, a random string selection protocol P is (μ, ϵ)-*resilient* if for any set S of size at most $\mu 2^n$, the probability that the output of P is in S is at most ϵ, even if one of the parties cheats.

In order to compare our results with previous work we state the following claim.

Lemma 3. *For a random selection protocol P the following holds.*

1. *If P is $(\mu, d\mu^c)$-resilient for some constants $c, d > 0$ and any $\mu > 0$ then P is $(n - O(1), n - O(1))$-good.*
2. *If P is $(n - O(1), n - O(1))$-good then for some constant d and any $\mu > 0$ it is $(\mu, d/\log(1/\mu))$-resilient.*

3   The Main Protocol

In this section we construct a protocol that is $(n - O(1), n - O(1))$-good. We start with the following protocol.

Lemma 4. *There is a $(n - 1, n - \log n)$-good protocol P_0 running in 3 rounds and communicating $n^2 + n + \log n$ bits. If Bob is honest then the outcome of P_0 has min-entropy at least $n - \log n$.*

Proof. The protocol $P_0(A, B)$ is as follows:

1. Player A picks $x_1, x_2, \ldots, x_n \in \{0, 1\}^n$ uniformly at random and sends them to Player B.
2. Player B picks $y \in \{0, 1\}^n$ uniformly at random and sends it to Player A.
3. Player A picks an index $j \in \{1, \ldots, n\}$ uniformly at random and sends it to B.
4. The outcome $\mathbf{R}$ of the protocol is $x_j \oplus y$, i.e., the bit-wise xor of x_j and y.

Note that the entropy bounds are tight as a cheating Bob can set $y = x_1$ in the protocol and then $H(\mathbf{R}) = n - 1$. Similarly, a cheating Alice can enforce the first $\log n$ bits of the outcome to be all zero bits so $H(\mathbf{R}) = n - \log n$ in that case.

1) It is easy to verify that the outcome $\mathbf{R}$ of the protocol P_0(Alice, Bob) is uniformly distributed if both Alice and Bob follow the protocol and hence it has entropy n.

2) Assume that Alice follows the protocol and Bob is trying to cheat. Hence, Alice picks uniformly at random $x_1, \ldots, x_n \in \{0, 1\}^n$. Bob picks y. Then Alice picks a random index $j \in \{1, \ldots n\}$ and they set $\mathbf{R} = x_j \oplus y$. Clearly, $H(x_1, \ldots, x_n) = n^2$, thus

$$n^2 = H(x_1, \ldots, x_n) \leq H(x_1, \ldots, x_n, y) \leq H(x_1 \oplus y, \ldots, x_n \oplus y) + H(y)$$
$$\leq H(x_1 \oplus y, \ldots, x_n \oplus y) + n.$$

Here the first inequality holds by (5), the middle one by (1) and (6), and the last one by (2). Therefore,

$$(n^2-n)/n \le H(x_1\oplus y,\ldots,x_n\oplus y)/n \le \sum_{i=1}^{n} H(x_i\oplus y)/n = H(x_j\oplus y|j) \le H(x_j\oplus y).$$

Here the second inequality holds by (6), the equality holds, as Alice chooses j uniformly, and the last inequality is true by (3).

3) Assume that Bob follows the protocol and Alice is trying to cheat. Hence, Alice carefully selects $x_1,\ldots,x_n$, Bob picks a random string $y \in \{0,1\}^n$ and Alice carefully chooses $j \in \{1,\ldots,n\}$. Thus $H(y|\langle x_1,\ldots,x_n\rangle) = n$ and hence

$$H(x_j \oplus y) \ge H(x_j \oplus y|\langle x_1,\ldots,x_n\rangle) \ge H(y|\langle x_1,\ldots,x_n\rangle) - H(j|\langle x_1,\ldots,x_n\rangle)$$
$$\ge H(y|\langle x_1,\ldots,x_n\rangle) - H(j) \ge n - \log n.$$

Here the second inequality holds by (1) and (6). Verifying the lower bound on the min-entropy is straightforward.

Our protocol achieves our goal of having entropy of the outcome close to n if Alice is honest. However if she is dishonest she can fix up-to $\log n$ bits of the outcome to her will. Clearly, Alice's cheating power comes from the fact that she can choose up-to $\log n$ bits in the last round of the protocol. If we would reduce the number of strings x_j she can choose from in the last round, her cheating ability would decrease as well. Unfortunately, that would increase cheating ability of Bob. Hence, there is a trade-off between cheating ability of Alice and Bob. To overcome this we will reduce the number of strings Alice can choose from but at the same time we will also limit Bob's cheating ability by replacing his y by an outcome of yet another run of the protocol played with Alice's and Bob's roles reversed. By iterating this several times we can obtain the following protocol.

Let $\log^* n$ stand for the number of times we can apply the function $\lceil \log x \rceil$ until we get 1 from n. For instance, $\log^* 100 = 4$.

Theorem 1. *There is a $(n-2, n-3)$-good protocol running in $2\log^* n + 1$ rounds and communicating $n^2 + O(n\log n)$ bits. Depending on n, either if Alice or Bob is honest then the min-entropy of the protocol is at least $n - O(\log n)$.*

Proof. Let $k = \log^* n - 1$. Define $\ell_0 = n$ and $\ell_i = \lceil \log \ell_{i-1} \rceil$, for $i = 1,\ldots,k$, so $\ell_{k-1} \in \{3,4\}$ and $\ell_k = 2$.

For $i = 1,\ldots,k$ we define protocol $P_i(A,B)$ as follows.

1. Player A picks $x_1, x_2,\ldots,x_{\ell_i} \in \{0,1\}^n$ uniformly at random and sends them to Player B.
2. Players A and B now run protocol $P_{i-1}(B,A)$ (note that players exchange their roles) and set y to the outcome of that protocol.
3. Player A picks an index $j \in \{1,\ldots,\ell_i\}$ uniformly at random and sends it to B.
4. The outcome $\mathbf{R}_i$ of this protocol is $x_j \oplus y$.

We claim that the protocols are $(n - 2, n - \log 4\ell_i)$-good:

Lemma 5. *For all $i = 0, 1, \ldots, k$ the following is true.*

1. *If both Alice and Bob follow the protocol $P_i(Alice, Bob)$ then its outcome $\mathbf{R}_i$ satisfies $H(\mathbf{R}_i) = n$.*
2. *If Alice follows the protocol $P_i(\text{Alice}, \text{Bob})$ then the outcome $\mathbf{R}_i$ satisfies $H(\mathbf{R}_i) \geq n - 2$.*
3. *If Bob follows the protocol $P_i(\text{Alice}, \text{Bob})$ then the outcome $\mathbf{R}_i$ of the protocol satisfies $H(\mathbf{R}_i) \geq n - \log 4\ell_i$.*

All the bounds on the entropy are valid also when conditioned on the tuple consisting of all strings communicated before running P_i. Furthermore, if i is even and Bob is honest or i is odd and Alice is honest then $H_\infty(\mathbf{R}_i) \geq n - \sum_{j=1}^{i+1} \ell_j$.

Proof. The first claim is straightforward to verify. We prove the other two simultaneously by an induction on i. For $i = 0$ the claims follow from Lemma 4. So assume that the claims are true for $i - 1$ and we will prove them for i.

If Alice follows the protocol $P_i(Alice, Bob)$ then she picks $x_1, \ldots, x_{\ell_i}$ uniformly at random. Then the protocol $P_{i-1}(Bob, Alice)$ is invoked to obtain $y = \mathbf{R}_{i-1}$. We can reason just as in the proof of Lemma 4. However this time we have a better lower bound for $H(x_1, \ldots, x_{\ell_i}, y)$. Indeed, by induction hypothesis, since Alice follows the protocol,

$$H(y|x_1, \ldots, x_{\ell_i}) \geq n - \log 4\ell_{i-1} \geq n - 2\ell_i.$$

Here the last inequality holds for all $i < k$ as $\ell_{i-1} > 4$ in this case and hence $2\ell_i \geq 2 \log \ell_{i-1} > \log 4\ell_{i-1}$. For $i = k$ we have $\ell_{i-1} \in \{3, 4\}$ and $\ell_i = 2$ and the inequality is evident.

Thus,

$$H(x_1, \ldots, x_{\ell_i}, y) = H(x_1, \ldots, x_{\ell_i}) + H(y|x_1, \ldots, x_{\ell_i}) \geq \ell_i n - 2\ell_i + n.$$

Just as in Lemma 4, this implies

$$H(x_j \oplus y) \geq H(x_j \oplus y|j) = \sum_{s=1}^{l_i} H(x_s \oplus y)/l_i$$
$$\geq (H(x_1, \ldots, x_{\ell_i}, y) - H(y))/\ell_i \geq (\ell_i n - 2\ell_i + n - n)/\ell_i = n - 2.$$

Assume that Bob follows the protocol $P_i(Alice, Bob)$ but Alice deviates from it by carefully choosing $x_1, \ldots, x_{\ell_i}$ and j. Then the protocol $P_{i-1}(Bob, Alice)$ is invoked to obtain $y = \mathbf{R}_{i-1}$. By induction hypothesis $H(y|x_1, \ldots, x_{\ell_i}) \geq n - 2$. Now Alice chooses $j \in \{1, \ldots, \ell_i\}$. Similarly as in the proof of Lemma 4, we have

$$H(x_j \oplus y) \geq H(x_j \oplus y|\langle x_1, \ldots, x_{\ell_i}\rangle) \geq H(y|\langle x_1, \ldots, x_{\ell_i}\rangle) - H(j|\langle x_1, \ldots, x_{\ell_i}\rangle)$$
$$\geq H(y|\langle x_1, \ldots, x_{\ell_i}\rangle) - H(j) \geq n - 2 - \log \ell_i.$$

The claim about min-entropy follows by induction.

By the lemma, the protocol P_k is $(n-2, n-3)$ good. It runs in $2k+3 = 2(\log^* n - 1) + 3$ rounds.

The number of communicated bits is equal to

$$n^2 + n + \log n + \sum_{i=1}^{k}(n\ell_i + \log \ell_i)$$

All ℓ_i's in the sum are at most $\log n$ and decrease faster than a geometric progression. Hence the sum is at most its largest term $(n \log n)$ times a constant.

4 Improving Communication Complexity

In the previous section we have shown a protocol for Alice and Bob that guarantees that the entropy of the selected string is at least $n - O(1)$. The protocol has an optimal (up-to a constant factor) number of rounds and communicates $O(n^2)$ bits. In this section we will address the possibility of reducing the amount of communication in the protocol.

We focus on the basic protocol $P_0(A, B)$ as that protocol contributes to the communication the most. The protocol can be viewed as follows.

1. Player A picks $x \in \{0,1\}^{m_A}$ uniformly at random and sends it to Player B.
2. Player B picks $y \in \{0,1\}^{m_B}$ uniformly at random and sends it to Player A.
3. Player A picks an index $j \in \{0,1\}^{m'_A}$ uniformly at random and sends it to B.
4. A fixed function $f : \{0,1\}^{m_A} \times \{0,1\}^{m_B} \times \{0,1\}^{m'_A} \to \{0,1\}^n$ is applied to x, y and j to obtain the outcome $f(x, y, j)$.

We will denote such a protocol by $P_0(A, B, f)$. In the basic protocol the parameters are: $m_A = n^2$, $m_B = n$ and $m'_A = \log n$. We would like to find another suitable function f with a smaller domain.

We note first that three rounds in the protocol are necessary in order to obtain the required guarantees on the output of the protocol. In any two round protocol at least one of the parties can force the output to have entropy at most $n/2 + O(\log n)$. (In a two round protocol, if for some x, the range of $f(x, \cdot)$ is smaller than $n2^{n/2}$ then Alice can enforce entropy $n/2 + \log n$ by picking this x. On the other hand if $f(x, \cdot)$ has a large range for all x, then Bob can cheat by almost always enforcing the output to lie in a set of size $2^{n/2}$. Bob's *cheating* set can be picked at random.).

4.1 Non-explicit Protocol

The following claim indicates that finding a suitable function f should be feasible.

Lemma 6. *If $f : \{0,1\}^n \times \{0,1\}^n \times \{0,1\}^{8 \log n} \to \{0,1\}^n$ is taken uniformly at random among all functions then with probability at least $1/2$, $P_0(A, B, f)$ satisfies:*

1. *If both Alice and Bob follow the protocol $P_0(Alice, Bob, f)$ then its outcome $\mathbf{R}$ satisfies $H(\mathbf{R}) = n - O(1)$.*
2. *If Alice follows the protocol $P_0(Alice, Bob, f)$ then the outcome $\mathbf{R}$ satisfies $H(\mathbf{R}) \geq n - O(1)$.*
3. *If Bob follows the protocol $P_0(Alice, Bob, f)$ then the outcome $\mathbf{R}$ of the protocol satisfies $H(\mathbf{R}) \geq n - O(\log n)$ and $H_\infty(\mathbf{R}) \geq n - O(\log n)$.*

The question is how to find an explicit function f of similar properties. We propose the following three functions that we believe have the required properties. We prove several results in that direction.

1. $f_{rot} : \{0,1\}^n \times \{0,1\}^n \times \{1,\ldots,n\} \rightarrow \{0,1\}^n$ defined by $f(x,y,j) = x^j \oplus y$, where x^j is the j-th rotation of x, $x^j = x_j x_{j+1} \cdots x_n x_1 \cdots x_{j-1}$. Here n is assumed to be a prime.
2. $f_{lin} : F^{k-1} \times F^k \times F \rightarrow F^k$, where $F = GF[2^{\log n}]$, $k = n/\log n$ and $f(d,y,j) = (1, d_1, \ldots, d_{k-1}) * j + (y_1, \ldots, y_k)$.
3. $f_{mul} : F \times F \times H \rightarrow F$, where $F = GF[2^n]$, $H \subseteq F$, $|H| = n$, and $f(x,y,j) = x * j + y$.

In particular the function f_{rot} is interesting as it would allow very efficient implementation. We conjecture that for $f \in \{f_{rot}, f_{lin}, f_{mul}\}$ protocol $P_0(A, B, f)$ is $(n - o(n), n - O(\log n))$-good.

Lemma 7. $P_0(A, B, f_{rot})$ is $(n/2 - 3/2, n - \log n)$-good when n is prime and the min-entropy of the outcome is at least $n - O(\log n)$ when Bob follows the protocol.

A similar lemma holds also for our other two candidate functions.

Averaging the Asymmetry. One of the interesting features of our protocols is the asymmetry of cheating power of the two parties. We used this asymmetry to build the protocol with entropy $n - O(1)$. One can also use this asymmetry for "*averaging*" their cheating powers in the following simple way. Given a protocol $Q_n(A, B)$ for selecting an n bit string, Alice and Bob first select the first $n/2$ bits of the string by running the protocol $Q_{n/2}(\text{Alice}, \text{Bob})$ and then they select the other half of the string by running the protocol $Q_{n/2}(\text{Bob}, \text{Alice})$. If the protocol Q_n is $(k(n), l(n))$-good then the *averaging* protocol is $(k(n/2) + l(n/2), k(n/2) + l(n/2))$-good. Similarly if the min-entropy when Alice follows the protocol is bounded from below by $k_\infty(n)$ and when Bob follows the protocol by $l_\infty(n)$, then the min-entropy of the outcome of the averaging protocol is at least $k_\infty(n/2) + l_\infty(n/2)$.

Hence from Lemma 7 we obtain the following corollary.

Corollary 2. *There is a 5-round protocol for random string selection that communicates $2n + O(\log n)$ bits, that is $(3n/4 - O(\log n), 3n/4 - O(\log n))$-good and that has min-entropy at least $n/2 - O(\log n)$ when at least one of the parties follows the protocol.*

In the next section we show for a variant of $P_0(A, B, f_{lin})$ a similar security guarantee.

4.2 Geometric Protocols and the Problem of Kakeya

We exhibit here a variant of the protocol $P_0(A, B, f_{lin})$ and show that it achieves entropy at least $3n/4 - O(1)$ if at least one party is honest. Fix a finite field F and a natural $m \geq 2$. Let $q = |F|$. We rephrase the protocol as follows:

1. Alice picks at random a vector $d = (1, d_2, \ldots, d_m) \in F^m$ and sends it to Bob.
2. Bob picks at random $x = (x_1, \ldots, x_m) \in F^m$ and sends it to Alice.
3. Alice picks at random $t \in F$ and sends it to Bob.
4. The output of the protocol is

$$y = x + td = (x_1 + t, x_2 + td_2, \ldots, x_m + td_m).$$

The geometric meaning of the protocol is as follows. Alice picks at random a direction of an affine line in the m-dimensional space F^m over F. Bob chooses a random affine line going in that direction. Alice outputs a random point lying on the line.

It is easy to lower bound the entropy of the output y of this protocol assuming that Bob is honest.

Lemma 8. *If Bob is honest then the outcome y of the protocol satisfies*

$$H(y) \geq H_\infty(y) \geq (m - 1) \log q.$$

Note that Alice can cheat this much. For example, Alice can force $y_1 = 0$ by choosing always $t = -x_1$.

In the case when Alice is honest we are able to prove the bound $H(y) \geq (m/2 + 1) \log q - O(1)$. We do not know whether Bob indeed can cheat this much. This question is related to the following problem known as Kakeya problem for finite fields.

Kakeya Problem. *Let L be a collection of affine lines in F^m such that for each direction there is exactly one line in L going in that direction. Let P_L denote points in lines from L. How small can be $|P_L|$?*

For a family L of lines let $\mathbf{X}_L$ denote the random variable in P_L that is a random point on a random line in L. That is, to generate an outcome of $\mathbf{X}_L$, we pick a random line ℓ in L (all the lines are equiprobable) and then pick a random point on ℓ (all the points on ℓ are equiprobable).

Call any set of lines L satisfying the conditions of Kakeya problem a Kakeya family and let $H(m, q)$ stand for the minimum $H(\mathbf{X}_L)$ over all Kakeya families L. Let $H_\infty(m, q)$ stand for the similar value for min-entropy.

Lemma 9. *Assume that Alice is honest. Then the outcome of the protocol always satisfies $H(y) \geq H(m, q)$ and there is Bob's strategy such that $H(y) = H(m, q)$. The same is true for min-entropy in place of entropy.*

Proof. Let $\mathbf{Y}_S$ stand for the outcome of the protocol provided Bob uses a deterministic strategy S. There is an onto function $S \mapsto L$ from deterministic Bob's strategies to Kakeya sets such that $\mathbf{X}_L$ coincides with $\mathbf{Y}_S$.

Indeed, assume that Bob uses a deterministic strategy S. That is, for each $d = (1, d_2, \ldots, d_m)$ Bob chooses $x = x(d)$ deterministically. Thus Bob defines a Kakeya family L consisting of all lines of the form

$$\{x(d) + td \mid t \in F\}.$$

Obviously $\mathbf{X}_L = \mathbf{Y}_S$.

Conversely, for every Kakeya set L there is Bob's strategy S mapped by this function to L (choose any point in the line in L going in direction d specified by Alice).

This implies the statement of the lemma for deterministic strategies. For randomized strategies it follows from Lemma 2.

Note that for every family of lines L we have $H(Y_L) \leq \log |P_L|$. Thus to prove that the entropy of the outcome is at least α (provided Alice is honest) we need to show the lower bound $|P_L| \geq 2^\alpha$ for Kakeya problem. The best known lower bound for $|P_L|$ is $\Omega(q^{m/2+1})$ [6,5] (and it is conjectured that $|P_L|$ must be close to q^m). Note that this bound does not immediately imply that $H(\mathbf{Y}_L) \geq (m/2 + 1) \log q$ for every Kakeya set L, as the entropy of a random variable can be much less than the log-cardinality of the set of outcomes. However, the key proposition from the proof of the bound $|P_L| = \Omega(q^{m/2+1})$ presented in [5] indeed allows to prove a slightly weaker inequality $H(\mathbf{Y}_L) \geq (m/2 + 1) \log q - O(1)$.

Proposition 1 ([5]). *Let L be a collection of affine lines in F^m such that every 2-dimensional plane has at most $q + 1$ lines from L. Let P be a subset of F^m. Then*

$$|\{(p, l) \mid l \in L, \, p \in P, \, p \in l\}| \leq C \cdot (|P|^{1/2}|L|^{3/4}|F|^{1/4} + |P| + |L|)$$

for some constant C.

This proposition allows to prove the following

Theorem 2. *If Alice is honest then the outcome of the geometric protocol satisfies $H(y) \geq (m/2+1) \log q - O(1)$ and $H_\infty(y) \geq \log q$ provided that $m^2 \log^2 q/q \in O(1)$.*

Proof. The second statement is obvious. Let us prove the first one. By Lemma 9 it suffices to show that $H(\mathbf{X}_L) \geq (m/2 + 1) \log q - O(1)$ for every Kakeya family L.

Let α stand for $q^{-m/2-1}c$ where $c \geq 1$ is a constant to be defined later. We will show that $H(\mathbf{X}_L) \geq -\log \alpha - O(1)$. For each $y \in P_L$ let p_y stand for the probability that $\mathbf{X}_L = y$: that is, p_y is equal to the number of lines in L containing y divided by q^m. We classify y's according to the value of p_y as follows.

Let Q denote the set of those $y \in P_L$ with

$$p_y \leq \alpha$$

and S_i for $i = 1, 2, \ldots, -\log \alpha$ the set of those $y \in P_L$ with

$$\alpha 2^{i-1} < p_y \leq \alpha 2^i.$$

The entropy of $\mathbf{X}_L$ is the average value of $-\log p_y$. For all y in Q we have $-\log p_y \geq -\log \alpha$. For all y in S_i we have $-\log p_y \geq -\log \alpha - i$. Thus $H(\mathbf{X}_L)$ can be lower bounded by

$$H(\mathbf{X}_L) \geq -\log \alpha - \sum_i i \cdot \left(\sum_{y \in S_i} p_y \right) \geq -\log \alpha - \sum_i i \cdot |S_i| \cdot \alpha 2^i.$$

Thus we need to show that

$$\sum_i i \cdot |S_i| \cdot \alpha 2^i = O(1). \tag{7}$$

To this end we need to upper bound $|S_i|$. We are able to show the following bound.

Lemma 10. *For all i we have $|S_i| \cdot \alpha 2^i = O(2^{-i})$ or $|S_i| \cdot \alpha 2^i = O(q^{-1})$.*

As the series $\sum_i i2^{-i}$ converges and $q^{-1} \cdot (-\log^2 \alpha) \in O(1)$ these bounds obviously imply (7).

Proof. Note that every 2-dimensional plane has at most $q + 1$ lines from L (the number of different directions in every plane is equal to $q + 1$). Apply Proposition 1 to L and $P = S_i$. We obtain

$$|\{(p,l) \mid l \in L, \ p \in S_i, \ p \in l\}| \leq C \cdot (|S_i|^{1/2}|L|^{3/4}q^{1/4} + |S_i| + |L|)$$
$$= C \cdot (|S_i|^{1/2}q^{(3m-2)/4} + |S_i| + q^{m-1}).$$

Every point in S_i belongs to more than $\alpha 2^{i-1}q^m$ lines in L hence

$$|\{(p,l) \mid l \in L, \ p \in S_i, \ p \in l\}| > |S_i|\alpha 2^{i-1}q^m.$$

Combining the inequalities we obtain

$$|S_i|\alpha 2^i q^m < C \cdot (|S_i|^{1/2}q^{(3m-2)/4} + |S_i| + q^{m-1}).$$

If the last term in the right hand side is greater than the other ones, we have

$$|S_i|\alpha 2^i < 3C \cdot q^{-1}.$$

If the second term in the right hand side is greater than the other ones, we have

$$\alpha 2^i q^m < 3C.$$

Note that, since $m \geq 2$, $i \geq 1$, we have $\alpha 2^i q^m = 2^i c q^{m/2-1} \geq 2c$. Therefore this cannot be the case, if we let $c \geq 1.5C$.

In the remaining case (the first term in the right hand side is greater than the other ones) we have

$$|S_i|^{1/2} < 3C2^{-i}\alpha^{-1}q^{-m/4-1/2} \Rightarrow |S_i| < 9C^2 2^{-2i}\alpha^{-2}q^{-m/2-1},$$

and

$$|S_i|\alpha 2^i < 9C^2 2^{-i}\alpha^{-1}q^{-m/2-1} = 9C^2 2^{-i}.$$

The last equality holds by the choice of α. Therefore,

$$|S_i|\alpha 2^i \leq 9C^2 2^{-i}.$$

If we choose $m = 4$ then the lower bounds for $H(y)$ in the cases when Alice cheats and Bob cheats coincide and are equal to $3 \log q - O(1)$. Thus we get:

Theorem 3. *There is a $(3n/4 - O(1), 3n/4 - O(1))$-good 3-round protocol that communicates $2n$ bits.*

Using averaging we obtain the following corollary:

Theorem 4. *There is a $(3n/4 - O(1), 3n/4 - O(1))$-good 6-round protocol that communicates $2n$ bits and guarantees the min-entropy at least $n/2 - O(1)$ for both players.*

Acknowledgments

We would like to thank to Troy Lee and John Tromp for useful discussions and Navin Goyal for pointing us to the problem of Kakeya. We also thank anonymous referees for valuable comments on the preliminary version of this paper. Part of the work was done while the second, third, fourth, and sixth author were visiting CWI, Amsterdam. H. Buhrman was supported by EU project QAP and BRICKS project AFM1. H. Buhrman and M. Koucký were supported in part by an NWO VICI grant (639.023.302). M. Koucký was supported in part by grant GA ČR 201/07/P276, 201/05/0124, project No. 1M0021620808 of MŠMT ČR and Institutional Research Plan No. AV0Z10190503. The work of N. Vereshchagin was supported in part by RFBR grant 06-01-00122.

References

1. Alon, N., Spencer, J.: The probabilistic method, 2nd edn. John Wiley & sons, Chichester (2000)
2. Ambainis, A., Buhrman, H., Dodis, Y., Röhrig, H.: Multiparty Quantum Coin Flipping. In: IEEE Conference on Computational Complexity 2004, pp. 250–259 (2004)
3. Goldreich, O., Goldwasser, S., Linial, N.: Fault-tolerant computation in the full information model. SIAM Journ. on Computing 27(2), 506–544 (1998)
4. Gradwohl, R., Vadhan, S., Zuckerman, D.: Random selection with an Adversarial Majority. In: Dwork, C. (ed.) CRYPTO 2006. LNCS, vol. 4117, pp. 409–426. Springer, Heidelberg (2006)
5. Mockenhaupt, G., Tao, T.: Restriction and Kakeya phenomena for finite fields. Duke Math. J. 121, 35–74 (2004)
6. Wolff, T.: Recent work connected with the Kakeya problem, in Prospects. In: Rossi, H. (ed.) Mathematics. AMS (1999)
7. Muchnik, A., Vereshchagin, N.: Shannon Entropy vs. Kolmogorov Complexity. In: Grigoriev, D., Harrison, J., Hirsch, E.A. (eds.) CSR 2006. LNCS, vol. 3967, pp. 281–291. Springer, Heidelberg (2006)
8. Sanghvi, S., Vadhan, S.: The Round Complexity of Two-Party Random Selection. In: Proceedings of Thirty-seventh Annual ACM Symposium on Theory of Computing, Baltimore, MD, USA, pp. 338–347

Testing st-Connectivity

Sourav Chakraborty[1], Eldar Fischer[2], Oded Lachish[2], Arie Matsliah[2],
and Ilan Newman[3],[*]

[1] Department of Computer Science, University of Chicago, Chicago, IL-60637 USA
sourav@cs.uchicago.edu
[2] Department of Computer Science, Technion, Haifa 3200, Israel
{eldar,loded,ariem}@cs.technion.ac.il
[3] Department of Computer Science, University of Haifa, Haifa 31905, Israel
ilan@cs.haifa.ac.il

Abstract. We continue the study, started in [9], of property testing of graphs in the orientation model. A major question which was left open in [9] is whether the property of st-connectivity can be tested with a constant number of queries. Here we answer this question on the affirmative. To this end we construct a non-trivial reduction of the st-connectivity problem to the problem of testing languages that are decidable by branching programs, which was solved in [11]. The reduction combines combinatorial arguments with a concentration type lemma that is proven for this purpose. Unlike many other property testing results, here the resulting testing algorithm is highly non-trivial itself, and not only its analysis.

1 Introduction

We continue the study, started in [9], of property testing of graphs in the orientation model. This is a model that combines information that has to be queried with information that is known in advance, and so does not readily yield to general techniques such as that of the regularity lemma used in [1] and [3].

Specifically, the information given in advance is an underlying undirected graph $G = (V, E)$ (that may have parallel edges). The input is then constrained to be an orientation of G, and the distances are measured relative to $|E|$ and not to any function of $|V|$. An orientation of G is simply an orientation of its edges. That is, for every edge $e = \{u, v\}$ in $E(G)$ an orientation of G specifies which of u and v is the source vertex of e, and which is the target vertex. Thus an orientation defines a directed graph $\vec{G}$ whose undirected skeleton is G. Given the undirected graph G, a property of orientations is just a partial set of all orientations of G.

We study orientation properties in the framework of *property testing*. The meta problem in the area of property testing is the following: Given a combinatorial structure S, distinguish between the case that S satisfies some property $\mathcal{P}$ and the case that S is ϵ-*far* from satisfying $\mathcal{P}$. Roughly speaking, a combinatorial

---

[*] Research supported in part by an Israel Science Foundation grant number 55/03.

M. Charikar et al. (Eds.): APPROX and RANDOM 2007, LNCS 4627, pp. 380–394, 2007.

structure is said to be ϵ-far from satisfying some property $\mathcal{P}$ if an ϵ-fraction of its representation has to be modified in order to make S satisfy $\mathcal{P}$. The main goal in property testing is to design randomized algorithms, which look at a very small portion of the input, and use this information to distinguish with high probability between the above two cases. Such algorithms are called *property testers* or simply *testers* for the property $\mathcal{P}$. Preferably, a tester should look at a portion of the input whose size is a function of ϵ only and is independent of the input size. A property that admits such a tester is called *testable*.

Blum, Luby and Rubinfeld [4] were the first to formulate a question of this type, and the general notion of property testing was first formulated by Rubinfeld and Sudan [14], who were interested in studying various algebraic properties such as the linearity of functions. The definitions and the first study of property testing as a framework were introduced in the seminal paper of Goldreich, Goldwasser and Ron [6]. Since then, an extensive amount of work has been done on various aspects of property testing as well as on studying particular properties. For comprehensive surveys see [13,5].

Here the relevant combinatorial structure is an orientation $\boldsymbol{G}$ of the underlying graph G, and the distance between two orientations $\boldsymbol{G}_1$, $\boldsymbol{G}_2$ is the number of edges that are oriented differently in $\boldsymbol{G}_1$ and $\boldsymbol{G}_2$. Thus an orientation $\boldsymbol{G}$ is ϵ-far from a given property P if at least $\epsilon|E(G)|$ edges have to be redirected in $\boldsymbol{G}$ to make it satisfy P. Ideally the number of queries that the tester makes depends only on ϵ and on nothing else (in particular it depends neither on $|E|$ nor the specific undirected graph G itself).

A major question that has remained open in [9] is whether connectivity properties admit such a test. For a fixed $s, t \in V(G)$, an orientation $\boldsymbol{G}$ is st-connected if there is a directed path from s to t in it. Connectivity and in particular st-connectivity is a very basic problem in graph theory which has been extensively studied in various models of computation.

Our main result is that the property of being st-connected is testable by a one-sided error algorithm with a number of queries depending only on ϵ. That is, we construct a randomized algorithm such that for any underlying graph G, on input of an unknown orientation the algorithm queries only $O(1)$ edges for their orientation and based on this decides with success probability $\frac{2}{3}$ (this of course could be amplified to any number smaller than 1) between the case that the orientation is st-connected and the case that it is ϵ-far from being st-connected. Our algorithm additionally has one-sided error, meaning that st-connected orientations are accepted with probability 1. Note that the algorithm knows the underlying graph G in advance and G is neither alterable nor part of the input to be queried. The dependence of the number of queries in our test on ϵ is triply exponential, but it is independent of the size of the graph.

To put our result in context with previous works in the area of property testing, we note that graph properties were extensively studied since the early beginning in the defining paper of Goldreich, Goldwasser and Ron [6]. The model that was mainly studied is the *dense graphs* model in which an input is a graph represented as a subgraph of the complete graph. As such, for n-vertex graphs, the input representation size is $\binom{n}{2}$ which is the number of all possible unordered pairs. Thus, any property that has $o(n^2)$ witness size, and in particular the property of undirected st-connectivity, is trivially testable as every input is close to the property. Similarly, properties of directed graphs were studied in the same context mostly by [1,3]. Inputs in this model are subgraphs of the complete directed graph (with or without anti parallel edges). In this case, directed st-connectivity is again trivial.

Other models in which graph properties were studied are the *bounded-degree* graph model of [7] in which a sparse representation of sparse graphs is considered (instead of the adjacency matrix as in the dense model), and the *general density* model (also called the *mixed* model) of [12] and [10]. In those models edges can be added as well, and so st-connectivity (directed or not) is again trivial as the single edge (s, t) can always be added and thus every graph is close to having the property. Testing general (all-pairs) connectivity is somewhat harder, and for this constant query testers are generally known, e.g. the one in [7] for the undirected bounded degree case.

Apart from [9], the most related work is that of [8] in which a graph $G = (V, E)$ is given and the properties are properties of boolean functions $f : E(G) \rightarrow \{0, 1\}$. In [8] the interpretation of such a function is as an assignment to certain formulae that are associated with the underlying graph G, and in particular can viewed as properties of orientations (although the results in [8] concentrate on properties that are somewhat more "local" than our "global" property of being st-connected). Hence, the results here should be viewed as moving along the lines of the investigation that was started in [9] and [8]. A common feature of the current work with this previous one, which distinguishes these results from results in many other areas of property testing and in particular the dense graph models, is that the algorithms themselves are rather non-trivial in construction, and not just in their analysis.

The algorithm that we present here for st-connectivity involves a preprocessing stage that is meant to reduce the problem to that of testing a branching program of bounded width. Once this is achieved, a randomized algorithm simulating the test for the constant width branching program from [11] is executed to conclude the result.

In general, the decision problem of st-connectivity of orientations of a given graph is not known to be reducible to constant width branching programs. In fact, it is most probably not the case as st-connectivity is complete for NL (non-deterministic LOG space) while deciding constant width branching programs is in L. In particular, it is not clear how to deal with high degree vertices or with large cuts. The purpose of the preprocessing stage is to get rid of these difficulties (here it will be crucial that we only want to distinguish between inputs that have the property and inputs that are quite far from having the property). This is done in several steps that constitute the main part of the paper. In particular we have an interim result in which most but not all edges of the graph are partitioned into constant width layers. This is proved using a weak concentration lemma for sequences of integers, which is formulated and proved for this purpose.

After the small portion of edges not in constant width layers is dealt with (using a reduction based on graph contraction), we can reduce the connectivity problem to a constant width read once branching program. Once such a branching program is obtained, the result of [11] can be used essentially in a black box manner.

Some interesting related open problems still remain. We still do not know if the property of being *strongly* st-connected is testable with a constant number of queries. The orientation G is *strongly* st-connected if there is a directed path in G from s to t as well as a directed path from t to s. A more general problem is whether in this model we can test the property of being *all-pairs strongly-connected* using a constant number of queries. Another related property is the property that for a given $s \in V(G)$ every vertex is reachable by a directed path from s. The complexity of these problems is unknown, although there are some indications that similar methods as those used here may help in this regard.

The rest of this paper is organized as follows. In Section 2 we introduce the needed notations and definitions. Section 3 contains the statement of the main result and an overview of the proof. In Section 4 we reduce the problem of testing st-connectivity in general graphs to the problem of testing st-connectivity in nicely structured bounded-width graphs (we later call them *st-connectivity programs*). Due to space considerations, most of the proofs from Section 4 are omitted. In Section 5 we reduce from testing st-connectivity programs to testing *clustered* branching programs. Converting these clustered branching programs into regular ones, to which we can apply the testing algorithm from [11], is straightforward. Finally in Section 6 we combine these ingredients to wrap up the proof.

2 Preliminaries

2.1 Notations

In what follows, our graphs are going to possibly include parallel edges, so we use 'named' pairs for edges, i.e. we use $e = e\{u, v\}$ for an undirected edge named e whose end points are u and v. Similarly, we use $e = e(u, v)$ to denote the directed edge named e that is directed from u to v. Let $G = (V, E)$ be an undirected multigraph (parallel edges are allowed), and denote by n the number of vertices in V. We say that a directed graph G is an *orientation* of the graph G, or in short a G-orientation, if we can derive G from G by replacing every undirected edge $e = e\{u, v\} \in E$ with either $e(u, v)$ or $e(v, u)$, but not both. We also call G the *underlying graph* of G.

Given an undirected graph G and a subset $W \subset V$ of G's vertices, we denote by $G(W)$ the induced subgraph of G on W, and we denote by $E(W) = E(G(W))$ the edge set of $G(W)$. The distance between two vertices u, v in G is denoted by $\mathrm{dist}_G(u, v)$ and is set to be the length of the shortest path between u and v. Similarly, for a directed graph G, $\mathrm{dist}_G(u, v)$ denotes the length of the shortest *directed* path from u to v. The distance of a vertex from itself is $\mathrm{dist}_G(v, v) = \mathrm{dist}_G(v, v) = 0$. In the case where there is no directed path from u to v in G, we set $\text{dist}_G(u,v) = \infty$. The diameter of an undirected graph G is defined as $\text{diam}(G) = \max_{u,v \in V}\{\text{dist}_G(u,v)\}$. Through this paper we always assume that the underlying graph G is connected, and therefore its diameter is finite.

For a graph G and a vertex $v \in V$, let $\Gamma_{in}(v) = \{u : \exists e(u,v) \in E\}$ and $\Gamma_{out}(v) = \{u : \exists e(v,u) \in E\}$ be the set of incoming and outgoing neighbors of v respectively, and let $\Gamma(v) = \Gamma_{in}(v) \cup \Gamma_{out}(v)$ be the set of neighbors in the underlying graph G. Let $\deg_{in}(v)$, $\deg_{out}(v)$ and $\deg(v)$ denote the sizes of $\Gamma_{in}(v)$, $\Gamma_{out}(v)$ and $\Gamma(v)$ respectively. We denote the i-neighborhood (in the underlying undirected graph G) of a vertex v by $N_i(v) = \{u : \text{dist}_G(u,v) \le i\}$. For example, $N_1(v) = \{v\} \cup \Gamma(v)$, and for all $v \in V$, $V = N_{\text{diam}(G)}(v)$.

2.2 Orientation Distance, Properties and Testers

Given two G-orientations G_1 and G_2, the distance between G_1 and G_2, denoted by $\Delta(G_1, G_2)$, is the number of edges in $E(G)$ having different directions in G_1 and G_2.

Given a graph G, a property $\mathcal{P}_G$ of G's orientations is a subset of all possible G-orientations. We say that an orientation G satisfies the property $\mathcal{P}_G$ if $G \in \mathcal{P}_G$. The distance of G_1 from the property $\mathcal{P}_G$ is defined by $\delta(G_1, \mathcal{P}_G) = \min_{G_2 \in \mathcal{P}_G} \frac{\Delta(G_1, G_2)}{|E(G)|}$. We say that G is ϵ-far from $\mathcal{P}_G$ if $\delta(G, \mathcal{P}_G) \ge \epsilon$, and otherwise we say that G is ϵ-close to $\mathcal{P}_G$. We omit the subscript G when it is obvious from the context.

Definition 1 ((ϵ, q)-tester). *Let G be a fixed undirected graph and let P be a property of G's orientations. An (ϵ, q)-tester T for the property P is a randomized algorithm, that for any G that is given via oracle access to the orientations of its edges operates as follows.*

- *The algorithm T makes at most q orientation queries to G (where on a query $e \in E(G)$ it receives as the answer the orientation of e in G).*
- *If $G \in P$, then T accepts it with probability 1 (here we define only one-sided error testers, since our testing algorithm will be one).*
- *If G is ϵ-far from P, then T rejects it with probability at least $2/3$.*

The query complexity of an (ϵ, q)-tester T is the maximal number of queries q that T makes on any input. We say that a property P is testable if for every $\epsilon > 0$ it has an $(\epsilon, q(\epsilon))$-test, where $q(\epsilon)$ is a function depending only on ϵ (and independent of the graph size n).

2.3 Connectivity Programs and Branching Programs

Our first sequence of reductions converts general st-connectivity instances to well structured bounded-width st-connectivity instances, as formalized in the next definition.

Definition 2 (st-**Connectivity Program**). *An st-Connectivity Program of width w and length m over n vertices (or $CP(w, m, n)$ in short), is a tuple*

$\langle G, \mathcal{L} \rangle$, where G is an undirected graph with n vertices and $\mathcal{L}$ is a partition of G's vertices into layers $L_0, \ldots, L_m$. There are two special vertices in G: $s \in L_0$ and $t \in L_m$, and the edges of G are going only between vertices in consecutive layers, or between the vertices of the same layer, i.e. for each $e = e\{u, v\} \in E(G)$ there exists $i \in [m]$ such that $u \in L_{i-1}$ and $v \in L_i$, or $u, v \in L_i$. The partition $\mathcal{L}$ induces a partition $E_1, \ldots, E_m$ of $E(G)$, where E_i is the set of edges that have both vertices in L_i, or one vertex in L_{i-1} and another in L_i. In this partition of the edges the following holds: $\max_i\{|E_i|\} \leq w$.

Any orientation of G's edges (that maps every edge $e\{u, v\} \in E(G)$ to either $e(u, v)$ or $e(v, u)$) defines a directed graph G in the natural way. An st-connectivity program $C = \langle G, \mathcal{L}, \rangle$ defines a property P_C of G's orientations in the following way: $G \in P_C$ if and only if in the directed graph G there is a directed path from s to t.

Next we define branching programs. These are the objects to which we can apply the testing algorithm of [11].

Definition 3 (Branching Program). A Read Once Branching Program of width w over an input of n bits (or $BP(w, n)$ in short), is a tuple $\langle G, \mathcal{L}, X \rangle$, where G is a directed graph with $0/1$-labeled edges, $\mathcal{L}$ is a partition of G's vertices into layers $L_0, \ldots, L_n$ such that $\max_i\{|L_i|\} \leq w$, and $X = \{x_0, \ldots, x_{n-1}\}$ is a set of n Boolean variables. In the graph G there is one special vertex s belonging to L_0, and a subset $T \subset L_n$ of accepting vertices. The edges of G are going only between vertices in consecutive layers, i.e. for each $e = e(u, v) \in E(G)$ there is $i \in [n]$ such that $u \in L_{i-1}$ and $v \in L_i$. Each vertex in G has at most two outgoing edges, one of which is labeled by '0' and the other is labeled by '1'. In addition, all edges between two consecutive layers are associated with one distinct member of $X = \{x_0, \ldots, x_{n-1}\}$. An assignment $\sigma : X \to \{0, 1\}$ to X defines a subgraph G_σ of G, which has the same vertex set as G, and for every layer L_{i-1} (whose outgoing edges are associated with the variable x_{j_i}), the subgraph G_σ has only the outgoing edges labeled by $\sigma(x_{j_i})$. A read once branching program $B = \langle G, \mathcal{L}, X \rangle$ defines a property $P_B \subset \{0, 1\}^n$ in the following way: $\sigma \in P_B$ if and only if in the subgraph G_σ there is a directed path from the starting vertex s to any of the accepting vertices in T.

Branching programs that comply with the above definition can be tested by the algorithm of [11]. However, as we will see in Section 5, the branching programs resulting from the reduction from our st-connectivity programs have a feature that they require reading more than one bit at a time to move between layers. The next definition describes these special branching programs formally.

Definition 4 (Clustered Branching Program). A c-clustered Read Once Branching Program of width w and length m over an input of n bits (or shortly $BP_c(w, m, n)$) is a tuple $\langle G, \mathcal{L}, X, \mathcal{I} \rangle$, where similarly to the previous definition, G is a directed graph with labeled edges (see below for the set of labels), $\mathcal{L} = (L_0, \ldots, L_m)$ is a partition of G's vertices into m layers such that $\max_i\{|L_i|\} \leq w$, and $X = \{x_0, \ldots, x_{n-1}\}$ is a set of n Boolean variables. Here too, G has one

*special vertex s belonging to L_0, and a subset $T \subset L_n$ of accepting vertices. The
additional element $\mathcal{I}$ is a partition $(I_1, \ldots, I_m)$ of X into m components, such
that $\max_i \{|I_i|\} \leq c$.*

*All edges in between two consecutive layers L_{i-1} and L_i are associated with
the component I_i of $\mathcal{I}$. Each vertex in L_{i-1} has $2^{|I_i|}$ outgoing edges, each of them
labeled by a distinct $\alpha \in \{0,1\}^{|I_i|}$.*

*An assignment $\sigma : X \to \{0,1\}$ to X defines a subgraph G_σ of G, which has
the same vertex set as G, and for every layer L_i (whose outgoing edges are
associated with the component I_i), the subgraph G_σ has only the edges labeled by
$\left(\sigma(x_i) \right)_{i \in I_i}$. A c-clustered read once branching program $B = \langle G, \mathcal{L}, X, \mathcal{I} \rangle$ defines
a property $P_B \subset \{0,1\}^n$ in the following way: $\sigma \in P_B$ if and only if in the
subgraph G_σ there is a directed path from the starting vertex s to one of the
accepting vertices in T.*

Observe that $BP(w,m)$ is equivalent to $BP_1(w,m,m)$.

3   The Main Result

For an undirected graph G and a pair $s, t \in V(G)$ of distinct vertices, let P_G^{st} be
a set of G-orientations under which there is a directed path from s to t. Formally,
$P_G^{st} = \{G : \text{dist}_G(s,t) < \infty\}$.

Theorem 1. *The property P_G^{st} is testable. In particular, for any undirected
graph G, two vertices $s, t \in V(G)$ and every $\epsilon > 0$, there is an (ϵ, q)-tester
T for P_G^{st} with query complexity $q = (2/\epsilon)^{2^{(1/\epsilon) \cdot 2^{O(\epsilon^{-2})}}}$.*

Note that the property P_G^{st} is trivially testable whenever the undirected distance
from s to t in G is less than $\epsilon|E(G)|$. In particular, our result is interesting only
for sparse graphs, i.e. graphs for which $|E(G)| \leq |V(G)|/\epsilon$.

We can slightly improve the query complexity to $q = (2/\epsilon)^{2^{O((1/\epsilon)^{(1/\epsilon)})}}$ by
proving a stronger version of the concentration argument (Lemma 4), but we
omit this proof from this extended abstract.

3.1   Proof Overview

The main idea of the proof is to reduce the problem of testing st-connectivity
in the orientation model to the problem of testing a Boolean function that is
represented by a *small width read once branching program*. For the latter we
have the result of [11] asserting that each such Boolean function is testable.

Theorem 2 ([11]). *Let $P \subseteq \{0,1\}^n$ be the language accepted by a read-once
branching program of width w. Then testing P with one-sided error requires at
most $\left(\frac{2^w}{\epsilon} \right)^{O(w)}$ queries.*

By the definition above of $BP(w, n)$, one could already notice that testing the acceptance of a branching program resembles testing st-connectivity, and that the two problems seem quite close. However, there are several significant differences, as noted here.

1. In branching programs, every layer is associated with a variable, querying which reveals all the edges going out from this layer. In st-connectivity instances, in order to discover the orientation of these edges we need to query each of them separately.
2. The length of the input in branching programs is the number of layers rather than the total number of edges.
3. The edges in branching program graphs are always directed from L_{i-1} to L_i for some $i \in [n]$. In our case, the graph is not layered, and a pair u, v of vertices might have any number of edges in both directions.
4. In branching programs the graphs have out-degree exactly 2, while an input graph of the st-connectivity problem might have vertices with unbounded out-degree.
5. The most significant difference is that the input graphs of the st-connectivity problem may have unbounded width. This means that the naive reduction to branching programs may result in an unbounded width BPs, which we cannot test with a constant number of queries.

We resolve these points in several steps. First, given an input graph G, we reduce it to a graph $G^{(1)}$ which has the following property: for every induced subgraph W of $G^{(1)}$, the diameter of W is larger than ϵ times the number of edges in W. Then we prove that $G^{(1)}$ can be layered such that most of its edges lie within bounded-width layers. In particular, the total number of edges in the "wide" layers is bounded by $\frac{\epsilon}{2}E(G^{(1)})$. For this we need a concentration type lemma which is stated and proven here for this purpose. Next we reduce $G^{(1)}$ to a graph $G^{(2)}$, which can be layered as above, but without wide layers at all. This in particular means that the number of edges in $G^{(2)}$ is of the order of the number of layers, similarly to bounded width branching programs. In addition, in this layering of $G^{(2)}$ the vertex t is in the last layer (while the vertex s remains in the first layer). Then we reduce $G^{(2)}$ (which might be thought of as the *undirected* analogue of a bounded width branching program) to a clustered read once bounded width branching program. Finally we show that these clustered branching programs can be converted into non-clustered branching programs, to which we can apply the test from [11].

4   Reducing General Graphs to Connectivity Programs

In this section we make our first step towards proving Theorem 1. We reduce the problem of testing st-connectivity in a general graph to the problem of testing st-connectivity of an st-Connectivity Program. First we define the notion of reducibility in our context, and then we describe a sequence of reductions that will eventually lead us to the problem of testing read once bounded width BPs. Due to space considerations, some of the proofs from this section are omitted.

4.1 Reducibility Between st-Connectivity Instances

Let $\mathcal{G}^{st}$ denote the class of undirected graphs having two distinct vertices s and t, and let $G, G' \in \mathcal{G}^{st}$. We say that G is (ϵ, η)-*reducible* to G' if there is a function ρ that maps orientations of G to orientations of G' (from now on we denote by $\boldsymbol{G'}$ the orientation $\rho(\boldsymbol{G})$) such that the following holds.

- If $\boldsymbol{G} \in P_G^{st}$ then $\boldsymbol{G'} \in P_{G'}^{st}$
- If $\delta(\boldsymbol{G}, P_G^{st}) \geq \epsilon$ then $\delta(\boldsymbol{G'}, P_{G'}^{st}) \geq \eta$
- Any orientation query to $\boldsymbol{G'}$ can be simulated by a single orientation query to $\boldsymbol{G}$.

We say that G is (ϵ)-*reducible* to G' if it is (ϵ, ϵ)-reducible to G'. Notice that whenever G is (ϵ, η)-reducible to G', any (η, q)-tester T' for $P_{G'}^{st}$ can be converted into an (ϵ, q)-tester T for P_G^{st}. Or in other words, (ϵ, q)-testing P_G^{st} is reducible to (η, q)-testing $P_{G'}^{st}$.

In the following section we introduce our first reduction, which is referred to as the reduction from G to $G^{(1)}$ in the proof overview.

4.2 Reduction to Graphs Having High-Diameter Subgraphs

An undirected graph G is called ϵ-*long* if for every subset of vertices $W \subset V(G)$, $\mathrm{diam}(G(W)) > \epsilon |E(W)|$.

Lemma 1. *Any graph $G \in \mathcal{G}^{st}$ is (ϵ)-reducible to a graph $G' \in \mathcal{G}^{st}$ which is ϵ-long.*

We first define a general contraction operator for graphs, and then use it for the proof. Given a graph G let $W \subset V$ be a subset of its vertices, and let $C_1, \ldots, C_r$ be the vertex sets of the connected components of $G(W)$. Namely, for each $C_i \subset W$, the induced subgraph $G(C_i)$ of the underlying graph G is connected, and for any pair $u \in C_i, v \in C_j$ $(i \neq j)$ of vertices, $e\{u, v\} \notin E(G)$. We define a graph G/W as follows. The graph G/W has the vertex set $V/W = (V \backslash W) \cup \{c_1, \ldots, c_r\}$ and its edges are

$$E/W = \Big\{ e\{u, v\} : (u, v \in V \setminus W) \wedge (e \in E) \Big\} \cup$$

$$\Big\{ e\{c_i, v\} : (v \in V \setminus W) \wedge \exists_{u \in C_i} (e\{u, v\} \in E) \Big\}$$

Intuitively, in G/W we contract every connected component C_i of $G(W)$ into a single (new) vertex c_i without changing the rest of the graph. Note that such a contraction might create parallel edges, but loops are removed. Whenever $s \in C_i$ for some $i \in [r]$, we rename the vertex c_i by s, and similarly if $t \in C_j$ then we rename c_j by t. In the following a connected component containing both s and t will not be contracted, so we can assume that the distinguished vertices s and t remain in the new graph G/W. Given an orientation $\boldsymbol{G}$ of G we define the orientation $\boldsymbol{G/W} = \rho(\boldsymbol{G})$ of G/W as the orientation induced from $\boldsymbol{G}$ in the natural way (note that there are no "new" edges in G/W).

Lemma 2. *Let $W \subset V$ be a subset of G's vertices, such that $\mathrm{diam}(G(W)) \leq \epsilon|E(W)|$. Then G is (ϵ)-reducible to the graph G/W.*

Proof. Fix an orientation $\boldsymbol{G}$ of G. It is clear that if $\boldsymbol{G} \in P_G^{st}$ then $\boldsymbol{G}/W \in P_{G/W}^{st}$. Now assume that $\delta(\boldsymbol{G}, P_G^{st}) \geq \epsilon$. Let d and d' denote $\delta(\boldsymbol{G}, P_G^{st}) \cdot |E(G)|$ and $\delta(\boldsymbol{G}/W, P_{G/W}^{st}) \cdot |E(G/W)|$ respectively. From the definition of the graph G/W it follows that $d' \geq d - \mathrm{diam}(G(W))$. This is true since any st-path in $\boldsymbol{G}/W$ can be extended to an st-path in $\boldsymbol{G}$ by reorienting at most $\mathrm{diam}(G(W))$ edges in W (by definition, $\mathrm{diam}(G(W))$ is an upper bound on the undirected distance from any "entry" vertex to any "exit" vertex in $G(W)$). From the condition on W we have $|E(\boldsymbol{G}/W)| = |E| - |E(W)| \leq |E| - \frac{\mathrm{diam}(G(W))}{\epsilon}$. Combining these two together we have

$$\delta(\boldsymbol{G}/W, P_{G/W}^{st}) = \frac{d'}{|E(\boldsymbol{G}/W)|} \geq \frac{d - \mathrm{diam}(G(W))}{|E| - \mathrm{diam}(G(W))/\epsilon} \geq$$
$$\frac{d - \mathrm{diam}(G(W))}{d/\epsilon - \mathrm{diam}(G(W))/\epsilon} = \epsilon$$

In addition, it is clear that we can simulate each query to $\boldsymbol{G}/W$ by making at most one query to $\boldsymbol{G}$. $\square$

Now the proof of Lemma 1 follows by applying this contraction (iteratively) for each "bad" subgraph W, until eventually we get a graph G' in which all vertex subsets W satisfy $\mathrm{diam}(G(W)) > \frac{|E(W)|}{\epsilon}$. If in some stage we have both s and t contained in the contracted subgraph W, then we just output a tester that accepts all inputs (since in this case all orientations are ϵ-close to being st-connected). Note that this process may result in several different graphs (depending on choices of the set W in each iteration), but we are only interested in one (arbitrary) graph G'. $\square$

4.3   Properties of ϵ-Long Graphs

Next we show that an ϵ-long graph G can be "layered" so that the total number of edges in the "wide" layers is bounded by $\frac{\epsilon}{2}E(G)$. We first define graph layering.

Definition 5 (Graph layering and width). *Given a graph G and a vertex $s \in V(G)$, let m denote the maximal (undirected) distance from s to any other vertex in G. We define a layering $\mathcal{L} = (L_0, L_1, \ldots, L_m)$ of G's vertices as follows. $L_0 = \{s\}$ and for every $i > 0$, $L_i = N_i(s) \setminus N_{i-1}(s)$. Namely L_i is the set of vertices which are at (undirected) distance exactly i from s. Note that for every edge $e\{u, v\}$ of G either both u and v are in the same layer, or $u \in L_{i-1}$ and $v \in L_i$ for some $i \in [m]$.*

We also denote by $E_i^{\mathcal{L}}$ the subset of G's edges that either have one vertex in L_i and the other in L_{i-1}, or edges that have both vertices in L_i. Alternatively, $E_i^{\mathcal{L}} = E(L_i \cup L_{i-1}) \setminus E(L_{i-1})$. We refer to the sets L_i and $E_i^{\mathcal{L}}$ as vertex-layers and edge-layers respectively. We might omit the superscript $\mathcal{L}$ from the edge-layers notation when it is clear from the context.

The vertex-width of a layering $\mathcal{L}$ is $\max_i\{|L_i|\}$, and the edge-width of $\mathcal{L}$ is $\max_i\{|E_i|\}$.

Bounding the Number of Edges Within Wide Edge-Layers. The following lemma states that in a layering of an ϵ-long graph most of the edges are captured in edge-layers of bounded width.

Lemma 3. *Consider the layering $\mathcal{L}$ of an ϵ-long graph G as defined above, and let $I = \{i : |E_i| > 2^{100/\epsilon^2}/\epsilon\}$ be the set of indices of the wide edge-layers. Then the following holds: $\sum_{i \in I} |E_i| \leq \frac{\epsilon}{2}|E|$.*

Proof. Denote by $\mathcal{A} = \langle a_0, a_1, \ldots, a_m \rangle$ a sequence of integers, where $a_0 = 1$ and for every $i \geq 1$, $a_i = |E_i|$.

Definition 6 (ϵ-good sequence). *Let $0 < \epsilon < 1$ be a positive constant. A sequence $1 = a_0, a_1, \ldots, a_m$ of positive natural numbers is ϵ-good if for every $0 \leq k < m$ and $\ell \in [m - k]$ we have that*

$$\sum_{i=k+1}^{k+\ell} a_i \leq \frac{\ell \cdot a_k}{\epsilon}.$$

Claim 1. *Let G be a graph in which for any induced subgraph W we have $|E(W)| < \mathrm{diam}(W)/\epsilon$. Then the sequence $\mathcal{A} = \langle a_0, a_1, \ldots, a_m \rangle$ defined above is $\epsilon/4$-good.*

Proof (Proof of Claim 1). Let us assume the contrary of the claim. Let k and ℓ be such that

$$\sum_{i=k+1}^{k+\ell} a_i > \frac{4\ell \cdot a_k}{\epsilon}$$

Consider the subgraph W defined by the vertices $\cup_{i=k}^{k+\ell} L_i$ and the edges $\cup_{i=k+1}^{k+\ell} E_i$. Now the number of edges in W is clearly $\sum_{i=k+1}^{k+\ell} a_i$. For each vertex v in L_k consider the neighborhood of distance $\ell + 1$ from v, $N_{\ell+1}(v)$, and denote the subgraph it spans by W_v. Notice that each W_v is of diameter at most $2(\ell+1) \leq 4\ell$, and that the union of the edge sets of the W_v includes the whole edge set of W, so we have $\sum_{v \in L_k} |E(W_v)| \geq \sum_{i=k+1}^{k+\ell} a_i$. The number of vertices in L_k is at most a_k, so by the pigeonhole principle we know that at least one of the vertices v in L_k has at least $\frac{1}{a_k} \sum_{i=k+1}^{k+\ell} a_i$ edges in W_v. By our assumption on $\sum_{i=k+1}^{k+\ell} a_i$ we have $|E(W_v)| > \frac{4\ell}{\epsilon} \geq \mathrm{diam}(W_v)/\epsilon$, a contradiction. $\square$

Proving Lemma 3 requires the following concentration type lemma, proof of which is omitted due to lack of space.

Lemma 4 (The weak concentration lemma). *Let $\langle a_0, \ldots, a_m \rangle$ be an ϵ-good sequence and let $B = \{i \mid a_i > 2^{5/\epsilon^2}/\epsilon\}$. Then $\sum_{i \in B} a_i \leq \epsilon \sum_{i=1}^{m} a_i$.* $\square$

Now the proof of Lemma 3 follows directly from Claim 1 and Lemma 4. $\square$

4.4   Reduction to Bounded Width Graphs

In this section we prove that ϵ-long graphs can be reduced to graphs that have bounded width. In terms of the proof overview, we are going to reduce the graph $G^{(1)}$ to the graph $G^{(2)}$.

Let $G = (V, E) \in \mathcal{G}^{st}$, and let $\mathcal{L} = (L_0, L_1, \ldots, L_m)$ be the layering of G as above. We call an edge-layer E_i wide if $|E_i| > \frac{1}{\epsilon} \cdot 2^{100/\epsilon^2}$. Let $\mathcal{W}$ be the set of all wide edge-layers.

Lemma 5. *If $G \in \mathcal{G}^{st}$ satisfies $\sum_{E_i \in \mathcal{W}} |E_i| \leq \frac{\epsilon}{2}|E|$ then G is $(\epsilon, \epsilon/2)$-reducible to a graph G' which has no wide edge-layers at all.*

The proof of Lemma 5 appears in omitted.

4.5   Reducing Bounded Width Graphs to st-Connectivity Programs

So far we reduced the original graph G to a graph G' which has a layering $\mathcal{L} = (L_0, L_1, \ldots, L_m)$ of edge-width at most $w = \frac{1}{\epsilon} \cdot 2^{100/\epsilon^2}$, and in which the source vertex s belongs to layer L_0. The remaining difference between G' and an st-connectivity program is that in G' the target vertex t might not be in the last vertex-layer L_m. The following lemma states that we can overcome this difference by another reduction.

Lemma 6. *Let G' be a graph as described above. Then G' is $(\epsilon, \epsilon/2)$-reducible to an st-connectivity program S of width at most $w + 1$.* □

5   Reducing st-Connectivity Programs to Branching Programs

We now show how to reduce an st-connectivity program to a clustered branching program (recall Definition 2 and Definition 4). First observe that we can assume without loss of generality that if an st-connectivity program has edge-width w, then its vertex-width is at most $2w$ (since removing vertices of degree 0 essentially does not affect the st-connectivity program, and a vertex in L_i with edges only between it and L_{i+1} can be safely moved to L_{i+1}).

Before moving to the formal description of the reduction, we start with a short intuitive overview. A branching program corresponds to a (space bounded) computation that moves from the start vertex s, which represents no information about the input at all, and proceeds (via the edges that are consistent with the input bits) along a path to one of the final vertices. Every vertex of the branching program represents some additional information gathered by reading more and more pieces of the input. Thus, the best way to understand the reduction is to understand the implicit meaning of each vertex in the constructed branching program.

Given a graph G of a bounded width st-connectivity program, and its layering $L_0, L_1, \ldots L_m$, we construct a graph G' (with layering $L'_0, L'_1, \ldots L'_m$) for the bounded-width branching program. The graph G' has the same number of layers as G. Each level L_i' in G' will represent the conditional connectivity of the vertices in the subgraph $G_i = G(\bigcup_{j=0}^{i} L_i)$ of G. To be specific, the knowledge we want to store in a typical vertex at layer L_i' of G' is the following.

- for every $u \in L_i$ whether it is reachable from s in G_i.
- for every $v, u \in L_i$ whether v is reachable from u in G_i.

Hence, the amount of information we store in each node $x \in L_i'$ has at most $2w + (2w)^2$ many bits, and so there will be at most 4^{2w^2+w} vertices in each L_i', meaning that the graph G' of the branching program is of bounded width as well.

Lemma 7. *Let $\epsilon > 0$ be a positive constant. Given a $CP(w, m, n)$ instance $C = \langle G, \mathcal{L} \rangle$, we can construct a $BP_w(4^{2w^2+w}, m, n)$ instance $B = \langle G', \mathcal{L}', X', \mathcal{I}' \rangle$ and a mapping ρ from G-orientations to assignments on X such that the following holds,*

- *if G satisfies P_C then $\sigma = \rho(G)$ satisfies P_B.*
- *if G is ϵ-far from satisfying P_C then $\sigma = \rho(G)$ is ϵ-far from satisfying P_B.*
- *any assignment query to σ can be simulated using a single orientation query to G.*

Proof. First we describe the construction, and then show that it satisfies the requirements above.

The vertices of G': We fix i and show, based on the layer L_i of G, how to construct the corresponding layer L_i' of G'. Each vertex in L_i' corresponds to a possible value of a pair (S_i, R_i) of sets. The first set $S_i \subseteq L_i$ contains vertices $v \in L_i$ for which there is a directed path from s to v in the subgraph of G induced on $\bigcup_{j=0}^{i} L_j$. The second set $R_i \subseteq L_i \times L_i$ is a set of ordered pairs of vertices, such that every ordered pair (u, v) is in R_i if there is a directed path from u to v in the subgraph of G induced on $\bigcup_{j=0}^{i} L_j$ (the path can be of length 0, meaning that the R_i's contain all pairs (v, v), $v \in L_i$). Notice that $|L_i'| = 2^{|L_i|^2+|L_i|} \leq 4^{2w^2+w}$ for all i.

The edges of G': Now we construct the edges of G'. Recall that E_{i+1}' denotes the set of (labeled) edges having one vertex in L_{i+1}' and the other in L_i'. Fix i and a vertex $v \in L_i'$. Let (S, R) be the pair of sets that correspond to v. Let S'' be the set $L_{i+1} \cap \Gamma_{out}(S)$, namely the neighbors of vertices from S that are in L_{i+1}, and set $R'' = \{(v, v) : v \in L_{i+1}\} \cup \{(u, v) : (u, v \in L_{i+1}) \wedge (v \in \Gamma_{out}(u))\} \cup \{(u, v) : (u, v \in L_{i+1}) \wedge ((\Gamma_{out}(u) \times \Gamma_{in}(v)) \cap R \neq \emptyset)\}$. Now define R' as the transitive closure of R'', and set $S' = S'' \cup \{v \in L_{i+1} : \exists_{u \in S''}(u, v) \in R'\}$. Let $v' \in L_{i+1}'$ be the vertex corresponding to the pair (S', R') that we defined above. Then the edges of E_{i+1}' are given by all such pairs of vertices (v, v').

The variables in X': Each variable $x_i' \in X'$ is associated with an edge $e_i \in E(G)$. This association is actually the mapping ρ above, i.e. every orientation $\mathbf{G}$ of G defines an assignment σ on X'.

The partition $\mathcal{I}'$ of X': Recall that E_i denotes the set of edges of G having either one vertex in L_{i-1} and the other in L_i, or both vertices in L_i. The partition $\mathcal{I}'$ of X' is induced by the partition $\mathcal{L}$ of $V(G)$. Namely, the component I_i of $\mathcal{I}$ contains the set of variables in X' that are associated with edges in E_i. Thus w is also a bound on the sizes of the components in $\mathcal{I}$.

The set $T' \subset L_m'$ of accepting vertices: The set T' is simply the subset of vertices in L_m' whose corresponding set S contains the target vertex t of G.

Note that each value of a variable in X' corresponds exactly to an orientation of an edge in G. This immediately implies the third assertion in the statement of the lemma. Distances between inputs are clearly preserved, so to prove the other assertions it is enough to show that the branching program accepts exactly those assignments that correspond to orientations accepted by the connectivity program. It is straightforward (and left to the reader) to see that the vertex reached in each L_i' indeed fits the description of the sets R and S, and so an assignment is accepted if and only if it corresponds to a connecting orientation. □

The branching programs resulting from the above reduction have a feature that they require reading more than one bit at a time to move between layers. Specifically, they conform to Definition 4. The result in [11], however, deals with standard branching programs (see Definition 3), which in relation to the above are a special case in which essentially $m = n$ and all the I_i's have size 1. Going from here to a standard (non-clustered) branching program (in which the edges between two layers depend on just one Boolean variable) is easy.

6   Wrapping Up – Proof of Theorem 1

We started with a graph G and wanted to construct an (ϵ, q)-testing algorithm for st-connectivity of orientations of G. In Section 4.1, Section 4.2 and Section 4.3 we constructed a graph G_1 such that if we have an (ϵ, q)-test for st-connectivity in G_1, then we have an (ϵ, q)-test for G. Additionally G_1 has the property that most of the edge-layers in G_1 are of size at most $w = \frac{1}{\epsilon} \cdot 2^{100/\epsilon^2}$. Then in Section 4.4 we constructed a graph G_2 such that if we have an $(\frac{\epsilon}{2}, q)$-test for st-connectivity in G_2 then we have an (ϵ, q)-test for G_1 and hence we have one for G. Moreover G_2 has all its edge-layers of size at most w. Finally in Section 4.5 we built a graph G_3 which has all its edge-layers of width at most $w + 1$, and in addition, the vertices s and t are in the first and the last vertex-layers of G_3 respectively. We also showed that having an $(\frac{\epsilon}{4}, q)$-test for st-connectivity in G_3 implies an $(\frac{\epsilon}{2}, q)$-test for G_2, and hence an (ϵ, q)-test for G. This ends the first part of the proof, which reduces general graphs to st-connectivity programs.

Then in Section 5 from G_3 we constructed a read once $(w + 1)$-clustered Branching Program that has width $4^{2(w+1)^2 + w + 1}$ so that an $(\frac{\epsilon}{4}, q)$-test for this

BP gives an $(\frac{\epsilon}{4}, q)$-test for st-connectivity in G_3. Then we converted the $(w+1)$-clustered Branching Program to a non-clustered Branching Program which has width $w_1 = 4^{2(w+1)^2 + (w+1)} 2^{(w+1)}$. Once we have our read once bounded width branching program then by applying the algorithm of [11] for testing branching programs we get an $(\frac{\epsilon}{4}, q)$-test with $q = (\frac{2^{w_1}}{\epsilon/4})^{O(w_1)}$ queries for our problem. Hence by combining all of the above, we get an (ϵ, q) testing algorithm for our original st-connectivity problem, where $q = (2/\epsilon)^{2^{O((1/\epsilon) \cdot 2^{(100/\epsilon^2)})}}$.   □

References

1. Alon, N., Fischer, E., Newman, I., Shapira, A.: A Combinatorial Characterization of the Testable Graph Properties: It's All About Regularity. In: Proceedings of the 38th ACM STOC, pp. 251–260. ACM, New York (2006)
2. Alon, N., Kaufman, T., Krivilevich, M., Ron, D.: Testing triangle-freeness in general graphs. In: Proceedings of the 17th SODA, pp. 279–288 (2006)
3. Alon, N., Shapira, A.: A Characterization of the (natural) Graph Properties Testable with One-Sided Error. In: Proceedings of the 46th IEEE FOCS, pp. 429–438 (2005) Also SIAM Journal on Computing (to appear)
4. Blum, M., Luby, M., Rubinfeld, R.: Self-testing/correcting with applications to numerical problems. Journal of Computer and System Sciences 47, 549–595 (1993) (a preliminary version appeared in Proc. 22nd STOC, 1990).
5. Fischer, E.: The art of uninformed decisions: A primer to property testing. In: Paun, G., Rozenberg, G., Salomaa, A. (eds.) Current Trends in Theoretical Computer Science: The Challenge of the New Century, vol. I, pp. 229–264. World Scientific Publishing (2004)
6. Goldreich, O., Goldwasser, S., Ron, D.: Property testing and its connection to learning and approximation. JACM 45(4), 653–750 (1998)
7. Goldreich, O., Ron, D.: Property testing in bounded degree graphs. Algorithmica 32(2), 302–343 (2002)
8. Halevy, S., Lachish, O., Newman, I., Tsur, D.: Testing Properties of Constraint-Graphs. In: Halevy, S., Lachish, O., Newman, I., Tsur, D. (eds.) Proceedings of the 22nd IEEE Annual Conference on Computational Complexity (CCC 2007) (to appear)
9. Halevy, S., Lachish, O., Newman, I., Tsur, D.: Testing Orientation Properties. In: Electronic Colloquium on Computational Complexity (ECCC) (2005)
10. Kaufman, T., Krivelevich, M., Ron, D.: Tight bounds for testing bipartiteness in general graphs. SICOMP 33(6), 1441–1483 (2004)
11. Newman, I.: Testing Membership in Languages that Have Small Width Branching Programs. SIAM Journal on Computing 31(5), 1557–1570 (2002)
12. Parnas, M., Ron, D.: Testing the diameter of graphs. Random Structures and Algorithms 20(2), 165–183 (2002)
13. Ron, D.: Property testing (a tutorial). In: Pardalos, P.M., Reif, J.H., Rolim, J.D.P. (eds.) Handbook of Randomized Computing, vol. II, ch. 15, Kluwer Press, Dordrecht (2001)
14. Rubinfeld, R., Sudan, M.: Robust characterization of polynomials with applications to program testing. SIAM Journal on Computing 25, 252–271 (1996) (first appeared as a technical report, Cornell University, 1993).

Properly 2-Colouring Linear Hypergraphs*

Arkadev Chattopadhyay[1] and Bruce A. Reed[1,2]

[1] School of Computer Science, McGill University, Montreal, Canada
{achatt3,breed}@cs.mcgill.ca
[2] Projet Mascotte, Laboratoire I3S, CNRS Sophia-Antipolis, France

Abstract. Using the symmetric form of the Lovász Local Lemma, one can conclude that a k-uniform hypergraph $\mathcal{H}$ admits a proper 2-colouring if the maximum degree (denoted by Δ) of $\mathcal{H}$ is at most $\frac{2^k}{8k}$ independently of the size of the hypergraph. However, this argument does not give us an algorithm to find a proper 2-colouring of such hypergraphs. We call a hypergraph *linear* if no two hyperedges have more than one vertex in common.

In this paper, we present a deterministic polynomial time algorithm for 2-colouring every k-uniform linear hypergraph with $\Delta \leq 2^{k-k^{\epsilon}}$, where $1/2 < \epsilon < 1$ is any arbitrary constant and k is larger than a certain constant that depends on ϵ. The previous best algorithm for 2-colouring linear hypergraphs is due to Beck and Lodha [4]. They showed that for every $\delta > 0$ there exists a $c > 0$ such that every linear hypergraph with $\Delta \leq 2^{k-\delta k}$ and $k > c \log \log(|E(\mathcal{H})|)$, can be properly 2-coloured deterministically in polynomial time.

1 Introduction

The probabilistic method [2] is widely used in theoretical computer science and discrete mathematics to guarantee the existence of a combinatorial structure with certain desired properties. However, these techniques are often nonconstructive. Efficient construction of structures with desired properties is an important research theme in various areas like Ramsey theory, graph colouring and coding theory.

In many applications of the method, there are N events (typically "bad") in some probability space, and one is interested in showing that the probability that none of the events happen is positive. This will be true if the events are *mutually independent* and each event occurs with probability less than one. In practice, however one often finds events that are not independent of each other. The Lovász Local Lemma is a powerful sieve method that can be used in this context when the events have limited dependence.

* Research of the first author is supported by a NSERC graduate scholarship and research grants of Prof. D. Thérien, the second author is supported by a Canada Research Chair in graph theory. We would like to thank an anonymous referee for pointing out the reference [10].

M. Charikar et al. (Eds.): APPROX and RANDOM 2007, LNCS 4627, pp. 395–408, 2007.
© Springer-Verlag Berlin Heidelberg 2007

The power of the Lemma comes from the fact that it allows one to conclude a global result by analysing the local property of random combinatorial structures. In particular, the symmetric form of the Lemma implies that a random 2-colouring of the vertices of a k-uniform hypergraph of maximum degree $\frac{2^k}{8k}$, generates no monochromatic edges with non-zero probability[1]. But it provides no clue as to how to find such a structure efficiently. The Local Lemma merely guarantees the existence of, what has been called a "needle in a haystack". It can be easily verified that if the total number of edges m in the hypergraph is much larger than 4^k, then with extremely small probability the colouring is proper. The question that we are interested in is the following :
"Is there an efficient algorithm to find such a 'rare' colouring?"

Beck[3] and Alon[1] developed algorithmic versions of the Lemma which evoked its symmetric form. Beck showed that if the maximum degree Δ of the hypergraph is reduced to $2^{\alpha k}$, where $\alpha = 1/48$, then indeed there is a positive answer to the question above.

An interesting and natural class to consider is that of *linear* (also called 'almost disjoint') hypergraphs. It is known that such hypergraphs allow significantly higher degree while remaining 2-colourable i.e. Erdös and Lovász [5] showed that they are 2-colourable even for $\Delta \leq (4 - o(1))^k$. Beck and Lodha[4] obtained a deterministic polytime algorithm to find a proper 2-colouring for such hypergraphs under the restrictions that $\Delta < 2^{(1-\delta)k}$ and that the total number of hyperedges (denoted by m) satisfies $(2m)^{1+\beta} < 2^{(k-2k')2^{k/4}}$, where $\beta > 0$ is a real constant and $3 \leq k' \leq 2k/3$ is an integer. Their method builds upon the ideas in [3] and uses involved combinatorial analysis that runs into several cases.

1.1 Our Result

We build upon the general method outlined in [9] to get an algorithm for 2-colouring k-uniform linear hypergraphs with larger maximum degree. Further, our algorithm does not require any conditions to be imposed on the relationship between total size of the hypergraph and the number of vertices in a hyperedge.

Theorem 1. *There exists a deterministic polytime algorithm that for every constant $\epsilon > 1/2$, finds a proper 2-colouring of any k-uniform linear hypergraph $\mathcal{H}$, whose maximum degree is bounded by 2^{k-k^ϵ}, provided $k \geq k_\epsilon$, where k_ϵ is a positive integer that depends only on ϵ.*

Our deterministic algorithm is constructed in two steps. First, we develop a randomized algorithm that is conceptually fairly simple. The random algorithm generates the colouring iteratively in phases. It employs novel *freezing* techniques that may be useful for dealing with other problems. In particular, our freezing is guided by the crucial use of the asymmetric form of the Local Lemma as opposed to the method of Beck and Lodha [4] that uses the symmetric form. This allows us to minimize freezing enabling the random colouring to work more effectively.

---

[1] It is known that k-uniform hypergraphs remain 2-colourable even for $\Delta \leq c * 2^k/\sqrt{k \log k}$, for some constant c and k larger than a certain constant (see [10]).

We finish off by derandomizing our algorithm by a simple application of the method of conditional expectations due to Erdös and Selfridge [6].

In Section 2, we introduce the basic terminology and the needed background on the Local Lemma. Section 3 introduces a randomized algorithm, Section 4 provides its analysis and Section 5 derandomizes the algorithm proving Theorem 1. We remark that the technique in this paper can be suitably modified to make it work for the more general case of hypergraphs having constant co-degree (co-degree of any pair of vertices being the number of edges containing that pair). Note that linear hypergraphs have co-degree at most one.

2   Basic Notions

A *hypergraph* $\mathcal{H}$ is given by a set of vertices (denoted by V) and a set of non-empty subsets of V (denoted by E). Each subset in E is called a *hyperedge*. In this paper, we will often call a hyperedge simply an edge. The *size* of $\mathcal{H}$ is simply the sum of the cardinalities of V and E. $\mathcal{H}$ is called *k-uniform* if each of its hyperedges has cardinality k. The *degree* of a vertex v in V is the number of hyperedges containing v and is denoted by d_v. We denote the *maximum degree* of $\mathcal{H}$ by Δ i.e $\Delta = max_{v \in V} d_v$. Two hyperedges *intersect* at a vertex v if v is contained in both of them. A hypergraph is called *linear* if every pair of hyperedges intersect at most at one vertex. We associate an undirected graph $G_{\mathcal{H}}$ with every hypergraph $\mathcal{H}$ in the following way: $V(G_{\mathcal{H}}) = V(\mathcal{H}) \cup E(\mathcal{H})$, and $E(G_{\mathcal{H}}) = \{(x,y)|x,y \in V(\mathcal{H}), x, y \in e \text{ for some } e \in E(\mathcal{H}) \text{ OR } x \in V(\mathcal{H}), y \in E(\mathcal{H}), x \in y \text{ OR } x, y \in E(\mathcal{H}), x \cap y \neq \emptyset\}$. A *path* from one vertex/edge x of $\mathcal{H}$ to another vertex/edge y of $\mathcal{H}$ is just a simple path in $G_{\mathcal{H}}$ from x to y. The *distance* from x to y is the length of the shortest path in $G_{\mathcal{H}}$ from x to y. We say that vertex x and y are *reachable* from each other if there exists a path from one to the other. A *(connected) component* of $\mathcal{H}$ is a hypergraph C such that $V(C)$ is a maximal set of vertices of $\mathcal{H}$ that are reachable from each other and $E(C) = \{e \in E(\mathcal{H})|e \subseteq V(C)\}$.

A *2-colouring* χ of $\mathcal{H}$ is any assignment of colours from the set of 2-colours (we use red and blue) to its vertices i.e $\chi : V \to \{red, blue\}$. We say a hyperedge e is *monochromatic* under the colouring χ if every vertex contained in e receives the same colour. χ is called a *proper 2-colouring* precisely if it does not generate any monochromatic hyperedge. In this work, we generate a 2-colouring iteratively, and each phase of our algorithm generates a partial colouring of vertices. For such a partial colouring, we call an edge *uni-coloured* if its vertices have not received two different colours. An uncoloured vertex is called *nice* if it lies in exactly one uni-coloured edge. For any uni-coloured edge e, its *restriction* to the uncoloured vertices is $e' = \{v \in e : v \text{ is not coloured}\}$. A partial colouring χ of $\mathcal{H}$ *induces* the hypergraph $\mathcal{H}'$ with its vertex set consisting of all uncoloured vertices and its edge set consisting of the restriction of all its uni-coloured edges. The following simple remarks will be useful:

Remark 2. A hypergraph $\mathcal{H}$ has a proper 2-colouring if and only if each of its components has one.

Remark 3. A partial colouring χ of $\mathcal{H}$ extends to a proper colouring if the induced hypergraph $\mathcal{H}'$ has a proper colouring.

Remark 4. Let χ be a partial colouring of $\mathcal{H}$ and let e be a uni-coloured edge that has a pair of nice vertices v_1 and v_2. Extend χ to χ' by colouring v_1 and v_2 such that $\chi'(v_1) \neq \chi'(v_2)$. Then the hypergraph induced by χ' (denoted by $\mathcal{H}''$) has the set of vertices $V(\mathcal{H}') - v_1 - v_2$ and its set of edges is $E(\mathcal{H}') - e$, where $\mathcal{H}'$ is the hypergraph induced by χ. Further, it has a proper 2-colouring iff $\mathcal{H}'$ has one.

We denote the probability of an event X by $\mathbf{Pr}[X]$ and the expected value of a random variable y by $\mathbf{E}[y]$. We now state the Asymmetric form of the Lovász Local Lemma [5] that is one of the most powerful tools of the probabilistic method.

Lemma 5 (Asymmetric Local Lemma). *Let $\mathcal{E} = A_1, \ldots, A_m$ be m (typically bad) events in a probability space, where event A_i occurs with probability p_i and is mutually independent of events in $\mathcal{E} - \mathcal{D}_i$, for some $\mathcal{D}_i \subseteq \mathcal{E}$. Then, the probability that none of these m events occur is positive provided that*

$$- \sum_{k:A_k \in \mathcal{D}_i} p_k \leq \tfrac{1}{4}. \text{ for all } i.$$

In the application of the Local Lemma to colouring a k-uniform hypergraph, we consider a random 2-colouring of its vertices. With each hyperedge e, we associate the event (denoted by A_e) that e is monochromatic. One can easily verify that A_e is mutually independent of all other events except those in $\mathcal{D}_e = \{A_f | f \cap e \neq \emptyset\}$. Clearly, $|\mathcal{D}_e| \leq 1 + (\Delta - 1)k$. Since $\mathbf{Pr}[A_e] = 1/2^{k-1}$ for all edges e, the local lemma implies that the probability of not having any monochromatic edge is positive if $\Delta \leq \frac{2^k}{8k}$.

We will denote by $\mathcal{BIN}(n, p)$ the sum of n independent Bernoulli random variables, each equal to 1 with probability p and 0 otherwise. The following inequality, called Chernoff's bound (see [2]), is used to bound the probability of the sum deviating from its expected value np:

$$\mathbf{Pr}(|\mathcal{BIN}(n, p) - np| > t) < 2e^{-((1 + \frac{t}{np})ln(1 + \frac{t}{np}) - \frac{t}{np})np} \tag{1}$$

3 Randomized Algorithm

We assume that the vertices are labelled $\{1, \ldots, n\}$. We first provide a randomized algorithm to obtain a proper 2-colouring and then show that it can be easily derandomized. The randomized algorithm works in three phases. In the first phase, we randomly colour a vertex v_i, unless it is frozen using one of the rules described below. These freezing rules are imposed to ensure that the Local Lemma can be applied to prove that the partial colouring obtained in the first phase extends to a proper colouring. We shall also show that with probability near 1, the connected components of the induced hypergraph are small in size. We post-process the random colouring in the following way : identify those uni-coloured edges that contain at least two nice vertices. For every such uni-coloured edge, we choose a pair of nice vertices and assign them different colours. By Remark 4, the hypergraph induced by this partial colouring (denoted by $\mathcal{H}_1$) is still properly 2-colourable and we shall show that it has small components, with high probability.

In the second phase, we reapply the procedure of the first phase to each component of $\mathcal{H}_1$ separately. We use very similar freezing rules and do identical post-processing to that in Phase 1. We show again that the partial colouring extends to proper colouring using the Local Lemma. We denote the induced hypergraph at the end of phase 2 by $\mathcal{H}_2$ and as before, each of its component is very small with high probability.

Finally, in the third phase, we apply brute force to explicitly compute a proper completion for the partial colouring of each component of $\mathcal{H}_2$. This is possible in poly-time as the size of each component is very small.

Forthwith the details. To ensure that we can apply the Local Lemma to show that our partial colouring extends to a proper colouring, we introduce a random variable H_e for each edge e as given below:

$$H_e = \sum_{f:e \cap f \neq \emptyset} \mathbf{Pr}[A_f | partial\ colouring]$$

Note that if H_e lies below $1/4$ at the end of a phase for every edge e, then the asymmetric form of the Local Lemma can be applied to conclude that the partial colouring extends to a proper 2-colouring. It will be more convenient to think of H_e as being a sum of k random variables - one for each vertex v contained in edge e. Formally, $H_e = \sum_{v \in e} H_v - (|e| - 1)\mathbf{Pr}[A_e | partial\ colouring]$, where H_v is given by

$$H_v = \sum_{f:v \in f} \mathbf{Pr}[A_f | partial\ colouring]$$

Thus, $H_v \leq 1/(4k)$ ensures that our partial colouring extends to a proper colouring. The key technique in our procedure is to control H_v, for each vertex v. A vertex v is considered *bad* if H_v exceeds a prescribed upper bound b_i for Phase i. Whenever a vertex v turns bad, we freeze all uncoloured vertices in each edge containing v. If this were the only freezing done, then the probability of v turning bad would be substantial and our freezing would be widespread. Indeed, the probability that a specific edge e containing v is monochromatic is $1/2^{k-1}$. So, the probability that v turns bad is at least $1/2^{k-1}$. To deal with this, recall that H_v is a sum of conditional probabilities. The intuition is if each of these probabilities in the sum is small, then H_v is concentrated and the probability that it turns bad is extremely small provided the threshold b_i is appropriately chosen. In particular, for each of the first two phases, we specify a lower bound on the number of vertices remaining uncoloured in a uni-coloured edge. If edge e attains this lower bound, we call e *naughty* and freeze all of its uncoloured vertices. This freezing of naughty edges ensures that the conditional probabilities that appear in the sum for H_v remain small, and reduces the probability of a vertex becoming bad to well below 2^{-k}. Unfortunately, there is considerable freezing done due to naughty edges. However, as we shall see, the post-processing step allows us to ignore all but a vanishingly small proportion of this second kind of freezing. This is the basic structure of our random colouring procedure in each phase. We fix constants β, α_1 and α_2 such that $1/2 < \alpha_2 < \beta < \alpha_1 < \epsilon$ for reasons that will become clear from subsequent discussion. For Phase 1, $b_1 = 2^{-k^\beta - 1}$ and for Phase 2 $b_2 = 1/(8k)$. The lower bound used for freezing uncoloured vertices in naughty edges in Phase i will be k^{α_i}, $i = 1, 2$.

With this intuitive description behind us, we are ready to give the formal description of the basic random colouring procedure parameterized with t_1, the threshold for detecting naughty edges and t_2, the threshold for detecting bad vertices:

RANDCOLOUR$(\mathcal{F}, t_1, t_2)$;
Input: Hypergraph $\mathcal{F}$ s.t. $\forall e \in E(\mathcal{F})\ |e| > t_1, \forall v \in V(\mathcal{F})\ H_v \leq t_2$.
Returns a partial colouring and the induced hypergraph $\mathcal{F}'$ such that $\forall e \in E(\mathcal{F}')\ |e| \geq t_1, \forall v \in V(\mathcal{F}')\ H_v \leq 2t_2$.

- **Main loop:** For each vertex $v_i \in V$.
  - If v_i is frozen, skip colouring it and go back to the main loop
    * Colour v_i uniformly at random and then do the following:
      1. If some uni-coloured edge e has only t_1 uncoloured vertices remaining, freeze every uncoloured vertex in e.
      2. If for some vertex v, H_v exceeds t_2, freeze all uncoloured vertices in uni-coloured edges containing v.
- **Post-processing:** For every uni-coloured edge containing two nice vertices, colour one of them red and the other blue.

Note that the main loop of RANDCOLOUR runs $|V(\mathcal{H})|$ times. To determine if an edge is naughty or a vertex is bad, we need to only look at edges containing the vertex that got coloured in that iteration. Thus, each loop takes $O(k.\Delta^2)$ time. Summing this up,

Remark 6. The running time of **RANDCOLOUR** is $O(k\Delta^2 \cdot |V(\mathcal{H})|)$.

We make another remark that is useful to get the intuition behind our post-processing step.

Remark 7. Every uni-coloured edge at the end of Phase i contains at least $k^{\alpha_i} > \sqrt{k}$ uncoloured vertices and satisfies at least one of the following conditions:

1. it is naughty
2. for every uncoloured vertex v in it, $v \in g$ for some edge g that is naughty or contains a bad vertex.

Let us call a uni-coloured edge *bad* if it intersects with at least $\sqrt{k}$ other naughty edges. Intuitively, we expect bad edges to occur with small probability. The following fact points out that our post-processing step deals effectively with naughty edges that are not bad and are far from both bad edges and vertices.

Proposition 8. *Consider a uni-coloured edge e. If e is at distance at least 4 from every bad edge and bad vertex, then e has two nice vertices, provided k is larger than a certain constant.*

Proof. Consider a uni-coloured edge f that intersects e. Remark 7 implies the following: if f is not naughty then it must be bad or it is at distance at most 2 from a bad vertex. As e is at distance at least 4 from every bad edge/vertex, f must be naughty. This shows that every uni-coloured edge that intersects e is naughty. Since e cannot be bad, there are at most $\sqrt{k}$ uni-coloured edges intersecting e. Hence for large k satisfying $k^{\alpha_i} - \sqrt{k} \geq 2$, we conclude that e has 2 nice vertices.

Thus, Remark 4 implies that all uni-coloured edges that are not within distance 3 of any bad vertex or bad edge, do not remain uni-coloured after post-processing and hence are not part of the induced hypergraph returned by RANDCOLOUR. We will see later in the analysis of our algorithm in Section 4 that this observation is very helpful in bounding the size of components of the hypergraph returned by RANDCOLOUR. Lemma 12 and 14 in the next section imply that the component size at the end of Phase i is s_i with high probability, where $s_1 = 2^{6k}(c_\epsilon/k^{\epsilon+1/2}) \cdot \log(n + m)$, $s_2 = 2^{6k}(c_\epsilon/k^{\beta+1/2}) \cdot \log(s_1)$ and n, m are respectively the number of vertices and edges of the hypergraph. The constant c_ϵ depends on ϵ and is chosen according to Lemma 14.

Using RANDCOLOUR as our key sub-routine, the entire 3-phase algorithm is described below.

2-COLOUR($\mathcal{H}$, α_1, α_2, β, ϵ)

- *Phase 1*: $\mathcal{H}_1$ = RANDCOLOUR($\mathcal{H}$, k^{α_1}, $2^{-k^\beta-1}$). If the largest component of $\mathcal{H}_1$ is larger than s_1 then repeat.
- *Phase 2*: Enumerate the components of $\mathcal{H}_1$ as $C_1, \ldots, C_j$.
  - For every connected component C_i
    * $\mathcal{H}_2^i$ = RANDCOLOUR(C_i, k^{α_2}, $1/16k$).
      · *Case 1.* $2^{6k} \leq \log(n+m)/\log\log(n+m)$. If the largest component of $\mathcal{H}_2^i$ is larger than s_2, then repeat call to RANDCOLOUR.
      · *Case 2.* $2^{6k} > \log(n+m)/\log\log(n+m)$. If there are any uni-coloured edges remaining, repeat call to RANDCOLOUR.
- *Phase 3* Let $\mathcal{H}_2 = \cup_{i=1}^j \mathcal{H}_2^i$.
  - If there are uni-coloured edges left in $\mathcal{H}_2$, then
    * Colour each connected component C of $\mathcal{H}_2$ in turn, by considering all $2^{|V(C)|}$ colourings.

Proposition 9. *Every partial colouring produced at the end of Phase 1 or Phase 2 can be completed to a proper 2-colouring.*

Proof. Let $\mathcal{H}_1'$ be the hypergraph induced at the end of random colouring in Phase 1 just before the post-processing step. Consider any edge f and any vertex u in f. It is simple to verify that colouring u at most doubles the probability that f becomes monochromatic. Hence, freezing done due to bad vertices ensures that for every vertex v, H_v is at most 2^{-k^β} which is less than $1/4k$ for large k.

Hence, $H_e < 1/4$ for all edges e in $\mathcal{H}'_1$. This ensures that the second condition of the Local Lemma is satisfied for each event A_e. Thus $\mathcal{H}'_1$ has a proper colouring. Recalling Remark 4, we see colouring a pair of nice vertices in a uni-coloured edge differently, does not increase the probability of any event A_e. Thus, $\mathcal{H}_1$ has a proper colouring. Using Remark 3, the partial colouring at the end of Phase 1 extends to a proper 2-colouring. A very similar argument applied to each connected component of $\mathcal{H}_1$ shows that each partial colouring obtained at the end of Phase-2 also has a proper completion.

Our aim in the next section is to show the result given below:

Theorem 10. *With high probability,* **2-COLOUR** *produces a proper 2-colouring of hypergraph* $\mathcal{H}$ *in time polynomial in the size of* $\mathcal{H}$.

4   Analysis of 2-COLOUR

We have already noted in Remark 6 that RANDCOLOUR runs in poly-time. Further Case 1 and Case 2 handled in Phase 2 of 2-COLOUR ensures that either the size of every component of $\mathcal{H}_2$ is at most s_2 or no uni-coloured edge is left at end of Phase 2. Thus Phase 3 of 2-COLOUR also runs in poly-time. Hence, to prove Theorem 10 it is sufficient to show the following:

Lemma 11. *Every call from 2-COLOUR to RANDCOLOUR succeeds with probability at least* $1/2$.

To prove Lemma 11, we simply need to show that the probability that the induced hypergraph returned by RANDCOLOUR has large components is less than a half. One way to do this would be to establish the following:

Claim. $\mathbf{E}[X_i] < 1/2$, for $i = 1, 2$.

where, X_i denotes the number of components of size at least s_i of the hypergraph $\mathcal{H}_i$. Lemma 11 follows from the above claim by a simple application of Markov's inequality.

Instead of bounding directly the expected value of X_i, we will define another integer valued random variable Y_i, such that $X_i \leq Y_i$. Bounding the expected value of Y_i turns out to be more convenient. But we need to introduce a few more notions before we can define Y_i.

Consider an auxiliary graph G_i associated with $\mathcal{H}_i$ in the following way. Each node of G_i corresponds to either a vertex or a hyperedge of hypergraph $\mathcal{H}_i$. Two nodes of G_i are connected by an arc if their distance from each other in $\mathcal{H}_i$ is at least 4 and at most 10. Every rooted tree T_i of G_i is called a $(4, 10)$-tree of $\mathcal{H}_i$. We call a $(4, 10)$-tree *bad* if every node of the tree corresponds to either a bad edge or a bad vertex generated in Phase i. Let Y_i be the number of bad $(4, 10)$ trees of $\mathcal{H}_i$ of size $s_i \cdot 2^{-6k}$. The following lemma establishes the desired relationship between X_i and Y_i (i.e. $X_i \leq Y_i$).

Lemma 12. *If $\mathcal{H}_i$ has a component C of size s, then there is a bad $(4, 10)$-tree of size at least $s/2^{6k}$, whose set of nodes is contained in $V(C) \cup E(C)$.*

Proof. Let T be a maximal bad $(4, 10)$-tree formed from vertices and edges of C. Let C_1 be the sub-hypergraph in the component that includes all edges that are within distance 6 from T. We show that there cannot be any vertex or edge outside of C_1 in C and the result follows from that by recalling that the degree of any vertex is at most 2^{k-k^ϵ}.

Suppose the contrary is true. Then, let b be a edge that is at minimal distance (at least 7) in C from T. Consider a minimal path P in C from b to T. Consider a node u in P at distance 7 from T. We have following possibilities:

- u is a bad edge, in which case one can add u to T.
- u is not bad. Recalling Proposition 8 and our post-processing step in RAND-COLOUR, there is a node u' that corresponds to either a bad vertex or bad edge at distance at most 3 from u. Hence u' is at distance at least 4 and at most 10 from T. We can grow T by adding u'.

Thus, in each case our assumption that T is maximal is contradicted if there exists any edge outside of C_1.

Now, to estimate the expected value of Y_i we need to calculate the probabilites of certain events. We denote the events of any edge e becoming naughty and bad in phase i by N_e^i and B_e^i respectively. Let the event that a vertex v becomes bad in phase i be denoted by B_v^i. We now estimate the probabilities of these events below (detailed calculations are in the appendix).

Proposition 13. *Let $1/2 < \alpha_2 < \beta < \alpha_1 < \epsilon < 1$. Then for sufficiently large k, the following probability bounds hold:*

1. $\mathbf{Pr}[B_e^1] \leq 2^{-k^{\epsilon+1/2}+k^{\alpha_1+1/2}+\sqrt{k}\log k+1}$ *for any edge e.*
2. $\mathbf{Pr}[B_v^1] \leq 2c^{-2^{k^{\alpha_1}-k^\beta-\log k-1}} \leq 2^{-2^{0.5k^{\alpha_1}}}$, *for any vertex v.*
3. $\mathbf{Pr}[B_e^2] \leq 2^{-k^{\beta+1/2}+k^{\alpha_2+1/2}+\sqrt{k}\log k+1}$ *for any edge e.*
4. $\mathbf{Pr}[B_v^2] \leq 2e^{-2^{k^{\alpha_2}-3\log k-6}} \leq 2^{-2^{0.5k^{\alpha_2}}}$, *for any vertex v.*

Using the bounds on probability of relevant events, we calculate the expected number of bad $(4, 10)$-trees in each phase of 2-COLOUR in the lemma below (see Appendix for a proof).

Lemma 14. *There exists a constant c_ϵ for $i = \{1, 2\}$ such that the expected number of bad $(4, 10)$-trees in $\mathcal{H}_i$ of size $(c_\epsilon/k^{\delta_i+1/2}) \cdot \log(n_i + m_i)$ is less than $\frac{1}{2}$, where $\delta_1 = \epsilon$ and $\delta_2 = \beta$ and n_i, m_i are respectively the number of vertices and edges in the largest component of $\mathcal{H}_i$.*

Proof (of Lemma 11 and Theorem 10). Combining Lemma 12 and Lemma 14 with Markov's inequality yields the bound on the size of the components of the hypergraph induced at the end of Phase 1 (i.e. $\mathcal{H}_1$). The argument for analyzing Phase 2 when *Case 1* holds is very similar, just noting that each component C_i of $\mathcal{H}_1$ now has size at most $2^{6k}c_\epsilon \log(n + m)$. In *Case 2*, we have

$$2^{6k} > \log(n + m)/\log\log(n + m) \qquad (2)$$

Consider any component C of $\mathcal{H}_1$. Let X_B be the random variable representing the total number of bad vertices and bad edges generated by phase 2 for C. Clearly, (2) and bounds from proposition 13 imply

$$\mathbf{E}[X_B] \leq |C| \cdot 2^{-0.5k^{\beta+1/2}} \leq 2^{6k}c_\epsilon \log(n+m) \times 2^{-0.5k^{\beta+1/2}} = o(1) \quad (3)$$

since $\beta + 1/2 > 1$. Markov's inequality implies that with probability tending to one, the random colouring procedure of Phase 2 will not generate bad vertices and bad edges. The post-processing done at the end of Phase 2 thus ensures that no uni-coloured edges remain at the end of Phase 2.

Hence, the expected number of times that Phase 1 is run is at most 2 and that Phase 2 is run is at most $2(n + m)$. Moreover, each run of Phase 1 and Phase 2 takes poly-time. One can verify that in *Case 1*, the size of each component is small enough to ensure that brute force search can be completed in polynomial time. In *Case 2*, it follows from the previous discussion *any* completion of the partial colouring of Phase 2 is proper.

5   Derandomization

Proof (of Theorem 1). We adapt the method of conditional expectations (due to Erdös and Selfridge [6]). In Phase 1 of 2-COLOUR, when we consider assigning a colour to a vertex v, we compute the colour that minimizes the expected number of bad $(4, 10)$-trees of size $\ell_1 = (c_\epsilon/k^{\epsilon+1/2}) \log(n+m)$ that would arise if the colouring is randomly completed as prescribed in Phase 1. This colour is then assigned to v. The number of such bad trees that we need to take into consideration is bounded from above by $(n+m) \cdot 2^{10(k-k^\epsilon)\ell_1} \cdot 4^{\ell_1}$. Since $\epsilon > 1/2$, it gets easily verified that the number of such bad trees is $O((n+m)^{1+\delta})$ for every $\delta < 1$ when k is growing function of n. Otherwise it is $O(n^c)$ for some constant c. As shown in Lemma 14, the expected number of such bad $(4, 10)$-trees at the beginning of Phase 1 is less than a half. The deterministic variant of Phase 1 hence produces no bad trees and so $\mathcal{H}_1$ has no large components.

If $2^{6k} \leq \log(n + m)/\log\log(n + m)$, we apply a similar procedure to derandomize Phase 2 by considering trees of size $\ell_2 = (c_\epsilon/k^{\beta+1/2}) \log\log(n+m)$. The rest of the method and argument is same as above. For larger k, when colouring a vertex v in a component C of $\mathcal{H}_1$, we assign v the colour that minimizes the expected number of bad vertices and edges of C. Using the argument given in the proof of Lemma 11, we conclude that Phase 2 generates no bad vertices or edges. Thus, no edge at the end of Phase 2 is uni-coloured.

References

1. Alon, N.: A parallel algorithmic version of the Local Lemma. Random Structures and Algorithms 2, 367–379 (1991)
2. Alon, N., Spencer, J.: The Probabilistic Method. Wiley, New York (1992)
3. Beck, J.: An algorithmic approach to the Lovász Local Lemma. Random Structures and Algorithms 2(4), 343–365 (1991)

4. Beck, J., Lodha, S.: Efficient proper 2-coloring of almost disjoint hypergraphs. In: SODA, pp. 598–605 (2002)
5. Erdös, P., Lovász, L.: Problems and results on 3-chromatic hypergraphs and some related questions. In: H., A., et al. (eds.) Infinite and Finite Sets. Colloq. Math. Soc. J. Bolyai, vol. 11, pp. 609–627 (1975)
6. Erdös, P., Selfridge, J.: On a combinatorial game. Journal of Combinatorial Theory (A) 14, 298–301 (1973)
7. Harary, F., Palmer, E.: Graphical Enumeration, 1st edn. Academic Press, London (1973)
8. Molloy, M., Reed, B.: Further algorithmic aspects of the Local Lemma. In: STOC, pp. 524–529 (1998)
9. Molloy, M., Reed, B.: Graph Colouring and the Probabilistic Method. Springer, Heidelberg (2002)
10. Radhakrishnan, J., Srinivasan, A.: Improved bounds and algorithms for hypergraph 2-coloring. Random Structures and Algorithms 16, 4–32 (2000)

Appendix

Proof of Proposition 13

Proof. We consider a set S_e of $\sqrt{k}$ edges $\{e_1, \ldots, e_{\sqrt{k}}\}$ that intersect an edge e. Let V_{e_i} be the set of those vertices of e_i that are not contained in any edge in $S_e \setminus \{e_i\}$. Since $\mathcal{H}$ is linear, $|V_{e_i}| \geq k - \sqrt{k} + 1$, for all i. If e_i is to become naughty, the first $k - k^{\alpha_i} - \sqrt{k}$ vertices in V_{e_i} that get coloured, receive the same colour. The actual vertices in V_{e_i} that get coloured in any instance by RANDCOLOUR may depend on the freezing caused by colouring in the rest of the hypergraph. Nevertheless, every instance of RANDCOLOUR that makes e_i naughty, makes a sequence L_i of $k - k^{\alpha_i} - \sqrt{k}$ independent colour assignments to some vertices in V_{e_i}. Since $V_{e_i} \cap V_{e_j} = \emptyset$ for $i \neq j$, every call to RANDCOLOUR makes all edges in S_e naughty with probability at most $2^{(-k+k^{\alpha_1}+\sqrt{k})\sqrt{k}}$. We have at most $(\Delta \cdot k)^{\sqrt{k}} = (2^{k-k^\epsilon} \cdot k)^{\sqrt{k}}$-many choices for S_e. Combining everything yields the first bound of our proposition.

We now compute $\mathbf{Pr}[B_v^1]$. Let $n_{v,i}$ represent the number of edges containing v and having i vertices uncoloured with $k - i$ vertices coloured the same. Let $n'_{v,i}$ be a random variable that is defined in the following way:

$$n'_{v,i} = \sum_{e: v \in e} x_e^i$$

where, for every edge e, x_e^i is a Bernoulli random variable that takes value 1 with probability 2^{-k+i+1}. Since our hypergraph is linear, the edges containing v only intersect each other at v. Thus, using sequences of colour assignments for each edge as before in the first part, one can show that $\mathbf{Pr}[n_{v,i} \geq N] \leq \mathbf{Pr}[n'_{v,i} \geq N]$ for every number N. Further, one can easily verify $\mathbf{E}[n'_{v,i}] = \Delta \cdot 2^{-k+i+1} = 2^{i+1-k^\epsilon}$. Since $n'_{v,i}$ is the sum of independent and identically distributed random variables, the classical result of Chernoff says that it is concentrated around its mean value. We re-write H_v in the following way:

$$H_v = \sum_{i:k^{\alpha_1} \leq i \leq k} H_{v,i}$$

where

$$H_{v,i} = 2^{-i+1} \times n_{v,i}$$

If $H_v \geq 2^{-k^\beta}$, then for some i, we have $H_{v,i} \geq \frac{2^{-k^\beta}}{k}$ and so

$$n_{v,i} \geq \frac{2^{i-1-k^\beta}}{k} = \mathbf{E}[n'_{v,i}] \cdot 2^{k^\epsilon - k^\beta - \log k - 2}$$

Since, $n'_{v,i} = \mathcal{BIN}(\Delta, p)$ for $p = 2^{-k+i+1}$, we get

$$\mathbf{Pr}\left[n_{v,i} \geq \frac{2^{i-1-k^\beta}}{k}\right] \leq \mathbf{Pr}\left[|\mathcal{BIN}(\Delta, p) - \Delta p| \geq \Delta p \cdot \left(2^{k^\epsilon - k^\beta - \log k - 2} - 1\right)\right] \tag{4}$$

Recalling that $\Delta p = 2^{i+1-k^\epsilon}$, we apply Chernoff's bound from (1) to the RHS of (4)

$$\mathbf{Pr}\left[n_{v,i} \geq \frac{2^{i-1-k^\beta}}{k}\right] \leq 2\exp\left(-2^{k^\epsilon - k^\beta - \log k - 2}(k^\epsilon - k^\beta - 2)\left(2^{i+1-k^\epsilon}\right)\right) \tag{5}$$

Noting that our freezing ensures that $i \geq k^{\alpha_1}$, we get finally

$$\text{RHS of (5)} \leq 2\exp\left(-(k^\epsilon - k^\beta - 2)2^{k^{\alpha_1} - k^\beta - \log k - 2}\right) \tag{6}$$

Recalling our assumption that $\beta < \alpha_1 < \epsilon$, we see that (6) directly yields the second bound of our Proposition for large k.

To obtain the probability bounds for events in the second phase of the algorithm, we introduce another notation. Let $\mathcal{H}_i$ represent the subgraph induced by those monochromatic edges at the end of Phase 1 that have exactly i uncoloured vertices left. Note that the degree of $\mathcal{H}_i$ denoted by Δ_i is at most 2^{i-k^β}, since $H_v \leq 2^{-k^\beta}$ for all vertices v at the end of phase 1.

Consider $\sqrt{k}$ second phase naughty edges $e_1, \ldots, e_{\sqrt{k}}$ intersecting edge e, where edge e_j had i_j uncoloured vertices at the beginning of phase 2. Applying a similar argument as before for computing $\mathbf{Pr}[B_e^1]$ and observing that there are at most $k^{\sqrt{k}}$ ways of choosing the $\sqrt{k}$-tuple $(i_1, \ldots, i_{\sqrt{k}})$ where $k^{\alpha_1} \leq i_j \leq k$ for each j, we get

$$\mathbf{Pr}[B_e^2] \leq k^{\sqrt{k}} \times \prod_{j=1}^{\sqrt{k}} 2^{-i_j + k^{\alpha_2} + \sqrt{k}} \cdot (\Delta_{i_j} k) \leq 2^{(-k^\beta + k^{\alpha_2} + 2\log k + 1)\sqrt{k}}$$

giving us our third bound.

Let $n_{v,j}^i$ be the number of monochromatic edges containing vertex v that have i uncoloured vertices at the end of Phase 1 and have $j \leq i$ uncoloured vertices at the end of Phase 2. As in the argument for bounding $\mathbf{Pr}[B_v^1]$, we introduce a random variable $n_{v,j}^{',i}$ for each $n_{v,j}^i$ such that $\mathbf{Pr}[n_{v,j}^i \geq N] \leq \mathbf{Pr}[n_{v,j}^{',i} \geq N]$ for every number N. Like before, each $n_{v,j}^{',i}$ is a sum of identical and independent Bernoulli random variables and hence, is concentrated around its mean. Clearly, $\mathbf{E}[n_{v,j}^{',i}] = \Delta_i \cdot 2^{-i+j} = 2^{j-k^\beta}$. Let $n_{v,j,2} = \sum_{i=k^{\alpha_1}}^k n_{v,j}^i$ i.e. the number of edges through v that have j uncoloured vertices at end of phase 2. As before, define $H_{v,j}^2 = 2^{-j+1} \cdot n_{v,j,2}$, so that H_v at end of phase 2 is simply $\sum_{j=k^{\alpha_2}}^k H_{v,j}^2$. Thus, $H_v \geq 1/8k$ implies that for some i, j,

$$n_{v,j}^i \geq \frac{2^{j-1}}{8k^3} = \mathbf{E}[n_{v,j}^{',i}] \cdot \frac{2^{k^\beta - 3}}{8k^3}$$

Applying Chernoff's bound as we did for computing the second bound of the proposition,

$$\mathbf{Pr}[n_{v,i}^j \geq \frac{2^{j-1}}{8k^3}] \leq 2e^{-2^{k^\beta - 3 \log k - 6} \cdot \mathbf{E}[n_{v,i}^{',j}]}$$

Plugging $\mathbf{E}[n_{v,i}^{',j}] \geq 2^{k^{\alpha_2} - k^\beta}$, we get our desired bound.

Proof of Lemma 14

Proof. Let X_ℓ^i denote the random variable that is equal to the number of bad $(4, 10)$-trees of size ℓ at end of Phase i. In the following discussion, we shall drop the superscript i denoting the Phase in which events occur. Consider T to be a $(4, 10)$-tree of size ℓ with nodes $n_1, \ldots, n_\ell$. Further, let B_T denote the event that T turns bad at the end of Phase i. We first want to compute $\mathbf{Pr}[B_T]$. For T to turn bad, each edge and vertex of the hypergraph comprising it has to become bad. These events are not independent as freezing done due to a vertex (an edge) becoming bad (naughty) affects the probabilities of other vertex/edge becoming bad/naughty. But event B_v (B_e) is determined by just exposing colours assigned to a set of vertices at distance 1 (2) from v (e) in $\mathcal{H}_i$. Let this set be denoted by V_v (V_e). For two edges e and f, it is easily verified that sets $V_e \cap V_f \neq \emptyset$ implies that the distance between e and f is less than 4. Similarly one can verify that sets V_{n_j} and V_{n_k} are disjoint for every pair of nodes n_j, n_k in T as distance between n_j and n_k is at least 4 in $\mathcal{H}_i$. For each node n in T, let B_n' denote the auxiliary event that n turns bad when just vertices in V_n are randomly coloured. Our previous observation implies that the auxiliary events associated with nodes of a $(4, 10)$-tree are independent of each other. Further, $\mathbf{Pr}[B_T] \leq \mathbf{Pr}[B_{n_1}' \cap B_{n_2}' \cap \cdots \cap B_{n_\ell}']$. Thus, using the bounds obtained for probabilities in Proposition 13, one gets the following:

$$\mathbf{Pr}[B_T^i] \leq 2^{-0.5\ell(k^{\delta_i + 1/2})} \tag{7}$$

where $\delta_1 = \epsilon$ and $\delta_2 = \beta$. We find out an upper bound on N_ℓ i.e. the possible number of such trees of size ℓ. There are 4^ℓ ways of choosing an unlabelled tree of size ℓ (see [7]). The root of our bad tree can be chosen in $n_i + m_i$ ways. We fix an unlabelled tree and the root of our bad tree and then choose the remaining nodes of the tree in a breadth-first fashion. Each node can then be chosen in at most $(2^{k-k^\epsilon})^{10} \cdot k$ ways. Thus, the total number of choices for a $(4, 10)$-bad tree of size ℓ is at most

$$N_\ell = (n_i + m_i) \times 4^\ell \times (2^{10(k-k^\epsilon)} \cdot k^{10})^{\ell-1} \tag{8}$$

Combining (7) and (8) for sufficiently large k we have,

$$\mathbf{E}[X_\ell^i] \leq (n_i + m_i) \cdot 2^{(-0.5k^{\delta_i+1/2}+10k+2)\ell} \tag{9}$$

Taking $\ell = c_\epsilon \cdot \log(n_i + m_i))$, for large k one gets that $\mathbf{E}[X_\ell^i] < 1/2$, since $\delta_i + 1/2 > 1$ and c_ϵ is a constant that just depends on ϵ.

Random Subsets of the Interval and P2P Protocols*

Jacek Cichoń, Marek Klonowski, Łukasz Krzywiecki, Bartłomiej Różański, and Paweł Zieliński

Institute of Mathematics and Computer Science
Wrocław University of Technology
Poland
{Jacek.Cichon,Marek.Klonowski,Lukasz.Krzywiecki,
Pawel.Zielinski}@pwr.wroc.pl

Abstract. In this paper we compare two methods for generating finite families of random subsets according to some sequence of independent random variables $\zeta_1, \ldots, \zeta_n$ distributed uniformly over the interval $[0, 1]$. The first method called *uniform split* uses ζ_i values straightforwardly to determine points of division of $[0, 1]$ into subintervals. The second method called *binary split* uses ζ_i only to perform subsequent divisions of already existing subintervals into exact halves. We show that the variance of lengthes of obtained intervals in the first method is approximately $\frac{1}{n^2}$ and that the variance of lengthes of obtained intervals in the second method is approximately $\frac{1}{n^2}(\frac{1}{\ln 2} - 1)$.

The uniform split is used in the Chord peer-to-peer protocol while the binary split is used in the CAN protocol. Therefore our analysis applies to this protocols and shows that CAN has a better probabilistic properties than Chord. We propose also a simple modification of the Chord protocol which improves its statistical properties.

1 Introduction

We investigate the problem of splitting a given interval into a finite number of nonoverlapping subintervals that appears in some peer-to-peer protocols. Splitting is done according to a sequence of random values $\zeta_1, \ldots, \zeta_n$ distributed uniformly in $[0, 1]$, and some fixed split method.

In this paper we present an analysis of two split methods. The first among them is rather straightforward. The family of subintervals is composed of all nonoverlapping intervals defined by the set $\{0, 1\} \cup \{\zeta_1, \ldots, \zeta_n\}$. This method corresponds to the sequential splitting by adding points – for a new point ζ_i we select an interval $(\zeta_j, \zeta_k]$ such that $\zeta_j < \zeta_i \leq \zeta_k$ and divide it into two parts: $(\zeta_j, \zeta_i]$ and $(\zeta_i, \zeta_k]$. We call this method *uniform split*. It is well known that this method has significant flaws in terms of subset length uniformity. Note that the uniform split corresponds to the process of adding a nodes in the Chord

---

* Partially supported by the EU within the 6th Framework Programme under contract 001907 (DELIS).

M. Charikar et al. (Eds.): APPROX and RANDOM 2007, LNCS 4627, pp. 409–421, 2007.
© Springer-Verlag Berlin Heidelberg 2007

peer-to-peer protocol (see [12] and [9]). It is well known, and our calculation confirms this fact, that the capacity of Chord nodes' areas (intervals) is a random variable with large variation. Now let us recall that amount of data and the number of requests passed via node in Chord is proportional to the length of its area. Hence, large variation of area size introduces a discrepancy between nodes' workload.

The second method called *binary split* is based on the following idea: if ζ_i values are used only to determine which existing interval is to be split, the splitting point is chosen always in the middle of selected interval. We show in Section 2 that the uniformity of interval lengths is significantly better than in the uniform split case. Binary split corresponds to a sequential process where each ζ_i determines an existing interval to split in two halves. The process starts with whole interval $[0, 1]$. ζ_1 obviously splits it into $[0, 0.5]$ and $(0.5, 1]$, but ζ_2 may either make it $[0, 0.25]$, $(0.25, 0.5]$ and $(0.5, 1]$ or $[0, 0.5]$, $(0.5, 0.75]$ and $(0.75, 1]$– depending on which initial interval ζ_2 falls in; and so on for all ζ_i for $1 < i \leq n$. The resulting family consists of $n + 1$ nonoverlapping intervals with lengths from the set $\{\frac{1}{2^k} : k \leq n + 1\}$. The binary split corresponds to the process of adding nodes in CAN peer-to-peer protocol (see [11]). In the last section of this paper we shall propose a small modification of the classical Chord protocol which is based on the binary split and which has better probabilistic properties than the original one.

The authors wish to express thanks to referees for their helpful suggestions concerning the presentation of this paper.

1.1   Notation

We denote the real numbers and integers by $\mathbb{R}$ and $\mathbb{Z}$, respectively. Let X be a random variable. We denote its expected value, variance and standard deviation by $\mathbf{E}[X]$, $\mathbf{var}[X]$ and $\mathbf{std}[X]$, respectively.

Let f be a complex function. We denote the residuum of the function $f(z)$ at the point a by $\mathrm{Res}\,[f(z)|z = a]$ (see [1]). The imaginary unit is denoted by $\mathbf{i}$, the real and imaginary parts of the complex number z are denoted by $\Re(z)$ and $\Im(z)$, respectively.

1.2   Arbitrary Split Method

Let $\mathcal{P}_n$ be any randomized method of generating a random set of n points from the interval $[0, 1]$. The set $\mathcal{P}_n(\omega)$ defines a sequence $(x_1^{\mathcal{P}_n(\omega)}, \ldots, x_{n+1}^{\mathcal{P}_n(\omega)})$ of lengths of consecutive intervals. By definition $x_1^{\mathcal{P}_n(\omega)} + \ldots + x_{n+1}^{\mathcal{P}_n(\omega)} = 1$, hence $\frac{1}{n+1}(x_1^{\mathcal{P}_n(\omega)} + \ldots + x_{n+1}^{\mathcal{P}_n(\omega)}) = \frac{1}{n+1}$. Let

$$\mathbf{var}[\mathcal{P}_n] = \mathbf{E}\left[\frac{1}{n+1}\sum_{i=1}^{n+1}(x_i^{\mathcal{P}_n} - \frac{1}{n+1})^2\right]$$

and $\mathbf{std}[\mathcal{P}_n] = \sqrt{\mathbf{var}[\mathcal{P}_n]}$. We may treat the number $\mathbf{std}[\mathcal{P}_n]$ as a measure of non-uniformity of distribution of points from a random set of cardinality n generated by process $\mathcal{P}_n$. It is easy to check that

$$\mathbf{var}[\mathcal{P}] = \frac{1}{n+1} \left(\mathbf{E} \left[\sum_{i=1}^{n+1} (x_i^{\mathcal{P}_n})^2 \right] - \frac{1}{n+1} \right). \tag{1}$$

Let us fix some subset $a = \{a_1, \ldots, a_n\}$ of $[0,1]$, and let us now choose some random point $\zeta \in [0,1]$ according to the uniform distribution in $[0,1]$. Then there exists an unique subinterval I generated by points of a such that $\zeta \in I$. We call this interval a *randomly uniformly chosen interval*. Let us recall the following basic fact:

Theorem 1. *Let $\mathcal{P}_n$ be an arbitrary method of generation of a random subset $\{a_1, \ldots, a_n\}$ of the interval $[0,1]$. Then the number*

$$\mathbf{ELRI}[\mathcal{P}_n] = \mathbf{E} \left[\sum_{i=1}^{n+1} (x_i^{\mathcal{P}})^2 \right] \tag{2}$$

is the expected value of the length of randomly chosen interval.

Proof. Let $I_1(\omega), \ldots I_{n+1}(\omega)$ be the sequence of intervals generated by the set $\mathcal{P}(\omega)$. Let ζ be a random number from the uniform distribution on $[0,1]$ and let $L(\omega, \zeta)$ be the length of this interval $I_i(\omega)$ that $\zeta \in I_i(\omega)$. Then we have

$$\mathbf{ELRI}[\mathcal{P}_n] = \int_{\Omega \times [0,1]} L(\omega, x)(dP \times d\lambda)(\omega, x)$$

$$= \int_\Omega \left(\int_0^1 L(\omega, x) dx \right) dP(\omega) = \int_\Omega \left(\sum_{i=1}^{n+1} |I_i(\omega)| \cdot \Pr(x \in I_i(\omega)) \right) dP(\omega)$$

$$= \int_\Omega \left(\sum_{i=1}^{n+1} |I_i(\omega)|^2 \right) dP(\omega) = \mathbf{E} \left[\sum_{i=0}^{n+1} (x_i^{\mathcal{P}})^2 \right]. \qquad \square$$

1.3   The Uniform Split

Let us consider a sequence $X_1, \ldots, X_n$ of independent uniformly distributed in $[0,1]$ random variables. They generate a random subset $\{X_1, \ldots, X_n\}$ of the interval $[0,1]$ and we denote this method by $unif_n$ and call it *uniform split* (see [8]). The set $\{X_1, \ldots, X_n\}$ induces a partition of $[0,1]$ into n subintervals whose lengths, taken in proper order, will be denoted by $x_1, \ldots, x_{n+1}$. Then for any $t_1 \geq 0, \ldots, t_{n+1} \geq 0$ we have

$$\Pr(x_1 \geq t_1, \ldots, x_{n+1} \geq t_{n+1}) = (1 - (t_1 + \ldots + t_{n+1}))_+^n, \tag{3}$$

where $(a)_+ = \max\{a, 0\}$ (see Feller [3]).

412   J. Cichoń et al.

Let us consider the random variable x_1, i.e. the length of the first interval. From equation (3) we see that $\Pr(x_1 \geq t) = (1-t)^n$ for $t \in [0,1]$. Therefore the density of the variable x_1 equals $\varphi(t) = (1-(1-t)^n)' = n(1-t)^{n-1}$. Notice also that the remaining variables $x_2, \ldots, x_{n+1}$ have the same density as x_1.

Theorem 2. $\mathbf{ELRI}[unif_n] = \frac{2}{n+2}$

Proof. The result follows from the following direct calculations:

$$\mathbf{ELRI}[unif_n] = \mathbf{E}\left[\sum_{i=1}^{n+1} x_i^2\right] = (n+1)\mathbf{E}\left[x_1^2\right] = (n+1)\int_0^1 x^2\varphi(x)dx =$$

$$(n+1)n\int_0^1 x^2(1-x)^{n-1}dx = \frac{2}{n+2}\ .$$

Remark 1. We used the identity $\int_0^1 x^2(1-x)^{n-1}dx = \frac{2}{n(n+1)(n+2)}$ which can be proved by induction on n or can be evaluated by the use of the Euler beta function $\int_0^1 x^{a-1}(1-x)^{b-1}dx = \frac{\Gamma(b)\Gamma(b)}{\Gamma(a+b)}$.

From Theorem 1, Theorem 2 and Equation (1) we get:

Corollary 1. $\mathbf{var}[unif_n] = \frac{n}{(1+n)^2(2+n)}$.

2   The Binary Split

Let us fix a natural number n. Let us consider the following method of generation of a random subset of $[0,1]$ of cardinality n. We start from an empty set of points. Suppose we already have set A_k of points from $[0,1]$ and a new point a_{k+1} is to be added. We choose a random point $y \in [0,1]$ and select an interval I generated by points from the set A_k such that $y \in I$. Then we define a_{k+1} as the the middle point of the interval I and put $A_{k+1} = A_k \cup \{a_{k+1}\}$. We stop this process after n steps. We call this method the *binary split* and we denote this method by bin_n.

Our goal is to calculate the value of $\mathbf{var}[bin_n]$. Let us start with putting $f_0 = 1$ and $f_n = \mathbf{ELRI}[bin_n]$ for $n > 0$.

Lemma 1. *For all $n \in \mathbb{N}$ we have*

$$f_{n+1} = \frac{1}{2^{n+1}} \sum_{k=0}^{n} \binom{n}{k} f_k. \tag{4}$$

Proof. Let us consider the sequence $(\xi_1, \ldots, \xi_n, \xi_{n+1})$ of independent random variables uniformly distributed in the interval $[0,1]$ defined on a probabilistic space Ω and let $\omega \in \Omega$. At the beginning the number $\xi_1(\omega)$ splits $[0,1]$ into two equal parts: $[0,0.5]$ and $(0.5,1]$. Let us now define $A = \{i > 1 : \xi_i(\omega) \leq 0.5\}$ and $B = \{i > 1 : \xi_i(\omega) > 0.5\}$. Then the variables $\{\xi_i : i \in A\}$ can only split the $[0,0.5]$ interval while variables from the set $\{\xi_i : i \in B\}$ can only split the

$(0.5, 1]$. Note that $\{2\xi_i : i \in A\}$ split the interval $[0, 1]$, hence $\mathbf{E}[(2\xi_i)_{i \in A}] = f_{|A|}$. A similar observation is true for the sequence $(2\xi_i - 1)_{i \in B}$. Therefore we have

$$f_{n+1} = \sum_{A \subseteq \{2,\ldots,n+1\}} \left(\frac{1}{4} f_{|A|} + \frac{1}{4} f_{|B|}\right) \left(\frac{1}{2}\right)^n$$

$$= \frac{1}{2^{n+2}} \sum_{k=0}^{n} \binom{n}{k}(f_k + f_{n-k}) = \frac{1}{2^{n+1}} \sum_{k=0}^{n} \binom{n}{k} f_k. \qquad \square$$

Let

$$L_n = \prod_{j=n}^{\infty} \left(1 - \frac{1}{2^j}\right).$$

It is easy to calculate that $L_1 \simeq 0.2888$ and easy estimations shows that the inequalities $1 - \frac{4}{2^n} < L_n < 1 - \frac{1}{2^n}$ holds for each $n \geq 1$. We shall express numbers f_n in terms of numbers L_n.

Lemma 2. $f_n = \sum_{m \geq 0} (\frac{1}{2})^m (1 - (\frac{1}{2})^m)^n L_{m+1}$.

Proof. Let us consider the exponential generating function

$$x(t) = \sum_{n \geq 0} f_n \frac{t^n}{n!}$$

of the sequence $(f_n)_{n \geq 0}$. From equation (4) we get

$$x'(t) = \sum_{n \geq 0} f_{n+1} \frac{t^n}{n!} = \sum_{n \geq 0} \frac{1}{2^{n+1}} \sum_{k=0}^{n} \binom{n}{k} f_k \frac{t^n}{n!} \qquad (5)$$

$$= \frac{1}{2} \sum_{n \geq 0} \left(\sum_{k=0}^{n} \binom{n}{k} f_k\right) \frac{(t/2)^n}{n!}, \qquad (6)$$

hence the function $x(t)$ satisfies the following functional equation

$$2x'(2t) = x(t)e^t,$$

i.e. $x'(t) = \frac{1}{2}x(\frac{t}{2})e^{\frac{t}{2}}$. If we put $X(t) = x(t)e^{-t}$ then we obtain a slightly simpler equation

$$X'(t) = \frac{1}{2}X\left(\frac{t}{2}\right) - X(t),$$

which can be solved explicitly. Namely we have

$$X(t) = \sum_{n \geq 0} \frac{t^n}{n!}(-1)^n \prod_{k=1}^{n}\left(1 - \left(\frac{1}{2}\right)^k\right).$$

Since $x(t) = X(t)e^t$ we obtain

$$f_n = \sum_{k=0}^{n} \binom{n}{k}(-1)^k \prod_{j=1}^{k}\left(1 - \left(\frac{1}{2}\right)^j\right).$$

The above formula is hard to be calculated accurately because it contains large coefficients with alternating signs. Therefore we need to transform it into a more suitable form. We put into the Euler partition formula (see [7])

$$\prod_{k=0}^{\infty} \frac{1}{1-q^k z} = \sum_{n \geq 0} \frac{z^n}{\prod_{k=1}^{n}(1-q^k)}$$

values $z = q^{a+1}$ and $q = \frac{1}{2}$, and get

$$\frac{1}{\prod_{k=a+1}^{\infty}(1-(\frac{1}{2})^k)} = \sum_{n \geq 0} \frac{(\frac{1}{2})^{(a+1)n}}{\prod_{k=1}^{n}(1-(\frac{1}{2})^k)}.$$

After multiplying both sides of this equality by L_1 we get

$$\prod_{j=1}^{a}\left(1-\frac{1}{2^j}\right) = \sum_{n \geq 0}\left(\frac{1}{2}\right)^{(a+1)n} L_{n+1},$$

and hence we obtain

$$f_n = \sum_{k=0}^{n}\binom{n}{k}(-1)^k \sum_{m \geq 0}\left(\frac{1}{2}\right)^{(k+1)m} L_{m+1} =$$

$$= \sum_{m \geq 0} L_{m+1}\left(\frac{1}{2}\right)^m \sum_{k=0}^{n}\binom{n}{k}(-1)^k\left(\left(\frac{1}{2}\right)^m\right)^k =$$

$$= \sum_{m \geq 0}\left(\frac{1}{2}\right)^m\left(1-\left(\frac{1}{2}\right)^m\right)^n L_{m+1},$$

which proves the lemma. □

Remark 2. From Lemma 2 we may deduce that $f_{n+1} < f_n$ for each n.

Let us consider now the following function

$$\varphi_n(x) = \left(\frac{1}{2}\right)^x\left(1-\left(\frac{1}{2}\right)^x\right)^n$$

defined on the interval $[0,\infty)$. The function φ_n has the global maximum at point $\log_2(n+1)$ and $\varphi_n(\log_2(n+1)) = \frac{1}{ne}+o(\frac{1}{n})$. Notice that $\sum_{m \geq 0}(\frac{1}{2})^m(1-(\frac{1}{2})^m)^n = \sum_{m \geq 0}\varphi_n(m)$. Moreover, $\int_0^{\infty}\varphi_n(x)dx = \frac{1}{(1+n)\ln 2}$. From these observations we deduce that

$$\sum_{m \geq 0}(\frac{1}{2})^m(1-(\frac{1}{2})^m)^n = O(\frac{1}{n}). \tag{7}$$

Lemma 3. $f_n = \sum_{m \geq 0}\frac{1}{2^m}(1-\frac{1}{2^m})^n + o(\frac{1}{n})$.

Proof. The proof is done by a simple estimation. Let us first show the following approximation

$$\sum_{m=0}^{\log_2 \sqrt{n}} \frac{1}{2^m}(1 - \frac{1}{2^m})^n = o(\frac{1}{n}).$$

This fact follows immediately from monotonicity of the function φ_n on the interval $[0, \log(n+1)]$. Namely,

$$\varphi_n(\log_2 \sqrt{n}) < \frac{1}{\sqrt{n}e^{\sqrt{n}}},$$

so

$$\sum_{m=0}^{\log_2 \sqrt{n}} \frac{1}{2^m}(1 - \frac{1}{2^m})^n \le \frac{\log_2 \sqrt{n}}{\sqrt{n}e^{\sqrt{n}}} \le \frac{1}{e^{\sqrt{n}}}$$

and $\frac{1}{e^{\sqrt{n}}} = o(\frac{1}{n})$.

Observe that if $k > \log_2 \sqrt{n}$ then $L_k > 1 - \frac{4}{\sqrt{n}}$, so we have

$$(1 - \frac{4}{\sqrt{n}}) \sum_{m>\log_2 \sqrt{n}} \frac{1}{2^m}(1 - \frac{1}{2^m})^n \le \sum_{m>\log_2 \sqrt{n}} \frac{1}{2^m}(1 - \frac{1}{2^m})^n L_{m+1}$$

and from equation (7) we obtain

$$\left| \sum_{m>\log_2 \sqrt{n}} \frac{1}{2^m}(1 - \frac{1}{2^m})^n - \sum_{m>\log_2 \sqrt{n}} \frac{1}{2^m}(1 - \frac{1}{2^m})^n L_{m+1} \right| \le \frac{C}{n\sqrt{n}} = o(\frac{1}{n}),$$

which proves the lemma. $\square$

Lemma 4. $\sum_{k\ge0} \frac{1}{2^k}(1 - \frac{1}{2^k})^n = \sum_{k=0}^{n} \binom{n}{k}(-1)^k \frac{1}{1-(\frac{1}{2})^{1+k}}$

Proof. The proof follows from following transformations:

$$\sum_{k\ge0} \frac{1}{2^k}(1 - \frac{1}{2^k})^n = \sum_{k\ge0}(\frac{1}{2^k}) \sum_{l=0}^{n} \binom{n}{l}(-1)^l \frac{1}{2^{kl}} =$$

$$\sum_{l=0}^{n} \binom{n}{l}(-1)^l \sum_{k\ge0}(\frac{1}{2})^{kl+k} = \sum_{l=0}^{n} \binom{n}{l}(-1)^l \frac{1}{1 - (\frac{1}{2})^{1+l}}. \qquad \square$$

Theorem 3. *The sequence f_n satisfies*

$$f_n = \frac{1}{n+1}\left(\frac{1}{\ln 2} + \omega(\log_2(n+1)) + \eta(n)\right) + O(\frac{1}{n^2}),$$

where ω is a periodic function with period 1 such that $|\omega(x)| < 1.42602 \cdot 10^{-5}$ and $|\eta(x)| < 6.72 \cdot 10^{-11}$.

In the proof of this theorem we use a method of the treatment of oscillating sums attributed to S.O. Rice by D.E. Knuth (see [7]).

Proof. Let us, for simplicity, denote

$$s_n = \sum_{k=0}^{n} \binom{n}{k} (-1)^k \frac{1}{1 - (\frac{1}{2})^{1+k}}$$

and let us consider the following function

$$f(z) = \frac{1}{1 - (\frac{1}{2})^{1-z}}$$

with complex argument z. Then f is a meromorphic function with single poles at points

$$\mathfrak{z}_k = 1 + \frac{2\pi k}{\ln 2} \mathbf{i},$$

where $k \in \mathbb{Z}$, $\mathbf{i}$ is the imaginary unit and

$$s_n = \sum_{k=0}^{n} \binom{n}{k} (-1)^k f(-k).$$

Let $B(x, y) = \Gamma(x)\Gamma(y)/\Gamma(x + y)$ be the Euler beta function. The function $B(n + 1, z)$ has single poles at points $0, -1, \ldots, -n$ and

$$\text{Res}\,[B(n + 1, z)|z = -k] = (-1)^k \binom{n}{k}.$$

(see [7] for details). Notice that the function f is holomorphic on the half-plane $\Re(z) < 0.5$. Therefore

$$s_n = \sum_{k=0}^{n} \text{Res}\,[B(n + 1, z)f(z)|z = -k]. \tag{8}$$

Let us consider a big rectangle C_k with end-point $\pm k \pm (2k+1)\pi \mathbf{i}/\ln 2$, where k is a natural number. It is quite easy to check that

$$\lim_{k \to \infty} \oint_{C_k} B(n + 1, z)f(z)dz = 0$$

(in proof of this fact the equality $f(1 + (2k + 1)\pi \mathbf{i}/\ln 2) = 1/2$ plays a crucial role). Therefore, using Cauchy Residue Theorem (see [1]), we get

$$s_n = -\sum \{\text{Res}\,[B(n + 1, z)f(z)|z = \mathfrak{z}_k] : k \in \mathbb{Z}\}. \tag{9}$$

Further, it can be easily checked that

$$\text{Res}\,[B(n + 1, z)f(z)|z = 1] = -\frac{1}{(n + 1)\ln 2}. \tag{10}$$

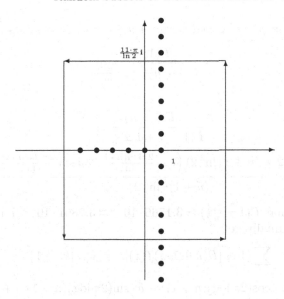

Fig. 1. Singular points of the function $B(4, z)f(z)$ and the contour of integration C_5

This first residue gives us the first part of the approximation of the number s_n. Generally, we have

$$\text{Res}\left[B(n+1, z)f(z)|z = 3k\right] = -\frac{\Gamma(1+n)\Gamma(1 + \frac{2k\pi i}{\ln 2})}{\Gamma(2+n+\frac{2k\pi i}{\ln 2})\ln 2}.$$

Notice that if $x \in \mathbb{R}$ then

$$\left|\frac{\Gamma(n+1)\Gamma(1+ix)}{\Gamma(2+n+ix)}\right| = \frac{n!}{\sqrt{(1^2+x^2)(2^2+x^2)\cdots((n+1)^2+x^2)}} \leq$$

$$\frac{1}{n+1} \cdot \frac{1}{\sqrt{(1+\frac{x^2}{1^2})(1+\frac{x^2}{2^2})\cdots(1+\frac{x^2}{n^2})}}.$$

Let $a = \frac{2\pi}{\ln 2}$. It can be checked numerically that

$$\sum_{k=2}^{\infty} \frac{1}{\sqrt{\prod_{m=1}^{100}(1+(ka/m)^2)}} \approx 2.32781 \cdot 10^{-11}.$$

Therefore

$$\left|\sum\{\text{Res}\left[B(n+1, z)f(z)|z = 3k\right] : |k| \geq 2\}\right| \leq \frac{6.72 \cdot 10^{-11}}{n+1} \qquad (11)$$

for all $n \geq 100$. Next, we use the following well known approximation formula

$$\frac{\Gamma(z+a)}{\Gamma(z+b)} = z^{a-b}\left(1 + \frac{(a-b)(a+b-1)}{2z} + O(\frac{1}{z^2})\right),$$

which holds when $|z| \to \infty$ and $|\arg(z + a)| < \pi$ to expressions

$$\frac{\Gamma(1+n)}{\Gamma(1+n+1\pm\frac{2\pi i}{\ln 2})},$$

and we obtain

$$\frac{\Gamma(1+n)}{\Gamma(1+n+1\pm\frac{2\pi i}{\ln 2})} =$$

$$\frac{(2\pi^2 + i\pi \ln 2 + (n+1)\ln^2 2)\left(\cos\frac{2\pi i \ln(n+1)}{\ln 2} \mp i\sin\frac{2\pi i \ln(n+1)}{\ln 2}\right)}{(n+1)^2\ln^2 2} + O(\frac{1}{n^2}).$$

After noticing that $\Gamma(1\pm\frac{2\pi i}{\ln 2}) \approx 3.1766 \cdot 10^{-6} \mp 3.7861 \cdot 10^{-6} \cdot i$ and some simple calculations we finally get

$$\sum\{\operatorname{Res}\left[B(n+1,z)f(z)|z=\mathfrak{z}_k\right] : |k| = 1\} = \qquad (12)$$

$$\frac{10^{-6}}{n+1}(a \cdot \cos(2\pi \log_2(n+1)) - b \cdot \sin(2\pi \log_2(n+1))) + O(\frac{1}{n^2}),$$

where $a \approx 9.166$ and $b \approx 10.924$. Putting together Equations (9), (10), (11) and (12) we obtain the thesis of the theorem. $\qquad\qquad\qquad\qquad\qquad\square$

From the last Theorem and Equation (1) we obtain

Corollary 2

$$\mathbf{var}[bin_n] = \frac{1}{(n+1)^2}\left(\frac{1}{\ln 2} - 1 + \omega(\log_2(n+1)) + \eta(n)\right) + O(\frac{1}{n^3}),$$

where ω and η are the function from Theorem 3.

Notice that $\frac{1}{\ln 2} - 1 \approx 0.4427$, therefore $\mathbf{std}[bin_n] \approx \frac{0.665}{n}$, hence $\mathbf{std}[bin_n]$ is significantly smaller than the value $\mathbf{std}[unif_n] \approx \frac{1}{n}$ (see Corollary 1). From this we conclude that the binary split process generates a more uniform distribution of random points in the interval $[0,1]$ than the uniform one.

Remark 3. The main influence on the asymptotic behavior of the sequence (f_n) is played by the three main poles of the function $f(z) = (1 - (\frac{1}{2})^{1-z})^{-1}$: $\mathfrak{z}_0 = 1$, $\mathfrak{z}_1 = 1 + \frac{2\pi i}{\ln 2}$ and $\mathfrak{z}_{-1} = 1 - \frac{2\pi i}{\ln 2}$. The first one, located at point 1, is responsible for the component $\frac{1}{(n+1)\ln 2}$. The next two poles are responsible for relatively small oscillations of the sequence f_n. The size of oscillations is relatively small because $\Im(1 + 2\pi i/\ln 2) \approx 10$ and the function Γ decreases rapidly when the imaginary part of a number grows.

Remark 4. Some aspects of the binary split model, namely the properties of the length of the first node, were investigated P. Flajolet (see [4]) and later by P. Kirchenhofer and H. Prodinger (see [6]) in their analyses of properties of the R. Morris probabilistic counter (see [10]). In their investigations a fluctuation factor of the form $\omega(\log_2 n)$ appears, too.

3 Discussion

Let $\mathcal{P}_n$ be any randomized method of generating random subsets of interval $[0, 1]$ of cardinality n. Let

$$\mathbf{CV}[\mathcal{P}_n] = \frac{\mathrm{std}[\mathcal{P}_n]}{\mathbf{E}\,[\mathcal{P}_n]}$$

denotes the coefficient of variation of $\mathcal{P}_n$. The number $\mathbf{CV}[\mathcal{P}_n]$ is a measure of dispersion of lengths of intervals generated by the method $\mathcal{P}_n$. We have proved that $\mathbf{CV}[unif_n] \simeq 1$ and $\mathbf{CV}[bin_n] \simeq 0.665$. Therefore the random subsets generated by the binary split methods has smaller dispersion than subsets generated by the uniform method.

The length of intervals in the uniform split with n elements varies between $\frac{1}{n^2}$ and $\frac{\ln(n)}{n}$. To be more precise let us consider the following two random variables $\min(\mathcal{P}_n) = \min\{x_i^{\mathcal{P}_n}, \ldots, x_{n+1}^{\mathcal{P}_n}\}$, $\max(\mathcal{P}_n) = \max\{x_i^{\mathcal{P}_n}, \ldots, x_{n+1}^{\mathcal{P}_n}\}$. It follows almost directly from Equation (3) that $\mathbf{E}\,[\min(unif_n)] = \frac{1}{n^2}$. Moreover $\mathbf{E}\,[\max(unif_n)] \sim \frac{\ln n}{n} + \frac{\gamma}{n}$ (see [2]). These observations were done by several authors working with P2P protocols–see e.g. [5].

Let $m_o = M_0 = 1$ and $m_n = \mathbf{E}\,[\min(bin_n)]$ and $M_n = \mathbf{E}\,[\max(bin_n)]$ for $n > 0$. Using similar arguments as in Lemma 1 we may show that

$$m_{n+1} = \frac{1}{2^{n+1}} \sum_{i=0}^{n} \binom{n}{i} \min(m_i, m_{n-i}),$$

$$M_{n+1} = \frac{1}{2^{n+1}} \sum_{i=0}^{n} \binom{n}{i} \max(M_i, M_{n-i}).$$

Numerical calculations show that $\frac{0.4}{\ln \ln n} \frac{1}{n} \leq m_n < M_n \leq \frac{2.2 \ln \ln n}{n}$ for each $n \geq 5$, however we do not have a precise mathematical proof of this fact. This shows that the length of intervals in the binary split is much better concentrated near its medium value $\frac{1}{n}$ than in the uniform split.

4 Conclusion

Our mathematical analysis was motivated by the problems arising in computer science. It is well known that the capacity of Chord (see [12] and [9]) nodes' areas (intervals) is a random variable with large variation. Our calculation of $\sigma(unif_n)$ confirms this fact. Now let us recall that amount of data and number of requests passed via node in Chord is proportional to the length of its area. Hence, large variation of area size introduces a discrepancy between nodes' workload.

This problem can be partially solved by a very simple modification of the original Chord protocol. The modification is based on *binary split* method which can be easily embedded into Chord's protocol. Namely, we need to modify only one of Chord's procedure, namely the procedure *join*. The original „join" procedure accepts a new node at an arbitrarily chosen position ζ_i. In the modification we can use ζ_i only to determine which interval the new node P will split upon arrival. Then the target interval is slitted into two halves and the node currently responsible for the interval will keep roughly half of the resources and P takes over the responsibility for the rest.

In other words, instead of using random protocol address, the new node randomly and uniformly picks a point and joins the Chord protocol precisely in the middle of the interval that is controlled by the node responsible for chosen point. The method described above can be treated as Chord protocol with a one dimensional CAN's split method (see [11]).

In reality Chord is a dynamic structure; nodes both leave and join the network. It is possible to modify the structure of remaining nodes after a single node leaves the system in such a way that after this modification we shall obtain a division generated by the binary split method, however, this is a quite complicated procedure. Our proposition is to ignore this fact.

We have made a lot of numerical experiments for checking what happens when we use the binary split only in the „join" procedure. Figure 2 contains a summary of one experiment. In this experiment we have build a Chord structure based on the binary split method with 10^5 nodes and later we successively removed one randomly chosen node in the „normal way" and add one node using the binary split regime $2 \cdot 10^5$ times. We observed that afer the initialization phase the variable $\mathbf{CV}[\mathcal{P}]$ increase, but later its value stabilize and in the stable regime

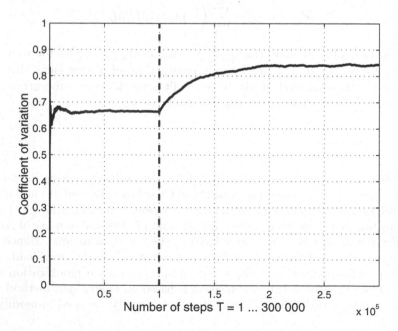

Fig. 2. Experiment with 10^5 nodes

we have $\mathbf{CV}[\mathcal{P}] \approx 0.85$. This coefficient of variation is bigger than the coefficient of variation of the binary split but it is less than the coefficient of variation of the uniform split. However, this behavior requires and still awaits for precise theoretical explanation.

References

1. Cartan, H.: Elementary Theory of Analytic Functions of One or Several Complex Variables. Herman, Paris (1973)
2. Devroye, L.: Laws of the iterated logarithm for order statistics of uniform spacings. The Annals of Probability 9(5), 860–867 (1981)
3. Feller, W.: An Introduction to Probability Theory and Its Applications, vol. II. John Wiley and Sons Inc, New York (1965)
4. Flajolet, P.: Approximate counting: A detailed analysis. BIT 25, 113–134 (1985)
5. King, V., Saia, J.: Choosing a random peer. In: Proceedings of the 23rd Annual ACM Symposium on Principles of Distributed Computing, pp. 125–130 (2004)
6. Kirschenhofer, P., Prodinger, H.: Approximate counting: an alternative approach. Informatique Theorique et Applications 25, 43–48 (1991)
7. Knuth, D.E. (ed.): Sorting and Searching. In: Knuth, D.E. (ed.) The art of computer programming, 3rd edn. The art of computer programming, Addison-Wesley, Reading, Massachusetts (1997)
8. Kopociński, B.: A random split of the interval [0,1]. Aplicationes Mathematicae 31, 97–106 (2004)
9. Liben-Nowell, D., Balakrishnan, H., Karger, D.: Analysis of the evolution of peer-to-peer systems. In: 21st ACM Symposium on Principles of Distributed Computing (PODC), Monterey, CA, July 2002 (2002)
10. Morris, R.: Counting large numbers of events in small registers. Communications of The ACM 21, 161–172 (1978)
11. Ratnasamy, S., Francis, P., Handley, M., Karp, R., Shenker, S.: A scalable content-addressable network. In: Proceedings of the ACM SIGCOMM '01 Conference, San Diego, California, USA, August 2001 (2001)
12. Stoica, I., Morris, R., Karger, D., Kaashoek, M.F., Balakrishnan, H.: Chord: A scalable peer-to-peer lookup service for internet applications. In: Proceedings of the ACM SIGCOMM '01 Conference, San Diego, California, USA, August 2001 (2001)

The Cover Time of Random Digraphs

Colin Cooper[1] and Alan Frieze[2],[*]

[1] Department of Computer Science, King's College, University of London,
London WC2R 2LS, UK
[2] Department of Mathematical Sciences, Carnegie Mellon University,
Pittsburgh PA15213, USA

Abstract. We study the cover time of a random walk on the random digraph $D_{n,p}$ when $p = \frac{d \log n}{n}, d > 1$. We prove that **whp** the cover time is asymptotic to $d \log \left(\frac{d}{d-1} \right) n \log n$.

1 Introduction

Let $D = (V, E)$ be a strongly connected digraph with $|V| = n$, and $|E| = m$. For $v \in V$ let C_v be the expected time taken for a simple random walk $\mathcal{W}_v = (\mathcal{W}_v(t), t = 0, 1, \dots$ on D starting at v, to visit every vertex of D. The *cover time* C_D of D is defined as $C_D = \max_{v \in V} C_v$.

For connected undirected graphs, the cover time is well understood, and has been intensely studied. It is an old result of Aleliunas, Karp, Lipton, Lovász and Rackoff [2] that $C_G \leq 2m(n-1)$. It was shown by Feige [8], [9], that for any connected graph G, the cover time satisfies $(1 - o(1))n \ln n \leq C_G \leq (1+o(1))\frac{4}{27}n^3$. As an example of a graph achieving the lower bound, the complete graph K_n has cover time determined by the Coupon Collector problem. The *lollipop* graph consisting of a path of length $n/3$ joined to a clique of size $2n/3$ gives the asymptotic upper bound for the cover time.

In a sequence of papers we have investigated the cover time of various classes of random graph. The main results of these papers can be summarised as follows:

- [4] If $p = d \ln n / n$ and $d > 1$ then **whp** $C_{G_{n,p}} \sim d \ln \left(\frac{d}{d-1} \right) n \ln n$.
- [7] Let $d > 1$ and let x denote the solution in $(0, 1)$ of $x = 1 - e^{-dx}$. Let X_g be the giant component of $G_{n,p}$, $p = d/n$. Then **whp** $C_{X_g} \sim \frac{dx(2-x)}{4(dx-\ln d)}n(\ln n)^2$.
- [5] If $G_{n,r}$ denotes a random r-regular graph on vertex set $[n]$ with $r \geq 3$ then **whp**
$$C_{G_{n,r}} \sim \frac{r-1}{r-2}n \ln n.$$
- [6] If G_m denotes a *preferential attachment graph* of average degree $2m$ then **whp**
$$C_{G_m} \sim \frac{2m}{m-1}n \ln n.$$

In this paper we turn our attention to random directed graphs. Let $D_{n,p}$ be the random digraph with vertex set $V = [n]$ where each possible directed edge

[*] Supported in part by NSF grant CCF0502793.

M. Charikar et al. (Eds.): APPROX and RANDOM 2007, LNCS 4627, pp. 422–435, 2007.
© Springer-Verlag Berlin Heidelberg 2007

(i,j), $i \neq j$ is included with probability p. It is known that if $np = d \ln n + \omega$ where $\omega = (d-1) \ln n \to \infty$ then $D_{n,p}$ is strongly connected **whp**. (If ω as defined tends to $-\infty$ then **whp** $D_{n,p}$ is not strongly connected). We discuss the covertime of $D_{n,p}$ for p at or above the strong connectivity threshold.

Theorem 1. *Suppose that $np = d \ln n$ where $d-1$ is at least a positive constant. Then* **whp**

$$C_{D_{n,p}} \sim d \ln \left(\frac{d}{d-1} \right) n \ln n.$$

Note that if $d = d(n) \to \infty$ with n then we have $C_{D_{n,p}} \sim n \ln n$.

Our analysis is based on Lemma 1 below, which is proved in its current state in [7]. In order to apply this lemma we need to have estimates of (i) the steady state $\pi(v)$, $v \in V$ of our walk and (ii) the mixing time. For an undirected graph G, (i) is trivial, we just take $\pi(v) = \deg(v)/2m$ where deg denotes degree and m is the number of edges in G. There is no such simple formula for digraphs and the main technical task for us is to find good estimates. We summarise our result concerning the steady state as follows:

Theorem 2. *Suppose that $np = d \ln n$ where $d-1$ is at least a positive constant. Then* **whp**

$$\pi_v \sim \frac{deg^-(v)}{m} \qquad \forall v \in V$$

where $deg^-(v)$ is the in-degree of v and m is the number of edges of $D_{n,p}$.

Note that if $d = d(n) \to \infty$ with n then the steady state distribution is asymptotically uniform.

Once we know the steady state distribution, it is basically plain sailing. We can plug into the program from the previously cited papers.

The next section describes Lemma 1. Section 3 deals with estimating the steady state distribution. We then estimate the cover time asymptotically in Section 4.

There is not room for complete proofs of every statement within the page limit. A full version of the paper can be obtained from http://www.math.cmu.edu/ af1p/Dnp.pdf.

2   Main Lemma

In this section D denotes a fixed strongly connected digraph with n vertices. A random walk $\mathcal{W}_u$ is started from a vertex u. Let $\mathcal{W}_u(t)$ be the vertex reached at step t, let P be the matrix of transition probabilities of the walk and let $P_u^{(t)}(v) = \mathbf{Pr}(\mathcal{W}_u(t) = v)$. We assume the random walk $\mathcal{W}_u$ on G is ergodic with steady state distribution π.

Let

$$d(t) = \max_{u,x \in V} |P_u^{(t)}(x) - \pi_x|,$$

and let T be such that, for $t \geq T$

$$\max_{u,x \in V} |P_u^{(t)}(x) - \pi_x| \leq n^{-3}. \tag{1}$$

Fix two vertices u, v. Considering the walk $\mathcal{W}_v$, starting at v, let $r_t = \mathbf{Pr}(\mathcal{W}_v(t) = v)$ be the probability that this walk returns to v at step $t = 0, 1, \dots$. Let

$$R_T(z) = \sum_{j=0}^{T-1} r_j z^j \tag{2}$$

and

$$\lambda = \frac{1}{KT} \tag{3}$$

for some sufficiently large constant K.

An almost identical lemma was first proved in [5]. For $t \geq T$ let $A_t(v)$ be the event that $\mathcal{W}_u$ does not visit v in steps $T, T+1, \dots, t$.

Lemma 1. *Suppose that*

(a) *For some constant $\theta > 0$, we have*

$$\min_{|z| \leq 1+\lambda} |R_T(z)| \geq \theta.$$

(b) $T^2 \pi_v = o(1)$ *and* $T\pi_v = \Omega(n^{-2})$.

Let

$$p_v = \frac{\pi_v}{R_T(1)(1 + O(T\pi_v))}, \tag{4}$$

where $R_T(1)$ is from (2).
Then for all $t \geq T$,

$$\mathbf{Pr}(A_t(v)) = \frac{(1 + O(T\pi_v))}{(1 + p_v)^t} + o(e^{-\lambda t/2}). \tag{5}$$

3   Estimating the Steady State

3.1   Two Useful Lemmas

We first give a simple lemma concerning the degree sequence of $D_{n,p}$. It can easily be proven by the use of the first and second moment method. Let $\deg^+$ denote out-degree and $\deg^-$ denote in-degree.

We deal with two cases. In Case (a) we have $d = \omega(n) \to \infty$ and in Case (b) we have $d > 1$ a fixed constant.

Lemma 2. *Let $\epsilon_1 = \omega^{-1/3}$, $c_a = 1 - \epsilon_1$, $C_a = 1 + \epsilon_1$, $c_b = d'/2$ and $C_b = d'' + 1$ where $d' < d''$ are the two roots of $x \ln(e/x) = 1 - \ln n/np$. Then,*

(a) *If $np = \omega \ln n$ then for all $v \in V$, $deg^+(v), deg^-(v) \in I_a = [c_a np, C_a np]$, with probability $\rho_a = 1 - O(n^{-K})$ for any $K > 0$.*

(b) *If $np = d \ln n$ where $d = O(1)$ then with probability $1 - n^{-\psi}$ where $\psi = \psi(d) > 0$:*

1. *Simultaneously, for all $v \in V$, $deg^+(v), deg^-(v) \in I_b = [c_b np, C_b np]$*
2. *For $k \in I_b$ there are $\leq 2(nep/k)^k n^{1-d}$ vertices v with $deg^-(v) = k$.*
3. *Let $k^* = (d-1) \ln n$. There are $\sim (nep/k^*)^{k^*} n^{1-d} = n^{\gamma_d + o(1)}$ vertices v with $deg^-(v) = k^*$, where $\gamma_d = (d-1) \ln(d/(d-1))$.*

Let $\mathcal{D}^\pm$ be the events that in- or out-degrees are within the bounds specified in Lemma 2. Let $\mathcal{D} = \mathcal{D}^+ \cap \mathcal{D}^-$. Refining this, for a set of vertices S, or for cases (a), (b) of Lemma 2 we let $\mathcal{D}^\pm(S), \mathcal{D}_a^\pm(S), \mathcal{D}_b^\pm(S)$ be the same conditions restricted to the vertices in S or the relevant case. Also, we will often drop suffices and use the constants c, C.

Lemma 3. *Let B_I denote the binomial random $B(n-1, p) |_{B(n-1,p) \in I}$[1] where $I = [cnp, Cnp]$, where $c < 1 < C$ are constants such that ρ, the probability that $B(n-1, p) \in I = 1 - o(1/n)$. For $\xi = X, Y, Z$ let $W_0^\xi = 1$ and let*

$$W_t^X = \sum_{i=1}^{B_0-\theta} \frac{A_i^X}{B_0}, \qquad W_t^Y = \sum_{i=1}^{B_0-\theta} \frac{A_i^Y}{B_i}, \qquad W_t^Z = \sum_{i=1}^{B_0-\theta} \frac{A_i^Z}{B_i} + \sum_{j=1}^{\theta} \frac{A_j^Z}{cnp}.$$

where the $B_i, i \geq 0$ are independent copies of B_I, the $A_i, i \geq 1$ are independent copies of W_{t-1}^ξ, and θ is a fixed non-negative integer. Then for $|\lambda| \leq M$, $M > 1$, $Mt = o(np)$, sufficiently large n, and $\xi = X, Y, Z$,

$$\mathbf{E}(e^{\lambda W_t^\xi}) \leq \exp\left\{ \lambda + \frac{L_\xi(5+\theta)|\lambda| t}{cnp} \right\}, \tag{6}$$

where $L_X = 1, L_Y = L_Z = M$.

Proof is omitted. ∎

We will apply Lemma 3 as follows. Let (6) be written as $\mathbf{E}(e^{\lambda W_t}) \leq e^{\lambda + \gamma |\lambda|}$, and recall that $|\lambda| \leq M$. By the Markov Inequality,

$$\mathbf{Pr}(W_t \leq 1 - A) \leq e^{-M(A-\gamma)} \tag{7}$$
$$\mathbf{Pr}(W_t \geq 1 + B) \leq e^{-M(B-\gamma)}. \tag{8}$$

We next give a brief outline of our approach to approximating the stationary distribution π. Iterating the equation $\pi = \pi P$, k times gives $\pi = \pi P^k$. For fixed y this gives $\pi_y = \sum_{x \in V} \pi_x P_x^{(k)}(y)$. By bounding $P_x^{(k)}(y)$ from above and below by values independent of x, i.e. $P_x^{(k)}(y) \sim \theta_y$ we obtain $\pi_y \sim \theta_y$.

---

[1] We will use the notation $Z |_\mathcal{E}$ to denote random variable Z conditioned on the occurrence of the event $\mathcal{E}$.

3.2   Lower Bounds on the Stationary Distribution

To bound $P_x^{(k)}(y)$ from below, we consider random walks between x and y consisting of simple directed (x, y)-paths of length k. There are three cases;

Case I: $np = d \ln n$, where $d \geq 1$ constant or $d = \omega \to \infty$ and $\omega \leq n^{3/10}$.
Case II: $n^{3/10} \leq \omega \leq n^{3/5}$.
Case III: $\omega \geq n^{3/5}$.

Most of the work is involved in the proof of the first case.

Lower Bounds for Case I

Lemma 4. Whp,

$$\pi_v \geq (1 - o(1))\frac{deg^-(v)}{m} \text{ for all } v \in V,$$

where m is the number of edges in $D_{n,p}$.

Proof. Let

$$\ell = \left\lfloor \frac{2}{3} \log_{np} n \right\rfloor. \tag{9}$$

Fix $x \in V$ and using Breadth First Search construct the sets $X_0 = \{x\}$, $X_1, \ldots, X_\ell$ where $X_{i+1} = N^+(X_i) \setminus (X_0 \cup \cdots \cup X_i)$ for $0 \leq i < \ell$. Here $N^+(S)$ is the set of out-neighbours of set S.

Let $X = \bigcup_{i=0}^\ell X_i$ and let T_X denote the BFS tree constructed in this manner. If $w \in X_{i+1}$ is the out-neighbour of more than one vertex of X_i, we only keep the *first* edge into w for T_X.

Given X and y, ($y = x$ is allowed) we define $Y_0 = \{y\}, Y_1, \ldots, Y_\ell$ where $Y_{i+1} = N^-(Y_i) \setminus (X \cup Y_0 \cup \cdots \cup Y_i)$ for $0 \leq i < \ell$. If $w \in Y_{i-1}$ is the in-neighbour of more than one vertex of Y_i, we only keep the first edge into w for T_Y. Let $Y = \bigcup_{i=0}^\ell Y_i$ and let T_Y denote the BFS tree constructed in this manner.

Observe that

$$\mathcal{D}^+ \text{ implies } |X| \leq (Cnp)^\ell \leq n^{2/3+o(1)}. \tag{10}$$
$$\mathcal{D}^- \text{ implies } |Y| \leq (Cnp)^\ell \leq n^{2/3+o(1)}. \tag{11}$$

For $u \in X_i$ let P_u denote the path from x to u in T_X and

$$\alpha_{i,u} = \prod_{\substack{w \in P_u \\ w \neq u}} \frac{1}{deg^+(w)} \leq \mathbf{Pr}(\mathcal{W}_x(i) = u)$$

and for $v \in Y_i$ let Q_v denote the path from v to y in T_Y and

$$\beta_{i,v} = \prod_{\substack{w \in Q_v \\ w \neq y}} \frac{1}{deg^+(w)} \leq \mathbf{Pr}(\mathcal{W}_v(i) = y). \tag{12}$$

Given $D_{n,p}$ we have

$$P_x^{(2\ell+1)}(y) \geq Z = Z(x,y) = \sum_{\substack{u \in X_\ell \\ v \in Y_\ell}} \alpha_{\ell,u} \beta_{\ell,v} \frac{1_{uv}}{\deg^+(u)} \tag{13}$$

where 1_{uv} is the indicator for the existence of the edge (u,v) and we take $\frac{1_{uv}}{\deg^+(u)} = 0$ if $\deg^+(u) = 0$.

Let $\mathcal{C} = \mathcal{C}(x,y)$ denote $X_i, Y_i, 0 \leq i \leq \ell$ and the collection of edge sets $(X_{i-1} : X_i), 1 \leq i \leq \ell$ in T_X (resp. $(Y_i : Y_{i-1}), 1 \leq i \leq \ell$ in T_Y). We have, with $q = 1 - p$,

$$\mathbf{E}(Z \mid \mathcal{C}) = \sum_{\substack{u \in X_\ell \\ v \in Y_\ell}} \alpha_{\ell,u} \beta_{\ell,v} \sum_{k=1}^{n-1} \binom{n-1}{k} p^k q^{n-1-k} \frac{k}{n-1} \frac{1}{k}$$

$$= \frac{(1+o(1))}{n} \left(\sum_{u \in X_\ell} \alpha_{\ell,u} \right) \left(\sum_{v \in Y_\ell} \beta_{\ell,v} \right). \tag{14}$$

We will show that the distribution of $\mathcal{C}$ is such that for $\epsilon_X, \epsilon_Y = o(1)$,

$$\mathbf{Pr}\left(\sum_{u \in X_\ell} \alpha_{\ell,u} < 1 - \epsilon_X \Big| \mathcal{D}^+ \right) = o(n^{-2}), \tag{15}$$

$$\mathbf{Pr}\left(\sum_{v \in Y_\ell} \beta_{\ell,v} < (1 - \epsilon_Y) \frac{\deg^-(y)}{m} \Big| \mathcal{D}^- \right) = o(n^{-2}). \tag{16}$$

Let

$$\epsilon_X = \frac{1}{\ln \ln \ln n}.$$

Now for each $v \in X_i$ and $j > i$ there is a unique path $v = v_i, v_{i+1}, \ldots, v_j$ from v to X_j in T_X. For such a path, let

$$\gamma_{i,j;v,v_j} = \prod_{k=i}^{j-1} \frac{1}{\deg^+(v_k)}.$$

Now consider the equation

$$\sum_{u \in X_\ell} \alpha_{\ell,u} = \sum_{u \in X_\ell} \gamma_{0,\ell;x,u} = \sum_{w \in N_D^+(x)} \sum_{u \in X_{\ell,w}} \frac{\gamma_{1,\ell;w,u}}{\deg^+(x)} \tag{17}$$

where $X_{\ell,w} = \{u \in X_\ell : \text{the path from } x \text{ to } u \text{ in } T_X \text{ goes through } w \in X_1\}$ and $N_D^\pm(x)$ is the set of in/out-neighbours of x in T_X.

This leads us to claim that given $\mathcal{D}^+$,

$\sum_{u \in X_\ell} \gamma_{0,\ell;x,u}$ (and hence $\sum_{u \in X_\ell} \alpha_{\ell,u}$) dominates the random variable $\widehat{W}_\ell^X$ defined below.

$$\tag{18}$$

Let $\nu = n - n^{2/3+o(1)}$ (see (10)). Let $\widehat{W}_0^X = 1$ and let

$$\widehat{W}_t^X = \sum_{i=1}^{B_0-B_0'} \frac{A_i}{B_0}$$

where (i) B_0 has the distribution B_I, (ii) B_0' is the number of successes for B_0 in the first $n^{2/3+o(1)}$ trials ($o(1)$ as in (10)) and (iii) the A_i are independent copies of $\widehat{W}_{t-1}^X$.

We can use induction to verify (18): Going back to (17), given $\mathcal{D}^+$, $\deg^+(x)$ is distributed as B_0 and $\gamma_{\ell,1;w,u}$ dominates $\widehat{W}_{\ell-1}^X$. We sum from 1 to B_0-B_0' in order to account for the out-neighbours of $u \in X_{\ell-t}$ which are in $X_0 \cup X_1 \cup \cdots X_{\ell-t}$. We rely here on the tree structure and the fact that our BFS construction of T_X only checks edges of the form (a,b) where b has not yet been placed in X to give us the claimed independence.

Note that

$$\mathbf{Pr}(B_0' \geq 100) \leq \rho^{-1}(n^{2/3+o(1)})^{100}p^{100} \leq n^{-3}. \tag{19}$$

In generating $\widehat{W}_\ell^X$ we will $\mathbf{qs}^2$ sample B_0' at most $o(n^2)$ times and so if W_t^X is defined as in Lemma 3 with $\theta = 100$ then

$$\mathbf{Pr}(W_\ell^X \geq \widehat{W}_\ell^X) = o(n^{-2}).$$

Thus for any value σ,

$$\mathbf{Pr}\left(\sum_{u\in X_\ell} \alpha_{\ell,u} \geq \sigma \mid \mathcal{D}^+\right) \geq \mathbf{Pr}(W_\ell^X \geq \sigma) - o(n^{-2}). \tag{20}$$

Using (7) with $M = 10/(\epsilon_X c)$ we have

$$\mathbf{Pr}\left(\sum_{u\in X_\ell} \alpha_{\ell,u} \leq 1 - \epsilon_X/2 \Big| \mathcal{D}^+\right)$$

$$\leq \mathbf{Pr}(W_\ell^X \leq 1 - \epsilon_X/2)$$

$$\leq \rho^{-1}\sum_{k\in I}\binom{n-1}{k}p^k q^{n-1-k}\mathbf{Pr}\left(\sum_{i=1}^{k-100} A_i \leq k(1-\epsilon_X/2)\right)$$

$$= o(n^{-2}).$$

So,

$$\mathbf{Pr}\left(\exists x,y: \sum_{u\in X_\ell}\alpha_{\ell,u} \leq 1 - \epsilon_X/2\right) \leq$$

$$n^2\mathbf{Pr}\left(\sum_{u\in X_\ell}\alpha_{\ell,u} \leq 1 - \epsilon_X/2\Big|\mathcal{D}^+, x=1, y=2\right) + \mathbf{Pr}(\mathcal{D}^+) = o(1). \tag{21}$$

This completes the proof of (15).

---

[2] An event $\mathcal{E}_n$ occurs *quite surely* (**qs**) if $\mathbf{Pr}(\neg\mathcal{E}_n) = O(n^{-K})$ for any constant $K > 0$.

We next consider the proof of (16). Now for each $v \in Y_i$ and $j < i$ with unique path $v = v_i, v_{i-1}, \ldots, v_j$ from v to Y_j in T_Y, let

$$\gamma_{i,j;v,v_j} = \prod_{k=j}^{i-1} \frac{1}{deg^+(v_k)}.$$

Consider the equation

$$\sum_{v \in Y_\ell} \beta_{\ell,v} = \sum_{v \in Y_\ell} \gamma_{\ell,0;v,y} = \sum_{w \in N_D^-(y) \cap Y_1} \sum_{v \in Y_{\ell,w}} \frac{\gamma_{\ell,1;v,w}}{deg^+(w)} \qquad (22)$$

where $Y_{\ell,w} = \{v \in Y_\ell :$ the path from v to y in T_Y goes through $w \in Y_1\}$. This leads us to claim that given $\mathcal{D}^-$ and $\mathcal{D}^+(X)$,

$$\sum_{v \in Y_\ell} \gamma_{\ell,0;v,y} \text{ (and hence } \sum_{v \in Y_\ell} \beta_{\ell,v}) \text{ dominates the random variable } \widehat{W}_\ell^Y \qquad (23)$$

which is defined as follows: Let $\widehat{W}_0^Y = 1$ and for $1 \le t \le \ell$,

$$\widehat{W}_t^Y = \sum_{i=1}^{B_0 - B_0'} \frac{A_i}{B_i}, \qquad (24)$$

where (i) B_0 has the distribution B_I, (ii) B_0' is the number of successes for B_0 in the first $n^{2/3+o(1)}$ trials (see (11)), (iii) B_i, $i = 1, 2, \ldots B_0 - B_0'$ are independent with distribution B_I, and (iii) the A_i are independent copies of $\widehat{W}_{t-1}^Y$.

We can use induction to verify (23): Going back to (22), $deg^+(w)$ is distributed as B_I and $\gamma_{\ell,1;v,w}$ dominates $\widehat{W}_{\ell-1}^Y$. B_0' accounts for in-neighbours of $u \in Y_{\ell-t}$ that are in $X \cup Y_0 \cup \cdots \cup Y_{\ell-t}$. To justify the independence of the B_i, given $\mathcal{D}^- \cap \mathcal{D}^+(X)$, we observe that given $Y_0, Y_1, \ldots, Y_{\ell-t}$, the out-degrees of the vertices in $Y_{\ell-t-1}$ are independent and distributed as B_I.

We can replace B_0' by 100 as we did for W^X. In the case where $np = d \ln n$, d constant, we make the following adjustment. Let $\zeta = 1/\ln \ln n$ and let a vertex y be *normal* if at most $\zeta_0 = \lceil 12/(\zeta^2 d) \rceil$ of its in-neighbours have out-degrees which are not in the range $[(1 - \zeta)np, (1 + \zeta)np]$. We show next that

$$\mathbf{Pr}(\exists \text{ a vertex which is not normal}) = o(1). \qquad (25)$$

Indeed

$$\mathbf{Pr}(\exists \text{ a vertex which is not normal}) \le o(1) + n \sum_{s=c_b np}^{C_b np} \binom{s}{\zeta_0} (2e^{-\zeta^2 np/3})^{\zeta_0}$$

$$= o(1) + O((\ln n)n^{-3}).$$

Let $\mathcal{N}_y$ be the event that vertex y is normal.

We continue by proving, that conditional on $\mathcal{D}^- \cap \mathcal{N}_y$, that with probability $1 - o(n^{-2})$

$$\sum_{v \in Y_\ell} \beta_{\ell,v} \geq (1 - \epsilon_Y) \frac{\deg^-(y)}{np}. \tag{26}$$

where ϵ_Y is defined below.

For this calculation we will condition on a fixed value s for $B_0 = |N^-(y) \cap Y_1|$ and that $B_i, i = 1, 2, \ldots, B_0$ are the out-degrees of the in-neighbours $w_1, w_2 \ldots, w_s$ of y in Y_1.

Furthermore, $\deg^-(y) - s$ is dominated by $B(n^{2/3+o(1)}, p)$ and so with probability $1 - O(n^{-3})$ we have $\deg^-(y) \leq s + 100$ and so we can replace $\deg^-(y)$ by s in (26).

For $np = d \ln n$, d constant, let $\epsilon_Y = 2\zeta$ and $M = 10/(\epsilon_Y c)$. For $np = \omega \ln n$, $\omega \to \infty$ let $\epsilon_Y = 1/\omega^{1/3}$ and $M = 1$. Then,

$$\mathbf{Pr}\left(\sum_{v \in Y_\ell} \beta_{\ell,v} \leq (1 - \epsilon_Y)\frac{s}{np} \bigg| \mathcal{D}^-, \mathcal{N}_y\right)$$

$$\leq 2\mathbf{Pr}\left(\sum_{i=1}^{s-\zeta_0} \frac{A_i}{(1+\zeta)np} \leq (1 - \epsilon_Y)\frac{s}{np}\right)$$

$$\leq 2\mathbf{Pr}\left(\sum_{i=1}^{s-\zeta_0} A_i \leq (1 - \epsilon_Y/2)s\right)$$

$$\leq 2\exp\left\{-Ms\left(\left(1 - \frac{105M(\ell-1)}{cnp}\right)(1 - \zeta_0/s) - (1 - \epsilon_Y/2)\right)\right\}$$

$$\leq 2e^{-M\epsilon_Y s/3}$$

$$= o(n^{-3}).$$

So,

$$\mathbf{Pr}\left(\exists x, y : \sum_{v \in Y_\ell} \beta_{\ell,v} \leq (1 - \epsilon_Y/2)\frac{\deg^-(y)}{np}\right) \leq$$

$$n^2\mathbf{Pr}\left(\sum_{v \in Y_\ell} \alpha_{\ell,v} \leq 1 - \epsilon_Y/2 \bigg| \mathcal{D}^+, x = 1, y = 2\right) + \mathbf{Pr}(\mathcal{D}^+) + \mathbf{Pr}(\mathcal{D}^-) + n\mathbf{Pr}(\neg\mathcal{N}_2) = o(1). \tag{27}$$

Using (14), (21) and (27) we see that with **whp**, $\forall x, y$, $\mathcal{C}(x, y)$ is such that

$$\mathbf{E}(Z \mid \mathcal{C}) \geq (1 - o(1))\frac{1}{n}\frac{\deg^-(y)}{np} = (1 - o(1))\frac{\deg^-(y)}{m}.$$

We next consider the concentration of $(Z \mid \mathcal{C})$.

For $u \in X_\ell$, and $|Y_\ell| = n^{2/3+o(1)}$, from (19) we have,

$$\mathbf{Pr}(|N^+(u) \cap Y_\ell| \geq 100 \mid \mathcal{D}^-(Y_i), i < \ell) \leq n^{-3}.$$

We write $Z = \sum_{u \in X_\ell} Z_u$ where

$$Z_u = \frac{\alpha_{\ell,u}}{\deg^+(u)} \sum_{v \in Y_\ell} \beta_{\ell,v} 1_{uv}.$$

Conditional on $\mathcal{D}^-(Y_i)$, $i < \ell$ and $|N^+(u) \cap Y_\ell| \leq 100$ we have $Z_u \leq 100/(cnp)^{2\ell+1}$. Let $\widehat{Z}_u = (cnp)^{2\ell+1} Z_u/100$, then for $u \in X_\ell$ the $\widehat{Z}_u$ are independent random variables, and $0 \leq \widehat{Z}_u \leq 1$. Let $\widehat{Z} = \sum_{u \in X_\ell} \widehat{Z}_u$. By the Hoeffding inequality we see that,

$$\mathbf{Pr}(|\widehat{Z} - \mathbf{E}(\widehat{Z})| \geq 4\sqrt{np\mathbf{E}(\widehat{Z})}) = o(n^{-4}).$$

However $Z = 100\widehat{Z}/(cnp)^{2\ell+1}$, and we conclude that, with probability $1 - o(1/n)$, for all $x, y \in V$,

$$|Z - \mathbf{E}(Z)| \leq \frac{400\sqrt{np\,\mathbf{E}(\widehat{Z})}}{(cnp)^{2\ell+1}} = O\left(\frac{1}{n^{7/6+o(1)}}\right). \tag{28}$$

It follows that **whp**

$$P_x^{(2\ell+1)}(y) \geq (1 - o(1)) \frac{\deg^-(y)}{m} \qquad \forall x, y. \tag{29}$$

Finally, for any $y \in V$ we have

$$\pi_y = \sum_{x \in V} \pi_x P_x^{(2\ell+1)}(y) \geq (1 - o(1)) \frac{\deg^-(y)}{m} \sum_{x \in V} \pi_x = (1 - o(1)) \frac{\deg^-(y)}{m}, \tag{30}$$

completing the proof of Lemma 4 $\qquad\qquad\qquad\qquad\qquad\qquad\qquad\qquad$ □

Lower Bounds for Cases II, III

Lemma 5. Whp,

$$\pi_v \geq \frac{1 - o(1)}{n} \ \text{ for all } v \in V.$$

Proof is omitted. $\qquad\qquad\qquad\qquad\qquad\qquad\qquad\qquad\qquad\qquad\qquad\qquad\qquad\qquad$ □

At this point we have proved that the expression in Theorem 2 is a lower bound for the steady state.

Upper Bound on Mixing Time. We next show that the mixing time T as defined in (1) satisfies

$$T = o(\ell \ln n) = o((\ln n)^2). \tag{31}$$

Define

$$\bar{d}(t) = \max_{x, x' \in V} |P_x^{(t)} - P_{x'}^{(t)}| \tag{32}$$

be the maximum over x, x' of the variation distance between $P_x^{(t)}$ and $P_{x'}^{(t)}$. Equation (29) implies that

$$\bar{d}(2\ell + 1) = o(1). \tag{33}$$

Lemma 20 of Chapter 2 of Aldous and Fill [1] proves that

$$\bar{d}(s+t) \le \bar{d}(s)\bar{d}(t) \quad \text{and} \quad \max_x |P_x^{(t)} - \pi_x| \le \bar{d}(t)$$

and so (31) follows immediately from (33).

3.3   Upper Bounds on the Stationary Distribution

We will use the following values: Here $\eta > 0$ is a sufficiently small constant and $\Lambda = \log_{np} n$.

$$\ell_0 = (1+\eta)\Lambda, \quad \ell_1 = (1-10\eta)\Lambda, \ \ell_2 = 11\eta\Lambda,$$
$$\ell_3 = (1-\eta/10)\Lambda, \ \ell_4 = \eta\Lambda/20, \quad \ell_5 = 9\eta\Lambda/10.$$

Case 1: $np \le n^\delta$ where $0 < \delta \ll \eta$ is a Positive Constant. We begin with a lemma that will help simplify our calculations.

Lemma 6. *Suppose that $np \le n^\delta$ where $\delta \ll 1$. Then* **whp** $S \subseteq V, |S| \le s_0 = \frac{1}{2}\log_{np} n$ *implies that S contains at most $|S|$ edges.*

Proof. The expected number of sets S with more than $|S|$ edges can be bounded by

$$\sum_{s=3}^{s_0} \binom{n}{s}\binom{s^2}{s+1}p^{s+1} \le \sum_{s=3}^{s_0}(e^2 np)^s sep = o(1). \qquad \square$$

Fix x, y and let X_i, $0 \le i \le \ell_3$ be the set of vertices that are reachable from x by a walk of length i. These sets are slightly larger than the X_i of Lemma 3.2 etc. in that we allow them to overlap. Let

$$X^*(x) = \bigcup_{i=0}^{\ell_3} X_i.$$

For $1 \le i \le \ell_1$ let

$$\tilde{X}_i = \left\{ a \in X_i \setminus \bigcup_{k=0}^{i-1} X_k : \exists b \in X_j \text{ and } j \le i \text{ such that } (a,b) \text{ is an edge} \right\}$$

and let $\tilde{X} = \bigcup_{i=\ell_4}^{\ell_1} \tilde{X}_i$.

Lemma 7. Whp,

(a) *For all $x \in V$, we have that for each $z \in X^*$ has at most $100/\eta$ in-neighbours in X^*.*
(b) *For all $x \in V$, we have $|\tilde{X}_i| \le n^{2\delta - 10\eta}(Cnp)^i + O(\ln n)$ for $\ell_4 \le i \le \ell_1$.*

Proof

(a) Let $T = \min\{t \le \ell_3 : |X_{t+1}| > Cnp|X_t|\}$ if such t exist and equal to $\ell_3 + 1$ otherwise. Let ζ be the number of in-neighbours of z in $\bigcup_{i=1}^{T} X_i$. ζ is stochastically dominated by $B((Cnp)^{\ell_3}, p)$. Hence if $r = 100/\eta$

$$\mathbf{Pr}(\zeta \ge r) \le (Cnp)^{r\ell_3} p^r \le n^{r(\delta - \eta/10)} \le n^{-4}.$$

Now $T = \ell_3 + 1$ with probability $1 - o(n^{-1})$ and part (a) of the lemma follows.

(b) Assuming that $i \le T$ we have $|X_i| \le (Cnp)^i$. Given this and $\ell_1 \le T$ we see that $|\tilde{X}_i|$ is dominated by $B(n^{1-10\eta+o(1)}(Cnp)^i, p)$ and the result follows from Chernoff bounds. □

We say that a vertex $z \in X_i$, $i \le \ell_4$ is *special* if it has two in-neighbours in $\bigcup_{j<i} X_j$ or if it has x as an out-neighbour. It follows from Lemma 6 that **whp** there can be at most one special vertex for a given x.

Lemma 8. Whp,

$$\pi_v \le (1 + o(1)) \frac{deg^-(v)}{m} \ for \ all \ v \in V.$$

Proof is omitted. □

4   The Cover Time

We see immediately from (31) that Condition (b) of Lemma 1 is satisfied.

We will show shortly that if $\mathcal{D}$ holds then

$$R_T(1) = 1 + o(1). \tag{34}$$

If $|z| \le 1 + \lambda$, then as $\lambda = 1/KT$ we have

$$R_T(z) \ge 1 - \sum_{t=1}^{T} r_t |z|^t \ge 1 - (1 + \lambda)^T \sum_{t=1}^{T} r_t = 1 - o(1)$$

and Condition (a) of the Lemma is also satisfied. So for $v \in V_0$,

$$p_v = (1 + o(1)) \frac{deg^-(v)}{n}.$$

Proof of (34): We first observe that because the minimum out-degree of $D_{n,p}$ is $\Omega(d \ln n)$ we have for any x, y

$$\mathbf{Pr}(\mathcal{W}_v(t) = y \mid \mathcal{W}_v(t-1) = x) = O\left(\frac{1}{d \ln n}\right). \tag{35}$$

The expected number of returns to $v \in V$ by $\mathcal{W}_v$ is therefore $O(T/d \ln n)$. Thus if $d \ge (\ln n)^2$ we are done immediately.

If $d \le (\ln n)^2$ then a simple first moment calculation shows that **whp** for every vertex $v \in V$, there is at most one edge from a vertex in $N^+(v)$ to $\{v\} \cup N^+(v)$ or from a vertex in $N^+(N^+(v))$ to $\{v\} \cup N^+(v) \cup N^+(N^+(v))$. Thus with probability $1 - O(1/(\ln n)^2)$, $x = \mathcal{W}_v(2)$ satisfies $dist(x, v) \ge 3$ and then the probability of a return to v is $O(T/(\ln n)^3)$.

4.1   Upper Bound on the Cover Time

For $np = d \ln n$, d constant, let $t_0 = (1 + \epsilon) d \ln \left(\frac{d}{d-1} \right) n \ln n$. For $np = d \ln n$
$d = d(n) \to \infty$ let $t_0 = (1 + \epsilon) n \ln n$. Here $\epsilon \to 0$ sufficiently slowly so that all
claimed inequalities below are valid.

Let $T_G(u)$ be the time taken by the random walk $\mathcal{W}_u$ to visit every vertex of
D. Let U_t be the number of vertices of G which have not been visited by $\mathcal{W}_u$ at
step t. We note the following:

$$C_u = \mathbf{E}(T_G(u)) = \sum_{t>0} \mathbf{Pr}(T_G(u) \geq t), \tag{36}$$

$$\mathbf{Pr}(T_G(u) \geq t) = \mathbf{Pr}(T_G(u) > t - 1) = \mathbf{Pr}(U_{t-1} > 0) \leq \min\{1, \mathbf{E}(U_{t-1})\} \tag{37}$$

Recall that $A_v(t)$, $t \geq T$ denotes the event that $\mathcal{W}_u(t)$ did not visit v in the
interval $[T, t]$. It follows from (36), (37) that for all $t \geq T$,

$$C_u \leq t + 1 + \sum_{s \geq t} \mathbf{E}(U_s) \leq t + 1 + \sum_v \sum_{s \geq t} \mathbf{Pr}(A_s(v)). \tag{38}$$

Assume first that $d(n) \to \infty$. If $v \in V$ and $t \gg T$ then

$$\mathbf{Pr}(A_s(v)) \leq (1 + o(1)) \exp \left\{ -\frac{(1 - o(1))s}{n} \right\} + o(e^{-s/T}). \tag{39}$$

Plugging (39) into (38) we get

$$C_u \leq t_0 + 1 + 2n \sum_{s \geq t_0} \left(\exp \left\{ -\frac{(1 - o(1))s}{n} \right\} + o(e^{-s/T}) \right) \tag{40}$$

$$\leq t_0 + 1 + 3n^2 \exp \left\{ -\frac{(1 - o(1))t_0}{n} \right\} + o(nTe^{-t_0/T})$$

$$= (1 + o(1))t_0.$$

Now assume that d is constant. For $v \in V$ we have

$$p_v \geq (1 - o(1)) \frac{\deg^-(v)}{m}.$$

Then in place of (40) we find, using the bounds in Lemma 2 (b1),

$$C_u \leq t_0 + 1 + \frac{2 + o(1)}{n^{d-1}} \sum_{k=cnp}^{Cnp} \left(\frac{nep}{k} \right)^k \sum_{s \geq t_0} \left(\exp \left\{ -\frac{(1 - o(1))ks}{m} \right\} + o(e^{-s/T}) \right)$$

$$= (1 + o(1))t_0,$$

where we have used the fact that $(nep(d - 1))/(kd))^k$ is maximized at $k = np(d - 1)/d$.

4.2 Lower Bound on the Cover Time

Let $t_1 = (1 - \epsilon)d\ln\left(\frac{d}{d-1}\right)n\ln n$. Let $V^* = \{v : \deg^-(v) = k^*\}$ where $k^* = (d-1)\ln n$. Let $V^\dagger$ denote the set of vertices in V^{**} that have not been visited by $\mathcal{W}_u$ by time t_1. We show that $\mathbf{E}(|V^\dagger|) \to \infty$ and $\mathbf{E}(|V^\dagger|^2) \sim \mathbf{E}(|V^\dagger|)^2$ and then use the Chebychev inequality to show that $V^\dagger \neq \emptyset$ **whp**.

References

1. Aldous, D., Fill, J.: Reversible Markov Chains and Random Walks on Graphs, http://stat-www.berkeley.edu/pub/users/aldous/RWG/book.html
2. Aleliunas, R., Karp, R.M., Lipton, R.J., Lovász, L., Rackoff, C.: Random Walks, Universal Traversal Sequences, and the Complexity of Maze Problems. In: Proceedings of the 20th Annual IEEE Symposium on Foundations of Computer Science, pp. 218–223. IEEE Computer Society Press, Los Alamitos (1979)
3. Alon, N., Spencer, J.: The Probabilistic Method, 2nd edn. Wiley-Interscience, Chichester (2000)
4. Cooper, C., Frieze, A.M.: The cover time of sparse random graphs. In: Proceedings of the 14th ACM-SIAM Symposium on Discrete Algorithms, pp. 140–147 (2003)
5. Cooper, C., Frieze, A.M.: The cover time of random regular graphs. SIAM Journal on Discrete Mathematics 18, 728–740 (2005)
6. Cooper, C., Frieze, A.M.: The cover time of the preferential attachment graph. Journal of Combinatorial Theory Series B (to appear)
7. Cooper, C., Frieze, A.M.: The cover time of the giant component of a random graph
8. Feige, U.: A tight upper bound for the cover time of random walks on graphs. Random Structures and Algorithms 6, 51–54 (1995)
9. Feige, U.: A tight lower bound for the cover time of random walks on graphs. Random Structures and Algorithms 6, 433–438 (1995)
10. Hoeffding, W.: Probability inequalities for sums of bounded random variables. Journal of the American Statistical Association 58, 13–30 (1963)
11. Kim, J.H., Vu, V.: Concentration of multivariate polynomials and its applications. Combinatorica 20, 417–434 (2000)

Eigenvectors of Random Graphs: Nodal Domains

Yael Dekel[1], James R. Lee[2,*], and Nathan Linial[1]

[1] Hebrew University
[2] University of Washington

Abstract. We initiate a systematic study of eigenvectors of random graphs. Whereas much is known about eigenvalues of graphs and how they reflect properties of the underlying graph, relatively little is known about the corresponding eigenvectors. Our main focus in this paper is on the *nodal domains* associated with the different eigenfunctions. In the analogous realm of Laplacians of Riemannian manifolds, nodal domains have been the subject of intensive research for well over a hundred years. Graphical nodal domains turn out to have interesting and unexpected properties. Our main theorem asserts that there is a constant c such that for almost every graph G, each eigenfunction of G has at most 2 nodal domains, together with at most c exceptional vertices falling outside these primary domains. We also discuss variations of these questions and briefly report on some numerical experiments which, in particular, suggest that there are almost surely no exceptional vertices.

1   Introduction

Let G be a graph and let A be its adjacency matrix. The eigenvalues of A turn out to encode a good deal of interesting information about the graph G. Such phenomena have been intensively investigated for over half a century. We refer the reader to the book [13, Ch. 11] for a general discussion of this subject and to the survey article [9] for the connection between eigenvalues and expansion. Strangely, perhaps, not much is known about the *eigenvectors* of A and how they are related to the properties of G. However, in many application areas such as machine learning and computer vision, eigenvectors of graphs are being used with great success in various computational tasks such as partitioning and clustering, for example see the work of Shi and Malik [17], Coifman, et al. [4,5], Pothen, Simon and Liou [14] and others. For all we know, the success of these methods has not yet been given a satisfactory theoretical explanation and we hope that our investigations will help to shed some light on these issues as well.

There is, on the other hand, a rich mathematical theory dealing with the spectrum and eigenfunctions of Laplacians. We will not go into this area at any depth and refer the reader who wants to get an impression of the size and depth of this theory to Chapter 8 in Marcel Berger's monumental panorama of Riemannian Geometry [2]. Suffices it to say that the adjacency matrix of a graph is a discrete analogue of the Laplacian (we mention a bit more about

* Research supported by NSF CAREER award CCF-0644037.

M. Charikar et al. (Eds.): APPROX and RANDOM 2007, LNCS 4627, pp. 436–448, 2007.

other analogues below). The only facts the reader should know are that the (geometric) Laplacian has a discrete spectrum and that its first eigenfunction is the constant function. This is analogous to the fact that the first eigenvector of a finite connected graph is a positive vector and in particular, if the graph at hand is d-regular, then its first eigenvalue is d with the all-ones vector as the corresponding eigenvector.

One of the classical questions about eigenfunctions of the Laplacian concerns their *nodal domains*. We will only discuss this concept in the realm of graphs and refer the interested reader to [2] for further information about the geometric seting. So what are nodal domains? Let G be a finite connected graph. It is well-known that every eigenfunction f but the first takes both positive and negative values. These values induce a partition of the vertex set $V(G)$ into maximal connected components on which f does not change its sign. These are f's nodal domains. Let us stay a bit inaccurate here and state that a classical theorem of Courant (see, for example, [3]) says that if we arrange the eigenvalues of G as $\lambda_1 \geq \lambda_2 \geq \ldots$ with corresponding eigenfunctions $f_1, f_2, \ldots$, then f_k has at most k nodal domains. We refer the reader to [7] for a full account of Courant's theorem for graphs.

Definition 1. *Let $G = (V, E)$ be an n-vertex graph and let $f : V \to \mathbb{R}$ be an eigenfunction of G. We say that f has a* zero crossing *on the edge $xy \in E$ if $f(x)f(y) \leq 0$. Let $Z_f \subset E$ be the set of all edges on which f has a zero crossing. The connected components of the graph $G_f = (V \setminus \{x : f(x) = 0\}, E \setminus Z_f)$ are called the* nodal domains *of f.*

We remark that one of the basic techniques for spectral partitioning involves splitting a graph according to its nodal domains (see, e.g. [20]) as a special case.

The main focus of our research is the following problem.

Question 1. How many nodal domains do the eigenfunctions of G tend to have for G that comes from various random graph models?

To get some initial idea, we started our research with a numerical experiment the outcomes of which were quite unexpected. It turns out that for graphs sampled from the random graph space $G(n, \frac{1}{2})$ *all* eigenfunctions have *exactly two nodal domains*. The same experimental phenomenon was observed for several smaller values of $p > 0$ in the random graph model $G(n, p)$, provided that n is large enough. Even more unexpected were the results obtained for *random regular graphs*. Some of these results are shown in Fig. 1. We found that quite a few of the first eigenvectors have just two nodal domains, and only then the number of nodal domains starts to grow.

It is also of interest to investigate similar questions for the so-called *combinatorial Laplacian* of random graphs. In the regular case there is no difference, of course, but for $G(n, p)$ it turns out that the picture changes slightly. The typical picture is that all eigenfunctions except for a small number at the very end of the spectrum still have exactly two nodal domains. However, a small number of eigenfunctions at the low end of the spectrum have three nodal domains. In Fig. 2

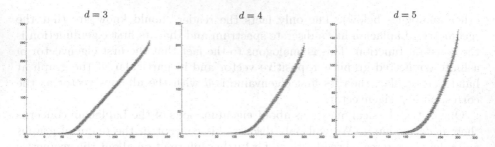

Fig. 1. The number of nodal domains in a random d-regular 300 vertex graph. There are y nodal domains corresponding to the x'th eigenvector (eigenvalues are sorted). For each d we show the average and standard deviation of 100 random graphs.

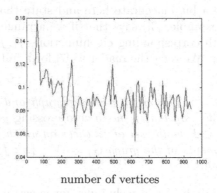

number of vertices

Fig. 2. The probability in $G(n, \frac{1}{2})$ that the last eigenvector of the Laplacian has three nodal domains. For each n 500 experiments were carried out.

we see that in a constant fraction of the graphs in $G\left(n, \frac{1}{2}\right)$ the last eigenvector has three nodal domains.

Our main theorem for $G(n, p)$ partly establishes this observed phenomenon.

Theorem 1. *For every $p \in (0, \frac{1}{2}]$, if $G \sim G(n, p)$, then almost surely it holds that every eigenfunction of G has at most two primary nodal domains, along with at most $O(1/\text{poly}(p))$ other vertices. In particular, almost surely every eigenfunction of G has at most $O(1/\text{poly}(p))$ nodal domains.*

1.1 Overview of Our Approach

Our basic approach is fairly straightforward, although it requires a subtle reduction to some nontrivial machinery. Given $G \sim G(n, p)$ and an eigenfunction f of G, we first partition $V(G) = \mathcal{P}_f \cup \mathcal{N}_f \cup \mathcal{E}_f$ where $\mathcal{P}_f, \mathcal{N}_f$ are the largest positive and negative nodal domains, respectively, and $\mathcal{E}_f$ is a set of exceptional vertices.

Next, we show that if $\mathcal{E}_f$ is large, then we can use the eigenvalue condition, combined with upper estimates on the eigenvalues of $G(n, p)$ to find a large subset

S of coordinates for which $\|f|_S\|_2$ is smaller than one would expect for a random unit vector $f \in S^{n-1}$. The last step is to show that the probability that some eigenfunction has small 2-norm on a large set of coordinates is exponentially small, and then to take a union bound over all such subsets.

The problem is that the final step seems to require finding very strong upper bounds on $\|A^{-1}\|$ (i.e. lower bounds on the smallest singular value of A) for a random discrete matrix A. Although there has been a great deal of progress in this direction [12,11,15,18,16,6], (see also the survey [19]), when A is symmetric, the known bounds are far too weak. Thus it is crucial that we reduce to proving upper bounds on $\|A^{-1}\|$ when A is a random *non-symmetric, rectangular* ± 1 Bernoulli matrix. In this case, we can employ the optimal bounds of [11,12] which also yield the exponential failure probability that we require. This reduction takes place in Theorem 6.

In Section 3.1, we show that it is possible to get significantly better control on the nodal domains of $G(n,p)$ if we could get slightly non-trivial bounds on the ℓ_∞ norms of eigenfunctions. This is a possible avenue for future research.

1.2 Preliminaries

Notation 1.1. *We denote by $G(n,p)$ a random graph with n vertices, where each edge is chosen independently with probability p. For any vertex $x \in V(G)$ we denote the set of its neighbors by $\Gamma(x)$.*

Notation 1.2. *For any eigenvector f of a graph G and any subset $S \subseteq V(G)$ we denote $f(S) = \sum_{y \in S} f(y)$, and particularly, for every $x \in V$ we denote $f(\Gamma(x)) = \sum_{y \sim x} f(y)$.*

Definition 2. *For $p \in [0,1]$, we define the random variable X_p by*

$$
X_p = \begin{cases} p - 1 & \text{with probability } p \\ p & \text{with probability } 1 - p. \end{cases}
$$

In particular, $\mathbb{E}X_p = 0$.

Definition 3. *Let $M_{m \times k}(p)$ be the $m \times k$ matrix whose entries are independent copies of X_p, and let $M_k^{\text{sym}}(p)$ be the symmetric $k \times k$ matrix whose off diagonal entries are independent copies of X_p, and and whose diagonal entries are p.*

Unless otherwise stated, throughout the manuscript, all eigenvectors are assumed to be normalized in ℓ_2.

2 Spectral Properties of Random Matrices

We now review some relevant properties of random matrices.

Theorem 2 (Tail bound for symmetric matrices). *If $A \sim M_k^{\text{sym}}(p)$, then for every $\xi > 0$*

$$\Pr\left(\|A\| \ge (2\sqrt{p(1-p)} + \xi)\sqrt{k}\right) \le 4e^{-\xi^2 k/8}$$

Here $\|A\|$ stands for the ℓ_2 operator norm of A.

Proof. Füredi and Komlós prove in [8] that the expected value of the largest magnitude eigenvalue of A is $2\sqrt{p(1-p)}\sqrt{k}$. Alon, Krivelevich and Vu prove in [1] (see also [10]) that the probability that the largest eigenvalue of A exceeds its median by $\xi\sqrt{k}$ is at most $2e^{-\xi^2 k/8}$, and so is the probability that the smallest eigenvalue of A is smaller than its median by $\xi\sqrt{k}$. As usual, the expected value of the first and last eigenvalues differs from the median by at most $O(1)$, hence

$$\Pr\left(\|A\| \ge (2\sqrt{p(1-p)} + \xi)\sqrt{k}\right) \le 4e^{-(1-o(1))\xi^2 k/8}$$

We also need a similar bound for $A \sim M_{m \times k}(p)$ whose proof is standard.

Theorem 3 (Tail bound for non-symmetric matrices). *For any $m \ge k$, if $A \sim M_{m \times k}(p)$ then*

$$\Pr\left(\|A\| \ge a_1\sqrt{m}\right) \le e^{-a_2 m}$$

where a_1, a_2 are some absolute constants.

First, we need a subgaussian tail bound.

Lemma 1. *For every $0 \le p \le \frac{1}{2}$,*

$$\mathbb{E}e^{tX_p} \le e^{(1-p)^2 t^2/2} .$$

Proof. Let us look at the Taylor expansion of both sides of the above inequality. The left hand side:

$$\mathbb{E}e^{tX_p} = pe^{-t(1-p)} + (1-p)e^{tp}$$

$$= \sum_{i=0}^{\infty}\left[\left(p(1-p)^{2i} + (1-p)p^{2i}\right)\frac{t^{2i}}{(2i)!} + \left((1-p)p^{2i+1} - p(1-p)^{2i+1}\right)\frac{t^{2i+1}}{(2i+1)!}\right]$$

$$\le \sum_{i=1}^{\infty}\left[\frac{(1-p)^{2i}t^{2i}}{(2i)!} + p(1-p)\left(p^{2i} - (1-p)^{2i}\right)\frac{t^{2i+1}}{(2i+1)!}\right] \le \sum_{i=0}^{\infty}\frac{(1-p)^{2i}t^{2i}}{(2i)!}$$

and the right hand side:

$$e^{(1-p)^2 t^2/2} = \sum_{i=0}^{\infty}\frac{(1-p)^{2i}t^{2i}}{2^i i!} .$$

The inequality follows from noticing that for every $i \ge 0$, $(2i)! \ge 2^i i!$.

Proof (Proof of Theorem 3). Take a $\frac{1}{3}$-net $N_1 \subseteq S^{k-1}$ and a $\frac{1}{3}$-net $N_2 \subseteq S^{m-1}$. We can assume that the nets are symmetric in the sense that if $x \in N_1$ (resp. $y \in N_2$) then $-x \in N_1$ (resp. $-y \in N_2$), since this only increases their size by at most a factor of 2. Let $x \in N_1$ and $y \in N_2$, and calculate $\Pr\left(\langle y, Ax \rangle > t\sqrt{m}\right)$. Using Markov's inequality we get

$$\Pr\left(\langle y, Ax \rangle > t\sqrt{m}\right) = \Pr\left(e^{\lambda \langle y, Ax \rangle} > e^{\lambda t \sqrt{m}}\right) \leq \frac{\mathbb{E}e^{\lambda \langle y, Ax \rangle}}{e^{\lambda t \sqrt{m}}}$$

and by Lemma 1 we have

$$\mathbb{E}e^{\lambda \langle y, Ax \rangle} = \prod_{i=1}^{k}\prod_{j=1}^{m} \mathbb{E}e^{\lambda A_{ij} x_i y_j} \leq \prod_{i=1}^{k}\prod_{j=1}^{m} e^{(1-p)^2 \lambda^2 x_i^2 y_j^2/2} = e^{(1-p)^2 \lambda^2/2}$$

since $\sum_{i,j} x_i^2 y_j^2 = 1$. Writing $\lambda = \frac{t\sqrt{m}}{(1-p)^2}$ yields

$$\Pr\left(\langle y, Ax \rangle > t\sqrt{m}\right) \leq e^{-\frac{t^2 m}{2(1-p)^2}}$$

and therefore

$$\Pr\left(\exists x \in N_1, y \in N_2 : |\langle y, Ax \rangle| > t\sqrt{m}\right) < e^{-\frac{t^2 m}{2(1-p)^2}} |N_1||N_2| \leq e^{-\frac{t^2 m}{2(1-p)^2}} 9^{k+m}$$

(this is true since by the choice of N_1, N_2, it holds that $\exists x \in N_1, y \in N_2 :$ $|\langle y, Ax \rangle| > t\sqrt{m} \iff \exists x \in N_1, y \in N_2 : \langle y, Ax \rangle > t\sqrt{m}$).

Let $x \in S^{k-1}$ and $y \in S^{m-1}$ be any two vectors. By successive approximation, we can find sequences $\{x_i\} \subseteq N_1$ and $\{y_i\} \subseteq N_2$ such that $x = \sum_{i \geq 1} \alpha_i x_i$ and $y = \sum_{i \geq 1} \beta_i x_i$ where $|\alpha_i|, |\beta_i| \leq 3^{-i+1}$. Hence if $|\langle y_i, Ax_j \rangle| \leq t\sqrt{m}$ for all $x_j \in N_1, y_i \in N_2$, then

$$|\langle y, Ax \rangle| \leq \sum_{i,j \geq 1} 3^{-i-j+2} |\langle y_i, Ax_i \rangle| \leq O(t)\sqrt{m}$$

as well. Thus choosing $t = O(1)$ large enough finishes the proof.

Theorem 4 (Tail bound for eigenvalues of $G(n,p)$). *For any $p \in (0, \frac{1}{2}]$, let A be the adjacency matrix of $G(n,p)$ and $\lambda_1, \ldots, \lambda_n$ its eigenvalues. Then for every $i \geq 2$ and every $\xi > 0$*

$$\Pr\left(|\lambda_i| \geq (2\sqrt{p(1-p)} + \xi)\sqrt{n}\right) \leq \exp(-\xi^2 n/32)$$

Proof. Füredi and Komlós prove in [8] that $\mathbb{E}\left(\max_{i \geq 2} |\lambda_i|\right) = 2\sqrt{p(1-p)}\sqrt{n}(1 + o(1))$, and here too we get a tail bound from [1], so that for every $\xi > 0$

$$\Pr\left(|\lambda_i| \geq (2\sqrt{p(1-p)} + \xi)\sqrt{n}\right) \leq \exp(-\xi^2 n/32)$$

We now state the following theorem from [12] (Theorem 3.3; a slight generalization of [11]), specialized to the case of the sub-gaussian random variables X_p.

Theorem 5. *[12] For any $p \in (0, \frac{1}{2}]$, $\delta > 0$, $n \geq 1$, and $N = (1 + \delta)n$, suppose that $Q \sim M_{N \times n}(p)$. Then there exist constants $\alpha = \alpha(p, \delta), \beta = \beta(p) > 0$ such that for all sufficiently large n, and every fixed $w \in \mathbb{R}^N$, we have*

$$\Pr\left[\exists v \in S^{n-1} \text{ s.t. } \|Qv - w\|_2 \leq \alpha\sqrt{N}\right] \leq \exp(-\beta N)$$

Remark 1. Strictly, speaking the results in [12] only apply for symmetric random variables (and thus only to $X_{\frac{1}{2}}$ in our setting). But the stated bound requires three components in [12]: (1) An exponential tail bound on the operator norm, which we prove in Theorem 3, (2) a Paley-Zygmund derived small-ball probability estimate, and (3) a Berry-Esséen-type small-ball probability estimate. Both (2) and (3) can be proved by hand for the variables X_p; these are standard estimates which we expound upon in the full version.

The next result follows from taking a union bound in the preceding theorem.

Corollary 1. *Under the setup of Theorem 5, there exist $\alpha = \alpha(p, \delta), \beta = \beta(p) > 0$ such that for all sufficiently large n, and every fixed $w \in \mathbb{R}^N$, we have*

$$\Pr\left[\exists c \in \mathbb{R}, \exists v \in S^{n-1} \text{ s.t. } \|Qv - cw\|_2 \leq \alpha\sqrt{N}\right] \leq \exp(-\beta N)$$

Proof. We may assume that $\|w\|_2 = 1$. Since $\|Qv\|_2 \leq N$ always, it suffices to prove the bound for $c \in [-2N, 2N]$. Take a 1-net M on $[-2N, 2N]$ with $|M| \leq 4N$, so applying Theorem 5 for every point in M and taking a union bound we get that

$$\Pr\left(\exists c \in M \text{ s.t. } \exists v \in S^{n-1} \text{ s.t. } \|Qv - cw\|_2 \leq \alpha\sqrt{N}\right) \leq \exp\left(-\beta N + \ln(4N)\right)$$

But now for any $c' \in [-2N, 2N]$, let $c \in M$ be such that $|c - c'| \leq 1$. For any matrix Q and $v \in S^{n-1}$

$$\|Qv - c'w\|_2 \leq \|Qv - cw\|_2 + \|(c' - c)w\|_2 \leq \|Qv - cw\|_2 + 1 .$$

3 Nodal Domains

First, we will show that the ℓ_2 mass of an eigenvector cannot become very small on a large subset of vertices.

Theorem 6 (Large mass on large subsets). *For every $p \in (0, \frac{1}{2}]$ and every $\varepsilon > 0$, there exist values $\alpha = \alpha(\varepsilon, p) > 0$ and $\beta = \beta(p) > 0$ such that if $G \sim G(n, p)$, and $S \subseteq [n]$ is a fixed subset with $|S| \geq (\frac{1}{2} + \varepsilon)n$, then for n large enough,*

$$\Pr\left[\exists a \text{ non-first eigenfunction } f \text{ of } G \text{ satisfying } \|f|_S\|_2 < \alpha\right] \leq \exp(-\beta n).$$

Proof. Let A be the adjacency matrix of $G = (V, E)$, and let $f : V \to \mathbb{R}$ be a non-first eigenfunction of G with eigenvalue λ. Assume that $\alpha \leq \frac{1}{2}$, and fix a subset $S \subseteq V$ with $|S| \geq (\frac{1}{2} + \varepsilon)n$.

For every $x \in S$, the eigenvalue condition $\lambda f(x) = f(\Gamma(x))$ implies that

$$\left| \sum_{y \in V} A_{xy} f(y) \right| \leq |\lambda f(x)|.$$

Or equivalently,

$$\left| \sum_{y \in V} (p - A_{xy}) f(y) - p \sum_{y \in V} f(y) \right| \leq |\lambda f(x)|.$$

Squaring the preceding inequality and summing over all $x \in S$ yields

$$\sum_{x \in S} \left| \sum_{y \in V} (p - A_{xy}) f(y) - p \sum_{y \in V} f(y) \right|^2 \leq |\lambda|^2 \cdot \|f|_S\|_2^2. \tag{1}$$

In particular, if we define $M = pJ - A$, where J is the $n \times n$ all ones matrix, and let B be the $|S| \times n$ sub-matrix of M consisting of rows corresponding to vertices in S, then (1) implies that $\|Bf - p\bar{f}\mathbf{1}\|_2 \leq |\lambda| \cdot \|f|_S\|_2$, where $\bar{f} = \langle f, \mathbf{1} \rangle$.

Furthermore, if we decompose $B = [P \ Q]$ where P contains the columns corresponding to vertices in S, and Q the others, then clearly $P \sim M_{|S|}^{\text{sym}}(p)$ and $Q \sim M_{|S| \times (n - |S|)}(p)$. Because $Bf = P(f|_S) + Q(f|_{\bar{S}})$, we can write

$$\|Q(f|_{\bar{S}}) - p\bar{f}\mathbf{1}\|_2 \leq \|Bf - p\bar{f}\mathbf{1}\|_2 + \|Pf|_S\|_2 \leq |\lambda| \cdot \|f|_S\|_2 + \|P(f|_S)\|_2 \tag{2}$$

Now, since f is a non-first eigenfunction, by Theorems 4 and 2, there exist constants $C = C(p), \beta' = \beta'(p) > 0$ such that

$$\Pr\left[|\lambda| + \|P\| \geq C\sqrt{n} \right] \leq \exp(-\beta' n)$$

for some $\beta' = \beta'(p) > 0$.

If we assume that $|\lambda| + \|P\| \leq C\sqrt{n}$ and also $\|f|_S\|_2 < \alpha \leq \frac{1}{2}$, then $a = \|f|_{\bar{S}}\|_2 \geq \frac{1}{2}$. In this case, (2) implies that

$$\|Q(\tfrac{1}{a} f|_{\bar{S}}) - \tfrac{p}{a}\bar{f}\mathbf{1}\|_2 \leq 2C\sqrt{n}\alpha.$$

Thus, letting $k = n - |S|$, if we can show that for any $\delta \geq \frac{4\varepsilon}{1 - 2\varepsilon} > 0$, there exists $\alpha, \beta > 0$ (with β depending only on p) such that for n large enough,

$$\Pr_{Q \sim M_{k(1+\delta) \times k}(p)} \left[\exists v \in S^{k-1}, \exists c \in \mathbb{R} \text{ s.t. } \|Qv - c\mathbf{1}\|_2 \leq 2C\alpha\sqrt{n} \right] \leq \exp(-\beta n),$$

then we are done. But this is precisely the content of Corollary 1.

We record the following simple corollary.

Corollary 2. *For every $p \in (0, \frac{1}{2}]$, there exist values $\alpha = \alpha(p) > 0$ and $\varepsilon = \varepsilon(p)$ with $0 < \varepsilon < \frac{1}{2}$ such that if $G \sim G(n,p)$, then almost surely, for every subset $S \subseteq V(G)$ with $|S| \geq (\frac{1}{2} + \varepsilon)n$ we have $\|f|_S\|_2 \geq \alpha$.*

Proof. We need to take a union bound over all subsets $S \subseteq [n]$ with $|S| \geq (\frac{1}{2} + \varepsilon)n$. But for every value $\beta = \beta(p)$ from Theorem 6, there exists an $\varepsilon < \frac{1}{2}$ such that the number of such subsets is $o(\exp(-\beta n))$, hence the union bound applies.

We fix now some $p \in (0, \frac{1}{2}]$ and consider $G(V, E) \sim G(n, p)$ and any non-first eigenfunction $f : V \to \mathbb{R}$ of G with eigenvalue λ. By Theorem 4, we have $|\lambda| = O_p(\sqrt{n})$ almost surely, and thus we assume this bound holds for the remainder of this section. Let $\mathcal{P}_f$ and $\mathcal{N}_f$ be the largest positive and negative nodal domains of f. We refer to the set $\mathcal{E}_f = V \setminus (\mathcal{P}_f \cup \mathcal{N}_f)$ as the set of *exceptional vertices*.

Lemma 2. *If $D_1, \ldots, D_m$ are the nodal domains in $\mathcal{E}_f$, then almost surely $m = O(\log n)$ and $|D_i| = O(\log n)$ for every $i \in [m]$.*

Proof. Consider e.g. the positive nodal domains $P_1, P_2, \ldots, P_s$ in $\mathcal{E}_f$. Then selecting one element from every P_i yields an independent set of size s, hence almost surely $s = O(\log n)$. Furthermore, since each P_i has no edges to $\mathcal{P}_f$, almost surely one of the two sets has size at most $O(\log n)$, hence in particular $|P_i| = O(\log n)$ for each $i \in [s]$. Similar statements hold for the negative nodal domains in $\mathcal{E}_f$.

Next we show that it is impossible to find too many exceptional vertices of small magnitude.

Lemma 3. *There exists some values $k \geq 1$ and $\alpha > 0$ such that almost surely there do not exist $x_1, \ldots, x_k \in \mathcal{E}_f$ with $|f(x_i)| \leq \alpha$ for every $i \in [k]$.*

Proof. Suppose, to the contrary, that such $x_1, \ldots, x_k \in \mathcal{E}_f$ exist, and let $\Gamma = \left(\bigcup_{i=1}^{k} \Gamma(x_i) \right) \setminus \mathcal{E}_f$. Then by Lemma 2, and properties of random graphs, we have $|\Gamma| \geq [1 - (1-p)^k]n - O_{k,p}(\sqrt{n})$ almost surely. Choose k large enough so that $|\Gamma| \geq (\frac{1}{2} + \varepsilon + \varepsilon')n$ where $\varepsilon = \varepsilon(p)$ is the constant from Corollary 2 and $\varepsilon + \varepsilon' < 1/2$.

For each $i \in [k]$, let D_i be the nodal domain of x_i with respect to f. Using the eigenvalue condition, for every $i \in [k]$, we have $|f(\Gamma(x_i))| \leq \alpha|\lambda| = \alpha \cdot O_p(\sqrt{n})$. Every neighborhood $\Gamma(x_i)$ has non-trivial intersection with only of $\mathcal{P}_f$ or $\mathcal{N}_f$, hence grouping terms by sign, we have

$$\sum_{x \in \Gamma(x_i)} |f(x)| \leq \left| f\left(\Gamma(x_i) \cap (\mathcal{P}_f \cup \mathcal{N}_f) \right) + f\left(\Gamma(x_i) \cap \mathcal{E}_f \setminus D_i \right) \right| + |f(D_i)|$$

$$\leq |f(\Gamma(x_i))| + |D_i|$$

$$\leq \alpha \cdot O_p(\sqrt{n}) + O(\log n),$$

where we have used the estimate on $|D_i|$ from Lemma 2 which holds almost surely. In particular,

$$\sum_{i=1}^{k} \sum_{x \in \Gamma(x_i)} |f(x)| \leq k\alpha \cdot O_p(\sqrt{n}) + O(k \log n). \tag{3}$$

Now let $\Gamma' = \{x \in \Gamma : |f(x)| \leq \frac{c}{\sqrt{n}}\}$ for some $c > 0$ to be chosen momentarily. Using (3), we see that $|\Gamma \setminus \Gamma'| \leq \frac{k\alpha \cdot O_p(n)}{c}$, hence choosing $c = O(k\alpha/\varepsilon')$ large enough, we have $|\Gamma'| \geq (\frac{1}{2} + \varepsilon)n$. On the other hand, we have $\|f|_{\Gamma'}\|_\infty \leq \frac{c}{\sqrt{n}}$, hence $\|f|_{\Gamma'}\|_2 \leq c = O(k\alpha/\varepsilon')$. Now choosing $\alpha > 0$ small enough, by Corollary 2, such a set Γ' and eigenvector f having small 2-norm on Γ' almost surely do not exist, completing the proof.

Corollary 3. *Almost surely every eigenfunction f of G has $|\mathcal{E}_f| = O_p(1)$.*

Proof. From the Perron-Frobenius theorem, we know that the first eigenvector has constant (positive) sign, and hence exactly one nodal domain. So consider a non-first eigenvector f. If $|\mathcal{E}_f|$ is sufficiently large, then since $\|f\|_2 = 1$, we can find $x_1, \ldots, x_k \in \mathcal{E}_f$ with $|f(x_i)| \leq \alpha$ for every $i \in [k]$, where k and α are as in the lemma. Hence $|\mathcal{E}_f| = O_p(1)$ almost surely.

3.1 ℓ_∞ Bounds and Nodal Domains

A more careful analysis of the preceding arguments shows that almost surely every eigenfunction of $G(n, p)$ has at most $O(1/\text{poly}(p))$ exceptional vertices. In this section, we show that if one can obtain a slightly non-trivial bound on $\|f\|_\infty$ for eigenfunctions f of $G(n, p)$, then almost surely all eigenfunctions of such graphs have at most $O(\log \frac{1}{p})$ exceptional vertices, and e.g. at most one exceptional vertex for $p \in [0.21, 0.5]$. First, we pose the following.

Question 2. Is it true that, almost surely, every eigenfunction f of $G(n, p)$ has $\|f\|_\infty = o(1)$?

A positive answer to this question yields more precise control on the nodal domains of $G(n, p)$.

Theorem 7. *Suppose it holds almost surely that, for every eigenvector f of $G \sim G(n, p)$, we have $\|f\|_\infty = o(1)$ as $n \to \infty$, then almost surely every eigenvector has at most $k_p = \lfloor \frac{1}{\log_2(1/(1-p))} \rfloor$ exceptional vertices.*

In order to prove the preceding theorem, we need the following strengthening of Theorem 6. In particular, we require that the subset of vertices S is allowed to depend, in a limited way, on the randomness of $G(n, p)$.

Theorem 8. *For every $p \in (0, \frac{1}{2}]$, and $\varepsilon > 0$, there exist values $\alpha = \alpha(\varepsilon, p) > 0$ and $\beta = \beta(p) > 0$ such that the following holds. Suppose $G \sim G(n, p)$ and A is the adjacency matrix of G. Suppose further that $S \subseteq [n]$ is a (possibly) random*

*subset of the vertices of G such that S is fixed by conditioning on the rows of A
indexed by a set $T \subseteq [n]$ with $|T| = o(n)$. Then for all sufficiently large n, with
probability at most $\exp(-\beta n)$ there exists a non-first eigenfunction f of G with
$\|f|_S\|_2 < \alpha\|f|_{V\setminus T}\|_2$ for some $S \subseteq V$ with $|S| \geq (\frac{1}{2} + \varepsilon)n$.*

Proof. Let A be the adjacency matrix of $G = (V, E)$, and let $f : V \to \mathbb{R}$ be a
non-first eigenfunction of G with eigenvalue λ. Let $S \subseteq V$ be a possibly random
subset with $|S| \geq (\frac{1}{2} + \varepsilon)n$. Let $T \subseteq [n]$ be such that $|T| \leq o(n)$ and S is
determined after conditioning on the values of A in the rows indexed by T.
Assume, furthermore, that $\|f|_S\|_2 < \alpha\|f|_T\|_2$.

Again, for every $x \in S$, the eigenvalue condition $\lambda f(x) = f(\Gamma(x))$ implies
that

$$\left| \sum_{y \in V\setminus T} A_{xy} f(y) + \sum_{y \in T} A_{xy} f(y) \right| \leq |\lambda f(x)|.$$

Or equivalently,

$$\left| \sum_{y \in V\setminus T} (p - A_{xy}) f(y) - p \sum_{y \in V\setminus T} f(y) + \sum_{y \in T} A_{xy} f(y) \right| \leq |\lambda f(x)|.$$

Squaring the preceding inequality and summing over all $x \in S \setminus T$ yields

$$\sum_{x \in S\setminus T} \left| \sum_{y \in V\setminus T} (p - A_{xy}) f(y) - p \sum_{y \in V\setminus T} f(y) + \sum_{y \in T} A_{xy} f(y) \right|^2 \leq |\lambda|^2 \cdot \|f|_S\|_2^2 \quad (4)$$

If we define $M = pJ - A$, let B be the $|S\setminus T| \times |V\setminus T|$ sub-matrix of M consisting
of rows corresponding to vertices in $S\setminus T$ and columns corresponding to vertices
in $V \setminus T$, and let $g = \frac{f|_{V\setminus T}}{\|f|_{V\setminus T}\|_2}$, then (4) implies that

$$\|Bg - c_f w_T\|_2 \leq 2|\lambda| \cdot \alpha$$

where $w_T \in \mathbb{R}^{|S\setminus T|}$ is a vector which depends only on the rows of A indexed by T
and $c_f \in \mathbb{R}$ is some constant depending on f. Note that we have used the fact that
$\|f|_S\|_2 \leq \alpha\|f|_{V\setminus T}\|_2$. Furthermore, B and S are independent random variables
conditioned on T. From this point, the proof concludes just as in Theorem 6.

Proof (Proof of Theorem 7). Our goal is to show that almost surely $|\mathcal{E}_f| \leq k_p$
for every non-first eigenfunction f with associated eigenvalue λ. Suppose, to
the contrary, that $|\mathcal{E}_f| > k_p$, and let $U \subseteq \mathcal{E}_f$ have $|U| = k_p + 1$. Consider
$\Gamma = \bigcup_{x \in U} \Gamma(x)$. By properties of random graphs, it holds that

$$|\Gamma| \geq [1 - (1-p)^{k_p+1}]n - O_p(k_p\sqrt{n \log n}) \geq \left(\frac{1}{2} + \varepsilon_p\right) n - O_p(\sqrt{n \log n}),$$

for some $\varepsilon_p > 0$. Thus for n large enough, we may assume that indeed $|\Gamma| \geq \left(\frac{1}{2} + \varepsilon_p\right) n$. Furthermore, as in Lemma 3, we have.

$$\sum_{x \in U} \sum_{y \in \Gamma(x)} |f(y)| \leq (k_p + 1)|\lambda| \cdot \|f\|_\infty + O(k_p \log n). \tag{5}$$

By Theorem 4, we have $|\lambda| = O_p(\sqrt{n})$ almost surely, and thus we assume this bound holds for the remainder of the proof.

Now let $\Gamma' = \{x \in \Gamma : |f(x)| \leq \frac{c}{\sqrt{n}}\}$ for some $c = c(n) > 0$ to be chosen momentarily. Using (5), we see that $|\Gamma \setminus \Gamma'| \leq \frac{O(k_p \sqrt{n})|\lambda| \cdot \|f\|_\infty}{c}$. Under the assumption $|\lambda| \cdot \|f\|_\infty = o(\sqrt{n})$, we can choose $c = o(1)$ so that $|\Gamma \setminus \Gamma'| = o(n)$, in which case we may assume that for n large enough, $|\Gamma'| \geq \left(\frac{1}{2} + \varepsilon_p'\right) n$ for some $\varepsilon_p' > 0$. We also have $\|f|_{\Gamma'}\|_2 = o(1)$, hence also $\|f|_{\Gamma'}\|_2 = o(1) \cdot \|f|_{V \setminus U}\|_2$, since $\|f|_U\|_2 \leq |U| \cdot \|f\|_\infty = o(1)$.

Let $y(n) = o(n)$ be an upper bound on the size of $|\Gamma \setminus \Gamma'|$, and consider

$$\mathcal{U} = \{W \subseteq \Gamma : |W| \geq |\Gamma| - y(n)\}.$$

Then, we have $|\mathcal{U}| \leq \binom{n}{y(n)}$, $\Gamma' \in U$, and the members of $\mathcal{U}$ can be enumerated $\mathcal{U} = U_1, U_2, \ldots$ in such a way that U_i is completely determined by the rows of the adjacency matrix of G corresponding to the vertices in U. We can thus apply Theorem 8 to each U_i to obtain, for some $\beta = \beta(p) > 0$,

$$\Pr[\exists U \text{ s.t. } \|f|_{\Gamma'}\|_2 = o(1) \cdot \|f|_{V \setminus U}\|_2$$

and

$$|\Gamma'| \geq (\tfrac{1}{2} + \varepsilon_p')n] < \binom{n}{k_p + 1}\binom{n}{y(n)} \exp(-\beta n),$$

and the latter quantity is $o(1)$ since $y(n) = o(n)$.

References

1. Alon, N., Krivelevich, M., Vu., V.H: On the concetration of eigenvalues of random symmetric matrices. Israel J. Math. 131, 259–267 (2002)
2. Berger, M.: A Panoramic View of Riemannian Geometry. Springer-Verlag (2003)
3. Chavel, I.: Eigenvalues in Riemannian Geometry. Academic Press (1984)
4. Coifman, R.R., Lafon, S., Lee, A.B., Maggioni, M., Nadler, B., Warner, F., Zucker, S.: Geometric Diffusion as a tool for harmonic analysis and structure definition of data, part i: Diffusion maps. Proc. Natl. Acad. Sci. 102(21), 7426–7431 (2005)
5. Coifman, R.R., Lafon, S., Lee, A.B., Maggioni, M., Nadler, B., Warner, F., Zucker, S.: Geometric Diffusion as a tool for harmonic analysis and structure definition of data, part i: Multiscale methods. Proc. Natl. Acad. Sci. 102(21), 7432–7437 (2005)
6. Costello, K., Tao, T., Vu, V.: Random symmetric matrices are almost surely non-singular. Duke Math. J (to appear 2007)
7. Davies, E.B., Gladwell, G.M.L., Leydold, J., Stadler, P.F.: Discrete nodal domain theorems. Linear Algebra and its Applications 336(1–3), 51–60 (2001)

8. Füredi, Z., Komlós, J.: The eigenvalues of random symmetric matrices. Combinatorica 1(3), 233–241 (1981)
9. Hoory, S., Linial, N., Wigderson, A.: Expander graphs and their applications. Bull. AMS 43(4), 439–561 (2006)
10. Krivelevich, M., Vu., V.H.: Approximating the independence number and the chromatic number in expected polynomial time. J. Comb. Optim. 6(2), 143–155 (2002)
11. Litvak, A.E., Pajor, A., Rudelson, M., Tomczak-Jaegermann, N.: Smallest singular value of random matrices and geometry of random polytopes. Adv. Math. 195(2), 491–523 (2005)
12. Litvak, A.E., Pajor, A., Rudelson, M., Tomczak-Jaegermann, N., Vershynin, R.: Euclidean embeddings in spaces of finite volume ratio via random matrices. J. Reine Angew. Math. 589, 1–19 (2005)
13. Lovász, L.: Combinatorial Problems and Exercises. Elsevier, Amsterdam (1979)
14. Pothen, A., Simon, H.D., Liou, K.-P.: Partitioning sparse matrices with eigenvectors of graphs. SIAM J. Matrix Anal. Appl. 11(3), 430–452 (1990)
15. Rudelson, M.: Invertibility of random matrices: norm of the inverse. Ann. Math (to appear 2007)
16. Rudelson, M., Vershynin, R.: The littlewood-offord problem and invertibility of random matrices (submitted)
17. Shi, J., Malik, J.: Normalized cuts and image segmentation. IEEE Transactions on Pattern Analysis and Machine Intelligence 22(8), 888–905 (2000)
18. Tao, T., Vu, V.: Inverse Littlewood-Offord theorems and the condition number of random discrete matrices. Ann. Math. (to appear 2007)
19. Vu, V.: Random discrete matrices (2007)
20. Weiss, Y.: Segmentation using eigenvectors: A unifying view. In: Proc. of the IEEE conf. on Computer Vision and Pattern Recognition, pp. 520–527 (1999)

Lower Bounds for Swapping Arthur and Merlin

Scott Diehl[*]

University of Wisconsin-Madison
sfdiehl@cs.wisc.edu

Abstract. We prove a lower bound for swapping the order of Arthur and Merlin in two-round Merlin-Arthur games using black-box techniques. Namely, we show that any AM-game requires time $\Omega(t^2)$ to black-box simulate MA-games running in time t. Thus, the known simulations of MA by AM with quadratic overhead, dating back to Babai's original paper on Arthur-Merlin games, are tight within this setting. The black-box lower bound also yields an oracle relative to which MA-TIME$[n] \not\subseteq$ AM-TIME$[o(n^2)]$.

Complementing our lower bounds for swapping Merlin in MA-games, we prove a time-space lower bound for simulations that drop Merlin entirely. We show that for any $c < \sqrt{2}$, there exists a positive d such that there is a language recognized by linear-time MA-games with one-sided error but not by probabilistic random-access machines with two-sided error that run in time n^c and space n^d. This improves recent results that give such lower bounds for problems in the second level of the polynomial-time hierarchy.

1 Introduction

Interactive protocols extend the traditional, static notion of a proof by allowing the verifier to enter into a conversation with the prover, wherein the goal is to convince the computationally limited verifier of a claim's validity in a probabilistic sense. Such an allowance affords the power to prove membership in what we believe to be significantly more difficult languages: Traditional polynomial-time proof systems (with a deterministic verifier) capture the class of languages recognizable in nondeterministic polynomial time, whereas interactive protocols have been shown to capture the entire class of languages recognizable in polynomial space [1, 2].

A particularly interesting subclass of interactive protocols is the class of public-coin protocols in which the number of communication rounds between the prover and the verifier is bounded by a constant— We know such protocols as *Arthur-Merlin games*. For example, graph nonisomorphism, which is not known to be in NP, has a two-round Arthur-Merlin game [3]. In fact, any language that has an Arthur-Merlin game with a constant number of rounds can be decided by an Arthur-Merlin game with only two rounds [4, 5]. Thus, there are at most

---

[*] Supported by NSF Career award CCR-0133693 and Cisco Systems Distinguished Graduate Fellowship.

M. Charikar et al. (Eds.): APPROX and RANDOM 2007, LNCS 4627, pp. 449–463, 2007.

two classes of languages recognized by Arthur-Merlin games, determined by the order in which the parties act: Those recognized by two-round games where the prover (Merlin) acts first, MA, and those recognized by two-round games where the verifier (Arthur) acts first, AM.

We know that the latter type of game is at least as powerful as the former: Any MA-game can be transformed into an AM-game recognizing the same language. Thus, we can prove at least as many languages when Arthur goes first as when Merlin goes first. This transformation does come at a cost, though: We must allow for some polynomial factor more time to verify the prover's claim in the resulting AM-game than in the original MA-game. For example, in Babai's transformation [4], one first reduces the error probability of the MA-game to exponentially small (by taking the majority vote of parallel trials) and then applies a union bound to show that, since the probability of proving an invalid claim is so small, switching the parties' order does not give the prover an unfair advantage. Since the error reduction necessitates verifying a number of trials of the MA-game that is linear in the number of bits sent by Merlin, this transformation incurs a quadratic-time overhead in general.

1.1   Time Overhead and MA Versus AM

Motivated by recent results in time-space lower bounds for probabilistic machines [6, 7], the main goal of this paper is to investigate the polynomial-time overhead needed in any such MA-to-AM transformation, and, in particular, whether or not it can be made subquadratic. We answer this question in the negative for a broad class of simulations, encompassing all currently known transformations: *black-box* simulations. Informally, a simulation is black-box if its outcomes are determined only by some number of calls to a given subroutine; In the case where we wish to simulate MA, this subroutine is the underlying predicate of an arbitrary MA-game.

Theorem 1 (Main Result). *For any time-constructible t, there is no black-box simulation of* MA*-games running in time t by* AM*-games running in time $o(t^2)$.*

Babai's simulation of MA by AM is indeed black-box, since it makes its decision by the majority vote of a linear number of trials of the MA-game. The same holds for all other known simulations, such as those using pseudorandom generators [8] and those using extractors [9]. Thus, Theorem 1 implies that those with a quadratic overhead (such as Babai's) are optimal within this setting; No tweaking or optimization of other known techniques—even by improvements to extractor or pseudorandom generator parameters—can possibly yield better performance.

We point out that the black-box restriction on the lower bound is the best one can hope for without solving a (seemingly) much more difficult open problem: An unrestricted (non-black-box) version of Theorem 1 would imply a time hierarchy for both MA and AM. For example, the known simulation of MA by AM guarantees that any language with a linear-time MA-game also has a quadratic-time AM-game; On the other hand, a non-black-box lower bound would provide such a language that has no subquadratic-time AM-game, witnessing a hierarchy for AM. Therefore, removing the black-box restriction would be at least as hard as solving the long-standing open problem of proving a fully-uniform time hierarchy for Arthur-Merlin games [10].

Theorem 1 allows the construction of a relativized world in which linear-time MA-games can solve a language that subquadratic-time AM-games cannot.

Corollary 2. *There exists an oracle relative to which there is a language recognizable by a linear-time* MA-*game but not by any subquadratic-time* AM-*game.*

As far as we know, these results are the first to compare the running-time of MA-games to equivalent AM-games. Santha compared the power of MA and AM by showing that there is an oracle relative to which AM is not contained in MA [11], but this result does not concern the overhead of the opposite inclusion. Canetti et al. gave a lower bound for black-box samplers [12], which implies that the error reduction in Babai's simulation (or in Goldreich and Zuckerman's simulation via extractors [9]) cannot be done more efficiently in this setting; However, amplification is not the only black-box technique through which we can simulate MA—for example, one can use pseudorandom generators [8]—so this does not even suffice to lower bound all *known* techniques (while our result does).

1.2   Time-Space Lower Bounds and MA Versus BPP

As mentioned in Sect. 1.1, the lower bound of Theorem 1 is motivated by our investigation of lower bounds on the time required by a probabilistic, polynomial-time verifier to solve problems *without* the help of the prover (i.e., BPP). Since Boolean satisfiability is in NP but is widely held to require exponential time to solve probabilistically (with bounded error), we strongly expect the verifier to be severely handicapped by the prover's absence. Despite this, a nontrivial lower bound for satisfiability on probabilistic machines has eluded the community so far.

However, progress has been made in proving concrete lower bounds for such hard problems by restricting the machines solving them to use a small amount of space. For example, a recent line of work gives a time lower bound of $n^{1.759}$ for *deterministic* machines to solve satisfiability in subpolynomial ($n^{o(1)}$) space [6, 13, 14]. However, for *probabilistic* machines, we only know of such lower bounds for problems in the second level of the polynomial-time hierarchy: Diehl and Van Melkebeek proved a time lower bound of $n^{2-o(1)}$ for subpolynomial-space probabilistic machines to solve $QSAT_2$, the problem of verifying the validity of a quantified Boolean formula with one quantifier alternation [6]. We bring these lower bounds closer to satisfiability by deriving time-space lower bounds for simulations of MA by probabilistic machines.[1]

---

[1] This was independently discovered by Emanuele Viola and Thomas Watson [personal communication].

Theorem 3. *For every constant $c < \sqrt{2}$, there exists a constant $d > 0$ such that there is a language recognized by a linear-time MA-game with one-sided error that cannot be recognized by probabilistic machines with two-sided error running in time $O(n^c)$ and space $O(n^d)$.*

Theorem 3 follows as a corollary to the quadratic time-space lower bound for $QSAT_2$ on probabilistic machines by Diehl and Van Melkebeek. This is because a time-n^c probabilistic algorithm for linear-time MA-games yields a time-n^{c^2} probabilistic algorithm for linear-time in the second level of the polynomial-time hierarchy; Thus, a lower bound of $n^{2-o(1)}$ for $QSAT_2$ implies one of $n^{\sqrt{2}-o(1)}$ for the class linear-time MA-games. The argument requires some additional technical steps to achieve the lower bound for one-sided error MA-games.

At first glance, it may seem strange that we don't already have a lower bound for SAT; After all, the lower bound for $QSAT_2$ seems as though it should imply a lower bound for SAT since the two are intimately connected. Indeed, this is true in the *deterministic* setting: Similar to the above case of MA, a deterministic lower bound of n^2 for $QSAT_2$ implies one of $n^{\sqrt{2}}$ for SAT. However, this is not necessarily the case when we consider *probabilistic* machines. The main culprit is that the known arguments one uses to construct an algorithm for $QSAT_2$ from one for SAT utilize the simulation of MA by AM; By Theorem 1, this carries at least a quadratic-time overhead by black-box techniques. In this manner, we only know how to translate a probabilistic algorithm for SAT running in time n^c into one for $QSAT_2$ running in time n^{2c^2}; Thus, an n^2 lower bound for $QSAT_2$ gives nothing for SAT. In order to obtain a lower bound for SAT on probabilistic machines by this line of argument, we require either a stronger lower bound for $QSAT_2$ or a subquadratic simulation of MA by AM. Recalling Theorem 1, the latter is impossible to achieve with any known simulation. As such, the class MA is currently as close to NP as we can stretch the lower bound for $QSAT_2$.

1.3 Techniques

The proof of Theorem 1 is inspired by recent work of Viola [7], where he proves a quadratic-time lower bound for black-box simulations of BPP in the second level of the polynomial-time hierarchy. Viola's result involves a technique inspired by a switching lemma of Segerlind, Buss, and Impagliazzo [15]—later improved upon by Razborov [16]—which uses random restrictions that set very few variables.

The lower bound for BPP yields the interesting corollaries that black-box simulations of standard two-sided error AM- or MA-games by one-sided error games of the respective type require a quadratic-time overhead. This is because the two-sided error games contain BPP, while the one-sided error games are (by definition) contained in the second level of the polynomial-time hierarchy. However, since the transformation to the second level requires a quadratic-time overhead for *both* classes, this does not rule out a linear-overhead, black-box simulation of MA by AM in the standard two-sided error model. Therefore, something more is needed.

Our proof can be summed up as follows: First, we transform the setting from AM-games into systems of disjunctive normal-form (DNF) formulas. Notice that we can view the behavior of an AM-game after the verifier sends his message as a nondeterministic computation where each guess corresponds to a possible prover's message. The black-box setting enforces that the outcome after the nondeterministic guess is determined by some number of trials $k(n)$ of the underlying predicate of the MA-game being simulated. Thus, by the standard polynomial-time hierarchy-to-circuit connection [17], the result of the game after the verifier's move can be viewed as the output of an exponential-size, depth-two OR-of-ANDs circuit with bottom fan-in k— i.e., a k-DNF.

Next, we show that the ideas underlying the switching lemma of Segerlind et al. [15, 16] and its adaptation by Viola [7] present us with a dichotomy for any such k-DNF: Either (i) its terms are spread out over the input variables and it accepts very often on random inputs, or (ii) its terms are focused on a small set of variables whose influence is therefore elevated above the rest. Our proof takes advantage of this dichotomy by constructing two distributions on the outcomes of MA-games: One that generates mostly "no" instances and one that generates mostly "yes" instances. We show that, for small k, if a k-DNF is of type (i), then it accepts on most instances of the "no" distribution; If the k-DNF is of type (ii), then it depends (with non-negligible probability) on too few variables to discern the difference between the "yes" and "no" distributions.

We conclude by showing that this means the verifier's actions either put the prover in a position to wrongfully convince the verifier of false claims or prevent the prover from helping the verifier separate fact from fallacy with bounded error— either case is a failure. Therefore, the number of trials k made by the AM-game cannot be small, which leads to the lower bound.

2 Black-Box Lower Bound

This section begins by precisely defining the models and promise problems to which our lower bound arguments apply.

2.1 Black-Box Model and the ∃-ApprMaj Problem

We define the class of languages recognizable by a time-bounded AM-game. For any time bound t, we say that a language $L \in$ AM-TIME$[t]$ if there exists a deterministic Turing machine V that takes three inputs and runs in time $t(n)$ such that for any x,

$$x \in L \Rightarrow \Pr_{|z|=t(n)} [\exists y \in \{0,1\}^{t(n)} \text{ such that } V(x,y,z) = 1] \geq 2/3 \qquad (1)$$

$$x \notin L \Rightarrow \Pr_{|z|=t(n)} [\exists y \in \{0,1\}^{t(n)} \text{ such that } V(x,y,z) = 1] \leq 1/3, \qquad (2)$$

where $n = |x|$ and the probabilities are taken over the uniform distribution.

Similarly, the class of languages recognized by MA-games bounded by time t, MA-TIME$[t]$, are the languages L where there is a deterministic, time $t(n)$ Turing machine V such that for any x,

$$x \in L \Rightarrow \exists y \in \{0,1\}^{t(n)} : \Pr_{|z|=t(n)} [V(x,y,z) = 1] \geq 2/3 \qquad (3)$$

$$x \notin L \Rightarrow \forall y \in \{0,1\}^{t(n)} : \Pr_{|z|=t(n)} [V(x,y,z) = 1] \leq 1/3. \qquad (4)$$

An AM- or MA-game has *one-sided error* if the acceptance probability in the case that $x \in L$ (i.e., the right-hand side of (1) or, respectively, (3)) is replaced by 1 in the above definitions. The resulting classes are referred to as AM-TIME$_1[t]$ and MA-TIME$_1[t]$, respectively.

We often refer to the Turing machine V as the underlying predicate, the string x as the problem input, y as the prover's message, and z as the verifier's message. The machine model we consider is that which allows *random-access* to the input and worktape (although our lower bounds hold for any fixed reasonable model), and our results assume that any time-bounds involved are constructible.

We now precisely define the notion of a black-box simulation. This is meant to capture any method of showing that MA is contained in AM by the design of a "modular" AM-game, into which one can plug *any* MA-game's underlying predicate V as a subroutine (i.e., it is oblivious to the MA-games it simulates[2]). Such a simulation uses only the results of queries to V to decide an answer and is indifferent to the computational details of how these answers are obtained: It should arrive at the right answer provided only that V satisfies the promise of an MA-game given by conditions (3) and (4). Therefore, a black-box AM-simulation of MA is an AM-game that solves the *promise problem* of whether or not a given subroutine satisfies (3) or (4) on input x.

We cast this promise problem as one on the (exponentially long) characteristic vector of the given subroutine on a fixed x. To do so, we first define the promise problem ApprMaj corresponding to the acceptance condition of a probabilistic computation with bounded error. This allows us to define the desired promise problem $\exists$-ApprMaj as one on matrices, where each row corresponds to a possible prover message, and each column corresponds to a possible verifier message.

Definition 4. ApprMaj *is the promise problem on Boolean strings where*

$$\text{ApprMaj}_Y = \{r \mid \text{at least } 2|r|/3 \text{ bits of } r \text{ are set to 1}\},$$
$$\text{ApprMaj}_N = \{r \mid \text{at most } |r|/3 \text{ bits of } r \text{ are set to 1}\}.$$

$\exists$-ApprMaj *is the promise problem on Boolean matrices where*

$$\exists\text{-ApprMaj}_Y = \{M \mid \text{at least one row of } M \text{ belongs to ApprMaj}_Y\},$$
$$\exists\text{-ApprMaj}_N = \{M \mid \text{every row of } M \text{ belongs to ApprMaj}_N\}.$$

Thus, the definition of an AM-game that efficiently black-box simulates MA fitting our intuition is one that solves $\exists$-ApprMaj on appropriately-sized matrices with few queries. We charge the simulation t time units for each query, where t

---

[2] We point out the difference to a relativizing inclusion, where we can have a different simulation for each oracle.

is the running-time of the simulated MA-game, to capture the fact that a black-box simulation must calculate the bits of the $\exists$-ApprMaj instance on the fly by running trials of the MA-game.

Definition 5. *We say that an* AM-*game black-box q-simulates* MA-*games running in time $t(n)$ if there is a polynomial $p(n)$ and an oracle machine S such that for any input $M \in \{0,1\}^{2^t \times 2^t}$,*

$$M \in \exists\text{-ApprMaj}_Y \Rightarrow \Pr_{|z|=p(n)}[\exists y \in \{0,1\}^{p(n)} \text{ such that } S^M(y,z) = 1] \geq 2/3$$

$$M \in \exists\text{-ApprMaj}_N \Rightarrow \Pr_{|z|=p(n)}[\exists y \in \{0,1\}^{p(n)} \text{ such that } S^M(y,z) = 1] \leq 1/3,$$

where S queries M in at most $q(n)$ locations.

We say that an AM-*game running in time $t'(n)$ black-box simulates* MA-*games running in time $t(n)$ if there is an* AM-*game that black-box q-simulates such* MA-*games for $q = t'/t$ and $p = t'$.*

We prove the lower bound of Theorem 1 by showing that any AM-game black-box q-simulating MA-games running in time t must have $q = \Omega(t)$; Therefore, the running time of such a simulation must be at least $\Omega(t^2)$. In fact, the other computational aspects of the AM-simulation, such as the bounds on the message lengths or the running-time of the underlying predicate V, turn out to be irrelevant to the number of required queries. Therefore, we simplify the situation by proving the query lower bound via a lower bound on the *bottom fan-in* of depth-3 circuits for $\exists$-ApprMaj (where by bottom fan-in we mean the fan-in of the gates adjacent to the input literals). In particular, the circuits we consider have the same correspondence to AM-games as standard constant-depth circuits have to the polynomial-time hierarchy [17].

Definition 6. *An* AM-*circuit is a depth-3 circuit with literals as inputs (variables or their complements) whose output gate is an oracle gate for* ApprMaj *and each depth-2 subcircuit is an* OR *of* AND*'s (i.e., a* DNF*); Furthermore, we are guaranteed that the input to the* ApprMaj *gate formed by the depth-2 subcircuits on any input X is either in* ApprMaj$_Y$ *or* ApprMaj$_N$.

The aforementioned connection to AM-games is given by the following proposition.

Proposition 7. *If an* AM-*game black-box q-simulates* MA-*games running in time t, then there is a family of* AM-*circuits with bottom fan-in q computing $\exists$-ApprMaj on matrices of size $2^t \times 2^t$.*

2.2   Proof of the Main Result

Proposition 7 allows us to prove Theorem 1 by showing a lower bound on the bottom fan-in of AM-circuits for $\exists$-ApprMaj.

Theorem 8. *Any* AM-*circuit solving* $\exists$-ApprMaj *on* $2^t \times 2^t$ *matrices has bottom fan-in* $\Omega(t)$.

As outlined in Sect. 1.3, the main idea behind the proof of Theorem 8 stems from an analysis of the k-DNF subcircuits of an AM-circuit on different input distributions. For an AM-circuit with bottom fan-in k to compute $\exists$-ApprMaj, Definition 6 requires most of its k-DNF subcircuits to accept when the input is drawn from a distribution producing mostly "yes" instances of $\exists$-ApprMaj and most to reject when the input is drawn from a distribution producing mostly "no" instances. By a union bound, a nontrivial fraction of the subcircuits output the correct answer most of the time for *both* the "yes" and "no" distributions. We construct specific "yes" and "no" distributions such that any k-DNF with $k = o(t)$ cannot do both at the same time; Therefore, such an AM-circuit for $\exists$-ApprMaj cannot exist.

We begin by defining the distributions producing mostly "yes" instances and "no" instances of $\exists$-ApprMaj that we use to prove the lower bound.

Definition 9. *Let* $\mathcal{D}_N^{n \times m}$ *be the distribution on* $n \times m$ *matrices obtained by assigning each entry independently to 1 with probability 1/6 and 0 with probability 5/6.*

Furthermore, let $\mathcal{D}_Y^{n \times m}$ *be the distribution on* $n \times m$ *matrices obtained by choosing one row uniformly at random and setting each entry within this row to 1, while the remaining entries are set as in* $\mathcal{D}_N^{n \times m}$.

When n *and* m *are clear from context, we simply use* $\mathcal{D}_Y$ *and* $\mathcal{D}_N$.

Notice that the distribution $\mathcal{D}_Y$ always generates instances in $\exists$-ApprMaj$_Y$, while $\mathcal{D}_N$ generates instances in $\exists$-ApprMaj$_N$ with high probability when $m = \Omega(\log n)$ (by the Chernoff bound).

Proposition 10. *There exists a constant* $\alpha > 0$ *such that for large enough* n *and* $m \geq \alpha \log n$,

$$\Pr_{M \in \mathcal{D}_Y^{n \times m}} [M \in \exists\text{-ApprMaj}_Y] = 1 \text{ and } \Pr_{M \in \mathcal{D}_N^{n \times m}} [M \in \exists\text{-ApprMaj}_N] \geq (1 - 1/n).$$

To analyze the relationship between the probability that a given k-DNF φ accepts on $\mathcal{D}_Y$ and the probability that it accepts on $\mathcal{D}_N$, we introduce a "lazy" procedure, Eval, that tries to evaluate $\varphi(M)$ (see Fig. 1). Given φ, input M, and a parameter s, Eval attempts to reduce the problem of evaluating a k-DNF to that of evaluating a $(k-1)$-DNF by querying the values of a set of variables that *cover* the terms of φ (a cover, or hitting set, of a DNF ψ is a subset Γ of the variables such that each term of ψ contains at least one variable in Γ). Eval then restricts φ appropriately to obtain a $(k-1)$-DNF φ', whose evaluation on the remaining variables is the same as the evaluation of φ, and recurses. However, Eval is "lazy" in the sense that it refuses to query more than s variables at each step; If at any step it becomes impossible to achieve the above term-size reduction in such few queries, Eval simply gives up and outputs the current subformula. Thus, Eval has three possible outputs: 0, 1, or a restriction of φ that has a large minimum cover. We point out that if Eval outputs 0, then $\varphi(M) = 0$, and if Eval outputs 1, then $\varphi(M) = 1$; However, the converse does not hold in either case due to Eval's laziness.

Procedure **Eval**(φ, M, s)
If $\varphi \equiv 0$, **output** 0.
Else if $\varphi \equiv 1$, **output** 1.
Else if every cover of φ has size greater than s, **output** φ.
Else
$\Gamma \leftarrow$ a cover of φ of size at most s.
Query the variables in Γ to obtain the partial assignment $\Gamma(M)$.
$\varphi' \leftarrow \varphi|_{\Gamma(M)}$.
Eval(φ', M, s).

Fig. 1. The "lazy" procedure, Eval

Our proof requires the cover size s to balance two desires: (i) We need s large enough so that if Eval outputs a formula ψ, then ψ is almost certainly satisfied by an input from $\mathcal{D}_N$, and (ii) we need s small enough so that Eval has not queried many variables when it arrives at a 0/1 answer. Property (i) ensures that the non-Boolean output case of Eval cannot happen very often for φ a k-DNF in an AM-circuit computing $\exists$-ApprMaj on $\mathcal{D}_N$, so the output of Eval(φ, M, s) almost always matches the evaluation of $\varphi(M)$ when $M \in \mathcal{D}_N$. Property (ii) guarantees that the event of Eval(φ, M, s) answering either 0 or 1 depends very weakly on most rows, which is not enough for Eval to distinguish the one-row difference between samples from $\mathcal{D}_Y$ and $\mathcal{D}_N$. Therefore, if φ behaves well on $\mathcal{D}_N$ and outputs 0 most of the time, then Eval(φ, M, s) not only outputs 0 most of the time for $M \in \mathcal{D}_N$ by (i), but also for $M \in \mathcal{D}_Y$ by (ii), so φ rejects too often on $\mathcal{D}_Y$.

We first quantify the effect of property (i) on the correctness of Eval on $\mathcal{D}_N$.

Lemma 11. *If φ is a k-DNF, then*

$$\Pr_{M \in \mathcal{D}_N}[\text{Eval}(\varphi, M, s) \neq 0] \leq \frac{\Pr_{M \in \mathcal{D}_N}[\varphi(M) = 1]}{1 - e^{-\frac{s}{k6^k}}}.$$

The proof begins by showing that any DNF ψ output by Eval must have many variable-disjoint terms due to its large cover size. These terms present too many independent events that must align in order for ψ to reject reliably on $\mathcal{D}_N$.

Lemma 12. *Let φ be a k-DNF. Then*

$$\Pr_{M \in \mathcal{D}_N}[\varphi(M) = 0 | \text{Eval}(\varphi, M, s) \notin \{0, 1\}] \leq e^{-\frac{s}{k6^k}}.$$

Proof. Suppose that Eval outputs some formula ψ. Notice that $\psi \not\equiv 0$ is a restriction of φ, so it is a k'-DNF for $k' \leq k$. Thus, a term of ψ is satisfied by $\mathcal{D}_N$ with probability at least $1/6^k$. We claim that ψ must have more than s/k

variable-disjoint terms Γ. The lemma follows if this is the case, since the events that a term in Γ is satisfied are independent, so

$$\Pr_{M \in \mathcal{D}_N}[\psi(M) = 0] \leq \Pr_{M \in \mathcal{D}_N}[\text{Each clause in } \Gamma \text{ is not satisfied}]$$
$$\leq (1 - 1/6^k)^{s/k} \leq e^{-\frac{s}{k6^k}}.$$

To prove the claim, consider a maximal set Γ of variable-disjoint terms of ψ. There are at most $|\Gamma|k$ variables in these terms, and they must form a cover of ψ; If they do not, then there is a term t that is not covered by the variables of Γ, i.e., t does not contain any of the variables in Γ. Thus, $\Gamma \cup \{t\}$ is a set of variable-disjoint terms, which contradicts the maximality of Γ. Furthermore, since ψ is a formula output by Eval, it cannot have a cover of size at most s. Therefore, we have that $|\Gamma|k > s$, so $|\Gamma| > s/k$ as claimed. □

Thus, since the formulas on which Eval does not give a 0/1 output are very likely to accept $\mathcal{D}_N$ by Lemma 12, Eval cannot give such an output often on a formula φ that largely rejects $\mathcal{D}_N$. Therefore, Eval must correctly output 0 for φ on most of these instances despite its laziness, yielding Lemma 11.

We now turn to quantify the effect of property (ii) on the relationship between Eval's behavior on $\mathcal{D}_Y$ and $\mathcal{D}_N$.

Lemma 13. *Let φ be a k-DNF. Then for any $\delta \geq \frac{sk}{n}$,*

$$\Pr_{M \in \mathcal{D}_Y}[\text{Eval}(\varphi, M, s) = 0] \geq (1 - \frac{sk}{\delta n}) \left(\Pr_{M \in \mathcal{D}_N}[\text{Eval}(\varphi, M, s) = 0] - \delta \right).$$

Proof. Recall that when the formula φ passed to Eval has small cover size, Eval reduces the term size of φ by one to obtain φ' (since Γ is a cover of φ). Therefore, Eval can make at most k recursive calls on a k-DNF. At each such call, Eval queries at most s variables, for a total of at most sk variables queried during the entire run. Different outcomes for the sample M cause Eval to form different subformulas φ' with (possibly) different covers, so the variables queried by Eval follow some probability distribution determined by the input distribution. We show that most variables, and in fact most rows, are queried with very small probability.

For a fixed k-DNF φ, s, and $\delta > 0$, define B as

$$B \doteq \{i | \Pr_{M \in \mathcal{D}_N}[\text{Eval}(\varphi, M, s) \text{ queries row } i] \geq \delta\}, \tag{5}$$

the set of rows queried by Eval with probability at least δ. Since any individual run of Eval queries at most sk variables, we have that

$$\sum_i \Pr_{M \in \mathcal{D}_N}[\text{row } i \text{ is queried by Eval}(\varphi, M, s)] \leq sk.$$

Therefore, B cannot be too large:

$$|B| \leq \frac{sk}{\delta}. \tag{6}$$

So (5) and (6) tell us that, for most rows i, Eval queries a variable in row i with very small probability when $\delta > \frac{sk}{n}$.

We would like to isolate the cases of $\mathcal{D}_Y$ that choose a row outside of B to set to 1. Towards this end, define $\mathcal{D}_Y^i$ to be the distribution on $n \times m$ matrices where each entry in row i is assigned 1 and each other entry is assigned 0 with probability 5/6 (and 1 otherwise). Then sampling M from $\mathcal{D}_Y$ is equivalent to choosing $i \in \{1, \ldots, n\}$ at random and sampling M from $\mathcal{D}_Y^i$. This allows us to divide the probability that Eval rejects on inputs from $\mathcal{D}_Y$ into cases by row. Discarding the rows in B, we have

$$\Pr_{M \in \mathcal{D}_Y}[\text{Eval}(\varphi, M, s) = 0] \geq \sum_{i \notin B} \Pr_{M \in \mathcal{D}_Y^i}[\text{Eval}(\varphi, M, s) = 0]/n. \tag{7}$$

Now consider how $\text{Eval}(\varphi, M, s)$ differs when M is drawn from $\mathcal{D}_N$ and when it is drawn from $\mathcal{D}_Y^i$ where $i \notin B$. Notice that the only time that Eval queries variables which are distributed differently in $\mathcal{D}_N$ and $\mathcal{D}_Y^i$ is when it queries a variable in row i. We know that this happens with probability less than δ under $\mathcal{D}_N$ by (5); This is also the case under $\mathcal{D}_Y^i$, since the distribution is identical to $\mathcal{D}_N$ outside of row i. Thus, the outcome of Eval on $\mathcal{D}_N$ and $\mathcal{D}_Y^i$ can be different on at most a δ fraction of inputs for $i \notin B$,

$$\Pr_{M \in \mathcal{D}_Y^i}[\text{Eval}(\varphi, M, s) = 0] \geq \Pr_{M \in \mathcal{D}_N}[\text{Eval}(\varphi, M, s) = 0] - \delta. \tag{8}$$

The lemma follows by combining (6), (7), and (8).   □

The two properties captured by Lemma 11 and Lemma 13 formalize the contradictory dichotomy that allows us to prove Theorem 8.

Proof (of Theorem 8). Let $n = m = 2^t$ and C be an AM-circuit with bottom fan-in k purported to solve $\exists$-ApprMaj. Without loss of generality, we assume that $9/10$ of C's DNF subcircuits accept on $M \in \exists$-ApprMaj$_Y$ while $9/10$ reject on $M \in \exists$-ApprMaj$_N$—this can be achieved by naïve amplification while increasing the bottom fan-in by only a constant factor.

Consider sampling M from $\mathcal{D}_Y$ or $\mathcal{D}_N$ and a random depth-2 subcircuit φ connected to the output gate of C. Then by Proposition 10,

$$\Pr_{M \in \mathcal{D}_Y, \varphi \in_U C}[\varphi(M) = 1] \geq \frac{9}{10} \text{ and } \Pr_{M \in \mathcal{D}_N, \varphi \in_U C}[\varphi(M) = 0] \geq \frac{9}{10}(1 - \frac{1}{n}) > 5/6$$

for large enough n. Therefore, the probability that a random subcircuit is correct on most instances of $\mathcal{D}_Y$ is greater than $1/2$ (and similarly for $\mathcal{D}_N$):

$$\Pr_{\varphi \in_U C}[\Pr_{M \in \mathcal{D}_Y}[\varphi(M) = 1] > 2/3] > 1/2 \text{ and } \Pr_{\varphi \in_U C}[\Pr_{M \in \mathcal{D}_N}[\varphi(M) = 0] > 2/3] > 1/2.$$

Thus, by a union bound, there is a k-DNF φ^* where

$$\Pr_{M \in \mathcal{D}_Y}[\varphi^*(M) = 0] \leq 1/3 \text{ and } \Pr_{M \in \mathcal{D}_N}[\varphi^*(M) = 1] \leq 1/3. \tag{9}$$

Our proof proceeds by showing that such a k-DNF $\varphi*$ cannot exist when $k = o(\log n)$. By Lemma 11, we have that

$$\Pr_{M \in \mathcal{D}_N}[\mathrm{Eval}(\varphi^*, M, s) = 0] \geq 1 - 1/\left(3(1 - e^{-\frac{s}{k6^k}})\right). \tag{10}$$

Plugging (10) into Lemma 13 gives

$$\Pr_{M \in \mathcal{D}_Y}[\mathrm{Eval}(\varphi^*, M, s) = 0] \geq \underbrace{(1 - \frac{sk}{\delta n})}_{(*)}\left(1 - 1/\left(3\underbrace{(1 - e^{-\frac{s}{k6^k}})}_{(**)}\right) - \delta\right). \tag{11}$$

We claim that for small enough k, we can set δ and s so that the right-hand side of (11) is at least $1/2$. Since Eval outputs 0 no more often than $\varphi^*(M) = 0$, this contradicts that $\Pr_{M \in \mathcal{D}_Y}[\varphi^*(M) = 0] \leq 1/3$ (from (9)), completing the lower bound for such k.

All that remains is to find the largest k so that we can choose δ and s appropriately. In order for the right-hand side of (11) to be large enough, we need $(*)$ and $(**)$ to be close to 1 and δ to be small. The condition on $(*)$ requires $s = O(\frac{\delta n}{k})$, while the condition on $(**)$ requires $s = \Omega(k6^k)$. Choosing δ to be a small constant, say, $\delta = 1/100$, we see that such a setting is possible when $k^2 6^k = O(n)$— i.e., when $k = c\log n$ for small enough c.   □

2.3   Remarks on Theorem 1

We now highlight a few details of the preceding analysis. First, notice that the time complexity of the simulated MA-game's verification procedure V enters into the running-time of the AM-simulation only when accounting for the resources needed to compute each query in the black-box setting of Definition 5— The circuit lower bound of Theorem 8 depends solely on the dimensions of the instance of ∃-ApprMaj, which correspond to the number of bits sent by the prover and verifier in the MA-game.

Additionally, notice that the only place that the number of columns m in the instance of ∃-ApprMaj (corresponding to the number of bits flipped by the verifier in an MA-game) enters into the proof is in Proposition 10, where a lower bound on m is needed to ensure that the distribution $\mathcal{D}_N$ behaves properly. Thus, Theorem 8 actually gives a lower bound of $\Omega(n)$ for $n \times m$ instances of ∃-ApprMaj, where $m \geq \alpha \log n$ for α as defined in Proposition 10. This implies that the black-box lower bound of Theorem 1 holds even when simulating MA-games where the verifier flips very few bits.

The above two observations show that the crucial factor in determining the number of queries needed by an AM-game black-box simulating MA is the *number of bits sent by the prover*. We point out that this behavior is shared by the known simulations of MA by AM (such as Babai's).

Furthermore, the instances produced by $\mathcal{D}_Y$ satisfy an even stricter promise than ∃-ApprMaj$_Y$, since *every* bit in some row is set to 1. This corresponds to the promise on the "yes" side fulfilled by MA-games with *one-sided error*. Therefore, Theorem 1 holds even for simulations of the weaker class of time-t MA-games with one-sided error.[3]

Finally, we take a moment to compare our techniques to those of Viola's [7]. Our approach is similar in that it uses ideas inspired by recent small-restriction switching lemmas [15, 16] to show that k-DNF's behave similarly enough on certain "yes" and "no" instances of the problem in question. In the setting of normal depth-3 circuits that Viola considers (for the problem ApprMaj), it is sufficient to give an argument that explicitly constructs a "yes" instance rejected by one particular depth-2 subcircuit (in the case where the top gate is AND) assuming that the subcircuit behaves correctly on "no" instances. However, one incorrect subcircuit is insufficient to fool an AM-circuit (as required by our setting), since the top gate only rejects if *most* subcircuits reject. Our approach addresses this issue via the aforementioned *probabilistic* construction of the bad "yes" instances. By producing instances that are independent of the specific subcircuits they are designed to fool, we simultaneously violate many more of them than we could by an explicit construction.

The bounded error of the AM-circuit setting also leads to a quantitatively better statement of the lower bound than in the Viola's standard depth-3 setting: The latter result states that no depth-3 circuit can compute ApprMaj with small bottom fan-in unless it is very large; In contrast, Theorem 8 states that no AM-circuits with small bottom fan-in can compute ∃-ApprMaj, *regardless of their size*.

3   Future Work

The most compelling open problem related to this work is determining whether or not a subquadratic-overhead simulation of MA by AM is at all possible— i.e., to either prove a general quadratic lower bound for the overhead of *any* simulation of MA by AM or to use some non-black-box technique to actually achieve a subquadratic AM-simulation. Either case yields something interesting: In the former case, one gets a hierarchy for MA and AM (see Sect. 1.1); In the latter case, one makes progress towards time-space lower bounds for SAT on probabilistic machines (see Sect. 1.2).

Furthermore, we ask if there are any common restrictions on the verifier that a non-black-box approach could take advantage of to allow a subquadratic simulation of a restricted class of MA-games. Especially interesting is the case of a space-bounded verifier: A subquadratic AM-simulation that exploits the verifier's space bound could also be used to achieve time-space lower bounds for SAT on probabilistic machines.

Another direction is to extend the black-box lower bound to relationships between other complexity class inclusions where the best known simulation has

---

[3] Since the usual definition of an MA-game has two-sided error and applies the same bound to the message bits and the time bound of the verification procedure, we chose not to draw any of these distinctions in Theorem 1.

quadratic overhead. In particular, inclusions that proceed by amplifying the error of a probabilistic computation to exponentially small seem particularly amenable, since we already know that the amplification cannot be done faster in a black-box manner [12]. We propose finding black-box lower bounds for the overhead of inclusions such as MA $\subseteq$ PP, $\oplus \cdot$ BPP $\subseteq$ BP $\cdot \oplus$P, etc. The latter is particularly interesting, since it applies to the time-overhead of the collapse of the polynomial-time hierarchy to BP $\cdot \oplus$P in Toda's Theorem [18]; It is currently unknown how to achieve such a collapse with sublinear factor increase in time per alternation, which could yield new time-space lower bounds for counting problems.

Acknowledgments

The author would like to thank Bess Berg, Jeff Kinne, and Emanuele Viola for their helpful discussions, and especially Dieter van Melkebeek for his extensive advice, support, and enthusiasm. We would also like to thank the anonymous referees for their comments.

References

[1] Lund, C., Fortnow, L., Karloff, H., Nisan, N.: Algebraic methods for interactive proof systems. Journal of the ACM 39, 859–868 (1992)
[2] Shamir, A.: IP = PSPACE. Journal of the ACM 39, 869–877 (1992)
[3] Goldreich, O., Micali, S., Wigderson, A.: Proofs that yield nothing but their validity or all languages in NP have zero-knowledge proof systems. Journal of the ACM 38, 691–729 (1991)
[4] Babai, L.: Trading group theory for randomness. In: Proceedings of the 17th ACM Symposium on the Theory of Computing, pp. 421–429. ACM Press, New York (1985)
[5] Babai, L., Moran, S.: Arthur-Merlin games: a randomized proof system, and a hierarchy of complexity classes. Journal of Computer and System Sciences 36, 254–276 (1988)
[6] Diehl, S., van Melkebeek, D.: Time-space lower bounds for the polynomial-time hierarchy on randomized machines. SIAM Journal on Computing 36(3), 563–594 (2006)
[7] Viola, E.: On probabilistic time versus alternating time. Technical Report TR-05-137, Electronic Colloquium on Computational Complexity (2005)
[8] Nisan, N., Wigderson, A.: Hardness vs. randomness. Journal of Computer and System Sciences 49, 149–167 (1994)
[9] Goldreich, O., Zuckerman, D.: Another proof that BPP $\subseteq$ PH (and more). Technical Report TR-97-045, Electronic Colloquium on Computational Complexity (1997)
[10] van Melkebeek, D., Pervyshev, K.: A generic time hierarchy for semantic models with one bit of advice. In: Proceedings of the 21st IEEE Conference on Computational Complexity, pp. 129–142. IEEE, Los Alamitos (2006)
[11] Santha, M.: Relativized Arthur-Merlin versus Merlin-Arthur games. Information and Computation 80(1), 44–49 (1990)

[12] Canetti, R., Even, G., Goldreich, O.: Lower bounds for sampling algorithms for estimating the average. Information Processing Letters 53, 17–25 (1995)

[13] Fortnow, L., Lipton, R., van Melkebeek, D., Viglas, A.: Time-space lower bounds for satisfiability. Journal of the ACM 52, 835–865 (2005)

[14] Williams, R.: Better time-space lower bounds for SAT and related problems. In: Proceedings of the 20th IEEE Conference on Computational Complexity, pp. 40–49. IEEE, Los Alamitos (2005)

[15] Segerlind, N., Buss, S., Impagliazzo, R.: A switching lemma for small restrictions and lower bounds for k-DNF resolution. SIAM Journal on Computing 33(5), 1171–1200 (2004)

[16] Razborov, A.: Pseudorandom generators hard for k-DNF resolution and polynomial calculus resolution. Unpublished manuscript (2002–2003)

[17] Furst, M.L., Saxe, J.B., Sipser, M.: Parity, circuits, and the polynomial-time hierarchy. Mathematical Systems Theory 17(1), 13–27 (1984)

[18] Toda, S.: PP is as hard as the polynomial-time hierarchy. SIAM Journal on Computing 20(5), 865–877 (1991)

Lower Bounds for Testing Forbidden Induced Substructures in Bipartite-Graph-Like Combinatorial Objects

Eldar Fischer and Eyal Rozenberg

Department of Computer Science, Technion, Haifa, Israel
{eldar,eyal}@cs.technion.ac.il

Abstract. We investigate the property of k-uniform k-partite (directed) hypergraphs with colored edges of being free of a fixed family of forbidden induced substructures. We show that this property is not testable with a number of queries polynomial in $1/\varepsilon$, presenting proofs for the case of two colors and $k = 3$, as well as the case of three colors and $k = 2$ (edge-colored bipartite graphs). This settles an open question from [1], implying that the polynomial testability proof for two colors and $k = 2$ cannot be extended to these structures.

1 Introduction

The domain of combinatorial property testing deals with promise problems which are relaxed versions of decision problems: One is given a property of combinatorial structures of a certain type, and for such a structure, must determine using as few queries as possible whether the structure satisfies the property or is far from satisfying it. The distance of structures from satisfying a property is measured by the fraction ε of bits in the representation of the structure that one must modify in order to reach a structure satisfying the property. For example, a graph with n vertices is ε-far from a property of graphs if one must add or remove at least $\varepsilon\binom{n}{2}$ edges to obtain a graph satisfying the property. A property is said to be *testable* if one can solve the above-mentioned promise problem with probability at least $2/3$ by querying only a constant number of places in the input structure, which is independent of the size of the structure and may depend only on the distance parameter ε of the relaxed decision problem.

The study of property testing began with the paper of Blum, Luby and Rubinfeld [2]; the notion of property testing was first given a formal definition by Rubinfeld and Sudan in [3]. Goldreich, Goldwasser and Ron first investigated properties of combinatorial structures, and specifically testing properties in the dense model for graphs, in [4]. Surveys of results in the field can be found in [5] and [6].

When considering a testable property, one wishes to tightly bound its query complexity $q(\varepsilon)$, that is the dependence of the number of queries necessary for testing an input for the property, on the distance parameter ε.

M. Charikar et al. (Eds.): APPROX and RANDOM 2007, LNCS 4627, pp. 464–478, 2007.

Many graph properties have been shown to be testable using the celebrated Szemerédi Regularity Lemma ([7]; see [8, Chapter 7] for a detailed exposition of the lemma, its proof and an example of its use). Applying it, one can partition large enough graphs in a way conductive to proving the testability of many types of properties (see [9] and more recently [10]); these upper bounds apply, essentially, to bipartite graphs as well. The drawback of using the lemma is the severely high dependence on the distance parameter ε of the number of queries necessary for testers – a tower function in $1/\varepsilon$ or worse. Recently, the Regularity Lemma has been extended to the case of k-uniform hypergraphs, implying similar upper bounds for testing non-induced sub-hypergraph freeness and (see [11], [12], [13] and the discussion in [14]); Ishigami, in the yet-unpublished [15], presents a different variant of a hypergraph regularity lemma which may allow for the application of the bounds also to testing induced sub-hypergraph freeness. There is a significant gap between the upper bounds on the number of queries in these results, and known lower bounds on testing forbidden substructures, which have a mild super-polynomial dependence on $1/\varepsilon$ (see below for details).

Less abundant for properties of this type are hardness results; thus there is often a significant gap between the best known lower and upper bounds for the query complexity of a property (see below).

In [1], Alon, Fischer and Newman prove that every property of bipartite graphs characterized by a set of forbidden induced subgraphs is polynomially testable (improving upon a result from [16]). Two open questions put forth at the conclusion of [1] are whether similar results may be obtained for graphs with colored edges, or for structures of higher dimensions.

In this article we will add to the known lower bound results on testing by considering the case of edge-colored bipartite graphs and the case of k-partite k-hypergraphs. We will prove a lower bound result for testing properties of such structures characterizable by finite sets of forbidden substructures.

These types of properties, which are closed under the taking of induced substructures, have known positive results regarding their testing in graphs (e.g. [10], although [9] suffices for the types of properties we consider); such results seem to extend naturally to colored graphs, and with the use of the versions of the regularity lemma for hypergraphs (mentioned above) extend to hypergraphs as well. Thus it is of interest to provide lower bounds for testing such properties, even if the bounds are not known to be tight.

Alon proved in [17] that it is impossible to test a graph for containing a copy of some (not necessarily induced) subgraph with polynomially-many queries in $1/\varepsilon$, if this subgraph is non-bipartite, e.g. an odd cycle. Alon and Shapira extended this result in [14], finding super-polynomial lower bounds for testing digraphs for containing copies of certain other digraphs (e.g. directed cycles of length 3 or more). In [18], Alon and Shapira proved that the property of general (non-k-partite) k-uniform hypergraphs of being free of a forbidden induced hypergraph is not, in most cases, polynomially testable. The ideas of the constructions in [17] and [14] will be of use to us in proving our two lower bounds.

This paper is organized as follows. In Section 2 we provide exact definitions of our terms, discuss the relations between matrices, bipartite graphs, k-graphs and tensors, and recall results regarding canonical testing. In Section 3 we state our two lower-bound results, for colored bipartite graphs and for k-graphs, and give a sketch of their proof. The two theorems, Theorem 1 and Theorem 2, are proven in Section 4 and Section 5 respectively.

2   Preliminaries

2.1   'Generalizations' of Bipartite Graphs

A matrix over $\{0,1\}$ can be thought of as a bipartite graph $G = (U, V, E)$, with $U = \{u_1, \ldots, u_n\}$, $V = \{v_1, \ldots, v_n\}$ and $E(G) = \{(u_i, v_j) \in U \times V \mid M(i,j) = 1\}$ — that is, a bipartite graph whose vertices are ordered, with edges between the pairs whose corresponding matrix cell has value 1.

A *colored bipartite graph* is a tuple $G = (U, V, \mathbf{col})$, where $\mathbf{col}$ is a function assigning to each edge one out of a finite set of colors $\{0, \ldots, \sigma - 1\}$. We assume w.l.o.g. that colored bipartite graphs have the complete set of edges $U \times V$, since we can interpret one of the edge colors as 'no edge'. We shall refer to graphs for which the set of colors is of cardinality σ as σ-*colored*; specifically, simple bipartite graphs are considered to be 2-colored.

A *k-uniform k-partite (uncolored) hypergraph* is a tuple $G = (V_1, \ldots, V_k, E)$ with the V_i being sets of vertices and $E \subseteq \prod_{i=1}^{k} V_i$ the set of the hyperedges. We shall refer to k-uniform k-partite hypergraphs as k-*graphs* for short.

A σ-colored bipartite graph G can also be thought of as an adjacency matrix M of dimensions $(|V_1|, |V_2|)$, with cell values $M(i,j) = \mathbf{col}(i,j) \in \{0, \ldots, \sigma - 1\}$. Similarly, a k-graph can be thought of as a k-dimensional adjacency tensor T over $\{0,1\}$, in which $T(v_1, \ldots, v_k) = 1$ if $(v_1, \ldots, v_k) \in E(G)$ and 0 otherwise.

The conceptual similarity between colored matrices and colored bipartite graphs, as well as between binary k-dimensional tensors and k-graphs will be used implicitly throughout the sections discussing these structures. However, these are not equivalent structures: The coordinates of matrices and tensors are ordered. Thus, when we refer to 'submatrices' of a bipartite graph's adjacency matrix, we are still referring to subgraphs — the submatrix coordinates may be selected irrespectively of the order of coordinates in the adjacency matrix, i.e. we allow row and column permutations of the submatrices. The same applies to k-graphs and their 'subtensors'.

2.2   Properties and Testers

A *property* $\mathcal{P}$ of (colored) bipartite graphs is a set of such graphs which is closed under graph isomorphism. Similarly, a property of matrices, considered as (colored) bipartite graph adjacency matrices is a set of matrices closed under permutations of the coordinates along each of the two axes.

A bipartite graph G with n vertices in each part is ε-*far* from satisfying a property $\mathcal{P}$ if G differs from every $G' \in \mathcal{P}$ with n vertices in each part by at least εn^2 edges, i.e. One needs to modify the values of at least εn^2 cells in its corresponding colored matrix to reach the colored matrix corresponding to G'.

The notions of a property and of ε-distance are defined in a similar fashion also for (uncolored) k-graphs — we consider k-graphs of *leg* n, that is, with dimensions $(n, n, \ldots, n)$, and the k-graph or the tensor is said to be ε-*far* from satisfying the property if one must modify (in this case, add or remove) more than εn^k of the possible hypergraph edges or of the tensor cells.

An ε-*tester* for a property $\mathcal{P}$ of colored bipartite graphs is an algorithm which, for a graph G with n vertices, is given the leg n, makes a certain number of queries of the colors of edges — queries of the type $\mathbf{col}\,(v_1, v_2)$ for some pair (v_1, v_2) — and distinguishes with high probability (e.g. $2/3$) between the case of G being in $\mathcal{P}$ and the case of G being ε-far from $\mathcal{P}$. The number of queries may only depend on ε. Testers for k-graph properties are defined similarly — a tester is given the leg n (the size of each of the k vertex sets), is able to make queries of whether $(v_1, \ldots, v_k) \in E$ or not, and must distinguish as described above between k-graphs in $\mathcal{P}$ and k-graphs which are ε-far from $\mathcal{P}$.

An ε-tester is *one-sided* if it identifies all members of $\mathcal{P}$ as such with probability 1, and all ε-far graphs (resp. matrices) as such at least with some constant non-zero probability (e.g. $2/3$).

A property $\mathcal{P}$ is *polynomially testable* if there exists some polynomial $p(x)$ such that for every ε, there exists an ε-tester for the property $\mathcal{P}$ making $p(1/\varepsilon)$ queries.

3   Overview of the Results

3.1   A Lower Bound for Colored Bipartite Graphs

Theorem 1. *There exists a 2-colored bipartite graph H with two vertices per part, such that an ε-tester of 3-colored bipartite graphs for being free of having H as an induced subgraph, performs no less than $(c/\varepsilon)^{c \cdot \log(c/\varepsilon)}$ queries for some global constant c — provided the tester is either one-sided, or works independently of the size of the input.*

We note that failure of the bound to apply to two-sided error testers whose operation might depend on the size of the input is not merely a technical issue specific to our results; see Subsection 5.2 below for some more discussion regarding this point.

Our argument for proving this theorem is based on the construction of a hard-to-test bipartite graph, as per the following:

Lemma 1. *There exists a $(2, 2)$ bipartite graph F such that for every ε and for every $n > 16(c/\varepsilon)^{-c \cdot \log(c/\varepsilon)}$, there exists a 3-colored bipartite graph G which is ε-far from being free of F, and yet the fraction of $(2, 2)$ subgraphs of G which are copies of F is no more than $(c/\varepsilon)^{-c \cdot \log(c/\varepsilon)}$ for some global constant c.*

This lemma will be proven by a sequence of constructions described in the following:

Construction Task List

T1. A colored matrix can represent 4-partite digraphs, with the representation preserving the distributions of induced substructures in the digraph. Specifically, submatrices of the representation will correspond to induced subgraphs of the digraphs on certain vertices.

T2. One may construct 4-partite digraphs which have only four ordered pairs of vertex sets between which edges are present (four 'edge layers'), and in which induced directed 4-cycles are super-polynomially rare.

These constitute a proof of a rougher, weaker version of Lemma 1: For one, we will have used many more than 3 colors — the representation of a digraph will not be at all terse; also, we will have used numerous forbidden submatrices, as a submatrix of the digraph representation will also contain information about edges other than those constituting a 4-cycle. We will then proceed to refine our construction:

T3. One may construct digraphs as described in Task List Item 2, with the additional constraint that the first three edge layers are identical.

T4. One may construct 4-partite digraphs as described in Task List Item 3, with the additional constraint that the edge layers are symmetric with respect to a relevant ordering of the vertices in each part.

T5. The construction for Task List Item 4 can be shown to satisfy the additional constraint that no two vertices are connected in all four edge layers.

These refinements of the construction will reduce the number of possible edge configurations of two vertex indices j_1, j_2 to just three; consequently, we will only need 3 colors for the matrix representation of the digraph. It will be shown that the existence of a 4-cycle involving vertices (j_1, j_2, j_3, j_4) determines precisely the edge configuration for each of the pairs (j_1, j_2), (j_2, j_3), (j_3, j_4) and (j_4, j_1). This in turn implies that we do not require more than one forbidden matrix (up to permutations) to constrain the represented 4-partite digraphs to be cycle-free. Lemma 1 will follow.

To use the construction of Lemma 1 in the proof of Theorem 1, we will finally argue that a tester cannot essentially do much better than sample submatrices of the matrix and check them for forbidden submatrices within the sample.

3.2 A Lower Bound for k-Graphs and Tensors

Theorem 2. *There exists a 3-graph H with leg 2, such that every ε-tester of 3-graphs for being free of copies of H as an induced 3-graph, performs no less than $(c/\varepsilon)^{c \cdot \log(c/\varepsilon)}$ queries for some global constant c — provided that the tester is either one-sided, or works independently of the size of the input.*

4 The Lower Bound for Colored Bipartite Graphs

In this section we present the details of the proof of Theorem 1. We will be limiting our digraphs to have edges only between certain pairs of subsets of the vertices. Specifically,

Definition 1. *A cyclic k-partite digraph $G = (V_1, \ldots, V_k, E)$ is a k-partite digraph in which every edge in E extends from V_i to $V_{(i \bmod k)+1}$ for $1 \leq i \leq k$.*

4.1 Task List Item 1: Representing Cyclic k-Partite Digraphs by Colored Matrices

The edges of a cyclic k-partite digraph can be decomposed into k bipartite graphs between the ordered pairs of cycle-consecutive parts. The edge relation between each of these ordered pairs can be seen as an $n \times n$ binary adjacency matrix (with n being the number of vertices in each part), in which a cell is set to 1 when an edge is present and to 0 otherwise. This leads to the following representation:

Definition 2. *Let $G = (V_1, \ldots, V_k, E)$ be a cyclic k-partite digraph, with k vertex sets of size n each, where $V_i = \{v_{i,1}, \ldots, v_{i,n}\}$. The colored matrix representation of G, denoted $\mathcal{CM}(G)$, is the matrix of leg n, each of whose cells has one of 2^{2k} colors, corresponding to all possible combinations of the following 2k binary values: For $M = \mathcal{CM}(G)$, each cell $M(j_1, j_2)$ has a distinct color bit for each one of the k edges $(v_{1,j_1}, v_{2,j_2}), \ldots, (v_{k-1,j_1}, v_{k,j_2}), (v_{k,j_1}, v_{1,j_2})$, and another bit for each one of the k edges $(v_{1,j_2}, v_{2,j_1}), \ldots, (v_{k-1,j_2}, v_{k,j_1}), (v_{k,j_2}, v_{1,j_1})$. Each bit is set to 1 if its respective edge exists, and to 0 otherwise.*

The forbidden submatrices. Our lower bound construction will utilize cyclic 4-partite digraphs which are far from not containing a directed square (a 4-cycle), yet have few copies of it. We shall choose a directed cycle of length four, and consequently set henceforth $k = 4$. The reason for this choice is that four is the first even length of such a hard-to-test subgraph, as has been described in [14]. Our matrix representations $\mathcal{CM}(\cdot)$ therefore have cells in $2^{2k} = 256$ possible colors. We demonstrate later how the number of colors can be progressively reduced to the minimum of 3.

A choice of a cell (j_1, j_2) yields information about the edges in all four layers; choosing a 2×2 submatrix with coordinates $(j_1, j_3) \times (j_2, j_4)$ yields information about several directed 4-cycles, one of which is $C = (v_{1,j_1}, v_{2,j_2}, v_{3,j_3}, v_{4,j_4})$. We note that for every 4-cycle of G there is a choice of $j_1, \ldots, j_4$ such that C as defined above corresponds to that cycle. Thus we only need to forbid 2×2 matrices that witness the existence of the four edges of the directed cycle C associated with a given submatrix. There are many possible colored 2×2 submatrices witnessing the existence of their C, since the existence of any of the rest of the $(k/2)^2 \cdot 2k - k = 28$ edges represented in the submatrix cells does not affect the presence of C. Hence, the forbidden submatrices are the 2^{28} matrices in which the four color bits for the edges of C are set.

Note that in some cycles of G it may be the case that $j_1 = j_3$ and/or $j_2 = j_4$. We refer to such cycles as *degenerate*; our construction and our arguments below will only involve graphs with no degenerate cycles.

For every copy of a (non-degenerate) 4-cycle in G, there exists exactly one leg-2 forbidden submatrix in $\mathcal{CM}(G)$ (recall that we are disregarding row and column permutations). This is true despite the fact that it is possible to *infer* the existence of a 4-cycle also from other submatrices of $\mathcal{CM}(G)$. In other words, a selection of a leg-2 submatrix of $\mathcal{CM}(G)$, and a check of whether its C exists, corresponds to a selection of four vertices in the four parts of G and a check of whether they form a (non-degenerate) cycle. With $n = |V_i|$ as the size of each V_i, There are $\binom{n}{2}^2$ such possible choices.

4.2   Task List Item 2: Construction of a Hard-to-Test Matrix

Definition 3. *The trivial integer solutions to the equation $x_1 + x_2 + \ldots + x_r = r \cdot x_{r+1}$ are those in which all of $x_1, \ldots, x_r$ are equal.*

Lemma 2 ([17, Lemma 3.1] and [14, Lemma 6.1]). *For every fixed integer $r \geq 2$, and for every positive integer m, there exists a subset $X \subseteq \{1, \ldots, m\}$, with size at least $m / \exp\left(10\sqrt{\log(m)\log(r)}\right)$, with no non-trivial solution to the equation $x_1 + x_2 + \ldots + x_r = r \cdot x_{r+1}$.*

Fix $r = 3$ and $\varepsilon' = 8\varepsilon$. Let m maximize $\varepsilon' < 4^{-8}\exp\left(-10\sqrt{\log(m)\log(3)}\right)$, obtaining, for an appropriate constant c, the bound $m \geq (c/\varepsilon')^{c \cdot \log(c/\varepsilon')}$.

Using the set X from Lemma 2, a cyclic 4-partite digraph T is constructed: the four parts of T's vertex set, $V_1, \ldots, V_4$, have cardinalities $m, 2m, 3m, 4m$ respectively. For every $1 \leq i \leq 3, 1 \leq j \leq im$ and $x \in X$, the edge (v_j, v_{j+x}) exists between V_i and V_{i+1}; the edges between V_4 and V_1 are the pairs (v_{j+3x}, v_j) for every $x \in X$ and $1 \leq j \leq m$.

As can be observed ([14, Lemma 6.2]), $E(T)$ consists of $4m|X|$ edges forming $m|X|$ edge-disjoint copies of the directed 4-cycle; T has more than $4^8 \varepsilon' m^2$ edges. For our use we would like all parts V_i to have the same cardinality, so we add isolated vertices making every V_i of size exactly $4m$. Let T_1 be the graph resulting from this addition. We note that all the cycles of T_1 are non-degenerate.

Definition 4. *A blow-up of a digraph $G = (V, E)$ by a factor s is a graph $G' = (V', E')$, with $V' = V \times [s]$, and $E' = \{((v, j), (u, j)) \mid (v, u) \in E \wedge j_1, j_2 \in [s]\}$. That is, every vertex is copied s times and every edge is replaced with the complete bipartite graph $K_{s,s}$ between the corresponding sets of vertex copies.*

Lemma 3 (special case of [14, Lemma 6.3]). *Let $K = (V(K), E(K))$ be a digraph and let $T = (V(T), E(T))$ be an s-blow-up of K. Let $R \subseteq E(T)$ be a subset of the set of edges of T, and suppose that each copy of K in T contains at least one edge of R. Then $|R| > |E(T)|/|V(K)|^4$.*

The graph T_1 is now blown-up by a factor of $s = \lfloor n/(4^2 m)\rfloor$; let G_1 denote the resulting blown-up graph. We have $|E(G_1)| = s^2 \cdot 4^8 \varepsilon' m^2 > 4^4 \varepsilon' n^2$. Since $E(T_1)$ is a set of edge-disjoint 4-cycles, $E(G_1)$ is a set of edge-disjoint s-blown-up 4-cycles. By the lemma above, at least a $1/4^4$-fraction of the edges of each of these edge sets must be removed so as to remove all 4-cycles; G_1 is thus ε'-far from being 4-cycle-free. On the other hand, G_1 only has $m^2 s^4 < c_1 n^4/m$ copies of the 4-cycle, for some constant c_1. One can also verify that all cycles of G_1 are non-degenerate.

We must now transform the argument regarding the testing hardness of the graph G_1 into an argument regarding $\mathcal{CM}(G_1)$.

Proposition 1. *For $\sigma = 2^8$ there exists a finite set $\mathcal{F}$ of σ-colored leg-2 matrices, such that for every ε and $n > (c/\varepsilon)^{c \cdot \log(c/\varepsilon)}$, there exists a σ-colored matrix M which is ε-far from being free of members of $\mathcal{F}$, and yet, the fraction of leg-2 submatrices of M which are copies of a member of $\mathcal{F}$ is no more than $(c/\varepsilon)^{-c \cdot \log(c/\varepsilon)}$ for some global constant c.*

Proof. Let $M = \mathcal{CM}(G_1)$, and set the family of forbidden matrices to be the 2^{28} matrices defined above. To prove the second part of the claim we recall that there is only one copy of a forbidden matrix in $\mathcal{CM}(G_1)$ for every copy of a 4-cycle in G. Only $c_1 n^4/m$ of the $\binom{n}{2}^2$ possible directed non-degenerate 4-cycles with vertices in consecutive parts appear in G, so no more than an $8c_1/m$ fraction of the $\binom{n}{2}^2$ submatrices of $\mathcal{CM}(G_1)$ of leg 2 are copies of forbidden matrices.

For the first part of the claim, we note that by modifying a matrix cell one affects the representation of at most 8 edges of G_1. Thus, unless at least $\varepsilon' n^2/8 = \varepsilon n^2$ cells are modified, more than $(1 - \varepsilon') n^2$ of the edges of G have their two representing color bits (i.e. in both the cells $\mathcal{CM}(G_1)(i,j)$ and $\mathcal{CM}(G_1)(j,i)$) unmodified. Consequently, the cycles contained in this unmodified subset of $E(G_1)$ still have their representing forbidden submatrix intact. G_1 has a non-degenerate directed 4-cycle remaining for every choice of less than $\varepsilon' n^2$ edges, so $\mathcal{CM}(G_1)$ still contains a copy of a forbidden matrix. $\square$

4.3   Reducing the Number of Colors

As mentioned above, 256 colors are more than necessary to construct a hard to test matrix. We now reduce this number, in several stages.

Note that by reducing the number of colors, we lose the ability to represent many of the graphs that were previously representable; we maintain, however, the ability to represent the graphs we construct for proving the lower bound.

Task List Item 3: Making most edge layers identical. We note that in the graph T_1, the edge sets in the 'first' 3 layers, those between V_i and V_{i+1} for $1 \leq i \leq 3$, are quite similar: $v_{i,j}$ is connected to $v_{i,j+x}$. The difference is that in each of the V_i's, only the first im vertices are connected onwards to vertices in V_{i+1}. We now consider the consequences of connecting all possible vertices of V_i onwards – that is, making $(v_{i,j}, v_{i+1,j+x})$ an edge whenever $j + x \leq 4m$ and $x \in X$.

Denote this new graph by T_2. As with the graph T_1, every directed 4-cycle $(v_{1,j_1}, v_{2,j_2}, v_{3,j_3}, v_{4,j_4})$ in T_2 satisfies

$$(j_2 - j_1) + (j_3 - j_2) + (j_4 - j_3) = (j_4 - j_1)$$

so if we denote

$$x_1 = j_2 - j_1 \quad x_2 = j_3 - j_2 \quad x_3 = j_4 - j_3 \quad x_4 = (j_4 - j_1)/3$$

the equation becomes $x_1 + x_2 + x_3 = 3x_4$, and since $x_1, \ldots, x_4 \in X$, all four values must be equal. Also, if such a cycle begins with $j_1 > m$ in V_1, then

$$(j_1 - m\lfloor j_1/m \rfloor, j_2 - m\lfloor j_1/m \rfloor, j_3 - m\lfloor j_1/m \rfloor, j_4 - m\lfloor j_1/m \rfloor)$$

is another cycle in T_2 (the vertex indices all remain positive), which begins with $j_1 \leq m$, i.e. it corresponds to a cycle in the original T_1. It follows that the total number of cycles has increased by no more than a factor of 4, and that all cycles are still non-degenerate.

Since all cycles are edge-disjoint in T_2 as well, the number of cycles increases with the s-blow-up of T_2 into G_2 by a factor of s^4, as in the case of G_1. G_2 has the same vertex sets as G_1, and a superset of the edges of G_1, making it at least as far from being 4-cycle free as G_1. As for the number of cycles, T_1 had $m|X| < m^2$ 4-cycles, T_2 has at most $4m|X| < 4m^2$ 4-cycles, and G_2 has at most $4m^2 s^4 < c_2 n^4/m$ 4-cycles, for some constant c_2.

We can now use our different construction of T_2 to reduce the number of colors necessary for its representation: As the bits for the three $V_i \to V_{i+1}$ edge layers are the same, we only need two bits for each type of layer (one for the $j_1 \to j_2$ edge and one for the 'flip' edge $j_2 \to j_1$), times two types of layers ($V_i \to V_{i+1}$ and $V_4 \to V_1$): in total we now use only $2^4 = 16$ colors. This property of T_2's first three layers carries over to G with the blow-up.

Our observations thus lead us to conclude that Proposition 1 also holds for $\sigma = 2^4$, with a different choice of the constants.

Task List Item 4: Making the edge layers symmetrical. The number of color bits may be further reduced – halved – if we could ensure that whenever $(v_{i_1,j_1}, v_{i_2,j_2})$ is an edge, so is $(v_{i_1,j_2}, v_{i_2,j_1})$. This can be affected by simply adding all 'flip' edges to T_2 – in addition to the edge $(j_1, j_1 + x)$ between V_i and V_{i+1}, and the edge $(j_1 + 3x, j_1)$ between V_4 and V_1, we also add $(j_1 + x, j_1)$ between V_i and V_{i+1}, and $(j_1, j_1 + 3x)$ between V_4 and V_1 respectively. The addition of the 'flip' edges may result in an excessive increase in the number of cycles, and possibly also results in intersections of the edges of different cycles. To avoid this, we again modify our pre-blowup graph T. Let us first consider a replacement of T_2 by the following T_3': Each of the four vertex sets is now $\{1, \ldots, 4^2 \cdot m\}$. The edges in the first three layers (we continue to maintain the uniformity of these layers) are $(j_1, j_1 + x + 3m)$ for all $x \in X$ and $j_1 \in \{1, \ldots, 4^2 m - x - 3m\}$; the edges between V_4 and V_1 are $(j_1 + 3(x + 3m), j_1)$ for all $x \in X$ and $j_1 \in \{1, \ldots, 4^2 m - 3(x + 3m)\}$. Each directed 4-cycle $(v_{1,j_1}, v_{2,j_2}, v_{3,j_3}, v_{4,j_4})$ must still satisfy

$$(j_2 - j_1) + (j_3 - j_2) + (j_4 - j_3) = (j_4 - j_1)$$

We denote

$$x_1 = j_2 - j_1 - 3m \quad x_2 = j_3 - j_2 - 3m \quad x_3 = j_4 - j_3 - 3m \quad x_4 = (j_4 - j_1 - 9m)/3$$

and this yields again the equation $x_1 + x_2 + x_3 = 3x_4$. Thus as in the case of T_2 above, cycles only exist when the edge x-values are all equal, i.e. T_3' has no more than $4^2 m|X|$ copies of a 4-cycle.

We now add all flip edges to T_3': the edges of the form $(v_{i,j_1+x+3m}, v_{i+1,j_1})$ are added in the first three layers, and the edges of the form $(v_{4,j_1}, v_{1,j_1+3(x+3m)})$ are added in the fourth layer. Let T_3 denote the resulting graph.

Lemma 4. *Every cycle in T_3 is either a cycle in T_3' (a no-flip-edge cycle) or a cycle consisting only of flip edges.*

Proof. Consider first some tuple (j_1, j_2, j_3, j_4) of vertex indices in the four parts where the first two edges are non-flip while the third one is a flip edge. In this case, we find that j_4 cannot be very far from j_1:

$$|j_1 - j_4| = (j_1 - j_2) + (j_3 - j_2) - (j_3 - j_4) \le 2 \cdot (3m + m) - 3m < 9m$$

however, for (j_4, j_1) to be an edge in the fourth layer (either a non-flip or a flip edge), we must have $|j_4 - j_1| = 9m + 3x$ for some $x \in X$. No such edges exist, proving that such a cycle is impossible. The remaining cases where one of three $V_i \to V_{i+1}$ edges is in the direction opposite to the other two edges are similarly impossible, implying that the edges in the first three layers are in the same direction for every cycle of T_3. If these three edges are non-flip edges, the j's are an increasing sequence, and so the fourth edge must have $j_4 > j_1$, i.e. it must also be a non-flip; if the edges in first three layers are flip edges, the j's are a decreasing sequence, and $j_4 < j_1$, i.e. the fourth edge must also be a flip edge. □

As for the number of cycles with all-flip or all-non-flip edges: If $(v_{i_1,j_1}, v_{i_2,j_2})$ is a non-flip edge, then $(v_{i_1,4^2m-j_1+1}, v_{i_2,4^2m-j_2+1})$ is a flip edge as well, and $(v_{i_1,4^2m-j_2+1}, v_{i_2,4^2m-j_1+1})$ is a non-flip edge. Thus if $(v_{1,j_1}, v_{2,j_2}, v_{3,j_3}, v_{4,j_4})$ is a cycle with no flip edges, $(v_{1,4^2m-j_1+1}, v_{2,4^2m-j_2+1}, v_{3,4^2m-j_3+1}, v_{4,4^2m-j_4+1})$ is an all-flip-edge cycle, and vice-versa. This one-to-one correspondence, together with the lemma above, bring us to conclude that there are exactly twice as many cycles in T_3 as there are in T_3', and that they are all edge-disjoint. Furthermore, the necessity of the first three edges to be in the same direction means that $j_1 \ne j_3$ and $j_2 \ne j_4$, so all cycles are still non-degenerate.

T_3 is a graph with $4^3 m$ vertices and no more than $2 \cdot 4^2 m|X|$ 4-cycles, all edge-disjoint. Blowing it up by a factor of $s = n/(4^3 m)$, we obtain a graph G_3 with n vertices and $2 \cdot 4^2 m|X| \cdot s^4 \le c_3 n^4/m$ cycles for an appropriate constant c_3. G_3 is also ε'-far from being cycle-free, by an argument similar to the case of G_1, with a proper choice of $m(\varepsilon') = (c/\varepsilon)^{c \cdot \log(c/\varepsilon)}$ reflecting the change in the constants used in the construction of T and the blow-up.

To represent G_3, we only need two bits of color: One bit for the first three layers (a single bit now suffices for both the 'non-flip' and the 'flip' edge), and one bit for the V_4-to-V_1 layer. We have thus brought down σ, the number of cell colors of a matrix for which Proposition 1 holds, to $2^2 = 4$ (again, with a different choice of a constant c).

Task List Item 5: Mutual exclusion between the edge layers. Obviously, the result of [1] implies it is impossible to reduce the number of color bits from two to one without making the matrix easy to test for the presence of forbidden submatrices. There is still the possibility of reducing the number of colors from four to three. In fact, if we review the construction of T_3 and G_3 carefully, we notice that for any (j_1, j_2), we only have three edge combinations represented for (j_1, j_2) (and the now-symmetrical j_2, j_1):

1. (j_1, j_2) is an edge in the $V_i \to V_{i+1}$ layers, but not in $V_4 \to V_1$.
2. (j_1, j_2) is not an edge in $V_i \to V_{i+1}$ layers, but is an edge in $V_4 \to V_1$.
3. (j_1, j_2) is not an edge in any layer.

No (j_1, j_2) can be an edge in all four layers, since edges in $V_4 \to V_1$ correspond to index differences $|j_1 - j_2|$ of at least $9m + 1$ (before the blowup of T_3 into G_3), while edges $V_i \to V_{i+1}$ correspond to differences of at most $4m$. Thus Proposition 1 holds for $\mathcal{CM}(G_3)$ as a 3-colored matrix as well. In fact, we are now able to prove Lemma 1:

Proof of Lemma 1. G is $\mathcal{CM}(G_3)$ constructed above. Indeed, there is now only one possible leg-2 submatrix (up to permutations) of $\mathcal{CM}(G_3)$ witnessing the presence of its corresponding cycle C in G_3: $M_F = \begin{pmatrix} 1 & 2 \\ 1 & 1 \end{pmatrix}$ (this is a matrix over $\{0, 1, 2\}$). One may verify that in all other leg-2 submatrices, at least one of the cycle edges must be missing. Thus F is the subgraph with adjacency matrix M_F. $\square$

4.4   Canonical Testing and Proof of Theorem 1

Definition 5. *A property $\mathcal{P}$ of (colored) bipartite graphs is hereditary if it is closed also under taking induced subgraphs.*

One can readily see that the properties we investigate here indeed fall under this definition.

Lemma 5. *If $\mathcal{P}$ is a hereditary property of colored bipartite graphs, and for some ε there exists a tester making $q(\varepsilon)$ queries, which works independently of the size of the input graph, then there exists a one-sided tester whose querying is done by uniformly sampling $O(q(\varepsilon))$ vertices from each one of the two parts of the graph, querying the subgraph they form and accepting if and only if this subgraph itself satisfies $\mathcal{P}$.*

As was done in [1, Lemma 5.5], the above lemma can be proven by almost word-by-word the same proof as that of [19, Appendix D, due to Alon].

Note, however, that these lemmata refer only to testers whose operation does not depend on the size of the input graph; this fact is essential to the proof in [19] of their lemma as well as the similar proof of Lemma 5. It is not trivial to show (and in fact not known at the moment) that a two-sided-error tester, which works differently for different values of n, implies the existence of a one-sided tester whose query complexity is not much worse than that of the original tester. See the errata [20] for [19] for further discussion of this point.

We now have everything necessary for proving our first main result:

Proof of Theorem 1. Consider an ε-tester of 3-colored bipartite graphs for being free of a forbidden subgraph which makes at most $q(\varepsilon)$ queries. If the tester works independently of the size n of the input graph, then by Lemma 5 there exists a one-sided tester for the property, which uniformly samples $\mathcal{O}(q)$ vertices from each of the two parts of the graph, queries the subgraph they form and accepts if and only if it is free of the forbidden matrix. If the original tester is one-sided, it is possible to show that w.l.o.g.it works independently of n; we omit the details.

By Lemma 1, there exists (for any sufficiently high n) a graph G and a forbidden subgraph F, such that G is ε-far from being free of F, but only a $(c'/\varepsilon)^{-c' \cdot \log(c'/\varepsilon)}$ fraction of its leg-2 subgraphs are copies of F, for some global constant c'.

The expected number of copies of the forbidden subgraph in the uniformly sampled subgraph of G is no more than $\mathcal{O}(q^4) \big/ (c'/\varepsilon)^{c' \cdot \log(c'/\varepsilon)}$ – the expected number of copies of $\mathcal{CM}(F)$ in a submatrix of $\mathcal{CM}(G)$ of leg $\mathcal{O}(q)$. If the original tester makes $q(\varepsilon)$ queries, then the number of queries of the uniform-subgraph-sampling tester is at most $\mathcal{O}(q^2)$. Thus if $q(\varepsilon) < (c/\varepsilon)^{c \cdot \log(c/\varepsilon)}$, for an appropriate constant c, then the expected number of forbidden subgraphs discovered is $o(1)$; the tester therefore accepts G with probability $1 - o(1)$, a contradiction. $\square$

5 The Lower Bound for 3-Graphs and Binary Tensors

5.1 Construction of a Hard-to-Test Tensor

Fix ε. Let M be as in the proof of Lemma 1, but with distance parameter $\varepsilon' = 2\varepsilon$. Let us consider the 3-colored matrix M as having two color bit layers: One bit-layer for the first three edge-layers of the 4-cycle (V_i to V_{i+1}), and another bit-layer for the 4th edge-layer (V_4 to V_1); it is still the case that no matrix cell $M(j_1, j_2)$ has both of its bits set.

Let us separate M into two binary matrices M' and M'', with $M'(j_1, j_2)$ being the first color bit of $M(j_1, j_2)$ and $M''(j_1, j_2)$ being the second color bit. Using these two matrices, we construct a 3-dimensional tensor T of leg n:

$$T(x, y, z) = \begin{cases} M'(x, y) & 1 \leq z \leq n/2 \\ M''(x, y) & n/2 < z \leq n \end{cases}$$

We split the forbidden leg-2 matrix M_F of Lemma 1 in a similar fashion, to obtain a forbidden leg-2 subtensor T_F:

$$\left[\begin{pmatrix} 1 & 0 \\ 1 & 1 \end{pmatrix}, \begin{pmatrix} 0 & 1 \\ 0 & 0 \end{pmatrix} \right]$$

(the two matrices are the layers for the two values in the z coordinate).

Lemma 6. *Let T' be a subtensor of T with coordinates $(j_1, j_3) \times (j_2, j_4) \times (z_1, z_2)$. $T' = T_F$ if and only if the following holds: (j_1, j_2, j_3, j_4) are vertex indices of a cycle in G_3, $z_1 \in \{1, \ldots, \frac{n}{2}\}$ and $z_2 \in \{\frac{n}{2} + 1, \ldots, n\}$.*

Proof. If $z_1, z_2 \leq n/2$ or $z_1, z_2 > \frac{n}{2}$, then T' is invariant along the z-axis and is therefore not a copy of T_F. Now suppose that $z_2 \in \{1, \ldots, \frac{n}{2}\}$ and $z_1 \in \{\frac{n}{2} + 1, \ldots, n\}$; in this case, all of (v_{j_1}, v_{j_2}), (v_{j_3}, v_{j_2}) and (v_{j_3}, v_{j_4}) are edges in the fourth edge layer of G_3 and (v_{j_4}, v_{j_1}) is an edge in the first three edge layers. We recall that G_3 is a blow-up of T_3, thus there exist vertices $v_{j'_1}, \ldots, v_{j'_4} \in T_3$ such that $(v_{j'_1}, v_{j'_2}), (v_{j'_3}, v_{j'_2}), (v_{j'_3}, v_{j'_4})$ are edges in T_3's fourth edge layer, and $(v_{j'_4}, v_{j'_1})$ an edge in its first three edge layers. Now, the edges in the fourth layer correspond to index differences $|j'_1 - j'_2|, |j'_3 - j'_2|$ and $|j'_3 - j'_4|$ of at least $9m + 1$. Thus either $j_1 < 5m$ or $j_1 > 9m + 1$. In the first case, $j_2 > 9m + 1$, $j_3 < 5m$ and $j_4 > 9m + 1$, thus $|j'_1 - j'_4| > 4m$, which makes it impossible for (j'_4, j'_1) to be an edge in the first three layers. The second case is similar. Thus whenever $z_2 \in \{1, \ldots, \frac{n}{2}\}$ and $z_1 \in \{\frac{n}{2} + 1, \ldots, n\}$, it is impossible that $T' = T_F$.

Finally, suppose $(z_1, z_2) \in \{1 \ldots \frac{n}{2}\} \times \{\frac{n}{2} + 1, \ldots, n\}$. In this case $T'(\cdot, \cdot, z_1)$ is the first color bit of a leg-2 submatrix of M, and $T'(\cdot, \cdot, z_2)$ is the second color bit thereof. If (j_1, j_2, j_3, j_4) are not vertex indices of a cycle of G_4, then at least one of the four '1' bits of T_F must be missing from T', so again $T' \neq T_F$.

For the second direction of the lemma, let (j_1, j_2, j_3, j_4) be vertex indices of a cycle in G_3, and let $(z_1, z_2) \in \{1, \ldots, \frac{n}{2}\} \times \{\frac{n}{2} + 1, \ldots, n\}$. The existence of the cycle constrains the four subtensor cells corresponding to the four edges to be 1, and the fact that no edge can exist both in the first three edge layers of G_3 and in its fourth layer constrains the other four bits to 0, so indeed $T' = T_F$.   □

5.2   Hardness of the Tensor and Proof of Theorem 2

Proposition 2. *There exists a single 3-dimensional binary tensor T_F of leg 2, such that for every n, ε there exists a tensor T which is ε-far from being free of that subtensor, and yet, the fraction of leg-2 subtensors of T which are copies of T_F is no more than $(c/\varepsilon)^{-c \cdot \log(c/\varepsilon)}$ for some global constant c.*

Proof. Let T, T_F and ε' be as in Subsection 5.1. By Lemma 6, we have that for every choice of z_1, z_2, either no choices of $(j_1, j_3) \times (j_2, j_4)$ yield a copy of T_F (for the case of $z_1 > \frac{n}{2}$ or $z_2 \leq \frac{n}{2}$), or at most a $(c'/\varepsilon')^{-c' \cdot \log(c'/\varepsilon')}$ fraction of these choices yields such a copy (due to the properties of M). Setting $c = c'/2$ we conclude that at most a $1/4 \cdot (c'/\varepsilon')^{-c' \cdot \log(c'/\varepsilon')} < (c/\varepsilon)^{-c \cdot \log(c/\varepsilon)}$ fraction of the leg-2 subtensors of T are copies of the forbidden subtensor.

As for the distance from being T_F-free, for every $z_1 \in \{1, \ldots, n/2\}$, one must modify enough cells of $T(\cdot, \cdot, z_1) = M'$ and $T(\cdot, \cdot, z_1 + \frac{n}{2}) = M''$ to affect all copies of T_F located in this pair of layers. These copies are in 1-1 correspondence with the copies of the forbidden leg-2 matrix in M, and the number of x, y coordinate pairs in which M has to be changed to remove all copies of the forbidden submatrix is at least $\varepsilon' n^2$; thus at least $2\varepsilon n^2$ changes are necessary to remove all copies of T_F in $T(\cdot, \cdot, z_1), T(\cdot, \cdot, z_1 + \frac{n}{2})$. There are $\frac{n}{2}$ disjoint pairs of such layers, so at least εn^3 changes are needed in total. T is therefore ε-far from being T_F-free. $\qquad\square$

We now state another lemma very similar to [19, Appendix D, due to Alon] and Lemma 5, again with a virtually identical proof, which will be necessary for achieving our second lower bound.

Lemma 7. *If $\mathcal{P}$ is a hereditary property of k-graphs, and for some ε there exists a tester which works independently of the size of the input graph, and makes $q(\varepsilon)$ queries, then there exists a one-sided tester whose querying is done by uniformly sampling $\mathcal{O}(q(\varepsilon))$ vertices from each one of the k parts, querying the k-subgraph they form and accepting if and only if this k-subgraph itself satisfies $\mathcal{P}$.*

Proof of Theorem 2. Consider a one-sided ε-tester of 3-graphs for being free of a some forbidden 3-subgraph which makes at most $q(\varepsilon)$ queries. If the tester works independently of the size of the input 3-graph, then by Lemma 7, there exists a one-sided tester which uniformly samples $\mathcal{O}(q)$ vertices from each one of the three parts, querying the 3-subgraph they form and accepting if and only if it is free of the forbidden 3-subgraph. If the original tester is one-sided, it is possible to show that w.l.o.g.it works independently of n.

By Proposition 2, there exists a 3-dimensional tensor T of leg n that is ε-far from being free of a leg-2 subtensor T_F, but only a $(c'/\varepsilon)^{c' \cdot \log(c'/\varepsilon)}$ fraction of its leg-2 subtensors are copies of T_F, for some global constant c'. We use T_F as the adjacency tensors for the forbidden 3-subgraph H, and T as the adjacency tensor for a 3-graph G to be tested.

The expected number of copies of H from the forbidden set in a uniformly sampled 3-subgraph of G is no more than $\mathcal{O}(q^6)\big/(c'/\varepsilon)^{c' \cdot \log(c'/\varepsilon)}$ – the expected number of copies of T_F in a uniformly sampled subtensor of T of leg $\mathcal{O}(q)$. If the original tester makes $q(\varepsilon)$ queries, the number of queries of the uniform-3-subgraph-sampling tester is at most $\mathcal{O}(q^3)$. Thus if $q(\varepsilon) < (c/\varepsilon)^{c \cdot \log(c/\varepsilon)}$, for an appropriate constant c, then the expected number of copies of H discovered is $o(1)$; the tester therefore accepts G with probability $1 - o(1)$, a contradiction. $\qquad\square$

References

1. Alon, N., Fischer, E., Newman, I.: Testing of bipartite graphs. SIAM Journal on Computing (2006) (to appear), available at the following URL, http://www.math.tau.ac.il/~nogaa/PDFS/afn2.pdf
2. Blum, M., Luby, M., Rubinfeld, R.: Self-testing/correcting with applications to numerical problems. In: STOC 1990: Proceedings of the twenty-second annual ACM symposium on Theory of computing, New York, NY, USA, pp. 73–83. ACM Press, New York (1990)
3. Rubinfeld, R., Sudan, M.: Robust characterizations of polynomials with applications to program testing. SIAM Journal on Computing 25(2), 252–271 (1996)
4. Goldreich, O., Goldwasser, S., Ron, D.: Property testing and its connection to learning and approximation. J. ACM 45(4), 653–750 (1998)

5. Ron, D.: Property testing (a tutorial). In: Rajasekaran, S., Pardalos, P.M., Reif, J.H., Rolim, J.D.P. (eds.) Handbook of Randomized Computing, Kluwer Academic Publishers, Dordrecht (2001)
6. Fischer, E.: The art of uninformed decisions: A primer to property testing. In: Paun, G., Rozenberg, G., Salomaa, A. (eds.) Current Trends in Theoretical Computer Science: The Challenge of the New Century, vol. 1, pp. 229–264. World Scientific Publishing (2004)
7. Szemerédi, E.: Regular partitions of graphs. In: Bermond, J.C., Fournier, J.C., M.L.V., Sotteau, D.(eds.) Proc. Colloque Inter. CNRS, pp. 399–401 (1978)
8. Diestel, R.: Graph Theory, 3rd edn. Springer, Heidelberg (2005)
9. Alon, N., Fischer, E., Krivelevich, M., Szegedy, M.: Efficient testing of large graphs. Combinatorica 20, 451–476 (2000)
10. Alon, N., Shapira, A.: A characterization of the (natural) graph properties testable with one-sided error. In: Proceedings of 46th Annual IEEE Symposium on Foundations of Computer Science (FOCS 2005), 23-25 October 2005, Pittsburgh, PA, USA, pp. 429–438 (2005) Will also appear in the SIAM Journal on Computing (special issue of FOCS 2005)
11. Rödl, V., Skokan, J.: Regularity lemma for k-uniform hypergraphs. Random Structures and Algorithms 25(1), 1–42 (2004)
12. Nagle, B., Rödl, V., Schacht, M.: The counting lemma for regular k-uniform hypegraphs (manuscript)
13. Gowers, W.T.: The counting lemma for regular k-uniform hypegraphs (manuscript 2005), available at the following URL, http://www.dpmms.cam.ac.uk/~wtg10/hypersimple4.pdf
14. Alon, N., Shapira, A.: Testing subgraphs in directed graphs. Journal of Computer Systems Science 69(3), 354–382 (2004)
15. Ishigami, Y.: A simple regularization of hypergraphs (2006), available at, http://www.citebase.org/abstract?id=oai:arXiv.org:math/0612838
16. Fischer, E., Newman, I.: Testing of matrix properties. In: STOC 2001: Proceedings of the 33rd annual ACM symposium on Theory of computing, New York, NY, USA, pp. 286–295. ACM Press, New York (2001)
17. Alon, N.: Testing subgraphs in large graphs. Random Structures and Algorithms 21(3-4), 359–370 (2002)
18. Alon, N., Shapira, A.: Linear equations, arithmetic progressions and hypergraph property testing. In: SODA: Proceedings of the sixteenth annual ACM-SIAM symposium on Discrete algorithms, Philadelphia, PA, USA, pp. 708–717. Society for Industrial and Applied Mathematics (2005)
19. Goldreich, O., Trevisan, L.: Three theorems regarding testing graph properties. Random Structures and Algorithms 23(1), 23–57 (2003)
20. Goldreich, O., Trevisan, L.: Errata for [19] (2005), available at the following URL, http://www.wisdom.weizmann.ac.il/~oded/PS/tt-err.ps

On Estimating Frequency Moments of Data Streams

Sumit Ganguly[1] and Graham Cormode[2]

[1] Indian Institute of Technology, Kanpur
sganguly@iitk.ac.in
[2] AT&T Labs–Research
graham@research.att.com

Abstract. Space-economical estimation of the pth frequency moments, defined as $F_p = \sum_{i=1}^{n} |f_i|^p$, for $p > 0$, are of interest in estimating all-pairs distances in a large data matrix [14], machine learning, and in data stream computation. Random sketches formed by the inner product of the frequency vector $f_1, \ldots, f_n$ with a suitably chosen random vector were pioneered by Alon, Matias and Szegedy [1], and have since played a central role in estimating F_p and for data stream computations in general. The concept of p-stable sketches formed by the inner product of the frequency vector with a random vector whose components are drawn from a p-stable distribution, was proposed by Indyk [11] for estimating F_p, for $0 < p < 2$, and has been further studied in Li [13].

In this paper, we consider the problem of estimating F_p, for $0 < p < 2$. A disadvantage of the stable sketches technique and its variants is that they require $O(\frac{1}{\epsilon^2})$ inner-products of the frequency vector with dense vectors of stable (or nearly stable [14,13]) random variables to be maintained. This means that each stream update can be quite time-consuming. We present algorithms for estimating F_p, for $0 < p < 2$, that does not require the use of stable sketches or its approximations. Our technique is elementary in nature, in that, it uses simple randomization in conjunction with well-known summary structures for data streams, such as the COUNT-MIN sketch [7] and the COUNTSKETCH structure [5]. Our algorithms require space $\tilde{O}(\frac{1}{\epsilon^{2+p}})$ [1] to estimate F_p to within $1 \pm \epsilon$ factors and requires expected time $O(\log F_1 \log \frac{1}{\delta})$ to process each update. Thus, our technique trades an $O(\frac{1}{\epsilon^p})$ factor in space for much more efficient processing of stream updates. We also present a stand-alone iterative estimator for F_1.

1 Introduction

Recently, there has been an emergence of *monitoring applications* in diverse areas including network traffic monitoring, network topology monitoring, sensor networks, financial market monitoring, and web-log monitoring. In these applications, data is generated rapidly and continuously, and must be analyzed efficiently, in real-time and in a single-pass over the data to identify large trends, anomalies, user-defined exception conditions, and so on. In many of these applications, it is often required to continuously track the "big picture", or an aggregate view of the data, as opposed to a detailed

---

[1] Following standard convention, we say that $f(n)$ is $\tilde{O}(g(n))$ if $f(n) = O\left(\left(\frac{1}{\epsilon}\right)^{O(1)} (\log m)^{O(1)} (\log n)^{O(1)} g(n) \right)$.

M. Charikar et al. (Eds.): APPROX and RANDOM 2007, LNCS 4627, pp. 479–493, 2007.
© Springer-Verlag Berlin Heidelberg 2007

view of the data. In such scenarios, efficient approximate computation is often acceptable. The data streaming model has gained popularity as a computational model for such applications—where incoming data (or updates) are processed very efficiently and in an online fashion using space, much less than that needed to store the data in its entirety.

A data stream S is viewed as a sequence of arrivals of the form (i, v), where i is the identity of an item that is a member of the domain $[n] = \{1, \ldots, n\}$ and v is the *update* to the frequency of the item. $v > 0$ indicates an insertion of multiplicity v, while $v < 0$ indicates a corresponding deletion. The frequency of an item i, denoted by f_i, is the sum of the updates to i since the inception of the stream, that is, $f_i = \sum_{(i,v) \text{ appears in } S} v$.

In this paper, we consider the problem of estimating the pth frequency moment of a data stream, defined as $F_p = \sum_{i=1}^{n} |f_i|^p$, for $0 < p < 2$. Equivalently, this can be interpreted as the pth power of the L_p norm of a vector defined by the stream. The techniques used to design algorithms and lower bounds for the frequency moment problem have been influential in the design of algorithmic and lower bound techniques for data stream computation. We briefly review the current state of the art in estimating F_p, with particular emphasis to the range $0 < p < 2$.

1.1 Review

Alon, Matias and Szegedy [1] present the seminal technique of AMS sketches for estimating F_2. An (atomic) AMS sketch is a random integer $X = \sum_{i=1}^{n} f_i \xi_i$, where $\{\xi_i\}_{i=1,2,\ldots,n}$ is a family of four-wise independent random variables assuming the values 1 or -1 with equal probability. An AMS sketch is easily maintained over a stream: for each update (i, v), the sketch X is updated as $X := X + v\xi_i$. Moreover, since the family $\{\xi_i\}$ is only 4-wise random, for each i, ξ_i can be obtained from a randomly chosen cubic polynomial h over a field F that contains the domain of items $[n]$, ($\xi_i = 1$ if the last bit of $h(i)$ is 1 and $\xi_i = -1$ otherwise). It then follows that $\mathsf{E}[X^2] = F_2$ and $\mathsf{Var}[X^2] \leq 2F_2^2$ [1]. An estimate of F_2 that is accurate to within $1 \pm \epsilon$ with confidence $\frac{7}{8}$ can therefore be obtained as the average of the squares of $O(\frac{1}{\epsilon^2})$ independent sketch values.

There has also been significant study of the case $p = 0$, also known as the distinct elements problem. Alon, Matias and Szegedy [1] gave a constant factor approximation in small space. Gibbons and Tirthapura [10] showed a $(1 \pm \epsilon)$ factor approximation space $\tilde{O}(\frac{1}{\epsilon^2})$; subsequent work has improved the (hidden) logarithmic factors [2].

p-Stable Sketches. The use of p-stable sketches was pioneered by Indyk [11] for estimating F_p, with $0 < p < 2$. A stable sketch S is defined as $Y = \sum_{i=1}^{n} f_i s_i$, where s_i is drawn at random from a p-stable distribution, denoted $S(p, 1)$ (the second parameter of $S(\cdot, \cdot)$ is the scale factor). By the defining property of p-stable distribution, Y is distributed as $S(p, (\sum_{i=1}^{n} |f_i|^p)^{1/p})$. In other words, Y is p-stable distributed, with scale factor $F_p^{1/p}$. Indyk gives a technique to estimate F_p by keeping $O(\frac{1}{\epsilon^2})$ independent p-stable sketches and returning the median of the these observations [11]. Woodruff [18] presents an $\Omega(\frac{1}{\epsilon^2})$ space lower bound for the problem of estimating F_p, for all $p \geq 0$, implying that the stable sketches technique is space optimal up to logarithmic factors.

Li [13] further analyses of stable sketches and suggests the use of the geometric mean estimator, that is,

$$\hat{F}_p = c \cdot |Y_1|^{1/k} |Y_2|^{1/k} \cdots |Y_k|^{1/k}$$

where $Y_1, Y_2, \ldots, Y_k$ are k independent p-stable sketches of the data stream. Li shows the above estimator is unbiased, that is, $\mathsf{E}[\hat{F}_p] = F_p$ and $\mathsf{Var}[\hat{F}_p] \approx \frac{\pi^2 F_p^2}{6kp^2}$. It follows (by Chebychev's inequality) that returning the geometric mean of $O(\frac{1}{\epsilon^2 p^2})$ sketches returns an estimate for F_p that is accurate to within factors of $(1 \pm \epsilon)$ with probability $\frac{7}{8}$. Li also shows that the geometric means estimator has lower variance than the median estimator proposed originally by Indyk [11].

Very Sparse sketches. The "very sparse sketch" method due to Li *et al.* aims to maintain the same space and accuracy bounds, but reduce the time cost to process each update [14,13]. Note that this technique applies only when the data satisfies some uniformity properties, whereas the preceeding techniques need no such assumptions. A very sparse (nearly) p-stable sketch is a linear combination of the form $W = \sum_{i=1}^{n} f_i w_i$, where w_i is P_p with probability $\beta/2$, $-P_p$ with probability $\beta/2$, and 0 otherwise. Here, P_p is the p-Pareto distribution with probability tail function $\Pr\{P_p > t\} = \frac{1}{t^p}$, $t \geq 1$. Pareto distributions are proposed since they are much simpler to sample from as compared to stable distributions. Further, Li shows that W is asymptotically p-stable provided $\frac{F_\infty}{F_p^{1/p}} \to 0$. Thus, very sparse sketches provide for a principled way of reducing the data stream processing time provided the data satisfies certain uniformity properties.

Drawbacks of Stable-Based Methods. A drawback of the original technique of stable sketches is that, in general, for each stream update all of the $O(\frac{1}{\epsilon^2})$ stable sketches must be updated. Each sketch update requires a pseudo-random generation of a random variable drawn from a p-stable distribution, making it time-consuming to process each stream update. The very sparse stable sketches somewhat alleviates this problem by speeding up the processing time by a factor of approximately $1/\beta$, although the data must now satisfy certain uniformity conditions. In general, it is not possible to a-priori guarantee or verify whether the data stream satisfies the desired properties. We therefore advocate that in the absence of knowledge of the data distribution, the geometric means estimator over p-stable sketches is the most reliable of the known estimators—which is quite expensive.

Contributions. In this paper, we present a technique for estimating F_p, for $0 < p < 2$. Our technique requires space $O(\frac{1}{\epsilon^{2+p}} \log^2 F_1 \log n)$ to estimate F_p to within relative error $(1 \pm \epsilon)$ with probability $7/8$. Further, it requires $O(\log n \log F_1)$ expected time (and $O(\log F_1 \log^2 n)$ worst-case time) to process each stream update. Thus, our technique trades a factor of $O(\frac{1}{\epsilon^p})$ space for improved processing time per stream update. From an algorithm design viewpoint, perhaps the most salient feature of the technique is that it does not recourse to stable distributions. Our technique is elementary in nature and uses simple randomization in conjunction with well-known summary structures for data streams, such as the COUNT-MIN sketch [7] and the COUNTSKETCH structure [5]. It is based on making some crucial and subtle modifications to the HSS technique [3].

Organization. The remainder of the paper is organized as follows. In Section 2, we review the HSS technique for estimating a class of data stream metrics. Sections 3 and 4 respectively, present a family of algorithms for estimating F_p and a recursive estimator for F_1, respectively. Finally, we conclude in Section 5.

2 Review of HSS Technique

In this section, we briefly review the HSS (for "Hierarchical Sampling from Sketches") procedure [3] for estimating F_p, $p > 2$ over data streams. Appendix A reviews the COUNTSKETCH and the COUNT-MIN algorithms for finding frequent items in a data stream and algorithms to estimate the residual first and second moments respectively of a data stream [9].

The HSS method is a general technique for estimating a class of metrics over data streams of the form:

$$\Psi(\mathcal{S}) = \sum_{i: f_i > 0} \psi(f_i). \tag{1}$$

From the input stream $\mathcal{S}$, sub-streams $\mathcal{S}_0 \ldots \mathcal{S}_L$ are created by successive sub-sampling, that is, $\mathcal{S}_0 = \mathcal{S}$ and for $1 \leq l \leq L$, $\mathcal{S}_l$ is obtained from $\mathcal{S}_{l-1}$ by sub-sampling each distinct item appearing in $\mathcal{S}_{l-1}$ independently with probability $\frac{1}{2}$ (hence $L = O(\log n)$). Let k be a space parameter. At each level l, we keep a frequent items data-structure, denoted by $\mathcal{D}_l(k)$, that takes as input the sub-stream $\mathcal{S}_l$, and returns an approximation to the $top(k)$ items of its input stream and their frequencies. $\mathcal{D}_l(k)$ is instantiated by either the COUNT-MIN or COUNTSKETCH data structures. At level l, suppose that the frequent items structure at this level has an additive error of $\Delta_l(k)$ (with high probability), that is, $|\hat{f}_i - f_i| \leq \Delta_l(k)$ with probability $1 - 2^{-t}$ where t is a parameter. Define $F_1^{res}(k, l)$ to be (the random variable denoting) the value of F_1 of the sub-stream $\mathcal{S}_l$ after removing the k largest absolute frequencies; and $F_2^{res}(k, l)$ to be the corresponding value of F_2. The (non-random value) $F_1^{res}(k, 0)$ (respectively, $F_2^{res}(k, 0)$) is written as $F_1^{res}(k)$ (resp. $F_2^{res}(k)$).

Lemma 1 (Lemma 1 from [3])

1. *For $l \geq 1$ and $k \geq 48$, $F_1^{res}(k, l) \leq \frac{F_1^{res}(k)}{2^l}$ with probability $\geq 1 - 2^{-\frac{k}{24}+1}$.*
2. *For $l \geq 1$ and $k \geq 48$, $F_2^{res}(k, l) \leq \frac{F_2^{res}(k)}{2^l}$ with probability $\geq 1 - 2^{-\frac{k}{24}+1}$.*

By the above lemma, let $\Delta_0 = \frac{F_1^{res}(k)}{k}$ or $\Delta_0 = \left(\frac{F_2^{res}(k)}{k}\right)^{1/2}$, depending on whether the COUNT-MIN or the COUNTSKETCH structure is being used as the frequent items structure at each level. Let $\bar{\epsilon} = \frac{\epsilon}{16}$, $T_0 = \frac{2\Delta_0}{\bar{\epsilon}}$ and $T_l = \frac{T_0}{2^l}$, $l = 0, 1, 2\ldots, \log T_0$. The items are grouped into groups $G_0, G_1, \ldots, G_L$ as follows: $G_0 = \{i \in \mathcal{S} : f_i \geq T_0\}$ and $G_l = \{i \in \mathcal{S} : T_l \leq f_i < T_{l-1}\}$, $1 \leq l \leq L$. It follows that, with high probability, for all items of G_l that are present in the random sub-stream $\mathcal{S}_l$, $\hat{f}_i \geq \frac{\Delta_l}{\bar{\epsilon}}$ and $|\hat{f}_i - f_i| \leq \epsilon f_i$.

Corresponding to every stream update (i, v), we use a hash-function $h : [n] \to [n]$ to map the item i to a level $u = lsb(h(i))$, where, $lsb(x)$ is the position of the least significant "1" in binary representation of x. The stream update (i, v) is then propagated to the frequent items data structures $\mathcal{D}_l$ for $0 \leq l \leq u$, so in effect, i is included in the sub-streams from level 0 to level u. The hash function is assumed to be chosen randomly from a fully independent family; later we reduce the number of random bits required by a standard data streaming argument.

At the time of inference, the algorithm collects samples as follows. From each level l, the set of items whose estimated frequency crosses the threshold $\frac{\Delta_0}{\epsilon 2^l}$ are identified, using the frequent items structure $\mathcal{D}_l$. It is possible for the estimate $\hat{f}_{i,l}$ of an item i obtained from the sub-stream $\mathcal{S}_l$ to exceed this threshold for multiple levels l. We therefore apply the "disambiguation-rule" of using the estimate obtained from the *lowest level* at which it crosses the threshold for that level. The estimated frequency after disambiguation is denoted as $\hat{f}_i$. Based on their disambiguated frequencies, the sampled items are sorted into their respective groups, $\bar{G}_0, \bar{G}_1, \ldots, \bar{G}_L$, as follows:

$$\bar{G}_0 = \{i | \hat{f}_i \geq T_0\} \text{ and } \bar{G}_l = \{i | T_{l-1} < \hat{f}_i \leq T_l \text{ and } i \in \mathcal{S}_l\}, 1 \leq l \leq L.$$

We define the estimator $\hat{\Psi}$ and a second idealized estimator $\bar{\Psi}$ which is used for analysis only.

$$\hat{\Psi} = \sum_{l=0}^{L} \sum_{i \in \bar{G}_l} \psi(\hat{f}_i) \cdot 2^l \qquad \bar{\Psi} = \sum_{l=0}^{L} \sum_{i \in \bar{G}_l} \psi(f_i) \cdot 2^l \qquad (2)$$

We now briefly review the salient points in the error analysis. Lemma 2 shows that the expected value of $\bar{\Psi}$ is close to Ψ.

Lemma 2 (Lemma 2 from [3]). *Suppose that for $0 \leq i \leq N - 1$ and $0 \leq l \leq L$, $|\hat{f}_{i,l} - f_i| \leq \epsilon f_i$ with probability $\geq 1 - 2^{-t}$. Then $|\mathsf{E}[\bar{\Psi}] - \Psi| \leq \Psi \cdot 2^{-t+\log L}$.*

We now present a bound on the variance of the idealized estimator. The frequency group G_l is partitioned into three sub-groups, namely, $\mathrm{lmargin}(G_l) = [T_l, T_l(1 + \bar{\epsilon}/2)]$, $\mathrm{rmargin}(G_l) = [T_{l-1}(1 - \bar{\epsilon}), T_{l-1}]$ and $\mathrm{midregion}(G_l) = [T_l(1 + \bar{\epsilon}/2), T_{l-1}(1 - \bar{\epsilon})]$, that respectively denote the lmargin (left-margin), rmargin (right-margin) and midregion of the group G_l. An item i is said to belong to one of these regions if its true frequency lies in that region. For any item i with non-zero frequency, we denote by $l(i)$ the group index l such that $i \in G_l$. For any subset $T \subset [n]$, denote by $\psi(T)$ the expression $\sum_{i \in T} \psi(f_i)$. Let $\Psi^2 = \Psi^2(\mathcal{S})$ denote $\sum_{i=1}^{n} \psi^2(f_i)$.

Lemma 3 (Lemma 3 from [3]). *Suppose that for all $0 \leq i \leq N - 1$ and $0 \leq l \leq L$, $|\hat{f}_{i,l} - f_i| \leq \epsilon f_i$ with probability $\geq 1 - 2^{-t}$. Then,*

$$\mathrm{Var}[\bar{\Psi}] \leq 2^{-t+L+2} \cdot \Psi^2 + \sum_{i \notin (G_0 - \mathrm{lmargin}(G_0))} \psi^2(f_i) \cdot 2^{l(i)+1}.$$

Corollary 4. *If the function $\psi(\cdot)$ is non-decreasing in the interval $[0 \ldots T_0 + \Delta_0]$, then, choosing $t = L + \log \frac{1}{\epsilon^2} + 2$, we get*

$$\mathrm{Var}[\bar{\Psi}] \leq \epsilon^2 \Psi^2 + \sum_{l=1}^{L} \psi(T_{l-1}) \psi(G_l) 2^{l+1} + 2\psi(T_0 + \Delta_0)\psi(\mathrm{lmargin}(G_0)) \qquad (3)$$

The error incurred by the estimate $\hat{\Psi}$ is $|\hat{\Psi} - \Psi|$, and can be written as the sum of two error components using the triangle inequality.

$$|\hat{\Psi} - \Psi| \le |\bar{\Psi} - \Psi| + |\hat{\Psi} - \bar{\Psi}| = \mathcal{E}_1 + \mathcal{E}_2$$

Here, $\mathcal{E}_1 = |\Psi - \bar{\Psi}|$ is the error due to sampling and $\mathcal{E}_2 = |\hat{\Psi} - \bar{\Psi}|$ is the error due to the approximate estimation of the frequencies. By Chebychev's inequality, $\mathcal{E}_1 = |\Psi - \bar{\Psi}| \le |E[\bar{\Psi}] - \Psi| + 3\sqrt{\text{Var}[\bar{\Psi}]}$ with probability $\frac{8}{9}$. Using Lemma 2 and Corollary 4, and choosing $t = L + \log \frac{1}{\epsilon^2} + 2$, the expression for $\mathcal{E}_1$ can be simplified as follows:

$$\mathcal{E}_1 \le \frac{\epsilon^2 L \Psi}{2^L} + 3\left(\epsilon^2 \Psi^2 + \sum_{l=1}^{L} \psi(T_{l-1})\psi(G_l)2^{l+1} + 2\psi(T_0 + \Delta_0)\psi(\text{lmargin}(G_0))\right)^{1/2}$$

(4)

with probability $\frac{8}{9}$. We now present an upper bound on $\mathcal{E}_2$.

Lemma 5. *Suppose that for $1 \le i \le n$ and $0 \le l \le L$, $|\hat{f}_{i,l} - f_i| \le \epsilon f_i$ with probability $\ge 1 - 2^{-t}$. Then $\mathcal{E}_2 \le \Delta_0 \sum_{l=0}^{L} \sum_{i \in G_l} \frac{|\psi'(\xi_i)|}{2^l}$ with probability $\ge \frac{9}{10} - 2^{-t}$, where for $i \in G_l$, ξ_i lies between f_i and $\hat{f}_i$, and maximizes $\psi'()$.*

The analysis assumes that the hash function mapping items to levels is completely independent. We adopt a standard technique of reducing the required randomness by using a pseudo-random generator (*PRG*) of Nisan [15] along the lines of Indyk in [11] and Indyk and Woodruff in [12]. More details are provided in Appendix B.

3 Estimating F_p

In this section, we use the Hss technique with some subtle but vital modifications to estimate F_p for $0 < p < 2$. We use the COUNTSKETCH structure as the frequent items structure at each level l.

We observe that a direct application of the Hss technique does not present an $\tilde{O}(1)$ space procedure, and so we need some novel analysis. To see this, suppose that k is the space parameter of the COUNTSKETCH procedure at each level. Firstly, observe that

$$\mathcal{E}_2 \le \Delta_0 \sum_{l=0}^{L} \sum_{i \in G_l} \frac{|\psi'(\xi_i)|}{2^l} = \sum_{l=0}^{L} \sum_{i \in G_l} \frac{\Delta_0}{2^l} \cdot p \cdot |f_i|^{p-1} \le 2\epsilon \sum_{l=0}^{L} \sum_{i \in G_l} |f_i|^p \le 2\epsilon^{1+p/2} F_p \le 2\epsilon F_p$$

as required, since $p < 2$ and $\Delta_0 2^{-l} \le \epsilon |f_i|$ for $i \in G_l$. Now, by equation (4) and using $t = L + 2\log \frac{1}{\epsilon} + 2$, we have

$$\mathcal{E}_1 \le \epsilon^2 F_{2p} + 3\left(\epsilon^2 F_{2p} + \sum_{l=1}^{L} \left(\frac{F_2^{res}(k)}{2^l k}\right)^{\frac{p}{2}} F_p(G_l)2^{l+1} + 2(1+\epsilon)\left(\frac{F_2^{res}(k)}{k}\right)^{\frac{p}{2}} F_p(\text{lmargin}(G_0))\right)^{\frac{1}{2}}.$$

Further, if we write $f_{rank(r)}$ to denote the rth largest frequency (in absolute value)

$$F_2^{res}(k) = \sum_{r>k} f_{rank(r)}^2 \le \sum_{r>k} f_{rank(k+1)}^{2-p} f_{rank(r)}^p \le \left(\frac{F_p}{k}\right)^{2/p-1} \cdot F_p \le k \cdot \left(\frac{F_p}{k}\right)^{2/p}, \text{ for } p > 0$$

and hence the expression for $\mathcal{E}_1$ simplifies to

$$\mathcal{E}_1 \leq \epsilon^2 F_{2p} + 3 \left(\epsilon^2 F_{2p} + F_p \sum_{l=1}^{L} 2^{l+1-lp/2} F_p(G_l) + \frac{2(1+\epsilon)F_p}{k} \cdot F_p(\mathrm{lmargin}(G_0)) \right)^{1/2}$$

The main problem arises with the the middle term in the above expression, namely, $F_p \sum_{l=1}^{L} 2^{l+1-lp/2} F_p(G_l)$, which can be quite large. Our approach relies on altering the group definitions to make them depend on the (randomized) quantities $F_2^{res}(k, l)$ so that the resulting expression for $\mathcal{E}_1$ becomes bounded by $O(\epsilon F_p)$. We also note that the expression for $\mathcal{E}_2$ derived above remains bounded by ϵF_p.

Altered Group Definitions and Estimator. We use the COUNTSKETCH data structure as the frequent items structure at each level $l = 0, 1, \ldots, L$, with $k = O(\frac{1}{\epsilon^{2+p}})$ buckets per hash function and $s = O(\log F_1)$ hash functions per sketch. We first observe that Lemma 1 can be slightly strengthened as follows:

Lemma 6 (A slightly stronger version of Lemma 1). *For $l \geq 1$ and $k \geq 48$*

1. $\qquad F_1^{res}(k, l) \leq \dfrac{F_1^{res}(2^{l-1}k)}{2^l}$ *with probability* $\geq 1 - 2^{-\frac{k}{24}+1}$.

2. $\qquad F_2^{res}(k, l) \leq \dfrac{F_2^{res}(2^{l-1}k)}{2^l}$ *with probability* $\geq 1 - 2^{-\frac{k}{24}+1}$

Proof. The result follows from the proof given for Lemma 1 [3].   □

At each level l, we use the COUNTSKETCH structure to estimate $F_2^{res}(k, l)$ to within an accuracy of $(1 \pm \frac{1}{4})$ with probability $1 - \frac{1}{16L}$, where, $\bar{\epsilon} = \frac{\epsilon}{32}$. Let $\bar{F}_2^{res}(k, l)$ denote $\frac{5}{4} \cdot F_2^{res}(k, l)$. We redefine the thresholds $T_0, T_1, \ldots$, as follows:

$$\Delta_l = \left(\frac{\bar{F}_2^{res}(k, l)}{k \cdot 2^l} \right)^{1/2}, \quad T_l = \frac{\Delta_l}{\epsilon}, \quad l = 0, 1, 2, \ldots, L.$$

The groups $G_0, G_1, G_2, \ldots$, are set in the usual way, using the new thresholds:

$$G_0 = \{i \mid f_i \geq T_0\} \text{ and } G_l = \{i \mid T_l \leq f_i \leq T_{l-1}\}$$

The estimator is defined by (1) as before.

Lemma 7. *Suppose $k \geq \frac{16}{\epsilon^{2+p}}$. Then, $\mathcal{E} \leq 8\epsilon F_p$ with probability at least $\frac{3}{4}$.*

Proof. We use the property of $F_2^{res}(t)$ derived above, that for any $1 \leq t \leq F_0$.

$$F_2^{res}(t) \leq t \left(\frac{F_p}{t} \right)^{2/p} = \frac{F_p^{2/p}}{t^{2/p-1}}, \text{ for } p > 0. \tag{5}$$

We therefore have,

$$T_l = 2\left(\frac{\bar{F}_2^{res}(k,l)}{\epsilon^2 k \cdot 2^l}\right)^{1/2} \leq 2\left(\frac{5F_2^{res}(k \cdot 2^{l-1})}{4\epsilon^2 k \cdot 2^l}\right)^{1/2}, \qquad \text{by Lemma 6}$$

$$\leq 2\left(\frac{5F_p^{2/p}}{4(k \cdot 2^{l-1})^{2/p-1}\epsilon^2 k \cdot 2^l}\right)^{1/2}, \qquad \text{by (5)}$$

$$= \frac{1}{\epsilon}\sqrt{\frac{5}{2}}\left(\frac{2F_p}{k2^l}\right)^{1/p} \tag{6}$$

By equation (4), one component of $\mathcal{E}_1$ can be simplified as follows:

$$\mathcal{E}_{1,1} \leq \sum_{l=1}^{L} T_l^p F_p(G_l) \cdot 2^{l+1}$$

$$\leq \epsilon^{-p}\left(\frac{5}{2}\right)^{p/2} \sum_{l=1}^{L} \frac{2F_p}{k \cdot 2^l} F_p(G_l) 2^{l+1} \qquad \text{substituting (6)}$$

$$\leq \frac{10}{k\epsilon^p} \cdot F_p \sum_{l=1}^{L} F_p(G_l) \qquad \text{since } p < 2$$

$$= \frac{10}{k\epsilon^p} \cdot F_p(F_p - F_p(G_0))$$

$$\leq \frac{5}{8}\epsilon^2 F_p^2 \qquad \text{since } k \geq \frac{16}{\epsilon^{2+p}}$$

The other component of $\mathcal{E}_1$ is

$$\mathcal{E}_{1,2} = 2T_0^p(1+\epsilon)F_p(\text{lmargin}(G_0)) \leq 2\epsilon^{-p}(5/2)^{p/2}2\frac{F_p}{k}(1+\epsilon)F_p \leq \frac{10}{k\epsilon^p}(1+\epsilon)F_p^2 \leq \epsilon^2 F_p^2,$$

also using $k \geq \frac{16}{\epsilon^{2+p}}$ and $\epsilon < \frac{1}{2}$. Substituting in (4), we have,

$$\mathcal{E}_1 \leq \epsilon F_p + 3(\epsilon^2 F_p^2 + \mathcal{E}_{1,1} + \mathcal{E}_{1,2})^{1/2} < 6\epsilon F_p. \tag{7}$$

Adding, the total error is bounded by

$$\mathcal{E} \leq \mathcal{E}_1 + \mathcal{E}_2 \leq 8\epsilon F_p \qquad \qquad \Box$$

We summarize this section in the following theorem.

Theorem 8. *There exists an algorithm that returns $\hat{F}_p$ satisfying $|\hat{F}_p - F_p| \leq \epsilon F_p$ with probability $\frac{3}{4}$ using space $O(\frac{1}{\epsilon^{2+p}}(\log^2 n)(\log F_1))$ and processes each stream update in expected time $O(\log n \log F_1)$ and worst case time $O(\log^2 n \log F_1)$ standard arithmetic operations on words of size $\log F_1$ bits.* $\qquad \Box$

Remarks

1. We note that for $0 < p < 1$, an estimator for F_p with similar properties may be designed in an exactly analogous fashion by using COUNT-MIN instead of COUNTSKETCH as the frequent items structure at each level. Such an estimator would require an ϵ-accurate estimation of F_1 (which would imply estimation of F_1^{res} using standard techniques), which could either be done using Cauchy-sketches [11,13] or using the stand-alone technique presented in Section 4. However, using Cauchy-sketches means that, in general, $O(\frac{1}{\epsilon^2})$ time is required to process each stream update. In order to maintain poly-logarithmic processing time per stream update, the technique of Section 4 may be used.

2. The space requirement of the stable sketches estimator grows as $\tilde{O}(\frac{1}{\epsilon^2 p^2})$ as a function of p [13], whereas, the HSS-based technique requires space $\tilde{O}(\frac{1}{\epsilon^{2+p}})$. For small values of p, i.e. $p = O\left(\frac{1}{\log \epsilon^{-1}(1 + \log\log \epsilon^{-1})}\right)$, the HSS technique can be asymptotically more space-efficient.

4 An Iterative Estimator for F_1

In this section, we use the HSS technique to present a stand-alone, iterative estimator for $F_1 = \sum_{i=1}^{n} |f_i|$. The previous section presents an estimator for F_p that uses, as a sub-routine, an estimator for $F_2^{res}(k)$ at each level of the structure. In this section, we present a stand-alone estimator that uses only COUNT-MIN sketch to estimate F_1. The technique may be of independent interest.

The estimator uses two separate instantiations of the HSS structure. The first instantiation uses COUNT-MIN sketch structure with $k = \frac{8}{\bar{\epsilon}^3}$ buckets per hash function, and $s = O(\log G)$ hash functions, where, $G = O(F_2)$ and $\bar{\epsilon} = \frac{\epsilon}{8}$. A collection of $s_2 = O(\log \frac{1}{\delta})$ independent copies of the structure are kept for the purpose of boosting the confidence of an estimate by taking the median. The second instantiation of the HSS structure uses $k' = \frac{128}{\bar{\epsilon}^3}$ buckets per hash function (so $k' = 16k$) and $s = O(\log G)$ hash functions. For estimating F_1, we use a two-step procedure, namely, (1) first, we obtain an estimate of F_1 that is correct to within a factor of 16 using the first HSS instantiation and (2) then, we use the second instantiation to obtain an ϵ-accurate estimation of F_1.

The first step of the estimation is done using the first instantiation of the HSS structure as follows. We set the threshold T_0 to a parameter t, $T_l = \frac{T_0}{2^l}$ and the threshold frequency for finding in group l to be $\frac{T_l}{2}$. The group definitions are as defined earlier: $G_0 = [t, F_1]$, $G_l = [\frac{t}{2^l}, \frac{t}{2^{l-1}})$, $1 \le l \le L$. The disambiguation rule for the estimated frequency is as follows: if $\hat{f}_{i,l} > T_l$, then, $\hat{f}_i$ is set to the estimate obtained from the lowest of these levels. The sampled groups $\bar{G}_l$ are defined as follows.

$$\bar{G}_0 = \{i \mid \hat{f}_i \ge T_0\}, \qquad \bar{G}_l = \left\{i \mid \frac{T_0}{2^l} \le \hat{f}_i < \frac{T_0}{2^{l-1}} \text{ and } i \in S_l\right\}, 1 \le i \le L.$$

The estimators $\hat{F}_1$ and $\bar{F}_1$ are defined as before—these are now functions of t.

$$\hat{F}_1(t) = \sum_{l=0}^{L} \sum_{i \in \bar{G}_l} |\hat{f}_i| 2^l \qquad\qquad \bar{F}_1(t) = \sum_{l=0}^{L} \sum_{i \in \bar{G}_l} |f_i| 2^l.$$

Estimator. Let t iterate over values $1, 2, 2^2, \ldots, G$ and for each value of t let $\hat{F}_1^{\mathrm{med}}(t)$ denote the median of the estimates $\hat{F}_1$ returned from the $s_1 = O(\log \frac{1}{\delta})$ copies of the HSS structure, each using independent random bits and the same value of t. Let $t_{\max}$ denote the largest value of t satisfying

$$\hat{F}_1^{\mathrm{med}}(t) \geq \frac{16t}{1.01\epsilon^2}.$$

The final estimate returned is $\hat{F}_1^{\mathrm{med}}(t_{\max})$ using the second HSS instantiation.

Analysis. We first note that Lemmas 2 and 3 hold for all choices of t. Lemma 5 gets modified as follows.

Lemma 9. *Suppose that for $1 \leq i \leq n$ and $0 \leq l \leq L$, $|\hat{f}_{i,l} - f_i| \leq \frac{\Delta_0}{2^l}$ with probability $\geq 1 - 2^{-t}$, where, $\Delta_0 = \frac{F_1^{res}(k)}{k}$. Then, $\mathcal{E}_2 \leq 16 \cdot \Delta_0 \sum_{l=0}^L \sum_{i \in G_l} \frac{|\psi'(\xi_i)|}{2^l}$ with probability $\geq \frac{9}{10} - 2^{-t}$, where for an $i \in G_l$, ξ_i lies between f_i and $\hat{f}_i$, and maximizes $\psi'()$.* □

Lemma 10. *Let $\bar{\epsilon} = \frac{\epsilon}{8}$, $k = \frac{8}{\bar{\epsilon}^3}$ and $\epsilon \leq \frac{1}{8}$. Then, with probability $1 - \delta$ each,*

1. *For $\frac{4F_1}{\bar{\epsilon}k} \leq t \leq \frac{8F_1}{\bar{\epsilon}k}$*

$$|\hat{F}_1^{med}(t) - F_1| \leq \frac{1.01\epsilon F_1}{2} \quad \text{and} \quad \hat{F}_1^{med}(t) \geq \frac{16t}{1.02\epsilon^2}$$

2. *For any $t \geq \frac{64F_1}{\bar{\epsilon}k}$, $\hat{F}_1^{med}(t) < \frac{16t}{1.02\epsilon^2}$ with probability $1 - \delta$.*

Proof. We consider the two summation terms in error term $\mathcal{E}_1$ given by equation (4) separately.

$$\mathcal{E}_{1,1} = \sum_{l=1}^L \frac{t}{2^l} F_1(G_l) 2^{l+1} \leq 2t(F_1 - F_0), \quad \text{and} \quad \mathcal{E}_{1,2} = 2tF_1(\mathrm{lmargin}(G_0)).$$

Adding, $\mathcal{E}_1 \leq (2(t + \Delta_0)F_1)^{1/2}$.

We ignore the term $2^{-t+L+2}F_2$ in $\mathcal{E}_1$, since, by choosing $t = O(L)$, this term can be made arbitrarily small in comparison with the other two terms. Since, $|G_l| \leq \frac{F_1 \cdot 2^l}{t}$, our bound on $\mathcal{E}_2$ becomes

$$\mathcal{E}_2 \leq 16\Delta_0 \sum_{l=0}^L \frac{|G_l|}{2^l} \leq \frac{16F_1^2}{kt}.$$

Therefore, the expression for total error is $\mathcal{E}(t) = \mathcal{E}_1 + \mathcal{E}_2$

$$\mathcal{E}(t) \leq 2F_1 \left(\frac{t}{F_1} + \frac{1}{k} \right)^{1/2} + \frac{16F_1^2}{kt}. \tag{8}$$

Suppose $k = \frac{128}{\bar{\epsilon}^3}$ and $\frac{4F_1}{\bar{\epsilon}k} \leq t \leq \frac{8F_1}{\bar{\epsilon}k}$. Using $\epsilon \leq \frac{1}{8}$ and $\bar{\epsilon} = \frac{\epsilon}{8}$, we have

$$\mathcal{E}(t) \leq 2F_1 \left(\frac{t}{F_1} + \frac{1}{k} \right)^{1/2} + \frac{16F_1^2}{kt} \leq \frac{1.01\epsilon F_1}{2}. \tag{9}$$

We therefore have,

$$|\hat{F}_1^{\text{med}}(t) - F_1| \leq \mathcal{E}(t) \leq \frac{1.01\epsilon F_1}{2}, \quad \text{for } \frac{4F_1}{\bar{\epsilon}k} \leq t \leq \frac{8F_1}{\bar{\epsilon}k} \text{ with probability } 1 - \delta.$$

Therefore, for $\frac{4F_1}{\bar{\epsilon}k} \leq t \leq \frac{8F_1}{\bar{\epsilon}k}$, with probability $1 - \delta$, we have from (9) that

$$\frac{1.01\epsilon F_1}{2} \geq 2\sqrt{tF_1} \text{ and so } t \leq \frac{(1.01)^2\epsilon^2 F_1}{16} \leq \frac{1.01\epsilon^2 \hat{F}_1^{\text{med}}}{16}(1 + 0.505\epsilon) \text{ so } \hat{F}_1^{\text{med}}(t) \geq \frac{16t}{1.02\epsilon^2}.$$
$$(10)$$

Let $t = \frac{2^{j+2}F_1}{\bar{\epsilon}k}$ for some $j \geq 0$, and suppose that (10) is satisfied. Then by (8)

$$\mathcal{E}(t) \leq 2^{(j-1)/2-2}\epsilon(1.01)F_1 + 2^{-j-2}\epsilon F_1 \leq 2^{j/2}\epsilon F_1 \quad \text{with probability } 1 - \delta.$$

With probability $1 - \delta$, $|\hat{F}_1^{\text{med}}(t) - F_1| \leq \mathcal{E}$ and so, using $\epsilon \geq \frac{1}{8}$,

$$2^{j/2-3}F_1 \geq 2^{j/2}\epsilon F_1 \geq \mathcal{E}(t) \geq |\hat{F}_1^{\text{med}}(t) - F_1| \geq \frac{16t}{1.02\epsilon^2} - F_1 \quad \text{by (10)}$$

$$\geq \frac{2^{j-3}F_1}{1.02} - F_1 \quad \text{using } \bar{\epsilon} - \frac{\epsilon}{8} \text{ and } k = \frac{8}{\bar{\epsilon}^3}$$

which is a contradiction for $j \geq 4$, proving claim 2. $\qquad\square$

The correctness of the algorithm follows from the above Lemma. The space requirement of the algorithm is $O(\frac{1}{\bar{\epsilon}^3}(\log^3 n)(\log^2 F_1))$ bits and the expected time taken to process each stream update is $O(\log F_1 \log \frac{1}{\delta})$ standard arithmetic operations on words of size $O(\log n)$.

5   Conclusion

We present a family of algorithms for the randomized estimation of F_p for $0 < p < 2$ and another family of algorithms for estimating F_p for $0 < p < 1$. The first algorithm family estimates F_p by using the COUNTSKETCH structure and F_2 estimation as sub-routines. The second algorithm family estimates F_p by using the COUNT-MIN sketch structure and F_1 estimation as a sub-routines. The space required by these algorithms are $O(\frac{1}{\epsilon^{2+p}}(\log^2 n)(\log^2 F_1)(\log \frac{1}{\delta})$ and the expected time required to process each stream update is $O(\log n \log F_1 \log \frac{1}{\delta})$. Finally, we also present a stand-alone iterative estimator for F_1 that only uses the COUNT-MIN sketch structure as a sub-routine.

Prior approaches to the problem of estimating F_p [11,13] used sketches of the frequency vector with random variables drawn from a symmetric p-stable distribution. An interesting feature of the above algorithms is that they do not require the use of stable distributions. The proposed algorithms trade an extra factor of $O(\epsilon^{-p})$ factor of space for improved procesing time (with no polynomial dependency on ϵ) per stream update.

References

1. Alon, N., Matias, Y., Szegedy, M.: The space complexity of approximating frequency moments. Journal of Computer and System Sciences 58(1), 137–147 (1998)
2. Bar-Yossef, Z., Jayram, T.S., Kumar, R., Sivakumar, D., Trevisan, L.: Counting distinct elements in a data stream. In: Rolim, J.D.P., Vadhan, S.P. (eds.) RANDOM 2002. LNCS, vol. 2483, Springer, Heidelberg (2002)
3. Bhuvanagiri, L., Ganguly, S.: Estimating Entropy over Data Streams. In: Azar, Y., Erlebach, T. (eds.) ESA 2006. LNCS, vol. 4168, Springer, Heidelberg (2006)
4. Carney, D., Cetintemel, U., Cherniack, M., Convey, C., Lee, S., Seidman, G., Stonebraker, M., Tatbul, N., Zdonik, S.: Monitoring streams – a new class of data management applications. In: Bressan, S., Chaudhri, A.B., Lee, M.L., Yu, J.X., Lacroix, Z. (eds.) CAiSE 2002 and VLDB 2002. LNCS, vol. 2590, Springer, Heidelberg (2003)
5. Charikar, M., Chen, K., Farach-Colton, M.: Finding frequent items in data streams. In: Widmayer, P., Triguero, F., Morales, R., Hennessy, M., Eidenbenz, S., Conejo, R. (eds.) ICALP 2002. LNCS, vol. 2380, pp. 693–703. Springer, Heidelberg (2002)
6. Cormode, G., Muthukrishnan, S.: What's New: Finding Significant Differences in Network Data Streams. In: IEEE INFOCOM, IEEE Computer Society Press, Los Alamitos (2004)
7. Cormode, G., Muthukrishnan, S.: An Improved Data Stream Summary: The Count-Min Sketch and its Applications. J. Algorithms 55(1), 58–75 (2005)
8. Flajolet, P., Martin, G.N.: Probabilistic Counting Algorithms for Database Applications. Journal of Computer and System Sciences 31(2), 182–209 (1985)
9. Ganguly, S., Kesh, D., Saha, C.: Practical Algorithms for Tracking Database Join Sizes. In: Ramanujam, R., Sen, S. (eds.) FSTTCS 2005. LNCS, vol. 3821, Springer, Heidelberg (2005)
10. Gibbons, P., Tirthapura, S.: Estimating simple functions on the union of data streams. In: Proc. SPAA (2001)
11. Indyk, P.: Stable Distributions, Pseudo Random Generators, Embeddings and Data Stream Computation. In: Proc. IEEE FOCS, IEEE Computer Society Press, Los Alamitos (2000)
12. Indyk, P., Woodruff, D.: Optimal Approximations of the Frequency Moments. In: Proc. ACM STOC, ACM Press, New York (2005)
13. Li, P.: Very Sparse Stable Random Projections, Estimators and Tail Bounds for Stable Random Projections. (Manuscript, 2006)
14. Li, P., Hastie, T.J., Church, K.W.: Very Sparse Random Projections. In: Proc. ACM SIGKDD, ACM Press, New York (2006)
15. Nisan, N.: Pseudo-Random Generators for Space Bounded Computation. In: Proc. ACM STOC, ACM Press, New York (1990)
16. Thorup, M., Zhang, Y.: Tabulation based 4-universal hashing with applications to second moment estimation. In: Proc. ACM SODA, January 2004, pp. 615–624. ACM Press, New York (2004)
17. Wegman, M.N., Carter, J.L.: New Hash Functions and their Use in Authentication and Set Equality. Journal of Computer and System Sciences 22, 265–279 (1981)
18. Woodruff, D.P.: Optimal approximations of all frequency moments. In: Proc. ACM SODA, January 2004, pp. 167–175. ACM Press, New York (2004)

A   COUNT-MIN and COUNTSKETCH Summaries

Given a data stream defining a set of item frequencies, $rank(r)$ returns an item with the rth largest absolute value of the frequency (ties are broken arbitrarily). We say that an item i has rank r if $rank(r) = i$. For a given value of k, $1 \leq k \leq n$, the set $top(k)$ is the set of items with rank $\leq k$. The residual second moment [5] of a data stream, denoted by $F_2^{res}(k)$, is defined as the second moment of the stream after the top-k frequencies have been removed. Then, $F_2^{res}(k) = \sum_{r>k} f_{rank(r)}^2$. The residual first moment [7] of a data stream, denoted by F_1^{res}, is analogously defined as the F_1 norm of the data stream after the top-k frequencies have been removed, that is, $F_1^{res} = \sum_{r>k} |f_{rank(r)}|$.

A *sketch* [1] is a random integer $X = \sum_i f_i \cdot x_i$, where, $x_i \in \{-1, +1\}$, for $i \in \mathcal{D}$ and the family of variables $\{x_i\}_{i \in \mathcal{D}}$ with certain independence properties. The family of random variables $\{x_i\}_{i \in \mathcal{D}}$ is referred to as the *sketch basis*. For any $d \geq 2$, a d-wise independent sketch basis can be constructed in a pseudo-random manner from a truly random seed of size $O(d \log n)$ bits as follows. Let F be field of characteristic 2 and of size at least $n + 1$. Choose a degree $d - 1$ polynomial $g : F \to F$ with coefficients that are randomly chosen from F [17]. Define x_i to be 1 if the first bit (i.e., the least significant position) of $g(i)$ is 1, and define x_i to be -1 otherwise. The d-wise independence of the x_i's follows from an application of Wegman and Carter's universal hash functions [17].

Pair-wise independent sketches are used in [5] to design the COUNTSKETCH algorithm for finding the top-k frequent items in an insert-only stream. The data structure consists of a collection of $s = O(\log \frac{1}{\delta})$ independent hash tables $U_1, U_2, \ldots, U_s$ each consisting of $8k$ buckets. A pair-wise independent hash function $h_j : [n] \to \{1, 2, \ldots, 8k\}$ is associated with each hash table that maps items randomly to one of the $8k$ buckets, where, k is a space parameter. Additionally, for each table index $j = 1, 2, \ldots, s$, we keep a pair-wise independent family of random variables $\{x_{ij}\}_{i \in [n]}$, where, each $x_{ij} \in \{-1, +1\}$ with equal probability. Each bucket keeps a sketch of the sub-stream that maps to it, that is, $U_j[r] = \sum_{i:h_j(i)=r} f_i x_{ij}$, $1 \leq j \leq s, 1 \leq r \leq 8k$. An estimate $\hat{f}_i$ is returned as follows: $\hat{f}_i = \text{median}_{j=1}^s U_j[h_j(i)] x_{ij}$. The accuracy of estimation is stated as a function Δ of the residual second moment given parameters k and b is defined as [5]

$$\Delta(b, k) = 8 \left(\frac{F_2^{res}(k)}{b} \right)^{1/2}.$$

The space versus accuracy guarantees of the COUNTSKETCH algorithm is presented in Theorem 11.

Theorem 11 ([5]). *Let $\Delta = \Delta(k, 8k)$. Then, for any given $i \in [n]$, $\Pr\{|\hat{f}_i - f_i| \leq \Delta\} \geq 1 - \delta$. The space used is $O(k \cdot \log \frac{1}{\delta} \cdot (\log F_1))$ bits, and the time taken to process a stream update is $O(\log \frac{1}{\delta})$.*

The COUNTSKETCH algorithm can be adapted to return approximate frequent items and their frequencies. The original algorithm [5] uses a heap for maintaining the current top-k items in terms of their estimated frequencies. After processing each arriving stream record of the form (i, v), where, v is assumed to be non-negative, an estimate for $\hat{f}_i$ is calculated using the scheme outline above. If i is already in the current estimated top-k heap then its frequency is correspondingly increased. If i is not in the heap but $\hat{f}_i$ is larger than the current smallest frequency in the heap, then it replaces that element in the heap. This scheme is applicable to insert-only streams. A generalization of this method for strict update streams is presented in [6] and returns, with probability $1 - \delta$, (a) all items with frequency at least $(\frac{F_2^{res}(k)}{k})^{1/2}$ and,

(b) does not return any item with frequency less than $(1 - \epsilon)(\frac{F_2^{res}(k)}{k})^{1/2}$ using space $O\left(k\epsilon^{-2}\log n \log(k\epsilon^{-1}\log(k\epsilon^{-1}))\log F_1\right)$ bits. For general update streams, a variation of this technique can be used for retrieving items satisfying properties (a) and (b) above using space $O(\epsilon^{-2}k\log(\delta^{-1}n)\log F_1)$ bits.

The COUNT-MIN algorithm [7] for finding approximate frequent items keeps a collection of $s = O(\log \frac{1}{\delta})$ independent hash tables $T_1, T_2, \ldots, T_s$, where each hash table T_j is of size $b = 2k$ buckets and uses a pair-wise independent hash function $h_j : [n] \rightarrow \{1, \ldots, 2k\}$, for $j = 1, 2, \ldots, s$. The bucket $T_j[r]$ is an integer counter that maintains the following sum $T_j[r] = \sum_{i:h_j(i)=r} f_i$. The estimated frequency $\hat{f}_i$ is obtained as $\hat{f}_i = \mathrm{median}_{r=1}^s T_j[h_j(i)]$. The space versus accuracy guarantees for the COUNT-MIN algorithm is given in terms of the quantity $F_1^{res}(k) = \sum_{r>k} |f_{rank(r)}|$.

Theorem 12 ([7]). $\Pr\{|\hat{f}_i - f_i| \le \frac{F_1^{res}(k)}{k}\} \ge 1 - \delta$ with probability using space $O(k\log \frac{1}{\delta} \log F_1)$ bits and time $O(\log \frac{1}{\delta})$ to process each stream update.

Estimating F_1^{res} and F_2^{res}. [9] presents an algorithm to estimate $F_2^{res}(k)$ to within an accuracy of $(1 \pm \epsilon)$ with confidence $1 - \delta$ using space $O(\frac{k}{\epsilon^2}\log(F_1)\log(\frac{n}{\delta}))$ bits. The data structure used is identical to the COUNTSKETCH structure. The algorithm basically removes the top-k estimated frequencies from the COUNTSKETCH structure and then estimates F_2. Let $\hat{f}_{\tau_1}, \ldots, \hat{f}_{\tau_k}$ denote the top-k estimated frequencies from the COUNTSKETCH structure. Next, the contributions of these estimates are removed from the structure, that is, $U_j[r]:=U_j[r] - \sum_{t:h_j(\tau_t)=r} f_{\tau_t}x_{j\tau_t}$. Subsequently, the *Fast-AMS* algorithm [16], a variant of the original sketch algorithm [1], is used to estimate the second moment as $\hat{F}_2^{res} = \mathrm{median}_{j=1}^s \sum_{r=1}^{8k} (U_j[r])^2$. Formally, we can state:

Lemma 13 ([9]). *For a given integer $k \ge 1$ and $0 < \epsilon < 1$, there exists an algorithm for update streams that returns an estimate $\hat{F}_2^{res}(k)$ satisfying $|\hat{F}_2^{res}(k) - F_2^{res}(k)| \le \epsilon F_2^{res}(k)$ with probability $1 - \delta$ using space $O(\frac{k}{\epsilon^2}(\log \frac{F_1}{\delta})(\log F_1))$ bits.*

A similar argument can be applied to estimate $F_1^{res}(s)$, where, instead of using the COUNTSKETCH algorithm, we use the COUNT-MIN algorithm for retrieving the top-k estimated absolute frequencies. In parallel, a set of $s = O(\frac{1}{\epsilon^2})$ sketches based on a 1-stable distribution [11] (i.e., $Y_j = \sum_i f_i z_{ji}$, where z_{ji} is drawn from a 1-stable distribution). After retrieving the top-k frequencies $f_{\tau_1}, \ldots, f_{\tau_k}$ with respect to their absolute values, we reduce the sketches $Y_j := Y_j - \sum_{r=1}^k f_{\tau_r} z_{j\tau_r}$ and estimate $F_1^{res}(k)$ as $\mathrm{median}_{j=1}^s |Y_j|$. We summarize this in Lemma 14.

Lemma 14. *For a given integer $k \ge 1$ and $0 < \epsilon < 1$, there exists an algorithm for update streams that returns an estimate $\hat{F}_1^{res}(k)$ satisfying $|\hat{F}_1^{res}(k) - F_1^{res}(k)| \le \epsilon F_1^{res}(k)$ with probability $1 - \delta$ using $O(\frac{1}{\epsilon}(k + \frac{1}{\epsilon})(\log \frac{k}{\delta})(\log F_1))$ bits.*

B   Reducing Random Bits by Using a PRG

We use a standard technique of reducing the randomness by using a pseudo-random generator (*PRG*) of Nisan [15] along the lines of Indyk in [11] and Indyk and Woodruff in [12].

Notation. Let M be a finite state machine that uses S bits and has running time R. Assume that M uses the random bits in k segments, each segment consisting of kb bits. Let U^r be a uniform distribution over $\{0,1\}^r$ and for a discrete random variable X, let $\mathcal{F}[X]$ denote the probability distribution of X, treated as a vector. Let $M(x)$ denote the state of M after using the random bits in x. The generator $G : \{0,1\}^u \to \{0,1\}^{kb}$ expands a "small" number of u bits that are truly random to a sequence of kb bits that "appear" random to M. G is said to be a pseudo-random generator for a class $\mathcal{C}$ of finite state machines with parameter ϵ, provided, for every $M \in \mathcal{C}$

$$\left| \mathcal{F}[M_{x \in U^{kb}}(x)] - \mathcal{F}[M_{x \in U^m}(G(x))] \right|_1 \leq \epsilon$$

where, $|y|_1$ denotes the F_1 norm of the vector y. Nisan [15] shows the following property (this version is from [11]).

Theorem 15 ([15]). *There exists a* PRG G *for Space(S) and Time(R) with parameter* $\epsilon = 2^{-O(S)}$ *that requires $O(S)$ bits such that G expands $O(S \log R)$ bits into $O(R)$ bits.*

This is sufficient due to the fact that we can compute the frequency moments by considering each (aggregate) frequency f_i in turn and use only segments of $O(\log F_1)$ bits to store and process it.

Distribution-Free Testing Lower Bounds for Basic Boolean Functions

Dana Glasner* and Rocco A. Servedio**

Department of Computer Science
Columbia University
New York, NY 10027, USA
{dglasner,rocco}@cs.columbia.edu

Abstract. In the *distribution-free* property testing model, the distance between functions is measured with respect to an arbitrary and unknown probability distribution $\mathcal{D}$ over the input domain. We consider distribution-free testing of several basic Boolean function classes over $\{0,1\}^n$, namely monotone conjunctions, general conjunctions, decision lists, and linear threshold functions. We prove that for each of these function classes, $\Omega((n/\log n)^{1/5})$ oracle calls are required for any distribution-free testing algorithm. Since each of these function classes is known to be distribution-free properly learnable (and hence testable) using $\Theta(n)$ oracle calls, our lower bounds are within a polynomial factor of the best possible.

1   Introduction

The field of property testing deals with algorithms that decide whether an input object has a certain property or is far from having the property after reading only a small fraction of the object. Property testing was introduced in [22] and has evolved into a rich field of study (see [3,7,11,20,21] for some surveys). A standard approach in property testing is to view the input to the testing algorithm as a function over some finite domain; the testing algorithm is required to distinguish functions that have a certain property P from functions that are ϵ-far from having property P. In the most commonly considered property testing scenario, a function f is ϵ-far from having a property P if f disagrees with every function g that has property P on at least an ϵ fraction of the points in the input domain; equivalently, the distance between functions f and g is measured with respect to the uniform distribution over the domain. The testing algorithm "reads" f by adaptively querying a black-box oracle for f at points x of the algorithm's choosing (such oracle calls are often referred to as "membership queries" in computational learning theory). The main goal in designing property testing algorithms is to use as few queries as possible to distinguish the two types of functions; ideally the number of queries should depend only on ϵ and should be independent of the size of f's domain.

One can of course view any property P as a class of functions (the class of those functions that have property P). In recent years there has been considerable work in

---

* Supported in part by an FFSEAS Presidential Fellowship.
** Supported in part by NSF award CCF-0347282, by NSF award CCF-0523664, and by a Sloan Foundation Fellowship.

M. Charikar et al. (Eds.): APPROX and RANDOM 2007, LNCS 4627, pp. 494–508, 2007.

the standard "uniform distribution" property testing scenario on testing various natural properties of Boolean functions $f : \{0, 1\}^n \rightarrow \{0, 1\}$, i.e. testing various Boolean function classes. Some classes for which uniform distribution testing results have been obtained are monotone functions [6,9,13]; Boolean literals, monotone conjunctions, general conjunctions and s-term monotone DNFs [19]; J-juntas [8]; parity functions (which are equivalent to degree-1 polynomials) [4]; degree-d polynomials [2]; decision lists, s-term DNFs, size-s decision trees and s-sparse polynomials [5]; and linear threshold functions [18].

Distribution-Free Property Testing. A natural generalization of property testing is to consider a broader notion of the distance between functions. Given a probability distribution $\mathcal{D}$ over the domain, we may define the distance between f and g as the probability that an input x drawn from $\mathcal{D}$ has $f(x) \neq g(x)$; the "standard" notion of property testing described above corresponds to the case where $\mathcal{D}$ is the uniform distribution. *Distribution-free property testing* is the study of property testers in a setting where distance is measured with respect to a *fixed but unknown and arbitrary* probability distribution $\mathcal{D}$. Since the distribution $\mathcal{D}$ is unknown, in this scenario the testing algorithm is allowed to draw random samples from $\mathcal{D}$ in addition to querying a black-box oracle for the value of the function.

Distribution free property testing is well-motivated by very similar models in computational learning theory (namely the model of distribution-free PAC learning with membership queries, which is closely related to the well-studied model of exact learning from equivalence and membership queries), and by the fact that in various settings the uniform distribution may not be the best way to measure distances. Distribution-free property testing has been considered by several authors [1,12,14,15,16]; we briefly describe some of the most relevant prior work below.

Goldreich et al. [12] introduced the model of distribution-free property testing, and observed that any *proper* distribution-free PAC learning algorithm (such a learning algorithm for a class of functions always outputs a hypothesis function that itself belongs to the class) can be used as a distribution-free property testing algorithm. They also showed that several graph properties that have testing algorithms with query complexity independent of input size in the uniform-distribution model (such as bipartiteness, k-colorability, ρ-clique, ρ-cut and ρ-bisection) do not have distribution-free testing algorithms with query complexity independent of input size. In contrast, Halevy and Kushilevitz [15] gave a distribution-free algorithm for testing connectivity in sparse graphs that has $\text{poly}(1/\epsilon)$ query complexity independent of input size.

A range of positive and negative results have been established for distribution-free testing of Boolean functions over $\{0, 1\}^n$. [16] showed that any distribution-free monotonicity testing algorithm over $\{0, 1\}^n$ must make $2^{\Omega(n)}$ queries; this is in contrast with the uniform distribution setting, where monotonicity testing algorithms are known that have query complexity $\text{poly}(n, 1/\epsilon)$ [6,9,13]. On the other hand, [14] showed that for several important function classes over $\{0, 1\}^n$ such as juntas, parities, low-degree polynomials and Boolean literals, there exist distribution-free testing algorithms with query complexity $\text{poly}(1/\epsilon)$ independent of n; these distribution-free results match the query bounds of uniform distribution testing algorithms for these classes.

To sum up, the current landscape of distribution-free property testing is intriguingly varied. For some testing problems (juntas, parities, Boolean literals, low-degree polynomials, connectivity in sparse graphs) the complexity of distribution-free testing is known to be essentially the same as the complexity of uniform-distribution testing; but for other natural testing problems (monotonicity, bipartiteness, k-colorability, ρ-clique, ρ-cut, ρ-bisection), distribution-free testing provably requires many more queries than uniform-distribution testing.

This Work. Our work is motivated by the fact that for many Boolean function classes over $\{0,1\}^n$ that are of fundamental interest, a very large gap exists between the query complexities of the best known distribution-free property testing algorithms (which typically follow trivially from learning algorithms and have query complexity $\Omega(n)$) and the best known uniform distribution property testing algorithms (which typically have query complexity $\text{poly}(1/\epsilon)$ independent of n). A natural goal is to try to close this gap, either by developing efficient distribution-free testing algorithms or by proving lower bounds for distribution-free testing for these classes.

We study distribution-free testability of several fundamental classes of Boolean functions that have been previously considered in the uniform distribution testing framework, and have been extensively studied in various distribution-free learning models. More precisely, we consider the following classes (in order of increasing generality): monotone conjunctions, arbitrary conjunctions, decision lists, and linear threshold functions. Each of these four classes is known to be testable in the uniform distribution setting using $\text{poly}(1/\epsilon)$ many queries, independent of n (see [19] for monotone and general conjunctions, [5] for decision lists, and [18] for linear threshold functions). On the other hand, for each of these classes the most efficient known distribution-free testing algorithm is simply to use a proper learning algorithm. Using the fact that each of these classes has Vapnik-Chervonenkis dimension $\Theta(n)$, standard results in learning theory yield well-known algorithms that use $O(n/\epsilon)$ random examples and no membership queries (see e.g. Chapter 3 of [17]), and known results also imply that any learning algorithm must make $\Omega(n)$ oracle calls (see [23]).

Our main results are strong distribution-free lower bounds for testing each of these four function classes:

Theorem 1. *Let T be any algorithm which, given oracle access to an unknown f : $\{0,1\}^n \to \{0,1\}$ and (sampling) oracle access to an unknown distribution $\mathcal{D}$ over $\{0,1\}^n$, tests whether f is a monotone conjunction versus $\Theta(1)$-far from every monotone conjunction with respect to $\mathcal{D}$. Then T must make $\Omega((n/\log n)^{1/5})$ oracle calls in total. The same lower bound holds for testing general conjunctions, testing decision lists, and testing linear threshold functions.*

These results show that for these function classes, distribution-free testing is nearly as difficult (from a query perspective) as distribution-free learning, and is much more difficult than uniform-distribution testing.

Organization. After giving preliminaries in Section 2, in Section 3 we present our construction of "yes" and "no" (function, distribution) pairs that are used in the lower bound for monotone conjunctions. The actual lower bound proof is given in Section 4. In Section 5 we give a simple argument that extends the result to a lower bound for arbitrary conjunctions and for decision lists. Because of space limitations we do not present the proof for linear threshold functions here; it can be found in [10].

2   Preliminaries

Throughout the paper we deal with Boolean functions over n input variables.

Definition 1. *Let $\mathcal{D}$ be a probability distribution over $\{0,1\}^n$. Given Boolean functions $f, g : \{0,1\}^n \rightarrow \{0,1\}$, the* distance between f and g with respect to $\mathcal{D}$ *is defined by* $dist_\mathcal{D}(f, g) \stackrel{def}{=} \mathbf{Pr}_{x \sim \mathcal{D}}[f(x) \neq g(x)]$.

If C is a class of Boolean functions over $\{0,1\}^n$, we define the distance between f and C with respect to $\mathcal{D}$ *to be $dist_\mathcal{D}(f, C) \stackrel{def}{=} \min_{g \in C} dist_\mathcal{D}(f, g)$.*

We say that f is ϵ-far from C with respect to $\mathcal{D}$ if $dist_\mathcal{D}(f, C) \geq \epsilon$.

Now we can define the notion of a distribution-free tester for a class of functions C:

Definition 2. *A* distribution-free tester *for class C is a probabilistic oracle machine T which takes as input a distance parameter $\epsilon > 0$, is given access to*

- *a* black-box oracle *to a fixed (but unknown and arbitrary) function $h : \{0,1\}^n \rightarrow \{0,1\}$ (when invoked with input x, the oracle returns the value $h(x)$); and*
- *a* sampling oracle *for a fixed (but unknown and arbitrary) distribution $\mathcal{D}$ over $\{0,1\}^n$ (each time it is invoked this oracle returns a pair $(x, h(x))$ where x is independently drawn from $\mathcal{D}$),*

and satisfies the following two conditions: for any $h : \{0,1\}^n \rightarrow \{0,1\}$ and any distribution $\mathcal{D}$,

- *If h belongs to C, then $\mathbf{Pr}[T^{h,\mathcal{D}} = Accept] \geq \frac{2}{3}$; and*
- *If h is ϵ-far from C w.r.t. $\mathcal{D}$, then $\mathbf{Pr}[T^{h,\mathcal{D}} = Accept] \leq \frac{1}{3}$.*

This definition allows the tester to be adaptive and to have two-sided error; this is of course the strongest version for proving lower bounds.

The Classes We Consider. For completeness we define here all the classes of functions that we will consider: these are (in order of increasing generality) monotone conjunctions, general conjunctions, decision lists, and linear threshold functions. We note that each of these function classes is quite basic and natural and has been studied intensively in fields such as computational learning theory.

The class $MCONJ$ consists of all monotone conjunctions of any set of Boolean variables from $x_1, \ldots, x_n$, i.e. all ANDs of (unnegated) Boolean variables.

The class $CONJ$ consists of all conjunctions of any set of Boolean literals over $\{0,1\}^n$ (a literal is a Boolean variable or the negation of a variable).

A *decision list* L of length k over the Boolean variables $x_1, \ldots, x_n$ is defined by a list of k pairs and a bit $(\ell_1, \beta_1), (\ell_2, \beta_2), \ldots, (\ell_k, \beta_k), \beta_{k+1}$ where each ℓ_i is a Boolean literal and each β_i is either 0 or 1. Given any $x \in \{0,1\}^n$, the value of $L(x)$ is β_i if i is the smallest index such that ℓ_i is made true by x; if no ℓ_i is true then $L(x) = \beta_{k+1}$. Let DL denote the class of all decision lists of arbitrary length $k \geq 0$ over $\{0,1\}^n$.

A *linear threshold function* is defined by a list of $n+1$ real values $w_1, \ldots, w_n, \theta$. The value of the function on input $x \in \{0,1\}^n$ is 1 if $w_1 x_1 + \cdots + w_n x_n \geq \theta$ and is 0 if $w_1 x_1 + \cdots + w_n x_n < \theta$. We write LTF to denote the class of all linear threshold functions over $\{0,1\}^n$.

It is well known and easy to see that $MCONJ \subsetneq CONJ \subsetneq DL \subsetneq LTF$.

Notation. For a string $x \in \{0,1\}^n$ we write x_i to denote the i-th bit of x. For $x, y \in \{0,1\}^n$ we write $x \wedge y$ to denote the n-bit string z which is the bitwise AND of x and y, i.e. $z_i = x_i \wedge y_i$ for all i. The string $x \vee y$ is defined similarly to be the bitwise OR of x and y.

Recall that the *total variation distance*, or *statistical distance*, between two random variables X and Y that take values in a finite set S is $d_{TV}(X,Y) \overset{\text{def}}{=} \frac{1}{2} \sum_{\zeta \in S} |\Pr[X = \zeta] - \Pr[Y = \zeta]|$.

3   The Two Distributions for Monotone Conjunctions

In this section we define two distributions, $\mathcal{YES}$ and $\mathcal{NO}$, over pairs $(h, \mathcal{D})$ where $h : \{0,1\}^n \to \{0,1\}$ is a Boolean function and $\mathcal{D}$ is a distribution over the domain $\{0,1\}^n$. We will prove that these distributions have the following properties:

1. For every pair $(g, \mathcal{D}_g)$ in the support of $\mathcal{YES}$, the function g is a monotone conjunction (and hence any tester for $MCONJ$ must accept every such pair with probability at least $2/3$).
2. For every pair $(f, \mathcal{D}_f)$ in the support of $\mathcal{NO}$, the function f is $1/3$-far from $MCONJ$ with respect to $\mathcal{D}_f$ (and hence any tester for $MCONJ$ must accept every such pair with probability at most $1/3$).

Our constructions are parameterized by three values ℓ, m and s. As we will see the optimal setting of these parameters (up to multiplicative constants) for our purposes is

$$\ell \overset{\text{def}}{=} n^{2/5}(\log n)^{3/5}, \quad m \overset{\text{def}}{=} (n/\log n)^{2/5}, \quad s \overset{\text{def}}{=} \log n. \tag{1}$$

To keep the different roles of these parameters clear in our exposition we will present our constructions and analyses in terms of "ℓ," "m" and "s" as much as possible and only plug in the values from (1) toward the end of our analysis.

3.1   The $\mathcal{YES}$ Distribution

A draw from the distribution $\mathcal{YES}$ over $(g, \mathcal{D}_g)$ pairs is obtained as follows:

- Let $R \subset [n]$ be a set of size $2\ell m$ selected uniformly at random. Randomly partition the set R into $2m$ subsets $A_1, B_1, \ldots, A_m, B_m$, each of size ℓ. Let $a^i \in \{0,1\}^n$ be the string whose j-th bit is 0 iff $j \in A_i$. The string b^i is defined similarly. The string c^i is defined to be $a^i \wedge b^i$, and similarly we define the set $C_i = A_i \cup B_i$. We sometimes refer to a^i, b^i, c^i as the "points of the i-th block."
- Let g_1 be the conjunction of all variables in $[n] \setminus R$.

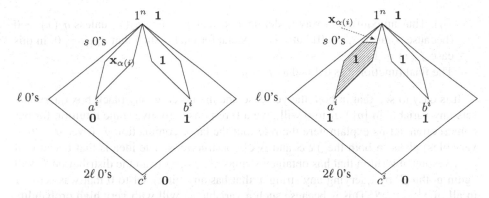

Fig. 1. The left figure shows how a yes-function g labels c^i and the points above it (including a^i and b^i). Bold print indicate the label that g assigns. Note that every point above b^i is labeled **1** by g, and points above a^i are labeled according to $\mathbf{x}_{\alpha(i)}$. The right figure shows how a no-function f labels c^i and the points above it (including a^i and b^i). Again, bold print indicates the label that f assigns. Note that every point above b^i is labeled **1** by f, and points above a^i with less than s 0's are labeled according to $\mathbf{x}_{\alpha(i)}$. The i-special points for block i are shaded and are labeled **1** by f.

– For each $i = 1, \ldots, m$ let $\alpha(i)$ be an element chosen uniformly at random from the set A_i; we say that $\alpha(i)$ is a *representative* of A_i. Let g_2 be a conjunction of length m formed by taking $g_2 = x_{\alpha(1)} \wedge \cdots \wedge x_{\alpha(m)}$, i.e. g_2 is an AND of the representatives from each of $A_1, \ldots, A_m$.

– The function g is taken to be $g = g_1 \wedge g_2$. For each $i = 1, \ldots, m$ the distribution $\mathcal{D}_g$ puts weight $2/(3m)$ on b^i and puts weight $1/(3m)$ on c^i.

It is clear that for every $(g, \mathcal{D}_g)$ in the support of $\mathcal{YES}$, the function g is a monotone conjunction that contains exactly $n - 2m\ell + m$ variables, so Property (1) indeed holds.

3.2 The $\mathcal{NO}$ Distribution

A draw from the distribution $\mathcal{NO}$ of $(f, \mathcal{D}_f)$ pairs is obtained as follows:

– As in the yes-case, let $R \subset [n]$ be a randomly selected set of size $2\ell m$, and randomly partition the set R into $2m$ subsets $A_1, B_1, \ldots, A_m, B_m$, each of size ℓ. The points a^i, b^i, c^i and sets A_i, B_i, C_i are defined as in the yes-case. The distribution $\mathcal{D}_f$ is uniform over the $3m$ points $a^1, \ldots, c^m$.

– Construct the conjunctions g_1 and g_2 exactly as in the yes-case: g_1 is the conjunction of all variables in $[n] \setminus R$ and g_2 is $x_{\alpha(1)} \wedge \cdots \wedge x_{\alpha(m)}$ where each $\alpha(i)$ is a representative chosen uniformly from A_i.

– Define the function f' as follows: $f'(x) = 0$ if there exists some $i \in [m]$ such that both the following conditions hold:
  - $x_{\alpha(i)} = 0$ and
  - (fewer than s of the elements $j \in A_i$ have $x_j = 0$) or ($x_j = 0$ for some $j \in B_i$).

The following terminology will be useful: we say that an input $x \in \{0,1\}^n$ is *i-special* if (at least s elements $j \in A_i$ have $x_j = 0$) and ($x_j = 1$ for all $j \in$

B_i). Thus an equivalent way to define f' is that $f'(x) = g_2(x)$ unless $g_2(x) = 0$ (because some $x_{\alpha(i)} = 0$) but x is i-special for each i such that $x_{\alpha(i)} = 0$; in this case $f'(x) = 1$.
- The final function f is defined as $f = g_1 \wedge f'$.

It is easy to see that in both the yes-case and the no-case, any black-box query that sets any variable in $[n] \setminus R$ to 0 will give a 0 response. To give some intuition for our construction, let us explain here the role that the large conjunction g_1 (over $n - 2\ell m$ variables) plays in both the $\mathcal{YES}$ and $\mathcal{NO}$ constructions. The idea is that because of g_1, a testing algorithm that has obtained strings $z^1, \ldots, z^q$ from the distribution $\mathcal{D}$ will "gain nothing" by querying any string x that has any bit x_i set to 0 that was set to 1 in all of $z^1, \ldots, z^q$. This is because such a variable x_i will with very high probability (over a random choice of $(f, \mathcal{D}_f)$ from $\mathcal{NO}$ or a random choice of $(g, \mathcal{D}_g)$ from $\mathcal{YES}$) be contained in g_1, so in both the "yes" and "no" cases the query will yield an answer of 0 with very high probability. Consequently there is no point in making such a query in the first place. (We give a rigorous version of this argument in Section 4.2).

For any $(f, \mathcal{D}_f)$ drawn from $\mathcal{NO}$, we have $f(a^i) = f(b^i) = 1$ and $f(c^i) = 0$ for each $i = 1, \ldots, m$. It is noted in [19] (and is easy to check) that any monotone conjunction h must satisfy $h(x) \wedge h(y) = h(x \wedge y)$ for all $x, y \in \{0, 1\}^n$, and thus must satisfy $h(c^i) = h(a^i) \wedge h(b^i)$. Thus any monotone conjunction h must disagree with f on at least one of a_i, b_i, c_i for all i, and consequently f is $1/3$-far from any monotone conjunction with respect to $\mathcal{D}_f$.

Thus we have established properties (1) and (2) stated at the beginning of this section. These give:

Lemma 1. *Any distribution-free tester for $MCONJ$ that is run with distance parameter $\epsilon = 1/3$ must accept a random pair $(g, \mathcal{D}_g)$ drawn from $\mathcal{YES}$ with probability at least $2/3$, and must accept a random pair $(f, \mathcal{D}_f)$ drawn from $\mathcal{NO}$ with probability at most $1/3$.*

4   The Lower Bound for Monotone Conjunctions

In this section we will prove the following theorem:

Theorem 2. *Let $q \stackrel{def}{=} \frac{1}{20}(\frac{n}{\log n})^{1/5}$. Let T be any probabilistic oracle algorithm that, given a pair $(h, \mathcal{D})$, makes at most q black-box queries to h and samples $\mathcal{D}$ at most q times. Then we have*

$$\left| \mathbf{Pr}_{(g,\mathcal{D}_g) \sim \mathcal{YES}}[T^{g,\mathcal{D}_g} = Accept] - \mathbf{Pr}_{(f,\mathcal{D}_f) \sim \mathcal{NO}}[T^{f,\mathcal{D}_f} = Accept] \right| \leq \frac{1}{4}.$$

Note that in the above theorem each probability is taken over the draw of the (function,distribution) pair from the appropriate distribution $\mathcal{YES}$ or $\mathcal{NO}$, over the random draws from the distribution $\mathcal{D}_f$ or $\mathcal{D}_g$, and over any internal randomness of algorithm T. Lemma 1 and Theorem 2 together immediately imply the first part of Theorem 1, our lower bound for monotone conjunctions.

4.1   The Idea

Here is some high-level intuition for the proof. If T could find a^i, b^i and c^i for some i then T would know which case it is in (yes versus no), because $h(a^i) \land h(b^i) = h(c^i)$ if and only if T is in the yes-case. Since T can only make $q \ll \sqrt{m}$ draws from $\mathcal{D}$, the birthday paradox tells us that with high probability the random sample that T draws contains at most one of a^i, b^i and c^i for each i. The c^i-type points (with $n - 2\ell$ ones) are labeled negative in both the yes- and no- cases, so these "look the same" to T in both cases. And since the distributions $\mathcal{D}_g$ (in the yes-case) and $\mathcal{D}_f$ (in the no-case) put weight only on the positive a^i and b^i-type points (with $n - \ell$ ones), these points "look the same" to T as well in both cases. So with high probability T cannot distinguish between yes-pairs and no-pairs on the basis of the first q random draws alone. (Corollary 1 formalizes this intuition).

Of course, though, T can also make q queries. Can T perhaps identify a triple (a^i, b^i, c^i) through these queries, or perhaps T can otherwise determine which case it is in even without finding a triple? The crux of the proof is to show that in fact queries actually cannot help T much; we now sketch the idea.

Consider a single fixed block $i \in [m]$. If none of a^i, b^i or c^i are drawn in the initial sample, then by the argument of Section 3.2 the tester will get no useful information about which case (s)he is in from this block. By the birthday paradox we can assume that at most one of a^i, b^i and c^i is drawn in the initial sample; we consider the three cases in turn.

If b^i is drawn, then by the Section 3.2 argument all query points will have all the bits in A_i set to 1; such queries will "look the same" in both the yes- and no- cases as far as the i-th block is concerned.

If a^i is drawn (so we are in the no-case), then by the Section 3.2 argument all query points will have all the bits in B_i set to 1. Using the definition of f', as far as the i-th block is concerned with high probability it will "look like" the initial a^i point was a b^i-point from the yes-case. This is because the only way the tester can tell that it is in the no-case is if it manages to query a point which has fewer than s bits from A_i set to 0 but the representative $\alpha(i)$ is one of those bits. Such points are hard to find since $\alpha(i)$ is randomly selected from A_i. (See the "a-witness" case in the proof of Lemma 6).

Finally, suppose that c^i is drawn. The only way a tester can distinguish between the yes- and no- cases is by finding an i-special point (or determining that no such point exists), but to find such a point it must make a query with at least s 0's in C_i, all of which lie in A_i. This is hard to do since the tester does not know how the elements of C_i are divided into the sets A_i and B_i. (See the "c-witness" case in the proof of Lemma 6).

4.2   Proof of Theorem 2

Fix any probabilistic oracle algorithm T that makes at most q black-box queries to h and samples $\mathcal{D}$ at most q times. Without loss of generality we may assume that T first makes exactly q draws from distribution $\mathcal{D}$, and then makes exactly q (adaptive) queries to the black-box oracle for h.

It will be convenient for us to assume that algorithm T is actually given "extra information" on certain draws from the distribution $\mathcal{D}$. More precisely, we suppose that each time T calls the oracle for $\mathcal{D}$,

- If a "c^i-type" labeled example $(c^i, h(c^i))$ is generated by the oracle, algorithm T receives the triple $(c^i, h(c^i), \alpha(i))$ (recall that $\alpha(i)$ is the index of the variable from C_i that belongs to the conjunction g_2);
- If a "non-c^i-type" labeled example $(x, h(x))$ is generated by the oracle where $x \neq c^i$ for all $i = 1, \ldots, m$, algorithm T receives the triple $(x, h(x), 0)$. (Thus there is no "extra information" given on non-c^i points).

It is clear that proving Theorem 2 for an arbitrary algorithm T that receives this extra information establishes the original theorem as stated (for algorithms that do not receive the extra information).

Following [16], we now define a *knowledge sequence* to precisely capture the notion of "what an algorithm learns from its queries." A knowledge sequence is a sequence of elements corresponding to the interactions that an algorithm has with each of the two oracles. The first q elements of a knowledge sequence are triples as described above; each corresponds to an independent draw from the distribution $\mathcal{D}$. The remaining elements of the knowledge sequence are input-output pairs corresponding to the algorithm's calls to the black-box oracle for h. (Recall that these later oracle calls are adaptive, i.e. each query point can depend on the answers received from previous oracle calls).

Notation. For any oracle algorithm ALG, let $\mathcal{P}_{yes}^{ALG}$ denote the distribution over knowledge sequences induced by running ALG on a pair $(g, \mathcal{D}_g)$ randomly drawn from $\mathcal{YES}$. Similarly, let $\mathcal{P}_{no}^{ALG}$ denote the distribution over knowledge sequences induced by running ALG on a pair $(f, \mathcal{D}_f)$ randomly drawn from $\mathcal{NO}$. For $0 \leq i \leq q$ we write $\mathcal{P}_{yes,i}^{ALG}$ to denote the length-$(q+i)$ prefix of $\mathcal{P}_{yes}^{ALG}$, and similarly for $\mathcal{P}_{no,i}^{ALG}$.

We will prove Theorem 2 by showing that the statistical distance $d_{TV}(\mathcal{P}_{yes}^T, \mathcal{P}_{no}^T)$ between distributions $\mathcal{P}_{yes}^T$ and $\mathcal{P}_{no}^T$ is at most $1/4$. Because of space constraints some proofs are omitted from the following presentation; all omitted proofs can be found in [10].

Most Sequences of Draws Are "Clean" in Both the Yes- and No- Cases. The main result of this subsection is Corollary 1; intuitively, this corollary shows that given only q draws from the distribution and no black-box queries, it is impossible to distinguish between the yes- and no- cases with high accuracy. This is achieved via a notion of a "clean" sequence of draws from the distribution, which we now explain.

Let $S = (x^1, y_1), \ldots, (x^q, y_q)$ be a sequence of q labeled examples drawn from distribution $\mathcal{D}$, where $\mathcal{D}$ is either $\mathcal{D}_f$ for some $(f, \mathcal{D}_f) \in \mathcal{NO}$ or $\mathcal{D}_g$ for some $(g, \mathcal{D}_g) \in \mathcal{YES}$. In either case there is a corresponding set of points $a^1, b^1, c^1, \ldots, a^m, b^m, c^m$ as described in Section 3. We say that S is *clean* if S does not hit any block $1, \ldots, m$ twice, i.e. if the number of different blocks from $1, \ldots, m$ for which S contains some point a^i, b^i or c^i is exactly q. With this definition, we have the following claim and its easy corollary:

Claim. We have $\mathbf{Pr}[\mathcal{P}_{yes,0}^T \text{ is clean}] = \mathbf{Pr}[\mathcal{P}_{no,0}^T \text{ is clean}] \geq 1 - q^2/m$. Furthermore, the conditional random variables $(\mathcal{P}_{yes,0}^T \mid \mathcal{P}_{yes,0}^T \text{ is clean})$ and $(\mathcal{P}_{no,0}^T \mid \mathcal{P}_{no,0}^T \text{ is clean})$ are identically distributed.

Corollary 1. *The statistical distance $d_{TV}(\mathcal{P}_{yes,0}^T, \mathcal{P}_{no,0}^T)$ is at most q^2/m.*

Eliminating Foolhardy Queries. Let T' denote a modified version of algorithm T which works as follows: like T, it starts out by making q draws from the distribution. Let Q be the set of all indices i such that all q draws from the distribution have the i-th bit set to 1. We say that any query string $x \in \{0,1\}^n$ that has $x_j = 0$ for some $j \in Q$ is *foolhardy*. After making its q draws from $\mathcal{D}$, algorithm T' simulates algorithm T for q black-box queries, except that for any foolhardy query that T makes, T' "fakes" the query in the following sense: it does not actually make the query but instead proceeds as T would proceed if it made the query and received the response 0.

Our goal in this subsection is to show that in both the yes- and no- cases, the executions of T and T' are statistically close. (Intuitively, this means that we can w.l.o.g. assume that the testing algorithm T does not make any foolhardy queries.) To analyze algorithm T' it will be useful to consider some other algorithms that are intermediate between T and T', which we now describe.

For each value $1 \le k \le q$, let U_k denote the algorithm which works as follows: U_k first makes q draws from the distribution $\mathcal{D}$, then simulates algorithm T for k queries, except that for each of the first $k - 1$ queries that T makes, if the query is foolhardy then U_k "fakes" the query as described above. Let U_k' denote the algorithm which works exactly like U_k, except that if the k-th query made by U_k is foolhardy then U_k' fakes that query as well. We have the following:

Lemma 2. *For all $k \in [q]$, the statistical distance*

$$d_{TV}((\mathcal{P}_{yes}^{U_k} \mid \mathcal{P}_{yes,0}^{U_k} \text{ is clean}), (\mathcal{P}_{yes}^{U_k'} \mid \mathcal{P}_{yes,0}^{U_k'} \text{ is clean}))$$

is at most $2\ell m/n$, and similarly $d_{TV}((\mathcal{P}_{no}^{U_k} \mid \mathcal{P}_{no,0}^{U_k} \text{ is clean}), (\mathcal{P}_{no}^{U_k'} \mid \mathcal{P}_{no,0}^{U_k'} \text{ is clean}))$ is also at most $2\ell m/n$.

Now a hybrid argument using Lemma 2 lets us bound the statistical distance between the executions of T and T'.

Lemma 3. *The statistical distance $d_{TV}(\mathcal{P}_{yes}^{T'}, \mathcal{P}_{yes}^{T})$ is at most $2\ell mq/n + q^2/m$, and the same bound holds for $d_{TV}(\mathcal{P}_{no}^{T'}, \mathcal{P}_{no}^{T})$.*

Bounding the Probability of Finding a Witness. Let T'' denote an algorithm that is a variant of T', modified as follows. T'' simulates T' except that T'' does not actually make queries on non-foolhardy strings; instead T'' simulates the answers to those queries "in the obvious way" that they should be answered if the target function were a yes-function and hence all of the draws from $\mathcal{D}$ that yielded strings with ℓ zeros were in fact b^i-type points. More precisely, assume that there are r distinct c^i-type points in the initial sequence of q draws from the distribution. Since for each c^i-type point the algorithm is given $\alpha(i)$, the algorithm "knows" r variables $x_{\alpha(i)}$ that are in the conjunction. To simulate an answer to a non-foolhardy query $x \in \{0,1\}^n$, T'' answers with 0 if any of the r $x_{\alpha(i)}$ variables are set to 0 in x, and answers with 1 otherwise. Note that consequently T'' does not actually make any black-box queries at all.

In this subsection we will show that in both the yes- and no- cases, the executions of T' and T'' are statistically close; once we have this it is not difficult to complete the proof of Theorem 2. In the yes-case these distributions are in fact identical (Lemma 4), but in the no-case these distributions are not identical; we will argue that they are close using properties of the function f' from Section 3.2.

We first address the easier yes-case:

Lemma 4. *The statistical distance* $d_{TV}(\mathcal{P}_{yes}^{T'}, \mathcal{P}_{yes}^{T''})$ *is zero.*

Proof. We argue that T' and T'' answer all queries in exactly the same way. Fix any $1 \le i \le q$ and let z denote the ith query made by T.

If z is a foolhardy query then both T' and T'' answer z with 0. So suppose that z is not a foolhardy query. Then any 0's that z contains must be in positions from points that were sampled in the first stage. Consequently the only variables that can be set to 0 that are in the conjunction g are the $x_{\alpha(i)}$ variables from the C_i sets corresponding to the c^i points in the draws. All the other variables that were "seen" are not in the conjunction so setting them to 0 or 1 will not affect the value of $g(z)$. Therefore, $g(z)$ (and hence T'''s response) is 0 if any of the $x_{\alpha(i)}$ variables are set to 0 in z, and is 1 otherwise. This is exactly how T'' answers non-foolhardy queries as well. □

To handle the no-case, we introduce the notion of a "witness" that the black-box function is a no-function.

Definition 3. *We say that a knowledge sequence contains a* witness *for* $(f, \mathcal{D}_f)$ *if elements* $q+1, \ldots$ *of the sequence (the black-box queries) contain either of the following:*

1. *A point* $z \in \{0,1\}^n$ *such that for some* $1 \le i \le m$ *for which* a^i *was sampled in the first q draws, the bit* $z_{\alpha(i)}$ *is 0 but fewer than s of the elements* $j \in A_i$ *have* $z_j = 0$. *We refer to such a point as an* a-witness *for block* i.
2. *A point* $z \in \{0,1\}^n$ *such that for some* $1 \le i \le m$ *for which* c^i *was sampled in the first q draws, z is i-special. We refer to such a point as a* c-witness *for block* i.

The following lemma implies that it is enough to bound the probability that $\mathcal{P}_{no}^T$ contains a witness:

Lemma 5. *The statistical distance* $d_{TV}((\mathcal{P}_{no}^{T'} \mid \mathcal{P}_{no}^{T'}$ *does not contain a witness and* $\mathcal{P}_{no,0}^{T'}$ *is clean)*, $(\mathcal{P}_{no}^{T''} \mid \mathcal{P}_{no}^{T''}$ *does not contain a witness and* $\mathcal{P}_{no,0}^{T''}$ *is clean))* *is zero.*

Proof. Claim 4.2 implies that $(\mathcal{P}_{no,0}^{T'} \mid \mathcal{P}_{no,0}^{T'}$ is clean) and $(\mathcal{P}_{no,0}^{T''} \mid \mathcal{P}_{no,0}^{T''}$ is clean) are identically distributed. We show that if there is no witness then T' and T'' answer all queries in exactly the same way; this gives the lemma. Fix any $1 \le i \le q$ and let z denote the ith query.

If z is a foolhardy query, then both T' and T'' answer z with 0. So suppose that z is not a foolhardy query and not a witness. Then any 0's that z contains must be in positions from points that were sampled in the first stage.

First suppose that one of the $x_{\alpha(i)}$ variables from some c^i that was sampled is set to 0 in z. Since z is not a witness, either z has fewer than s zeros from A_i or some variable from B_i is set to zero in z. So in this case we have $f(x_i) = g_2(x_i) = 0$.

Now suppose that none of the $x_{\alpha(i)}$ variables from the c^i's that were sampled are set to 0 in z. If no variable $x_{\alpha(i)}$ from any a^i that was sampled is set to 0 in z, then clearly

$f(z) = g(z) = 1$. If any variable $x_{\alpha(i)}$ from an a^i that was sampled is set to 0 in z, then since z is not a witness there must be at least s elements of A_i set to 0 and every element of B_i set to 1 for each such $x_{\alpha(i)}$. Therefore, $f(z) = 1$.

Thus $f(z)$ evaluates to 0 if any of the $x_{\alpha(i)}$ variables from the c^i's that were sampled is set to 0 and evaluates to 1 otherwise. This is exactly how T'' answers queries as well.   □

Let us consider a sequence of algorithms that hybridize between T' and T'', similar to the previous section. For each value $1 \leq k \leq q$, let V_k denote the algorithm which works as follows: V_k first makes q draws from the distribution $\mathcal{D}$, then simulates algorithm T' for k queries, except that each of the first $k - 1$ queries is faked (foolhardy queries are faked as described in the previous subsection, and non-foolhardy queries are faked as described at the start of this subsection). Thus algorithm V_k actually makes at most one query to the black-box oracle, the k-th one (if this is a foolhardy query then this one is faked as well). Let V_k' denote the algorithm which works exactly like V_k, except that if the k-th query made by V_k is non-foolhardy then V_k' fakes that query as well as described at the start of this subsection.

Lemma 6. *For each value* $1 \leq k \leq q$, *the statistical distance* $d_{TV}((\mathcal{P}_{no}^{V_k} \mid \mathcal{P}_{no,0}^{V_k}$ *is clean*), $(\mathcal{P}_{no}^{V_k'} \mid \mathcal{P}_{no,0}^{V_k'}$ *is clean*)) *is at most* $\max\{\frac{qs}{\ell}, \frac{q}{2^s}\} = qs/\ell$.

Proof. By Lemma 5, the executions of V_k and V_k' are identically distributed unless the k-th query string (which we denote z) is a witness for $(f, \mathcal{D}_f)$. Since neither V_k nor V_k' makes any black-box query prior to z, the variation distance between $\mathcal{P}_{no}^{V_k}$ and $\mathcal{P}_{no}^{V_k'}$ is at most $\mathbf{Pr}[z$ is a witness] where the probability is taken over a random draw of $(f, \mathcal{D}_f)$ from $\mathcal{NO}$ conditioned on $(f, \mathcal{D}_f)$ being consistent with the q draws from the distribution and with those first q draws being clean. We bound the probability that z is a witness by considering both possibilities for z (an a-witness or a c-witness) in turn.

- We first bound the probability that z is an a-witness. So fix some $i \in [m]$ and let us suppose that a^i was sampled in the first stage of the algorithm. We will bound the probability that z is an a-witness for block i; once we have done this, a union bound over the (at most q) blocks such that a^i is sampled in the first stage gives a bound on the overall probability that z is an a-witness.

  Fix any possible outcome for z. In order for z to be an a-witness for block i, it must be the case that fewer than s of the ℓ elements in A_i are set to 0 in z, but the bit $z_{\alpha(i)}$ is set to 0. For a random choice of $(f, \mathcal{D}_f)$ as described above, since we are conditioning on the q draws from the distribution being clean, the only information that these q draws reveal about the index $\alpha(i)$ is that it is some member of the set A_i. Consequently for a random $(f, \mathcal{D}_f)$ as described above, each bit in A_i is equally likely to be chosen as $\alpha(i)$, so the probability that $\alpha(i)$ is chosen to be one of the at most s bits in A_i that are set to 0 in z is at most s/ℓ. Consequently the probability that z is an a-witness for block i is at most s/ℓ, and a union bound gives that the overall probability that z is an a-witness is at most qs/ℓ.

- Now we bound the probability that z is a c-witness. Fix some $i \in [m]$ and let us suppose that c^i was sampled in the first stage of the algorithm. We will bound the probability that z is a c-witness for block i and then use a union bound as above.

Fix any possible outcome for z; let r denote the number of 0's that z has in the bit positions in C_i. In order for z to be a c-witness for block i it must be the case that z is i-special, i.e. $r \geq s$ and all r of these 0's in fact belong to A_i. For a random choice of $(f, \mathcal{D}_f)$ conditioned on being consistent with the q samples from the distribution and with those q samples being clean, the distribution over possible choices of A_i is uniform over all $\binom{2\ell}{\ell}$ possibilities for selecting a size-ℓ subset of C_i. Consequently the probability that all r 0's belong to A_i is at most

$$\frac{\binom{2\ell-r}{\ell-r}}{\binom{2\ell}{\ell}} = \frac{\ell(\ell-1)\cdots(\ell-r+1)}{2\ell(2\ell-1)\cdots(2\ell-r+1)} < \frac{1}{2^r} \leq \frac{1}{2^s}.$$

So the probability that z is a c-witness for block i is at most $1/2^s$, and by a union bound the overall probability that z is a c-witness is at most $q/2^s$.

So the overall probability that z is a witness is at most $\max\{\frac{qs}{\ell}, \frac{q}{2^s}\}$. Using (1) we have that the maximum is qs/ℓ, and the lemma is proved.   □

Now similar to Section 4.2, a hybrid argument using Lemma 6 lets us bound the statistical distance between the executions of T' and T''. The proof of the following lemma is entirely similar to that of Lemma 3 so we omit it.

Lemma 7. *The statistical distance $d_{TV}(\mathcal{P}_{no}^{T'}, \mathcal{P}_{no}^{T''})$ is at most $q^2 s/\ell + q^2/m$.*

Putting the Pieces Together. At this stage, we have that T'' is an algorithm that only makes draws from the distribution and makes no queries. It follows that the statistical distance $d_{TV}(\mathcal{P}_{yes}^{T''}, \mathcal{P}_{no}^{T''})$ is at most $d_{TV}(\mathcal{P}_{yes,0}^{T}, \mathcal{P}_{no,0}^{T})$. So we can bound $d_{TV}(\mathcal{P}_{yes}^{T}, \mathcal{P}_{no}^{T})$ as follows (we write "d" in place of "d_{TV}" for brevity):

$$d(\mathcal{P}_{yes}^{T}, \mathcal{P}_{yes}^{T'}) + d(\mathcal{P}_{yes}^{T'}, \mathcal{P}_{yes}^{T''}) + d(\mathcal{P}_{yes}^{T''}, \mathcal{P}_{no}^{T''}) + d(\mathcal{P}_{no}^{T''}, \mathcal{P}_{no}^{T'}) + d(\mathcal{P}_{no}^{T'}, \mathcal{P}_{no}^{T})$$

$$\leq d(\mathcal{P}_{yes}^{T}, \mathcal{P}_{yes}^{T'}) + d(\mathcal{P}_{yes}^{T'}, \mathcal{P}_{yes}^{T''}) + d(\mathcal{P}_{yes,0}^{T}, \mathcal{P}_{no,0}^{T}) + d(\mathcal{P}_{no}^{T''}, \mathcal{P}_{no}^{T'}) + d(\mathcal{P}_{no}^{T'}, \mathcal{P}_{no}^{T})$$

$$\leq 4q^2/m + 4q\ell m/n + q^2 s/\ell$$

where the final bound follows by combining Corollary 1, Lemma 3, Lemma 4 and Lemma 7. Recalling the parameter settings $\ell = n^{2/5}(\log n)^{3/5}$, $m = (n/\log n)^{2/5}$, and $s = \log n$ from (1) and the fact that $q = \frac{1}{20}(\frac{n}{\log n})^{1/5}$, this bound is less than $1/4$. This concludes the proof of Theorem 2.   □

5   Extending the Lower Bound to Conjunctions and Decision Lists

The construction and analysis from the previous sections easily give a lower bound for testing decision lists via the following lemma:

Lemma 8. *For any pair $(f, \mathcal{D}_f)$ in the support of $\mathcal{NO}$ and any decision list h, the function f is at least $1/6$-far from h w.r.t. $\mathcal{D}_f$.*

Proof. Fix any $(f, \mathcal{D}_f)$ in the support of $\mathcal{NO}$ and any decision list $h = (\ell_1, \beta_1)$, $(\ell_2, \beta_2), \ldots, (\ell_k, \beta_k)$, β_{k+1}. We will show that at least one of the six points a^1, b^1, c^1, a^2, b^2, c^2 is labeled differently by h and f. Grouping all m blocks into pairs and applying the same argument to each pair gives the lemma.

Let ℓ_{a_1} be the first literal in h that is satisfied by a^1, so the value $h(a^1)$ equals β_{a_1}. Define ℓ_{b_1}, ℓ_{c_1}, ℓ_{a_2}, ℓ_{b_2}, and ℓ_{c_2} similarly. We will assume that h and f agree on all six points, i.e. that $\beta_{a_1} = \beta_{b_1} = \beta_{a_2} = \beta_{b_2} = 1$ and $\beta_{c_1} = \beta_{c_2} = 0$, and derive a contradiction.

We may suppose w.l.o.g. that $a_1 = \min\{a_1, b_1, a_2, b_2\}$. We now consider two cases depending on whether or not $c_1 < a_1$. (Note that a_1 cannot equal c_1 since $f(a^1) = 1$ but $f(c^1) = 0$.)

Suppose first that $c_1 < a_1$. No matter what literal ℓ_{c_1} is, since c^1 satisfies ℓ_{c_1} at least one of a^1, b^1 must satisfy it as well. But this means that $\min\{a_1, b_1\} \leq c_1$, which is impossible since $c_1 < a_1$ and $a_1 \leq \min\{a_1, b_1\}$.

Now suppose that $a_1 < c_1$; then it must be the case that ℓ_{a_1} is a literal "x_j" for some $j \in B_1$. (The only other possibilities are that ℓ_{a_1} is "$\overline{x}_j$" for some $j \in A_i$ or is "x_j" for some $j \in ([n] \setminus C_1)$; in either case, this would imply that $f(c^1) = 1$, which does not hold.) Since $f(c^2) = 0$ and $(c^2)_j = 1$, it must be the case that $c_2 < a_1$. But no matter what literal ℓ_{c_2} is, since c^2 satisfies it at least one of a^2, b^2 must satisfy it as well. This means that $\min\{a_2, b_2\} \leq c_2 < a_1 \leq \min\{a_2, b_2\}$, which is a contradiction. $\square$

Since monotone conjunctions are a subclass of decision lists, for every $(g, \mathcal{D}_g)$ in the support of $\mathcal{YES}$ we have that g is computed by a decision list. We thus have the obvious analogue of Lemma 1 for decision lists; together with Theorem 2, this gives the $\Omega((n/\log n)^{1/5})$ lower bound for decision lists that is claimed in Theorem 1.

Since any conjunction (not necessarily monotone) can be expressed as a decision list, we immediately have an analogue of Lemma 8 for general conjunctions. The same line of reasoning described above now gives the $\Omega((n/\log n)^{1/5})$ lower bound for general conjunctions that is claimed in Theorem 1.

While the construction from Section 3 went through unchanged to give us lower bounds on general conjunctions and decision lists, it is not suited for a lower bound on LTF. To see this, observe that for any $(f, \mathcal{D}_f)$ in the support of $\mathcal{NO}$ the function f is 0-far from the linear threshold function $x_1 + \cdots + \cdots x_n \geq n - 3\ell/2$ with respect to $\mathcal{D}_f$. Consequently we need a somewhat different approach; in [10] a modified version of our construction from Section 3 is used to obtain a lower bound for LTF.

References

1. Ailon, N., Chazelle, B.: Information theory in property testing and monotonicity testing in higher dimension. Information and Computation 204, 1704–1717 (2006)
2. Alon, N., Kaufman, T., Krivelevich, M., Litsyn, S., Ron, D.: Testing low-degree polynomials over GF(2). In: Proceedings of RANDOM-APPROX, pp. 188–199 (2003)
3. Alon, N., Shapira, A.: Homomorphisms in Graph Property Testing - A Survey. Topics in Discrete Mathematics (to appear, 2007) available at
http://www.math.tau.ac.il/~asafico/nesetril.pdf
4. Blum, M., Luby, M., Rubinfeld, R.: Self-testing/correcting with applications to numerical problems. J. Comp. Sys. Sci. 47, 549–595, Earlier version in STOC'90 (1993)

5. Diakonikolas, I., Lee, H., Matulef, K., Onak, K., Rubinfeld, R., Servedio, R., Wan, A.: Testing for concise representations. Submitted for publication (2007)
6. Dodis, Y., Goldreich, O., Lehman, E., Raskhodnikova, S., Ron, D., Samorodnitsky, A.: Improved testing algorithms for monotonocity. In: Hochbaum, D.S., Jansen, K., Rolim, J.D.P., Sinclair, A. (eds.) RANDOM 1999 and APPROX 1999. LNCS, vol. 1671, pp. 97–108. Springer, Heidelberg (1999)
7. Fischer, E.: The art of uninformed decisions: A primer to property testing. Computational Complexity Column of The Bulletin of the European Association for Theoretical Computer Science 75, 97–126 (2001)
8. Fischer, E., Kindler, G., Ron, D., Safra, S., Samorodnitsky, A.: Testing juntas. In: Proceedings of the 43rd IEEE Symposium on Foundations of Computer Science, pp. 103–112 (2002)
9. Fischer, E., Lehman, E., Newman, I., Raskhodnikova, S., Rubinfeld, R., Samrodnitsky, A.: Monotonicity testing over general poset domains. In: Proc. 34th Annual ACM Symposium on the Theory of Computing, pp. 474–483. ACM Press, New York (2002)
10. Glasner, D., Servedio, R.: Distribution-free testing lower bounds for basic boolean functions (2007), http://www.cs.columbia.edu/rocco/papers/random07.html
11. Goldreich, O.: Combinatorial property testing – a survey. In: "Randomized Methods in Algorithms Design", AMS-DIMACS, pp. 45–61 (1998)
12. Goldreich, O., Goldwaser, S., Ron, D.: Property testing and its connection to learning and approximation. Journal of the ACM 45, 653–750 (1998)
13. Goldreich, O., Goldwasser, S., Lehman, E., Ron, D., Samordinsky, A.: Testing monotonicity. Combinatorica 20(3), 301–337 (2000)
14. Halevy, S., Kushilevitz, E.: Distribution-Free Property Testing. In: Proceedings of the Seventh International Workshop on Randomization and Computation, pp. 302–317 (2003)
15. Halevy, S., Kushilevitz, E.: Distribution-Free Connectivity Testing. In: Proceedings of the Eighth International Workshop on Randomization and Computation, pp. 393–404 (2004)
16. Halevy, S., Kushilevitz, E.: A lower bound for distribution-free monotonicity testing. In: Proceedings of the Ninth International Workshop on Randomization and Computation, pp. 330–341 (2005)
17. Kearns, M., Vazirani, U.: An introduction to computational learning theory. MIT Press, Cambridge, MA (1994)
18. Matulef, K., O'Donnell, R., Rubinfeld, R., Servedio, R.: Testing Halfspaces. (Manuscript, 2007)
19. Parnas, M., Ron, D., Samorodnitsky, A.: Testing basic boolean formulae. SIAM J. Disc. Math. 16, 20–46 (2002)
20. Ron, D.: Property testing (a tutorial). In: Rajasekaran, S., Pardalos, P.M., Reif, J.H., Rolim, J.D.P. (eds.) "Handbook of Randomized Computing, Volume II", Kluwer, Dordrecht (2001)
21. Rubinfeld, R.: Sublinear time algorithms (2006), available at http://theory.csail.mit.edu/~ronitt/papers/icm.ps
22. Rubinfeld, R., Sudan, M.: Robust characterizations of polynomials with applications to program testing. SIAM J. on Comput. 25, 252–271 (1996)
23. Turán, G.: Lower bounds for PAC learning with queries. In: COLT '93: Proc. 6th Annual Conference on Computational Learning Theory, pp. 384–391 (2002)

On the Randomness Complexity of Property Testing*

Testing*

Oded Goldreich and Or Sheffet**

Weizmann Institute of Science, Rehovot, Israel

Abstract. We initiate a general study of the randomness complexity of property testing, aimed at reducing the randomness complexity of testers without (significantly) increasing their query complexity. One concrete motivation for this study is provided by the observation that the product of the randomness and query complexity of a tester determine the actual query complexity of implementing a version of this tester that utilizes a weak source of randomness (through a randomness-extractor). We present rather generic upper- and lower-bounds on the randomness complexity of property testing and study in depth the special case of testing bipartiteness in two standard property testing models.

Due to space limitations, several proofs and various other details have been omitted from this version. A full version is available from

    http://www.wisdom.weizmann.ac.il/~oded/p_ors.html

1 Introduction

Property testing [RS, GGR] is concerned with a relaxed type of decision problems; specifically, for a fixed property (resp., a set) Π, the task is to distinguish between objects that have property Π (resp., are in Π) and objects that are "far" from have property Π (resp., are "far" from any object in Π). The focus of property testing is on sublinear-time algorithms, which in particular cannot examine the entire object. Instead, these algorithms, called *testers*, may obtain bits in the representation of the object by issuing adequate queries. Indeed, in this case, the query complexity of testers becomes a measure of central interest.

For natural properties, testers of sublinear query-complexity must be randomized (see articulation in Section 2.1). This is a qualitative assertion, and the corresponding quantitative question arises naturally: for any fixed property Π and a sublinear function q, *what is the randomness-complexity of testers for Π that have query-complexity q?*

---

* This work is based on the M.Sc. thesis of the second author, which was completed under the supervision of the first author. This research was partially supported by the Israel Science Foundation (grant No. 460/05).

** Work done while Or was a graduate student at the Weizmann Institute of Science.

M. Charikar et al. (Eds.): APPROX and RANDOM 2007, LNCS 4627, pp. 509–524, 2007.

In addition to the natural appeal of the foregoing question, there are concrete reasons to care about it. Firstly, the randomness-complexity of a tester determines the length of PCPs that are constructed on top of this tester. Indeed, this was the motivation for the interest of [GS, BSVW] in reducing the randomness complexity of low-degree testing. Secondly, the randomness-complexity of a tester affects the complexity of implementing a version of this tester while utilizing a weak source of randomness. This motivation is further discussed in Section 1.2.

Indeed, the randomness-complexity of testers was considered in some prior work, starting in [GS]. This subject is the pivot of [BSVW] and the main topic of [SW]. However, all these works refer to specific (algebraic) tasks (i.e., testing low-degree polynomials and group homomorphisms). In contrast, our focus in this paper is either on general properties (see Section 1.4) or on specific combinatorial properties (see Section 1.3).

1.1   The Perspective of Average-Estimation

Property testing is a vast generalization of the task of estimating the average value of a function. Specifically, consider the task of distinguishing between functions $f : \{0, 1\}^n \to \{0, 1\}$ having average value exceeding 0.5 and functions that are ϵ-far from having this property (i.e., functions having average value below $0.5 - \epsilon$). Clearly, this task can be solved by a randomized algorithm that queries the function at $O(1/\epsilon^2)$ (random) points. This query-complexity is optimal and any algorithm achieving it, called a *sampler*, must be randomized (see [CEG]). Furthermore, a quantitative study of the randomness-complexity of samplers in terms of their query-complexity was also carried out in [CEG]. The current paper may be viewed as extending this study to the domain of general property testing.

Note that estimating the average value of a function corresponds to very restricted properties of functions. In particular, these properties are *symmetric* (i.e., are invariant under any relabeling of the inputs to the function). In contrast, most of the study of property testing refers to properties that are not symmetric (e.g., being a low-degree polynomial, monotonicity, representing a graph that has a certain graph property, etc). Furthermore, while all symmetric properties of Boolean functions are easily testable by straightforward sampling, this cannot be said about property testing in general (nor about the numerous special cases that were studied in the last decade [F, R]).

1.2   A Concrete Motivation: Using Weak Sources of Randomness

In the standard context of randomized algorithms, a concrete motivation for minimizing the randomness-complexity is provided by the exponential effect of the latter measure on the time-complexity of a possible derandomization. In contrast, in the context of property testing, derandomization is typically infeasible, because (as noted above) deterministic testers cannot have sublinear query complexity. Instead, a different motivation (advocated in [G]), becomes very relevant in this context.

We refer to the effect of the randomness-complexity on the overhead involved in implementing the tester when using only weak sources of randomness (rather than perfect ones). Specifically, we refer to the paradigm of implementing randomized algorithms by using (a single sample from) such a weak source, and trying all possible seeds to an adequate randomness extractor (see below). We shall see that the overhead created by this method is determined by the randomness-complexity of the original algorithm.

Recall that a *randomness extractor* is a function $E : \{0,1\}^s \times \{0,1\}^n \rightarrow \{0,1\}^r$ that uses an s-bit long random seed in order to transform an n-bit long (outcome of a) weak source of randomness into an r-bit long string that is almost uniformly distributed in $\{0,1\}^r$. Specifically, we consider arbitrary weak sources that are restricted (only) in the sense that, for a parameter k, no string appears as the source outcome with probability that exceeds 2^{-k}. Such sources are called (n,k)-sources (and k is called the min-entropy). Now, E is called a (k,ϵ)-extractor if for any (n,k)-source X it holds that $E(U_s, X)$ is ϵ-close to U_r, where U_m denotes the uniform distribution over m-bit strings (and the term 'close' refers to the statistical distance between the two distributions). For further details about (k,ϵ)-extractors, the interested reader is referred to Shaltiel's survey [Shal].

Next we recall the standard paradigm of implementing randomized algorithms while using sources of weak randomness. Suppose that the algorithm A has time-complexity t and randomness-complexity $r \leq t$. Recall that, typically, the analysis of algorithm A refers to what happens when A obtains its randomness from a perfect random source (i.e., for each possible input α, we consider the behavior of $A(\alpha, U_r)$, where $A(\alpha, \omega)$ denotes the output of A on input α when given randomness ω). Now, suppose that we have at our disposal only a weak source of randomness; specifically, a (n,k)-source for $n \gg k \gg r$ (e.g., $n = 10k$ and $k = 2r$). Then, using a (k,ϵ)-extractor $E : \{0,1\}^s \times \{0,1\}^n \rightarrow \{0,1\}^r$, we can transform the n-bit long outcome of the weak source into 2^s strings, each of length r, and use the resulting 2^s strings (which are "random on the average") in 2^s corresponding invocations of the algorithm A. That is, upon obtaining the outcome $x \in \{0,1\}^n$ from the source, we invoke the algorithm A for 2^s times such that in the ith invocation we provide A with randomness $E(i,x)$. The results of these 2^s invocations are processed in the natural manner. For example, if A is a decision procedure, then we output the majority vote obtained in the 2^s invocations (i.e., when given the input α, we output the majority vote of the sequence $\langle A(\alpha, E(i,x)) \rangle_{i=1,\ldots,2^s}$).

Let us consider the cost of the foregoing implementation. We assume, for simplicity, that the running-time of the randomness extractor is dominated by the running-time of A. Then, algorithm A can be implemented using a weak source, while incurring an overhead factor of 2^s. Recalling that $s > \log_2(n - k)$ and $n > k > r - s$ must hold (cf. [Shal]), it follows that for $k = n - \Omega(n)$ the aforementioned overhead is at least linear in r. On the other hand, for $n = O(k) = O(r)$ (resp., $n = \text{poly}(k) = \text{poly}(r)$) efficient randomness-extractors using $s = (1 + o(1))\log_2 n$ (resp., $s = O(\log n)$) are known; see [Shal]. This establishes our claim that the time-complexity of implementing randomized algorithms when using weak sources is related to the randomness-complexity of these algorithms. The same applies to the query complexity of testers. Specifically, for $n = O(k) = O(r)$ (resp., $n = \text{poly}(k) = \text{poly}(r)$) the query-complexity of implementing a tester is almost linear in $r \cdot q$ (resp., is $\text{poly}(r) \cdot q$), where q is the query-complexity of the original tester (which uses a perfect source of randomness).

1.3   Specific Algorithms

The motivation discussed in Section 1.2 is best illustrated by our results regarding testing bipartiteness *in the bounded-degree model* of [GR1]. Specifically, fixing a degree bound d, the task is to distinguish (N-vertex) bipartite graphs of maximum degree d from (N-vertex) graphs of maximum degree d that are ϵ-far from bipartite (for some parameter ϵ), where ϵ-far means that $\epsilon \cdot dN$ edges have to be omitted from the graph in order to yield a bipartite graph. It is easy to see that no deterministic algorithm of $o(N)$ time-complexity can solve this problem. Yet, there exists a probabilistic algorithm of time-complexity $\tilde{O}(\sqrt{N}\text{poly}(1/\epsilon))$ that solves this problem correctly (with probability $2/3$). This algorithm makes $q \overset{\text{def}}{=} \tilde{O}(\sqrt{N}\text{poly}(1/\epsilon))$ incidence-queries to the graph, and (as described in the work [GR2]) has randomness-complexity $r > q > \sqrt{N}$ (yet $r < q \cdot \log_2 N$).[1]

Let us now turn to the question of implementing the foregoing tester in a setting where we have access only to a weak source of randomness. In this case, the implementation calls for invoking the original tester $\tilde{O}(r)$ times, which yields a total running time of $\tilde{O}(r) \cdot \tilde{O}(\sqrt{N}\text{poly}(1/\epsilon)) > N$ (and the same bound holds for its query-complexity). But in such a case we better use the standard (deterministic) decision procedure for bipartiteness!

Fortunately, a randomness-efficient implementation of the original tester of [GR2] is possible. This implementation (presented in Section 3.2) has randomness-complexity $r' = \text{poly}(\epsilon^{-1} \log N)$ (rather than $r = \text{poly}(\epsilon^{-1} \log N) \cdot \sqrt{N}$). Thus, the cost of the implementation that uses a weak source of randomness is related to $r' \cdot s = \tilde{O}(\sqrt{N}\text{poly}(1/\epsilon))$, which matches the original bound (up to differences hidden in the $\tilde{O}()$ and $\text{poly}()$ notation).

The randomness-efficient implementation of the [GR2]-tester presented in Section 3.2 is based on pin-pointing the "random features" used in the original analysis, and providing an alternative implementation that satisfies the same features. In contrast, the randomness-efficient tester presented in Section 3.1 is based on new ideas.

In Section 3.1 we consider testers for graph properties *in the adjacency matrix model* of [GGR]. Specifically, we consider the task of testing bipartiteness. Recall that the tester presented in [GGR] works by selecting a random set of $\tilde{O}(\epsilon^{-2})$ vertices and inspecting the (corresponding) induced subgraph. In fact, as shown [GGR], it suffices to make $\tilde{O}(\epsilon^{-3})$ queries. A randomness-efficient implementation of the "random features" used in the original analysis, allows

---

[1] We comment that $\Omega(\sqrt{N})$ is a lower-bound on the query-complexity of any property tester of bipartiteness (in the bounded-degree model; see [GR1]).

reducing the randomness-complexity to $\tilde{O}(\epsilon^{-1}) \cdot \log_2 N$, where N denotes the number of vertices. In contrast, using an alternative approach, we present a tester of randomness-complexity $O(\log(1/\epsilon)) \cdot \log_2 N$, while maintaining a query-complexity bound of $\tilde{O}(\epsilon^{-3})$. The latter randomness-efficient tester is the main technical contribution of this work.

1.4   Generic Bounds

In contrast to the specific algorithms described in Section 1.3, we now consider quite generic lower- and upper-bounds on the randomness-complexity of property testers as a function of their query-complexity. We stress that these results do not refer to the time-complexity of the testers, which makes the lower-bounds stronger (and the upper-bound weaker).

Loosely speaking, we show that, for a wide class of properties of functions defined over a domain of size D, *the randomness-complexity of testing with q queries is essentially* $\log_2(D/q)$. Needless to say, the dependence on the query-complexity is essential, because deterministic testers of query-complexity D exist for any property. Furthermore, the randomness-complexity of any tester can be decreased by an additive term of t while increasing the query complexity by a factor of 2^t.

The lower-bounds established in Section 2.1 are exactly of the foregoing form, and they apply to two general and natural classes of properties. In particular, these lower-bounds apply to testing low-degree polynomials (cf., e.g., [BLR, RS]), locally-testable codes (cf., e.g., [GS]), testing graph properties (in both the adjacency matrix and incidence-list models, see [GGR, GR1], resp.), testing monotonicity (cf., e.g., [GGLRS]), and testing of clustering (cf., e.g., [ADPR]). The upper-bound established in Section 2.2 refers to any property but is actually of the form $\log_2 D + \log_2 \log_2 R + O(1)$ (rather than $\log_2(D/q)$), where R is the size of the range of the functions we refer to.

2   Generic Bounds

We consider testing properties of functions from D to R. Fixing a set of such functions Π, we say that a randomized oracle machine T is an ϵ-tester for Π if the following two conditions hold:

1. For every $f \in \Pi$ it holds that $\Pr[T^f = 1] \geq 2/3$.
2. For every f that is ϵ-far from Π it holds that $\Pr[T^f = 1] \leq 1/3$, where f is ϵ-far from Π if for every $g \in \Pi$ it holds that $\Pr_{x \in D}[f(x) \neq g(x)] > \epsilon$.

In case the first condition holds with probability 1, we say that T has one-sided error. The query and randomness complexities of T are defined in the natural manner. A tester is called non-adaptive if it determines its queries based solely on its internal coin-tosses (and independently of the answers to prior queries).

Note that we have defined property testers for finite properties and a fixed value of the proximity parameter ϵ. The definition may be extended to infinite properties and varying ϵ, by providing the tester with $|D|, |R|$ and ϵ as inputs (and assuming $D = [D]$).[2]

2.1 Lower Bounds

We provide lower-bounds on the randomness complexity of testing two general classes of properties.

Strongly Evasive Properties. We first consider properties that are "strongly evasive" in the sense that determining the values of some function at a constant fraction of the domain leaves the promise problem (of distinguishing between yes-instances and "far from yes"-instances) undetermined.[3] That is, for fixed parameters ϵ and ρ, the property Π is called strongly evasive if there exists a function $f : D \to R$ such that for every $D' \subset D$ of density ρ (i.e., $D' = \rho \cdot D$), there exists $f_1 \in \Pi$ and $f_0 : D \to R$ that is ϵ-far from Π such that for every $x \in D'$ it holds that $f_1(x) = f_0(x) = f(x)$. Many natural properties are strongly evasive (with respect to various pairs of parameters); see examples below. The following result can be easily proved by extending a similar result regarding samplers (which is presented in [CEG]).

Theorem 1. *Let Π be strongly evasive with respect to ϵ and ρ. Then any ϵ-tester for Π that has query complexity q, must have randomness complexity greater than $\log_2(\rho|D|/q)$.*

Some Applications. Many graph properties are strongly evasive, but since such properties will be at the focus of the next subsection, we mention first a few examples that refer to different types of properties.

1. *Multi-variate polynomial.* For every m and d, the set of m-variate polynomial of total degree d (over a finite field F) is strongly evasive with respect to density $\rho = 1/2$ and distance $\epsilon = 1/4$.
2. *Codes of linear distance.* A binary code $C \subset \{0,1\}^n$ of distance $d = \Omega(n)$, is viewed as a set of functions of the form $f : [n] \to \{0,1\}$, where each function corresponds to a codeword. The set of these codewords is strongly evasive with respect to density $\rho = 1 - (d/2n)$ and distance $\epsilon = d/2n$.
3. *Monotone functions.* A function $f : \{0,1\}^n \to \{0,1\}$ is said to be monotone if $f(x) \leq f(y)$ for every $x \prec y$, where $\prec$ denotes the natural partial order among strings (i.e., $x_1 \cdots x_n \prec y_1 \cdots y_n$ if $x_i \leq y_i$ for every i and $x_i < y_i$ for some i). The set of monotone functions is strongly evasive with respect to density $\rho = 1/4$ and distance $\epsilon = 1/4$.

Turning back to graph properties, we focus on the bounded incidence lists model (of [GR1]) because the results of the following subsection do not apply to it. We mention a few properties of bounded-degree graphs that are strongly evasive in

---

[2] Occasionally, we shall assume that $\epsilon \geq |D|^{-1}$; otherwise, ϵ-testers coincide with standard decision procedures.

[3] This notion of "strongly evasive" is incomparable to the standard definition of evasiveness (cf. [LY]).

the (bounded) incidence lists model. Examples include connectivity, planarity, bipartiteness, and being Eulerian (or Hamiltonian).

Relabeling-Invariant Properties. We now consider properties that are invariant under some "nice" relabeling of D. Specifically, for any set S_D of permutations over D, we say that the property Π is S_D-invariant if for every $f : D \to R$ and every $\pi \in S_D$ it holds that $f \in \Pi$ if and only $(f \circ \pi) \in \Pi$, where $(f \circ \pi)(x) = f(\pi(x))$. We consider only sets S_D that correspond to a transitive group of permutations over D; that is, S_D is permutation group and for every $x, y \in D$ there exists a permutation $\pi \in S_D$ such that $\pi(x) = y$. Needless to say, the set of all permutations is a transitive group of permutations, but so are also many other permutation groups (e.g., the group of all cyclic permutations).

Theorem 2. *Let S_D be a transitive group of permutations over D, and Π be a non-empty and S_D-invariant property of functions from D to R. Suppose that, for some $\sigma \in R$, the all-σ function is 2ϵ-far from Π. Then any non-adaptive ϵ-tester for Π that has query complexity q, must have randomness complexity at least $\log_2(|D|/q) - 1$.*

Main Application. As hinted before, the most appealing application of Theorem 2 is to testing graph properties in the adjacency matrix model (of [GGR]). In this model, N-vertex graphs are represented by Boolean functions defined over $[N] \times [N]$. For technical reasons, we prefer to represent such graphs as Boolean functions defined over the set of the $\binom{N}{2}$ (unordered) vertex-pairs, which is actually more natural (as well as non-redundant). Note that the set of all permutations over $[N]$ induces a transitive group of permutations over these pairs, where the permutation $\pi : [N] \to [N]$ induces a permutation that maps pairs of the form $\{i, j\}$ to $\{\pi(i), \pi(j)\}$. Indeed, any graph property is invariant under this group, and Theorem 2 can be applied whenever either the empty graph or the complete graph is far from the property. We note that all the graph properties considered in [GGR] fall into the latter category (and that the testers of [GGR] are all non-adaptive).[4]

Other Applications. We note that any property that refers to sets of objects (e.g., sets of points as in [ADPR]) is invariant under the group of all permutations. Another application domain consists of matrix-properties that are preserved under row and column permutations.

Generalizations. Theorem 2 can be generalized to properties that are S_D-invariant under a set of permutations that is "sufficiently mixing" in the sense that a permutation selected uniformly in S_D maps each element of the domain to a distribution that has high min-entropy. For example, for a parameter $\alpha \geq 1$, it suffices that for every $x \in D$ and $y \in R$ it holds that

---

[4] Note that q adaptive Boolean queries can always be replaced by 2^q non-adaptive Boolean queries. We warn that the more query-efficient transformation provided in [GT] is inapplicable here, because this transformation does not preserve the randomness-complexity.

$\Pr_{\pi \in S_D}[\pi(x) = y] \leq \alpha/|D|$. In this case, we shall prove that $|D'| > |D|/2\alpha$, and a lower-bound of $\log_2(|D|/q) - \log_2(2\alpha)$ on the randomness-complexity follows. A different generalization is obtained by replacing σ with a set of values $S \subset R$ and referring to properties for which every function $f : D \rightarrow S$ is 2ϵ-far from the property.

2.2 Upper Bounds

We start with a totally generic bound, and later focus on graph properties.

A generic bound. Recall that we refer to properties of functions from D to R. The following result can be easily proved by extending a similar result regarding samplers (presented in [CEG]), which in turn is proved using well-known techniques.

Theorem 3. *If Π has an ϵ-tester that makes q queries then it has an ϵ-tester that makes $O(q)$ queries and tosses $\log_2 |D| + \log_2 \log_2 |R| + O(1)$ coins. Furthermore, one-sided error and/or non-adaptivity are preserved.*

Corollary. Applying Theorem 3 to testers of graph properties in the adjacency matrix model (of [GGR]), we conclude that *if a property of N-vertex graphs is ϵ-testable using q queries then it has an ϵ-tester that makes $O(q)$ queries and tosses $2 \log_2 N + O(1)$ coins.* We further discuss this model in the following subsection.

Bounds for Canonical Testers of Graph Properties. The proof of Theorem 3 shows that for every tester T (of randomness complexity r) there exists a small set of coin-sequences Ω_T ($\subset \{0,1\}^r$) that is essentially as good as the original set of coin-sequences used by this tester (i.e., $\{0,1\}^r$). This raises the question of whether there may exists a universal set Ω that is good for all testers (of randomness complexity r). Needless to say, the latter formulation is too general and is doomed to yield a negative answer (e.g., by considering, for any Ω, a pathological tester that behaves badly when fed with any sequence in Ω). Still such universal sets may exist for naturally restricted classes of testers.

One adequate class of testers was suggested in [GT], and it refers to testing graph properties in the adjacency matrix model. A canonical ϵ-tester for a property Π of N-vertex graphs is determined by an integer k and a property Π' of k-vertex graphs. Such a tester, sometimes referred to as k-canonical, selects uniformly a set of k vertices in the input graph G and accepts G if and only if the corresponding induced (k-vertex) subgraph has the property Π'. It was shown in [GT] that if Π is ϵ-testable with query complexity q then Π has a k-canonical ϵ-tester with $k = O(q)$. Thus, it is natural to consider the notion of a "universal set" of k-subsets of $[N]$ that is good for all k-canonical testers.

Definition 4. *A set $\Omega \subseteq \{S \subset [N] : |S| = k\}$ is called (ϵ, k)-universal if for every property Π of N-vertex graphs and for every k-canonical ϵ-tester for Π, denoted T, the following holds:*

1. If G has property Π, then $\text{Pr}_{\omega \in \Omega}[T^G(\omega) = 1] \geq 3/5$, where $T^G(\omega)$ denotes the execution of T when given the coin-sequence ω and oracle access to G.
2. If G is ϵ-far from property Π, then $\text{Pr}_{\omega \in \Omega}[T^G(\omega) = 1] \leq 2/5$.

Using an (ϵ, k)-universal set, we can reduce the randomness complexity of any k-canonical ϵ-tester T by selecting uniformly $\omega \in \Omega$ and emulating $T(\omega)$. The residual oracle machine, denoted T', is essentially an ϵ-tester for the same property, except that T' may err with probability at most $2/5$ (rather than $1/3$). Needless to say, T' has randomness complexity $\log_2 |\Omega|$ and query complexity $\binom{k}{2}$. Furthermore, T' preserves the possible one-sided error of T.

Clearly, the set of all k-subsets is (ϵ, k)-universal, because using this set coincides with the definition of a k-canonical ϵ-tester. We seek (ϵ, k)-universal sets that are much smaller; specifically, by prior results we may hope to have (ϵ, k)-universal sets of size $O(N^2)$. By extending the proof of Theorem 3, we can prove the following result.

Theorem 5. There exist (ϵ, k)-universal sets (for $[N]$) having size $2^{k^2} \cdot N^2$.

The randomness complexity of the derived ϵ-tester is $k^2 + 2 \log_2 N$. For relatively small k and in particular for k that only depends on ϵ (as in [GGR, AFKS, AFNS]), this is much smaller than the randomness complexity of the k-canonical ϵ-tester (i.e., $k \log_2 N$).

Open Problems. Can the upper-bound of Theorem 5 be improved; in particular, do there exist (ϵ, k)-universal sets (of subsets of $[N]$) having size $O(\text{poly}(k) \cdot N^2)$ or even $O(N^2)$? Can universal sets of small size (e.g., as in Theorem 5) be efficiently constructed?

3 Specific Algorithms: The Case of Bipartiteness

We consider two standard models for testing graph properties: the adjacency matrix model (introduced in [GGR]) and the bounded-degree model (introduced in [GR1]). We focus on the problem of testing bipartiteness in these models. Further details and additional testers are provided in [Shef]. We make extensive use of randomness-efficient hitters.[5]

3.1 In the Adjacency Matrix Model

In the adjacency matrix model an N-vertex graph $G = (V, E)$ is represented by the Boolean function $g : [N] \times [N] \to \{0, 1\}$ such that $g(u, v) = 1$ if and only if u and v are adjacent in G (i.e., $\{u, v\} \in E$). In this section we present a randomness-efficient Bipartite Tester for graphs in the adjacency matrix model.

---

[5] A hitter is a randomized algorithm that, on input parameters n, ϵ and δ, outputs a list of strings such that, for *any* function $f : \{0, 1\}^n \to \{0, 1\}$ that satisfies $|f^{-1}(1)| \geq \epsilon 2^n$, with probability at least $1 - \delta$ it holds that the list contains an f-preimage of 1 (i.e., an element of $f^{-1}(1)$).

This tester is strongly influenced by the tester of [GGR], but differs from it in significant ways. Still, it is instructive to start with a description of the tester of [GGR].

The Tester of [GGR]. Essentially, the bipartite tester of [GGR] selects a random set of $\tilde{O}(\epsilon^{-2})$ vertices, inspects the subgraph of G induces by this set, and accepts if and only if this induced subgraph is bipartite. The analysis in [GGR] actually refers to the following description, which also has a lower query-complexity.

Algorithm 6. *On input parameters N and ϵ, and oracle access to an adjacency predicate of an N-vertex graph, $G = (V, E)$, proceed as follows:*

1. *Uniformly select a sample U of $\tilde{O}(\epsilon^{-1})$ vertices.*
2. *Uniformly select a sample S of $\tilde{O}(\epsilon^{-2})$ vertex-pairs.*
3. *For each $u \in U$ and $(v_1, v_2) \in S$, check whether $\{u, v_1\}, \{u, v_2\}$ and $\{v_1, v_2\}$ are edges.*
4. *Accept if and only if the subgraph viewed in Step 3 is bipartite.*

Clearly, this algorithm never rejects a bipartite graph, and thus its analysis focuses on the case that G is ϵ-far from being bipartite. One key observation is that each 2-partition, (U_1, U_2), of U induces a 2-partition of the entire graph in which all neighbors of U_1 are on one side and all the other vertices are on the other side. A pair of vertices (v_1, v_2) detects that the latter partition is not a valid 2-coloring of G if there exists $u_1, u_2 \in U_1$ (resp., $u_1, u_2 \in U_2$) such that $\{u_1, v_1\}, \{v_1, v_2\}$ and $\{v_2, u_2\}$ are all edges of G. In such a case, we call the pair (v_1, v_2) a witness against (U_1, U_2). The analysis in [GGR] shows that if G is ϵ-far from being bipartite then, with high probability, for every 2-partition of U there exists a pair in S that is a witness against this 2-partition. Let us briefly recall how this is done.

The first step is proving that, with high probability (say, with probability at least $5/6$), the set U dominates[6] all but an $\epsilon/8$ fraction of the vertices of G that have degree at least $\epsilon N/8$. This step is quite straightforward. The next step is proving that this implies that *for every 2-partition of U there exists at least $\epsilon N^2/2$ (ordered) vertex-pairs that are each a witness against this 2-partition.* The implication is proved by confronting the following two facts:

1. Since G is ϵ-far from being bipartite, the 2-partition of V induced by any 2-partition of U has at least ϵN^2 (ordered) vertex-pairs that reside on the same side of the partition and yet are connected by an edge.
2. The number of (ordered) vertex-pairs (v_1, v_2) such that $\{v_1, v_2\} \in E$ but either v_1 or v_2 is not dominated by U is at most $\epsilon N^2/2$, because each low-degree vertex contributes at most $\epsilon N/4$ such (ordered) pairs and there are at most $\epsilon N/8$ high-degree vertices that are not dominated by U.

---

[6] We say that a set U **dominates** a vertex v in the graph G if v is adjacent to some vertex in U.

Having established the existence of at least $\epsilon N^2/2$ vertex-pairs that constitute a witness against any fixed 2-partition of U, it is clear that each random pair of vertices will be a witness with probability at least $\epsilon/2$, and selecting enough random pairs will do the job. The point, however, is that we need to rule out each of the $2^{|U|}$ possible 2-partitions of U. Thus, the number of selected pairs is set such that the probability that we do not find a witness against any specific 2-partition is smaller than $2^{-|U|}$. Indeed, setting $|S| = O(|U|/\epsilon)$ will do. This completes our review of [GGR].

As stated in Section 1.3, the problem with the foregoing approach is that it is impossible to implement it using randomness-complexity below $|U|$, which in turn is $\Omega(\epsilon^{-1})$. However, our aim is to obtain randomness-complexity that is linearly related to $O(\log(1/\epsilon))$.

A Warm-Up: Randomness-Efficient Tester of Query Complexity $\tilde{O}(\epsilon^{-4})$. A closer look at the foregoing argument reveals that a pair (v_1, v_2) such that $\{u_1, v_1\}, \{v_1, v_2\}$ and $\{v_2, u_2\}$ are all edges of G is not merely a witness against a *specific* 2-partition of U that places u_1 and u_2 on the same side. It is actually a witness against *any* 2-partition of U that places u_1 and u_2 on the same side. Viewed from a different perspective, such a pair (v_1, v_2) imposes a constraint on the "relevant" 2-partition of U; the constraint being that u_1 and u_2 should not be placed on the same side. It will be useful to consider the graph of these constraints, which has the vertex-set U and edges between each pair of vertices to which such a constraint is applied (i.e., there is an edge between u_1 and u_2 if there exists a pair $(v_1, v_2) \in V \times V$ that imposes a constraint on the pair (u_1, u_2)). Indeed, the 2-partitions of U that satisfy the set of these constraints are exactly the 2-colorings of this auxiliary graph.

The foregoing perspective suggests that it may be useful to try to accumulate constraints. At the very extreme, the graph of constraints will not be bipartite, which definitely allows us to reject (because it indicates that there are witnesses against each 2-partition of U). Discarding this case, we consider another extreme case in which the graph of constraints is connected, leaving us with a single allowed 2-partition of U (i.e., a single 2-coloring of the constraint graph), which can be checked as in Algorithm 6. The point, however, is that in this case it will suffice to set $|S| = O(\epsilon^{-1})$ and more importantly to have a sample that rules out the remaining partition with constant probability (rather than with probability $2^{-|U|}$). This opens the door to a randomness-efficient implementation.

But what if the graph of constraints that we found is not connected? Unless this event is due to sheer lack of luck, it indicates that there are few pairs in $V \times V$ that impose constraints regarding vertex-pairs in $U \times U$ that are in different connected components of the constraint graph. This implies that, for every 2-partition of U that is consistent with the constraint graph (i.e., every 2-coloring of this graph), there are many pairs in $V \times V$ that constitute a witness against the 2-partition of some of the connected components. That is, each such pair imposes a constraint that refers to vertices that reside in the same connected component, and furthermore this constraint contradicts the constraints that are already present regarding this connected component.

Needless to say, for the foregoing to work, we should determine adequate thresholds for the notion of "*few* pairs in $V \times V$ that impose a constraint regarding vertex-pairs" (in $U \times U$). Let us start by spelling out the notion of imposing (or rather forcing) a constraint. We say that the pair $(v_1, v_2) \in V \times V$ **constrains** the pair $(u_1, u_2) \in U \times U$ if $\{u_1, v_1\}, \{v_1, v_2\}$ and $\{v_2, u_2\}$ are all edges of G. Next, we say that a pair $(u_1, u_2) \in U \times U$ is ρ-**constrained** if there are at least $\rho \cdot N^2$ vertex-pairs in $V \times V$ that constrain (u_1, u_2). Leaving ρ unspecified for a moment, we make the following observations:

1. Using a sample of $O(\rho^{-1} \cdot \log |U|)$ vertex-pairs in $V \times V$, with high probability, it holds that *for every ρ-constrained pair $(u_1, u_2) \in U \times U$, the sample contains a pair that constrains (u_1, u_2)*. This holds even if the sample is generated using a randomness-efficient hitter (which hits any set of density ρ with probability at least $1 - (|U|^{-2}/10)$, using randomness-complexity $O(\log |V| + \log |U|) = O(\log |V|))$. The point is that there are at most $|U|^2$ relevant pairs (i.e., pairs that are ρ-constrained), and we may apply a Union Bound as long as we fail on each such pair with probability at most $|U|^{-2}/10$ (or so).

2. Consider the graph $G_{U,\rho}$ consisting of the vertex-set U and edges corresponding to the ρ-constrained pairs of vertices. Then, the number of vertex-pairs in $V \times V$ that constrain some pair of vertices (in U) that does not belong to the same connected component of $G_{U,\rho}$ is at most $|U|^2 \cdot \rho N^2$.

   Recall that if G is ϵ-far from bipartite and U is good (i.e., U dominates almost all high-degree vertices) then, for every 2-partition of U, there are at least $\epsilon N^2/2$ pairs that constrain some pair of vertices that are on the same side of this 2-partition. It follows that at least $((\epsilon/2) - |U|^2\rho) \cdot N^2$ of these pairs constrain pairs that are in the same connected component of $G_{U,\rho}$. Setting $\rho = \epsilon/(4|U|^2)$, we need to hit a set of density $\epsilon/4$, which is easy to do using a randomness-efficient hitter.

This analysis lead to an algorithm that resembles Algorithm 6, except that it uses a secondary sample S that has different features than in the original version. In Algorithm 6 the set S had to hit any fix set of density $\epsilon/2$ with probability at least $1 - 2^{-|U|}$. Here the set S needs to hit any fix set of density $\rho = \epsilon/(4|U|^2) < \epsilon^{-3}$ with probability at least $1 - (|U|^{-2}/10)$. Thus, while in Algorithm 6 we used $|S| = O(|U|/\epsilon)$ but generating the set S required at least $|U|$ random bits, here $|S| = O(|U|^2/\epsilon) = \tilde{O}(\epsilon^{-3})$ but generating the set S can be done using $O(\log N)$ random bits. (The set U is generated with the same aim as in Algorithm 6; that is, hitting a set of density ϵ with probability at least $1 - \epsilon^{-1}$. Such a set can be generated using $O(\log N)$ random bits).

Thus, we obtain a (computational efficient) ϵ-tester with randomness-complexity $O(\log N)$ and query-complexity $O(|U| \cdot |S|) = \tilde{O}(\epsilon^{-4})$. Our aim in the next section is to reduce the query-complexity to $\tilde{O}(\epsilon^{-3})$ while essentially maintaining the randomness-complexity.

The Actual Algorithm: Randomness-Efficient Tester of Query Complexity $\tilde{O}(\epsilon^{-3})$. The query-complexity bottleneck in the previous subsection is due to the size of S, which in turn needs to hit sets of density $\rho = O(\epsilon^3)$. Our improvement will follow by using a larger value of the threshold ρ (essentially $\rho = O(\epsilon^2)$). Recall that in the previous subsection we used $\rho = O(\epsilon^3)$ in order to bound the total number of pairs that constrain pairs that are not ρ-constrained. Thus, using $\rho = O(\epsilon^3)$ seems inherent to an analysis that refers to each pair separately, and indeed we shall deviate from that paradigm in this section.

The planned deviation is quite natural. After all, we not not care about having specific edges in our constraint graph, but rather care about the connected components of that graph. For example, looking at any vertex $u \in U$, any pair in $V \times V$ that constrains any pair (u, u'), where $u' \in U \setminus \{u\}$, increases the connected component in which u resides. That is, let $\gamma(u_1, u_2)$ denote the fraction of vertex-pairs in $V \times V$ that constrain (u_1, u_2), and recall that a pair (u_1, u_2) was called ρ-constrained if $\gamma(u_1, u_2) \geq \rho$. Thus, we (tentatively) say that $u \in U$ is ρ-constrained if $\sum_{u' \in U \setminus \{u\}} \gamma(u, u') \geq \rho$. Let us now see what happens.

1. Using a sample of $O(\rho^{-1} \cdot \log |U|)$ vertex-pairs in $V \times V$, with high probability, it holds that for every ρ-constrained vertex $u \in U$, the sample contains a pair that constrains (u, u'), for some $u' \in U \setminus \{u\}$. Again, this holds even if the sample is generated using a randomness-efficient hitter.
2. The number of vertex-pairs in $V \times V$ that constrain some pair of vertices $(u_1, u_2) \in U \times U$ such that either u_1 or u_2 is not ρ-constrained is at most $2|U| \cdot \rho N^2$. This means that we can ignore such vertex-pairs (in $V \times V$) even when setting $\rho = O(\epsilon/|U|)$ or so.

Thus, taking a sample S' as in Item 1, will result in having a constraint graph $G_{U,S'}$ in which each ρ-constrained vertex resides in non-singleton connected components. In particular, the number of non-singleton connected components is at most $|U|/2$.

Note, however, that unlike in the previous subsection, the foregoing facts do not yield an upper-bound on the number of vertex-pairs in $V \times V$ that constrain some pair of vertices (in U) that does not belong to the same connected component of $G_{U,S'}$. Loosely speaking, we shall iterate the same process on the non-singleton connected components of $G_{U,S'}$, while recalling that the only vertices that form singleton connected components in $G_{U,S'}$ are not ρ-constrained (and thus can be ignored). This suggests an iterative process, which will halt after at most $\log_2 |U|$ iterations in a situation analogous to having no ρ-constrained vertices. At this point we may proceed with a final sample of pairs that, with high probability, will yield a constraint that conflicts with the existing ones.

Clarifying the foregoing iterative process requires generalizing the notion of ρ-constrained vertices such that it will apply to the connected components determined in the previous iteration. Consider a partition of U, denoted $\overline{U} = (U^{(0)}, U^{(1)}, ..., U^{(k)})$, where $U^{(0)}$ may be empty and k may equal 0, but for every $i \in [k]$ it holds that $U^{(i)} \neq \emptyset$. In the first iteration, we use $\overline{U} = (\emptyset, \{u_1\}, ..., \{u_t\})$, where $U = \{u_1, ..., u_t\}$. In later iterations, $U^{(1)}, ..., U^{(k)}$ will correspond to connected components of the current constraint graph and $U^{(0)}$ will contain vertices that were cast aside at some point.

Definition 7 (being constrained w.r.t a partition). *For $i \in \{0, 1, ..., k\}$, we say that $u \in U^{(i)}$ is ρ-constrained w.r.t $\overline{U}$ if $\sum_{u' \in U'} \gamma(u, u') \geq \rho$, where $U' = \cup_{j \in [k] \setminus \{i\}} U^{(j)}$. Recall that $\gamma(u_1, u_2)$ denote the fraction of vertex-pairs in $V \times V$ that constrain (u_1, u_2), where the pair $(v_1, v_2) \in V \times V$ constrains the pair $(u_1, u_2) \in U \times U$ if $\{u_1, v_1\}, \{v_1, v_2\}$ and $\{v_2, u_2\}$ are all edges of G.*

We stress that the foregoing sum does not include vertices in either $U^{(0)}$ or $U^{(i)}$. Our analysis will refer to the following algorithm, which *can be implemented within randomness-complexity $O(\log(1/\epsilon)) \cdot \log_2 N$ and query-complexity $\tilde{O}(\epsilon^{-3})$.*

Algorithm 8 (The Bipartite Tester, revised)

1. Select a sample U of $\tilde{O}(\epsilon^{-1})$ vertices by using a hitter that hits any set of density $\epsilon/8$ with probability at least $1 - (\epsilon/100)$.
2. For $i = 1, ..., \ell + 1$, where $\ell = \log_2 |U|$, select a sample S_i of $\tilde{O}(\epsilon^{-2})$ vertex-pairs by using a hitter that hits any set of density $\rho = \epsilon/\tilde{O}(|U|)$ with probability at least $1 - \tilde{O}(|U|)^{-1}$. (This hitter has randomness-complexity $O(\log N + \log |U|) = O(\log N)$.) Let $S = \cup_{i=1}^{\ell+1} S_i$.
3. For each $u \in U$ and $(v_1, v_2) \in S$, check whether $\{u, v_1\}, \{u, v_2\}$ and $\{v_1, v_2\}$ are edges.
4. Accept if and only if the subgraph viewed in Step 3 is bipartite.

Needless to say, the peculiar way in which S is selected is aimed to support the analysis.

Lemma 9. *If G is ϵ-far from being bipartite then Algorithm 8 rejects with probability at least $2/3$.*

Proof Outline: We may assume that U is good in the sense that it dominates all but $\epsilon N/8$ of the vertices that have degree at least $\epsilon N/8$. As argued above (and shown in [GGR]), there are at most $\epsilon N^2/2$ vertex pairs that have an endpoint that is not dominated by $U = \{u_1, ..., u_t\}$. Starting with $\overline{U} = (\emptyset, \{u_1\}, ..., \{u_t\})$, the proof proceeds in iterations while establishing that in each iteration one of the following two events occur:

1. There are $\Omega(\epsilon N^2)$ vertex pairs that form constraints that contradicts the existing constraints. In this case, with very high probability, the algorithm will select such a pair and will reject (because the subgraph that it sees is not 2-colorable).
2. There exist ρ-constrained vertices with respect to the current partition $\overline{U} = (U^{(0)}, U^{(1)}, ..., U^{(k)})$, where $U^{(1)}, ..., U^{(k)}$ are connected components of the current constraint graph and $U^{(0)}$ contains vertices that were cast aside in previous iterations. It is also shown that ρ-constrained (w.r.t $\overline{U}$) vertices cannot be in $U^{(0)}$. In this case, with very high probability, the algorithm will find new constraints and in particular it will find such a constraint between every ρ-constrained (w.r.t $\overline{U}$) vertex and some vertex that is in one of the other k connected components.

Noting that the second case (i.e., Case 2) becomes impossible once $k = 1$ holds (in the current iteration), we conclude that at this point the algorithm must reject due to the first case (i.e., Case 1).

Open Problem. Needless to say, we are aware of the Bipartite Tester of [AK], which has better query-complexity than the tester of [GGR] (as well as ours). Specifically, the query-complexity of the tester of [AK] is $\tilde{O}(\epsilon^{-2})$ rather than $\tilde{O}(\epsilon^{-3})$. Theorem 3 implies that the tester of [AK] has a randomness-efficient implementation, but it does not provide an explicit one. We conjecture that there exists a randomness-efficient bipartite tester that has query-complexity $\tilde{O}(\epsilon^{-2})$ and time-complexity $\text{poly}(\epsilon^{-1} \log N)$.

3.2 In the Bounded-Degree Model

The bounded-degree model refers to a fixed degree bound, denoted d. An N-vertex graph $G = (V, E)$ (of maximum degree d) is represented in this model by a function $g : [N] \times [d] \to \{0, 1, ..., N\}$ such that $g(v, i) = u \in [N]$ if u is the ith neighbor of v and $g(v, i) = 0$ if v has less than i neighbors. In this section we outline a randomness-efficient implementation of the Bipartite Tester of [GR2], which refers to the bounded-degree model.

Recall that the tester of [GR2] proceeds in $T = O(1/\epsilon)$ iterations, where in each iteration a random start vertex, denoted s, is selected (uniformly in V), and $K \overset{\text{def}}{=} \text{poly}((\log N)/\epsilon) \cdot \sqrt{N}$ random walks starting from vertex s are performed, where each walk is of length $L \overset{\text{def}}{=} \text{poly}((\log N)/\epsilon)$. Focusing on a single iteration, we observe that the analysis of [GR2] refers to events that correspond to pairs of paths, and that it only relies on the variances and covariances of random variables corresponding to these events. Thus, the original analysis applies also when the walks are four-wise independent (so that pairs of pairs of paths are "covered"). Consequently, each iteration can be implemented using a randomness-efficient construction of K four-wise independent random strings, each specifying a random walk of length L (i.e., each being a string of length $L \log_2 d$). It follows that a single iteration can be implemented using $(\log_2 N) + 4 \cdot L \log_2 d = \text{poly}((\log N)/\epsilon)$ random coins (rather than $(\log_2 N) + K \cdot L \log_2 d = \Omega(\sqrt{N})$ random coins).

References

[ADPR] Alon, N., Dar, S., Parnas, M., Ron, D.: Testing of Clustering. SIDMA 16(3), 393–417 (2003)

[AFKS] Alon, N., Fischer, E., Krivelevich, M., Szegedy, M.: Efficient testing of large graphs. In: 40th FOCS, 1999, pp. 645–655 (1999)

[AFNS] Alon, N., Fischer, E., Newman, I., Shapira, A.: A Combinatorial Characterization of the Testable Graph Properties: It's All About Regularity. In: 38th STOC, 2006, pp. 251–260 (2006)

[AK] Alon, N., Krivelevich, M.: Testing k-Colorability. SIDMA 15(2), 211–227 (2002)

524   O. Goldreich and O. Sheffet

[BSVW]   Ben-Sasson, E., Sudan, M., Vadhan, S., Wigderson, A.: Randomness-
          efficient low degree tests and short PCPs via epsilon-biased sets. In: Proc.
          35th STOC, pp. 612–621 (June 2003)
[BLR]   Blum, M., Luby, M., Rubinfeld, R.: Self-testing/correcting with applications
          to numerical problems. JCSS 47, 549–595 (1993)
[CEG]   Canetti, R., Even, G., Goldreich, O.: Lower Bounds for Sampling Algo-
          rithms for Estimating the Average. IPL 53, 17–25 (1995)
[F]   Fischer, E.: The art of uninformed decisions: A primer to property testing.
          Bulletin of the EATCS 75, 97–126 (2001)
[G]   Goldreich, O.: Another motivation for reducing the randomness complexity
          of algorithms. Position paper. ECCC (2006)
[GGR]   Goldreich, O., Goldwasser, S., Ron, D.: Property testing and its connection
          to learning and approximation. JACM, 653–750 (July 1998)
[GGLRS]   Goldreich, O., Goldwasser, S., Lehman, E., Ron, D., Samorodnitsky, A.:
          Testing Monotonicity. Combinatorica 20(3), 301–337 (2000)
[GR1]   Goldreich, O., Ron, D.: Property testing in bounded degree graphs. Algo-
          rithmica, 302–343 (2002)
[GR2]   Goldreich, O., Ron, D.: A sublinear bipartite tester for bounded degree
          graphs. Combinatorica 19(3), 335–373 (1999)
[GS]   Goldreich, O., Sudan, M.: Locally testable codes and PCPs of almost linear
          length. JACM 53(4), 558–655 (2006)
[GT]   Goldreich, O., Trevisan, L.: Three theorems regarding testing graph prop-
          erties. Random Structures and Algorithms 23(1), 23–57 (2003)
[LY]   Lovász, L., Young, N.: Lecture Notes on Evasiveness of Graph Properties.
          In: TR-317-91, Computer Science Dept. Princeton University (1991)
[R]   Ron, D.: Property testing. In: Handbook on Randomization, pp. 597–649
          (2001)
[RS]   Rubinfeld, R., Sudan, M.: Robust characterization of polynomials with ap-
          plications to program testing. SICOMP 25(2), 252–271 (1996)
[Shal]   Shaltiel, R.: Recent Developments in Explicit Constructions of Extractors.
          Bulletin of the EATCS 77, 67–95 (2002)
[Shef]   Sheffet, O.: Reducing the Randomness Complexity of Property Testing,
          with an Emphasis on Testing Bipartiteness. M.Sc. Thesis, Weizmann Insti-
          tute of Science (December 2006) Available from
          http://www.wisdom.weizmann.ac.il/~oded/p_ors.html
[SW]   Shpilka, A., Wigderson, A.: Derandomizing Homomorphism Testing in Gen-
          eral Groups. SICOMP, Vol. 36-4, pp.1215–1230, 2006

On the Benefits of Adaptivity in Property Testing of Dense Graphs

Mira Gonen and Dana Ron*

School of Electrical Engineering, Tel Aviv University, Tel Aviv Israel
gonenmir@post.tau.ac.il, danar@eng.tau.ac.il

Abstract. We consider the question of whether adaptivity can improve the complexity of property testing algorithms in the *dense graphs* model. It is known that there can be at most a quadratic gap between adaptive and non-adaptive testers in this model, but it was not known whether any gap indeed exists. In this work we reveal such a gap.

Specifically, we focus on the well studied property of *bipartiteness*. Bogdanov and Trevisan (*IEEE Symposium on Computational Complexity, 2004*) proved a lower bound of $\Omega(1/\epsilon^2)$ on the query complexity of *non-adaptive* testing algorithms for bipartiteness. This lower bound holds for graphs with maximum degree $O(\epsilon n)$. Our main result is an *adaptive* testing algorithm for bipartiteness of graphs with maximum degree $O(\epsilon n)$ whose query complexity is[1] $\tilde{O}(1/\epsilon^{3/2})$. A slightly modified version of our algorithm can be used to test the combined property of being bipartite and having maximum degree $O(\epsilon n)$. Thus we demonstrate that adaptive testers are stronger than non-adaptive testers in the dense graphs model.

We note that the upper bound we obtain is tight up-to polylogarithmic factors, in view of the $\Omega(1/\epsilon^{3/2})$ lower bound of Bogdanov and Trevisan for *adaptive* testers. In addition we show that $\tilde{O}(1/\epsilon^{3/2})$ queries also suffice when (almost) all vertices have degree $\Omega(\sqrt{\epsilon} \cdot n)$. In this case adaptivity is not necessary.

1 Introduction

We consider the question of whether adaptivity can benefit property testing algorithms in the *dense graphs model* [10]. In this model a testing algorithm (tester) is given a distance parameter $\epsilon > 0$ and query access to the $n \times n$ adjacency matrix of a graph $G = (V, E)$ where $n = |V|$ (that is, the tester can perform *vertex-pair queries*). For a predetermined property $\mathcal{P}$ (e.g., bipartiteness), the tester should distinguish, with high constant probability, between the case that G has the property $\mathcal{P}$ and the case that more than ϵn^2 edges should be added to or removed from G in order to make it have the property. In the latter case

---

* This work was supported by the Israel Science Foundation (grant number 89/05).
[1] The notation $\tilde{O}(g(k))$ for a function g of a parameter k means $O(g(k) \cdot \text{polylog}(g(k)))$ where $\text{polylog}(g(k)) = \log^c(g(k))$ for some constant c.

M. Charikar et al. (Eds.): APPROX and RANDOM 2007, LNCS 4627, pp. 525–539, 2007.
© Springer-Verlag Berlin Heidelberg 2007

we say that the graph is ϵ-*far* from having $\mathcal{P}$. A tester is *adaptive* if its queries depend on answers to previously asked queries.

It is known that there is at most a quadratic gap in the query complexity between adaptive and non-adaptive testers in the dense graphs model [1,13]. Namely, if there exists any (possibly adaptive) tester for a particular property that has query complexity $q(n, \epsilon)$, then there exists a non-adaptive tester with query complexity $O((q(n, \epsilon))^2)$. This non-adaptive tester, referred to as *canonical*, takes a uniformly selected sample of $O(q(n, \epsilon))$ vertices and queries all pairs of vertices in the sample. It accepts or rejects depending on some property of the resulting induced subgraph. In fact, all known testers in the dense graphs model are non-adaptive.

In this work we focus on the basic and well studied graph property of *bipartiteness*. Goldreich et. al. [10] showed that $\tilde{O}(1/\epsilon^4)$ queries are sufficient for testing bipartiteness using a canonical tester, which accepts or rejects depending on whether the subgraph induced by the sampled vertices is bipartite. In fact, Goldreich et. al. showed that it is not necessary to perform all queries between pairs of selected vertices, as done by the canonical tester, and that $\tilde{O}(1/\epsilon^3)$ queries are sufficient, where the choice of queries is non-adaptive. By performing a more refined and sophisticated analysis, Alon and Krivelevich [2] improved on this result and showed that it is sufficient to randomly select $\tilde{O}(1/\epsilon)$ vertices and apply the canonical tester, thus reducing the query complexity to $\tilde{O}(1/\epsilon^2)$.

The result of Alon and Krivelevich is optimal in terms of the *number of vertices* that the tester inspects [2]. A natural question addressed by Bogdanov and Trevisan [5] is whether $1/\epsilon^2$ *queries on pairs of vertices* are necessary. Bogdanov and Trevisan showed that $\Omega(1/\epsilon^2)$ queries are indeed necessary for any *non-adaptive* tester. For *adaptive* testers they showed that $\Omega(1/\epsilon^{3/2})$ queries are necessary. Both lower bounds hold for graphs of small degree, that is, the degree of every vertex is $\Theta(\epsilon n)$. Bogdanov and Trevisan raised the question whether there exists a natural example where an actual gap between adaptive and non-adaptive testers occurs.

Our Results. Our main result is a proof of a gap between the complexity of adaptive and non-adaptive testers in the dense graphs model. Specifically, we describe an adaptive bipartiteness tester for graphs in which all vertices have degree $O(\epsilon n)$ that performs $\tilde{O}(1/\epsilon^{3/2})$ queries. Thus we demonstrate that adaptive testers are stronger than non-adaptive testers in this model, since the $\Omega(1/\epsilon^2)$ lower bound of [5] for non-adaptive testers holds in this case, which we refer to as the *low-degree* case. In other words, the gap between the complexity of adaptive and non-adaptive testers in the dense-graphs model is manifested in the problem of testing bipartitness under the promise that the graph has degree $O(\epsilon n)$.

We next describe an alternative formulation of our result, which removes the need for a promise. By slightly modifying our algorithm we can test for the combined property of being bipartite and having degree at most $c\epsilon n$ for any given constant c. The complexity of the algorithm remains $\tilde{O}(1/\epsilon^{3/2})$. Since the lower bound of [5] still holds when we add the degree bound restriction, we obtain a gap between the complexity of adaptive and non-adaptive testers in the dense-graphs model for the combined property.

In the above two formulations, we either relied on a promise or we combined the bipartiteness property with another property, where in both cases there is a dependence on the distance parameter ϵ (in the promise or in the combined property). A natural question is whether the complexity of testing bipartiteness in *general graphs* when allowing adaptivity is $o(1/\epsilon^2)$. A positive answer would establish a gap between adaptive and non-adaptive testers without relying on a promise and without allowing the property to depend on ϵ. While we do not answer this question, we provide another result that may be of use in further research on the problem. We show that $\tilde{O}(1/\epsilon^{3/2})$ queries suffice when (almost) all vertices have degree $\Omega(\sqrt{\epsilon} \cdot n)$, that is, relatively high degree. In this case adaptivity is not necessary. We conjecture that $\tilde{O}(1/\epsilon^{3/2})$ queries are sufficient for all graphs but leave this as an open question. Note that if our conjecture is not correct then one would obtain a surprising non-monotonic behavior of the complexity of testing bipartiteness as a function of the degree of the graph.

Techniques. Interestingly, though our work is strictly in the aforementioned model of dense graphs, we use techniques not only from the dense graphs model, but also from the *bounded-degree graphs* model. In the bounded-degree model there is a an upper bound, d, on the degree of vertices in the graph. A testing algorithm may perform queries of the form: "who is the i'th neighbor of vertex v" (*neighbor queries*) and is required to distinguish with high constant probability between the case that the graph is bipartite and the case that more than $\epsilon \cdot d \cdot n$ edges should be removed from the graph in order to make it bipartite. Thus, both the type of queries and the distance measure are different from those in the dense model.

Specifically, by extending techniques from [10] and [2] and combining them we prove the following lemma that is a building block in proving our main result.

Lemma 1. *If a graph G is ϵ-far from being bipartite and has maximum degree $\rho \cdot \epsilon n$ for some $\rho \geq 1$, then with high constant probability over a random choice of $s = \tilde{O}(\rho/\epsilon)$ vertices in G, the subgraph induced by the selected vertices is ϵ'-far from being bipartite (according to the distance measure for dense graphs) for*
$$\epsilon' = \Omega\left(\frac{\epsilon}{\rho^2 \log(1/\epsilon)}\right).$$

That is, it is not merely that (with high constant probability) the subgraph induced by the small sample is *not* bipartite (as shown in [2]), but rather that at least $\epsilon' \cdot s^2$ edges should be removed from this subgraph in order to make it bipartite. We note that Lemma 1 can be strengthened in the case where almost all vertices have roughly the same degree (see the full version of this paper [14]).

From this point on (until the end of the section) we focus on the case that $\rho = O(1)$, that is, the maximum degree in the graph is $O(\epsilon n)$. As noted previously, the known lower bounds of $\Omega(1/\epsilon^2)$ for non-adaptive algorithms and $\Omega(1/\epsilon^{3/2})$ for adaptive algorithms were shown for this case. Given Lemma 1, consider the small subgraph G' that is induced by the $s = \tilde{O}(1/\epsilon)$ sampled vertices, and assume that in fact $\Omega\left(\frac{\epsilon}{\log(1/\epsilon)} \cdot s^2\right)$ edges should be removed from G' in order to make it bipartite. Observe that given the $O(\epsilon n)$ upper bound on the degree of vertices in G, and the number of vertices in G', with high constant probability, the degree of vertices in G' is upper bounded by $d = \mathrm{polylog}(1/\epsilon)$. This means that the number of edges that should be removed from G' is at least $\epsilon'' \cdot d \cdot s$ for $\epsilon'' = 1/\mathrm{polylog}(1/\epsilon)$. In other words, in terms of the distance measure *defined for the bounded degree model*, the subgraph G' is very far from being bipartite.

We hence turn to the result in [11] on testing bipartiteness in the bounded-degree model. The tester of [11] is based on performing random walks, where the query complexity grows like a square root of the number of vertices in the graph and depends polynomially on the inverse of the distance to bipartiteness (according to the distance measure for bounded degree graphs). The algorithm has no dependence on the degree bound. Recall that in our case, the graph G' contains $s = \tilde{O}(1/\epsilon)$ vertices (so that $\sqrt{s} = \tilde{O}(1/\sqrt{\epsilon})$), and the distance to bipartiteness is $1/\mathrm{polylog}(1/\epsilon)$. However, we cannot simply apply the [11] tester to G', because we do not have direct access to neighbor queries in G', as is assumed in the bounded-degree model. Yet, we can easily emulate such queries using vertex-pair queries, at a cost of $s = \mathrm{polylog}(1/\epsilon) \cdot \epsilon^{-1}$ vertex-pair queries per neighbor query. The total number of queries performed by our tester is therefore $\tilde{O}(\epsilon^{-3/2})$.

Perspective. As noted previously, all known algorithms in the dense graphs model are non-adaptive. On the other hand, algorithms in the bounded-degree model are almost "adaptive by nature" in the sense that non-adaptive tester in this model are quite limited. Recall that in the bounded-degree model a tester can perform neighbor queries. That is, it can ask for names of neighbors of vertices of its choice. Therefore, in the bounded-degree model, essentially all that a non-adaptive tester can do is ask for neighbors of a predetermined set of vertices. If a tester wants to "go beyond" these neighbors, and, say, perform random walks, or a (bounded-depth) breadth first search, then the tester must use answers to previous queries (names of neighbors) in its new queries. More formally, Raskhodnikova and Smith [17] prove that for any *non-trivial* property, that is, a property that is not determined by the distribution of degrees in the graph, any non-adaptive tester must perform $\Omega(\sqrt{n}/d)$ queries in graphs with degree bound d. Our tester can be viewed as "importing" this adaptivity to the dense model by emulating a tester from the bounded-degree model in the dense model.

The question of whether adaptivity can be beneficial arises in other areas of property testing. Two interesting families of properties for which it was proved that adaptivity *cannot* help are monotonicity in one dimension [8], and properties of subsets of strings that are defined by linear subspaces [4]. Fischer's result [8] implies that a known lower bound for testing monotonicity in one dimension [6] that holds for non-adaptive testers, also holds for adaptive testers. In [4] the fact that adaptive algorithms are no stronger than non-adaptive algorithms simplified a lower-bound proof. In contrast, Fischer [8, Sec. 4] gives several examples of properties for which there is a large (e.g., exponential) gap between the complexity of adaptive and non-adaptive testers.

Other Related Work. Other results on testing bipartiteness can be found in [12] and [16]. For tutorials on property testing in general, see [9,7,18].

2 Preliminaries

Let $G = (V, E)$ be an undirected graph with n vertices. For a vertex v let $\Gamma(v)$ denote the set of neighbors of v and let $\deg(v)$ denote the degree of v. For a set of vertices $R \subseteq V$ and an edge $(u, v) \in E$ we say that R *spans* (u, v) if $u, v \in R$. We let $E(R)$ denote the set of edges spanned by R. For a partition (V_1, V_2) of V, we say that an edge $(u, v) \in E$ is a *violating* edge with respect to (V_1, V_2) if either $(u, v) \in E(V_1)$ or $(u, v) \in E(V_2)$.

Recall that $G = (V, E)$ is said to be *bipartite* if there exists a partition (V_1, V_2) of V with respect to which there are no violating edges. It is well know that a graph is bipartite if and only if it contains no odd-length cycles. We say that G is ϵ-*far* from being bipartite if the number of edges that must be removed from G so as to make it bipartite is greater than ϵn^2. In other words, for every partition (V_1, V_2) there are more than ϵn^2 violating edges with respect to (V_1, V_2).

Definition 1. *A testing algorithm for bipartiteness is given as input a distance parameter $\epsilon > 0$ and may query the adjacency matrix of G. Namely, it can perform queries of the form: "is there an edge between u and v in G?" We refer to such queries as vertex-pair queries. The requirements from the algorithm are:*

- *If G is bipartite then the algorithm should accept;*
- *If G is ϵ-far from being bipartite then the algorithm should reject with probability at least $2/3$.*

Observe that in the above definition we required that the algorithm have *one-sided error*.

3 Some Useful Sampling Claims

In this section we prove a few claims that will be useful in our analysis. First we introduce one more definition.

Definition 2. *For $0 \leq \alpha, \gamma \leq 1$, we say that a subset $S \subseteq V$ is an (α, γ) dominating set for G if all but at most a γ-fraction of the vertices in G with degree at least αn have a neighbor in S.*

Lemma 2. *For every graph G, with probability at least $1 - \delta$ over the choice of a uniformly selected random subset S of $\frac{1}{\alpha} \cdot \ln \frac{1}{\gamma \delta}$ vertices in G, the subset S is an (α, γ) dominating set for G.*

Proof. Let S denote the set of selected vertices. We may select $S = \{u_1, \ldots, u_s\}$ in s trials, where in the i'th trial we uniformly, independently at random select a vertex from $V \setminus \{u_1, \ldots, u_{i-1}\}$. Consider any fixed vertex v with degree at least αn. The probability that S does not contain any neighbor of v is upper bounded by $(1 - \alpha)^s < e^{-\alpha \cdot s} = \gamma \cdot \delta$. Therefore, the expected fraction of vertices with degree at least αn that do not have a neighbor in S is at most $\gamma \cdot \delta$. By Markov's inequality, the probability that there are more than a γ-fraction such vertices (that is, $\frac{1}{\delta}$ times the expected value) is less than δ.   □

In our analysis we make use of one of Janson's inequalities [15] (see also [3, Chapter 8, Theorem 7.2]). In order to state it we shall need to introduce some notation. Let Ω be a finite set and let R be a random subset of Ω given by $\Pr[r \in R] = p_r$, where these events are mutually independent over $r \in \Omega$. Let A_i, $i \in I$, be subsets of Ω, where I is a finite index set. For each $i \in I$ let X_i be a 0/1 random variable which is 1 if $A_i \subseteq R$ and is 0 otherwise, and let $X = \sum_{i \in I} X_i$ be the number of subsets A_i that are contained in R. Define

$$\Delta \stackrel{\text{def}}{=} \sum_{i \neq j,\ A_i \cap A_j \neq \emptyset} \Pr[X_i = 1 \ \& \ X_j = 1] \qquad (1)$$

(note that the sum is over all *ordered* pairs $i \neq j$).

Theorem 1 ([15]). *For any $\gamma > 0$,*

$$\Pr[X \leq (1 - \gamma)\mathrm{Exp}[X]] \leq \exp\left(-\gamma^2 \frac{\mathrm{Exp}[X]}{2(1 + \Delta/\mathrm{Exp}[X])}\right).$$

As a corollary of Theorem 1 we get:

Corollary 3. *Let $0 < \beta \leq \alpha < 1$ where $\alpha \geq \beta$, and let $B \subseteq E$ be a set of edges such that $|B| \geq \beta n^2$ and the number of edges in B that are incident to any $u \in V$ is at most αn. If we uniformly select a random subset of vertices R where $|R| = 2c_R \cdot \frac{\alpha}{\beta}$ and $c_R \geq 1/\alpha$, then*

$$\Pr\left[|E(R) \cap B| \leq \frac{1}{8}\beta|R|^2\right] \leq 2e^{-c_R/24}$$

Proof. We first analyze the case that R is chosen by independently selecting each vertex $u \in V$ with probability $p_R = c_R \cdot \frac{\alpha}{\beta} \cdot \frac{1}{n}$, so that we may apply Theorem 1.

Let $\Omega = V$ and for each edge $(u_i, v_i) \in B$ let $A_i = \{u_i, v_i\}$ and let X_i be the 0/1 random variable indicating whether $A_i \subseteq R$. Thus $X_i = 1$ if and only if R spans (u_i, v_i) and $X = \sum_i X_i = |E(R) \cap B|$. It follows that $\Pr[X_i = 1] = p_R^2$ and

$$\mathrm{Exp}[X] = |B| \cdot p_R^2 \geq \beta n^2 \cdot p_R^2. \qquad (2)$$

Observe that for every $i \neq j$ such that $A_i \cap A_j \neq \emptyset$ we have that $\Pr[X_i = 1 \ \& \ X_j = 1] = p_R^3$. For any vertex u let $\deg_B(u)$ denote the number of edges in B that are incident to u. Then

$$\Delta = \sum_{(u_i, v_i) \in B} (\deg_B(u_i) + \deg_B(v_i) - 2) \cdot p_R^3. \qquad (3)$$

Since $\deg_B(u) \le \alpha n$ for every $u \in V$, we get that $\Delta \le 2 \cdot |B| \cdot \alpha n \cdot p_R^3$ Therefore,

$$\Delta/\mathrm{Exp}[X] \le 2 \cdot \alpha n \cdot p_R = 2 \cdot \frac{c_R \alpha^2}{\beta} . \tag{4}$$

Since $\alpha \ge \beta$ and $c_R \ge 1/\alpha$, we have that $\frac{c_R \alpha^2}{\beta} \ge 1$, and so $2(1 + \Delta/\mathrm{Exp}[X]) \le 6\alpha n \cdot p_R$. If we apply Theorem 1 with $\gamma = \frac{1}{2}$ then we get that

$$\Pr\left[|E(R) \cap B| \le \frac{1}{2}|B| \cdot p_R^2 \right] \le \exp\left(-(1/4) \cdot \frac{|B| \cdot p_R^2}{6\alpha n \cdot p_R} \right)$$
$$\le \exp\left(-(1/24)(\beta n/\alpha)p_R \right) = \exp(-c_R/24) . \tag{5}$$

Since $\mathrm{Exp}[|R|] = p_R \cdot n$, by a multiplicative Chernoff bound, the probability that $|R| \ge 2 \cdot p_R \cdot n$ is $\exp(-p_R n/3) < \exp(-c_R/24)$. In other words, with probability at least $1 - \exp(-c_R/24)$, $|R|^2 \le 4p_R^2 n^2$. Combining this with Equation (5), we get that

$$\Pr\left[|E(R) \cap B| \le \frac{1}{8}\beta|R|^2 \right] \le \exp(-c_R/24) + \Pr\left[|E(R) \cap B| \le \frac{1}{2}|B| \cdot p_R^2 \right]$$
$$\le 2\exp(-c_R/24) . \tag{6}$$

It remains to deal with the way R is chosen. Instead of choosing R by independently selecting each vertex $u \in V$ with probability p_R, we can equivalently first select the size r of R that is induced by this distribution, and then uniformly at random select a subset of r vertices. Furthermore, since we have already shown that the probability that $|R| \ge 2 \cdot p_R \cdot n$ is $\exp(-p_R n/3) < \exp(-c_R/24)$ (and have already taken the probability of this event into account), we can restrict the choice of r to $r \le 2 \cdot p_R \cdot n$. The final observation is that the probability that $|E(R) \cap B| \le \frac{1}{2}|B| \cdot p_R^2$ monotonically decreases with the size r of R. Therefore, we can take r to equal $2 \cdot p_R \cdot n = 2c_R \cdot \frac{\alpha}{\beta}$. ⊓

4   An Adaptive Algorithm for Graphs with Small Degree

In this section we describe an adaptive tester for bipartiteness with query complexity $\frac{1}{\epsilon^{3/2}} \cdot \mathrm{poly}\left(\frac{\alpha}{\epsilon} \log \frac{1}{\epsilon} \right)$ for graphs with maximum degree αn. In particular, when $\alpha = O(\epsilon)$, the query complexity is $\tilde{O}(1/\epsilon^{3/2})$. Recall that the lower bounds in [5] ($\Omega(1/\epsilon^2)$ for non-adaptive testers and $\Omega(1/\epsilon^{3/2})$ for adaptive testers), hold for graphs with degree bound $O(\epsilon n)$. At the end of the section we address the question of testing the combined property of being bipartite and having degree at most αn for $\alpha = O(\epsilon)$.

Let $\rho = \frac{\alpha}{\epsilon}$ denote the ratio between the maximum degree in the graph and ϵn. We may assume that $\rho \ge 1$ or else the graph is trivially ϵ-close to being bipartite. The algorithm selects (uniformly at random) two subsets of vertices, denoted S and R, where $|S| = \Theta\left(\frac{\log(1/\epsilon)}{\epsilon} \right)$ and $|R| = \Theta\left(\frac{\rho \cdot \log(1/\epsilon)}{\epsilon} \right)$, such that $|R|$ is always at least a sufficiently large constant factor larger than $\rho \cdot |S|$. [2]

---

[2] In particular, setting $|S| = \frac{4}{\epsilon} \ln \frac{64}{\epsilon}$ and $|R| = 128\rho|S| = \frac{512\rho}{\epsilon} \ln \frac{64}{\epsilon}$, suffices.

The algorithm then performs queries on only some of the pairs of vertices in $S \cup R$ (as is explained subsequently) in an attempt to find an odd-length cycle in the subgraph induced by $S \cup R$. We first describe several properties of the two subsets and show that they hold with high constant probability (given the sizes of the two sets as stated above). In all that follows, when we say "high constant probability" we mean probability at least $1 - \delta$ for a small constant δ. The constant δ is selected so that when we sum over all (constant number of) failure events, we get at most $1/3$.

Property 1. *The total number of edges that are incident to vertices that do not have a neighbor in S is at most $(\epsilon/2)n^2$.*

Lemma 4. *With high constant probability S has Property 1.*

Proof. By Lemma 2, with high constant probability S is an $(\epsilon/4, \epsilon/4)$ dominating set for G. Namely, there are at most $(\epsilon/4)n$ vertices with degree at least $(\epsilon/4)n$ that do not have a neighbor in S. The total number of edges incident to these vertices is upper bounded by $(\epsilon/4)n^2$. Since there are at most n vertices with degree less than $(\epsilon/4)n$ (that may not have a neighbor in S), there are at most $(\epsilon/4)n^2$ additional edges. The lemma follows.   □

Definition 3. *For a partition (S_1, S_2) of S, we say that an edge $(u, v) \in E$ is conflicting with (S_1, S_2) if u and v both have a neighbor in S_1 or both have a neighbor in S_2.*

By Definition 3, if for every partition (S_1, S_2) of S it holds that R spans at least one edge that conflicts with (S_1, S_2), then the subgraph induced by $S \cup R$ is not bipartite. If, as in previous works [10,2] our algorithm would query all $\Omega(1/\epsilon^2)$ pairs of vertices in $S \cup R$ (or even just all $\Omega(1/\epsilon^2)$ pairs of vertex u, v such that $u \in S, v \in R$ or $u, v \in R$), then it would suffice to show that with high constant probability R spans at least one conflicting edge for every partition of S. We however need something stronger, which will allow us to save in the number of queries when running an adaptive algorithm.

Property 2. *For every partition (S_1, S_2) of S, the subset R spans at least $(\epsilon/16)|R|^2$ edges that conflict with (S_1, S_2).*

Lemma 5. *Let G be a graph that is ϵ-far from being bipartite and has maximum degree $\alpha n = \rho \cdot \epsilon n$. Suppose that S has Property 1. Then with high constant probability R has Property 2.*

Proof. For each fixed partition (S_1, S_2) of S, let $B(S_1, S_2)$ denote the set of edges that conflict with (S_1, S_2). We first show that conditioned on S having Property 1, $|B(S_1, S_2)| \geq (\epsilon/2)n^2$. To see why this is true, consider the following partition (V_1, V_2) of V: $V_1 = \{u : \Gamma(u) \cap S_2 \neq \emptyset\}$ and $V_2 = V \setminus V_1$. Since G is ϵ-far from bipartite, there are at least ϵn^2 edges that are violating with respect to (V_1, V_2). Since S has Property 1, among these edges there are at least $(\epsilon/2)n^2$ edges whose endpoints both have at least one neighbor in S. Consider any such edge (u, v). If $u, v \in V_1$ then u and v both have a neighbor in S_2, and if $u, v \in V_2$ then u and v both have a neighbor in S_1. Therefore, all these violating edges are conflicting edges and so $|B(S_1, S_2)| \geq (\epsilon/2)n^2$ as claimed.

We now apply Corollary 3 with $B = B(S_1, S_2)$ so that $\beta = \epsilon/2$, and with $\alpha = \rho \cdot \epsilon$ and $c_R = \Theta(\log(1/\epsilon)/\epsilon)$ (so that $|R| = 2c_R \cdot \frac{\alpha}{\beta} = \Theta(\rho \cdot \log(1/\epsilon)/\epsilon)$). It follows that the probability that R spans less than $(\epsilon/16)|R|^2$ edges in $B(S_1, S_2)$ is $\exp(-\Theta(\log(1/\epsilon)/\epsilon))$. The lemma follows by taking a union bound over all partitions (S_1, S_2) of S.   □

Let $T = S \cup R$ and let $\bar{n} = |T|$, so that $\bar{n} = \Theta(\rho \cdot \log(1/\epsilon)/\epsilon)$. Let G_T denote the subgraph induced by T.

Property 3. *The maximum degree in G_T is at most $2\alpha\bar{n} = O(\rho^2 \log(1/\epsilon))$.*

Lemma 6. *With high constant probability G_T has Property 3.*

Proof. Recall that $\bar{n} = |T|$, and let $u_1, \ldots, u_{\bar{n}}$ be random variables where u_i is the i'th vertex selected to be in $T = S \cup R$. (To be precise, since R and S may intersect (if ϵ is very small), $\bar{n}$ is the number of vertices in T, counted with repetitions.) For every $i, j \in \{1, \ldots, \bar{n}\}$, $i \neq j$, let $Y_{i,j}$ be a random variable indicating that $(u_i, u_j) \in E$, so that $\sum_{j \neq i} Y_{i,j}$ is the number of neighbors that u_i has in T. For each fixed choice of i, consider selecting u_i first, and then selecting the other vertices. By the definition of $Y_{i,j}$ we have that $\Pr[Y_{i,j} = 1] \leq \frac{\deg(u_i)}{n}$ for every $j \neq i$ (where we have an inequality rather than equality since S and R are each selected without repetitions). Hence for each fixed choice of i we have that $\text{Exp}[\sum_{j \neq i} Y_{i,j}] \leq \deg(u_i) \cdot (\bar{n} - 1)/n < \rho\epsilon\bar{n}$. By a multiplicative Chernoff bound, $\Pr[\sum_{j \neq i} Y_{i,j} > 2\rho\epsilon\bar{n}] < e^{-\rho\epsilon\bar{n}/3}$. By applying the union bound, the probability that this event occurs for some $i \in \{1, \ldots, \bar{n}\}$ is upper bounded by $\bar{n} \cdot e^{-\rho\epsilon\bar{n}/3}$ which for $\bar{n} = \Theta(\rho \cdot \log(1/\epsilon)/\epsilon)$ is smaller than any constant, as required.   □

Lemma 7. *Let G be a graph that is ϵ-far from being bipartite and has maximum degree $\alpha n = \rho\epsilon n$. If R has Property 2 and G_T has Property 3 then G_T is ϵ'-far from being bipartite for $\epsilon' \geq \epsilon/(c_2\rho^2 \log(1/\epsilon))$ where c_2 is a constant.*

Note that Lemma 1, which was stated in the introduction, follows from Lemmas 4–7.

Proof. Consider any fixed partition $(S_1 \cup R_1, S_2 \cup R_2)$ of $T = S \cup R$. We shall show that there are at least $\epsilon''|R|^2$ violating edges with respect to $(S_1 \cup R_1, S_2 \cup R_2)$ where $\epsilon'' \geq \epsilon/(c_2'\rho^2 \log(1/\epsilon))$ for some constant c_2'. Since $|R| \geq (1/2)\bar{n}$ (as $|R| > |S|$), $\epsilon''|R|^2 \geq \frac{1}{4}\epsilon''\bar{n}^2$, the lemma follows for $\epsilon' = \frac{1}{4}\epsilon''$.

Recall that R has Property 2 and so there are at least $(\epsilon/16)|R|^2$ edges (u, v), $u, v \in R$, that conflict with (S_1, S_2). Note that if an edge (u, v) (where $u, v \in R$) conflicts with (S_1, S_2) then it is not necessarily a violating edge with respect to $(S_1 \cup R_1, S_2 \cup R_2)$, as u and v may be on different sides of the partition. However, we can define a mapping from edges that conflict with (S_1, S_2) to edges that are violating with respect to $(S_1 \cup R_1, S_2 \cup R_2)$, where the number of conflicting edges mapped to each violating edge is at most $c_1'\rho^2 \log(1/\epsilon)$ for

534 M. Gonen and D. Ron

some constant c_1'. We thus get that the number of violating edges is at least $\frac{(\epsilon/16)|R|^2}{c_1'\rho^2\log(1/\epsilon)} = \frac{\epsilon}{c_2'\rho^2\log(1/\epsilon)}|R|^2$ as claimed.

In what follows, whenever we say that an edge (u,v) is a conflicting edge, we mean that it conflicts with (S_1, S_2) and is spanned by R. Similarly, whenever we say that an edge is violating, then it is violating with respect to $(S_1 \cup R_1, S_2 \cup R_2)$.

Consider any conflicting edge (u,v). If it is violating (i.e., $u,v \in R_1$ or $u,v \in R_2$) then it is mapped to itself. Otherwise, its two end points belong to different sides of the partition $(S_1 \cup R_1, S_2 \cup R_2)$. Without loss of generality, $u \in R_1$ and $v \in R_2$. Since (u,v) is conflicting, either u has a neighbor $w \in S_1$ and v has a neighbor $z \in S_1$, or u has a neighbor $w \in S_2$ and v has a neighbor $z \in S_2$. In the former case we map (u,v) to the violating edge (u,w), and in the latter case we map (u,v) to the violating edge (v,z).

In the worst case, for every vertex u, all conflicting edges (u,v) are mapped to the same edge (u,w). Since (by Property 3) u has at most $2\rho\epsilon\overline{n} = c_1'\rho^2\log(1/\epsilon)$ neighbors in R, the maximum number of conflicting edges that are mapped to each violating edge is at most $c_1'\rho^2\log(1/\epsilon)$ as claimed. □

As we show in [14], we can strengthen Lemma 7 in the case where almost all vertices have *approximately the same degree* (where this degree may be large). Specifically, in that case we show that if the graph is ϵ-far from bipartite, then the subgraph induced by a sample of $\tilde{\Theta}(1/\epsilon)$ vertices is ϵ'-far from bipartite for $\epsilon' = \Omega(\epsilon/\log(1/\epsilon))$.

4.1 Emulating an Algorithm for Bounded Degree Graphs

An alternative model for testing graph properties is known as the "bounded-degree model". When we discuss this model we shall add an 'overline' (−) to all notation so as to distinguish it from the dense model. In this model there is an upper bound $\overline{d}$ on the degree of all vertices, and a graph $\overline{G}$ over $\overline{n}$ vertices is said to be $\overline{\epsilon}$-far from being bipartite if more than $\overline{\epsilon}\overline{d}\overline{n}$ edges must be removed in order to make the graph bipartite. A testing algorithm in this model can perform "neighbor queries", that is, for any vertex v and index $1 \leq i \leq \overline{d}$, the algorithm may query what is the i'th neighbor of v. Clearly this model is most appropriate for constant degree graphs, and more generally, for graphs in which the average degree is of the same order as the maximum degree $\overline{d}$ (so that the number of edges is $\Theta(\overline{d}\overline{n})$).

Goldreich and Ron [11] presented an algorithm for testing bipartiteness in the bounded-degree model whose query complexity and running time are $\sqrt{\overline{n}} \cdot \text{poly}(\log\overline{n}, 1/\overline{\epsilon})$. The running time of the algorithm has no dependence on the degree bound $\overline{d}$ and it works for every $\overline{\epsilon}$ and $\overline{d}$, which in particular may be functions of $\overline{n}$.

We now describe the algorithm in [11]: It uniformly and independently at random selects $O(1/\overline{\epsilon})$ *starting* vertices. From each starting vertex s it performs $\sqrt{\overline{n}} \cdot \text{poly}(\log\overline{n}, 1/\overline{\epsilon})$ random walks, each of length $\text{poly}(\log\overline{n}, 1/\overline{\epsilon})$. Each step in the random walk is performed in the following manner. Let $\deg(u)$ be the degree of the current vertex reached by the walk (where initially $u = s$). Then for each vertex $v \in \Gamma(u)$, the walk traverses to v with probability $\frac{1}{2d}$, and with probability $1 - \frac{\deg(u)}{2d}$ (which is at least $1/2$), the walk remains at u. If the algorithm detects an odd-length cycle in the course of all the walks then it rejects, otherwise it accepts.

We now return to our problem. Let $\overline{G} = G_T$ and $\overline{n} = |T|$. By Lemmas 4–7, with high constant probability over the choice of S and R, all vertices in the graph $\overline{G} = G_T$ have degree at most $\overline{d} = O(\rho^2 \log(1/\epsilon))$, and it is necessary to remove more than $\epsilon' \overline{n}^2$ edges in order to make it bipartite, where $\epsilon' = \Omega(\epsilon/(\rho^2 \log(1/\epsilon)))$. But this means that, in the bounded-degree model, $\overline{G}$ is $\overline{\epsilon}$-far from being bipartite for

$$\overline{\epsilon} = \frac{\epsilon' \overline{n}^2}{\overline{d}\overline{n}} = \Omega \left(\frac{1}{\rho^3 \log(1/\epsilon)} \right).$$

Given the aforementioned discussion, we would like to run the algorithm of [11] on $\overline{G}$. The only difficulty is that we cannot perform neighbor queries in $\overline{G}$, where these queries are necessary for performing random walks, and we can only perform vertex-pair queries. Note that the number of neighbor queries performed by the algorithm is $\sqrt{\overline{n}} \cdot \text{poly}(\log \overline{n}, 1/\overline{\epsilon})$, which for $\overline{G} = G_T$ is $\text{poly}(\rho \cdot \log(1/\epsilon))/\sqrt{\epsilon}$. In order to perform a random step from a vertex u, we simply perform all queries (u, v) for $v \in T$, and take a random neighbor. The total cost is $\text{poly}(\rho \cdot \log(1/\epsilon))/\epsilon^{3/2}$.

Algorithm 1 (*A Testing Algorithm for Graphs with degree at most $\rho\epsilon n$*)

1. *Uniformly at random select two subsets of vertices, S and R from V, where $|S| = \Theta(\log(1/\epsilon)/\epsilon)$ and $|R| = \Theta(\rho \cdot \log(1/\epsilon)/\epsilon)$, and let $T = S \cup R$.*

2. *Uniformly and independently at random select $\Theta(\rho^3 \log(1/\epsilon))$ vertices from T. Let the set of vertices selected be denoted by W.*

3. *For each vertex $s \in W$, perform $\text{poly}(\rho \cdot \log(1/\epsilon))/\sqrt{\epsilon}$ random walks in G_T, each of length $\text{poly}(\rho \cdot \log(1/\epsilon))$. To perform a random walk step from vertex u, the algorithm performs all vertex pair queries (u, v) for $v \in T$. For each $v \in \Gamma(v) \cap T$ the walk continues to v with probability $\frac{1}{2d}$ (where $\overline{d} = c\rho^2 \log(1/\epsilon)$ for a constant c), and with probability $1 - \frac{|\Gamma(v) \cap T|}{\overline{d}}$ the walk remains at u.*

4. *If an odd-length cycle is detected in the subgraph induced by all random walks then* reject, *otherwise* accept.

The next theorem follows from Lemmas 5, 6 and 7 and the correctness of the algorithm in [11].

Theorem 2. *Algorithm 1 is a testing algorithm for graphs with maximum degree $\alpha n = \rho\epsilon n$. Its query complexity and running time are $\text{poly}(\rho \cdot \log(1/\epsilon))/\epsilon^{3/2}$. In particular, when $\rho = O(1)$, the complexity of the algorithm is $\tilde{O}(1/\epsilon^{3/2})$.*

4.2 Testing Bipartiteness Combined with Small Degree

We have shown that is possible to test bipartitness of graphs with query complexity and running time $\text{poly}(\rho \cdot \log(1/\epsilon))/\epsilon^{3/2}$ under the promise that the graph has degree at most $\alpha n = \rho\epsilon n$. Here we sketch the idea of how to remove the promise and test the combined property of being bipartite and having degree at most $\rho\epsilon n$ for any given $\rho = O(1)$. For full details see [14].

For the sake of simplicity we set $\rho = 3$. We claim that it is possible to test whether a graph has degree at most $3\epsilon n$ or is (ϵ/c)-far from any such graph for any constant c, using $\tilde{O}(1/\epsilon)$ queries. We refer to this as the *degree test*. Our algorithm for the combined property will first run the degree test with the distance parameter set to $\epsilon/16$. If the degree test rejects, then our algorithm rejects. Otherwise it runs Algorithm 1 with the distance parameter set to $\epsilon/4$. If at any step in the algorithm it observes a vertex that has degree greater than $6\epsilon\bar{n}$ in G_T (the subgraph induced by the sample T where $|T| = \bar{n}$) then it stops and rejects. Otherwise its returns the output of Algorithm 1.

If the graph G has degree at most $3\epsilon n$ and is bipartite, then it will pass the degree test and be accepted by Algorithm 1 with high constant probability. On the other hand, if the graph is ϵ-far from the combined property, then it is either at least $(\epsilon/2)$-far from having degree at most $3\epsilon n$ or it is at least $(\epsilon/2)$-far from being bipartite (or possibly both). In the first case, it will be rejected with high constant probability by the degree test. In fact, we set the distance parameter for the degree test to be such that with high constant probability it only passes graphs that are $(\epsilon/16)$-close to having degree at most $3\epsilon n$.

The main observation is that if a graph is $(\epsilon/16)$-close to having degree at most $3\epsilon n$ but is $(\epsilon/2)$-far from being bipartite, then the following holds: The subgraph of G, denoted G', which contains all vertices of G but only the edges between vertices with degree at most $6\epsilon n$, is $(\epsilon/4)$-far from bipartite. When we run Algorithm 1, it either detects a vertex with high degree in the sample, in which case it has evidence that G does not have degree at most $3\epsilon n$, or it effectively runs on G', and rejects with high constant probability.

5 An Algorithm for Graphs with Large Degree

In this section we show that $\tilde{O}\left(\frac{1}{\epsilon\cdot\alpha}\right)$ queries suffice for testing bipartiteness of graphs with minimum degree αn. In particular this implies that $\tilde{O}(\epsilon^{-3/2})$ queries suffice for testing bipartiteness of graphs with minimum degree $\Omega(\sqrt{\epsilon} \cdot n)$. In fact, we allow there to be at most $(\epsilon/4)n^2$ edges that are incident to vertices with degree smaller than αn.

Algorithm 2 (*A Testing Algorithm for Graphs with Degree at least αn*)

1. *Uniformly and independently at random select* $\Theta\left(\frac{\log(1/\epsilon)}{\alpha}\right)$ *vertices in* V.
   Let the set of selected vertices be denoted by S.[3] *Uniformly at random select a subset* R *of* $\Theta\left(\frac{\log^2(1/\epsilon)}{\epsilon}\right)$ *vertices in* V.

---

[3] Here S is chosen a bit differently from the way it was chosen in Algorithm 1 so that we can directly apply a result from [2].

2. *Perform all vertex-pair queries between vertices in S and vertices in R.*
3. *Uniformly and independently at random select $\Theta\left(\frac{\log^3(1/\epsilon)}{\epsilon \cdot \alpha}\right)$ pairs of vertices in R and perform a vertex-pair query on each pair.*
4. *Consider the subgraph induced by all edges revealed in the vertex-pair queries. If the subgraph is bipartite then* accept, *otherwise* reject.

Theorem 3. *Algorithm 2 is a testing algorithm for graphs with minimum degree at least αn. Its query complexity and running time are $\frac{\text{polylog}(1/\epsilon)}{\epsilon \cdot \alpha}$.*

In our analysis we build on Alon and Krivelevich [2], where at a certain point we deviate from their analysis and exploit the minimum degree of the graph. The next lemma is a variant of Proposition 3.2 in [2].

Lemma 8. *Let $0 < \alpha < 1$, and let G be a graph in which the number of edges incident to vertices with degree smaller than αn is at most $(\epsilon/4)n^2$. Suppose we select $\Theta\left(\frac{\log(1/\epsilon)}{\alpha}\right)$ vertices, uniformly, independently at random, and denote the subset of vertices selected by S. Then with high constant probability over the choice of S, there exist subsets of S, denoted $S^1, \ldots, S^t$, and edge-disjoint subgraphs of G, denoted $\{G^i = (V, E^i)\}_{i=1}^{t=\ln \frac{1}{\alpha}}$ for which the following conditions hold:*

1. *For every $1 \le i \le t$: (a) $|S^i| \le t \cdot e^{i+2}$; (b) Every vertex in G^i has at least one neighbor in $\bigcup_{j=1}^{i} S^j$; (c) For every vertex v in G^i, the degree of v in G^i is at most $\frac{n}{e^{i-1}}$.*
2. *$\left| E(G) \setminus \bigcup_{i=1}^{t} E(G^i) \right| \le \frac{\epsilon}{2} n^2.$*

As in Section 4, we show that certain properties of the samples S and R hold with high constant probability.

For each $1 \le i \le t$ let $S^{\le i} = \bigcup_{j=1}^{i} S^j$, and for each partition (S_1, S_2) of S, and for $\ell \in \{1, 2\}$, let $S_\ell^i = S_\ell \cap S^i$ and $S_\ell^{\le i} = S_\ell \cap S^{\le i} = \bigcup_{j=1}^{i} S_\ell^j$. Let $B^i(S_1, S_2) \subseteq E(G^i)$ be the subset of edges in $E(G^i)$ that conflict with $(S_1^{\le i}, S_2^{\le i})$ (recall Definition 3).

Property 4. *For every partition (S_1, S_2) of S, there exists an index $1 \le i \le t$ such that $|B^i(S_1, S_2)| \ge \frac{\epsilon}{2t} n^2$.*

Lemma 9. *Let G be a graph that is ϵ-far from being bipartite. If S satisfies the conditions in Lemma 8 then S has Property 4.*

Proof. Similarly to the proof of Lemma 5, consider the following partition (V_1, V_2) of V: $V_1 = \{u : \Gamma(u) \cap S_2 \ne \emptyset\}$ and $V_2 = V \setminus V_1$. Since G is ϵ-far from bipartite, there are at least ϵn^2 edges that are violating with respect to (V_1, V_2). Since S satisfies Condition 2 in Lemma 8, among these edges there are at least $(\epsilon/2)n^2$ edges that belong to $\bigcup_{i=1}^{t} E(G^i)$. Therefore, there exists an index $1 \le i \le t$ such that there are at least $\frac{\epsilon}{2t} n^2$ edges in $E(G^i)$ that are violating with respect to (V_1, V_2).

Consider any such edge (u, v). Since S satisfies Condition 1b in Lemma 8, u and v each have a neighbor in $S^{\leq i}$. If $u, v \in V_1$ then u and v both have a neighbor in S_2, and if $u, v \in V_2$ then u and v both have a neighbor in S_1. Therefore, all these violating edges are conflicting edges with respect to $(S_1^{\leq i}, S_2^{\leq i})$ and hence $|B^i(S_1, S_2)| \geq (\epsilon/2t)n^2$ as claimed. □

Property 5. *For every partition (S_1, S_2) of S and for every index $1 \leq i \leq t$ such that $|B^i(S_1, S_2)| \geq \frac{\epsilon}{2t}n^2$, the subset R spans at least $\frac{\epsilon}{16t}|R|^2$ edges in $B^i(S_1, S_2)$.*

Lemma 10. *If S satisfies the conditions in Lemma 8 then with high constant probability over the choice of R it has Property 5.*

Proof. Let us fix i, and let $Z = S^{\leq i}$. For any partition (Z_1, Z_2) of Z, let $B(Z_1, Z_2)$ denote the subset of edges in $E(G^i)$ that conflict with (Z_1, Z_2). Consider a fixed partition (Z_1, Z_2) for which $|B(Z_1, Z_2)| \geq \frac{\epsilon}{2t}n^2$. By Condition 1c in Lemma 8, the degree of vertices in G^i is at most $\frac{n}{e^{i-1}}$. We apply Corollary 3 with $B = B(Z_1, Z_2)$, so that $\beta = \frac{\epsilon}{2t}$, and with $\alpha = e^{-(i-1)}$ and $c_R = \Omega(\log(1/\epsilon) \cdot e^i)$. The probability that R spans less than $\frac{\epsilon}{16t}|R|^2$ edges in $B(Z_1, Z_2)$ is $\exp(-\Omega(\log(1/\epsilon) \cdot e^i))$.

By Condition 1a in Lemma 8, $|Z| = O(te^i)$. By taking the union bound over all $\exp(O(te^i))$ partitions (Z_1, Z_2) of $Z = S^{\leq i}$ and the union bound over all $1 \leq i \leq t$, we get that with high constant probability, for every $1 \leq i \leq t$ and for every partition (Z_1, Z_2) of $Z = S^{\leq i}$ for which $|B(Z_1, Z_2)| \geq \frac{\epsilon}{2t}n^2$, R spans at least $\frac{\epsilon}{16t}|R|^2$ edges in $B(Z_1, Z_2)$.

Now consider any partition (S_1, S_2) of S and index $1 \leq i \leq t$ such that $|B^i(S_1, S_2)| \geq \frac{\epsilon}{2t}$. Since, by definition, $B^i(S_1, S_2) = B(S_1^{\leq i}, S_2^{\leq i})$ where $(S_1^{\leq i}, S_2^{\leq i})$ is the partition of $S^{\leq i}$ induced by (S_1, S_2), the lemma follows. □

Lemma 11. *Suppose that S has Property 4 and R has Property 5. Then with high constant probability, for every partition (S_1, S_2) of S, in Step 3 of Algorithm 2 the algorithm selects a pair $u, v \in R$ such that (u, v) conflicts with (S_1, S_2).*

Proof. Let us fix a partition (S_1, S_2) and let i be the minimal index such that $|B^i(S_1, S_2)| \geq \frac{\epsilon}{2t}n^2$, where such an index exists because S has Property 4. Since R has Property 5 we know that it spans at least $\frac{\epsilon}{16t}|R|^2$ edges in $B^i(S_1, S_2)$. If we select $\Theta\left(\frac{\log^3(1/\epsilon)}{\epsilon \cdot \alpha}\right)$ pairs of vertices from R, then the probability that no edge in $B^i(S_1, S_2)$ is selected is $\exp\left(-\Theta\left(\frac{\log^3(1/\epsilon)}{t\alpha}\right)\right)$. Since $\alpha \geq \sqrt{\epsilon}$ the above probability is $\exp\left(-\Theta\left(\frac{\log^2(1/\epsilon)}{\alpha}\right)\right)$. The lemma follows by taking a union bound over all partitions of S. □

Theorem 3 follows from Lemmas 8–11: If for every partition (S_1, S_2) of S the algorithm queries a conflicting edge (u, v), $u, v \in R$, then for every partition $(S_1 \cup R_1, S_2 \cup R_2)$ the algorithm queries a violating edge and so that subgraph induced by all edges queried is not bipartite.

Acknowledgments

We would like to thank the anonymous referees of RANDOM 2007 for their helpful comments.

References

1. Alon, N., Fischer, E., Krivelevich, M., Szegedy, M.: Efficient testing of large graphs. Combinatorica 20, 451–476 (2000)
2. Alon, N., Krivelevich, M.: Testing k-colorability. SIAM J. on Discrete Mathematics 15, 211–227 (2002)
3. Alon, N., Spencer, J.H.: The Probabilistic Method. Wiley, New York (2000)
4. Ben-Sasson, E., Harsha, P., Raskhodnikova, S.: 3CNF properties are hard to test. In: Proc. 35th Symposium on the Theory of Computing (STOC), pp. 345–354 (2003)
5. Bogdanov, A., Trevisan, L.: Lower bounds for testing bipartiteness in dense graphs. In: Proc. 19th IEEE Symp. Computational Complexity Conference, pp. 75–81 (2004)
6. Ergun, F., Kannan, S., Kumar, S.R., Rubinfeld, R., Viswanathan, M.: Spot-checkers. Journal of Computer and Systems Science 60(3), 717–751 (2000)
7. Fischer, E.: The art of uninformed decisions: A primer to property testing. Bulletin of the European Association for Theoretical Computer Science 75, 97–126 (2001)
8. Fischer, E.: On the strength of comparisons in property testing. Information and Computation 189, 107–116 (2004)
9. Goldreich, O.: Combinatorial property testing - a survey. In: Randomization Methods in Algorithm Design, pp. 45–60 (1998)
10. Goldreich, O., Goldwasser, S., Ron, D.: Property testing and its connections to learning and approximation. Journal of the ACM 45, 653–750 (1998)
11. Goldreich, O., Ron, D.: A sublinear bipartite tester for bounded degree graphs. Combinatorica 19(3), 335–373 (1999)
12. Goldreich, O., Ron, D.: Property testing in bounded degree graphs. Algorithmica, 302–343 (2002)
13. Goldreich, O., Trevisan, L.: Three theorems regarding testing graph properties. Random Structures and Algorithms 23(1), 23–57 (2003)
14. Gonen, M., Ron, D.: On the benefits of adaptivity in property testing of dense graphs. Available (2007) from http://www.eng.tau.ac.il/~danar
15. Janson, S.: Poisson approximaton for large deviations. Random Structures and Algorithms 1(2), 221–229 (1990)
16. Kaufman, T., Krivelevich, M., Ron, D.: Tight bounds for testing bipartiteness in general graphs. SIAM Journal on Computing 33(6), 1441–1483 (2004)
17. Raskhodnikova, S., Smith, A.: A note on adaptivity in testing properties of bounded degree graphs. Technical Report TR06-089, Electronic Colloquium on Computational Complexity (ECCC), Available (2006) from http://www.eccc.uni-trier.de/eccc/
18. Ron, D.: Property testing. In Handbook on Randomization II, 597–649 (2001)

Slow Mixing of Markov Chains Using Fault Lines and Fat Contours[*]

Sam Greenberg[1] and Dana Randall[2]

[1] School of Mathematics, Georgia Institute of Technology, Atlanta, GA 30332-0160
[2] College of Computing, Georgia Institute of Technology, Atlanta, GA 30332-0280

Abstract. We show that local dynamics require exponential time for two sampling problems: independent sets on the triangular lattice (the hard-core lattice gas model) and weighted even orientations of the Cartesian lattice (the 8-vertex model). For each problem, there is a parameter λ known as the fugacity such that local Markov chains are expected to be fast when λ is small and slow when λ is large. However, establishing slow mixing for these models has been a challenge because standard contour arguments typically used to show that a chain has small conductance do not seem sufficient. We modify this approach by introducing the notion of *fat contours* that can have nontrivial d-dimensional volume and use these to establish slow mixing of local chains defined for these models.

1 Introduction

Markov chains based on local moves, known as *Glauber dynamics,* are used extensively in practice to sample from large state spaces. An example is the following Markov chain used to sample independent sets on $\mathbb{Z}^d$, the so-called "hard-core lattice gas model." The Gibbs (or Boltzmann) distribution is parameterized by a "fugacity" $\lambda > 0$ and is defined as $\pi(I) = \lambda^{|I|}/Z$, where Z is the normalizing constant known as the partition function. Glauber dynamics start at any initial state, say the empty independent set, and repeatedly add and remove single vertices according to the correct conditional probabilities so that the chain converges to the Gibbs distribution. We are interested in characterizing when simple chains converge quickly to equilibrium so that they can be used for efficient Monte Carlo algorithms.

An interesting phenomenon occurs as the parameter λ is varied: for small values of λ, Glauber dynamics converge quickly to stationarity, while for large values of λ, this convergence will be prohibitively slow. When λ is sufficiently large, dense independent sets dominate the stationary distribution π and it will take a very long time to move from an independent set that lies mostly on the odd sublattice to one that lies mostly on the even sublattice. The key observation is that local moves require the chain to visit an independent set that has roughly half of its vertices on the even sublattice and half on the odd sublattice, and such configurations are forced to have substantially fewer vertices and are therefore

---

[*] Supported in part by NSF grants CCR-0515105 and DMS-0505505.

M. Charikar et al. (Eds.): APPROX and RANDOM 2007, LNCS 4627, pp. 540–553, 2007.
© Springer-Verlag Berlin Heidelberg 2007

very unlikely when λ is large. This phenomenon is well known in the statistical physics community and characterizes a *phase transition* in the underlying model. Physicists observe such a dichotomy in the context of identifying when there will be a unique limiting distribution on the infinite lattice, known as a Gibbs state; below some critical λ_c there is a unique Gibbs state, while above λ_c there are multiple Gibbs states.

In order to show that a Markov chain is slow, it suffices to show that it has exponentially small *conductance*, i.e., that the state space can be partitioned into three sets such that the middle set has exponentially small probability compared to the other two and yet it is necessary to pass through the middle set to move between configurations in the large sets. In the context of independent sets on the Cartesian lattice, the large sets consist of configurations that lie predominantly on one sublattice, and the middle set contains configurations that are more balanced. Balanced independent sets are forced to contain many fewer vertices because there must be a space between regions that are mostly even and regions that are mostly odd; this makes such configurations exponentially unlikely when λ is large. When λ is large enough, then the total weight of the middle set will have small probability, even though it may contain an exponential number of configurations. *Peierls arguments* allow us to formalize this intuition by defining "contours" between regions bounded by odd or even vertices in the independent set and constructing injections that map configurations in the middle set to ones with substantially larger stationary probability. For independent sets, the injection is constructed by shifting the interior of a contour and adding many new vertices to the set (see, e.g., [3, 6, 5, 11]). Likewise, for spin configurations such as the Ising model (where vertices are assigned + or - spins and neighboring vertices prefer to have the same spin) contours separate regions that are mostly - and mostly + and the injection flips the spins on the interior of a contour [13].

1.1 Non-bipartite Independent Sets and Weighted Even Orientations

The dichotomy observed for Glauber dynamics on independent sets defined on the Cartesian lattice is believed to persist when the underlying lattice is the 2-dimensional triangular lattice. For small λ, Glauber dynamics are known to be rapidly mixing [9]. However, for large enough λ, independent sets will tend to be quite dense and hence any local dynamics should be slow. This is because there are three maximal independent sets, arising from the natural tri-partition of the triangular lattice into black, white and gray vertices, and the most likely configurations will be largely monochromatic. Configurations with fewer than $n/2$ vertices of each color are expected to be exponentially unlikely, and yet it is necessary to pass through such configurations to move from, say, a mostly black configuration to a mostly white one.

Unfortunately, Peierls arguments that succeed on bipartite lattice do not seem to generalize readily to the triangular lattice. The problem is that a contour surrounding a region whose boundary is black might be adjacent to some vertices that are white and some that are gray. The map must significantly increase the

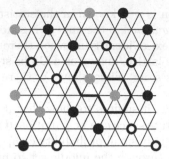 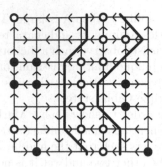

Fig. 1. An independent set of the triangle lattice and an even orientation of the Cartesian lattice

stationary probability for the argument to work, but there is no guarantee that shifting the black region in any direction will allow us to add enough new vertices to sufficiently increase the weight. An example of such a problematic balanced configuration is illustrated in the first half of Figure 1.

The second model we consider, *weighted even orientations*, also seems resistant to standard contour arguments. Given a rectangular region L in the Cartesian lattice $\mathbb{Z}^2$, the state space Ω_8 is the set of even orientations of $E(L)$, i.e., orientations of edges so that each vertex on the interior of L has even in-degree and out-degree. This is known as the *8-vertex model* in statistical physics, as there are 8 possible orientations of the edges incident to any vertex. We call vertices *sources* if they have in-degree 0 and *sinks* if they have out-degree 0; all other vertices are called *Eulerian* since their in-degree and out-degree are both 2. For $\sigma \in \Omega_8$, let $S(\sigma)$ be number of sources and sinks in σ. Given $\gamma > 0$, we assign configuration σ probability $\pi(\sigma) = \gamma^{S(\sigma)}/Z$, where Z is the normalizing constant.

The local Markov chain $\mathcal{M}_8$ is the Glauber dynamics defined on the set of even orientations. In each step, the chain chooses a cycle of length 4 in the lattice and either reverses the orientations of the edges around that cycle or keeps them unchanged according to the correct conditional probabilities given by the Gibbs distribution. This chain can be shown to connect the state space for all finite γ. When $\gamma = 0$, the only allowable configurations are Eulerian orientations (known as the *6-vertex model*) where every vertex has in-degree = out-degree $= 2$, and Glauber dynamics are known to be efficient [8, 12]. When γ is close to 1, we can use simple coupling arguments to show that Glauber dynamics are also fast. However, when γ is sufficiently large, we expect most vertices to be sources or sinks and it should take exponentially long to move from a configuration that has predominantly odd sources and even sinks to one with predominantly even sources and odd sinks. While it seems true that configurations that are "balanced" are exponentially unlikely, this does not seem to follow from any standard contour arguments. As demonstrated by the configuration in the second half of Figure 1, it is not always possible to define a map between valid configurations by flipping or shifting a single contour that is guaranteed to significantly increase the stationary probability, as required for the Peierls argument to work.

1.2 Our Results

We provide the first rigorous proofs that Glauber dynamics are slow for independent sets on the triangular lattice and for weighted even orientations on $\mathbb{Z}^2$. For even orientations, we show slow mixing of the local Markov chain on rectangular regions with fixed boundary conditions, while for independent sets we consider rhomboidal regions with periodic (toroidal) boundary conditions. These turn out to be the simplest regions for which our arguments can be made to work. Our two main theorems are as follows.

Theorem 1. *Let Λ be an $n \times n$ rhomboidal region of the triangular lattice with periodic boundary conditions, let Ω_{IS} be the set of independent sets on Λ, and let $\mathcal{M}_{IS}$ be Glauber dynamics on Ω_{IS}. There exists λ such that for all $\lambda' > \lambda$, the mixing time of $\mathcal{M}_{IS}(\lambda')$ is $\Omega(e^{kn})$ for some constant k.*

Theorem 2. *Let L be an $n \times n$ region in the Cartesian lattice, Ω_8 be the set of even orientations of L, and $\mathcal{M}_8$ be Glauber dynamics on Ω_8. There exists γ such that for all $\gamma' > \gamma$, the mixing time of $\mathcal{M}_8(\gamma')$ is $\Omega(e^{kn})$ for some constant k.*

Our proofs are based on several innovations. First, we abandon the approach of partitioning the state space into sets of configurations so that the middle sets are "balanced" in the sense described above. Instead we expand the approach in [11] of basing the partition of the state space on "topological obstructions." Roughly speaking, the middle set in our partition of the state space is defined by the presence of "fault lines," or paths from the top to bottom or left to right of the region that pass only through "unfavorable" vertices (in our cases, vacant vertices in the independent sets or Eulerian vertices in the even orientations). The absence of a fault line is characterized by the presence of a monochromatic blocking path of "favorable" vertices in each direction, and the color of this path determines which part of the state space a configuration lies in. To see why this is different from the standard approach, consider an independent set that has a single odd cross, and then includes all even vertices that can possibly be added; this independent set is considered "odd" even though it has $O(n^2)$ even vertices and only $O(n)$ odd ones. This partition of the state space was shown to greatly simplify the combinatorial methods underlying the Peierls argument for bipartite independent sets [11] and can be extended to the models we consider here as well.

It is still the case that the 1-dimensional contours used for independent sets and the Ising model on bipartite lattices do not readily generalize to our problems for the reasons outlined above. However, a generalized notion of contours that includes a larger 2-dimensional region can be made to work. Instead of defining a *minimal* connected set of unfavorable vertices, we define *fat contours* to be *maximal* connected sets of unfavorable vertices. We encode a fat contour by taking a depth-first search on the component, and define an injective map from configurations in the "middle set" of the state space by replacing the entire fat contour with favorable vertices (a maximal independent set or a maximal set of sources and sinks). This typically involves shifting parts of the independent set outside of the fat contour in two different directions. We then show that the gain is sufficient to outweigh the amount of information needed for the encoding of the fat contour.

Last, in order to show slow mixing in the context of independent sets on the triangular lattice with periodic boundary conditions, it is necessary to talk about multiple non-contractible fault lines, depending on the color of the boundary vertices. On the Cartesian lattice with periodic boundary conditions it was only necessary to find two non-contractible cycles and to shift (or flip) the configuration between these; on the triangular lattice it is sometimes necessary to find three non-contractible cycles since we are shifting regions in multiple directions.

In Section 2, we cover some background material for basic Markov chain mechanics more formally. In Sections 3 and 4, we prove Theorems 1 and 2. We note that we have not attempted to optimize constants throughout this version of the paper, but have focused on constructing simple arguments for the proofs.

2 Preliminaries

Let $\mathcal{M}$ be an ergodic (i.e., irreducible and aperiodic), reversible Markov chain with finite state space Ω, transition probability matrix P, and stationary distribution π. Let $P^t(x, y)$ be the t-step transition probability from x to y and let $||\cdot, \cdot||$ denote the *total variation distance*.

Definition 1. *For $\varepsilon > 0$, the* mixing time $\tau = \min\{t : \|P^{t'}, \pi\| \leq 1/4, \forall t' \geq t\}$.

We say a Markov chain is *rapidly mixing* if the mixing time is bounded by a polynomial in n and *slowly mixing* if the mixing time is exponential in n. Jerrum and Sinclair [7] defined the conductance of a chain and showed that it bounds mixing time.

Definition 2. *If a Markov chain has stationary distribution π, for any subset $S \subset \Omega$ of the state space with $\pi(S) \leq 1/2$, define*

$$\Phi_S = \frac{\sum_{x \in S, y \notin S} \pi(x) P(x, y)}{\pi(S)}$$

and define the conductance Φ *as*

$$\Phi = \min_{S : \pi(S) \leq 1/2} \Phi_S.$$

The conductance theorem due to Jerrum and Sinclair [7] reduces to the following that exactly characterizes when a Markov chain mixes rapidly or slowly.

Theorem 3. *An ergodic, reversible Markov chain with conductance Φ is rapidly mixing if $\Phi > 1/p(n)$ for some polynomial p and slowly mixing if $\Phi < c^{-n}$ for some $c > 1$.*

It follows that in order to show that a Markov chain mixes slowly, it suffices to identify a cut (S, S^C) in the state space such that $\Phi_S \leq c^{-n}$.

3   Weighted Even Orientations

Let L be an $n \times n$ region of the Cartesian lattice $\mathbb{Z}^2$, and let Ω_8 be the set of all orientations of the edges of L, so that all vertices in the interior of L have even in-degree and even out-degree. We are given a fixed constant $\gamma > 0$ representing the fugacity. Then for each $\sigma \in \Omega_8$, we define the Gibbs measure $\pi(\sigma) = \gamma^{S(\sigma)}/Z$, where $S(\sigma)$ is the number of sources and sinks in σ and Z is the normalizing constant. Let the Markov chain $\mathcal{M}_8$ be the Glauber dynamics on Ω_8 that chooses a face of L uniformly at random and flips the orientations of all edges incident to that face according to the correct conditional probabilities. More precisely, let s_1 be the number of sources and sinks among the four vertices defining the face in the current configuration and let s_2 be the number of sources and sinks that would surround that face if we were to flip the orientations of the four bounding edges. We flip the orientations of the four bounding edges with probability $\lambda^{s_2}/(\lambda^{s_1} + \lambda^{s_2})$ and we keep the orientation unchanged with probability $\lambda^{s_1}/(\lambda^{s_1} + \lambda^{s_2})$.

When $\lambda = 1$, all even orientations are equally likely. In this case, the probability of flipping the orientation of edges around any face is the same. If we define the distance between two configurations to be the number of edges in which their orientations differ, then it is easy to construct a coupling argument to show that the chain is rapidly mixing. (See [1] for the coupling theorem.) In fact, the distance function never increases during moves of the coupled chain. When $\lambda \neq 1$ the distance function can increase as well as decrease, but the coupling argument still works when λ is suffiently close to 1. We also know that the chain is rapidly mixing when $\lambda = 0$, since this corresponds to Eulerian orientations on Cartesian lattice regions [8].

However, when λ is large, Glauber dynamics behave quite differently than when λ is small and the chain converges rapidly to equilibrium. Before proceeding with our analysis of the mixing time, we present a reinterpretation of Ω_8 as an edge coloring. For every configuration, color an edge white if it points from an even vertex to an odd one, and color it black if it points from an odd vertex to an even one. An example of this transformation is shown in Figure 2. Now Ω_8 can be seen as the set of edge-colorings where every internal vertex has an even number of edges of each color. The sources and sinks are now monochromatic vertices (i.e., all incident edges are the same color). We call vertices that are incident to both black and white edges *bichromatic*. The cut in the state space will be characterized by configurations with many bichromatic vertices. Unlike the standard approach of partitioning the state space by relative numbers of black and white vertices, we instead use the approach of [11] and partition according to fault lines.

Call two vertices of L *edge-adjacent* if they share an edge of L and call two vertices of L *face-adjacent* if they lie on a common face of L. (Face-adjacent vertices can be edge-adjacent or diagonally opposite across a face.) Define a *vertical fault line* to be a connected path of bichromatic face-adjacent vertices from the top of L to the bottom. A *horizontal fault line* is defined similarly. Let $\mathcal{F} \subset \Omega_8$ be the set of all configurations containing a fault line.

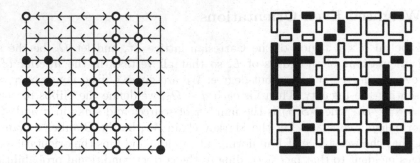

Fig. 2. A configuration $\sigma \in \Omega_8$ (with sources and sinks marked) and the corresponding edge-coloring

Now we define a *vertical bridge* to be a edge-connected path of monochromatic vertices which touches both the top and bottom of L. A *horizontal bridge* is defined similarly. We say that a configuration has a *cross* if it contains both a horizontal and a vertical bridge (of the same color). Let $\mathcal{W}$ be the set of configurations containing a white cross and $\mathcal{B}$ be the set of configurations containing a black cross. We now show that these three sets $\mathcal{F}$, $\mathcal{W}$, and $\mathcal{B}$ are disjoint and characterize all of Ω_8.

Lemma 1. *We may partition Ω_8 into $\mathcal{F}$, $\mathcal{W}$, and $\mathcal{B}$. That is, every configuration of Ω_8 has either a fault line, a white cross, or a black cross (but no two of these).*

Proof. The sets $\mathcal{W}$ and $\mathcal{B}$ are clearly disjoint, as every configuration in $\mathcal{W}$ has a vertical white bridge and every configuration in $\mathcal{B}$ has a horizontal black bridge; no configuration can have both. Similarly $\mathcal{F}$ is disjoint from $\mathcal{W}$ and $\mathcal{B}$ because fault lines obstruct crosses; if a configuration has a horizontal fault line it can not have a vertical bridge. What remains to be shown is that these three sets cover all of Ω_8, that any configuration without a cross must have a fault line.

Let σ be a configuration with no horizontal bridge. This means some vertices along the top of L must be bichromatic. Let T be the set of bichromatic vertices that have a face-connected path to the top of L. If T never reaches the bottom of L, then the boundary of T can be used to construct a horizontal bridge (and therefore a contradiction). Therefore T contains some part of the bottom of L, and so contains a vertical fault line.

Similarly, if σ has no vertical bridge then σ has a horizontal fault line.

We now show that for $\mathcal{M}_8$ to pass from $\mathcal{W}$ to $\mathcal{B}$, it must pass through $\mathcal{F}$, so $\mathcal{F}$ defines a cut in the state space.

Lemma 2. *For transition probability $P(\cdot, \cdot)$ of $\mathcal{M}_8$, we have $P(\sigma_W, \sigma_B) = 0$ for all $\sigma_W \in \mathcal{W}$ and $\sigma_B \in \mathcal{B}$.*

Proof. Assume we do have configurations $\sigma_W \in \mathcal{W}$ and $\sigma_B \in \mathcal{B}$ that differ by a single move of $\mathcal{M}_8$, say on face f. Configuration σ_W has both vertical and horizontal white bridges and σ_B has vertical and horizontal black bridges. Outside of f, $\sigma_W = \sigma_B$, so each contain both white and black paths from f to the top, bottom, left, and right of L. Moreover, for f to be critical to there being a black or a white cross, one edge e_B of f must be incident to only black edges outside of f, and the opposite edge e_W of f must be incident to only white edges outside of f. Coloring e_B black and e_W white, we must then have a left-right white crossing of the region and a top-bottom black crossing, a contradiction. We can conclude that $\mathcal{W}$ and $\mathcal{B}$ are not connected by a single move.

Next, we proceed to show that the stationary probability of $\mathcal{F}$ is exponentially small. We define a *fat contour* to be a maximally face-connected set of bichromatic vertices containing a fault line. Clearly any configuration with a fault line has a fat contour. To bound $\pi(\mathcal{F})$, we define a mapping $\psi : \mathcal{F} \to \Omega_8$ which takes $\sigma \in \mathcal{F}$ and recolors edges incident to a fat contour so that the contour contains only white (or black) vertices. Although ψ is not one-to-one, we will show that for every $\sigma' \in \mathrm{Img}(\psi)$, the stationary probability of the pre-image $\psi^{-1}(\sigma') = \{\sigma \in \mathcal{O} : \psi(\sigma) = \sigma'\}$ is exponentially less than $\pi(\sigma')$. First we prove a counting lemma that will help bound the size of the pre-image of ψ.

Lemma 3. *If F is a fat contour such that $|F| = \ell$, then there are fewer than $2n \cdot 64^\ell$ possible choices for the vertices of F and fewer than 6^ℓ colorings of the edges incident to those vertices.*

Proof. Without loss of generality, we assume F is a vertical fat contour. This at most halves the total choices for F. The fat contour must then include some vertex of the top of L. The location of this vertex together with a description of a DFS traversal of F starting at this vertex is sufficient to reconstruct the vertices in F. As there are at most eight choices at each step in the DFS (four adjacencies along edges and four diagonals across faces) there are at most $8^{2\ell}$ such traversals. With n choices for the starting vertex, we have the bound on F.

Finally, for each vertex in F, there are at most 6 possible Eulerian orientations of the incident edges, so 6^ℓ is an immediate (albeit weak) upper bound on the number of colorings of the edges defining F.

Our definition of $\psi(\sigma)$ proceeds as follows. First, choose an arbitrary fat contour of σ, F (e.g. farthest to the left or top). In the complement of F, each of the connected components (or "islands" of F) has a boundary which is either entirely white or entirely black. We reverse the color of every edge within the white islands (leaving black islands as they are). Now the edges incident to F are entirely black, and F can be recolored completely with black monochromatic vertices. An example of this modification is in Figure 3. The resulting configuration has $|F|$ more monochromatic vertices than the pre-image, corresponding to additional sources and sinks in the original even orientation.

Notice that to find the inverse of ψ, we need only the location of F and the colorings of the edges incident to vertices of F. The edges on the boundary of the fat contour will then define whether a island was originally white or black, and we can therefore recover the colors of edges in those islands accordingly.

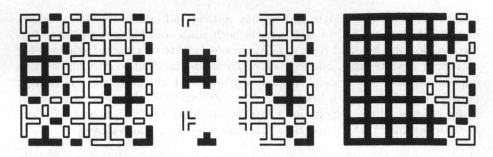

Fig. 3. A coloring σ, σ with the fat contour removed, and $\psi(\sigma)$

Lemma 4. *There exists constants $\gamma_0, n_0, c > 1$ such that, if $\gamma > \gamma_0$ and $n > n_0$, then $\pi(\mathcal{F}) < c^{-n}$.*

Proof. For each $\ell \in [n, n^2]$, let $\mathcal{F}_l$ be the edge-colorings in $\mathcal{F}$ where the fat contour chosen by ψ is of size ℓ. Then, for each $\sigma' \in \text{Img}(\psi)$,

$$
\pi(\psi^{-1}(\sigma')) = \sum_{\ell=n}^{n^2} \sum_{\sigma \in \mathcal{F}_l : \psi(\sigma) = \sigma'} \pi(\sigma)
$$

$$
= \sum_{\ell=n}^{n^2} \sum_{\sigma \in \mathcal{F}_l : \psi(\sigma) = \sigma'} \pi(\sigma') \cdot \gamma^{-\ell}
$$

$$
\leq \sum_{\ell=n}^{n^2} 2n 64^\ell 6^\ell \cdot \pi(\sigma') \cdot \gamma^{-\ell}
$$

$$
< 2n^3 \left(\frac{384}{\gamma} \right)^n \pi(\sigma').
$$

This bound on the pre-image allows us to bound $\pi(\mathcal{F})$ as follows:

$$
\pi(\mathcal{F}) = \sum_{\sigma' \in \text{Img}(\psi)} \pi(\psi^{-1}(\sigma'))
$$

$$
< \sum_{\sigma' \in \text{Img}(\psi)} 2n^3 \left(\frac{384}{\gamma} \right)^n \pi(\sigma')
$$

$$
= 2n^3 \left(\frac{384}{\gamma} \right)^n \pi(\text{Img}(\psi))
$$

$$
< 2n^3 \left(\frac{384}{\gamma} \right)^n.
$$

Taking $\lambda > 384$ yields the lemma.

The final step of the proof of Theorem 2 is to show that this lemma is sufficient to bound the conductance. Consider the cut defined by the set $\mathcal{W}$. By symmetry $\pi(\mathcal{W}) = \pi(\mathcal{B}) = (1 - \pi(\mathcal{F}))/2$. Therefore

$$\Phi \leq \Phi_{\mathcal{W}} \leq \frac{\pi(\mathcal{F})}{\pi(\mathcal{W})} \leq .49c^{-n},$$

where $c > 1$. Appealing to Theorem 3, this proves Theorem 2.

Using essentially the same arguments, we can extend the slow mixing result presented in this section to weighted even orientations on regions with periodic boundary conditions, as well as Cartesian lattice regions in higher dimensions. We leave these details for the full version of the paper.

4   Independent Sets on the Triangular Lattice

Let Λ be a $3n \times 3n$ rhomboidal region of the triangular lattice with periodic boundary conditions and let Ω_{IS} be the set of all independent sets on Λ. Given a constant $\lambda > 0$, define a probability distribution $\pi(I) = \lambda^{|I|}/Z$, where Z is the normalizing constant. The Glauber dynamics $\mathcal{M}_{IS}$ are as follows: choose a vertex of Λ uniformly at random and add or remove that vertex from I with the correct conditional probabilities, if possible. That is, if the vertex could be included without violating the independence condition, add it with probability $\lambda/(1 + \lambda)$ and remove it with probability $1/(1 + \lambda)$.

The lattice has a natural tri-partition, which we color black, white, and gray. Call a face of Λ empty if it is not incident to any vertex of I, as illustrated in Figure 4.

Call two faces of Λ adjacent if they share at least one vertex. We then define a fault line to be a non-contractible cycle of empty faces. We let $\mathcal{F} \subset \Omega_{IS}$ be the set of all independent sets with at least one fault line.

The obstruction preventing a fault line must be a set of tightly packed vertices of I. Call two vertices of an independent set touching if they are incident to faces which share an edge. Note that touching vertices must have the same color. We define a monochromatic bridge to be a non-contractible cycle of touching vertices of I. For any non-contractible cycle, the winding number is an ordered pair of

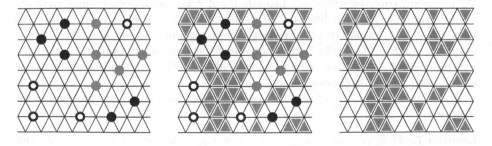

Fig. 4. An independent set on Λ and the corresponding empty faces

integers representing the number of times it winds around a fundamental domain in the horizontal and vertical directions before returning to the starting point. We say a configuration has a monochromatic *cross* if it contains two bridges with different winding numbers. Let $\mathcal{B}$, $\mathcal{W}$, and $\mathcal{G}$ be the set of independent sets containing black, white, and gray crosses, respectively.

We first show that these sets define a partition of Ω_{IS}.

Lemma 5. *Every independent set in Ω_{IS} has a fault line or a white, black, or gray cross, but no two of these.*

Proof. This proof is similar to the argument in [11] that shows that independent sets on the Cartesian lattice without fixed boundary conditions must have horizontal and vertical bridges of one color or there must be a fault line. A key point we use here is that any two non-contractible cycles of different winding numbers must intersect.

The sets $\mathcal{B}$, $\mathcal{W}$, and $\mathcal{G}$ are disjoint, as an independent set cannot have two crosses of different colors; that would involve two bridges of different winding numbers and different colors, whose intersection would lead to a contradiction. Similarly, $\mathcal{F}$ is disjoint from $\mathcal{B}$, $\mathcal{W}$, and $\mathcal{G}$, as no set can have both a cross and a fault line; the fault line must intersect at least one of the bridges, which is impossible.

To see that there must be either a cross or a fault line, examine the torus after the removal of a bridge. The remaining space is a non-contractible strip of the torus of the same winding number as the bridge. If there exists a path of touching vertices across this strip, we find a bridge of a different winding number and therefore a cross. However, if no such path exists, then there must exist a fault line along the strip.

We now show that removing $\mathcal{F}$ disconnects $\mathcal{W}$, $\mathcal{B}$, and $\mathcal{G}$. Let $P(\cdot, \cdot)$ be the transition probabilities of $\mathcal{M}_{IS}$.

Lemma 6. *Let $I_1 \in \mathcal{W}$, $I_2 \in \mathcal{B}$ and $I_3 \in \mathcal{G}$ be three independent sets. Then $P(I_i, I_j) = 0$ for all $i \neq j$.*

Proof. Individual moves of $\mathcal{M}_{IS}$ add or remove single vertices. Clearly it takes multiple moves to eliminate one cross and complete another.

We now show that the stationary probability of $\mathcal{F}$ is exponentially small. In doing so, we again extend fault lines to 2-dimensional regions. Define a *fat contour* to be a maximally connected set of empty faces containing a fault line. We define a mapping $\psi : \mathcal{F} \rightarrow \Omega_{IS}$ which eliminates at least one fat contour. Although the mapping is not one-to-one, we will show that each $I' \in \mathrm{Img}(\psi)$ has a pre-image whose total weight is exponentially smaller.

To bound the number of sets in this pre-image, we bound the number of fat contours.

Lemma 7. *If F is a fat contour with ℓ faces, then there are at most $2n^2 36^\ell$ choices for the locations of those faces.*

Fig. 5. Adjacent faces which are point-adjacent, edge-adjacent, and neither

Proof. First, we limit the notion of adjacencies in F. Define two adjacent faces to be *edge-adjacent* if they share an edge. Call them *point-adjacent* if they share a single vertex and yet are not both adjacent to a common face. As illustrated in Figure 5, not all pairs of adjacent faces are edge- or point-adjacent. However, note that edge- and point-adjacencies suffice to connect F, as a vertex of I removes a complete hexagon from F. We can therefore find a traversal of F that uses only edge- and point-adjacencies.

There are $2n^2$ choices for a face to start the DFS of F. Then each step of the DFS has six possible directions (three edge-adjacencies and three point-adjacencies), so there are at most 6^{2l} possible traversals starting at f.

Our definition of ψ is slightly more complicated than in Section 3 because we are considering toroidal regions. Suppose first that F contains two fault lines with different winding numbers. Then the complement of F contains only regions whose boundaries are contractible. By the maximality of F, each of these regions has a monochromatic boundary, so we may refer to these connected components (or "islands") by the colors of their boundaries. Note that if we shift all white islands one space East so that their boundaries become gray, and shift all black islands one space to the North-East so that their boundaries also become gray (and leave gray islands as they are), we form a new independent set of the same size. After this shift, all vertices incident to F are gray, so ψ may fill the entire fat contour with gray vertices. If F has ℓ faces, ψ adds at exactly $\ell/6$ vertices to I, each the center of a vacant hexagon.

Unfortunately, fat contours need not contain multiple fault lines with differing winding numbers and indeed the complement of F can contain regions whose boundaries are non-contractible. If this is the case, then F has a bridge on each side. Define a *bulge* to be a maximal set of touching vertices of I which contain a bridge. In this case F must be incident to two bulges. If these are of the same color, ψ can shift all islands within F to that color and fill F.

Complicating matters further, there may be no fat contour incident to two bulges of the same color. For instance, let F_1 be a fat contour incident to a white bulge on the left and a black one on the right. To the right of the black bulge there must be another fat contour, F_2. If F_2 is incident to a white bulge, ψ can shift the islands of F_1 and F_2 and the black bulge all to white. Then ψ can fill both fat contours with white vertices. If F_1 and F_2 contain ℓ_1 and ℓ_2 faces (respectively), then ψ adds $(\ell_1 + \ell_2)/6$ vertices to I.

In one final case, suppose we have no pair of neighboring fat contours bordered by bulges of the same color. Then there must then be a third fat contour F_3 which is incident to still another bulge. Luckily we only have three colors of bulges; at some point these colors must repeat. For example, if we have, in order, a black bulge, F_1, a white bulge, F_2, a gray bulge, F_3, and then a black bulge, then ψ can shift the white bulge, the gray bulge, and all islands of the fat contours to black. We may then fill all three fat contours with black vertices.

To find the inverse of ψ, note that we need only the faces of the fat contour(s); the colors of the neighboring vertices can be inferred from the shape of F, and these colors define the direction of the shift.

We may now bound the stationary probability of $\mathcal{F}$.

Lemma 8. *There exists constants* $\lambda_0, n_0, c > 1$ *such that, if* $\lambda > \lambda_0$ *and* $n > n_0$, *then* $\pi(\mathcal{F}) < c^{-n}$.

Proof. For each $\ell \in [n, 2n^2]$, let $\mathcal{F}_l \subset \mathcal{F}$ be the independent sets where the fat contours chosen by ψ contain a total of ℓ faces. Given ℓ_1, ℓ_2, ℓ_3 such that $\ell_1 + \ell_2 + \ell_3 = \ell$, Lemma 7 shows that there are at most $\prod_{i=1}^{3} 2n^2 36^{\ell_i} = 8n^6 36^\ell$ choices of faults such that $|F_i| = \ell_i$. Then, for each $I' \in \mathrm{Img}(\psi)$,

$$\pi(\psi^{-1}(I')) = \sum_{\ell=n}^{2n^2} \sum_{I \in \mathcal{F}_l : \psi(I) = I'} \pi(I)$$

$$= \sum_{\ell=n}^{2n^2} \sum_{I \in \mathcal{F}_l : \psi(I) = I'} \pi(I') \cdot \lambda^{-\frac{\ell_1 + \ell_2 + \ell_3}{6}}$$

$$< \sum_{\ell=n}^{2n^2} l^3 8n^6 36^\ell \cdot \pi(I') \cdot \lambda^{-\frac{\ell}{6}}$$

$$< 64n^{12} \left(\frac{36^6}{\lambda}\right)^{\frac{n}{6}} \pi(I').$$

The bound on the pre-image then allows us to bound $\pi(\mathcal{F})$ as follows:

$$\pi(\mathcal{F}) = \sum_{I' \in \mathrm{Img}(\psi)} \pi(\psi^{-1}(I'))$$

$$< \sum_{I' \in \mathrm{Img}(\psi)} 64n^{12} \left(\frac{36^6}{\lambda}\right)^{n} \pi(I')$$

$$= 64n^{12} \left(\frac{36^6}{\lambda}\right)^{\frac{n}{6}} \pi(\mathrm{Img}(\psi))$$

$$< 64n^{12} \left(\frac{36^6}{\lambda}\right)^{\frac{n}{6}}.$$

Taking $\lambda > 36^6$ completes the proof.

Note that by symmetry the sets $\mathcal{W}, \mathcal{B}$ and $\mathcal{G}$ have equal stationary probability. Observing now that their total weight is at least $1 - c^{-n}$, the conductance arguments of Sections 2 allow us to finish the proof of Theorem 1.

References

1. Aldous, D.: Random walks on finite groups and rapidly mixing Markov chains. Séminaire de Probabilités XVII, Lecture Notes in Mathematics vol. 986, pp. 243–297. Springer, Heidelberg (1981/82)
2. Baxter, R.J., Entig, I.G., Tsang, S.K.: Hard-square lattice gas. Journal of Statistical Physics 22, 465–489 (1989)
3. Borgs, C., Chayes, J.T., Frieze, A., Kim, J.H., Tetali, P., Vigoda, E., Vu, V.H.: Torpid mixing of some MCMC algorithms in statistical physics. In: Proceedings of the 40th IEEE Symposium on Foundations of Computer Science, 1999, pp. 218–229 (1999)
4. Dobrushin, R.L.: The problem of uniqueness of a Gibbsian random field and the problem of phase transitions. Functional Analysis and its Applications 2, 302–312 (1968)
5. Galvin, D., Kahn, J.: On phase transitions in the hard-core model on Z^d. Combinatorics, Probability and Computing 13, 137–164 (2004)
6. Galvin, D., Randall, D.: Sampling 3-colorings of the discrete torus. In: Proceedings of the 17th ACM/SIAM Symposium on Discrete Algorithms (SODA), 2007, pp. 376–384 (2007)
7. Jerrum, M.R., Sinclair, A.J.: The Markov chain Monte Carlo method: an approach to approximate counting and integration. In: Hochbaum, D.S. (ed.) Approximation Algorithms for NP-Hard Problems, pp. 482–520. PWS Publishing, Boston (1997)
8. Luby, M., Randall, D., Sinclair, A.J.: Markov Chains for Planar Lattice Structures. SIAM Journal on Computing 31, 167–192 (2001)
9. Luby, M., Vigoda, E.: Fast convergence of the Glauber dynamics for sampling independent sets. Random Structures and Algorithms, 229–241 (1999)
10. Metropolis, N., Rosenbluth, A.W., Rosenbluth, M.N., Teller, A.H., Teller, E.: Equation of state calculations by fast computing machines. Journal of Chemical Physics 21, 1087–1092 (1953)
11. Randall, D.: Slow Mixing of Glauber Dynamics via Topological Obstructions. In: Proceedings of the 16th ACM/SIAM Symposium of Discrete Algorithms (SODA), 2006, pp. 870–879 (2006)
12. Randall, D., Tetali, P.: Analyzing Glauber dynamics by comparison of Markov chains. Journal of Mathematical Physics 41, 1598–1615 (2000)
13. Thomas, L.E.: Bound on the mass gap for finite volume stochastic ising models at low temperature. Communications in Mathematical Physics 126, 1–11 (1989)

Better Binary List-Decodable Codes Via Multilevel Concatenation

Venkatesan Guruswami* and Atri Rudra**

Department of Computer Science and Engineering
University of Washington
Seattle, WA 98195
{venkat,atri}@cs.washington.edu

Abstract. We give a polynomial time construction of binary codes with the best currently known trade-off between rate and error-correction radius. Specifically, we obtain linear codes over fixed alphabets that can be list decoded in polynomial time up to the so called Blokh-Zyablov bound. Our work builds upon [7] where codes list decodable up to the Zyablov bound (the standard product bound on distance of concatenated codes) were constructed. Our codes are constructed via a (known) generalization of code concatenation called multilevel code concatenation. A probabilistic argument, which is also derandomized via conditional expectations, is used to show the existence of inner codes with a certain nested list decodability property that is appropriate for use in multilevel concatenated codes. A "level-by-level" decoding algorithm, which crucially uses the list recovery algorithm for folded Reed-Solomon codes from [7], enables list decoding up to the designed distance bound, aka the Blokh-Zyablov bound, for multilevel concatenated codes.

1 Introduction

1.1 Background and Context

A fundamental trade-off in the theory of error-correcting codes is the one between the proportion of redundancy built into codewords and the fraction of errors that can be corrected. Let us say we are interested in binary codes that can be used to recover the correct codeword even when up to a fraction ρ of its symbols could be corrupted by the channel. Such a channel can distort a codeword c (that is n bits long) into about $2^{H(\rho)n}$ possible received words, where $H(\rho) = -\rho \log_2 \rho - (1 - \rho) \log_2(1 - \rho)$ stands for the binary entropy function. Now for each of these words, the error-recovery procedure must identify c as a possibility for the true codeword. (In fact, even if the errors are random, the algorithm must identify c as a candidate codeword for *most* of these $2^{H(\rho)n}$ received words, if we seek a low decoding error probability.) To put it differently, if we require

---

* Supported in part by NSF CCF-0343672, a Sloan Research Fellowship and a David and Lucile Packard Foundation Fellowship.
** Supported in part by NSF CCF-0343672.

M. Charikar et al. (Eds.): APPROX and RANDOM 2007, LNCS 4627, pp. 554–568, 2007.

the error-recovery procedure to pin down a relatively small number of candidate codewords for all (or even most) received words, then there must be "nearly-disjoint" Hamming balls of size $2^{H(\rho)n}$ centered at each of the codewords. This implies that there can be at most about $2^{(1-H(\rho))n}$ codewords. Therefore the best rate of communication we can hope for when a fraction ρ of the bits can be corrupted is $1 - H(\rho)$.

If we could pack about $2^{(1-H(\rho))n}$ pairwise disjoint Hamming balls of radius ρn in $\{0,1\}^n$, then one can achieve a rate approaching $1 - H(\rho)$ while guaranteeing correct and unambiguous recovery of the codeword from an arbitrary fraction ρ of errors. Unfortunately, it is well known that such a "perfect" packing of Hamming balls in $\{0,1\}^n$ does not exist. Perhaps surprisingly (and fortunately), it turns out that it is possible to pack more than $2^{(1-H(\rho)-\varepsilon)n}$ such Hamming balls such that no $O(1/\varepsilon)$ of them intersect at a point. In fact a random packing has such a property with high probability.

In turn, this implies that for $0 < \rho < 1/2$ and any $\varepsilon > 0$, and all large enough n, there *exist* binary codes of rate $1 - H(\rho) - \varepsilon$ that enable correcting a fraction ρ of errors, outputting a list of at most $O(1/\varepsilon)$ answers in the worst-case (this error-recovery model is called "list decoding").[1] Therefore, one can approach the information-theoretically optimal rate of $1 - H(\rho)$. A similar result holds for codes over alphabet with q symbols – for correction of a fraction ρ, $0 < \rho < 1 - 1/q$, of errors, we can approach the optimal rate of $1 - H_q(\rho)$, where $H_q(\rho) = \rho \log_q(q-1) - \rho \log_q \rho - (1-\rho) \log_q(1-\rho)$ is the q-ary entropy function.

While the above pinpoints $R = 1 - H(\rho)$ as the optimal trade-off between the rate R of the code and the fraction ρ of errors that can corrected, it is a non-constructive result. The codes achieving this trade-off are shown to exist via a random coding argument and are not explicitly specified. Further, for a code to be useful, the decoding algorithm must be efficient, and for a random, unstructured code only brute-force decoders running in exponential time are known.

The big challenge then is to approach the above trade-off with explicit codes and polynomial time list decoding algorithms. Recently, in [7], we were able to achieve such a result for large alphabets. For large q, the optimal rate $1 - H_q(\rho)$ approaches $1 - \rho$, and in [7], we give explicit codes of rate $1 - \rho - \varepsilon$ over an alphabet of size $2^{(1/\varepsilon)^{O(1)}}$ with a polynomial time list decoding algorithm for a fraction ρ of errors (for any $0 < \rho < 1$). However, approaching the *list decoding capacity* of $1 - H_q(\rho)$ for any fixed small alphabet size q, such as $q = 2$, remains an important open question.

The best known tradeoff between R and ρ (from [7]) that can be achieved by an explicit binary code along with efficient list decoding algorithm is the so called Zyablov bound [12]. Figure 1 gives a pictorial comparison between the Zyablov bound and the list decoding capacity. As one can see, there is a still a huge gap between the nonconstructive results and what is known explicitly,

---

[1] The proof of Shannon's theorem for the binary symmetric channel also says that for most received words at most one codeword would be output.

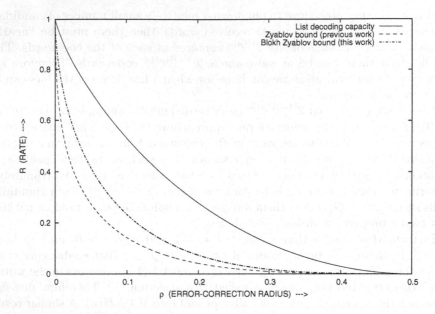

Fig. 1. Rate R of our binary codes plotted against the error-correction radius ρ of our algorithm. The best possible trade-off, i.e., capacity, is $\rho = H^{-1}(1 - R)$, and the Zyablov bound are also plotted.

Table 1. Values of rate at different error correction radius for List decoding capacity, Zyablov bound and Blokh Zyablov bound in the binary case. For rates above 0.4, the Blokh Zyablov bound is 0 up to 3 decimal places, hence we have not shown this.

ρ	0.01	0.02	0.03	0.05	0.10	0.15	0.20	0.25	0.30	0.35
Capacity rate	0.919	0.858	0.805	0.713	0.531	0.390	0.278	0.188	0.118	0.065
Zyablov rate	0.572	0.452	0.375	0.273	0.141	0.076	0.041	0.020	0.009	0.002
Blokh Zyablov rate	0.739	0.624	0.539	0.415	0.233	0.132	0.073	0.037	0.017	0.006

closing which is a challenging open problem. Narrowing this gap serves as the primary motivation for this work.

1.2 Our Results and Techniques

In this paper, we present linear codes over any fixed alphabet that can be constructed in polynomial time and can be efficiently list decoded up to the so called Blokh-Zyablov bound. This achieves a sizable improvement over the previous best known result (see Figure 1 and Table 1 for the binary case).

Our codes are constructed via multilevel concatenated codes. We will provide a formal definition later on — we just sketch the basic idea here. For an integer $s \geqslant 1$, a multilevel concatenated code C over $\mathbb{F}_q$ is obtained by combining s "outer" codes $C_{out}^0, C_{out}^1, \ldots, C_{out}^{s-1}$ of the same block length , say N, over large alphabets of size say $q^{a_0}, q^{a_1}, \ldots, q^{a_{s-1}}$, respectively, with a suitable "inner" code over $\mathbb{F}_q$. The inner code, say C_{in}, is of dimension $a_0 + a_1 \cdots + a_{s-1}$. Given messages $m_0, m_1, \ldots, m_{s-1}$ for the s outer codes, the encoding as per the multilevel generalized concatenation codes proceeds by first encoding each m_j as per C_{out}^j. Then for every $1 \leqslant i \leqslant N$, the collection of the ith symbols of $C_{out}^j(m_j)$ for $0 \leqslant j \leqslant s-1$, which can be viewed as a string over $\mathbb{F}_q$ of length $a_0 + a_1 + \cdots + a_{s-1}$, is encoded by the inner code. For $s = 1$ this reduces to the usual definition of code concatenation. In other words, this is like normal code concatenation with inner code C_{in} and outer code obtained by juxtaposing the symbols of codewords of $C_{out}^0, \ldots, C_{out}^{s-1}$.

We present a list decoding algorithm for C, given list recovering algorithms for the outer codes (list recovering is a generalization of list decoding that will be defined later) and list decoding algorithms for the inner code and some of its subcodes. What makes this part more interesting than the usual code concatenation, is that the inner code in addition to having list decodable properties, also needs to have good list decodable properties for certain subcodes. Specifically, the subcodes of dimension $a_j + a_{j+1} + \cdots + a_{s-1}$ of the inner code obtained by arbitrarily fixing the first $a_0 + \cdots + a_{j-1}$ symbols of the message, must have better list-decodability properties for increasing j (which is intuitively possible since they have lower rate). In turn, this allows the outer codes C_{out}^j to have rates increasing with j, leading to an overall improvement in the rate for a certain list-decoding radius.

To make effective use of the above approach, we also prove, via an application of the probabilistic method, that a random linear code over $\mathbb{F}_q$ has the required stronger condition on list decodability. By applying the method of conditional expectation ([1]), we can construct such a code deterministically in time singly exponential in the block length of the code (which is polynomial if the inner code encodes messages of length $O(\log n)$). Note that constructing such an inner code, given the existence of such codes, is easy in quasi-polynomial time by trying all possible generator matrices. The lower time complexity is essential for constructing the final code C in polynomial time.

1.3   Related Work

Our work can be seen as a generalization of the result of list decoding concatenated codes from [7]. The outer codes used in our work are the same as the ones used in [7]. However, the inner codes used in [7] are not sufficient for our purposes. Our proof of existence of the requisite inner codes (and in particular the derandomization of the construction of such codes using conditional expectation) is similar to the one used to establish list decodability properties of random "pseudolinear" codes in [6] (see also [5, Sec. 9.3]).

Concatenated codes were defined in the seminal thesis of Forney [4]. Its generalizations to linear multilevel concatenated codes were introduced by Blokh and Zyablov [2] and general multilevel concatenated codes were introduced by Zinoviev [10]. Our list decoding algorithm is inspired by the argument for "unequal error protection" property of multilevel concatenated codes [11].

1.4   Organization of the Paper

In Section 2, we start with some definitions and preliminaries. Section 3 presents a construction of a linear code that has good "nested" list decodable properties. In section 4, we present our algorithm for list decoding multilevel concatenated codes. Finally, in Section 5, we present the main result of the paper.

2   Preliminaries

2.1   Basic Coding Definitions

A code of *dimension* k and *block length* n over an alphabet Σ is a subset of Σ^n of size $|\Sigma|^k$. The *rate* of such a code equals k/n. Each vector in C is called a codeword. In this paper, we will focus on the case when Σ is a finite field. We will denote by $\mathbb{F}_q$ the field with q elements. A code C over $\mathbb{F}_q$ is called a linear code if C is a subspace of $\mathbb{F}_q^n$. In this case the dimension of the code coincides with the dimension of C as a vector space over $\mathbb{F}_q$. By abuse of notation we will also think of a code C as a map from elements in $\mathbb{F}_q^k$ to their corresponding codeword in $\mathbb{F}_q^n$. If C is linear, this map is a linear transformation, mapping a row vector $x \in \mathbb{F}_q^k$ to a vector $xG \in \mathbb{F}_q^n$ for a $k \times n$ matrix G over $\mathbb{F}_q$ called the generator matrix.

The Hamming distance between two vectors in Σ^n is the number of places they differ in. The (minimum) distance of a code C is the minimum hamming distance between any two pairs of distinct codewords from C. The relative distance is the ratio of the distance to the block length.

2.2   Multilevel Concatenated Codes

We will be working with multilevel concatenation coding schemes [3]. We start this section with the definition of multilevel concatenated codes. As the name suggests, these are generalizations of the well-studied concatenated codes. Recall that for a concatenated code, we start with a code C_{out} over a large alphabet (called the outer code). Then we need a code C_{in} that maps all symbols of the larger alphabet to strings over a smaller alphabet (called the inner code). The encoding for the concatenated code (denoted by $C_{out} \circ C_{in}$) is done as follows. We think of the message as being a string over the large alphabet and then encode it using C_{out}. Now we use C_{in} to encode each of the symbols in the codeword of C_{out} to get our codeword (in $C_{out} \circ C_{in}$) over the smaller alphabet. Most of the constructions of good binary codes are achieved via code concatenation. In particular, binary codes with the best known tradeoff (called the Zyablov bound) between rate and list decoding radius are constructed via code concatenation [7]. These codes have folded Reed-Solomon codes as outer codes and suitably chosen binary codes as inner codes, and can be list decoded up to the designed minimum distance, which is equal to the product of the outer and inner code distances.

Multilevel concatenation codes generalize the usual code concatenations in the following manner. Instead of there being one outer code, there are multiple outer codes. In particular, we "stack" codewords from these multiple outer codes and construct a matrix. The inner codes then act on the columns of these intermediate matrix. We now formally define multilevel concatenated codes (this will also contain the formal definition of the concatenated codes as a special case).

There are $s \geqslant 1$ outer codes, denoted by $C_{out}^0, C_{out}^1, \ldots, C_{out}^{s-1}$. For every $0 \leqslant i \leqslant s - 1$, C_{out}^i is a code of block length N and rate R_i and defined over a field $\mathbb{F}_{Q_i}$. The inner code C_{in} is code of block length n and rate r that maps tuples from $\mathbb{F}_{Q_0} \times \mathbb{F}_{Q_1} \times \cdots \times \mathbb{F}_{Q_{s-1}}$ to symbols in $\mathbb{F}_q$. In other words,

$$C_{out}^i : (\mathbb{F}_{Q_i})^{R_i N} \to (\mathbb{F}_{Q_i})^N,$$

$$C_{in} : \mathbb{F}_{Q_0} \times \mathbb{F}_{Q_1} \times \cdots \times \mathbb{F}_{Q_{s-1}} \to (\mathbb{F}_q)^n.$$

The multilevel concatenated code, denoted by $(C_{out}^0 \times C_{out}^1 \times \ldots C_{out}^{s-1}) \circ C_{in}$ is a map of the following form:

$$(C_{out}^0 \times C_{out}^1 \times \ldots C_{out}^{s-1}) \circ C_{in} : (\mathbb{F}_{Q_0})^{R_0 N} \times (\mathbb{F}_{Q_1})^{R_1 N} \times \cdots \times (\mathbb{F}_{Q_{s-1}})^{R_{s-1} N} \to \mathbb{F}_q^{nN}.$$

We now describe the encoding scheme. Given a message $(m_0, m_1, \ldots, m_{s-1}) \in (\mathbb{F}_{Q_0})^{R_0 N} \times (\mathbb{F}_{Q_1})^{R_1 N} \times \cdots \times (\mathbb{F}_{Q_{s-1}})^{R_{s-1} N}$, we first construct an $s \times N$ matrix M, whose i^{th} row is the codeword $C_{out}^i(m_i)$. Note that every column of M is an element from the set $\mathbb{F}_{Q_0} \times \mathbb{F}_{Q_1} \times \cdots \times \mathbb{F}_{Q_{s-1}}$. Let the jth column (for $1 \leqslant j \leqslant N$) be denoted by M_j. The codeword corresponding to the multilevel concatenated code ($C \stackrel{def}{=} (C_{out}^0 \times C_{out}^1 \times \ldots C_{out}^{s-1}) \circ C_{in}$) is defined as follows

$$C(m_0, m_1, \ldots, m_{s-1}) = (C_{in}(M_1), C_{in}(M_2), \cdots, C_{in}(M_N)).$$

(The codeword can be naturally be thought of as an $n \times N$ matrix, whose i'th column corresponds to the inner codeword encoding the i'th symbols of the s outer codewords.)

For the rest of the paper, we will only consider outer codes over the same alphabet, that is, $Q_0 = Q_1 = \cdots = Q_{s-1} = Q$. Further, $Q = q^a$ for some integer $a \geqslant 1$. Note that if $C_{out}^0, \ldots, C_{out}^{s-1}$ and C_{in} are all $\mathbb{F}_q$ linear, then so is $(C_{out}^0 \times C_{out}^1 \times \cdots \times C_{out}^{s-1}) \circ C_{in}$.

The gain from using multilevel concatenated codes comes from looking at the inner code C_{in} along with its subcodes. For the rest of the section, we will consider the case when C_{in} is linear (though the ideas can easily be generalized for general codes). Let $G \in \mathbb{F}_q^{as \times n}$ be the generator matrix for C_{in}. Let $r_0 = as/n$ denote the rate of C_{in}. For $0 \leqslant j \leqslant s - 1$, define $r_j = r_0(1 - j/s)$, and let G_j denote $r_j n \times n$ submatrix of G containing the last $r_j n$ rows of G. Denote the code generated by G_j by C_{in}^j; the rate of C_{in}^j is r_j. For our purposes we will actually look at the subcode of C_{in} where one fixes the first $0 \leqslant j \leqslant s - 1$ message symbols. Note that for every j these are just cosets of C_{in}^j. We will be looking at C_{in}, which in addition to having good list decoding properties as a "whole," also has good list decoding properties for each of its subcode C_{in}^j.

The multilevel concatenated code C ($= (C_{out}^0 \times \cdots \times C_{out}^{s-1}) \circ C_{in}$) has rate $R(C)$ that satisfies

$$R(C) = \frac{r_0}{s} \sum_{i=0}^{s-1} R_i . \tag{1}$$

The Blokh-Zyablov bound is the trade-off between rate and relative distance obtained when the outer codes meet the Singleton bound (i.e., C_{out}^j has relative distance $1 - R_j$), and the various subcodes C_{in}^j of the inner code, including the whole inner code $C_{in} = C_{in}^0$, lie on the Gilbert-Varshamov bound (i.e., have relative distance $\delta_j \geqslant H_q^{-1}(1 - r_j)$). The multilevel concatenated code then has relative distance at least $\min_{0 \leqslant j \leqslant s-1}(1 - R_j)H_q^{-1}(1 - r_j)$. Expressing the rate in terms of distance, the Blokh-Zyablov bound says that there exist multilevel concatenated C with relative distance at least δ with the following rate:

$$R_{BZ}^s(C) = \max_{0 < r < 1 - H_q(\delta)} r - \frac{r}{s}\sum_{i=0}^{s-1} \frac{\delta}{H_q^{-1}(1 - r + ri/s)}. \tag{2}$$

As s increases, the trade-off approaches the integral

$$R_{BZ}(C) = 1 - H_q(\delta) - \delta \int_0^{1-H_q(\delta)} \frac{dx}{H_q^{-1}(1 - x)}. \tag{3}$$

The convergence of $R_{BZ}^s(C)$ to $R_{BZ}(C)$ happens quite quickly even for small s such as $s = 10$.

2.3   List Decoding and List Recovering

Definition 1 (List decodable code). *For $0 < \rho < 1$ and an integer $L \geqslant 1$, a code $C \subseteq \mathbb{F}_q^n$ is said to be (ρ, L)-list decodable if for every $y \in \mathbb{F}_q^n$, the number of codewords in C that are within Hamming distance ρn from y is at most L.*

We will need to work with two different generalizations of list decoding. The first one is motivated by multilevel concatenation schemes. The definition looks more complicated than it really is.

Definition 2 (Nested linear list decodable code). *Given a linear code C in terms of some generator matrix $G \in \mathbb{F}_q^{k \times n}$, an integer s that divides k, a vector $\mathbf{L} = \langle L_0, L_1, \ldots, L_{s-1} \rangle$ of integers L_j $(0 \leqslant j \leqslant s - 1)$, a vector $\rho = \langle \rho_0, \rho_1 \ldots, \rho_{s-1} \rangle$ with $0 < \rho_j < 1$, and a vector $\mathbf{r} = \langle r_0, \ldots, r_{s-1} \rangle$ of reals where $r_0 = k/n$ and $0 \leqslant r_{s-1} < \cdots < r_i < r_0$, C is called an $(\mathbf{r}, \rho, \mathbf{L})$-nested list decodable if the following holds:*

For every $0 \leqslant j \leqslant s - 1$, C^j is a rate r_j code that is (ρ_j, L_j)-list decodable, where C^j is the subcode of C generated by the the last $r_j n$ rows of the generator matrix G.

The second generalization of list decoding called list recovering, a term first coined in [6] even though the notion existed before, has been extremely useful in list decoding concatenated codes. The input for list recovering is not a sequence of symbols but rather a sequence of lists (or more accurately sets, since the ordering of elements in the input lists does not matter).

Definition 3 (List recoverable code). *A code $C \subseteq \mathbb{F}_q^n$, is called (ρ, ℓ, L)-list recoverable if for every sequence of sets $S_1, S_2, \ldots, S_n$, where $S_i \subseteq \mathbb{F}_q$ and $|S_i| \leqslant \ell$ for every $1 \leqslant i \leqslant n$, there are at most L codewords in $c \in C$ such that $c_i \in S_i$ for at least $(1 - \rho)n$ positions i.*

The following simple folklore lemma shows how, for suitable parameters, a list recoverable outer code can be concatenated with a list decodable inner code to give a new list decodable code. The approach is simply to run the list decoding algorithm for each of the inner blocks, returning a list of possible symbols for each possible outer codeword symbol, which are then used as input to the list recovering algorithm for the outer code.

Lemma 1. *If C_{out} is a (ξ, ℓ, L)-list recoverable over an alphabet of size Q, and C_{in} is a (ρ, ℓ)-list decodable code with Q codewords, then the concatenated code $C_{out} \circ C_{in}$ is $(\xi \cdot \rho, L)$-list decodable.*

2.4   Known Result on List Recoverable Codes

We will use the following powerful result concerning good list recoverable codes from [7]; these codes will serve as the outer codes in our multilevel concatenation scheme.

Theorem 1. *For every integer $\ell \geqslant 1$, for all constants $\varepsilon > 0$, for all R, R'; $0 < R \leqslant R' < 1$, and for every prime p, there is an* explicit *family of folded Reed-Solomon codes, over fields of characteristic p that have rate at least R and which can be $(1 - R - \varepsilon, \ell, L(N))$-list recovered in polynomial time, where for codes of block length N, $L(N) = (N/\varepsilon^2)^{O(\varepsilon^{-1} \log(\ell/R))}$ and the code is defined over alphabet of size $(N/\varepsilon^2)^{O(\varepsilon^{-2} \log \ell/(1-R'))}$.*

We remark that the above theorem was stated with $R' = R$ in [7], though the above follows immediately from the proof for $R' = R$ and properties of the folded Reed-Solomon codes [9]. The proof for $R' > R$ uses folded Reed-Solomon codes with a larger "folding" parameter. A larger folding parameter increases the fraction of errors that can be tolerated at the expense of a larger alphabet size.

3   Linear Codes with Good Nested List Decodability

In this section, we will prove the following result concerning the existence (and constructibility) of linear codes over any fixed alphabet with good nested list decodability properties.

Theorem 2. *For any integer $s \geqslant 1$ and reals $0 < r_{s-1} < r_{s-2} < \cdots < r_1 < r_0 < 1$, $\varepsilon > 0$, let $\rho_j = H_q^{-1}(1 - r_j - 2\varepsilon)$ for every $0 \leqslant j \leqslant s-1$. Let $\mathbf{r} = \langle r_0, \ldots, r_{s-1} \rangle$, $\rho = \langle \rho_0, \rho_1, \ldots, \rho_{s-1} \rangle$ and $\mathbf{L} = \langle L_0, L_1, \ldots, L_{s-1} \rangle$, where $L_j = q^{1/\varepsilon}$. For large enough n, there exists a linear code (over fixed alphabet $\mathbb{F}_q$) that is $(\mathbf{r}, \rho, \mathbf{L})$-nested list decodable. Further, such a code can be constructed in time $q^{O(n/\varepsilon)}$.*

Proof. We will show the existence of the required codes via a simple use of the probabilistic method (in fact, we will show that a random linear code has the required properties with high probability). We will then use the method of conditional expectation ([1]) to derandomize the construction with the claimed time complexity.

Define $k_j = \lfloor r_i n \rfloor$ for every $0 \leqslant j \leqslant s-1$. We will pick a random $k_0 \times n$ matrix G with entries picked independently from $\mathbb{F}_q$. We will show that the linear code C generated by G has good nested list decodable properties with high probability. Let C_j, for $0 \leqslant j \leqslant s-1$ be the code generated by the "bottom" k_j rows of G. Recall that we have to show that with high probability C_j is $(\rho_j, q^{1/\varepsilon})$ list decodable for every $0 \leqslant j \leqslant s-1$ (C_j obviously has rate r_j). Finally for integers $J, k \geqslant 1$, and a prime power q, let $\mathrm{Ind}(q, k, J)$ denote the collection of subsets $\{x_1, x_2, \ldots, x_J\} \subseteq \mathbb{F}_q^k$ such that all vectors $x_1, \ldots, x_J$ are linearly independent over $\mathbb{F}_q$.

We recollect the following two straightforward facts: (i) Given any L distinct vectors from $\mathbb{F}_q^k$, for some $k \geqslant 1$, at least $\lceil \log_q L \rceil$ of them are linearly independent; (ii) Any set of linearly independent vectors in $\mathbb{F}_q^k$ are mapped to independent random vectors in $\mathbb{F}_q^n$ by a random $k \times n$ matrix over $\mathbb{F}_q$.

We now move on to the proof of existence of linear codes with good nested list decodability. We will actually do the proof in a manner that will facilitate the derandomization of the proof. Define $J = \lceil \log_q(q^{1/\varepsilon} + 1) \rceil$. For any vector $\mathbf{r} \in \mathbb{F}_q^n$, integer $0 \leqslant j \leqslant s-1$, subset $T = \{x_1, \ldots, x_J\} \in \mathrm{Ind}(q, k_j, J)$ and any collection $\mathcal{S}$ of subsets $S_1, S_2, \ldots, S_J \subseteq \{1, \ldots, n\}$ of size at most $\rho_j n$, define an indicator variable $I(j, \mathbf{r}, T, \mathcal{S})$ in the following manner. $I(j, \mathbf{r}, T, \mathcal{S}) = 1$ if and only if for every $1 \leqslant i \leqslant J$, $C(x_i)$ differs from $\mathbf{r}$ in exactly the set S_i. Note that if for some $0 \leqslant j \leqslant s-1$, there are $q^{1/\varepsilon} + 1$ codewords in C_j all of which differ from some received word $\mathbf{r}$ in at most $\rho_j n$ places, then this set of codewords is a "counter-example" that shows that C is not $(\mathbf{r}, \rho, \mathbf{L})$-nested list decodable. Since the $q^{1/\varepsilon} + 1$ codewords will have some set T of J linearly independent codewords, the counter example will imply that $I(j, \mathbf{r}, T, \mathcal{S}) = 1$ for some collection of subsets $\mathcal{S}$. In other words, the indicator variable captures the set of bad events we would like to avoid. Finally define the sum of all the indicator variables as follows:

$$S_C = \sum_{j=0}^{s-1} \sum_{\mathbf{r} \in \mathbb{F}_q^n} \sum_{T \in \mathrm{Ind}(q, k_j, J)} \sum_{\substack{\mathcal{S} = \{S_1, \ldots, S_J\}, \\ S_i \subseteq \{1, \ldots, n\}, |S_i| \leqslant \rho_j n}} I(j, \mathbf{r}, T, \mathcal{S}).$$

Note that if $S_C = 0$, then C is $(\mathbf{r}, \rho, \mathbf{L})$-nested list decodable as required. Thus, we can prove the existence of such a C if we can show that $\mathbf{E}_C[S_C] < 1$. By linearity of expectation, we have

$$\mathbf{E}[S_C] = \sum_{j=0}^{s-1} \sum_{\mathbf{r} \in \mathbb{F}_q^n} \sum_{T \in \mathrm{Ind}(q, k_j, J)} \sum_{\substack{\mathcal{S} = \{S_1, \ldots, S_J\}, \\ S_i \subseteq \{1, \ldots, n\}, |S_i| \leqslant \rho_j n}} \mathbf{E}[I(j, \mathbf{r}, T, \mathcal{S})]. \qquad (4)$$

Fix some arbitrary $j, \mathbf{r}, T = \{x_1, x_2, \ldots, x_J\}, \mathcal{S} = \{S_1, S_2, \ldots, S_J\}$ (in their corresponding domains). Then we have

$$\mathbf{E}[I(j,\mathbf{r},T,\mathcal{S})] = \Pr[I(j,\mathbf{r},T,\mathcal{S}) = 1]$$

$$= \prod_{x_i \in T} \Pr[C(x_i) \text{ differ from } \mathbf{r} \text{ in exactly the positions in } S_i]$$

$$= \prod_{i=1}^{J}\left(\frac{q-1}{q}\right)^{|S_i|}\left(\frac{1}{q}\right)^{n-|S_i|} = \prod_{i=1}^{J}\frac{(q-1)^{|S_i|}}{q^n}, \tag{5}$$

where the second and the third equality follow from the definition of the indicator variable, the fact that vectors in T are linearly independent and the fact that a random matrix maps linearly independent vectors to independent uniformly random vectors in $\mathbb{F}_q^n$. Using (5) in (4), we get

$$\mathbf{E}[S_C] = \sum_{j=0}^{s-1}\sum_{\mathbf{r}\in\mathbb{F}_q^n}\sum_{T\in\mathrm{Ind}(q,k_j,J)}\sum_{\substack{\mathcal{S}=\{S_1,\dots,S_J\},\\ S_i\subseteq\{1,\dots,n\},|S_i|\leqslant\rho_j n}}\prod_{i=1}^{J}\frac{(q-1)^{|S_i|}}{q^n}$$

$$= \sum_{j=0}^{s-1}\sum_{\mathbf{r}\in\mathbb{F}_q^n}\sum_{T\in\mathrm{Ind}(q,k_j,J)}\sum_{(\ell_1,\ell_2,\dots,\ell_J)\in\{0,1,\dots,\rho_j n\}^J}\prod_{i=1}^{J}\binom{n}{\ell_i}\frac{(q-1)^{\ell_i}}{q^n}$$

$$= \sum_{j=0}^{s-1}\sum_{\mathbf{r}\in\mathbb{F}_q^n}\sum_{T\in\mathrm{Ind}(q,k_j,J)}\left(\sum_{\ell=0}^{\rho_j n}\binom{n}{\ell}\frac{(q-1)^{\ell}}{q^n}\right)^J$$

$$\leqslant \sum_{j=0}^{s-1}\sum_{\mathbf{r}\in\mathbb{F}_q^n}\sum_{T\in\mathrm{Ind}(q,k_j,J)}q^{nJ(H_q(\rho_j)-1)} \leqslant \sum_{j=0}^{s-1}q^n\cdot q^{Jk_j}\cdot q^{nJ(H_q(\rho_j)-1)}$$

$$\leqslant \sum_{j=0}^{s-1}q^{nJ(1/J+r_j+1-r_j-2\varepsilon-1)} \leqslant sq^{-\varepsilon nJ}. \tag{6}$$

The first inequality follows the following known inequality for $p < 1 - 1/q$ ([8]): $\sum_{i=0}^{pn}\binom{n}{i}(q-1)^i \leqslant q^{H_q(p)n}$. The second inequality follows by upper bounding the number of J linearly independent vectors in $\mathbb{F}_q^{k_j}$ by q^{Jk_j}. The third inequality follows from the fact that $k_j = \lfloor r_j n\rfloor$ and $\rho_j = H_q^{-1}(1 - r_j - 2\varepsilon)$, The final inequality follows from the fact that $J = \lceil\log_q(q^{1/\varepsilon}+1)\rceil$.

Thus, (6) shows that there exists a code C (in fact with high probability) that is $(\mathbf{r}, \rho, \mathbf{L})$-nested list decodable. In fact, this could have been proved using a simpler argument. However, the advantage of the argument above is that we can now apply the method of conditional expectations to derandomize the above proof.

The algorithm to deterministically generate a linear code C that is $(\mathbf{r}, \rho, \mathbf{L})$-nested list decodable is as follows. The algorithm consists of n steps. At any step $1 \leqslant i \leqslant n$, we choose the ith column of the generator matrix to be the value $\mathbf{v}_i \in \mathbb{F}_q^{k_0}$ that minimizes the conditional expectation $\mathbf{E}[S_C|\mathbf{G}_1 = \mathbf{v}_1,\dots,\mathbf{G}_{i-1} = \mathbf{v}_{i-1}, \mathbf{G}_i = \mathbf{v}_i]$, where $\mathbf{G}_i$ denotes the ith column of $\mathbf{G}$ and $\mathbf{v}_1,\dots,\mathbf{v}_{i-1}$ are the column vectors chosen in the previous $i-1$ steps. This algorithm would work only if for any $1 \leqslant i \leqslant n$ and vectors $\mathbf{v}_1, \ldots, \mathbf{v}_i$, we can exactly compute $\mathbf{E}[S_C|\mathbf{G}_1 = \mathbf{v}_1, \ldots, \mathbf{G}_i = \mathbf{v}_i]$. Indeed from (4), we have $\mathbf{E}[S_C|\mathbf{G}_1 = \mathbf{v}_1, \ldots, \mathbf{G}_i = \mathbf{v}_i]$ is

$$
\sum_{j=0}^{s-1} \sum_{\mathbf{r} \in \mathbb{F}_q^n} \sum_{T \in \mathrm{Ind}(q, k_j, J)} \sum_{\substack{\mathcal{S}=\{S_1, \ldots, S_J\}, \\ S_i \subseteq \{1, \ldots, n\}, |S_i| \leqslant \rho_j n}} \mathbf{E}[I(j, \mathbf{r}, T, \mathcal{S})|\mathbf{G}_1 = \mathbf{v}_1, \ldots, \mathbf{G}_i = \mathbf{v}_i].
$$

Thus, we would be done if we can compute the following for every value of $j, \mathbf{r}, T = \{x_1, \ldots, x_J\}, \mathcal{S} = \{S_1, \ldots, S_J\}$: $\mathbf{E}[I(j, \mathbf{r}, T, \mathcal{S}) = 1|\mathbf{G}_1 = \mathbf{v}_1, \ldots, \mathbf{G}_i = \mathbf{v}_i]$. Note that fixing the first i columns of G implies fixing the value of the codewords in the first i positions. Thus, the indicator variable is 0 (or in other words, the conditional expectation we need to compute is 0) if for some message, the corresponding codeword does not disagree with $\mathbf{r}$ exactly as dictated by $\mathcal{S}$ in the first i positions. More formally, $I(j, \mathbf{r}, T, \mathcal{S}) = 0$ if the following is true for some $1 \leqslant \ell \leqslant i$ and $0 \leqslant i' \leqslant J$: $x_{i'} \cdot \mathbf{G}_\ell \neq \mathbf{r}_\ell$, if $\ell \notin S_{i'}$ and $x_{i'} \cdot \mathbf{G}_\ell = \mathbf{r}_\ell$ otherwise. However, if none of these conditions hold, then using argument similar to the ones used to obtain (5), one can show that

$$
\mathbf{E}[I(j, \mathbf{r}, T, \mathcal{S})|\mathbf{G}_1 = \mathbf{v}_1, \ldots, \mathbf{G}_i = \mathbf{v}_i] = \prod_{\ell=1}^{J} \left(\frac{q-1}{q} \right)^{|S'_\ell|} \left(\frac{1}{q} \right)^{n-i-|S'_\ell|},
$$

where $S'_\ell = S_\ell \setminus \{1, 2, \ldots, i\}$ for every $1 \leqslant \ell \leqslant J$.

To complete the proof, we need to estimate the time complexity of the above algorithm. There are n steps and at every step i, the algorithm has to consider $q^{k_0} \leqslant q^n$ different choices of $\mathbf{v}_i$. For every choice of $\mathbf{v}_i$, the algorithm has to compute the conditional expectation of the indicator variables for all possible values of $j, \mathbf{r}, T, \mathcal{S}$. It is easy to check that there are $\sum_{i=1}^{s} q^n \cdot q^{Jk_j} \cdot 2^{nJ} \leqslant sq^{n(1+2J)}$ possibilities. Finally, the computation of the conditional expected value of a fixed indicator variable takes time $O(snJ)$. Thus, in all the total time taken is $O(n \cdot q^n \cdot sq^{n(1+2J)} \cdot snJ) = q^{O(n/\varepsilon)}$, as required. ∎

4 List Decoding Multilevel Concatenated Codes

In this section, we will see how one can list decode multilevel concatenated codes, provided the outer codes have good list recoverability and the inner code has good nested list decodability. We have the following result, which generalizes Lemma 1 for regular concatenated codes (the case $s = 1$).

Theorem 3. *Let $s \geqslant 1$ and $\ell \geqslant 1$ be integers. Let $0 < R_0 < R_1 < \cdots < R_{s-1} < 1$, $0 < r_0 < 1$, $0 < \xi_0, \cdots, \xi_{s-1} < 1$, $0 < \rho_0, \cdots, \rho_{s-1} < 1$ and $\varepsilon > 0$ be reals. Let q be a prime power and let $Q = q^a$ for some integer $a > 1$. Further, let C_{out}^j $(0 \leqslant j \leqslant s-1)$ be an $\mathbb{F}_q$-linear code over $\mathbb{F}_Q$ of rate R_j and block length N that is (ξ_j, ℓ, L)-list recoverable. Finally, let C_{in} be a linear $(\mathbf{r}, \rho, \mathbf{L})$-nested list decodable code over $\mathbb{F}_q$ of rate r_0 and block length $n = as/r_0$, where $\mathbf{r} = \langle r_0, \cdots, r_{s-1} \rangle$ with $r_i = (1 - i/s)r_0$, $\rho = \langle \rho_0, \cdots, \rho_{s-1} \rangle$ and $\mathbf{L} = \langle \ell, \ell, \cdots, \ell \rangle$.*

Then $C = (C_{out}^0 \times \cdots \times C_{out}^{s-1}) \circ C_{in}$ is a linear $(\min_j \xi_j \cdot \rho_j, L^s)$-list decodable code. Further, if the outer code C_{out}^j can be list recovered in time $T_j(N)$ and the inner code C_{in} can be list decoded in time $t_j(n)$ (for the jth level), then C can be list decoded in time $O\left(\sum_{j=0}^{s-1} L^j \left(T_j(N) + N \cdot t_j(n)\right)\right)$.

Proof. Given list recovering algorithms for C_{out}^j and list decoding algorithms for C_{in} (and its subcodes C_{in}^j), we will design a list decoding algorithm for C. Recall that the received word is an $n \times N$ matrix over $\mathbb{F}_q$. Each consecutive "chunk" of n/s rows should be decoded to a codeword in C_{out}^j. The details follow.

Before we describe the algorithm, we will need to fix some notation. Define $\delta = \min_j \xi_j \rho_j$. Let $\mathbf{R} \in \mathbb{F}_q^{nN}$ be the received word, which we will think of as an $n \times N$ matrix over $\mathbb{F}_q$ (note that s divides n). For any $n \times N$ matrix M and for any $1 \leqslant i \leqslant N$, let $M_i \in \mathbb{F}_q^n$ denote the ith column of the matrix M. Finally, for every $0 \leqslant j \leqslant s-1$, let C_{in}^j denote the subcode of C_{in} generated by all but the first ja rows of the generator matrix of C_{in}. We are now ready to describe our algorithm.

Recall that the algorithm needs to output all codewords in C that differ from $\mathbf{R}$ in at most δ fraction of positions. For the ease of exposition, we will consider an algorithm that outputs matrices from $C_{out}^0 \times \cdots \times C_{out}^{s-1}$. The algorithm has s phases. At the end of phase j ($0 \leqslant j \leqslant s-1$), the algorithm will have a list of matrices (called $\mathcal{L}_j$) from $C_{out}^0 \times \cdots \times C_{out}^j$, where each matrix in $\mathcal{L}_j$ is a possible submatrix of some matrix that will be in the final list output by the algorithm. The following steps are performed in phase j (where we are assuming that the list decoding algorithm for C_{in}^j returns a list of messages while the list recovering algorithm for C_{out}^j returns a list of codewords).

1. Set $\mathcal{L}_j$ to be the empty set.
2. For every $\mathbf{c} = (c_0, \cdots, c_{j-1}) \in \mathcal{L}_{j-1}$ repeat the following steps (if this is the first phase, that is $j = 0$, then repeat the following steps once):
   (a) Let G_j be the first aj rows of the generator matrix of C_{in}. Let $\mathbf{X} = (G_j)^T \cdot \mathbf{c}$, where we think of $\mathbf{c}$ as an $ja \times N$ matrix over $\mathbb{F}_q$. Let $\mathbf{Y} = \mathbf{R} - \mathbf{X}$ (for $j = 0$ we use the convention that $\mathbf{X}$ is the all 0s matrix). For every $1 \leqslant i \leqslant N$, use the list decoding algorithm for C_{in}^j on column $\mathbf{Y}_i$ for up to ρ_j fraction of errors to obtain list $S_i^j \subseteq (\mathbb{F}_Q)^{s-j}$. Let $T_i^j \subseteq \mathbb{F}_Q$ be the projection of every vector in S_i^j on to its first component.
   (b) Run the list recovery algorithm for C_{out}^j on set of lists $\{T_i^j\}_i$ obtained from the previous step for up to ξ_j fraction of errors. Store the set of codewords returned in I_j.
   (c) Add $\{(\mathbf{c}, \mathbf{v}) | \mathbf{v} \in I_j\}$ to $\mathcal{L}_j$.

At the end, remove all the matrices $M \in \mathcal{L}_{s-1}$, for which the codeword $(C_{in}(M_1), C_{in}(M_2), \cdots, C_{in}(M_N))$ is at a distance more than δ from $\mathbf{R}$. Output the remaining matrices as the final answer.

We will first talk about the running time complexity of the algorithm. It is easy to check that each repetition of steps 2(a)-(c) takes time $O(T_j(N) + N \cdot t_j(n))$.

To compute the final running time, we need to get a bound on number of times step 2 is repeated in phase j. It is easy to check that the number of repetitions is exactly $|\mathcal{L}_{j-1}|$. Thus, we need to bound $|\mathcal{L}_{j-1}|$. By the list recoverability property of C_{out}^j, we can bound $|I_j|$ by L. This implies that $|\mathcal{L}_j| \leqslant L|\mathcal{L}_{j-1}|$, and therefore by induction we have

$$|\mathcal{L}_i| \leqslant L^{i+1} \quad \text{for } i = 0, 1, \ldots, s-1. \tag{7}$$

Thus, the overall running time and the size of the list output by the algorithm are as claimed in the statement of the theorem.

We now argue the correctness of the algorithm. That is, we have to show that for every $M \in C_{out}^0 \times \cdots \times C_{out}^{s-1}$, such that $(C_{in}(M_1), C_{in}(M_2) \cdots, C_{in}(M_N))$ is at a distance at most δ from $\mathbf{R}$ (call such an M a *good* matrix), $M \in \mathcal{L}_{s-1}$. In fact, we will prove a stronger claim: for every good matrix M and every $0 \leqslant j \leqslant s-1$, $M^j \in \mathcal{L}_j$, where M^j denotes the submatrix of M that lies in $C_{out}^0 \times \cdots \times C_{out}^j$ (that is the first j "rows" of M). For the rest of the argument fix an arbitrary good matrix M. Now assume that the stronger claim above holds for $j' - 1 (< s - 1)$. In other words, $M^{j'-1} \in \mathcal{L}_{j'-1}$. Now, we need to show that $M^{j'} \in \mathcal{L}_{j'}$.

For concreteness, let $M = (m_0, \cdots, m_{s-1})^T$. As M is a good matrix and $\delta \leqslant \xi_{j'}\rho_{j'}$, $C_{in}(M_i)$ can disagree with $\mathbf{R}_i$ on at least a fraction $\rho_{j'}$ of positions for at most $\xi_{j'}$ fraction of column indices i. The next crucial observation is that for any column index i, $C_{in}(M_i) = (G_{j'})^T \cdot (m_{0,i}, \cdots, m_{j'-1,i}) + (G \setminus G_{j'})^T \cdot (m_{j',i}, \cdots, m_{s-1,i})$, where $G_{j'}$ is as defined in step 2(a), $G \setminus G_{j'}$ is the submatrix of G obtained by "removing" $G_{j'}$ and $m_{j',i}$ is the ith component of the vector $m_{j'}$. The following might help the reader to visualize the different variables.

$$G^T \cdot M = \begin{pmatrix} (G_{j'})^T & (G \setminus G_{j'})^T \end{pmatrix} \cdot \begin{pmatrix} m_{0,1} & \cdots & m_{0,i} & \cdots & m_{0,N} \\ & & \vdots & & \\ m_{j'-1,1} & \cdots & m_{j'-1,i} & \cdots & m_{j'-1,N} \\ m_{j',1} & \cdots & m_{j',i} & \cdots & m_{j',N} \\ & & \vdots & & \\ m_{s-1,1} & \cdots & m_{s-1,i} & \cdots & m_{s-1,N} \end{pmatrix}$$

$$= \begin{pmatrix} \uparrow & & \uparrow & & \uparrow \\ C_{in}(M_1) & \cdots & C_{in}(M_i) & \cdots & C_{in}(M_N) \\ \downarrow & & \downarrow & & \downarrow \end{pmatrix}$$

Note that $G \setminus G_{j'}$ is the generator matrix of $C_{in}^{j'}$. Thus, for at most $\xi_{j'}$ fraction of column indices i, $(m_{j',i}, \cdots, m_{s-1,i}) \cdot (G \setminus G_{j'})$ disagrees with $\mathbf{R}_i - \mathbf{X}_i$ on at least $\rho_{j'}$ fraction of places, where $\mathbf{X}$ is as defined in Step 2(a), and $\mathbf{X}_i$ denotes the i'th column of $\mathbf{X}$. As $C_{in}^{j'}$ is $(\rho_{j'}, \ell)$-list decodable, for at least $1 - \xi_{j'}$ fraction of column index i, $M_i^{j'}$ will be in $S_i^{j'}$ (where $M_i^{j'}$ is M_i projected on it's last $s - j'$ co-ordinates and $S_i^{j'}$ is as defined in Step 2(a)). In other words, $m_{j',i}$ is in $T_i^{j'}$ for at least $1 - \xi_{j'}$ fraction of i's. Further, as $|S_i^{j'}| \leqslant \ell$, $|T_i^{j'}| \leqslant \ell$. This implies with the list recoverability property of $C_{out}^{j'}$ that $m_{j'} \in I_{j'}$, where $I_{j'}$ is as defined in step 2(b). Finally, step 2(c) implies that $M^{j'} \in \mathcal{L}_{j'}$ as required.

The proof of correctness of the algorithm along with (7) shows that C is (δ, L^s)-list decodable, which completes the proof. ∎

5   List Decoding Up to the Blokh-Zyablov Bound

We combine the results we have proved in the last couple of sections to get our main result.

Theorem 4 (Main). *For every fixed field* $\mathbb{F}_q$, *reals* $0 < \delta < 1, 0 < r \leqslant 1 - H_q(\delta), \varepsilon > 0$ *and integer* $s \geqslant 1$, *there exists linear codes* C *over* $\mathbb{F}_q$ *of block length* N *that are* $(\delta - \varepsilon, L(N))$-*list decodable with rate* R *such that*

$$R = r - \frac{r}{s} \sum_{i=0}^{s-1} \frac{\delta}{H_q^{-1}(1 - r + ri/s)}, \tag{8}$$

and $L(N) = (N/\varepsilon^2)^{O\left(s\varepsilon^{-3}\delta/(H_q^{-1}(1-r)-\delta)\right)}$. *Finally,* C *can be constructed in time* $(N/\varepsilon^2)^{O(s/(\varepsilon^6 r\delta))}$ *and list decoded in time polynomial in* N.

Proof. Let $\gamma > 0$ (we will define its value later). For every $0 \leqslant j \leqslant s - 1$ define $r_j = r(1 - j/s)$ and $R_j = 1 - \frac{\delta}{H_q^{-1}(1-r_j)}$. The code C is going to be a multilevel concatenated code $(C_{out}^0 \times \cdots \times C_{out}^{s-1}) \circ C_{in}$, where C_{out}^j is the code from Theorem 1 of rate R_j and block length N' (over $\mathbb{F}_{q^a}$) and C_{in} is an $(\langle r_0, \ldots, r_{s-1} \rangle, \rho, \mathbf{L})$-nested list decodable code as guaranteed by Theorem 2, where for $0 \leqslant j \leqslant s - 1$, $\rho_j = H_q^{-1}(1 - r_j - 2\gamma^2)$ and $L_j = q^{1/\gamma^2}$. Finally, we will use the property of C_{out}^j that it is $(1 - R - \gamma, q^{1/\gamma^2}, (N'/\gamma^2)^{O(\gamma^{-3}\log(1/R_j))})$-list recoverable. Theorem 1 implies that such codes exist with (where we apply Theorem 1 with $R' = \max_j R_j = 1 - \delta/H_q^{-1}(1 - r/s)$)

$$q^a = (N'/\gamma^2)^{O(\gamma^{-4}H_q^{-1}(1-r/s)/\delta)}. \tag{9}$$

Further, as codes from Theorem 1 are $\mathbb{F}_q$-linear [7], C is a linear code.

The claims on the list decodability of C follows from the choices of R_j and r_j and Theorems 1, 2 and 3. In particular, note that we invoke Theorem 3 with the following parameters: $\xi_j = 1 - R_j - \gamma$ and $\rho_j = H_q^{-1}(1 - r_j - 2\gamma^2)$ (which implies[2] that $\xi_j \rho_j \geqslant \delta - \varepsilon$ as long as $\gamma = \Theta(\varepsilon)$), $\ell = q^{1/\gamma^2}$ and $L = (N'/\gamma^2)^{O(\gamma^{-1}\log(\ell/R_j))}$. The choices of ℓ and γ imply that $L = (N/\varepsilon^2)^{O(\varepsilon^{-3}\log(1/R_j))}$. Now $\log(1/R_j) \leqslant \log(1/R_{min})$, where $R_{min} = \min_j R_j = 1 - \delta/H_q^{-1}(1-r)$. Finally, we use the fact that for any $0 < y < 1$, $\ln(1/y) \leqslant 1/y - 1$ to get that $\log(1/R_j) \leqslant O(1/R_{min} - 1) = O(\delta/(H_q^{-1}(1 - r) - \delta))$. The claimed upper bound of $L(N)$ follows as $L(N) \leqslant L^s$ (by Theorem 3).

---

[2] As for any $0 < x < 1$ and small enough $\alpha > 0$, $H_q^{-1}(x - \alpha^2) \geqslant H_q^{-1}(x) - \Theta(\alpha)$ [9].

By the choices of R_j and r_j and (1), the rate of C is as claimed. The construction time for C is the time required to construct C_{in}, which by Theorem 2 is $2^{O(n/\gamma^2)}$ where n is the block length of C_{in}. Note that $n = as/r$, which by (9) implies that the construction time is $(N/\varepsilon^2)^{O(\varepsilon^{-6}sH_q^{-1}(1-r/s)/(r\delta))}$. The claimed running time follows by using the bound $H_q^{-1}(1 - r/s) \leqslant 1$.

We finally consider the running time of the list decoding algorithm. We list decode the inner code(s) by brute force, which takes $2^{O(n)}$ time, that is, $t_j(n) = 2^{O(n)}$. Thus, Theorems 1, 3 and the bound on $L(N)$ implies the claimed running time complexity. ∎

Choosing the parameter r in the above theorem so as to maximize (8) gives us linear codes over any fixed field whose rate vs. list decoding radius tradeoff meets the Blokh-Zyablov bound (2). As s grows, the trade-off approaches the integral form (3) of the Blokh-Zyablov bound.

References

1. Alon, N., Spencer, J.: The Probabilistic Method. John Wiley and Sons, Inc. Chichester (1992)
2. Blokh, E.L., Zyablov, V.V.: Linear Concatenated Codes. (in Russian) Moscow: Nauka (1982)
3. Dumer, I.I.: Concatenated codes and their multilevel generalizations. In: Pless, V.S., Huffman, W.C. (eds.) Handbook of Coding Theory, vol. 2, pp. 1911–1988. North Holland (1998)
4. Forney, G.D.: Concatenated Codes. MIT Press, Cambridge, MA (1966)
5. Guruswami, V.: List decoding of error-correcting codes. LNCS, vol. 3282, Springer, Heidelberg (2004) (Winning Thesis of the 2002 ACM Doctorial Dissertation Competition)
6. Guruswami, V., Indyk, P.: Expander-based constructions of efficiently decodable codes. In: Proceedings of the 42nd Annual IEEE Symposium on Foundations of Computer Science, 2001, pp. 658–667. IEEE Computer Society Press, Los Alamitos (2001)
7. Guruswami, V., Rudra, A.: Explicit capacity-achieving list-decodable codes. In: Proceedings of the 38th Annual ACM Symposium on Theory of Computing, (May 2006) pp. 1–10. ACM Press, New York (2006)
8. MacWilliams, F.J., Sloane, N.J.A.: The Theory of Error-Correcting Codes. Elsevier/North-Holland, Amsterdam (1981)
9. Rudra, A.: List Decoding and Property Testing of Error Correcting Codes. PhD thesis, University of Washington (2007)
10. Zinoviev, V.A.: Generalized concatenated codes. Prob. Peredachi Inform. 12(1), 5–15 (1976)
11. Zinoviev, V.A., Zyablov, V.V.: Codes with unequal protection. Prob. Peredachi Inform. 15(4), 50–60 (1979)
12. Zyablov, V.V.: An estimate of the complexity of constructing binary linear cascade codes. Problems of Information Transmission 7(1), 3–10 (1971)

Worst-Case to Average-Case Reductions Revisited

Dan Gutfreund[1] and Amnon Ta-Shma[2]

[1] SEAS, Harvard University,
Cambridge, MA 02138
danny@eecs.harvard.edu
[2] Computer Science Department,
Tel-Aviv University, Israel, 69978
amnon@post.tau.ac.il

Abstract. A fundamental goal of computational complexity (and foundations of cryptography) is to find a polynomial-time samplable distribution (e.g., the uniform distribution) and a language in $\text{NTIME}(f(n))$ for some polynomial function f, such that the language is hard on the average with respect to this distribution, given that NP is worst-case hard (i.e. NP $\neq$ P, or NP $\not\subseteq$ BPP). Currently, no such result is known even if we relax the language to be in nondeterministic sub-exponential time. There has been a long line of research trying to explain our failure in proving such worst-case/average-case connections [FF93, Vio03, BT03, AGGM06]. The bottom line of this research is essentially that (under plausible assumptions) non-adaptive Turing reductions cannot prove such results.

In this paper we revisit the problem. Our first observation is that the above mentioned negative arguments extend to a non-standard notion of average-case complexity, in which the distribution on the inputs with respect to which we measure the average-case complexity of the language, is only samplable in super-polynomial time. The significance of this result stems from the fact that in this non-standard setting, [GSTS05] did show a worst-case/average-case connection. In other words, their techniques give a way to bypass the impossibility arguments. By taking a closer look at the proof of [GSTS05], we discover that the worst-case/average-case connection is proven by a reduction that "almost" falls under the category ruled out by the negative result. This gives rise to an intriguing new notion of (almost black-box) reductions.

After extending the negative results to the non-standard average-case setting of [GSTS05], we ask whether their positive result can be extended to the standard setting, to prove some new worst-case/average-case connections. While we can not do that unconditionally, we are able to show that under a mild derandomization assumption, the worst-case hardness of NP implies the average-case hardness of $\text{NTIME}(f(n))$ (under the uniform distribution) where f is computable in quasi-polynomial time.

1 Introduction

Proving that the worst-case hardness of NP implies the average-case hardness of NP, is a fundamental open problem in the fields of computational complexity

M. Charikar et al. (Eds.): APPROX and RANDOM 2007, LNCS 4627, pp. 569–583, 2007.
© Springer-Verlag Berlin Heidelberg 2007

and foundations of cryptography (as it is a necessary step towards basing the existence of one-way functions on worst-case NP hardness). Bogdanov and Trevisan [BT03] (building on Feigenbaum and Fortnow [FF93]), show that "it is impossible (using non-adaptive reductions) to base the average-case hardness of a problem in NP or the security of a one-way function on the worst-case complexity of an NP complete problem (unless the polynomial hierarchy collapses)". This result is taken as demonstrating a major obstacle for showing a worst-case/average-case equivalence within NP.

Our first observation is that the arguments of [BT03] can be extended to a non-standard notion of average-case complexity, in which hardness is measured with respect to distributions that are samplable in super-polynomial time (rather than in fixed polynomial time):

Theorem 1. *Suppose that there is a language $L \in NP$ and a distribution $\mathcal{D}$ samplable in time $n^{\log n}$ such that there is a non-adaptive reduction from solving SAT on the worst-case to solving L on the average with respect to $\mathcal{D}$. Then every language in coNP can be computed by a family of nondeterministic Boolean circuits of size $n^{polylog(n)}$.*

Similar to the original result of [BT03], this should be taken as demonstrating a major obstacle for showing a worst-case/average-case equivalence within NP, even for this non-standard notion of average-case complexity. Nevertheless, Gutfreund, Shaltiel and Ta-Shma [GSTS05] do prove exactly the worst-case/average-case connection ruled out in Theorem 1 (by virtue of non-adaptive reductions).

Theorem 2. *[GSTS05] There exists a distribution $\mathcal{D}$ samplable in time $n^{\log n}$, such that if there exists a BPP algorithm solving SAT on the average with respect to $\mathcal{D}$, then there exists a BPP algorithm solving SAT on the worst-case.*

This surprising state of affairs gives rise to two questions. First, what can we learn from the proof of [GSTS05] about worst-case to average-case reductions, given that their technique bypasses the limitations imposed by Theorem 1? Second, after showing that the negative arguments can be extended to the non-standard notion of [GSTS05], can we turn the argument around and show that their positive result be extended and serve as a basis to prove new worst-case/average-case connections under the standard notion of average-case complexity?

Let us start with the first question. Looking at the proof of [GSTS05], we observe that it follows by constructing a distribution $\mathcal{D}$ as in the statement, and a fixed polynomial time machine R, s.t. for every probabilistic polynomial-time machine A solving SAT well on the average with respect to $\mathcal{D}$, the machine R^A (i.e. R with oracle access to A) solves SAT well in the worst case. In fact, the machine R is easy to describe, and unexpectedly turns out to be the familiar search-to-decision reduction for SAT.[1] I.e., given a SAT formula ϕ, R^A runs a

---

[1] A search to decision reduction for SAT uses an oracle that decides SAT to find a satisfying assignment for a given formula if such an assignment exist. The known reductions are either sequential and deterministic or parallel, non-adaptive and randomized [BDCGL90].

search to decision reduction on ϕ, where for each SAT query in the reduction it queries A. At the end R^A holds an assignment, and it accepts ϕ iff the assignment satisfies ϕ. Since the proof works with any search to decision reduction, we can use the non-adaptive search-to-decision reduction of [BDCGL90], and obtain what seems to be a non-adaptive worst-case to average-case reduction that by Theorem 1 implies that every language in coNP can be computed by a family of nondeterministic Boolean circuits of size $n^{polylog(n)}$. So did we prove an unexpected collapse?

The answer is of course no. Instead we showed a way to bypass the limitation imposed by Theorem 1. But to understand how, we need to give a careful look at the seemingly innocent term "reduction".

1.1   What Is a Reduction?

The term "reduction" (or more precisely "Turing reduction") used in [BT03], and many other papers, is defined as follows. Suppose P and P' are two computational tasks. E.g., P might be solving SAT on the worst case, and P' might be solving SAT on the average with respect to the distribution $\mathcal{D}$. We say that P reduces to P' if there exists a probabilistic polynomial time oracle machine R such that for *every* oracle A that solves P', R^A solves P. So in our example, P reduces to P' if there exists one fixed polynomial time machine R s.t. for every A solving SAT well on the average with respect to $\mathcal{D}$, the machine R^A solves SAT well in the worst case. From such a reduction, one in particular can deduce that if P' is easy (e.g., for BPP) then so does P. Reversing it, one may conclude that if P is hard then so does P'.

So does the proof of [GSTS05] uses a reduction? On the one hand, R is indeed a probabilistic, polynomial time oracle machine. However, the proof shows something weaker than a reduction regarding R: for *every probabilistic, polynomial-time* oracle machine A that solves P', R^A solves P. Namely, instead of showing that for every A that solves P', R^A solves P, it is only shown that for every *efficient* A this is the case. Thus, the argument of [GSTS05] is *not* a reduction. Nevertheless, it is still *useful*. That is, we can still conclude that if efficient machines cannot solve SAT on the worst-case then efficient machines cannot solve SAT on the average with respect to $\mathcal{D}$. The fact that we restrict the proof of correctness only to apply to efficient machines, does not make a difference since efficient machines is all that we care about!

We believe that this state of affairs calls for some new notation. First, to make matters more precise, we believe what we simply called "a reduction" should be called "a black-box reduction". This is because the key property captured in the definition is the black-box use of A by R (i.e., the reduction is oblivious to the actual solution of P') and the black-box use of A in the correctness proof (i.e., R^A is correct whenever A is correct, regardless of what A is). We then suggest a new kind of reduction:

Definition 1 (Class-specific black-box reductions). *Let P, P' be two computational problems, and $\mathcal{C}$ a class of functions. We say that R is a $\mathcal{C}$-black-box* *reduction from P to P', if R is a probabilistic polynomial-time oracle machine such that for every oracle $A \in C$ that solves P', R^A solves P.*

If R queries A non-adaptively, we say that the reduction is non-adaptive. If C is the class of all *functions, we simply say that R is a black-box reduction.*

Later we will also consider reductions that run in super-polynomial time, and we will state the running time explicitly when this is the case. If we do not state the running time then it is polynomial. Also, unless stated otherwise, whenever we say reduction, we mean black-box reduction.

Note that Definition 1 is only meaningful when the class C is more powerful than R. Otherwise R can run the oracle by itself. This is indeed the case in [GSTS05] where R runs in linear time and A runs in arbitrary polynomial time.

In Definition 1 the reduction R is still black-box in A, but now the correctness proof is not black-box in A, and works only when A comes from a bounded class. As we said before, normally a reduction is used to prove that if some task P' is easy then so does P (or the contra-positive: if P is hard then so does P'). Notice that for this purpose class black-box reductions are as useful as general reductions, because the argument is that if an *efficient* A solves P', then an efficient algorithm R^A solves P. The big advantage is that class black-box reductions are more flexible, and as Theorems 1 and 2 show, we can construct a class black-box reduction that is impossible to achieve with a general reduction (assuming the hierarchy does not collapse).

1.2   Back to Average-Case Complexity

We now turn to the second question we raised, whether we can leverage the result (or the proof) of [GSTS05] to prove a new worst-case/average-case connection under the standard notion of average-case complexity. Indeed we are able to show such a result but only assuming an unproven derandomization assumption. Before we state and discuss our result, some background is needed.

Most cryptographic primitives require at least the existence of One-Way Functions (OWFs) [IL89]. A function $f : \{0,1\}^* \rightarrow \{0,1\}^*$ is one-way if f is computable in a fixed polynomial time, and, for every constant $c > 0$, for every polynomial time algorithm A, for all large enough input lengths, $\Pr_{y \in f(U_n)} [A(y) \in f^{-1}(y)] < \frac{1}{n^c}$. In words, f can be computed by an algorithm that runs in some fixed polynomial time, while f^{-1} is hard (on the average) for *all* polynomial-time algorithms.

It is a common belief that OWFs exist, and a standard assumption in cryptography. The question whether it can be based on the worst-case hardness of NP is a major open problem. There are several relevant notions of hardness involved, and we identify the following hierarchy of conjectures:

- (Worst-case) Some function in NP is worst-case hard for a class of adversaries (usually BPP).
- (Average-case) Some function in NP is average-case hard for a class of adversaries (usually BPP).

- (One-way) Some function in P is hard to invert on average by a class of adversaries (usually BPP).
- (Pseudo-randomness) Some function in P generates distributions of low entropy that are indistinguishable from uniform by a class of adversaries (usually BPP).

The only non-trivial relation among these assumptions that is known today is that the one-wayness assumption is equivalent to the pseudo-randomness assumption when the class of adversaries is BPP [HILL99].

In all the above assumptions we assume that some function "fools" some class of adversaries, where the meaning of "fooling" varies (being worst-case hardness, average-case hardness, one-wayness or pseudo-randomness). The usual choice in cryptography is that the function lies in P or NP (according to the assumption we work with), while the class of adversaries is BPP (or sometimes even BPTIME$(t(n))$ for some super-polynomial $t(n)$). Thus, while the function is computable in a fixed polynomial time, the adversary may run in unbounded polynomial time, and so *has more resources than the algorithm for the function itself*. We therefore call this setting the "weak-fooling-strong" setting.

Another setting that often arises in complexity, is almost identical to the above except that the class of adversaries is *weaker* than the class it tries to fool. E.g., a key observation of Nisan and Wigderson [NW94] is that generators that are used to derandomize probabilistic complexity classes, can run in time n^c while fooling algorithms running in time n^b for some $b \ll c$. We call this the "strong-fooling-weak" setting.

This difference is a crucial (though subtle) dividing line. The canonic example for this distinction is the Blum-Micali-Yao PRG v.s. the Nisan-Wigderson PRG. This distinction applies not only to PRGs, and in particular we believe it is a central issue when discussing worst-case to average-case reductions for NP. Indeed the techniques that we use, apply to the strong-fooling-weak setting, but not to the weak-fooling-strong setting.

So now we focus on the strong-fooling-weak setting, and review previous work on the problem. We begin by listing the exponential-time analogues of the assumptions we listed above:

- (Worst-case hardness in EXP) Some function in EXP is worst-case hard for BPP.
- (Average-case hardness in EXP) Some function in EXP is average-case hard for BPP.
- (Exponential-time pseudo-random generators) For every constant $\epsilon > 0$, there exists a pseudorandom generator $G : n^\epsilon \to n$ fooling BPP, and the generator is computable in time 2^{n^ϵ} (i.e. exponential in the seed length).

Note that each of the above statements is implied by the corresponding statement in the weak-fooling-strong setting. Impagliazzo and Wigderson [IW98] (see also [TV02]), building on a long line of works such as [NW94, BFNW93, IW97], show that all three assumptions are equivalent. Their work was done in the context of understanding the power of randomness in computation, and indeed the equivalence above easily extends to include the following statement about the ability to reduce the amount of randomness used by efficient probabilistic algorithms.

- (Subexponential-time derandomization) For every probabilistic polynomial-time TM A (that outputs one bit), and a constant $\epsilon > 0$, there exists another probabilistic TM A^ϵ, that runs in time 2^{n^ϵ}, uses at most n^ϵ random coins, and behaves essentially the same as A.[2]

These beautiful and clean connections shed light on some of the fundamental questions in complexity theory regarding randomness and computational hardness of functions. Unfortunately, no such connections are known below the exponential level. As an example consider the following question that lies in the strong-fooling-weak setting, yet in the sub-exponential regime.

Open Problem 1. *Does NP $\not\subseteq$ BPP imply the existence of a language L in $\widetilde{NP} = NTIME(n^{O(\log n)})$ such that L is hard on average for BPP with respect to the uniform distribution?*

We believe that solving Open Problem 1 will be a major breakthrough. Here we give an affirmative answer under a weak derandomization assumption. We begin with the derandomization assumption. Following Kabanets [Kab01], we say that two probabilistic TM's are δ-indistinguishable if no samplable distribution can output with δ probability an instance on which the answers of the machines differ significantly (the acceptance probabilities, averaging over their randomness, differ by at least δ). See Definition 3 for a formal definition. We can now formalize the derandomization hypothesis:

Hypothesis 1. *For every probabilistic polynomial-time decision algorithm A (that outputs one bit), and every constant $\epsilon > 0$, there exists another probabilistic polynomial-time algorithm A^ϵ that on inputs of length n, tosses at most n^ϵ coins, and A, A^ϵ are $\frac{1}{100}$-indistinguishable.*

We then prove:

Theorem 3. *(Informal) If NP is worst-case hard and weak derandomization of BPP is possible (i.e. Hypothesis 1 is true) then there exists a language in $\widetilde{NP}$ that is hard on average for BPP.*[3]

---

[2] Here we mean that A^ϵ maintains the functionality of A *on the average* in the sense of Kabanets [Kab01] (see Definition 3 and Hypothesis 1). This notion of derandomization is standard when working with hardness against uniform TM's (rather than circuits) and with generators that fool TM's (rather than circuits).

[3] Our result is actually slightly stronger than the one stated here. It gives a hard on average language in the class NTIME($n^{\omega(1)}$) with the additional constraint that membership witnesses are of *polynomial* length, i.e. only the verification takes super-polynomial time, making it closer to NP. In particular, this class is contained in EXP. Also, standard separation techniques (such as the nondeterministic time hierarchy) can not separate this class from NP.

The formal statement is given in Section 4 (Theorem 7). The weak derandomization assumption is the precise (polynomial-time) analogue of the subexponential-time derandomization assumption stated above. That is, in both assumptions, A^ϵ reduces the randomness of A by a polynomial factor, and the result is indistinguishable to polynomial time adversaries (see Definition 3). However, in the latter we let A^ϵ run in subexponential-time, while in Theorem 3 we demand it runs in polynomial time.

We remark that there is an almost trivial proof of the theorem if we replace the weak derandomization assumption by a strong derandomization assumption, namely that BPP = P. In our notation, this is like assuming that A^ϵ can replace the $poly(n)$ random coins of A by logarithmically many random coins. This assumption is much stronger than the assumption in Theorem 3, where we assume A^ϵ reduces the number of coins just by a polynomial factor. Indeed, under the strong assumption, we can apply standard hierarchy theorems to separate $\widetilde{\text{NP}}$ from BPP (which is now P) even under average-case complexity measure. Note however, that our weak derandomization does not imply that BPP is (strictly) contained in $\widetilde{\text{NP}}$ and therefore we cannot apply hierarchy theorems.

Another strong assumption that implies the average-case hardness in our conclusion, is the existence of ceratin pseudo-random generators. Again, our assumption is much weaker than that since it only states that derandomization is possible (not necessarily via a construction of pseudo-random generators).

We therefore believe that the relationship that we obtain between worst-case hardness, average-case hardness and derandomization, is intriguing as it shows that highly non-trivial connections between these notions do exist below the exponential level.

2 Preliminaries

BPTIME$(t(n), c(n))$ is the class of languages that can be decided by randomized Turing machines that run in time $t(n)$ and use $c(n)$ random coins. NTIME$(t(n), w(n))$ is the class of languages that can be decided by nondeterministic Turing machines that run in time $t(n)$, and take witnesses of length $w(n)$. BPTIME$(t(n))$ and NTIME$(t(n))$ stand for BPTIME$(t(n), t(n))$ and NTIME$(t(n), t(n))$ respectively. PPM denotes the class of probabilistic polynomial-time TM's.

An ensemble of distributions $\mathcal{D}$ is an infinite set of distributions $\{\mathcal{D}_n\}_{n \in N}$, where $\mathcal{D}_n$ is a distribution over $\{0,1\}^n$. We denote by $\mathcal{U}$ the uniform distribution. Let $A(\cdot; \cdot)$ be a probabilistic TM, using $m(n)$ bits of randomness on inputs of length n. We say that A is a sampler for the distribution $\mathcal{D} = \{\mathcal{D}_n\}_{n \in N}$, if for every n, the random variable $A(1^n, y)$ is distributed identically to $\mathcal{D}_n$, where the distribution is over the random string $y \in_R \{0,1\}^{m(n)}$. In particular, A always outputs strings of length n on input 1^n. If A runs in polynomial time we simply say $\mathcal{D}$ is *samplable*. A *distributional problem* is a pair $(L, \mathcal{D})$, where L is a language and $\mathcal{D}$ is an ensemble of distributions.

Definition 2 (Average BPP). *Let* $(L, \mathcal{D})$ *be a distributional problem, and* $s(n)$ *a function from* N *to* $[0, 1]$. *We say that* $(L, \mathcal{D})$ *can be* efficiently decided on average *with success* $s(n)$, *and denote it by* $(L, \mathcal{D}) \in Avg_{s(n)}BPP$, *if there is an algorithm* $A \in PPM$ *such that for every large enough* n, $\Pr[A(x) = L(x)] \geq s(n)$, *where the probability is over an instance* $x \in \{0, 1\}^n$ *sampled from* $\mathcal{D}_n$ *and the internal coin tosses of* A.

We mention that the average-case definition that we give here is from [Imp95] (there it is denoted Heur-BPP). It differs from the original definition of Levin [Lev86]. Generally speaking, all previous works about hardness amplification and worst-case to average-case reductions, and in particular those that we use here (e.g. [BT03, IL90, GSTS05]), hold under Definition 2.

We denote the computational problem of deciding $(L, \mathcal{D})$ on average with success s by $(L, \mathcal{D})_s$. For a language L defined by a binary relation $R \subseteq \{0, 1\}^* \times \{0, 1\}^*$ (i.e. $L = \{x : \exists y \text{ s.t. } (x, y) \in R\}$), the *search problem* associated with L is given x find y such $(x, y) \in R$ if such a y exist (i.e. if $x \in L$) and output 'no' otherwise. The average-case analogue of solving the search problem of $(L, \mathcal{D})$ with success s, is to solve the search problem of L with probability at least s, over an instance of L drawn from $\mathcal{D}$ (and the internal coins of the search process). We denote this computational problem by $(L, \mathcal{D})_{search, s}$.

We need to define a non-standard (and weaker) solution to search problems, by letting the searching procedure output a list of candidate witnesses rather than one, and not requiring that the algorithm recognize 'no' instances.[4] For a langauge L defined by a binary relation R, the *list-search problem* associated with L is given x find a list $y_1, \ldots, y_m$ (where $m = \text{poly}(|x|)$) such that $\exists i \in [m]$ for which $(x, y_i) \in R$, if $x \in L$. Note that for $x \notin L$ we are not required to answer 'no'. We denote by $(L, \mathcal{D})_{list-search, s}$ the average-case analogue.

Indistinguishability. Following Kabanets [Kab01], we say that two probabilistic TM's are indistinguishable if no samplable distribution can output with high probability an instance on which the answers of the machines differ significantly (averaging over their randomness). Below is the formal definition.

Definition 3. *Let* A_1 *and* A_2 *be two probabilistic TM's outputting* $0/1$, *such that on inputs of length* n A_1 *uses* $m_1(n)$ *random coins and* A_2 *uses* $m_2(n)$ *random coins. For* $\epsilon, \delta > 0$, *we say that* A_1 *and* A_2 *are* (ϵ, δ)-indistinguishable, *if for every samplable distribution* $\mathcal{D} = \{\mathcal{D}_n\}_{n \in N}$ *and every* $n \in N$,

$$\Pr_{x \in \mathcal{D}_n}\left[\left|\Pr_{r \in_R \{0,1\}^{m_1(n)}}[A_1(x, r) = 1] - \Pr_{r' \in_R \{0,1\}^{m_2(n)}}[A_2(x, r') = 1]\right| > \epsilon\right] \leq \delta$$

To save on parameters, we will sometimes take ϵ *to be equal to* δ *and then we will say that* A_1, A_2 *are* δ-indistinguishable *(meaning* (δ, δ)-indistinguishable*)*.

---

[4] The reason we need this is that we are going to apply search procedures on languages in $NTIME(n^{\omega(1)}, \text{poly}(n))$. In this case an efficient procedure cannot check whether a candidate witness is a satisfying one. We therefore cannot amplify the success probability of such procedures. On the other hand, when we only require a list that contains a witness we can apply standard amplification techniques.

3   Impossibility Results and How to Bypass Them

We begin with showing that under standard assumptions, the [GSTS05] result cannot be proven via black-box and non-adaptive reductions. We then take a closer look at the reduction of [GSTS05], in order to understand what enables it to bypass the impossibility result.

The following statement can be obtained by generalizing the proof of [BT03] to arbitrary time bounds.

Theorem 4 (Implicit in [BT03]). *Suppose that there is a language $L \in NTIME(n^{O(\log n)})$ and a reduction R from solving SAT on the worst-case, to solving $(L, \mathcal{U})$ on the average with success $1 - 1/n^{O(\log n)}$. Further suppose that R is non-adaptive and black-box, and is computable in time $n^{polylog(n)}$. Then every language in coNP can be computed by a family of nondeterministic Boolean circuits of size $n^{polylog(n)}$.*

From that we can conclude Theorem 1, which we now re-state more formally (the proof is omitted due to space limitation).

Theorem 5. *Suppose there is a language $L \in NP$ and a distribution D samplable in time $n^{\log n}$, such that there is a black-box and non-adaptive reduction from solving SAT on the worst-case, to solving (L, D) on the average with success $1 - 1/n$. Then every language in coNP can be computed by nondeterministic Boolean circuits of size $n^{polylog(n)}$.*

3.1   A Closer Look at the Reduction of [GSTS05]

There are two steps in the argument of [GSTS05]. First it is shown that assuming NP $\not\subseteq$ BPP, any probabilistic, polynomial-time algorithm BSAT for SAT has a hard polynomial-time samplable distribution $\mathcal{D}_{\text{BSAT}}$, and then they show one quasi-polynomial time distribution $\mathcal{D}$ that is hard for all polynomial-time, probabilistic algorithms.

We now recall how the first step is achieved. Given BSAT we define the probabilistic polynomial time algorithm SSAT that tries to solve the search problem of SAT using oracle calls to BSAT (via the downwards self-reducibility property of SAT), and answers "yes" if and only if it finds a satisfying assignment. We then define the SAT formula:

$$\exists_{x \in \{0,1\}^n} \ [\ \text{SAT}(x) = 1 \ \text{ and } \ \text{SSAT}(x) \neq \text{'yes'} \ ] \tag{1}$$

The assumption NP $\not\subseteq$ BPP implies that SSAT does not solve SAT and the sentence is true. We now ask SSAT to find a satisfying assignment to it. If it fails doing so, then BSAT is wrong on one of the queries made along the way. Otherwise, the search algorithm finds a SAT sentence x on which SSAT is wrong. This means that BSAT is wrong on one of the queries SSAT makes on input x. Things are somewhat more complicated because SSAT is a probabilistic algorithm and not a deterministic one, and so the above sentence is not really a

SAT formula, but we avoid these technical details and refer the interested reader to [GSTS05]. In any case, we produce a small set of queries such that on at least one of the sentences in the set, BSAT is wrong, i.e., we get a polynomial-time samplable distribution on which BSAT has a non-negligible error probability.

To implement the second step, [GSTS05] define the distribution $\mathcal{D}$ that on input 1^n picks at random a machine from the set of probabilistic machines with description size at most, say, $\log n$ and runs it up to, say, $n^{\log n}$ time. We know that for any probabilistic polynomial-time algorithm BSAT there is a hard distribution $\mathcal{D}_{\text{BSAT}}$, and this distribution is sampled by some polynomial-time algorithm with a fixed description size. Thus, for n large enough, we pick this machine with probability at least $1/n$ (because its description size is smaller than $\log n$) and then we output a bad input for BSAT with probability $1/n$. The sampling time for $\mathcal{D}$ is $n^{\log n}$ (or, in fact, any super-polynomial function).

We now ask: Can we interpret the [GSTS05] result as a worst-case to average-case reduction?

Indeed, given an algorithm BSAT for solving $(\text{SAT}, \mathcal{D})$, we define the algorithm $R^{\text{BSAT}} = \text{SSAT}$, where R is a search-to-decision reduction. The analysis shows that if BSAT indeed solves SAT well on the average with respect to $\mathcal{D}$, then it also solves it well on the average with respect to $\mathcal{D}_{\text{BSAT}}$. This implies that necessarily the sentence in Eq (1) is true, i.e., SSAT solves SAT in the worst-case. In other words, the standard search to decision reduction for SAT is also a worst-case to average-case reduction!

Another question is now in place: is the reduction black-box?

Looking at Definition 1 we see that a reduction is *black-box* if it has the following *two* properties:

1. (Property 1) R makes a black-box use of the adversary A (in our case BSAT). I.e., R may call A on inputs but is not allowed to look into the code of A.
2. (Property 2) The reduction is correct for any A that solves the problem (in our case SAT), putting no limitations on the nature of A. E.g. A may even be undecidable.

We see that in the reduction of [GSTS05], the first condition is satisfied. R is merely the standard search-to-decision reduction for SAT which queries the decision oracle on formulas along a path of the search tree. We can replace the standard search-to-decision reduction with the one by Ben-David et. al. [BDCGL90]. The latter makes only non-adaptive queries to the decision oracle. Thus we get a non-adaptive reduction. However, the second condition is violated. Indeed, the key point in the *analysis* of [GSTS05] is that it works only for *efficient* oracles BSAT. This is so because the analysis encodes the failure of R^{BSAT} as an NP statement. Here the analysis crucially uses the fact that BSAT (and therefore R^{BSAT}) is in BPP, and therefore its computation has a short description as a Boolean formula.

So let us now summarize this surprising situation: from the reduction R's point of view, it is black-box, i.e. Property 1 holds (R does not rely on the inner working of the oracle BSAT or its complexity), but for the *analysis* to work, the oracle BSAT has to be efficient, i.e. Property 2 is violated.[5] This motivates the definition of class-specific black-box reductions that we gave in the introduction.

Given this definition and the discussion about non-adaptivity above, we can restate the result of [GSTS05] as follows:

Theorem 6. *There exists a distribution $\mathcal{D}$ samplable in time $n^{\log n}$, such that there is a BPP-black-box and non-adaptive reduction from solving SAT on the worst-case to solving $(L, \mathcal{D})$ on average with success $1 - 1/n$.*

Theorem 6 is in sharp contrast to Theorem 5. Theorem 5, as well as [FF93, BT03, AGGM06], say that the requirement in black-box reductions that they succeed whenever they are given a "good" oracle, *regardless of its complexity*, is simply too strong, i.e. such reductions are unlikely to exist. Theorem 6, on the other hand, says that weakening the requirement to work only for efficient oracles (i.e. allowing to violate Property 2, but not property 1) is enough to bypass the limitation.

We mention that there are other cases of non-black-box reductions in complexity theory that bypass black-box limitations. For example, the fact that the polynomial-time hierarchy collapses under the assumption $NP = P$ is proven via a non-black-box reduction from solving SAT efficiently to solving $QSAT_2$ efficiently ($QSAT_2$ is the canonic Σ_2-complete problem of deciding the validity of a Boolean first-order formula with two quantifier alternations). Indeed if a black-box reduction between these tasks exists then the hierarchy collapses unconditionally. It is interesting, however, that in this argument both the reduction and the proof of correctness are non-black-box, because the reduction queries the NP-oracle on statements that are encodings of the computation of the oracle itself (assuming that this oracle can be realized efficiently). I.e. both Properties 1 and 2 are violated. Examples from the field of cryptography can be found in the work of Barak [Bar01] (and some following papers that use similar ideas). Again his proof (when considered as a reduction) violates both properties of black-boxness.

The only other BPP-black-box reduction that we are aware of appears in the work of Imapgliazzo and Wigderson [IW98].[6] Their proof shows that given an

---

[5] We mention that recently, Atserias [Ats06] gave an alternative proof to [GSTS05] where he shows that even the analysis can be done almost black-box. That is, it does not need to use the description of BSAT, it only needs to know the *running time* of BSAT. In contrast, the analysis in [GSTS05] does use the description of BSAT.

[6] Although some proofs of security in cryptography appear to use the fact that the adversary is efficient (e.g. in zero-knowledge proofs with black-box simulation), when written as reductions they are in fact black-box in the standard sense. That is, it is shown that *any* adversary that breaks the cryptographic primitive, implies breaking the security assumption (e.g. bit-commitments in the case of zero-knowledge). Of course, the contradiction with the security assumption is only true when the adversary is efficient. However, this is an artifact of the security assumption, not the way the proof is derived (or in other words, if we change the security assumption to hold against say, sub-exponential-time adversaries rather than polynomial-time adversaries, the same proof of security holds).

efficient oracle that breaks a ceratin pseudo-random generator, one can use it to compute an EXP-complete language. We do not know if this reduction can be done in a black-box way, nor do we have evidence that it cannot.

Finally, we want to mention that one should not confuse black-box limitations with non-relativizing arguments. The proof of [GSTS05] (as well as the collapse of the polynomial-time hierarchy) *can* be done in a relativizing way.

4   Top-Down Overview of Theorem 3

We re-state Theorem 3 in a formal way.

Theorem 7. *If Hypothesis 1 is true and $SAT \notin RP$ then there exists a language $L \in NTIME(t(n), poly(n))$, such that, $(L, \mathcal{U}) \notin Avg_{1/2+1/\log^\alpha n} BPP$, where $t(n) = n^{\omega(1)}$ is an arbitrary time-constructible super-polynomial function, and $\alpha > 0$ is a universal constant.*

We now explain the intuition and give a top-down overview of the proof. The full proof is omitted due to space limitations and will appear in the final version of the paper.

Our starting point is Theorem 2, which says that if SAT is worst-case hard, then there exist a single distribution on SAT instances, $\mathcal{D}_{hard}$, on which every probabilistic polynomial-time algorithm errs with relatively high probability, but the distribution is only samplable in quasi-polynomial-time (for this reason we denote it by $\mathcal{D}_{hard}$). Our goal is to somehow extract from $\mathcal{D}_{hard}$ a *simple* distribution on which the same holds. Ideally, the uniform distribution will be good enough. The key tool that we use is a reduction given by Impagliazzo and Levin [IL90] that shows that if there exists a *polynomial-time* samplable distribution $\mathcal{D}$ that is hard on average for some language $L \in NP$, then there exists another language $L' \in NP$ for which the *uniform distribution* $\mathcal{U}$ is hard on average. We would like to apply this reduction on SAT and the distribution $\mathcal{D}_{hard}$.

However we immediately run into a problem because $\mathcal{D}_{hard}$ is samplable in super-polynomial time, while the argument of [IL90] only applies to distributions that are samplable in polynomial time. To understand this, let us elaborate on how the complexity of the distribution influences the reduction of [IL90]. There are several different entities to consider in this reduction: The language L, The distribution $\mathcal{D}_{hard}$, The language L' we reduce to, and the reduction R itself that solves $(L, \mathcal{D}_{hard})$ on average given an oracle that solves $(L', \mathcal{U})$ on average.

We can expect that both the complexity of L' as well as the complexity of the reduction R depend on $\mathcal{D}_{hard}$. Indeed, using the [IL90] reduction, the non-deterministic procedure for the language L' involves checking membership in the language L as well as running the sampler for $\mathcal{D}_{hard}$. In our case, since $\mathcal{D}_{hard}$ is samplable in super-polynomial-time, this results in L' having a super-polynomial non-deterministic complexity (and so puts us in the strong-fooling-weak setting).

It seems that the same should hold for the reduction R. Indeed the reduction of [IL90] is from search problems to search problems, which means that R must handle the membership witnesses for the language L'. As we said above, this involves computing $\mathcal{D}_{hard}$ and in particular, the complexity of R is at least that of $\mathcal{D}_{hard}$. This however means that we can not deduce from the hardness on average of $(L, \mathcal{D}_{hard})$ the hardness on average of $(L', \mathcal{U})$, because the hardness of $(L, \mathcal{D}_{hard})$ is only against algorithms with complexity smaller than that of $\mathcal{D}_{hard}$, and the reduction's complexity is at least that of $\mathcal{D}_{hard}$. This makes the reduction of [IL90] useless for us (at least a direct use of it).

The surprising thing, and the main observation here, is that in the Impagliazzo-Levin argument the running time of the reduction does not depend on the time complexity of $\mathcal{D}_{hard}$! It only depends on the number of random coins the sampler for $\mathcal{D}_{hard}$ uses: while the reduction R does look at the membership witnesses for L', the size of these witnesses is only a function of the number of random coins the sampler uses. Furthermore, during this process, R is never required to verify the witnesses and therefore does not need to run the sampler. We formalize this observation in the following lemma.

Lemma 1. *For every Distribution $\mathcal{D}$ samplable in $BPTIME(t(n), c(n))$, a language $L \in NTIME(t_L(n), c_L(n))$ and $0 < \delta(n) < 1$, there exists $L' \in NTIME(t(n) + t_L(n) + poly(n, c(n)), c(n) + c_L(n))$ such that there is a probabilistic non-adaptive reduction, R, from $(L, \mathcal{D})_{search, 1-O(\delta(n) \cdot c^2(n))}$ to $(L', \mathcal{U})_{list\ search, 1-\delta(n)}$.*

Furthermore, the running time of R is $poly(n, c(n), t_L(n))$, and note that it is independent of $t(n)$.

The conclusion of the Lemma is that in order to keep the reduction efficient, we need to reduce the randomness complexity of the sampler for $\mathcal{D}_{hard}$ to a fixed polynomial, but we do not need to reduce its time complexity. This is fortunate because while in general there is no reason to believe that we can reduce the running time of algorithms, it is widely believed that randomness can be reduced without paying much penalty in running time. To that end we use Hypothesis 1, and prove:

Lemma 2. *Assume Hypothesis 1 is true. Let $t(n)$ be an arbitrary super-polynomial function. There is a distribution $\mathcal{D}$ samplable in $BPTIME(O(t(n), O(n^3))$ such that, $(SAT, \mathcal{D}) \in Avg_{1-1/n}BPP \Rightarrow SAT \in RP$.*

Note that Hypothesis 1 does not seem to derandomize general distributions that are samplable in super-polynomial-time. First, Hypothesis 1 only derandomizes polynomial-time algorithms and only by polynomial factors (while here we want to derandomize a sampler that runs in super-polynomial time and uses a super-polynomial number of coins). And second, Hypothesis 1 only applies to decision algorithms.[7] We show, however, that the specific distribution $\mathcal{D}_{hard}$ from [GSTS05] can be derandomized under Hypothesis 1. The proof is quite technical and involves getting into the details of [GSTS05].

---

[7] In general, standard derandomization results do not apply to probabilistic procedures that output many bits. We refer the reader to [DI06] for a discussion about derandomizing procedures that output many bits versus decision procedures (that output a single bit).

582   D. Gutfreund and A. Ta-Shma

The above two lemmas give us $1 - 1/\mathrm{poly}(n)$ hardness on the average for the list-search version of the problem (given the worst-case hardness and the derandomization assumptions). To get $1/2 + 1/\log^{\alpha} n$ hardness on the average for the decision problem, we use generalizations of known techniques in average-case complexity [BDCGL90, Tre05]. The tricky part is doing the hardness amplification using a reduction whose running time is $\mathrm{poly}(n, c(n))$ and, in particular, independent of $t(n)$. By using careful generalizations of [BDCGL90], Trevisan's amplification technique [Tre05] goes through, and we obtain Theorem 7.

Acknowledgements

We thank Ronen Shaltiel and Salil Vadhan for many helpful discussions. Andrej Bogdanov for pointing out to us that Theorem 5 follows directly from a generalization of [BT03] (i.e. Theorem 4). Previously we had a more complicated and weaker statement. Charlie Rackoff for suggesting the relevance of the argument that the polynomial-time hierarchy collapses under P = NP, Alex Healy for commenting on the manuscript, and Oded Goldreich for a helpful conversation.

The first author is supported by ONR grant N00014-04-1-0478 and NSF grant CNS-0430336. The second author is supported by the Israel Science Foundation grant 217/05, the Binational Science Foundation, and the EU Integrated Project QAP.

References

[AGGM06]   Akavia, A., Goldreich, O., Goldwasser, S., Moshkovitz, D.: On basing one-way functions on NP-hardness. In: Proceedings of the 38th Annual ACM Symposium on Theory of Computing, 2006, pp. 701–710. ACM Press, New York (2006)

[Ats06]   Atserias, A.: Distinguishing SAT from polynomial-size circuits through black-box queries. In: Proceedings of the 21th Annual IEEE Conference on Computational Complexity, 2006, pp. 88–95. IEEE Computer Society Press, Los Alamitos (2006)

[Bar01]   Barak, B.: How to go beyond black-box simulation barrier. In: Proceedings of the 43rd Annual IEEE Symposium on Foundations of Computer Science, 2006, pp. 106–115. IEEE Computer Society Press, Los Alamitos (2001)

[BDCGL90]   Ben-David, S., Chor, B., Goldreich, O., Luby, M.: On the theory of average case complexity. In: Proceedings of the 22nd Annual ACM Symposium on Theory of Computing,1990, pp. 379–386. ACM Press, New York (1990)

[BFNW93]   Babai, L., Fortnow, L., Nisan, N., Wigderson, A.: BPP has subexponential simulation unless Exptime has publishable proofs. Computational Complexity 3, 307–318 (1993)

[BT03]   Bogdanov, A., Trevisan, L.: On worst-case to average-case reductions for NP problems. In: Proceedings of the 44th Annual IEEE Symposium on Foundations of Computer Science, 2003, pp. 308–317. IEEE Computer Society Press, Los Alamitos (2003)

[DI06]   Dubrov, B., Ishai, Y.: On the randomness complexity of efficient sam-
             pling. In: Proceedings of the 38th Annual ACM Symposium on Theory
             of Computing, 2006, pp. 711–720. ACM Press, New York (2006)
[FF93]   Feigenbaum, J., Fortnow, L.: Random-self-reducibility of complete sets.
             SIAM Journal on Computing 22, 994–1005 (1993)
[GSTS05]   Gutfreund, D., Shaltiel, R., Ta-Shma, A.: if NP languages are hard in
             the worst-case then it is easy to find their hard instances. In: Proceed-
             ings of the 20th Annual IEEE Conference on Computational Complex-
             ity, pp. 243–257. IEEE Computer Society Press, Los Alamitos (2005)
[HILL99]   Håstad, J., Impagliazzo, R., Levin, L., Luby, M.: A pseudorandom gen-
             erator from any one-way function. SIAM Journal on Computing 28(4),
             1364–1396 (1999)
[IL89]   Impagliazzo, R., Luby, M.: One-way functions are essential for com-
             plexity based cryptography. In: Proceedings of the 30th Annual IEEE
             Symposium on Foundations of Computer Science,1989, pp. 230–235.
             IEEE Computer Society Press, Los Alamitos (1989)
[IL90]   Impagliazzo, R., Levin, L.: No better ways of finding hard NP-problems
             than picking uniformly at random. In: Proceedings of the 31st Annual
             IEEE Symposium on Foundations of Computer Science, 1990, pp. 812–
             821. IEEE Computer Society Press, Los Alamitos (1990)
[Imp95]   Impagliazzo, R.: A personal view of average-case complexity. In: Pro-
             ceedings of the 10th Annual Conference on Structure in Complexity
             Theory, 1995, pp. 134–147 (1995)
[IW97]   Impagliazzo, R., Wigderson, A.: P = BPP if E requires exponential
             circuits: Derandomizing the XOR lemma. In: Proceedings of the 29th
             Annual ACM Symposium on Theory of Computing, 1997, pp. 220–229.
             ACM Press, New York (1997)
[IW98]   Impagliazzo, R., Wigderson, A.: Randomness vs. time: de-
             randomization under a uniform assumption. In: Proceedings of
             the 39th Annual IEEE Symposium on Foundations of Computer Sci-
             ence, 1998, pp. 734–743. IEEE Computer Society Press, Los Alamitos
             (1998)
[Kab01]   Kabanets, V.: Easiness assumptions and hardness tests: Trading time
             for zero error. Journal of Computer and System Sciences 63(2), 236–252
             (2001)
[Lev86]   Levin, L.: Average case complete problems. SIAM Journal on Comput-
             ing 15(1), 285–286 (1986)
[NW94]   Nisan, N., Wigderson, A.: Hardness vs. randomness. Journal of Com-
             puter and System Sciences 49, 149–167 (1994)
[Tre05]   Trevisan, L.: On uniform amplification of hardness in NP. In: Proceed-
             ings of the 37th Annual ACM Symposium on Theory of Computing,
             2005, pp. 31–38. ACM Press, New York (2005)
[TV02]   Trevisan, L., Vadhan, S.: Pseudorandomness and average-case complex-
             ity via uniform reductions. In: Proceedings of the 17th Annual IEEE
             Conference on Computational Complexity, 2002, pp. 129–138. IEEE
             Computer Society Press, Los Alamitos (2002)
[Vio03]   Viola, E.: Hardness vs. randomness within alternating time. In: Pro-
             ceedings of the 18th Annual IEEE Conference on Computational Com-
             plexity, 2003, pp. 53–62. IEEE Computer Society Press, Los Alamitos
             (2003)

On Finding Frequent Elements in a Data Stream

Ravi Kumar[1] and Rina Panigrahy[2]

[1] Yahoo! Research, 701 First Ave, Sunnyvale, CA 94089, USA
ravikumar@yahoo-inc.com
[2] Microsoft Research, 1065 La Avenida, Mountain View, CA 94043, USA
rina@microsoft.com

Abstract. We consider the problem of finding the most frequent elements in the data stream model; this problem has a linear lower bound in terms of the input length. In this paper we obtain sharper space lower bounds for this problem, not in terms of the length of the input as is traditionally done, but in terms of the quantitative properties (in this case, distribution of the element frequencies) of the input *per se*; this lower bound matches the best known upper bound for this problem. These bounds suggest the study of data stream algorithms through an instance-specific lens.

1   Introduction

The data stream model of computation has evolved into an indispensable tool to cope with processing massive data sets. In this model, the input is processed as and when it arrives by a computational device that has only a limited storage space. The complexity in this model is typically measured in terms of the space required by the algorithm; this is often quantified as a function of the length of the input (n) and the size of the domain from which the elements of the inputs are drawn (m). Data stream algorithms are deemed interesting, efficient, and usable only if their space requirement is sublinear in n; this is especially so since n is assumed to be extremely large. Efficient data stream algorithms are out of question for most problems if neither randomization nor approximation is permitted. With randomization, sublinear space data stream algorithms have been developed for approximating several statistical data analysis primitives, ranging from frequency moments and L_p differences to histograms. For an excellent account of the important results in the area, see the recent monograph by Muthukrishnan [1] and the survey by Babcock et al. [2].

Motivation. Even though many problems are shown to have efficient data stream algorithms, it is provably impossible to obtain $o(n)$ space algorithms for many other interesting data processing problems, even with the allowances of approximation and randomization. This is a consequence of both the limitations of the data stream model itself and the inherent difficulty of the underlying problem.

   An important problem that falls in this doomed class is the *MostFrequent* problem: given an input sequence $X = x_1, \ldots, x_n$, output the element(s) that

M. Charikar et al. (Eds.): APPROX and RANDOM 2007, LNCS 4627, pp. 584–595, 2007.
© Springer-Verlag Berlin Heidelberg 2007

occur the most in the sequence X. This is an important and recurring problem in databases, information retrieval, and web analysis; for instance, search engines would like to quickly distill the frequently-issued queries from their daily query logs. Unfortunately, Alon, Matias, and Szegedy [3] showed an $\Omega(n)$ lower bound for estimating the frequency of the largest item given an arbitrary data stream. An approximation version of this problem involves finding an element whose frequency is close that of the optimum; unfortunately even this version needs linear space in the worst case. Throughout the rest of the paper, *MostFrequent* will always refer to the approximation version.

Despite this seemingly insurmountable lower bound for the *MostFrequent* problem, a wide variety of heuristics that work very well in practice have been proposed; for example, see [4,5,6,7]. These heuristics are successful because the lower bound assumes a certain (near uniform) distribution of the frequencies of the elements in the stream, where as, in practice one can exploit the fact that the frequencies are power law or Zipfian. Charikar, Chen, and Colton [7] obtained an algorithm along with a performance guarantee for a relaxed version of the *MostFrequent* problem — given a parameter ϵ they can find an element with frequency at least $(1 - \epsilon)$ times that of the most frequent element in space $\tilde{O}(F_2/F_\infty^2)$, where F_2 is the second frequency moment of the input and F_∞ is the frequency of the most frequent input element.[1] Although F_2/F_∞^2 may be linear in n in the worst case, for many distributions including the Zipfian or power law distributions encountered in practice, it is sublinear. A natural question arises: for such (and other) distributions in order to find the most frequent elements, do we need $\Omega(F_2/F_\infty^2)$ space or can we do better?

Our Contributions. For the *MostFrequent* problem, where we are given the frequency distribution of the elements in a stream and an adversary is allowed to choose the order in which the elements arrive, we show tight space bound (up to logarithmic factors) of $\tilde{\Theta}(F_2/F_\infty^2)$, where F_2 is the second frequency moment of the input and F_∞ is the frequency of the most frequent input element. The upper bound can be inferred from the algorithm of Charikar, Chen, and Colton [7]; see also [8]. We show a matching lower bound for multi-pass data stream algorithms that are even allowed to approximate the most frequent element.[2] The lower bound is established by information-theoretic arguments as in Bar-Yossef et al. [9] and consists of two steps. The first step involves obtaining an information complexity lower bound for a functional version of set disjointness. The second step constructs separate instances of the set disjointness problem, carefully interleaves them, and uses the lower bound from the first step together with additional information-theoretic arguments to obtain the final lower bound.

---

[1] For a sequence $X = x_1, \ldots, x_n$ where each $x_i \in Y$, let $f_y(X)$ denote the number of times $y \in Y$ appears in the sequence. Then, $F_2(X) = \sum_{y \in Y} f_y^2(X)$ and $F_\infty(X) = \max_{y \in Y} f_y(X)$.

[2] Since $F_0 \geq F_2/F_\infty^2$, it is tempting to exploit this relationship towards a lower bound in terms of F_0 via a reduction to the indexing problem. However, this approach does not seem to lead to tight bounds even for one-pass algorithms and certainly does not say much for multi-pass algorithms.

In particular, our bounds show the precise behavior of the space requirements as a function of the second and the infinity frequency moments of the input. In particular, it shows that the algorithm in [7] for Zipf distributions is essentially optimal.

At a philosophical level, our results suggest that the worst-case lower bounds for many data stream problems are perhaps way too pessimistic and stronger and better bounds could be stated if one were to take the properties of inputs themselves into account and parametrize the complexity of the algorithms differently. To quote [1]:

> "...the goal would be to find workable approximations under suitable assumptions on input data distributions."

Our results makes a modest progress towards this goal. In fact, we propose the study of data stream algorithms through an *instance-specific* lens. Roughly speaking, the plan is to obtain optimal space bounds, both upper and lower, for data stream algorithms parametrized in terms of the *actual* properties of the input sequence than just its length, n. We call these the *instance-specific data stream bounds*.

Notation. Let $X = x_1, \ldots, x_n$ denote the input sequence of length n. We assume that the input sequence X arrives in an arbitrary order. We work with randomized approximation algorithms, i.e., algorithms that given an error parameter $\epsilon > 0$ and a confidence parameter $\delta > 0$, output a value that is an $(1 \pm \epsilon)$ multiplicative approximation to the correct value, with probability at least $1 - \delta$. We use $\tilde{O}, \tilde{\Omega}, \tilde{\Theta}$ to suppress factors that are logarithmic in n.

2   Main Result

First, we formally state the *MostFrequent* problem. Let $F_\infty = F_\infty(X)$ denote the frequency of the most frequent element in the input sequence X. Let $F_2 = F_2(X)$ denote the second frequency moment of the input sequence X.

Problem 1 (MostFrequent). In the *MostFrequent* problem, we are given the distribution of the frequencies of the elements, i.e., we are told how many elements have a certain frequency but not the identity of those elements. The objective is to find an element with frequency at least $(1 - \epsilon) \cdot F_\infty$, in one or more passes over the input.

2.1   Upper Bound

For sake of completeness, we briefly discuss the known algorithm for the *MostFrequent* problem. For simplicity of exposition, we stated it as a two-pass algorithm; it is possible to make it run in a single pass.

The algorithm maintains h hash tables of size b buckets each. In the first pass, each input element is hashed into the h tables using pairwise-independent hash functions, keeping a counter per bucket. In the second pass, input elements that hash to a large number buckets with a high counter value — above a threshold τ — are reported as frequent elements.

Here is a glimpse of the analysis. Focusing on one hash table, if there are b buckets, then the expected count in a bucket cell is n/b. The variance of this count can be computed to be at most F_2/b. But the expected count of the bucket with the most frequent element is at least $n/b + F_\infty(1 - 1/b)$. So, as long as the gap between the expectations $F_\infty(1 - 1/b)$ is at least $c \cdot \sqrt{F_2/b}$ for sufficiently chosen constant $c > 0$, we can set the threshold τ so that there is a constant gap in the probability of a count exceeding the threshold for the frequent element versus any other element. This gives a bound on the number of buckets to be $b > \Omega(F_2/F_\infty^2)$. By using several (constant) such tables and using the median count, the probability gap can be amplified arbitrarily.

Theorem 1 ([7]). *There is a randomized data stream algorithm for the Most-Frequent problem that uses space $\tilde{O}(F_2/F_\infty^2)$.*

2.2 Tools for the Lower Bound

We assume the reader is familiar with communication complexity and information theory. See the book by Kushilevitz and Nisan [10] for the former and the book by Cover and Thomas [11] for the latter.

In our usage, $X \perp Y$ will denote the random variables X and Y are independent. We use $H(\cdot)$ to denote the entropy and $I(\cdot; \cdot)$ to denote the mutual information between two random variables.[3] Some basic properties of the information-theoretic quantities can be found in [11].

First, we need the following simple lemma that relates the entropies of two random variables to the mutual information with two other random variables that are independent.

Lemma 1. *For random variables A, B, X, Y such that $X \perp Y$, we have*

$$H(A) + H(B) \geq I(AB; XY) + I(A; B) \geq I(AB; X) + I(A; Y) + I(B; Y).$$

Proof. The first inequality follows easily since

$$H(A) + H(B) = H(AB) + I(A; B) \geq I(AB; XY) + I(A; B).$$

To show the second inequality, we first show

$$
\begin{aligned}
I(AB; XY) &= H(XY) - H(XY \mid AB) \\
&= H(X) + H(Y) - H(XY \mid AB) \quad \text{(by } X \perp Y\text{)} \\
&\geq H(X) + H(Y) - H(X \mid AB) - H(Y \mid AB) \quad \text{(by subadditivity)} \\
&= I(AB; X) + I(AB; Y).
\end{aligned}
\tag{1}
$$

---

[3] Let A, B be random variables. Then, the entropy $H(A) = -\sum_a \Pr[A = a] \log \Pr[A = a]$, the conditional entropy $H(A \mid B) = \sum_b H(A \mid B = b) \Pr[B = b]$ and the mutual information $I(A; B) = H(A) - H(A \mid B) = H(B) - H(B \mid A)$.

Next, we show that

$$I(B; Y \mid A) + I(A; B) = H(B \mid A) - H(B \mid A, Y) + H(B) - H(B \mid A)$$
$$= H(B) - H(B \mid A, Y)$$
$$\geq H(B) - H(B \mid Y) \quad \text{(by dropping condition)}$$
$$= I(B; Y). \tag{2}$$

Finally, the second inequality follows as

$$I(AB; XY) + I(A; B) \geq I(AB; X) + I(AB; Y) + I(A; B) \quad \text{by (1)}$$
$$= I(AB; X) + I(A; Y) + I(B; Y \mid A) + I(A; B)$$
$$\geq I(AB; X) + I(A; Y) + I(B; Y). \quad \text{by (2)}$$

Next, we generalize this to a collection of random variables defined on a tree.

Theorem 2. *Given random variables $U_0, \ldots, U_{2^k-1}$ and $k + 1$ random variables $X_0, \ldots, X_k$ that are independent. Consider a binary tree of height k, with $U_0, \ldots, U_{2^k-1}$ at the leaves in that order and each node is the labeled by the concatenation of all the variables in leaves of the subtree rooted at that node. That is, the parents of the leaves have labels $U_0 U_1, U_2 U_3, U_4 U_5, \ldots, U_{2^k-2} U_{2^k-1}$ in that order; the root has label $U_0 U_1 .. U_{2^k-1}$; and in general the j-th node from left at the i-th level (root is at level 0) is labeled by $V_{ij} = U_{j2^i} U_{j2^i+1} \cdots U_{(j+1)2^i-1}$ where $i \in \{0, \ldots, k\}$, $j \in j(k, i) = \{0, \ldots, 2^{(k-i)} - 1\}$.*
Then

$$\sum_{j=0}^{2^k-1} H(U_j) \geq \sum_{i \in [k], j \in j(k,i)} I(V_{ij}; X_i).$$

Note that the order in which the U's appear in V_{ij} is immaterial as the order does not affect the entropy and the mutual information. All that is used here is that each node is labeled with a concatenation of labels of its descendant leaves in any order.

Proof. We prove by induction on k. The base case for $k = 1$ is given by the Lemma 1. For notational simplicity, we first illustrate the proof for $k = 2$.

$$H(U_0) + H(U_1) + H(U_2) + H(U_3)$$
$$= H(U_0 U_1) + H(U_2 U_3) + I(U_0; U_1) + I(U_2; U_3)$$
$$\geq I(U_0 U_1 U_2 U_3; X_0) + I(U_0 U_1; X_1 X_2) + I(U_2 U_3; X_1 X_2) + I(U_0; U_1) + I(U_2; U_3)$$
$$\quad \text{(by Lemma 1)}$$
$$\geq I(U_0 U_1 U_2 U_3; X_0) + I(U_0 U_1; X_1) + I(U_0; X_2) + I(U_1; X_2) + I(U_2 U_3; X_1)$$
$$\quad + I(U_2; X_2) + I(U_3; X_2). \quad \text{(by Lemma 1 applied twice)}$$

This completes the proof for $k = 2$. For general $k > 2$, we proceed by induction. Let X' be a sequence of k independent random variables such that $X'_i = X_i$ for $i = 0, \ldots, k - 1$ and $X'_{k-1} = X_{k-1} X_k$.

$$\sum_{j=0}^{2^k-1} H(U_j)$$

$$= \sum_{j'=0}^{2^{k-1}-1} H(U_{2j'}U_{2j'+1}) + \sum_{j'=0}^{k} I(U_{2j'}; U_{2j'+1})$$

$$\geq \sum_{i\in[k-1], j\in j(k-1,i)} I(V_{ij}; X_i') + \sum_{j'=0}^{k} I(U_{2j'}; U_{2j'+1})$$

$$\text{(by induction hypothesis and using } X')$$

$$= \sum_{i\in[k-2], j\in j(k-2,i)} I(V_{ij}; X_i) + \sum_{j'=0}^{k} I(U_{2j'}U_{2j'+1}; X_{k-1}X_k) + \sum_{j'=0}^{k} I(U_{2j'}; U_{2j'+1})$$

$$\geq \sum_{i\in[k-2], j\in j(k-2,i)} I(V_{ij}; X_i) + \sum_{j'=0}^{k} I(U_{2j'}U_{2j'+1}; X_{k-1})$$

$$+ \sum_{j'=0}^{k} I(U_{2j'}; X_k) + I(U_{2j'+1}; X_k) \quad \text{(by Lemma 1 applied } k \text{ times)}$$

$$= \sum_{i\in[k], j\in j(k,i)} I(V_{ij}; X_i).$$

Note that the above theorem easily extends to a conditional version where all entropies and mutual information quantities are conditioned on some random variable. Such a conditional version will be used later.

We consider both multi-party communication protocols. A protocol specifies the rules of interaction between the players and a transcript $\Pi(X)$ contains the messages exchanged between the players upon the input X. If A is a random variable corresponding to a distribution on the inputs, then $I(A; \Pi(A))$ measures the mutual information between the input and the transcript; this is called *information cost*. The *information complexity* of a function is the minimum information cost of a protocol that correctly computes the function. Conditional version of these quantities are defined in an analogous manner; the reader is referred to [9] for more details.

2.3   The *IndexedAnd* Problem

We consider the following communication problem with t players, where each player has a bit vector. The objective is to find an index in the vector so that all t players have a '1' in that index.

Problem 2 (IndexedAnd). In the *IndexedAnd* problem, there are t players each with a binary vector of length n. Viewing the input to the players as a $n \times t$

binary matrix, the goal of the players is to output an index $i \in [n]$, if any, so that the i-th row in the matrix is all 1's; here a i-th row corresponds to the i-th bit of all players. If there is no such row with all 1's, then any output is valid.

Note that *IndexedAnd* is a functional version of the t-player set disjointness problem. Despite this close relationship, it is not clear if the communication lower bounds for set disjointness directly translate to the *IndexedAnd* problem.

We consider the following input instances for the problem.

(1) NO instance. In the inputs corresponding to the NO instance, each row has one 1 with probability $1/2$ at a random column. As stated before, any output is considered valid if the input is a NO instance.

(2) YES instance. The inputs corresponding to the YES instance is the same as above except that exactly one row has all 1's; the correct answer in this case would be the index of that row.

A random NO instance can be generated as follows. Let $D = (D_1, \ldots, D_n)$ where D_i denotes the column number in row i that is set to 1 with probability $1/2$; so each D_i takes a random value in the range $[t]$. Let $X \in \{0,1\}^{n \times t}$ be the random variable denoting the input and let $X_1, \ldots X_n$ denote the rows of X; so $X_{ij} = 1$ with probability $1/2$ if $D_i = j$ and always 0 otherwise. Note that $X_1, \ldots, X_n$ are independent random variables, even conditioned on D.

We consider protocols that work correctly on the above YES and NO promise instances; note that the protocol is allowed to output any answer on a NO instance. Let $\mathrm{IC}_{IndexedAnd} = \min_\Pi I(X; \Pi(X) \mid D)$ be the conditional information complexity over all protocols that solve the above promise instances of *IndexedAnd*. Note that we use the above distribution corresponding to the NO instances to measure the information complexity. Let Π be the (randomized) protocol that achieves the minimum conditional information cost.

We also need the one-bit version of the *IndexedAnd* problem.

Problem 3 (And). In the *And* problem, there are t players with one bit each. The inputs are restricted to the cases where either all the t players have 1's or at most one player as a 1. The objective is to decide which case.

2.4 Lower Bound for the *IndexedAnd* Problem

We show the main theorem in this section.

Theorem 3. *There is a constant c so that for $n > ct$, $\mathrm{IC}_{IndexedAnd} \geq \tilde{\Omega}(n/t)$. (Alternately, we may say $\mathrm{IC}_{IndexedAnd} \geq \tilde{\Omega}(n/t - c)$.)*

To show this theorem, we first argue that there is a row i about which the protocol does not reveal much information. For simplicity let $\mathrm{IC} = \mathrm{IC}_{IndexedAnd}$ and let $\Pi = \Pi(X)$.

Lemma 2. *There is a row $i \in [n]$ such that $\Pr[i$ is output in NO instance$] \leq 2/n$ and $I(X_i; \Pi \mid D) \leq (2/n) \cdot \mathrm{IC}$.*

Proof. Since $X_1, \ldots, X_n$ are independent given D, it follows (à la [9]) that

$$
\begin{aligned}
\mathrm{IC} &= I(X; \Pi \mid D) \\
&= H(X_1, \ldots, X_n \mid D) - H(X_1, \ldots, X_n \mid \Pi, D) \\
&\geq \left(\sum_{i=1}^{n} H(X_i \mid D) \right) - \left(\sum_{i=1}^{n} H(X_i \mid \Pi, D) \right) \\
&= \sum_{i=1}^{n} I(X_i; \Pi \mid D).
\end{aligned}
\tag{3}
$$

Since the protocol is allowed to output anything on a NO instance, by averaging over the rows, there are at least $n/2$ rows so that for such row i, $\Pr[i$ is output in NO instance$] \leq 2/n$. Using (3) and averaging once more over the rows, there are at least $n/2$ rows so that for each such row i, we have $I(X_i; \Pi \mid D) \leq 2 \cdot \mathrm{IC}/n$. Putting these together, the proof is complete.

Let D_{-i} denote the entries of D except for the entry i. We now obtain a conditional version of Lemma 2.

Lemma 3. *There is a row* $i \in [n]$ *and values* d_{-i} *for* D_{-i} *so that* $\Pr[i$ *is output in NO case* $\mid D_i, D_{-i} = d_{-i}] \leq 6/n$ *and* $I(X_i; \Pi \mid D_i, D_{-i} = d_{-i}) \leq 6 \cdot \mathrm{IC}/n$.

Proof. We use the i provided by Lemma 2. First, we have the guarantee $I(X_i; \Pi \mid D_i D_{-i}) \leq 2 \cdot \mathrm{IC}/n$. This is equivalent to $E_{d_{-i}}[I(X_i; \Pi \mid D_i, D_{-i} = d_{-i}] \leq 2 \cdot \mathrm{IC}/n$ over different values d_{-i} of D_{-i}. So by Markov's inequality, for at least $2/3$-fraction of values d_{-i} of D_{-i}, we have $I(X_i; \Pi \mid D_i, D_{-i} = d_{-i}) \leq 6 \cdot \mathrm{IC}/n$. Similarly, for at least $2/3$-fraction of values d_{-i} of D_{-i}, we have $\Pr[i$ is output in NO case $\mid D_i, D_{-i} = d_{-i}] \leq 6/n$. Together the two events must happen in at least $1/3$-fraction of the cases completing the proof.

To prove theorem 3 we provide a reduction from the *And* problem. A lower bound on the information complexity of the one-bit *And* problem was proved in [9,12].

Theorem 4 ([9,12]). $\mathrm{IC}_{And} \geq \tilde{\Omega}(1/t)$.

We are now ready to prove Theorem 3.

Proof (Proof of Theorem 3). To complete the proof, we provide a reduction. Given an instance of the one-bit *And* problem and a protocol for *IndexedAnd*, we embed the one-bit *And* problem in row i that is given by Lemma 3. We use the values of d_{-i} to fill rows other than i. Note that once the values of d_{-i} are known, these rows can be filled without any communication among players, but by just using the players' private randomness.

The protocol for *IndexedAnd* is run on this input. Answer YES if output of this protocol equals i and NO otherwise. Note that in the NO instance of one-bit AND chosen randomly by setting a random position to 1 with probability $1/2$, the probability of outputting YES is at most $6/n$ and in the YES instance it is 1. Out of the total $t + 1$ NO instances of the And problem each occurs in this random distribution with probability at least $1/(2t)$. So the probability of outputting YES when any of the $t + 1$ NO instances is implanted on row i is at most $12t/n$. For $n > ct$, this probability is small and results in a protocol for the And problem. Therefore, $6 * \mathrm{IC}/n \geq \mathrm{IC}_{And}$ and thus $\mathrm{IC} \geq \Omega(n) * \mathrm{IC}_{And}$.

2.5   Lower Bound for the $MostFrequent$ Problem

We now prove a lower bound on the space required for the $MostFrequent$ problem. The set up is that we are given a vector $f = \{(n_1, k_1), (n_2, k_2), \ldots, (n_s, k_s), (1, t)\}$ where n_i elements occur with frequency k_i ($i \in [s]$) and the most frequent element occurs with frequency t that is the greater than all other frequencies k_i.

For the purposes of proving a lower bound we will make the following relaxations on the frequency vectors and the input instance. We assume that all the frequencies k_i and t are powers of 2; for such distributions any hardness on finding the most frequent element will directly mean a hardness on finding an approximate most frequent element to within factor 2.

Allowing Randomness in Frequency Counts. Next, instead of producing a stream with frequency vector exactly f, we will allow some randomness in the number of elements with the different frequencies — the number of elements with frequency k_i will be sharply concentrated around n_i. Precisely, frequency counts are obtained by drawing elements from a universe U that is a disjoint union $U_1 \cup U_2 \cup \cdots \cup U_s$ where $|U_i| = 2n_i$. Each element of U_i occurs independently and randomly in the stream either 0 or k_i times with probability $1/2$. So the expected count of number of elements with frequency k_i is n_i; if n_i is large, this count is sharply concentrated around its n_i. Each element of the universe U occurs in the stream independently and randomly in this way, except that one element may occur t times.

If there is such an element in the stream with frequency t, then the algorithm is required to identify that element; if no such element occurs any output is allowed.

Theorem 5. *The space complexity for the above problem is $\tilde{\Omega}(\sum_i n_i k_i^2 / t^2)$.*

We prove the lower bound by reduction from the $IndexedAnd$ problem.

Reduction Idea. If all the frequencies k_i were the same, say k (i.e., $s = 1$), then the reduction is easy: given an instance of the $IndexedAnd$ problem with $2n$ rows and t/k players we can construct a sub-stream for each player where the element is present or absent based on the bit in the input. Further each element is replicated k times. The final stream is the concatenation of the sub-streams for each player, where each element occurs k times and the element corresponding to the all 1's row occurs t times. Recall that $IndexedAnd$ problem is hard even if we restrict to YES instances where one row has all 1's and each of the other rows have a 1 in a random column with probability $1/2$ and all 0's with probability $1/2$. So if a YES instance is translated into a stream, the count of the number of elements with frequency k is sharply concentrated around n. By Theorem 3, the total communication complexity is at least $\tilde{\Omega}(\frac{n}{(t/k)})$. So the average communication between successive players among the t/k players is $\tilde{\Omega}(\frac{n}{(t/k)^2}) = \tilde{\Omega}(nk^2/t^2)$.

Reduction for Multiple Frequencies. When the number of frequencies s is more than 1 the reduction is more complicated. Instead of working with one instance of the *IndexedAnd* problem we will look at s instances each with t/k_i players. We will then interleave the sub-streams corresponding the different players over all the instances in such as way as to be able to add up their individual communication complexities.

Formally, we prove the lower bound by assuming a data stream algorithm to solve the *MostFrequent* problem and use it to obtain a communication protocol for *IndexedAnd*. Consider a collection of s random NO instances of the *IndexedAnd* problem, where the i-th instance has $2n_i$ rows and t/k_i players. We will show how a solution to *MostFrequent* can be used to solve a random YES instance of *IndexedAnd* implanted on any of these instances. Construct an instance of *MostFrequent* as follows.

By assumption $t = 2^{t'}$. Also assume without loss of generality that all the frequencies between 1 and t that are powers of 2 are present — this is easily achieved by setting the n_i corresponding to the absent ones to 0. So, $s = t', k_i = 2^{i-1}, i = 1, \ldots, s$.

Interleaving the Different Inputs. A block in the stream is constructed for each player; the j-th player in the i-th instance (for $i \in \{1, .., t'\}$ and $j \in \{1, .., 2^{t'-i+1}\}$) constructs a block b_{ij} by looking at its input column and including k_i copies of the element if a bit 1 is present in the corresponding row and none otherwise. To construct the stream we first arrange the blocks in a binary tree of height $t' + 1$ with a dummy root; the blocks b_{1j} corresponding to the first instance are at the leaves in order of increasing j; the blocks b_{ij} corresponding to the i-th instance are at height i; and the children of the root are two blocks $b_{t',1}$ and $b_{t',2}$ corresponding to the $t' - 1$-th instance. We construct a data stream from these blocks by laying them down in the depth-first search (preorder) traversal order;

Observe that if one of the $s = t'$ instances of the *IndexedAnd* problem is a YES instance and all the others are NO instances then there is an element in the stream that occurs t times that corresponds to the solution to the YES instance. Further the number of elements in the stream with frequency k_i is sharply concentrated around n_i. The essential idea is that a protocol for the *MostFrequent* problem can be used to solve a YES instance implanted into one of the s instances and random NO instances for the others.

Let Y_i denote a random NO input for the i-th instance. Let $Y = (Y_1, \ldots, Y_s)$ denote the combination of all these inputs. Let D denote the union of all the conditioning variables for all the instances. The first instance has t blocks. Let

Π_r denote the output of the block b_{1r} in the streaming algorithm; by the output of a block we mean the content in memory after processing the block. We will lower bound $\sum_{r=1}^{t} H(\Pi_r \mid D)$.

Define $V_{1,1} = \Pi_1, \ldots, \Pi_t$. Define $V_{2,1} = \Pi_1 \Pi_3 \ldots \Pi_{t-1}$ and $V_{2,2} = \Pi_2 \Pi_4 \ldots \Pi_t$. Recursively define $V_{i,1}, V_{i,2}, \ldots, V_{i,2^i}$ where $V_{i,2j-1}$ and $V_{i,2j}$ are obtained from $V_{i-1,j}$ by taking all odd and even elements respectively (for example $V_{3,1} = \Pi_1 \Pi_5 \ldots \Pi_{t-3}, V_{3,2} = \Pi_3 \Pi_7 \ldots \Pi_{t-1}$).

Lemma 4. $I(V_{i,j}; Y_j \mid D) \geq \mathrm{IC}_{n_i,t/k_i}$ where $\mathrm{IC}_{n_i,t/k_i}$ is the conditional information complexity of the IndexedAnd problem with n_i rows and t/k_i players.

Proof. For simplicity let us prove first for $i = j = 1$. Assume that $I(V_{1,1}; Y_1 \mid D) < \mathrm{IC}_{n_1,t/k_1}$. Again let D_i denote the conditioning variables for the i-th instance and D_{-i} denote the conditioning variables for all the other instances. Then there is a value d_{-i} for D_{-i} so that $I(V_{1,1}; Y_1 \mid D_1, D_{-1} = d_{-1}) < \mathrm{IC}_{n_1,t/k_1}$. Then we have a protocol for the first instance of the IndexedAnd problem with improved conditional information complexity than the best possible which is a contradiction. Given an IndexedAnd instance with n_1 rows and t players, it can be implanted in the first instance of the reduction. The other instances can be generated using private random coin tosses based on d_{-i}.

Similarly for $V_{2,1} = \Pi_1 \Pi_3 \ldots \Pi_{t-1}$ and $V_{2,2} = \Pi_2 \Pi_4 \ldots \Pi_t$, observe that because of our specific ordering of blocks in the stream outputs $\Pi_1, \Pi_3, \ldots, \Pi_{t-1}$ partition the stream into $t/2$ parts and each block of the second instance resides in a different partition. So again $I(V_{2,1}; Y_2 \mid D) > \mathrm{IC}_{n_2,t/k_2}$ as otherwise we have a protocol for the second instance of the IndexedAnd problem with improved conditional information complexity. Similarly since the outputs $\Pi_2, \Pi_4, \ldots, \Pi_t$ also separate the $t/2$ blocks of the second instance into different partitions we have $I(V_{2,2}; Y_2 \mid D) > \mathrm{IC}_{n_2,t/k_2}$.

In this manner, the outputs in $V_{i,j}$ can be used to design a protocol for the i-th instance of the IndexedAnd problem giving $I(V_{i,j}; Y_i \mid D) > \mathrm{IC}_{n_i,t/k_i}$.

We now view the variables $V_{i,j}$ arranged in a binary tree of depth t' to invoke Theorem 2; $V_{i-1,j}$ is the j-th node at depth $i-1$ and has children $V_{i,2j-1}$ and $V_{i,2j}$. Each left (right) child has the even (odd) alternate variables of its parent. The leaves of this tree are $\Pi_1, \ldots, \Pi_t$ (not in that order). Further as required by Theorem 2, each node $V_{i,j}$ in the tree is a concatenation of the labels of the leaves of the subtree rooted at that node (note that although the order in which they are concatenated is different from one in the statement of Theorem 2; as noted earlier, this ordering does not make any difference in the entropy values). So Theorem 2 along with lemma 4 and Theorem 3 gives

$$\sum_{i=1}^{t} H(\Pi_i \mid D) \geq \sum_{i=1}^{t'} \sum_{j=1}^{2^i} I(V_{i,j}; Y_i \mid D)$$

$$\geq \sum_{i=1}^{t'} \sum_{j=1}^{2^i} \mathrm{IC}_{n_i,t/k_i}$$

$$\geq \sum_{i=1}^{t'} 2k_i \tilde{\Omega}(n_i/(t/k_i - c)))$$

$$= \sum_{i=1}^{t'} \tilde{\Omega}(n_i k_i^2/t) - O(t).$$

By averaging, for some $i \in [t]$, $H(\Pi_i \mid D) \geq \tilde{\Omega}(\sum n_i k_i^2/t^2) - O(1)$. Recall that the space complexity of the *MostFrequent* problem is at least $\max_i H(\Pi_i \mid D)$. So ignoring the constant additive term, the space complexity is at least $\tilde{\Omega}(\sum n_i k_i^2/t^2)$ completing the proof of Theorem 5.

References

1. Muthukrishnan, S.: Data Streams: Algorithms and Applications. Now Publishers (2005)
2. Babcock, B., Babu, S., Datar, M., Motwani, R., Widom, J.: Models and issues in data stream systems. In: Proceedings of the 21st ACM Symposium on Principles of Databases Systems, pp. 1–16. ACM Press, New York (2002)
3. Alon, N., Matias, Y., Szegedy, M.: The space complexity of approximating the frequency moments. Journal of Computer and System Sciences 58(1), 137–147 (1999)
4. Gibbons, P., Matias, Y.: New sampling-based summary statistics for improving approximate query answers. In: Proceedings of the ACM SIGMOD International Conference on Management of Data, pp. 331–342. ACM Press, New York (1998)
5. Gibbons, P., Matias, Y.: Synposis data structures for massive data sets. In: Proceedings of 10th Annual ACM-SIAM Symposium on Discrete Algorithms, pp. 909–910. ACM Press, New York (1999)
6. Fang, M., Shivakumar, N., Garcia-Molina, H., Motwani, R., Ullman, J.: Computing iceberg queries efficiently. In: Proceedings of 22nd International Conference on Very Large Data Bases, pp. 307–317 (1996)
7. Charikar, M., Chen, K., Farach-Colton, M.: Finding frequent items in data streams. In: Proceedings of the 29th International Colloquium on Automata, Languages, and Programming, pp. 693–703 (2002)
8. Indyk, P., Woodruff, D.P.: Optimal approximations of the frequency moments of data sets. In: Proceedings of the 37th Annual ACM Symposium on Theory of Computing, pp. 202–208. ACM Press, New York (2005)
9. Bar-Yossef, Z., Jayram, T.S., Kumar, R., Sivakumar, D.: An information statistics approach to data stream and communication complexity. Journal of Computer and System Sciences 68(4), 702–732 (2004)
10. Kushilevitz, E., Nisan, N.: Communication Complexity. Cambridge University Press, Cambridge (1997)
11. Cover, T.M., Thomas, J.A.: Elements of Information Theory. John Wiley & Sons, Inc, Chichester (1991)
12. Chakrabarti, A., Khot, S., Sun, X.: Near-optimal lower bounds on the multiparty communication complexity of set-disjointness. In: Proceedings of the 18th Annual IEEE Conference on Computational Complexity, pp. 107–117. IEEE Computer Society Press, Los Alamitos (2003)

Implementing Huge Sparse Random Graphs

Moni Naor[*] and Asaf Nussboim[**]

Department of Computer Science and Applied Mathematics
Weizmann Institute of Science, Rehovot, Israel
{moni.naor, asaf.nussbaum}@weizmann.ac.il

Abstract. Consider a scenario where one desires to simulate the execution of some graph algorithm on huge random $G(N, p)$ graphs, where $N = 2^n$ vertices are fixed and each edge independently appears with probability $p = p_n$. Sampling and storing these graphs is infeasible, yet Goldreich et al. [7], and Naor et al. [12] considered emulating dense $G(N, p)$ graphs by efficiently computable 'random looking' graphs. We emulate *sparse* $G(N, p)$ graphs - including the densities of the $G(N, p)$ threshold for containing a giant component ($p \sim 1/N$), and for achieving connectivity ($p' \sim \ln N/N$). The reasonable model for accessing sparse graphs is neighborhood queries where on query-vertex v, the entire neighbor-set $\Gamma(v)$ is efficiently retrieved (without sequentially deciding adjacency for each vertex). Our emulation is faithful in the sense that our graphs are indistinguishable from $G(N, p)$ graphs from the view of any efficient algorithm that inspects the graph by neighborhood queries of its choice. In particular, the $G(N, p)$ degree sequence is sufficiently well approximated.

1 Introduction

Consider a scenario where one desires to simulate the execution of some graph algorithm on random input graphs of huge size, perhaps even exponentially large in the input length, n, of the corresponding algorithms. Sampling and storing these huge random graphs is clearly infeasible, but can they be emulated by 'random looking' graphs that are efficiently computable?

This question of emulating huge random graphs continues a rich body of research regarding the implementation of huge random objects. Huge random objects can often be faithfully replaced by some 'random-looking' counterparts that are sampled from distributions of significantly smaller support, and can thus be efficiently utilized (namely, using polynomially bounded resources). In general, huge objects are not represented explicitly, but rather by an efficient procedure that evaluates queries regarding the object (e.g. input-output queries on functions). These queries are evaluated using a succinct ($poly(n)$-length) representation of the object called seed. Thus, random looking distributions are sampled by randomly picking the seed.

---

[*] Partly supported by a grant from the Israel Science Foundation.
[**] Partly supported by the Minerva Foundation 2-8495.

M. Charikar et al. (Eds.): APPROX and RANDOM 2007, LNCS 4627, pp. 596–608, 2007.
© Springer-Verlag Berlin Heidelberg 2007

Examples of highly influential random looking objects, which, in fact, underly the foundations of modern cryptography, are pseudorandom functions $f : \{0,1\}^n \to \{0,1\}^n$ and pseudorandom permutations over similar domains. The former were defined and constructed by Goldreich, Goldwasser and Micali [6] (under the necessary assumption that one-way functions exist[1]), and the latter were provided by Luby and Rackoff [10] (based on [6]). The criterion introduced in [6] for faithful emulation of random objects is computational indistinguishability. Namely, no efficient distinguishing algorithm that inspects the function (permutation, resp.) via a sequence of input-output queries of its choice, can tell with probability significantly better then $\frac{1}{2}$ whether the function (permutation, resp.) is sampled from the pseudorandom distribution or from the uniform distribution over functions (permutations, resp.) $f : \{0,1\}^n \to \{0,1\}^n$.

1.1 Implementing Huge Random Graphs

Recently, Goldreich et al. [7], and Naor et al. [12] studied the implementation of huge random graphs. They considered the canonical random graphs $G(N, p_n)$, where $N = 2^n$ labeled vertices are fixed and each edge appears with probability $p = p_n$ independently of all other edges. These works focused on relatively dense graphs, where the natural access model is via edge-queries that inquire whether a specific edge appears in the graph.

In contrast, the focus of this work is sparse graphs. The latter term refers to a wide range of densities p, including the Erdös-Rényi threshold density for containing a giant component ($p \sim 1/N$ [4]), and for achieving connectivity ($p' \sim \ln N/N$ [3]). As edge-queries rarely detect any adjacency in sparse graphs, the reasonable model for accessing them is by providing the entire neighborhood (namely, the entire list of adjacent vertices) $\Gamma(v)$ in response to a query-vertex v (this neighborhood-queries model is the common one in the context of sparse graphs, and in particular, in the field of testing graph properties without inspecting the entire graph [8]).

Supporting neighborhood-queries is far from trivial, as one has to specify which of the *exponentially* many potential vertices indeed appears in $\Gamma(v)$. In addition, since the graphs are undirected, consistency is required in the sense that $u \in \Gamma(w)$ iff $w \in \Gamma(u)$. In particular, previous works [7,12] implement a graph via a Boolean function f, where $f(e) = 1$ iff the edge e appears in the graph. Using pseudorandom Boolean functions [6] guarantees efficiency (in the usage of randomness and memory), but eventually fails in our context, as supporting even a single neighborhood-query requires the evaluation of f on exponentially many inputs.

To overcome this, a different approach was applied by Goldreich et al. to support neighborhood-queries [7]. They construct sparse d-regular graphs that achieve computational pseudo-randomness w.r.t. the uniform distribution over

---

[1] A one-way function is an efficiently computable function $h : \{0,1\}^* \to \{0,1\}^*$ that cannot be inverted efficiently on random inputs with non-negligible probability of success.

d-regular graphs. They rely on the fact that the views observed by efficient distinguishers (that examine sparse random regular graphs) are typically cycle-free and are symmetric (w.r.t. the role of different vertices in the view). Thus, randomly shuffling the vertex-names of any single large-girth d-regular graph will produce a random looking cycle-free view, as desired. Cycle-free views characterize *non-regular* sparse random graphs too. Yet our main challenge in the present work is (not only to construct graphs that produce cycle-free views, but) mainly to properly approximate the degree distribution of $G(N, p)$.

1.2   Our Contribution

We construct computationally pseudorandom graphs w.r.t. neighborhood-queries, under the necessary assumption that one-way functions (OWF) exist. Pseudorandomness is achieved (even) w.r.t. adaptive distinguishers, that may choose the next query depending on previous replies (in particular, the next query vertex may appear in a previous reply). Our results hold for the entire range of densities where typical degrees are $poly(n)$-bounded, namely, for arbitrary $p_n \leq \frac{n^{O(1)}}{N}$.

Standard pseudorandomness arguments imply the necessity of the OWF assumption whenever $p \geq \frac{1}{Nn^{O(1)}}$. Thus, the OWF assumption is necessary to capture the threshold densities for containing a giant component ($p \sim 1/N$), and for achieving connectivity ($p' \sim \ln N/N$. For smaller densities $p \leq \frac{1}{Nn^{\omega(1)}}$, OWFs are no longer needed since $G(N, p)$-views rarely include any adjacencies, and therefore the empty graph provides the desired pseudorandom implementation.

We remark that one may consider a weaker notion of efficiency, under which it is possible to handle (again, under the OWF assumption) higher densities: here neighborhoods are allowed to have super-polynomial size, yet each query is handled in polynomial time in the size of the reply. With respect to this weaker efficiency definition, our construction applies whenever the density is negligible. The latter roughly captures the entire range where the edge-queries model is no longer reasonable (so neighborhood-queries are used instead). We stress that subsequent definitions and results in this paper relate only to the stronger (and more standard) notion of efficiency.

Description of our Construction. We first provide a costly interim construction that emulates $G(N, p)$ well, and then 'de-randomize' our interim implementation to obtain an efficient one. In the interim construction, $G_{\Pi Bin}$, the degree of each specific vertex has Binomial distribution $BIN(N - 1, p)$ (just as in $G(N, p)$). However, unlike the $G(N, p)$ case, all the degrees in $G_{\Pi Bin}$ are independent of each other[2] (for instance, the sum of degrees in $G_{\Pi Bin}$ is allowed to be odd). Given the degrees, edges are assigned to the graph via the traditional configuration method (Bollobás [2]) where each vertex of degree d is associated with d unique 'ports'. A uniformly random matching over the ports decides the edges of the graph s.t. two

---

[2] Thus the notation ΠBin stands for the fact that the joint distribution of the degrees $(d_1, ..., d_n)$ is the product-distribution of N Binomial distributions.

vertices are adjacent iff any of their ports are matched (self-loops and multi-edges are ignored[3]).

The indistinguishability of $G_{\Pi Bin}$ from $G(N, p)$ is established by showing that the distribution of $G_{\Pi Bin}$-replies is statistically close[4] to the corresponding distribution in the $G(N, p)$ case - as long as the number of queries is $poly(n)$-bounded. To this end, it is observed that in the $G(N, p)$ case the size of the next reply Γ_j has Binomial distribution, with each specific vertex being equally likely to appear in Γ_j (this holds regardless of the previous replies $\Gamma_1, ..., \Gamma_{j-1}$). Thus, the main technical part of the proof is to analyze the distribution of the next reply in the $G_{\Pi Bin}$ model and to establish it's closeness (up to a negligible difference) to the $G(N, p)$ case.

The interim construction is 'de-randomized' as follows. Neighborhood-queries are handled by using random-looking functions that efficiently support interval-queries where on interval-query (α, β) the entire sum $\sum_{\alpha \leq x \leq \beta} f(x)$ is retrieved. Implementing such functions is due to Goldreich et al. [7] and to Naor and Reingold (in [5]). Specifically, we use a Boolean function f over $N(N-1)$ inputs that are partitioned into N blocks of size $N-1$. The sum of f over the v'th block corresponds to the number of ports d_v of vertex v. Thus, the integers $1, ..., d_1$ are the ports of the first vertex, $d_1 + 1, ..., d_1 + d_2$ are the ports of the second vertex, etc.. To retrieve the neighborhood $\Gamma(v)$, we use interval-queries over f to identify the exact set of ports that v possesses, and then apply the random matching on these ports. To resemble $G_{\Pi Bin}$, the aforementioned construction [7,5] guarantees our functions f to be computationally indistinguishable (w.r.t. interval-queries) from the (truly random) functions that correspond to $N(N-1)$ independent Bernoulli trials with success probability p.

Finally, as the total number of ports is typically exponentially large, the required random matching over the ports is formalized as a random involution (with no fixed-points) and Naor and Reingold's construction of pseudorandom involutions [13] is applied to efficiently match the ports and decide $\Gamma(v)$.

Achieving almost k-Wise Independence. We briefly discuss (k, ϵ)-wise independence, which is an alternative criterion (incomparable to computational pseudorandomness) for being a good 'random looking' graph. Meeting this criterion means that for any k vertices $u_1, ..., u_k$ (fixed in advance) the distribution of the neighborhoods $\Gamma(u_1), ..., \Gamma(u_k)$ is within ϵ statistical distance from the corresponding distribution in the $G(N, p)$ case.

Our construction can achieve (k, ϵ)-wise independence for any prescribed $poly(n)$ bounded value of k and any prescribed exponentially small ϵ. This is done by slightly modifying the implementation of the pseudorandom involutions and the implementation of the pseudorandom functions that support interval-queries. Rather than using computationally pseudorandom bits for the two latter imple-

---

[3] When the total number of ports is odd, a single port remains unmatched and thus induces no edge.

[4] The statistical distance between two distributions $\mathcal{D}_1, \mathcal{D}_2$ is defined as $\frac{1}{2} \sum_x |\Pr_{\mathcal{D}_1}[x] - \Pr_{\mathcal{D}_2}[x]|$ where the sum ranges over the entire sample space.

mentations (as originally done in [5,7,13]), k'-wise independent bits are used instead. The latter refer to distributions s.t. for any fixed k' bits, their joint distribution is precisely uniform. It can be shown that taking some $k' \leq poly(n,k)$ suffices for our resulting graphs to be (k,ϵ)-wise independent. Thus by applying known efficient constructions of k'-wise independent bits (cf. [1] chp. 15) the efficiency of our modified construction is retained. Constructions of bits that are both pseudorandom and k'-wise independent are easily given (by combining the original constructions for each type), thus providing graphs which are simultaneously pseudorandom as well as (k,ϵ)-wise independent.

Emulating Related Models of Random Graphs.   Consider the original Erdös-Rényi models $G(N,M)$ and $\{G(N,t)\}_t$ which are closely related to $G(N,p)$ graphs. In both models N labeled vertices are fixed (as before), with $G(N,M)$ being the uniform distribution over all graphs with precisely M edges, whereas $\{G(N,t)\}_t$ is the random graph process, where the initial graph $G(N,0)$ is empty and at each time step, $G(N,t+1)$ is obtained from $G(N,t)$ by adding a uniformly random edge (that hasn't been chosen before). Thus, $\{G(N,t)\}_t$ at time $t = M$ is identical to $G(N,M)$, and it is well known that the combinatorial properties of $G(N,p)$ and $G(N,M)$ graphs are very similar when $p \sim M/\binom{N}{2}$.

We demonstrate how to emulate $G(N,M)$ and $\{G(N,t)\}_t$ graphs where for $\{G(N,t)\}_t$ graphs the reasonable types of queries are: i) which edge was added at time step t, and ii) whether some specific edge already appears at time t. The efficiency and pseudorandomness of the following constructions is easy to establish given the main theorems of this paper.

The first construction (appropriate for dense graphs too) uses a pseudorandom bijection σ from the set of all possible time-steps $\{1, ..., \binom{N}{2}\}$ to the set of all $\binom{N}{2}$ potential edges. Thus, $\sigma(t)$ is the edge joined to the graph at time t, and the edge e appears in the graph at time t iff $\sigma^{-1}(e) \leq t$. Similarly, $G(N,M)$ is emulated by including in the graph precisely those edges s.t. $\sigma^{-1}(e) \leq M$.

For sparse graphs, the $\{G(N,t)\}_t$ process is bounded a-priori at some $T = n^{\Theta(1)}N$ time, and neighborhood-queries should be supported. To this end, the main construction of this paper is used with the following two adaptations. i) We first (trivially) modify the construction of the pseudorandom range-summable functions, s.t. precisely $2T$ ports are produced (instead of a Binomially distributed number of ports). Deciding the edge-set of the resulting graph, and supporting neighborhood-queries in a pseudorandom manner is done as before. ii) In addition, we use a pseudorandom bijection σ from the set of all possible time-steps $\{1, ..., T\}$ to the set of T edges that match the ports in our construction s.t. $\sigma(t)$ is the edge joined to the graph at time t. Deciding whether the edge $\{u,v\}$ already appears at time t, is done by enumerating all $poly(n)$ ports ρ_i of u, and for each of them checking whether ρ_i was matched to a port of v prior to time t. Unfortunately, this account of time steps fails to ignore double-edges and self loops, in contrast with $\{G(N,t)\}_t$, (but in line with the variant of $\{G(N,t)\}_t$ where at each step a uniformly random vertex-pair is added, with repetitions allowed).

2   Preliminaries

The main definitions provided are of the models $G_{\Pi\text{Bin}}$ and $G_{\Pi\text{Bin}}^{\text{OTF}}$ (under 'graphs and configurations'), and the definitions (derived from [7]) of pseudorandomness not only for graphs, but also for interval-summable functions and for involutions.

Basics, Arithmetics and Asymptotic Analysis. Efficient procedures are algorithms that run in worst-case polynomial time in their input length n. Throughout, all graphs have $N = 2^n$ vertices. Negligible terms $\epsilon(n)$ are ones that vanish faster than the reciprocal of any polynomial (for all j, $|\epsilon(n)| = o(n^{-j})$). The notation $X(1 \pm \delta)$ stands for some term E s.t. $X(1 - \delta) \le E \le X(1 + \delta)$, and the notation $A \sim B$ implies that $A = B(1 \pm \epsilon)$ for some negligible ϵ. We often use the fact that $\sim$ is a transitive relation that behaves well under summation, multiplication and division. Some of the inequalities used throughout hold only for sufficiently large (yet very reasonable) values of n. We let $[m] = \{1, ..., m\}$.

Graphs and Configurations. We consider only simple, undirected graphs (no self-loops or multi-edges allowed), over the vertex-set $\{1, ..., N\}$. A port sequence is any sequence $\boldsymbol{t} = (t_1, ..., t_N) \in \{0, ..., N - 1\}^N$, regardless of whether $\boldsymbol{t}$ is indeed a graphic-sequence (namely the degree sequence of some N-vertex graph). In particular we allow the sum $\sum_{v=1}^{N} t_v$ to be odd. The term 'degree' and the notation 'd_v' are abused to refer not only to degrees of vertices, but also to the number of ports that a vertex v has in the configurational model.

• The $G_{\Pi\text{Bin}}$ model. In this model each vertex v is associated with d_v unique ports where the number of ports has Binomial distribution $d_v \sim \text{BIN}(N - 1, p)$, and all the random variables d_v are independent of each other. If the sum $\sum_{v=1}^{N} d_v$ is odd, a single 'parity-vertex' with a single port is added (to obtain an even sum). A matching over the ports is later chosen uniformly at random (among all possible matchings), s.t. two vertices are adjacent iff any of their ports are matched. Self-loops, multi-edges, and (possibly) the edge connected to the parity vertex are all ignored in the final graph.

• The $G_{\Pi\text{Bin}}^{\text{OTF}}$ model. Here the port sequence is produced as in $G_{\Pi\text{Bin}}$, but the matching over the ports is decided 'on-the-fly', during a single interaction with some distinguishing algorithm C. On query-vertex v, the available ports associated with v are matched one after the other to a uniformly random available port. After the execution of C terminates, the remaining available ports are uniformly matched at random. It is not too hard to see, that the graph distribution produced by the two models is identical.

Computational Pseudorandomness. As in [7], the following definitions capture pseudorandomness not only for graphs (w.r.t. neighborhood-queries) but also for interval-summable functions and for involutions. To this end, we use query representation functions.

• Query representation functions (QRFs). Given the type of queries we wish to support, we represent each specific object o (e.g. a single graph) by a specific function f_o s.t. evaluating f_o on a single input corresponds to supporting a single (possibly complex) query over o. We call f_o the QRF of o. For instance to support neighborhood queries over N vertex graphs, each specific graph g is represented by a function $f_g : [N] \rightarrow 2^{[N]}$ s.t. for any vertex v, $f_g(v) = \Gamma(v)$. Similarly, to support interval-queries over Boolean functions with domain $[M]$, each specific function h is represented by another function $f'_h : [M] \times [M] \rightarrow [M]$ s.t. $f'_h(\alpha, \beta) = \sum_{\alpha \leq x \leq \beta} h(x)$ for all $\alpha, \beta \in [M]$. The QRF for involutions (w.r.t. input-output queries) is the original involution itself.

• Pseudorandomness w.r.t. complex queries. Note that any distribution $\mathcal{D}$ over the original objects induces a corresponding distribution $\mathcal{F}_\mathcal{D}$ over the QRFs. Consequently, pseudorandomness of the original objects $\mathcal{D}$ w.r.t. a given type of queries, reduces to the pseudorandomness of the QRFs $\mathcal{F}_\mathcal{D}$ w.r.t. simple input-output queries. We thus define pseudorandomness w.r.t. complex queries simply as pseudorandomness (in the classical sense of GGM [6]) of the corresponding QRFs. For a given sequence of densities $\{p_n\}_{n \in \mathbb{N}}\, p_n \in [0, 1]$ the following definitions are used:

i) Neighborhood queries pseudorandom graphs. Let $\{\mathcal{G}_n\}_{n \in \mathbb{N}}$ be a sequence of distributions, where each $\mathcal{G}_n$ is taken over N-vertex graphs. Then $\{\mathcal{G}_n\}_{n \in \mathbb{N}}$ is neighborhood-queries pseudorandom w.r.t. $\{G(N, p_n)\}_{n \in \mathbb{N}}$ if the neighborhood QRFs $\{\mathcal{F}_{\mathcal{G}_n}\}_{n \in \mathbb{N}}$ induced by $\{\mathcal{G}_n\}_{n \in \mathbb{N}}$ is pseudorandom w.r.t. the neighborhood QRFs induced by $\{G(N, p_n)\}_{n \in \mathbb{N}}$.

ii) Interval-sum queries pseudorandom functions. Consider a sequence of distributions $\{\mathcal{H}_n\}_{n \in \mathbb{N}}$ where each $\mathcal{H}_n$ is taken over Boolean functions with domain D_n. Then $\{\mathcal{H}_n\}_{n \in \mathbb{N}}$ is interval-sum query pseudorandom if its interval-sum QRFs are pseudorandom w.r.t. the interval-sum QRFs of the truly random functions. The latter refer to the distribution over Boolean functions h_n with domain D_n, where for each input x we have $h_n(x) = 1$ with probability p_n independently of the value of h_n over other inputs.

iii) Pseudorandom involutions. A sequence of integers $\{M_n\}_{n \in \mathbb{N}}$ forms proper involutions domains if all M_n are even and $M_n = n^{\omega(1)}$. A sequence of distributions $\{\Pi_n\}_{n \in \mathbb{N}}$ over involutions with no fixed-points $\pi_n : [M_n] \rightarrow [M_n]$ are pseudorandom involutions if it is pseudo-random w.r.t. the sequence of uniform distributions over all involutions with no fixed-points and same domains $[M_n]$.

Finally, recall that efficiently constructing pseudorandom functions is possible iff one-way functions (OWFs) exist, which are efficiently computable functions $f : \{0,1\}^* \rightarrow \{0,1\}^*$ that cannot be inverted efficiently on uniformly random inputs with non-negligible probability of success.

3   Our Main Construction

This section formally describes our construction of sparse pseudorandom graphs.

We first define the range of 'proper densities' p that our arguments handle. To ensure that all the degrees are $poly(n)$-bounded, proper densities are upper bounded. Our techniques also require a lower bound which guarantees that the total number of ports is (almost surely) (i) super-polynomial (in n) and (ii) extremely close to its expectation. These facts are (i) frequently used while proving the similarity of the models $G_{\Pi Bin}$ and $G(N,p)$, and (ii) used to validate the usage of the pseudorandom involutions. For densities too small to be proper, $G(N,p)$-views rarely include any adjacencies, so $G(N,p)$ graphs are emulated well by the empty graph.

Definition 1 (Proper density). *A sequence of densities* $\{p_n\}_{n\in\mathbb{N}}$ *is proper if for all* n, $0 < p_n < 1$ *and* $\frac{(\lg N)^{\omega(1)}}{N^2} \leq p_n \leq \frac{(\lg N)^{O(1)}}{N}$.

We next present the main constructions we use as sub-routines.

Theorem 1 ([7,5] Interval-summable functions). *Let* $\{p_n\}_{n\in\mathbb{N}}$ *be proper densities. Assuming the existence of one-way functions, then [7,5] provide interval-summable queries pseudorandom functions with domains* $D_n = [N] \times [N-1]$.

Theorem 2 ([13] Pseudorandom involutions). *Let* $\{M_n\}_{n\in\mathbb{N}}$ *be a proper involutions domains.[5] Assuming the existence of one-way functions, [13] provide pseudorandom involutions w.r.t. the domains* $\{[M_n]\}_{n\in\mathbb{N}}$.

Our main construction is given bellow. The underlying intuition is discussed in the introduction.

Construction 1 (Implementing sparse random graphs). *On input* $(1^n, p)$, *construct a graph on vertex-set* $[N]$ *as follows. If* $p \leq N^{-1.5}$ *pick the empty graph so each query is replied with the empty set. Otherwise,*

- **Sampling** - *Pick an Interval-summable function* f *over domain* $X = [N(N-1)]$ *with parameter* p, *as in Theorem 1. Set* $E = \sum_{x\in X} f(x)$ *to be the total number of ports. If* E *is odd increase it by 1 (adding the parity port). Sample a pseudorandom involution (with no fixed points)* $\pi : [E] \to [E]$, *as in Theorem 2.*
- **Supporting neighborhood-queries** - *On query-vertex* v:
  (i) *Compute* $S_v = \sum_{x=1}^{(v-1)(N-1)} f(x)$, $S'_v = \sum_{x=1}^{v(N-1)} f(x)$ *and* $d_v = S'_v - S_v$ *(thus* $S_v + 1, ..., S_v + d_v$ *are the ports associated with* v*).*
  (ii) *For* $i = 1, ..., d_v$ *compute* $T_i = \pi(S_v + i)$ *(*T_i *is the port matched with* S_i*). Unless* T_i *is the parity-port, conduct a binary search to decide to which vertex* u_i *the port* T_i *belongs (the space of the search is the vertex-set* $[N]$*. At each stage of the search step (i) is invoked to check whether* T_i *belongs to* u *or to a previous or consequent vertex* u'*).*
  (iii) *Output the set* $\{u_1, ..., u_{d_v}\} \setminus \{v\}$.

4   Pseudorandomness of the Main Construction

This section presents our main result (Theorem 3) that establishes the pseudorandomness of construction 1. It is proved by reducing it to our main technical result (Theorem 3.1) which asserts a negligible distance between views produced by the $G_{\Pi Bin}$ and the $G(N,p)$ models.

---

[5] The original [13] construction handles only domains of size which is a power of 2. However, the adaptation to the general case is not involved (cf.[11]).

Theorem 3 (Pseudorandomness of construction 1). *Let* $\{p_n\}_{n \in \mathbb{N}}$ *be arbitrary proper densities. Then, assuming the existence of one-way functions, the graphs distributions* $\{\mathcal{G}_n\}_{n \in \mathbb{N}}$ *produced by construction 1 on inputs* $(1^n, p_n)$ *are neighborhood-queries pseudorandom w.r.t.* $\{G(N, p_n)\}_{n \in \mathbb{N}}$.

Proof (Theorem 3). Correctness is trivial for $p_n \leq N^{-1.5}$, since the views of a $G(N, p)$ graph reveal adjacencies only with negligible probability. For larger densities, by Theorems 1 and 2 the interval-sum function sub-routine and the involution sub-routine used by our construction are efficient. Since the remaining computations executed by our construction are trivial, the entire construction is efficient. Pseudorandomness - Using the transitivity of indistinguishability, we establish the pseudorandomness of $\mathcal{G}_n$ by demonstrating (i) the indistinguishability of $\mathcal{G}_n$ from $G_{\Pi \text{Bin}}(N, p_n)$, and (ii) the indistinguishability of $G_{\Pi \text{Bin}}(N, p_n)$ from $G(N, p_n)$. Part (i) follows by a standard argument from the pseudorandomness of both the interval summable functions and the involutions used. Indeed, if (i) was false there would exist an efficient distinguisher D between $\mathcal{G}_n$ and $G_{\Pi \text{Bin}}(N, p_n)$. This means that by combining our construction with D, one gets an efficient procedure for distinguishing between truly random interval summable functions and involutions and the pseudorandom ones - a contradiction to their pseudorandomness. Part (ii) follows immediately from Theorem 3.1, since the statistical distance between views produced by $G_{\Pi \text{Bin}}$ and $G(N, p)$ upper-bounds the distinguishing advantage of the distinguisher. $\square$

Theorem 3.1. (Statistical distance between views produced by $\mathbf{G_{\Pi Bin}}$ and $\mathbf{G(N, p)}$). *Let* C *be an efficient distinguisher and let* $\underline{V}, \overline{V}$ *denote the distributions of the view of* C *as the input graphs are sampled from either* $G_{\Pi \text{Bin}}(N, p_n)$ *or* $G(N, p_n)$, *respectively. Then, the statistical distance between* $\underline{V}$ *and* $\overline{V}$ *is negligible.*

Proof (Theorem 3.1). Fix n and assume w.l.o.g. that the distinguisher C is a circuit.[6] We may further assume C to be deterministic, as the coin-tosses that produce the largest statistical distance between the views $\underline{V}, \overline{V}$ can be hard-wired into the circuit. Let $v_1, ..., v_q$ denote the vertex-queries of C, and let $R_1, ..., R_q$ denote the responses that C receives. Thus R_j is the entire neighbor-set of v_j, and $\boldsymbol{R} = \{R_1, ..., R_q\}$ denotes the entire view. Note that $v_j, R_j, \boldsymbol{R}$ are all random variables as probabilities are taken over the choice of the graph (from either $G_{\Pi \text{Bin}}$ or $G(N, p)$). Next, let u_j, Γ_j and Γ, respectively denote specific values of v_j, R_j and $\boldsymbol{R}$. As $\overline{\text{Pr}}\,[\cdot]$, $\underline{\text{Pr}}\,[\cdot]$ denote probabilities taken over $G(N, p_n)$ and $G_{\Pi \text{Bin}}(N, p_n)$, respectively, our goal is to establish a negligible upper bound on

$$\Sigma_\Gamma \left| \overline{\text{Pr}}\,[\boldsymbol{R} = \boldsymbol{\Gamma}] - \underline{\text{Pr}}\,[\boldsymbol{R} = \boldsymbol{\Gamma}] \right|. \tag{1}$$

The proof proceeds as follows. (i) We first separate the sum in equation 1 into 'likely' terms and into terms with either an unlikely port sequence or an unlikely view (formal definitions are given in the next paragraph). (ii) Next, the (un-surprising) negligible bound on the contribution of the unlikely terms is claimed in Lemma 2 (it is proved in [11]). (iii) Then, the negligible bound for the

---

[6] We thus strengthen the distinguisher *without* assuming our OWFs to be hard to invert even for circuits (and not only for Turing machines).

likely terms is claimed in Lemma 1. Lemma 1 (which considers the entire view) is proved by reducing it to Claim 1 which considers the distributions of the next reply in the view (given the previous replies) and establishes sufficient closeness of these distributions in the $G_{\Pi Bin}$ and the $G(N,p)$ models. (iv) Finally, claim 1 itself is proved by first observing that in the $G(N,p)$ case the size of the next neighborhood Γ_j has Binomial distribution, with each specific vertex being equally likely to appear in Γ_j. Thus, the main technical part of the proof is to analyze the distribution of the next reply in the $G_{\Pi Bin}$ model. It will turn out that some terms concerning the distribution of the random port sequence will cancel nicely with other terms concerning the distribution of the random matching (given the port sequence). This way an (almost) Binomial distribution is established for the size of the next reply in the $G_{\Pi Bin}$ case and claim 1 follows.

• **Definitions and notation.** An improper edge is either a self-edge, a multi-edge or an edge connected to the parity-vertex. A port is called proper if the edge it induces is proper, and a vertex is proper if all it's ports are. A degree t_v is likely if $0 \leq t_v \leq d_{\max}$ for $d_{\max} = \lg N \lceil p(N-1) \rceil$. An entire port sequence t is likely if all the degrees t_v are likely and in addition $\sum_{v=1}^{N} t_v = \mu(1 \pm \epsilon)$, where $\mu = N(N-1)p$ is the expected value of the sum, and the error-term is $\epsilon = \frac{\lg \lg N}{\sqrt{\bar{p}N}}$. We let $\mathbb{D}$ denote the set of all likely port sequences. The random variable that indicates the resulting port sequence is denoted d.

A view Γ is improper if some query-vertex v_j is improper. We let $\mathbb{V}$ denote the collection of all 'likely' views that are simultaneously (i) Reply-collision-free - for any $i < j$, $[\Gamma(v_i) \bigcap \Gamma(v_j)] \setminus \{v_1, ..., v_j\} = \emptyset$. Namely, the distinguisher detects no 'non-trivial' collisions, but may produce trivial collisions by choosing, say, some $v_2 \in \Gamma(v_1)$, and some $v_3 \in \Gamma(v_2)$ so $v_2 \in [\Gamma(v_1) \bigcap \Gamma(v_j)]$. (ii) Have small neighbor-sets - $|\Gamma_j| \leq d_{\max} = \lg N \lceil p(N-1) \rceil$ for all j. (iii) Contain only proper vertices (as defined above). Given that a (partial) view $\{\Gamma_1, ..., \Gamma_{j-1}\}$ is likely, we say that a following reply Γ_j is likely if $\{\Gamma_1, ..., \Gamma_j\}$ remains likely.

Separating the likely and unlikely terms. The triangle inequality gives

$$\Sigma_\Gamma \left| \overline{\Pr}[R = \Gamma] - \underline{\Pr}[R = \Gamma] \right| =$$

$$\Sigma_{\Gamma \in \mathbb{V}} \left| \overline{\Pr}[R = \Gamma] - (\underline{\Pr}[R = \Gamma, d \in \mathbb{D}] + \underline{\Pr}[R = \Gamma, d \notin \mathbb{D}]) \right| +$$

$$\Sigma_{\Gamma \notin \mathbb{V}} \left| \overline{\Pr}[R = \Gamma] - \underline{\Pr}[R = \Gamma] \right| \leq$$

$$\underbrace{\Sigma_{\Gamma \in \mathbb{V}} \left| \overline{\Pr}[R = \Gamma] - \underline{\Pr}[R = \Gamma, d \in \mathbb{D}] \right|}_{\overset{\text{def}}{=} T_1} +$$

$$\underbrace{\underline{\Pr}[R \in \mathbb{V}, d \notin \mathbb{D}] + \overline{\Pr}[R \notin \mathbb{V}] + \underline{\Pr}[R \notin \mathbb{V}]}_{\overset{\text{def}}{=} T_2}$$

(this a-symmetric separation of events is crucial to our argument). Thus, proving Theorem 3.1 reduces to establishing the following lemmata.

Lemma 1 (Statistical distance between *likely* views). *For* $C, \overline{\mathrm{Pr}}\left[\cdot\right], \underline{\mathrm{Pr}}\left[\cdot\right],$
$\boldsymbol{R}, \mathbb{V}, \mathbb{D}$ *and* T_1 *as above the term* T_1 *is negligible.*

Lemma 2 (Statistical distance between *unlikely* views). *For* $C, \overline{\mathrm{Pr}}\left[\cdot\right],$
$\underline{\mathrm{Pr}}\left[\cdot\right], \boldsymbol{R}, \mathbb{V},$ *and* $\mathbb{D}$ *as above the term* T_2 *is negligible.*

The unsurprising Lemma 2 is proved in [11]. We continue with Lemma 1.

Proof of Lemma 1. Assume w.l.o.g. that C performs the same number of queries
q on any N-vertex graph. Then for any likely view $\boldsymbol{\Gamma} = \{\Gamma_1, ..., \Gamma_q\} \in \mathbb{V}$ we have

$$\overline{P} \stackrel{\text{def}}{=} \overline{\mathrm{Pr}}\left[\boldsymbol{R} = \boldsymbol{\Gamma}\right] = \Pi_{j=1}^q \overline{P}_j,$$

$$\underline{P} \stackrel{\text{def}}{=} \underline{\mathrm{Pr}}\left[\boldsymbol{R} = \boldsymbol{\Gamma}, \boldsymbol{d} \in \mathbb{D}\right] = \underline{\mathrm{Pr}}\left[\boldsymbol{d} \in \mathbb{D}\right] \Pi_{j=1}^q \underline{P}_j,$$

where

$$\overline{P}_j = \overline{\mathrm{Pr}}\left[R_j = \Gamma_j \mid R_1 = \Gamma_1, ..., R_{j-1} = \Gamma_{j-1}\right]$$

and

$$\underline{P}_j = \underline{\mathrm{Pr}}\left[R_j = \Gamma_j \mid \boldsymbol{d} \in \mathbb{D}, R_1 = \Gamma_1, ..., R_{j-1} = \Gamma_{j-1}\right].$$

We will show (Claim 1) that for some negligible ϵ, and any $\boldsymbol{\Gamma} \in \mathbb{V}$ and $0 \leq j \leq q$
we have $\underline{P}_j = \overline{P}_j(1 \pm \epsilon)$, as well as $\underline{\mathrm{Pr}}\left[\boldsymbol{d} \in \mathbb{D}\right] \geq 1 - \epsilon$. As q is $poly(n)$-bounded and
ϵ is negligible, then $q\epsilon = o(1)$ so we get

$$\underline{P} = \overline{P}(1 \pm \epsilon)^{q+1} = \overline{P}(1 \pm \Theta(\epsilon q)).$$

Consequently,

$$T_1 = \Sigma_{\boldsymbol{\Gamma} \in \mathbb{V}} \Theta(\epsilon q) \overline{\mathrm{Pr}}\left[\boldsymbol{R} = \boldsymbol{\Gamma}\right] = \Theta(\epsilon q) \Sigma_{\boldsymbol{\Gamma} \in \mathbb{V}} \overline{\mathrm{Pr}}\left[\boldsymbol{R} = \boldsymbol{\Gamma}\right] = \Theta(\epsilon q),$$

which is negligible. Therefore Lemma 1 follows once we establish claim 1.   □

Claim 1. *For* $\overline{P}_j, \underline{P}_j$, *and* $\mathbb{D}$ *as above there exist a negligible* ϵ, *s.t. for all likely
views* $\boldsymbol{\Gamma} \in \mathbb{V}$ *and all* $0 \leq j \leq q$ *then* $\underline{P}_j = \overline{P}_j(1 \pm \epsilon)$, *and* $\underline{\mathrm{Pr}}\left[\boldsymbol{d} \in \mathbb{D}\right] \geq 1 - \epsilon$.

Proof (Claim 1). We focus on the $\underline{P}_j = \overline{P}_j(1 \pm \epsilon)$ part, as the $\underline{\mathrm{Pr}}\left[\boldsymbol{d} \in \mathbb{D}\right] \geq 1 - \epsilon$
part merely states that the port sequence is likely (for a complete proof see [11]).

• **Notation.** Let W denote the set of vertices $w \notin \left(\left(\bigcup_{i=1}^{j-1} \Gamma_i\right) \bigcup \{v_1, ..., v_j\}\right)$ that
haven't appeared in the view up to stage j, and let $N' = |W|$. Let A denote the event
that the next reply Γ_j is likely. Consider an arbitrary likely degree $k \leq d_{\max}$, and
let H denote the event that a specific (partial) view has been observed, namely that
$(R_1 = \Gamma_1, ..., R_{j-1} = \Gamma_{j-1})$. Similarly, let $H' = (\boldsymbol{d} \in \mathbb{D}) \bigcap (R_1 = \Gamma_1, ..., R_{j-1} = \Gamma_{j-1})$.

As the claim deals only with likely views, then the view is reply-collision-free (see
notations at page 605) so the next query vertex v_j appears in either one or none of
the previous replies $\Gamma_1, ..., \Gamma_{j-1}$. Assume w.l.o.g. that the first case holds (adapting
the proof to the complement case is trivial). Therefore, whenever A occurs, we have:
$(|R_j| = k)$ iff v_j has precisely $k - 1$ neighbors in W.

The G(N, p_n) case (Claim 1). By the above and by the total independence of edges in $G(N, p_n)$,

$$\overline{\Pr}\left[A, |R_j| = k \,\middle|\, H\right] = \binom{N'}{k-1} p^{k-1}(1-p)^{N'-(k-1)}$$

with all specific choices of vertices $w_1, ..., w_{k-1} \in W$ for R_j being equiprobable.

The $G_{\Pi Bin}$ case (Claim 1). It isn't too hard to see that whenever A holds, the symmetry of the $G_{\Pi Bin}$ model implies that all specific choices of vertices $w_1, ..., w_{k-1} \in W$ are equiprobable for R_j (just as in $G(N, p_n)$). Thus, it remains to show that (analogously to the $G(N, p)$ case) the next reply has approximately Binomially distributed size, that is, to prove the following claim:

Claim 1.1 (Establishing the (approximately) Binomial size of the next reply in the $G_{\Pi Bin}$ case). *For A, H', N' as above,*

$$\underline{\Pr}\left[A, |R_j| = k \,\middle|\, H'\right] \sim \binom{N'}{k-1} p^{k-1}(1-p)^{N'-(k-1)}.$$

Proof. The following intuitive argument and the formal proof given in [11] apply similar ideas.

The informal argument. Let $d_v^* = |\Gamma(v_j) \bigcap W|$. Again, since v_j appears in the current view (that is, in $\{\Gamma_1, ..., \Gamma_{j-1}\}$) precisely once, then whenever A occurs, we have: $(|R_j| = k)$ iff v_j has precisely $k-1$ neighbors in W. Hence,

$$\underline{\Pr}\left[A, |R_j| = k \,\middle|\, H'\right] = \underline{\Pr}\left[A, d_v^* = k - 1 \,\middle|\, H'\right]$$

$$= \frac{\Pr\left[A, d_v^* = k - 1, H'\right]}{\Pr\left[H'\right]}$$

$$= \frac{\Pr\left[A \,\middle|\, H', d_v^* = k - 1\right]\Pr\left[H' \,\middle|\, d_v^* = k - 1\right]\Pr\left[d_v^* = k - 1\right]}{\sum_t \Pr\left[H' \,\middle|\, d_v^* = t\right]\Pr\left[d_v^* = t\right]},$$

(the sum taken over all likely degrees t).

We will soon argue that $\underline{\Pr}\left[H' \,\middle|\, d_v^* = t'\right] \sim \underline{\Pr}\left[H' \,\middle|\, d_v^* = t''\right]$ holds for any pair of likely degrees t', t''. Assuming this, we may cancel out (up to negligible terms) the $\underline{\Pr}\left[H' \,\middle|\, d_v^* = k - 1\right]$ from the nominator with the terms $\underline{\Pr}\left[H' \,\middle|\, d_v^* = t\right]$ from the denominator. As the event A is extremely likely, the term $\underline{\Pr}\left[A \,\middle|\, H', d_v^* = k - 1\right]$ ~ 1 and cancels too. Thus,

$$\underline{\Pr}\left[A, |R_j| = k \,\middle|\, H'\right] \sim \frac{\Pr\left[d_v^* = k - 1\right]}{\sum_t \Pr\left[d_v^* = t\right]}.$$

Next, $\sum_t \underline{\Pr}\left[d_v^* = t\right] \sim 1$ since we sum over all likely degrees. Therefore,

$$\underline{\Pr}\left[A, |R_j| = k \,\middle|\, H'\right] \sim \underline{\Pr}\left[d_v^* = k - 1\right]$$

and the final term could be shown to have Binomial distribution, so our claim follows.

To demonstrate that $\underline{\Pr}\left[H' \,\middle|\, d_v^* = t'\right] \sim \underline{\Pr}\left[H' \,\middle|\, d_v^* = t''\right]$ (this is where our argument becomes informal), we first assume that all the degrees except d_{v_j} are fixed.

We consider the equivalent model $G_{\Pi \text{Bin}}^{\text{OTF}}$, where the view H is produced by repeatedly matching each of the available ports σ_i of the current query vertex with a random available port. Let $\sigma_1, ..., \sigma_z$ denote the ports of the first $j-1$ query vertices. Clearly, when σ_i is matched, any available port τ is chosen with probability precisely $1/E_i$ for $E_i = (\sum_v d_v) + 1 - 2i$. Thus, the difference between the cases $d_v = t'$ and $d_v = t''$ is that each choice (of matching some τ with σ_i) occurs w.p. $1/E_i'$ instead of $1/E_i''$. Since E_i', E_i'' are both super-polynomial (as the port sequence is likely), and since $E_i' - E_i'' = t' - t'' \le poly(n)$, this induces an insignificant difference between the resulting distributions. This difference remains insignificant even as diversities accumulate over the $poly(n)$ many ports σ_i that are matched to decide H'. Finally, as this holds for any choice of the degrees (excluding d_{v_j}), one can get $\underline{\Pr}[H' \mid d_v^* = t'] \sim \underline{\Pr}[H' \mid d_v^* = t'']$.

The formal proof of Claim 1.1 is given in [11]. Given Claim 1.1, then Claim 1 follows, so Lemma 1 and hence our main Theorems 3.1 and 3 follow as well. □

Acknowledgements. We thank Gil Segev, Noam Livne and the anonymous referees for carefully reading and commenting on a draft of this paper.

References

1. Alon, N., Spencer, J.: The Probabilistic Method. John Wiley, New York (1992)
2. Bollobás, B.: A probabilistic proof of an asymptotic formula for the number of labelled regular graphs. Preprint Series, Matematisk Institut, Aarhus Universitet (1979)
3. Erdös, P., Rényi, A.: On random graphs I. Publicationes Mathematicae 6, 290–297 (1959)
4. Erdös, P., Rényi, A.: On the evolution of random graphs. Publications of the Mathematical Institute of the Hungarian Academy of Sciences 5, 17–61 (1960)
5. Gilbert, A., Guha, S., Indyk, P., Kotidis, Y., Muthukrishnan, S., Strauss, M.: Fast, Small-Space Algorithms for Approximate Histogram Maintenance. In: 34'th annual ACM symposium on Theory of computing, pp. 389–398 (2002)
6. Goldreich, O., Goldwasser, S., Micali, S.: How to Construct Random Functions. Journal of the ACM 33(4), 276–288 (1985)
7. Goldreich, O., Goldwasser, S., Nussboim, A.: On the Implementation of Huge Random Objects. In: proc. 44'th FOCS IEEE Symp. on Foundations of Computer Science, pp. 68–79 (2003)
8. Goldreich, O., Ron, D.: Property Testing in Bounded Degree Graphs. In: Proceedings of the 29'th ACM Symposium on Theory of Computing, pp. 406–415 (1997)
9. Kaplan, E., Naor, M., Reingold, O.: Derandomized Constructions of k-Wise (Almost) Independent Permutations. In: APPROX-RANDOM 2005, 354–365 (2005)
10. Luby, M., Rackoff, C.: How to Construct Pseudo-Random Permutations from Pseudo-Random Functions. SIAM J. on Computing 17, 373–386 (1998)
11. Naor, M., Nussboim, A.: Implementing Huge Sparse Random Graphs, Available at http://www.wisdom.weizmann.ac.il/~asafn/PAPERS/sparseGnp.ps
12. Naor, M., Nussboim, A., Tromer, E.: Efficiently Constructible Huge Graphs that Preserve First Order Properties of Random Graphs. In: Proceedings of the 2'nd Theory of Cryptography Conference, pp. 66–85 (2005)
13. Naor, M., Reingold, O.: Constructing Pseudo-Random Permutations with a Prescribed Cyclic Structure. Journal of Crypto. 15(2), 97–102 (2002)

Sublinear Algorithms for Approximating String Compressibility*

Sofya Raskhodnikova[1,**], Dana Ron[2,***], Ronitt Rubinfeld[3], and Adam Smith[1,†]

[1] Pennsylvania State University, USA
{sofya,asmith}@cse.psu.edu
[2] Tel Aviv University, Israel
danar@eng.tau.ac.il
[3] MIT, Cambridge MA, USA
ronitt@csail.mit.edu

Abstract. We raise the question of approximating the compressibility of a string with respect to a fixed compression scheme, in sublinear time. We study this question in detail for two popular lossless compression schemes: run-length encoding (RLE) and Lempel-Ziv (LZ), and present sublinear algorithms for approximating compressibility with respect to both schemes. We also give several lower bounds that show that our algorithms for both schemes cannot be improved significantly.

Our investigation of LZ yields results whose interest goes beyond the initial questions we set out to study. In particular, we prove combinatorial structural lemmas that relate the compressibility of a string with respect to Lempel-Ziv to the number of distinct short substrings contained in it. In addition, we show that approximating the compressibility with respect to LZ is related to approximating the support size of a distribution.

1 Introduction

Given an extremely long string, it is natural to wonder how compressible it is. This fundamental question is of interest to a wide range of areas of study, including computational complexity theory, machine learning, storage systems, and communications. As massive data sets are now commonplace, the ability to estimate their compressibility with extremely efficient, even sublinear time, algorithms, is gaining in importance. The most general measure of compressibility, Kolmogorov complexity, is not computable (see [14] for a textbook treatment), nor even approximable. Even under restrictions which make it computable (such

---

* A full version of this paper is available [17]. These results appeared previously as part of a technical report [16].
** Research done while at the Hebrew University of Jerusalem, Israel, supported by the Lady Davis Fellowship, and while at the Weizmann Institute of Science, Israel.
*** Supported by the Israel Science Foundation (grant number 89/05).
† Research done while at the Weizmann Institute of Science, Israel, supported by the Louis L. and Anita M. Perlman Postdoctoral Fellowship.

M. Charikar et al. (Eds.): APPROX and RANDOM 2007, LNCS 4627, pp. 609–623, 2007.
© Springer-Verlag Berlin Heidelberg 2007

as a bound on the running time of decompression), it is probably hard to approximate in polynomial time, since an approximation would allow distinguishing random from pseudorandom strings and, hence, inverting one-way functions. However, the question of how compressible a large string is with respect to a *specific compression scheme* may be tractable, depending on the particular scheme.

We raise the question of approximating the compressibility of a string with respect to a fixed compression scheme, in sublinear time, and give algorithms and nearly matching lower bounds for several versions of the problem. While this question is new, for one compression scheme, answers follow from previous work. Namely, compressibility under Huffman encoding is determined by the entropy of the symbol frequencies. Batu *et al.* [3] and Brautbar and Samorodnitsky [5] study the problem of approximating the entropy of a distribution from a small number of samples, and their results immediately imply algorithms and lower bounds for approximating compressibility under Huffman encoding.

In this work we study the compressibility approximation question in detail for two popular lossless compression schemes: run-length encoding (RLE) and Lempel-Ziv (LZ) [19]. In the RLE scheme, each run, or a sequence of consecutive occurrences of the same character, is stored as a pair: the character, and the length of the run. Run-length encoding is used to compress black and white images, faxes, and other simple graphic images, such as icons and line drawings, which usually contain many long runs. In the LZ scheme[1], a left-to-right pass of the input string is performed and at each step, the longest sequence of characters that has started in the previous portion of the string is replaced with the pointer to the previous location and the length of the sequence (for a formal definition, see Section 4). The LZ scheme and its variants have been studied extensively in machine learning and information theory, in part because they compress strings generated by an ergodic source to the shortest possible representation (given by the entropy) in the asymptotic limit (cf. [10]). Many popular archivers, such as gzip, use variations on the LZ scheme. In this work we present sublinear algorithms and corresponding lower bounds for approximating compressibility with respect to both schemes, RLE and LZ.

Motivation. Computing the compressibility of a large string with respect to specific compression schemes may be done in order to decide whether or not to compress the file, to choose which compression method is the most suitable, or check whether a small modification to the file (e.g., a rotation of an image) will make it significantly more compressible[2]. Moreover, compression schemes are used as tools for measuring properties of strings such as similarity and entropy. As such, they are applied widely in data-mining, natural language processing and genomics (see, for example, Lowenstern *et al.* [15], Kukushkina *et al.* [11],

---

[1] We study the variant known as LZ77 [19], which achieves the best compressibility. There are several other variants that do not compress some inputs as well, but can be implemented more efficiently.

[2] For example, a variant of the RLE scheme, typically used to compress images, runs RLE on the concatenated rows of the image and on the concatenated columns of the image, and stores the shorter of the two compressed files.

Benedetto *et al.* [4], Li *et al.* [13] and Calibrasi and Vitányi [8,9]). In these applications, one typically needs only the *length* of the compressed version of a file, not the output itself. For example, in the clustering algorithm of [8], the distance between two objects x and y is given by a normalized version of the length of their compressed concatenation $x\|y$. The algorithm first computes all pairwise distances, and then analyzes the resulting distance matrix. This requires $\Theta(t^2)$ runs of a compression scheme, such as gzip, to cluster t objects. Even a weak approximation algorithm that can quickly rule out very incompressible strings would reduce the running time of the clustering computations dramatically.

Multiplicative and Additive Approximations. We consider three approximation notions: additive, multiplicative, and the combination of additive and multiplicative. On the input of length n, the quantities we approximate range from 1 to n. An *additive approximation* algorithm is allowed an additive error of ϵn, where $\epsilon \in (0,1)$ is a parameter. The output of a *multiplicative approximation* algorithm is within a factor $A > 1$ of the correct answer. The combined notion allows both types of error: the algorithm should output an estimate $\widehat{C}$ of the compression cost C such that $\frac{C}{A} - \epsilon n \leq \widehat{C} \leq A \cdot C + \epsilon n$. Our algorithms are randomized, and for all inputs the approximation guarantees hold with probability at least $\frac{2}{3}$.

We are interested in sublinear approximation algorithms, which read few positions of the input strings. For the schemes we study, purely multiplicative approximation algorithms must read almost the entire input. Nevertheless, algorithms with additive error guarantees, or a possibility of both multiplicative and additive error are often sufficient for distinguishing very compressible inputs from inputs that are not well compressible. For both the RLE and LZ schemes, we give algorithms with combined multiplicative and additive error that make few queries to the input. When it comes to additive approximations, however, the two schemes differ sharply: sublinear additive approximations are possible for the RLE compressibility, but not for LZ compressibility.

1.1   Results for Run-Length Encoding

For RLE, we present sublinear algorithms for all three approximation notions defined above, providing a trade-off between the quality of approximation and the running time. The algorithms that allow an additive approximation run in time independent of the input size. Specifically, an ϵn-additive estimate can be obtained in time[3] $\tilde{O}(1/\epsilon^3)$, and a combined estimate, with a multiplicative error of 3 and an additive error of ϵn, can be obtained in time $\tilde{O}(1/\epsilon)$. As for a strict multiplicative approximation, we give a simple 4-multiplicative approximation algorithm that runs in expected time $\tilde{O}(\frac{n}{C_{\text{rle}}(w)})$ where $C_{\text{rle}}(w)$ denotes the compression cost of the string w. For any $\gamma > 0$, the multiplicative error can be improved to $1 + \gamma$ at the cost of multiplying the running time by $\text{poly}(1/\gamma)$. Observe that the algorithm is more efficient when the string is less compressible,

---

[3] The notation $\tilde{O}(g(k))$ for a function g of a parameter k means $O(g(k) \cdot \text{polylog}(g(k)))$ where $\text{polylog}(g(k)) = \log^c(g(k))$ for some constant c.

and less efficient when the string is more compressible. One of our lower bounds justifies such a behavior and, in particular, shows that a constant factor approximation requires linear time for strings that are very compressible. We also give a lower bound of $\Omega(1/\epsilon^2)$ for ϵn-additive approximation.

1.2 Results for Lempel-Ziv

We prove that approximating compressibility with respect to LZ is closely related to the following problem, which we call COLORS: *Given access to a string τ of length n over alphabet Ψ, approximate the number of* distinct *symbols ("colors") in τ*. This is essentially equivalent to estimating the support size of a distribution [18]. Variants of this problem have been considered under various guises in the literature: in databases it is referred to as approximating distinct values (Charikar *et al.* [7]), in statistics as estimating the number of species in a population (see the over 800 references maintained by Bunge [6]), and in streaming as approximating the frequency moment F_0 (Alon *et al.* [1], Bar-Yossef *et al.* [2]). Most of these works, however, consider models different from ours. For our model, there is an A-multiplicative approximation algorithm of [7], that runs in time $O\left(\frac{n}{A^2}\right)$, matching the lower bound in [7,2]. There is also an almost linear lower bound for approximating COLORS with additive error [18].

We give a reduction from LZ compressibility to COLORS and vice versa. These reductions allow us to employ the known results on COLORS to give algorithms and lower bounds for this problem. Our approximation algorithm for LZ compressibility combines a multiplicative and additive error. The running time of the algorithm is $\tilde{O}\left(\frac{n}{A^3\epsilon}\right)$ where A is the multiplicative error and ϵn is the additive error. In particular, this implies that for any $\alpha > 0$, we can distinguish, in sublinear time $\tilde{O}(n^{1-\alpha})$, strings compressible to $O(n^{1-\alpha})$ symbols from strings only compressible to $\Omega(n)$ symbols.[4]

The main tool in the algorithm consists of two combinatorial structural lemmas that relate compressibility of the string to the number of distinct short substrings contained in it. Roughly, they say that a string is well compressible with respect to LZ if and only if it contains few distinct substrings of length ℓ for all small ℓ (when considering all $n - \ell + 1$ possible overlapping substrings). The simpler of the two lemmas was inspired by a structural lemma for grammars by Lehman and Shelat [12]. The combinatorial lemmas allow us to establish a reduction from LZ compressibility to COLORS and employ a (simple) algorithm for approximating COLORS in our algorithm for LZ.

Interestingly, we can show that there is also a reduction in the *opposite direction*: namely, approximating COLORS reduces to approximating LZ compressibility. The lower bound of [18], combined with the reduction from COLORS to LZ, implies that our algorithm for LZ cannot be improved significantly. In particular, our lower bound implies that for any $B = n^{o(1)}$, distinguishing strings compressible by LZ to $\tilde{O}(n/B)$ symbols from strings compressible to $\tilde{\Omega}(n)$ symbols requires $n^{1-o(1)}$ queries.

---

[4] To see this, set $A = o(n^{\alpha/2})$ and $\epsilon = o(n^{-\alpha/2})$.

1.3 Further Research

It would be interesting to extend our results for estimating the compressibility under LZ77 to other variants of LZ, such as dictionary-based LZ78 [20]. Compressibility under LZ78 can be drastically different from compressibility under LZ77: e.g., for 0^n they differ roughly by a factor of $\sqrt{n}$. Another open question is approximating compressibility for schemes other than RLE and LZ. In particular, it would be interesting to design approximation algorithms for lossy compression schemes such as JPEG, MPEG and MP3. One lossy compression scheme to which our results extend directly is Lossy RLE, where some characters, e.g., the ones that represent similar colors, are treated as the same character.

1.4 Organization

We start with some definitions in Section 2. Section 3 contains our results for RLE. Section 4 deals with the LZ scheme. All missing details (descriptions of algorithms and proofs of claims) can be found in [17].

2 Preliminaries

The input to our algorithms is usually a string w of length n over a finite alphabet Σ. The quantities we approximate, such as compression cost of w under a specific algorithm, range from 1 to n. We consider estimates to these quantities that have both multiplicative and additive error. We call $\widehat{C}$ an (λ, ϵ)-estimate for C if $\frac{C}{\lambda} - \epsilon n \leq \widehat{C} \leq \lambda \cdot C + \epsilon n$, and say an algorithm (λ, ϵ)-estimates C (or is an (λ, ϵ)-approximation algorithm for C) if for each input it produces an (λ, ϵ)-estimate for C with probability at least $\frac{2}{3}$.

When the error is purely additive or multiplicative, we use the following shorthand: ϵn-additive estimate stands for $(1, \epsilon)$-estimate and λ-multiplicative estimate, or λ-estimate, stands for $(\lambda, 0)$-estimate. An algorithm computing an ϵn-additive estimate with probability at least $\frac{2}{3}$ is an ϵn-additive approximation algorithm, and if it computes an λ-multiplicative estimate then it is an λ-multiplicative approximation algorithm, or λ-approximation algorithm.

For some settings of parameters, obtaining a valid estimate is trivial. For a quantity in $[1, n]$, for example, $\frac{n}{2}$ is an $\frac{n}{2}$-additive estimate, $\sqrt{n}$ is a $\sqrt{n}$-estimate and ϵn is an (λ, ϵ)-estimate whenever $\lambda \geq \frac{1}{2\epsilon}$.

3 Run-Length Encoding

Every n-character string w over alphabet Σ can be partitioned into maximal runs of identical characters of the form σ^ℓ, where σ is a symbol in Σ and ℓ is the length of the run, and consecutive runs are composed of different symbols. In the *Run-Length Encoding* of w, each such run is replaced by the pair (σ, ℓ). The number of bits needed to represent such a pair is $\lceil \log(\ell + 1) \rceil + \lceil \log |\Sigma| \rceil$ plus the overhead which depends on how the separation between the characters and the lengths is implemented. One way to implement it is to use prefix-free encoding for lengths. For simplicity we ignore the overhead in the above expression, but our analysis can be adapted to any implementation choice. The *cost of the run-length encoding*, denoted by $C_{\mathrm{rle}}(w)$, is the sum over all runs of $\lceil \log(\ell + 1) \rceil + \lceil \log |\Sigma| \rceil$.

3.1   An ϵn-Additive Estimate with $\tilde{O}(1/\epsilon^3)$ Queries

Our first algorithm for approximating the cost of RLE is very simple: it samples a few positions in the input string uniformly at random and bounds the lengths of the runs to which they belong by looking at the positions to the left and to the right of each sample. If the corresponding run is short, its length is established exactly; if it is long, we argue that it does not contribute much to the encoding cost. For each index $t \in [n]$, let $\ell(t)$ be the length of the run to which w_t belongs. The cost contribution of index t is defined as

$$c(t) = \frac{\lceil \log(\ell(t) + 1) \rceil + \lceil \log |\Sigma| \rceil}{\ell(t)}. \tag{1}$$

By definition, $\dfrac{C_{\mathrm{rle}}(w)}{n} = \underset{t \in [n]}{\mathsf{E}}[c(t)]$, where $\mathsf{E}_{t \in [n]}$ denotes expectation over a uniformly random choice of t. The algorithm, presented below, estimates the encoding cost by the average of the cost contributions of the sampled short runs, multiplied by n.

ALGORITHM I: AN ϵn-ADDITIVE APPROXIMATION FOR $C_{\mathrm{rle}}(w)$

1. Select $q = \Theta\left(\frac{1}{\epsilon^2}\right)$ indices $t_1, \ldots, t_q$ uniformly and independently at random.
2. For each $i \in [q]$:
   (a) Query t_i and up to $\ell_0 = \frac{8 \log(4|\Sigma|/\epsilon)}{\epsilon}$ positions in its vicinity to bound $\ell(t_i)$.
   (b) Set $\hat{c}(t_i) = c(t_i)$ if $\ell(t_i) < \ell_0$ and $\hat{c}(t_i) = 0$ otherwise.
3. Output $\widehat{C}_{\mathrm{rle}} = n \cdot \underset{i \in [q]}{\mathsf{E}}[\hat{c}(t_i)]$.

Correctness. We first prove that the algorithm is an ϵn-additive approximation. The error of the algorithm comes from two sources: from ignoring the contribution of long runs and from sampling. The ignored indices t, for which $\ell(t) \geq \ell_0$, do not contribute much to the cost. Since the cost assigned to the indices monotonically decreases with the length of the run to which they belong, for each such index,

$$c(t) \leq \frac{\lceil \log(\ell_0 + 1) \rceil + \lceil \log |\Sigma| \rceil}{\ell_0} \leq \frac{\epsilon}{2}. \tag{2}$$

Therefore,

$$\frac{C_{\mathrm{rle}}(w)}{n} - \frac{\epsilon}{2} \leq \frac{1}{n} \cdot \sum_{t:\, \ell(t) < \ell_0} c(t) \leq \frac{C_{\mathrm{rle}}(w)}{n}. \tag{3}$$

Equivalently, $\frac{C_{\mathrm{rle}}(w)}{n} - \frac{\epsilon}{2} \leq \mathsf{E}_{i \in [n]}[\hat{c}(t_i)] \leq \frac{C_{\mathrm{rle}}(w)}{n}$.

By an additive Chernoff bound, with high constant probability, the sampling error in estimating $\mathsf{E}[\hat{c}(t_i)]$ is at most $\epsilon/2$. Therefore, $\widehat{C}_{\mathrm{rle}}$ is an ϵn-additive estimate of $C_{\mathrm{rle}}(w)$, as desired.

Query and time complexity. (Assuming $|\Sigma|$ is constant.) Since the number of queries performed for each selected t_i is $O(\ell_0) = O(\log(1/\epsilon)/\epsilon)$, the total number of queries, as well as the running time, is $O(\log(1/\epsilon)/\epsilon^3)$.

3.2 Summary of Positive Results on RLE

After stating Theorem 1 that summarizes our positive results, we briefly discuss some of the ideas used in the algorithms omitted from this version of the paper.

Theorem 1. *Let $w \in \Sigma^n$ be a string to which we are given query access.*

1. *Algorithm I gives ϵn-additive approximation to $C_{\mathrm{rle}}(w)$ in time $\tilde{O}(1/\epsilon^3)$.*
2. *$C_{\mathrm{rle}}(w)$ can be $(3, \epsilon)$-estimated in time $\tilde{O}(1/\epsilon)$.*
3. *$C_{\mathrm{rle}}(w)$ can be 4-estimated in expected time $\tilde{O}\left(\frac{n}{C_{\mathrm{rle}}(w)}\right)$. A $(1 + \gamma)$-estimate of $C_{\mathrm{rle}}(w)$ can be obtained in expected time $\tilde{O}\left(\frac{n}{C_{\mathrm{rle}}(w)} \cdot \mathrm{poly}(1/\gamma)\right)$. The algorithm needs no prior knowledge of $C_{\mathrm{rle}}(w)$.*

Section 3.1 gives a complete proof of Item 1. The algorithm in Item 2 partitions the positions in the string into *buckets* according to the length of the runs they belong to. It estimates the sizes of different buckets with different precision, depending on the size of the bucket and the length of the runs it contains. The main idea in Item 3 is to search for $C_{\mathrm{rle}}(w)$, using the algorithm from Item 2 repeatedly (with different parameters) to establish successively better estimates.

3.3 Lower Bounds for RLE

We give two lower bounds, for multiplicative and additive approximation, respectively, which establish that the running times in Items 1 and 3 of Theorem 1 are essentially tight.

Theorem 2

1. *For all $A > 1$, any A-approximation algorithm for C_{rle} requires $\Omega\left(\frac{n}{A^2 \log n}\right)$ queries. Furthermore, if the input is restricted to strings with compression cost $C_{\mathrm{rle}}(w) \geq C$, then $\Omega\left(\frac{n}{CA^2 \log(n)}\right)$ queries are necessary.*
2. *For all $\epsilon \in (0, \frac{1}{2})$, any ϵn-additive approximation algorithm for C_{rle} requires $\Omega(1/\epsilon^2)$ queries.*

A Multiplicative Lower Bound (Proof of Theorem 2, Item 1): The claim follows from the next lemma:

Lemma 3. *For every $n \geq 2$ and every integer $1 \leq k \leq n/2$, there exists a family of strings, denoted W_k, for which the following holds: (1) $C_{\mathrm{rle}}(w) = \Theta\left(k \log(\frac{n}{k})\right)$ for every $w \in W_k$; (2) Distinguishing a uniformly random string in W_k from one in $W_{k'}$, where $k' > k$, requires $\Omega\left(\frac{n}{k'}\right)$ queries.*

Proof. Let $\Sigma = \{0, 1\}$ and assume for simplicity that n is divisible by k. Every string in W_k consists of k blocks, each of length $\frac{n}{k}$. Every odd block contains only 1s and every even block contains a single 0. The strings in W_k differ in the locations of the 0s within the even blocks. Every $w \in W_k$ contains $k/2$ isolated 0s and $k/2$ runs of 1s, each of length $\Theta(\frac{n}{k})$. Therefore, $C_{\mathrm{rle}}(w) = \Theta\left(k \log(\frac{n}{k})\right)$. To distinguish a random string in W_k from one in $W_{k'}$ with probability 2/3, one must make $\Omega(\frac{n}{\max(k,k')})$ queries since, in both cases, with asymptotically fewer queries the algorithm sees only 1's with high probability. ∎

Additive Lower Bound (Proof Theorem 2, Item 1): For any $p \in [0, 1]$ and sufficiently large n, let $\mathcal{D}_{n,p}$ be the following distribution over n-bit strings. For simplicity, consider n divisible by 3. The string is determined by $\frac{n}{3}$ independent coin flips, each with bias p. Each "heads" extends the string by three runs of length 1, and each "tails", by a run of length 3. Given the sequence of run lengths, dictated by the coin flips, output the unique binary string that starts with 0 and has this sequence of run lengths.[5]

Let W be a random variable drawn according to $\mathcal{D}_{n,1/2}$ and W', according to $\mathcal{D}_{n,1/2+\epsilon}$. The following facts are established in the full version [17]: (a) $\Omega(1/\epsilon^2)$ queries are necessary to reliably distinguish W from W', and (b) With high probability, the encoding costs of W and W' differ by $\Omega(\epsilon n)$. Together these facts imply the lower bound.   ∎

4   Lempel Ziv Compression

In this section we consider a variant of Lempel and Ziv's compression algorithm [19], which we refer to as LZ77. In all that follows we use the shorthand $[n]$ for $\{1, \ldots, n\}$. Let $w \in \Sigma^n$ be a string over an alphabet Σ. Each symbol of the compressed representation of w, denoted $LZ(w)$, is either a character $\sigma \in \Sigma$ or a pair (p, ℓ) where $p \in [n]$ is a pointer (index) to a location in the string w and ℓ is the length of the substring of w that this symbol represents. To compress w, the algorithm works as follows. Starting from $t = 1$, at each step the algorithm finds the longest substring $w_t \ldots w_{t+\ell-1}$ for which there exists an index $p < t$, such that $w_p \ldots w_{p+\ell-1} = w_t \ldots w_{t+\ell-1}$. (The substrings $w_p \ldots w_{p+\ell-1}$ and $w_t \ldots w_{t+\ell-1}$ may overlap.) If there is no such substring (that is, the character w_t has not appeared before) then the next symbol in $LZ(w)$ is w_t, and $t = t + 1$. Otherwise, the next symbol is (p, ℓ) and $t = t + \ell$. We refer to the substring $w_t \ldots w_{t+\ell-1}$ (or w_t when w_t is a new character) as a *compressed segment*.

Let $C_{\mathrm{LZ}}(w)$ denote the number of symbols in the compressed string $LZ(w)$. (We do not distinguish between symbols that are characters in Σ, and symbols that are pairs (p, ℓ).) Given query access to a string $w \in \Sigma^n$, we are interested in computing an estimate $\widehat{C}_{\mathrm{LZ}}$ of $C_{\mathrm{LZ}}(w)$. As we shall see, this task reduces to estimating the number of distinct substrings in w of different lengths, which in turn reduces to estimating the number of distinct characters ("colors") in a string. The actual length of the binary representation of the compressed substring is at most a factor of $2 \log n$ larger than $C_{\mathrm{LZ}}(w)$. This is relatively negligible given the quality of the estimates that we can achieve in sublinear time.

We begin by relating LZ compressibility to COLORS (§4.1), then use this relation to discuss algorithms (§4.2) and lower bounds (§4.3) for compressiblity.

4.1   Structural Lemmas

Our algorithm for approximating the compressibility of an input string with respect to LZ77 uses an approximation algorithm for COLORS (defined in the

---

[5] Let b_i be a boolean variable representing the outcome of the ith coin. Then the output is $0b_1 0 1 \overline{b_2} 1 0 b_3 0 1 \overline{b_4} 1 \ldots$.

introduction) as a subroutine. The main tool in the reduction from LZ77 to
COLORS is the relation between $C_{LZ}(w)$ and the number of distinct substrings
in w, formalized in the two structural lemmas. In what follows, $d_\ell(w)$ denotes
the *number of distinct substrings* of length ℓ in w. Unlike compressed segments
in w, which are disjoint, these substrings may overlap.

Lemma 4 (Structural Lemma 1). *For every $\ell \in [n]$, $C_{LZ}(w) \geq \frac{d_\ell(w)}{\ell}$.*

Lemma 5 (Structural lemma 2). *Let $\ell_0 \in [n]$. Suppose that for some integer
m and for every $\ell \in [\ell_0]$, $d_\ell(w) \leq m \cdot \ell$. Then $C_{LZ}(w) \leq 4(m \log \ell_0 + n/\ell_0)$.*

Proof of Lemma 4. This proof is similar to the proof of a related lemma
concerning grammars from [12]. First note that the lemma holds for $\ell = 1$, since
each character w_t in w that has not appeared previously (that is, $w_{t'} \neq w_t$ for
every $t' < t$) is copied by the compression algorithm to $LZ(w)$.

For the general case, fix $\ell > 1$. Recall that $w_t \ldots w_{t+k-1}$ of w is a *compressed
segment* if it is represented by one symbol (p, k) in $LZ(w)$. Any substring of lenth
ℓ that occurs *within* a compressed segment must have occurred previously in the
string. Such substrings can be ignored for our purposes: the number of *distinct*
length-ℓ substrings is bounded above by the number of length-ℓ substrings that
start inside one compressed segment and end in another. Each segment (except
the last) contributes $(\ell-1)$ such substrings. Therefore, $d_\ell(w) \leq (C_{LZ}(w)-1)(\ell-1) < C_{LZ}(w) \cdot \ell$ for every $\ell > 1$. ∎

Proof of Lemma 5. Let $n_\ell(w)$ denote the number of compressed segments of
length ℓ in w, not including the last compressed segment. We use the shorthand
n_ℓ for $n_\ell(w)$ and d_ℓ for $d_\ell(w)$. In order to prove the lemma we shall show that
for every $1 \leq \ell \leq \lfloor \ell_0/2 \rfloor$,

$$\sum_{k=1}^{\ell} n_k \leq 2(m+1) \cdot \sum_{k=1}^{\ell} \frac{1}{k}. \tag{4}$$

For all $\ell \geq 1$, since the compressed segments in w are disjoint, $\sum_{k=\ell+1}^{n} n_k \leq \frac{n}{\ell+1}$.
If we substitute $\ell = \lfloor \ell_0/2 \rfloor$ in the last two equations and sum them up, we get:

$$\sum_{k=1}^{n} n_k \leq 2(m+1) \cdot \sum_{k=1}^{\lfloor \ell_0/2 \rfloor} \frac{1}{k} + \frac{2n}{\ell_0} \leq 2(m+1)(\ln \ell_0 + 1) + \frac{2n}{\ell_0}. \tag{5}$$

Since $C_{LZ}(w) = \sum_{k=1}^{n} n_k + 1$, the lemma follows.

It remains to prove Equation (4). We do so below by induction on ℓ, using
the following claim.

Claim 6. *For every $1 \leq \ell \leq \lfloor \ell_0/2 \rfloor$, $\displaystyle\sum_{k=1}^{\ell} k \cdot n_k \leq 2\ell(m+1)$.*

Proof. We show that each position $j \in \{\ell, \ldots, n-\ell\}$ that participates in a
compressed substring of length at most ℓ in w can be mapped to a distinct

length-2ℓ substring of w. Since $\ell \leq \ell_0/2$, by the premise of the lemma, there are at most $2\ell \cdot m$ distinct length-2ℓ substrings. In addition, the first $\ell - 1$ and the last ℓ positions contribute less than 2ℓ symbols. The claim follows.

We call a substring *new* if no instance of it started in the previous portion of w. Namely, $w_t \ldots w_{t+\ell-1}$ is *new* if there is no $p < t$ such that $w_t \ldots w_{t+\ell-1} = w_p \ldots w_{p+\ell-1}$. Consider a compressed substring $w_t \ldots w_{t+k-1}$ of length $k \leq \ell$. The substrings of length greater than k that start at w_t must be *new*, since LZ77 finds the longest substring that appeared before. Furthermore, every substring that contains such a *new* substring is also *new*. That is, every substring $w_{t'} \ldots w_{t+k'}$ where $t' \leq t$ and $k' \geq k + (t' - t)$, is *new*.

Map each position $j \in \{\ell, \ldots, n-\ell\}$ in the compressed substring $w_t \ldots w_{t+k-1}$ to the length-2ℓ substring that ends at $w_{j+\ell}$. Then each position in $\{\ell, \ldots, n-\ell\}$ that appears in a compressed substring of length at most ℓ is mapped to a distinct length-2ℓ substring, as desired.   ∎ (Claim 6)

Establishing Equation (4). We prove Equation (4) by induction on ℓ. Claim 6 with ℓ set to 1 gives the base case, i.e., $n_1 \leq 2(m + 1)$. For the induction step, assume the induction hypothesis for every $j \in [\ell - 1]$. To prove it for ℓ, add the equation in Claim 6 to the sum of the induction hypothesis inequalities (Equation (4)) for every $j \in [\ell - 1]$. The left hand side of the resulting inequality is

$$\sum_{k=1}^{\ell} k \cdot n_k + \sum_{j=1}^{\ell-1} \sum_{k=1}^{j} n_k = \sum_{k=1}^{\ell} k \cdot n_k + \sum_{k=1}^{\ell-1} \sum_{j=1}^{\ell-k} n_k$$

$$= \sum_{k=1}^{\ell} k \cdot n_k + \sum_{k=1}^{\ell-1} (\ell - k) \cdot n_k = \ell \cdot \sum_{k=1}^{\ell} n_k.$$

The right hand side, divided by the factor $2(m + 1)$, which is common to all inequalities, is

$$\ell + \sum_{j=1}^{\ell-1} \sum_{k=1}^{j} \frac{1}{k} = \ell + \sum_{k=1}^{\ell-1} \sum_{j=1}^{\ell-k} \frac{1}{k} = \ell + \sum_{k=1}^{\ell-1} \frac{\ell-k}{k} = \ell + \ell \cdot \sum_{k=1}^{\ell-1} \frac{1}{k} - (\ell - 1) = \ell \cdot \sum_{k=1}^{\ell} \frac{1}{k}.$$

Dividing both sides by ℓ gives the inequality in Equation (4).   ∎ (Lemma 5)

4.2   An Algorithm for LZ77

This subsection describes an algorithm for approximating the compressibility of an input string with respect to LZ77, which uses an approximation algorithm for COLORS as a subroutine. The main tool in the reduction from LZ77 to COLORS consists of structural lemmas 4 and 5, summarized in the following corollary.

Corollary 7. *For any $\ell_0 \geq 1$, let $m = m(\ell_0) = \max_{\ell=1}^{\ell_0} \frac{d_\ell(w)}{\ell}$. Then*

$$m \leq C_{LZ}(w) \leq 4 \cdot \left(m \log \ell_0 + \frac{n}{\ell_0} \right).$$

The corollary allows us to approximate C_{LZ} from estimates for d_ℓ for all $\ell \in [\ell_0]$. To obtain these estimates, we use the algorithm of [7] for COLORS as a subroutine (in the full version [17] we also describe a simpler COLORS algorithm with the same provable guarantees). Recall that an algorithm for COLORS approximates the number of distinct colors in an input string, where the ith character represents the ith color. We denote the number of colors in an input string τ by $C_{\mathrm{COL}}(\tau)$. To approximate d_ℓ, the number of distinct length-ℓ substrings in w, using an algorithm for COLORS, view each length-ℓ substring as a separate color. Each query of the algorithm for COLORS can be implemented by ℓ queries to w.

Let $\mathrm{ESTIMATE}(\ell, B, \delta)$ be a procedure that, given access to w, an index $\ell \in [n]$, an approximation parameter $B = B(n, \ell) > 1$ and a confidence parameter $\delta \in [0,1]$, computes a B-estimate for d_ℓ with probability at least $1-\delta$. It can be implemented using an algorithm for COLORS, as described above, and employing standard amplification techniques to boost success probability from $\frac{2}{3}$ to $1-\delta$: running the basic algorithm $\Theta(\log \delta^{-1})$ times and outputting the median. Since the algorithm of [7] requires $O(n/B^2)$ queries, the query complexity of $\mathrm{ESTIMATE}(\ell, B, \delta)$ is $O\left(\frac{n}{B^2} \ell \log \delta^{-1}\right)$. Using $\mathrm{ESTIMATE}(\ell, B, \delta)$ as a subroutine, we get the following approximation algorithm for the cost of LZ77.

ALGORITHM II: AN (A, ϵ)-APPROXIMATION FOR $C_{\mathrm{LZ}}(w)$

1. Set $\ell_0 = \lceil \frac{2}{A\epsilon} \rceil$ and $B = \frac{A}{2\sqrt{\log(2/(A\epsilon))}}$.
2. For all ℓ in $[\ell_0]$, let $\hat{d}_\ell = \mathrm{ESTIMATE}(\ell, B, \frac{1}{3\ell_0})$.
3. Combine the estimates to get an approximation of m from Corollary 7:
   set $\hat{m} = \max_\ell \dfrac{\hat{d}_\ell}{\ell}$.
4. Output $\widehat{C}_{\mathrm{LZ}} = \hat{m} \cdot \frac{A}{B} + \epsilon n$.

Theorem 8. *Algorithm II (A, ϵ)-estimates $C_{LZ}(w)$. With a proper implementation that reuses queries and an appropriate data structure, its query and time complexity are $\tilde{O}\left(\frac{n}{A^3\epsilon}\right)$.*

Proof. By the Union Bound, with probability $\geq \frac{2}{3}$, all values $\hat{d}_\ell$ computed by the algorithm are B-estimates for the corresponding d_ℓ. When this holds, $\hat{m}$ is a B-estimate for m from Corollary 7, which implies that

$$\frac{\hat{m}}{B} \leq C_{\mathrm{LZ}}(w) \leq 4 \cdot \left(\hat{m} B \log \ell_0 + \frac{n}{\ell_0}\right).$$

Equivalently, $\dfrac{C_{\mathrm{LZ}} - 4(n/\ell_0)}{4B \log \ell_0} \leq \hat{m} \leq B \cdot C_{\mathrm{LZ}}$. Multiplying all three terms by $\frac{A}{B}$ and adding ϵn to them, and then substituting parameter settings for ℓ_0 and B, specified in the algorithm, shows that $\widehat{C}_{\mathrm{LZ}}$ is indeed an (A, ϵ)-estimate for C_{LZ}.

As explained before the algorithm statement, each call to $\mathrm{ESTIMATE}(\ell, B, \frac{1}{3\ell_0})$ costs $O\left(\frac{n}{B^2} \ell \log \ell_0\right)$ queries. Since the subroutine is called for all $\ell \in [\ell_0]$, the straightforward implementation of the algorithm would result in $O\left(\frac{n}{B^2}\ell_0^2\log\ell_0\right)$ queries. Our analysis of the algorithm, however, does not rely on independence of queries used in different calls to the subroutine, since we employ the Union Bound to calculate the error probability. It will still apply if we first run ESTIMATE to approximate d_{ℓ_0} and then reuse its queries for the remaining calls to the subroutine, as though it requested to query only the length-ℓ prefixes of the length-ℓ_0 substrings queried in the first call. With this implementation, the query complexity is $O\left(\frac{n}{B^2}\ell_0\log\ell_0\right) = O\left(\frac{n}{A^3\epsilon}\log^2\frac{1}{A\epsilon}\right)$. To get the same running time, one can maintain counters for all $\ell \in [\ell_0]$ for the number of distinct length-ℓ substrings seen so far and use a trie to keep the information about the queried substrings. Every time a new node at some depth ℓ is added to the trie, the ℓth counter is incremented. ∎

4.3 Lower Bounds: Reducing COLORS to LZ77

We have demonstrated that estimating the LZ77 compressibility of a string reduces to COLORS. As shown in [18], COLORS is quite hard, and it is not possible to improve much on the simple approximation algorithm in [7] , on which we base the LZ77 approximation algorithm in the previous subsection. A natural question is whether there is a better algorithm for the LZ77 estimation problem. That is, is the LZ77 estimation strictly easier than COLORS? As we shall see, it is not much easier in general.

Lemma 9 (Reduction from COLORS to LZ77). *Suppose there exists an algorithm $\mathcal{A}_{LZ}$ that, given access to a string w of length n over an alphabet Σ, performs $q = q(n, |\Sigma|, \alpha, \beta)$ queries and with probability at least $5/6$ distinguishes between the case that $C_{LZ}(w) \leq \alpha n$ and the case that $C_{LZ}(w) > \beta n$, for some $\alpha < \beta$.*

Then there is an algorithm for COLORS taking inputs of length $n' = \Theta(\alpha n)$ that performs q queries and, with probability at least $2/3$, distinguishes inputs with at most $\alpha'n'$ colors from those with at least $\beta'n'$ colors, $\alpha' = \alpha/2$ and $\beta' = \beta \cdot 2 \cdot \max\left\{1, \frac{4\log n'}{\log|\Sigma|}\right\}$.

Two notes are in place regarding the reduction. The first is that the gap between the parameters α' and β' that is required by the COLORS algorithm obtained in Lemma 9, is larger than the gap between the parameters α and β for which the LZ-compressibility algorithm works, by a factor of $4 \cdot \max\left\{1, \frac{4\log n'}{\log|\Sigma|}\right\}$. In particular, for binary strings $\frac{\beta'}{\alpha'} = O\left(\log n' \cdot \frac{\beta}{\alpha}\right)$, while if the alphabet is large, say, of size at least n', then $\frac{\beta'}{\alpha'} = O\left(\frac{\beta}{\alpha}\right)$. In general, the gap increases by at most $O(\log n')$. The second note is that the number of queries, q, is a function of the parameters of the LZ-compressibility problem and, in particular, of the length of the input strings, n. Hence, when writing q as a function of the parameters of COLORS and, in particular, as a function of $n' = \Theta(\alpha n)$, the complexity may be somewhat larger. It is an open question whether a reduction without such increase is possible.

Prior to proving the lemma , we discuss its implications. [18] give a strong lower bound on the sample complexity of approximation algorithms for COLORS. An interesting special case is that a subpolynomial-factor approximation for COLORS requires many queries even with a promise that the strings are only slightly compressible: for any $B = n^{o(1)}$, distinguishing inputs with $n/11$ colors from those with n/B colors requires $n^{1-o(1)}$ queries. Lemma 9 extends that bound to estimating LZ compressibility: *For any $B = n^{o(1)}$, and any alphabet Σ, distinguishing strings with LZ compression cost $\tilde{\Omega}(n)$ from strings with cost $\tilde{O}(n/B)$ requires $n^{1-o(1)}$ queries.*

The lower bound for COLORS in [18] applies to a broad range of parameters, and yields the following general statement when combined with Lemma 9:

Corollary 10 (LZ is Hard to Approximate with Few Samples). *For sufficiently large n, all alphabets Σ and all $B \leq n^{1/4}/(4 \log n^{3/2})$, there exist $\alpha, \beta \in (0,1)$ where $\beta = \Omega\left(\min\left\{1, \frac{\log|\Sigma|}{4\log n}\right\}\right)$ and $\alpha = O\left(\frac{\beta}{B}\right)$, such that every algorithm that distinguishes between the case that $C_{LZ}(w) \leq \alpha n$ and the case that $C_{LZ}(w) > \beta n$ for $w \in \Sigma^n$, must perform $\Omega\left(\left(\frac{n}{B'}\right)^{1-\frac{2}{k}}\right)$ queries for $B' = \Theta\left(B \cdot \max\left\{1, \frac{4\log n}{\log|\Sigma|}\right\}\right)$ and $k = \Theta\left(\sqrt{\frac{\log n}{\log B' + \frac{1}{2}\log\log n}}\right)$*

Proof of Lemma 9. Suppose we have an algorithm $\mathcal{A}_{LZ}$ for LZ-compressibility as specified in the premise of Lemma 9. Here we show how to transform a COLORS instance τ into an input for $\mathcal{A}_{LZ}$, and use the output of $\mathcal{A}_{LZ}$ to distinguish τ with at most $\alpha'n'$ colors from τ with at least $\beta'n'$ colors, where α' and β' are as specified in the lemma. We shall assume that $\beta'n'$ is bounded below by some sufficiently large constant. Recall that in the reduction from LZ77 to COLORS, we transformed substrings into colors. Here we perform the reverse operation.

Given a COLORS instance τ of length n', we transform it into a string of length $n = n' \cdot k$ over Σ, where $k = \lceil\frac{1}{\alpha}\rceil$. We then run $\mathcal{A}_{LZ}$ on w to obtain information about τ. We begin by replacing each color in τ with a uniformly selected substring in Σ^k. The string w is the concatenation of the corresponding substrings (which we call *blocks*). We show that:

1. If τ has at most $\alpha'n'$ colors, then $C_{LZ}(w) \leq 2\alpha'n$;
2. If τ has at least $\beta'n'$ colors, then $\Pr_w[C_{LZ}(w) \geq \frac{1}{2}\cdot\min\left\{1, \frac{\log|\Sigma|}{4\log n'}\right\}\cdot\beta'n] \geq \frac{7}{8}$.

That is, in the first case we get an input w for COLORS such that $C_{LZ}(w) \leq \alpha n$ for $\alpha = 2\alpha'$, and in the second case, with probability at least $7/8$, $C_{LZ}(w) \geq \beta n$ for $\beta = \frac{1}{2}\cdot\min\left\{1, \frac{\log|\Sigma|}{4\log n}\right\}\cdot\beta'$. Recall that the gap between α' and β' is assumed to be sufficiently large so that $\alpha < \beta$. To distinguish the case that $C_{COL}(\tau) \leq \alpha'n'$ from the case that $C_{COL}(\tau) > \beta'n'$, we can run $\mathcal{A}_{LZ}$ on w and output its answer. Taking into account the failure probability of $\mathcal{A}_{LZ}$ and the failure probability in Item 2 above, the Lemma follows.

We prove these two claims momentarily, but first observe that in order to run the algorithm $\mathcal{A}_{LZ}$, there is no need to generate the whole string w. Rather, upon each query of $\mathcal{A}_{LZ}$ to w, if the index of the query belongs to a block that has already been generated, the answer to $\mathcal{A}_{\mathrm{LZ}}$ is determined. Otherwise, we query the element (color) in τ that corresponds to the block. If this color was not yet observed, then we set the block to a uniformly selected substring in Σ^k. If this color was already observed in τ, then we set the block according to the substring that was already selected for the color. In either case, the query to w can now be answered. Thus, each query to w is answered by performing at most one query to τ.

It remains to prove the two items concerning the relation between the number of colors in τ and $C_{\mathrm{LZ}}(w)$. If τ has at most $\alpha' n'$ colors then w contains at most $\alpha' n'$ distinct blocks. Since each block is of length k, at most k compressed segments start in each new block. By definition of LZ77, at most one compressed segment starts in each repeated block. Hence,

$$C_{\mathrm{LZ}}(w) \leq \alpha' n' \cdot k + (1 - \alpha')n' \leq \alpha' n + n' \leq 2\alpha' n.$$

If τ contains $\beta' n'$ or more colors, w is generated using at least $\beta' n' \cdot \log(|\Sigma|^k) = \beta' n \log |\Sigma|$ random bits. Hence, with high probability (e.g., at least $7/8$) over the choice of these random bits, any lossless compression algorithm (and in particular LZ77) must use at least $\beta' n \log |\Sigma| - 3$ bits to compress w. Each symbol of the compressed version of w can be represented by $\max\{\lceil \log |\Sigma| \rceil, 2\lceil \log n \rceil\} + 1$ bits, since it is either an alphabet symbol or a pointer-length pair. Since $n = n'\lceil 1/\alpha' \rceil$, and $\alpha' > 1/n'$, each symbol takes at most $\max\{4 \log n', \log |\Sigma|\} + 2$ bits to represent. This means the number of symbols in the compressed version of w is

$$C_{\mathrm{LZ}}(w) \geq \frac{\beta' n \log |\Sigma| - 3}{\max\{4 \log n', \log |\Sigma|\}) + 2} \geq \frac{1}{2} \cdot \beta' n \cdot \min\left\{1, \frac{\log |\Sigma|}{4 \log n'}\right\}$$

where we have used the fact that $\beta' n'$, and hence $\beta' n$, is at least some sufficiently large constant. ∎

Acknowledgements. We would like to thank Amir Shpilka, who was involved in a related paper on distribution support testing [18] and whose comments greatly improved drafts of this article. We would also like to thank Eric Lehman for discussing his thesis material with us and Oded Goldreich and Omer Reingold for helpful comments.

References

1. Alon, N., Matias, Y., Szegedy, M.: The space complexity of approximating the frequency moments. J. Comput. Syst. Sci. 58(1), 137–147 (1999)
2. Bar-Yossef, Z., Kumar, R., Sivakumar, D.: Sampling algorithms: lower bounds and applications. In: Proceedings of the thirty-third annual ACM symposium on Theory of computing, New York, NY, USA, 2001, pp. 266–275. ACM Press, New York (2001)
3. Batu, T., Dasgupta, S., Kumar, R., Rubinfeld, R.: Tugkan Batu, Sanjoy Dasgupta, Ravi Kumar, and Ronitt Rubinfeld. SIAM Journal on Computing 35(1), 132–150 (2005)

4. Benedetto, D., Caglioti, E., Loreto, V.: Language trees and zipping. Phys. Rev. Lett. 88(4) (2002). (See comment by Khmelev DV, Teahan WJ, Phys Rev Lett. 90(8):089803, (2003) and the reply Phys Rev Lett. 90(8):089804, 2003)
5. Brautbar, M., Samorodnitsky, A.: Approximating the entropy of large alphabets. In: Proceedings of the Eighteenth Annual ACM-SIAM Symposium on Discrete Algorithms (2007)
6. Bunge, J.: Bibligraphy on estimating the number of classes in a population, http://www.stat.cornell.edu/~bunge/bibliography.htm
7. Charikar, M., Chaudhuri, S., Motwani, R., Narasayya, V.R.: Towards estimation error guarantees for distinct values. In: PODS, pp. 268–279. ACM, New York (2000)
8. Cilibrasi, R., Vitányi, P.M.B.: Clustering by compression. IEEE Transactions on Information Theory 51(4), 1523–1545 (2005)
9. Cilibrasi, R., Vitányi, P.M.B.: Similarity of objects and the meaning of words. In: Cai, J.-Y., Cooper, S.B., Li, A. (eds.) TAMC 2006. LNCS, vol. 3959, pp. 21–45. Springer, Heidelberg (2006)
10. Cover, T., Thomas, J.: Elements of Information Theory. Wiley & Sons, Chichester (1991)
11. Kukushkina, O.V., Polikarpov, A.A., Khmelev, D.V.: Using literal and grammatical statistics for authorship attribution. Prob. Peredachi Inf. 37(2), 96–98 (2000) [Probl. Inf. Transm. (Engl. Transl.) 37, 172–184 (2001)]
12. Lehman, E., Shelat, A.: Approximation algorithms for grammer-based compression. In: Proc. 18th Annual Symp. on Discrete Algorithms, 2002, pp. 205–212 (2002)
13. Li, M., Chen, X., Li, X., Ma, B., Vitányi, P.M.B.: The similarity metric. IEEE Transactions on Information Theory 50(12), 3250–3264 (2004) (Prelim. version in *SODA 2003*)
14. Li, M., Vitányi, P.: An Introduction to Kolmogorov Complexity and Its Applications. Springer, Heidelberg (1997)
15. Loewenstern, D., Hirsh, H., Noordewier, M., Yianilos, P.: DNA sequence classification using compression-based induction. Technical Report 95-04, Rutgers University, DIMACS (1995)
16. Raskhodnikova, S., Ron, D., Rubinfeld, R., Shpilka, A., Smith, A.: Sublinear algorithms for approximating string compressibility and the distribution support size. Electronic Colloquium on Computational Complexity, TR05-125 (2005)
17. Raskhodnikova, S., Ron, D., Rubinfeld, R., Smith, A.: Sublinear algorithms for approximating string compressibility. Full version of this paper, in preparation, Arxiv Report 0706.1084 [cs.DS] (June 2007)
18. Raskhodnikova, S., Ron, D., Shpilka, A., Smith, A.: On the difficulty of approximating the support size of a distribution. (Manuscript 2007)
19. Ziv, J., Lempel, A.: A universal algorithm for sequential data compression. IEEE Transactions on Information Theory 23, 337–343 (1977)
20. Ziv, J., Lempel, A.: Compression of individual sequences via variable-rate coding. IEEE Transactions on Information Theory 24, 530–536 (1978)

Author Index

Lecture Notes in Computer Science

For information about Vols. 1–4546

please contact your bookseller or Springer